David Hughes.

First Year Chemistry

J. M. Coxon, J. E. Fergusson
and L. F. Phillips

**Department of Chemistry, University of Canterbury,
Christchurch, New Zealand**

Edward Arnold

© J. M. Coxon, J. E. Fergusson and L. F. Phillips 1980

First published 1980
by Edward Arnold (Publishers) Ltd.
41 Bedford Square
London WC1B 3DQ

ISBN: 0 7131 2700 7

Typesetting by Parkway Group, London and Abingdon
Printed in Hong Kong by
Wing King Tong Co Ltd

Preface

This book is designed to suit first year university courses in general chemistry. We have included rather more material than would normally be covered in such a course in order to allow selection of topics by the teacher or student. Thus an exploration of the consequences of the Second Law of Thermodynamics is deferred. The student is assumed to have a passing acquaintance with the basic ideas of elements and compounds, chemical reactions, stoichiometry and the ideal gas laws, and with the elementary physics of heat, light and electricity. A few traditional topics have been left out; for example, colligative properties are not called by that name, osmosis is not discussed and there is no chapter on nuclear chemistry. Other topics, such as the extraction of the elements and their reaction chemistry, we have chosen to treat in more detail than usual. The emphasis in the organic section is on functional groups and their reactions, with applications to synthesis emphasized through problems. We welcome readers' opinions of the choices we have made and we are willing to entertain complaints about favourite topics that have been omitted.

One of our main aims in writing this book was to produce a text in which approximately equal weight would be given to physical, inorganic and organic chemistry, and in which the divisions between the three branches of chemistry would as far as possible be minimized. Philosophically and pedagogically we believe this to be the correct approach. Whether we have succeeded is for the reader to judge.

J. M. Coxon
J. E. Fergusson
L. F. Phillips
Christchurch, New Zealand
1979

Contents

1 The Nature and Composition of Matter

1.1 A General Survey

Chemistry can be defined as the study of matter, its structure, composition and transformations. In practice this amounts to a study of the behaviour of matter as it is encountered in the world around us or under conditions that can be realized in the laboratory, and the interpretation of this behaviour in terms of processes that go on in a very different world, the sub-microscopic world of atoms and molecules. We observe, for example, that sulphur burns in air with a blue flame and a choking smell, and we interpret our observation in terms of an attack by gaseous oxygen molecules on sulphur molecules at the surface of the molten mass to produce gaseous sulphur dioxide. Looking a little deeper, we may find there is something interesting to be learned about the transfers of electrons involved in the breaking of chemical bonds in the sulphur and oxygen molecules and formation of new bonds in the sulphur dioxide molecules, or we may become engrossed in the details of the process by which some of the energy released by the chemical reaction is transformed into the light that we see as a blue flame. Alternatively, we might decide to investigate the molecular processes that result in the sensation of a choking smell when we breathe air that contains sulphur dioxide. These are examples of the chemist's way of looking at things. It is quite a fundamental way, and has proven very successful both in discovering new theoretical concepts and in finding solutions to practical problems. Nevertheless, the chemist's viewpoint is far from being the only one, and before we begin to examine the details of the chemical picture it will be instructive to consider briefly some of the other faces which matter presents to the world.

The behaviour that is to be regarded as characteristic of a particular sample of matter depends very much on the scale of the phenomena which are being observed. Thus, on a cosmic scale matter takes the form of rather widely spaced agglomerations, called solar systems, dust clouds, star clusters, and so on, which interact with one another over distances of the order of light years* mainly through the extremely feeble, long-range force of gravitation. At the opposite extreme, on the smallest scale for which we have experimental evidence, matter exists in the form of elementary particles, such as protons, neutrons, and electrons, which interact with one another electromagnetically and also through what are known as the strong and weak nuclear interactions. The nuclear interactions are only effective at very short range—distances of 10^{-15} m or less—whereas the electromagnetic interaction is also effective at long range, which on this scale means at distances in excess of 10^{-14} m. At short range the strong interaction, which is the force that holds atomic nuclei together, is about 100 times as strong as the electromagnetic interaction and about 10^{14} times as strong as the weak nuclear interaction. The weak interaction just manages to manifest itself in some forms of radioactive decay of atomic nuclei. At short range the force which results from the strong interaction between particles in the nucleus of an atom is about 10^{39} times stronger than the gravitational force due to the masses of the particles. However, both kinds of nuclear interaction fall rapidly to zero at distances greater than 10^{-15} m, whereas the electromagnetic and gravitational interactions vary inversely with the square of the distance, in accordance with Coulomb's and Newton's Laws, and so extend essentially to infinity. In principle the most fundamental description we can give of the behaviour of matter is a description in terms of interacting elementary particles.

The two extremes of cosmic and sub-nuclear phenomena represent the happy hunting grounds of the astronomer and the high-energy physicist, respectively. Between these extremes are two other well-defined regions, namely the region of macroscopic phenomena in which we live, where distances are measured in centimetres or kilometres, and the regime of atomic and molecular phenomena, where distances are conveniently measured in nanometres or tenths of nanometres (sometimes called Angstrom units). The chemical world of atoms and molecules is governed entirely by electromagnetic interactions; on a macroscopic scale both electromagnetic and gravitational interactions are significant. In the macroscopic world we also encounter the distinction between living and non-living matter, which introduces still

* A light year is the distance (9.46×10^{15} m) travelled by light in one year.

another viewpoint, that of the biologist. Present indications are that most if not all of the behaviour which distinguishes living things can be understood in terms of the chemistry of the very complicated molecules of which living matter is composed, but our understanding clearly has a long way to go in this direction.

For many purposes macroscopic objects, including living things, can be regarded as made up of more or less massive particles which obey the laws of classical mechanics. These are laws, built around the concepts of position, velocity, force and momentum, which govern the motions of billiard balls, planets, suspension bridges and moon rockets. Because we live on a macroscopic scale there is a natural temptation to regard the appearance and behaviour of matter on this scale as being in some sense its true appearance and behaviour, and to attempt to understand phenomena on different scales in terms of mental concepts which are strictly applicable only to the scale on which we live. To take a particular example, it is remarkably difficult to think of electrons in an atom without calling up a mental picture of a miniature solar system, with tiny classical particles travelling in well-defined orbits around the central nucleus. This carrying over of concepts from one scale to another is often helpful, but it can also be very misleading, as we shall see.

1.2 The Chemical Structure of Matter

The results of chemical experiments encourage us to believe that matter is composed of entities which we call atoms, molecules, radicals and ions. As a brief definition we can say that an atom consists of a small (10^{-15} m), massive, positively charged nucleus surrounded by a cloud of negatively charged electrons, the atom as a whole being electrically neutral. A molecule is then a stable collection of atoms held together by the forces of electrical attraction between the oppositely charged electrons and nuclei. The number of atoms comprising a molecule can be small: one for argon and helium, two for hydrogen, three for water—or large: 20 000 or more for a typical protein molecule. A radical is a fragment of a molecule, and as such is usually very reactive, having a strong tendency to join up with another radical and make a stable molecule. An ion is an atom, molecule, or radical that has gained or lost one or more electrons and so carries an excess negative or positive charge.

Considering atoms in more detail, we find that they can be characterized by their atomic number and by their mass. The *atomic number Z* is equal to the charge on the nucleus, in multiples of the charge of a proton, and defines the particular element of which a single atom represents the smallest possible sample. Since the atom as a whole is neutral, Z must also equal the number of electrons which surround the nucleus. The *atomic mass A*, which is nearly all concentrated in the nucleus, is measured in terms of a unit which is almost exactly equal to the mass of a proton or of a hydrogen atom (which consists of a proton plus a single electron). Prior to 1961 there were two different units of atomic mass, one employed by chemists and the other by physicists. The difference amounted to 0.028%, the physicists' unit being the smaller of the two. In 1961 the two competing scales were replaced by a single scale based on a unit which is defined as one twelfth of the mass of the most abundant sort of carbon atom, the atom that is designated by the symbol ^{12}C. The form of this definition implies that atoms of a given element are not necessarily all of the same mass, and this is indeed the case, the atoms of different mass being referred to as different *isotopes* of the element in question. Some elements, for example fluorine, have only one isotope, which in this case is ^{19}F. Others, such as cadmium or xenon, have eight or more. The superscript which is conventionally used to indicate the mass number of an isotope may be written either before the chemical symbol, as we have done, or after, whichever is the more convenient. The atomic number is sometimes written as a subscript, e.g. $^{16}_{8}O$ is an oxygen atom with atomic mass $A = 16$ and atomic number $Z = 8$.

The average atomic mass of a representative sample of an element as it occurs in nature is termed its *relative atomic mass*, A_r (formerly the 'atomic weight'), and is not usually dependent on the source from which the sample is obtained. Where differences *do* occur they may convey interesting information about the previous history of the sample; for example, the relative abundance of the different oxygen isotopes ^{16}O and ^{18}O in limestone can provide quite precise information about the temperatures of the ocean from which the limestone was precipitated in the geological past.

Except in the case of the most abundant hydrogen isotope, ^{1}H, the atomic mass A is always greater than the atomic number Z, commonly by a factor of 2 or more. Thus a carbon atom of mass 12 has atomic number 6, an oxygen atom of mass 16 has atomic number 8, and a uranium atom of mass 238 has atomic number 92. To account for the difference it is usual to assume that the nucleus of an atom of atomic number Z and atomic mass A can be regarded as a cluster of Z protons, each of unit mass and unit charge, and $(A - Z)$ neutrons, each of unit mass and zero charge, all held together by the strong nuclear interaction. The heavier an atom is, the greater is the ratio of neutrons to protons that seems to be required to produce a stable nucleus. For atomic masses in excess of about 210 it appears that all nuclei are unstable and subject to radioactive decay.

The *relative molecular mass M_r* (formerly the 'molecular weight') of a molecule or radical is the sum of the relative atomic masses of the constituent atoms. For a species of specified isotopic composition, for example $^{12}C^{16}O$, we must use instead the sum of the appropriate atomic masses. The quantity of a substance which is equal to the relative molecular mass in grams is called a *mole*. More generally, a mole of *anything at all* contains as many individual units as there are atoms in 0.012 kg of ^{12}C. This number of units is called Avogradro's number and amounts to 6.02×10^{23}. The mole is defined as a number of units, not in terms of mass. Nevertheless, the connection between the mole and the molecular mass is of fundamental importance. The very great usefulness of the mole concept stems from the fact that when we write out the balanced chemical equation for a reaction between molecules, for example:

$$H_2O + SO_3 \rightarrow H_2SO_4 \qquad (1.1)$$

the equation tells us not only what happens to the individual molecules, but also what happens on a macroscopic scale in terms of moles of material. Thus Eq. 1.1 states not only that one molecule of water reacts with one molecule of sulphur trioxide to produce a molecule of sulphuric acid, but also that 18 g of water reacts with 80 g of sulphur trioxide to produce 98 g of sulphuric acid.

Our introductory definition of a molecule stated that molecules are held together by the same force which keeps atoms intact, namely the electrical attraction of the oppositely charged electrons and nuclei. Except when the nuclei are very close together this attraction, which results essentially from a sharing of electrons between different atoms, is sufficient to overcome the repulsion of electrons and nuclei for particles of their own kind. However, the ability of any atom to share electrons is strictly limited, so that between stable molecules we find only relatively weak, residual electrical forces. These residual forces, which are known collectively as *Van der Waals' forces,* cause molecules to aggregate into liquids or solids provided the temperature is not too high. As an illustration of the various points we have mentioned so far, Fig. 1.1 shows the changes that occur when steadily increasing amounts of energy in the form of heat are supplied to a solid, in this case a piece of ice.

Because electrons are very much lighter than nuclei (electron mass = proton mass/1840) the motion of a collection of electrons and nuclei consists of a relatively rapid motion of the electrons, like a swarm of gnats, around the tiny but ponderous nuclei. Within a molecule the nuclei tend to

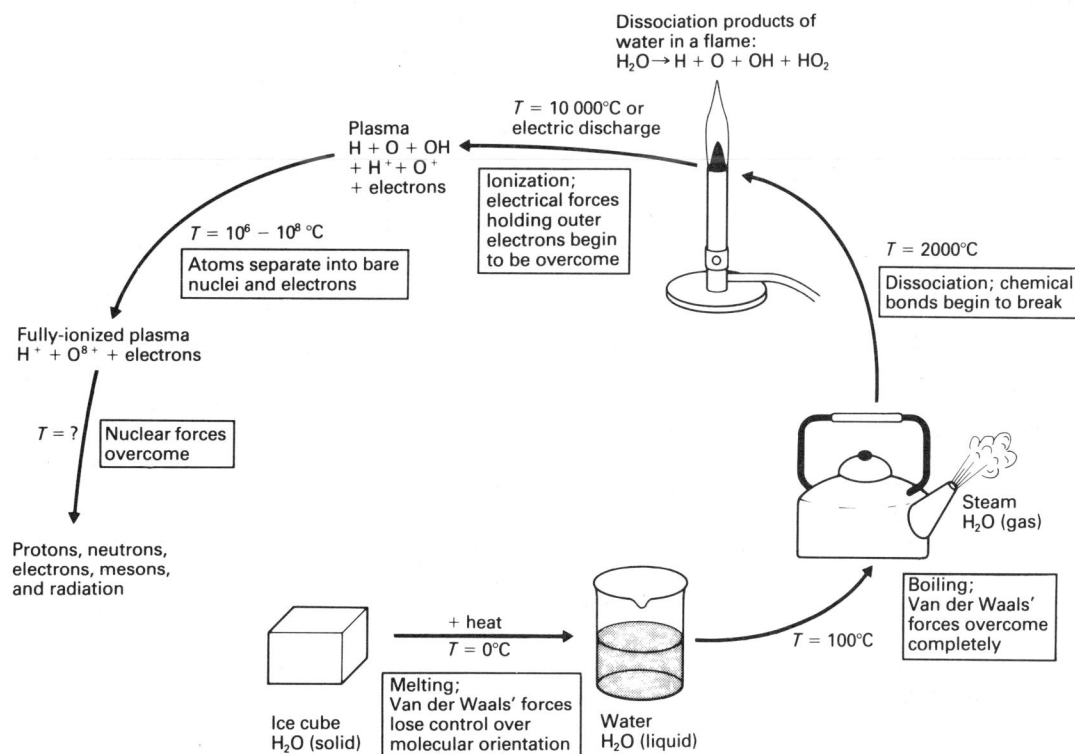

Fig. 1.1 Effect of applying increasing amounts of energy to a solid

remain in fixed positions relative to one another, only undergoing relatively slow oscillations about equilibrium positions which correspond to minimum potential energy of the whole system of electrons plus nuclei. These equilibrium positions give rise to well-defined distances, or bond lengths, between adjacent atoms in the same molecule, and to well-defined bond angles when two or more atoms are bound to a single atom. Some examples of experimentally determined bond lengths and angles are given in Fig. 1.2.

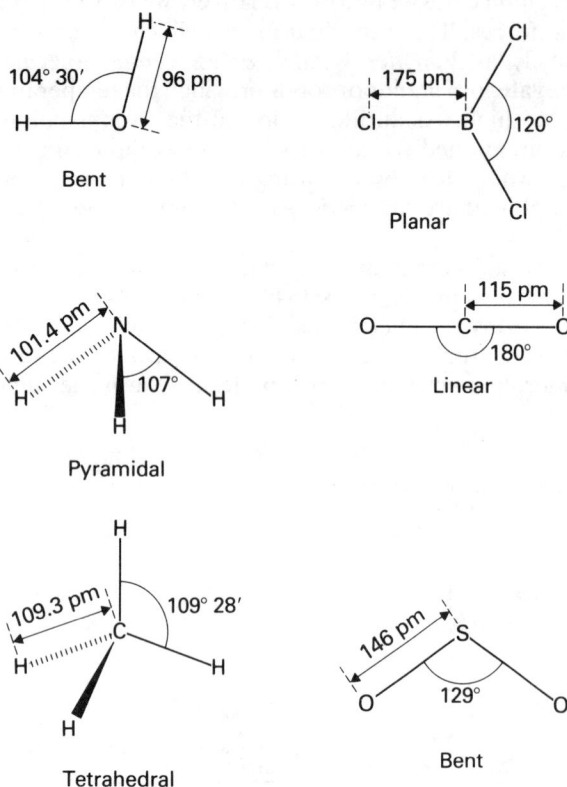

Fig. 1.2 Some experimentally determined bond lengths and angles in stable molecules

Under most circumstances atoms and molecules are sufficiently massive to be treated as particles in the sense of classical mechanics. Therefore it is permissible to consider their behaviour in terms of classical mechanics, and to make use of concepts such as atomic or ionic radius, internuclear distance, moment of inertia, and force constant (a measure of the resistance of a chemical bond to being stretched or compressed) which are easy to visualize in terms of analogous quantities in the macroscopic world. For electrons, on the other hand, the application of classical mechanics or the uncritical use of macroscopic concepts generally leads to predictions which experiments show to be wrong.

In order to deal successfully with electrons it has been necessary to devise the new form of mechanics known as *quantum mechanics*. The experimental results which were most influential in stimulating the development of quantum mechanics, and the kind of information which quantum theory provides, are discussed in the next section.

1.3 The Introduction of Quantum Theory

The first indication that something more than classical mechanics might be needed to describe small-scale phenomena came not from the study of matter, but from the study of light. By the end of the nineteenth century the wave theory of light was very firmly entrenched as a result of the observation of a whole host of phenomena which could only be understood in terms of wave motion. The most characteristic property of a wave is the ability to give rise to interference and diffraction effects (Fig. 1.3). These effects are observed when the crest of a wave from one source overlaps with the trough of a wave from another source, so that the two waves cancel one another at this place, while nearby two troughs or two crests coincide so that the waves reinforce one another. For light waves these effects were sufficiently well known to be used in practice to control the grinding of lenses and mirrors. The single disturbing feature at the time was the curious inability of wave theory to account for the familiar changes in colour, from dull red through orange-yellow to white, which occur when a solid is gradually heated to a high temperature. Experimentally the radiation from a hot body has a peak of intensity at some particular wavelength λ_{peak} and the intensity falls off rapidly at wavelengths less than that of the peak (Fig. 1.4). When the temperature of the body is raised the peak moves to shorter wavelengths in accordance with Wien's Law:

$$\lambda_{peak}\, T = \text{a constant} \qquad (1.2)$$

In contrast, a straightforward application of the wave theory by Rayleigh and Jeans led to the conclusion that the intensity of radiation from the hot body should increase steadily with decreasing wavelength, as shown by the dotted line in Fig. 1.4, and that the relative intensity at different wavelengths should be independent of temperature.

In 1900 Max Planck showed that in order to obtain a correct formula for the intensity distribution, by which we mean a formula that agreed with experiment, it was necessary to make the radical assumption that a hot body could emit energy as radiation only in discrete packets, or quanta, each containing an amount of energy E given by the formula

$$E = h\nu \qquad (1.3)$$

Here ν is the frequency of the radiation in s^{-1} and Planck's constant h has the value 6.627×10^{-34} joule seconds. Next,

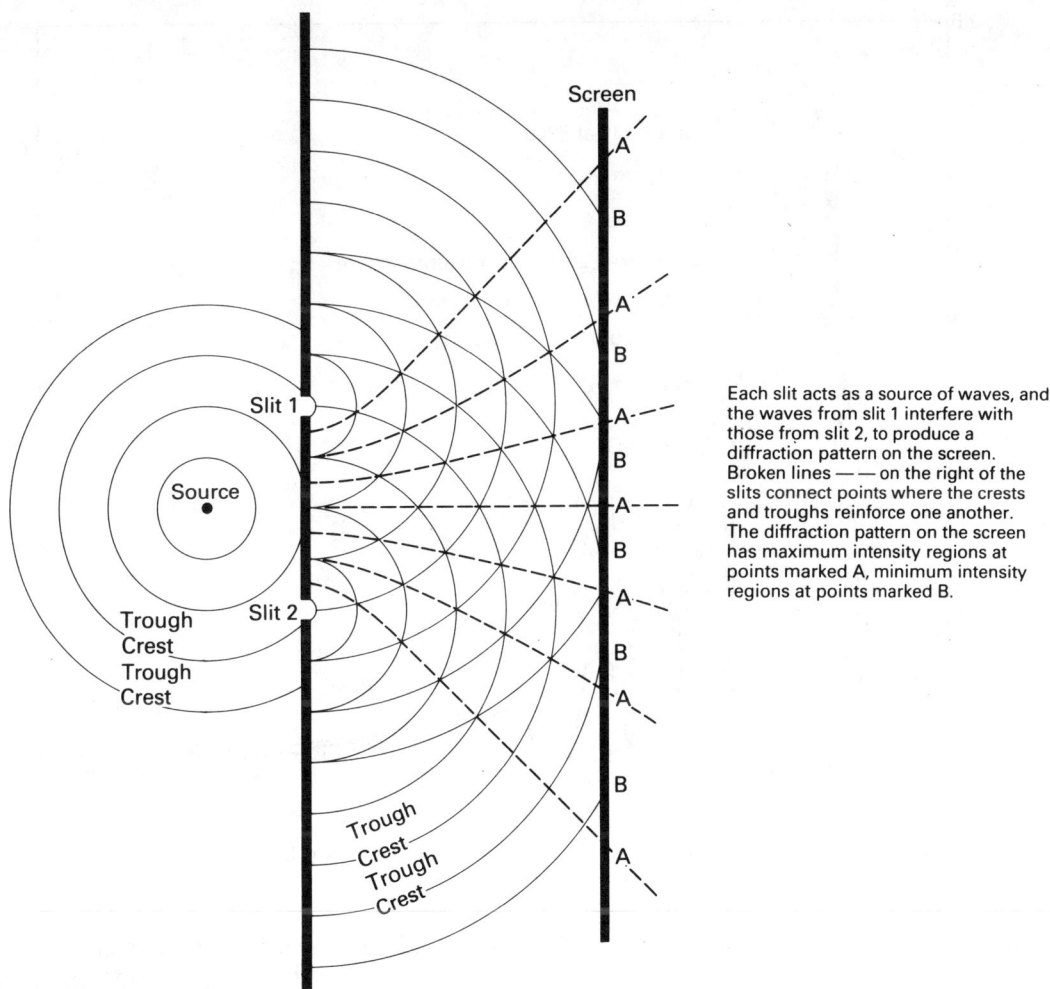

Each slit acts as a source of waves, and the waves from slit 1 interfere with those from slit 2, to produce a diffraction pattern on the screen. Broken lines —— on the right of the slits connect points where the crests and troughs reinforce one another. The diffraction pattern on the screen has maximum intensity regions at points marked A, minimum intensity regions at points marked B.

Fig. 1.3 A double-slit diffraction experiment

in 1904 Einstein showed that experimental data for the photoelectric effect (whereby light falling on a surface in vacuum causes the material of the surface to eject electrons) could be understood only if a body absorbs energy from a beam of radiation in the form of individual quanta, or *photons* as they are now usually called, whose energy is given by Eq. 1.3. Finally, in 1923 Compton demonstrated that a single photon could be observed to behave as a particle during collision with an electron, giving up energy to the electron in the process and going on its way as a photon of lower energy, i.e. longer wavelength. Since every interaction between light and matter reveals the existence of photons, we are driven to conclude that light consists entirely of photons whose motion must be such as to give rise to the observed interference and diffraction effects. In other words, photons are particles whose equation of motion is a wave equation.

The solution of a wave equation takes the form of values of amplitude of the wave—half the distance from trough to crest in the case of a water wave—at different points in space. The intensity of the wave at a given point is then proportional to the *square* of the amplitude at that point. For a light beam the intensity is also proportional to the number of photons arriving at a point in unit time. Hence, we can conclude that for any point in space the square of the amplitude, or *wave function,* that we obtain by solving the wave equation is proportional to the number of photons that arrive at that point in unit time. Therefore the square of the amplitude is proportional to the probability of finding a photon at the point at a particular instant. The equations of motion for a photon are seen to be rather uninformative —we can never predict exactly where the photon will be, but only the probability of finding one, not necessarily even the same one, at a given place.

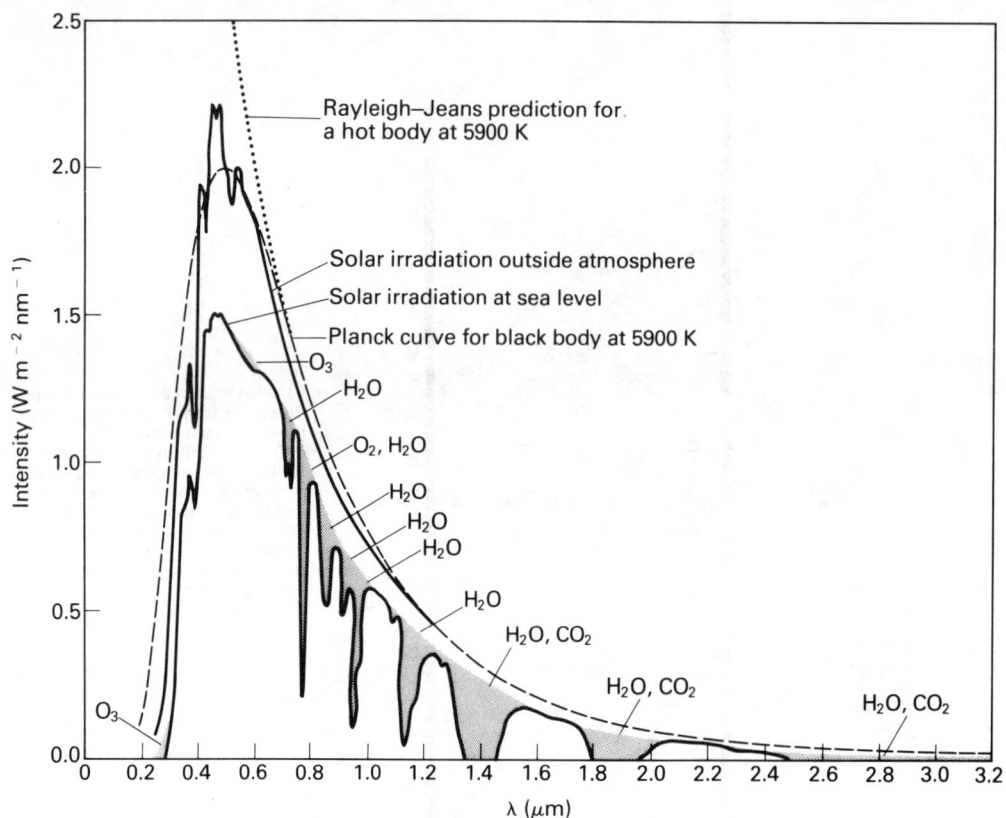

Fig. 1.4 A hot body—intensity of the Sun's radiation as a function of wavelength. (Shaded areas show light which is absorbed in the atmosphere by carbon dioxide, water, oxygen and ozone)

Now let us consider what this might have to do with particles of matter. Certainly the most famous result connected with Einstein's special theory of relativity is the equation which relates the total energy of a body to its mass, namely

$$E = mc^2 \qquad (1.4)$$

Here m is the mass of the body and c the velocity of light $(3 \times 10^8 \text{ m s}^{-1})$. For an ordinary particle whose mass can be measured when the particle is at rest, such as a proton or a billiard ball, the mass m is in general not constant, but increases without limit as the speed of the particle approaches the speed of light. For a particle to move with the velocity of light, as in the case of a photon, its rest mass (or *proper mass*) must be zero. For such a particle the linear momentum (the product of mass and linear velocity) is given by

$$p = mc \qquad (1.5)$$

If we substitute p for mc in Eq. 1.4, use Eq. 1.3, and remember that ν is equal to c/λ, we obtain the formula

$$\lambda = h/p \qquad (1.6)$$

which relates wavelength to momentum for a photon, or for any other particle of zero rest mass. In 1924 de Broglie proposed that this equation should apply to all particles, whatever their rest mass. His proposal was based on intuition rather than deduction; nevertheless it has proven to be correct. A direct test of Eq. 1.6 for particles of matter was reported in 1927, when two independent sets of workers showed that a beam of electrons could be diffracted by the regularly spaced planes of atoms in a crystal, and that the wavelength calculated from the results agreed precisely with de Broglie's prediction. The extraordinary feature of Eq. 1.6 is that it assigns a wavelength λ to anything that has momentum, whether it be an electron or a battleship. The saving feature of Eq. 1.6 is the presence of the factor h, which ensures that for all macroscopic particles the wavelength is so short that interference or diffraction effects are virtually impossible to detect. For such effects to be detectable the wavelength involved must be comparable with or greater than the dimensions of the system with which the wave interacts. For an electron in a molecule this is invariably the case, so that it is always necessary to solve a wave equation in order to determine the energy of the electron and the probability of finding it at any point.

Atoms and molecules are generally stable; they do not part with electrons very readily. Therefore the wave functions which describe their electrons must always remain concentrated in the vicinity of the atom or molecule, i.e. they must be *standing waves,* analogous to the waves in a violin string or in the membrane of a drum. The differential equation for a standing wave in one dimension* can be written

$$\frac{d^2\Psi}{dx^2} + \frac{4\pi^2}{\lambda^2}\Psi = 0 \qquad (1.7)$$

where Ψ (Greek 'psi') is the amplitude or *wave function,* x is, for example, the distance along the vibrating string, and λ is the wavelength (see Fig. 1.5). Since we are dealing with standing waves, only certain values of λ are allowed, namely those which correspond to nodes at the ends of the string.

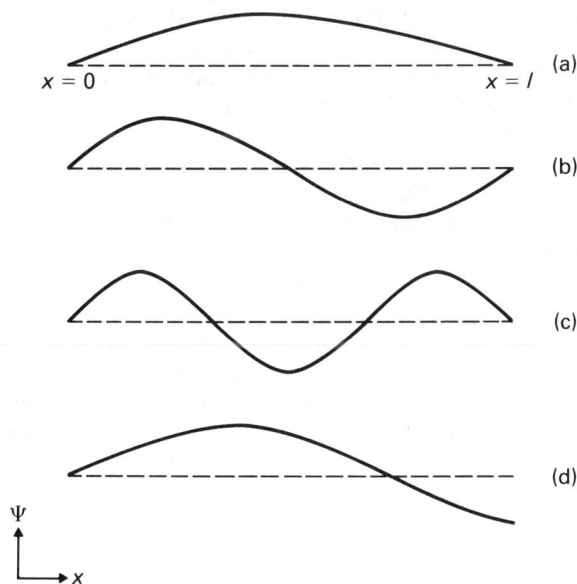

Fig. 1.5 Standing waves in a stretched string: (a) fundamental, $\lambda = 2l$; (b) first overtone, $\lambda = l$; (c) second overtone, $\lambda = 2l/3$; (d) unacceptable solution, $2l$ not an integral multiple of λ.
[$\Psi = a \sin 2\pi n/\lambda$. $\sin 2\pi\nu t$; λ = wavelength; ν = frequency; t = time; $l = n\lambda/2$, $n = 1,2,3,\ldots$]

The fundamental vibration has nodes only at the ends; the various harmonics have additional nodes distributed evenly along the length of the string. To obtain a wave equation for a particle it is plausible that we should simply substitute for λ from Eq. 1.6. While we are doing this it proves helpful first to use the expression

$$T = p^2/2m \qquad (1.8)$$

* To 'solve' such an equation, we seek a function Ψ whose dependence on x is such that the equation is obeyed. There may be many such functions; other criteria are then used to select the acceptable ones.

for the kinetic energy T of the particle, and then to use the fact that the total energy E is the sum of the particle's kinetic energy and its potential energy V, that is

$$T = E - V \qquad (1.9)$$

The outcome of these manipulations is that we obtain the equation

$$\frac{d^2\Psi}{dx^2} + \frac{8\pi^2 m}{h^2}(E - V)\Psi = 0 \qquad (1.10)$$

which happens, perhaps surprisingly, to be a correct representation of nature for the case of a particle constrained to move in one dimension. Eq. 1.10 can be generalized to deal with a three-dimensional system by replacing the differential operator d^2/dx^2 by its three-dimensional counterpart, the Laplacian operator:

$$\nabla^2 = \frac{\partial^2}{\partial x^2} + \frac{\partial^2}{\partial y^2} + \frac{\partial^2}{\partial z^2} \qquad (1.11)$$

and making the wave function Ψ a function of all three coordinates, x, y and z. Thus we obtain

$$\nabla^2\Psi + \frac{8\pi^2 m}{h^2}(E - V)\Psi = 0 \qquad (1.12)$$

Eq. 1.12 is called the *Schroedinger equation* for a stationary state. This equation is of central importance to modern theoretical chemistry. It can be written more succinctly in the form

$$H\Psi = E\Psi \qquad (1.13)$$

where the operator H, given by

$$H = -\frac{h^2}{8\pi^2 m}\nabla^2 + V \qquad (1.14)$$

is the celebrated *Hamiltonian operator.* Eq. 1.13 shows that when the Hamiltonian operator acts on the wave function the result is equal to the wave function multiplied by the energy. When this type of equation is obeyed the value of E is called an eigenvalue and the function Ψ is called an eigenfunction of the operator H.

In Chapters 2 and 3 we shall make use of some results that have been obtained by solving the Schroedinger equation for chemical systems. What is involved in each case is writing down the correct form of Hamiltonian operator for the system of interest and then solving Eq. 1.12 or its equivalent to determine the possible wave functions and energies. In general it happens that, as for the case of waves in a string, only certain values of λ correspond to acceptable wave functions. In view of Eq. 1.6 this means that for a particle only certain values of the energy E are allowable; these are the values which correspond to the energy levels of electrons in atoms and molecules. Unfortunately, for most systems of interest to chemists it turns out that the Schroedinger equation is impossibly difficult to solve, given the current state of knowledge, and we have to be content with the approximate solutions that can be obtained, for example, with the aid of a digital computer. For our present purpose it is sufficient to note that electrons in atoms and molecules have to be described by means of a wave equation, that the equation which is used for this purpose is called the Schroedinger equation, and that methods exist for obtaining more or less accurate solutions of the Schroedinger equation for molecular systems.

1.4 The States of Matter: Solids, Liquids and Gases

Solids and liquids are termed *condensed phases*. In them the constituent atoms, molecules or ions are packed closely together with very little free volume, and each is greatly affected by the behaviour of its nearest neighbours. A gas, by contrast, contains widely separated molecules, all in a state of rapid thermal motion independently of one another, and there is no such thing as a nearest neighbour.

Solids can be classified as either crystalline or glassy (an alternative term for glassy is *amorphous*), depending on the presence or absence or long-range order in the arrangement of the atoms, molecules or ions of which the solid is composed. A crystal consists essentially of a single basic unit, the *unit cell*, repeated many millions of times in a regular pattern throughout the bulk of the material. In a glass the extended regular pattern is absent. The glassy state is generally unstable relative to the crystalline state, but the rate of transformation of a glass to a crystal is likely to be immeasurably slow at ordinary temperatures. The difference between a crystal and a glass resembles the difference between a crystal and a liquid, and in fact glasses are normally obtained by cooling liquids so rapidly that their component molecules have too little time to untangle themselves into a regular array. The properties of a non-crystalline material often differ in desirable ways from the properties of the corresponding crystal, a fact which is exploited in certain alloys, as well as in ordinary window glass. Some common substances such as lamp-black or flowers of sulphur appear to lack crystalline form but are in fact micro-crystalline rather than truly amorphous. Two very interesting intermediate classes of solids are the plate-like materials, such as mica, which have a regular arrangement of atoms in two dimensions but tend to be disordered in the third dimension, and fibres, such as the fibrous proteins, which have a regular arrangement of atoms in one direction only.

Crystalline solids can themselves be classified into four main types, these being molecular crystals, giant molecules, ionic crystals, and metals. Representatives of each type are illustrated in Fig. 1.6. A molecular crystal, or Van der Waals' crystal, is formed by stacking individual molecules in a three-dimensional array which is held together only by the rather weak Van der Waals' forces. Such crystals are generally soft and have low melting points. Examples are solid carbon dioxide ('dry ice'), crystalline iodine, and the solid form of any substance which is gaseous at room temperature. Ordinary ice is a molecular crystal, although here the residual forces between the molecules ('hydrogen bonds') are considerably stronger than is the case for most substances.

Diamond, silicon dioxide (e.g. crystalline quartz), and silicon carbide (carborundum) are examples of materials which crystallize as giant molecules. The basic units of such crystals are individual atoms, and the forces which hold the crystal lattice together are ordinary covalent bonds, i.e. chemical bonds which result from the sharing of electrons between a pair of atoms. Such crystals are notably hard, diamond for example being the hardest substance known, and they have very high melting points.

In an ionic crystal, such as sodium chloride, the units which make up the crystal lattice are positive and negative ions, in this case Na^+ and Cl^-. For such an arrangement to be stable the neutral atoms from which the ions are derived must differ very markedly in their 'electronegativity', i.e. in their power to attract electrons. Therefore ionic solids tend to be formed by ions derived from the more reactive metals (Li^+, Na^+, K^+, Ca^{2+}, Mg^{2+}) and the more reactive non-metals (F^-, Cl^-, Br^-, O^{2-}) or groups of non-metals (NO_3^-, ClO_4^-, SO_4^{2-}). The electrostatic forces between oppositely charged ions are of similar strength to ordinary covalent bonds, so that ionic crystals tend to resemble the giant molecules in being hard and having fairly high melting points.

Metallic crystals are similar to those we have classified as giant molecules in that they are held together by chemical bonds which extend throughout the crystal lattice, and metals as a class are correspondingly hard and high-melting. However, they differ in that the shared electrons which make up the bonds are not localized in any one bond, but are relatively free to move throughout the crystal lattice under the influence of an externally applied voltage. Consequently metals are mostly good conductors of electricity, in contrast to materials like diamond or quartz which are excellent insulators. The high mobility of the electrons in a metal also makes for relatively easy breaking and reforming of bonds when the crystal lattice is distorted by an external force, which explains why metals tend to be malleable and ductile. For many purposes the properties of metallic solids

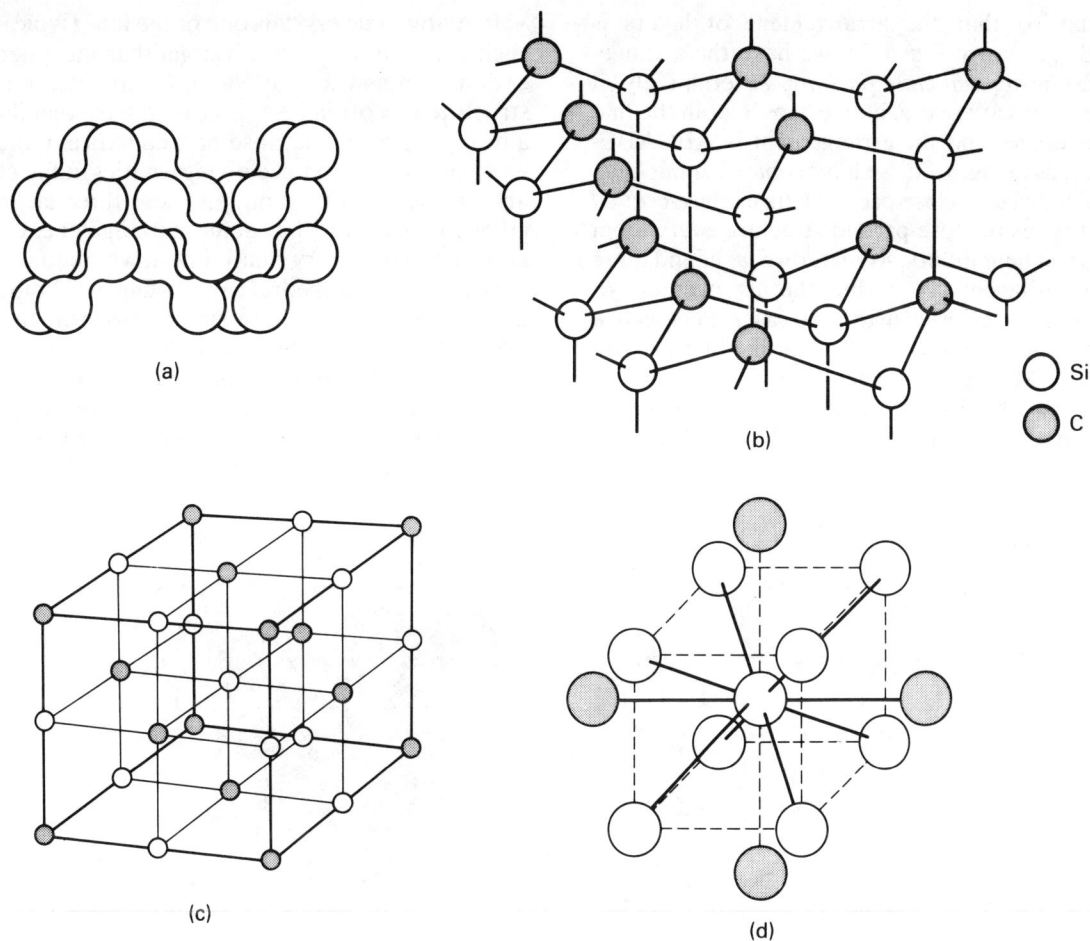

Fig. 1.6 (a) Molecular crystal (I_2). (b) Giant molecule (SiC). (c) Ionic crystal (NaCl structure). (d) Body-centre cubic structure of alkali metals. Four of the six second-nearest neighbours are shown

can be satisfactorily accounted for by a model which consists of a three-dimensional array of positive ions immersed in a sea of negative electrons, these being the non-localized bonding electrons, or conduction electrons.

The entities of which a crystal is composed—atoms, ions, or molecules—can be separated from one another with the expenditure of a greater or lesser amount of energy. The energy required to bring about this separation is called the *lattice energy*, and is a measure of the strength of cohesion of the crystal. Hard crystals have high lattice energies. For an ionic crystal the lattice energy is usually stated as the energy required to transform a mole of the crystal into a gas of separated positive and negative ions. During this separation, work is done against the electrostatic forces of attraction between the oppositely charged species. The same type of separation of ions occurs when the crystal dissolves in a solvent; hence the lattice energy is also a measure of the resistance of the crystal to going into solution. The forces between charged particles are greatly reduced if a medium of high dielectric constant is interposed between them. Consequently, ionic crystals dissolve most readily in solvents like water which have high dielectric constants.

The forces between atoms in a crystal fall off rapidly with increasing distance, so the most stable arrangement of atoms is likely to be one in which they are packed together as closely as possible. The closest possible packing in a plane is illustrated in Fig. 1.7a for spherical atoms of uniform size. To obtain a three-dimensional arrangement with the closest possible packing we can simply fit a second layer on top of the first, with atoms in the second layer located above gaps in the first as in Fig. 1.7b, and then a third on top of the second, and so on. However, a complication arises in that there are two possible ways of placing the third layer above the second. If the third layer is placed vertically above the first so that the arrangement of layers is ABABABAB...., as in Fig. 1.7c, we have the arrangement known as *hexagonal close packing*. Alternatively, if the third layer atoms are placed above gaps in both the first and second layers so that the arrangement is ABCABC-ABCABC...., as in Fig. 1.7d, we have *cubic close packing*. Another name for a cubic close-packed lattice is *face-centred cubic*. In both types of close-packed structure every atom has twelve nearest neighbours, six in its own layer and three in each of the adjacent layers, but the arrangement of nearest neighbours is different in the two cases. Perspective drawings of hexagonal and cubic close-packed lattices are given in Fig. 1.8. Most close-packed lattices are of one or other of these two types, but mixed lattices with layers arranged in a manner such as ABACABACABAC.... are also known.

In many ionic crystals one of the ions (typically an anion such as O^{2-} or F^-) is much larger than the other (typically a cation such as Ca^{2+} or Na^+). In this situation the crystal structure can often be appreciated most readily in terms of a hexagonal or cubic close-packed lattice of the larger ion, with the smaller ion occupying cavities in the close-packed arrangement. For example, where three large anions are adjacent in a plane they define a triangular cavity which can accommodate a very small cation; where four are adjacent they define a tetrahedral cavity which will hold a somewhat larger cation, and where six are adjacent they define an octahedral cavity which will accommodate a still larger cation. Many examples can be given; for the present we limit ourselves to three. Thus, in BeO we have a hexagonal close-packed lattice of O^{2-} ions with Be^{2+} ions in tetrahedral cavities at the positions marked B on Fig. 1.7b. In

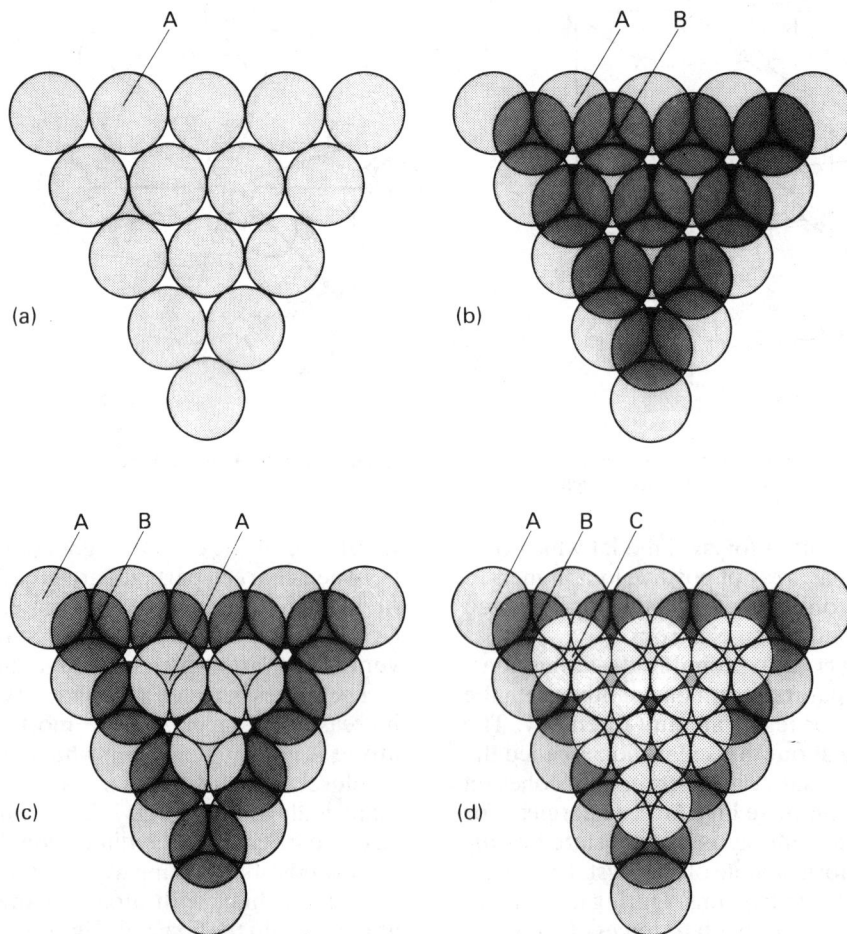

Fig. 1.7 (a) Close-packed layer of identical spheres in a plane. (b) Superposition of two close-packed layers. (c) Third layer added vertically above first, giving hexagonal closest packing. (d) Third layer added above gaps in first two layers, giving cubic closest packing

MgO, on the other hand, the Mg^{2+} ions occupy the unlabelled positions of Fig. 1.7b, i.e. they occupy octahedral cavities, in a cubic close-packed lattice of oxide ions. This is the 'rock salt' (NaCl) structure. In rock salt itself the Na^+ ions similarly occupy octahedral cavities in a cubic close-packed lattice of Cl^- ions. Other illustrations of these ideas will be given in later chapters.

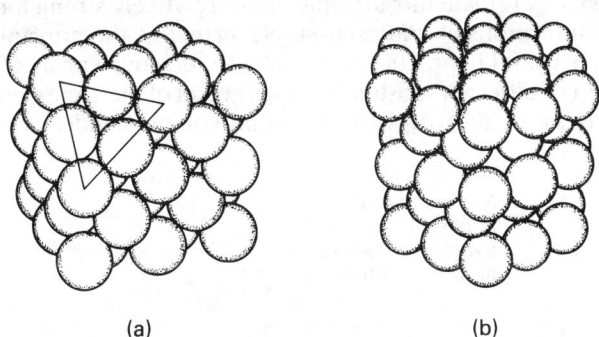

(a)   (b)

Fig. 1.8 (a) Cubic close packing, with corner atom removed to show part of a layer. (b) Hexagonal close packing

When a crystal melts the long-range order breaks down, but some short-range order persists in the liquid at temperatures right up to the boiling point. For a few substances a large amount of one- or two-dimensional order exists in the liquid. These are the fascinating materials known as liquid crystals. The breakdown of a lattice requires energy to be supplied in the form of the latent heat of fusion and usually leads to a less compact structure in the liquid, except for a few substances whose crystal lattice has a very open structure, notably water, antimony, and type-metal, for all of which the liquid is denser than the solid. The forces which formerly held the crystal together do not disappear in the liquid, but the thermal motion of the molecules causes these forces to lose their directional character. At the surface of the liquid there is an imbalance in the forces such that molecules in the surface are attracted towards the body of the liquid (Fig. 1.9). This is the origin of *surface tension*. The magnitude of the surface tension of a liquid is expressed by a coefficient, usually symbolized by the letter γ, which can be defined either as the energy required to form a unit area of surface, or the force exerted by the surface tension across a line of unit length located in the surface layer.

As a result of their thermal motion molecules have a tendency to escape from the surface of a liquid and enter the gas phase. Molecules which escape can be replaced by molecules from the gas phase which collide with the surface and so re-enter the liquid. If the number of molecules escaping exceeds the number returning there is a net evaporation of liquid; if the reverse holds there is a net condensation. The equilibrium gas pressure at which the rate of evaporation equals the rate of condensation is called the *vapour pressure* of the liquid. As the tendency for molecules in the surface to escape increases with increasing temperature, so does the vapour pressure. If a liquid is heated under a fixed external pressure the liquid as a whole boils at the temperature at which its vapour pressure equals the external pressure. At an external pressure of one atmosphere this temperature is the normal boiling point of the liquid. The molecules which escape tend to be those with the greatest amount of thermal energy; to replace this energy and keep the liquid boiling, heat has to be supplied. This is termed the latent heat of vaporization. If the external pressure is increased the temperature can be raised further without the liquid boiling; all that happens is that the density and surface tension of the liquid steadily decrease. However, this process cannot continue indefinitely, and at a certain well-defined temperature and pressure the density of the liquid and vapour become equal to one another and the surface tension falls to zero. This is the *critical point*, defined by the critical temperature T_c and pressure P_c. Above the critical point no liquid phase exists, so that the critical temperature of a gas represents an upper limit above which it is impossible to liquefy the gas by compression.

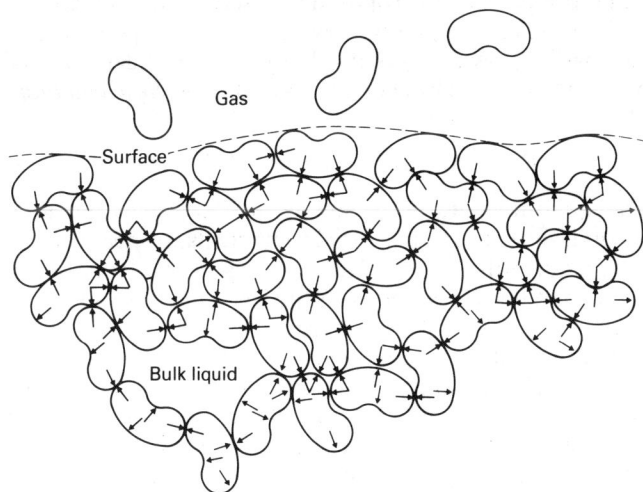

Fig. 1.9 Forces at the surface of a liquid; small arrows indicate the direction of attractive forces between molecules. Molecules in the surface experience a resultant force towards the bulk of the liquid, which gives rise to surface tension

In an ideal gas the molecules are so far apart that they collide only with the walls of the containing vessel, and both the molecular size and the forces of attraction and repulsion between molecules have no effect on the physical properties of the gas. Such a gas obeys the ideal gas law

$$PV = nRT \qquad (1.15)$$

where P is the pressure, V the volume, n the number of moles of gas, R the universal gas constant, and T the absolute temperature. For P measured in newtons per square metre, V measured in cubic metres, n in moles and T in kelvin, R has the value $8.318 \, J \, K^{-1} \, mol^{-1}$. All real gases liquefy at a sufficiently low temperature, so that at some point deviations from Eq. 1.15 must begin to appear. At low temperatures the Van der Waals' forces of attraction begin to add their effect to the applied pressure P. At the same time, as the volume decreases, the finite volume taken up by the molecules themselves begins to be comparable with the total volume of the container, V. An equation which makes specific allowance for these effects is Van der Waals' equation:

$$(P + a/V^2)(V - b) = nRT \qquad (1.16)$$

where the constant a is included to take account of the intermolecular forces and the constant b takes account of the volume which is excluded from V because it is occupied by molecules. The constants a and b can be related to the experimentally determined critical constants P_c, V_c and T_c of the gas. (V_c is the volume of a mole of gas at the critical temperature and pressure.) Many other semi-empirical equations have been proposed to describe the behaviour of real gases, each equation having its own proponents and each with its own region of validity. The most general type of equation which has been proposed is the *virial equation*, of which one form is

$$PV = A + B/V + C/V^2 + D/V^3 + \ldots \qquad (1.17)$$

The quantities A, B, C and D are known as the first, second, third and fourth virial coefficients, respectively. In practice a sufficient number of terms in the expansion is used to make PV agree with the experimental data. The right hand side of Eq. 1.17 reduces to just the term A, which must therefore be equal to nRT, when V is large. Surprisingly, this apparently quite empirical equation has a very adequate theoretical basis in terms of a model in which individual gas molecules are in equilibrium with short-lived clusters of two, three, or more molecules held together by Van der Waals' or other forces. The coefficient B then represents essentially the contribution to PV from binary clusters, C contains the contribution from clusters of three molecules, and so on. The virial equation is particularly useful at moderate gas densities, where the series in Eq. 1.17 converges rapidly, but becomes less useful as the liquid state is approached.

One further approach, which provides a convenient, rough guide for predicting the properties of dense gases and liquids, is through the *principle of corresponding states*. To use this principle we first define the *reduced* variables P/P_c, T/T_c, and V/V_c, i.e. we express the pressure, tempera-

ture and volume as fractions of their values at the critical point. The principle of corresponding states then says that all substances obey the same equation of state in terms of the reduced variables. If this principle were perfectly true the critical ratio, $P_c V_c/RT_c$, should have the same value for all substances; in fact it is found to vary from 0.30 for the inert gases to about 0.20 for substances such as water or CH_3CN (ethanenitrile) which have relatively strong forces of intermolecular attraction. Nevertheless the principle is very useful in practice, especially for dealing with a collection of different substances which are all of the same general type, e.g. all hydrocarbons, or all volatile halides of non-metals.

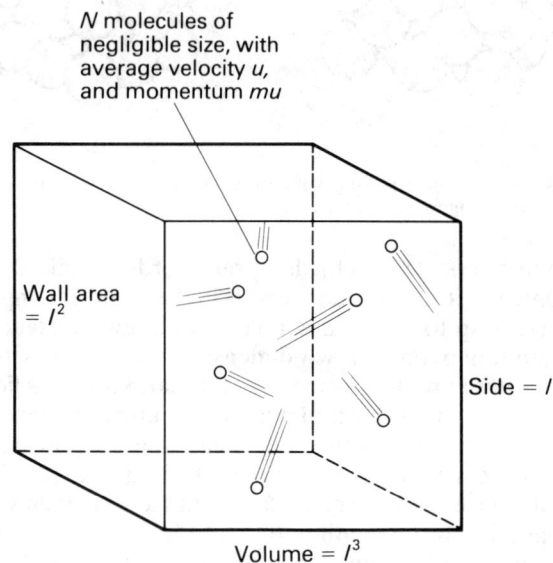

Fig. 1.10 Model used to discuss the properties of an ideal gas

Before we leave the topic of gases temporarily, it will be interesting to show how the equation of state for an ideal gas, Eq. 1.15, can be predicted on the basis of a model which represents the gas as a collection of molecules in rapid thermal motion, the molecules being considered so small that they collide only with the walls of the containing vessel. For a cubic vessel of side l metre the volume V is l^3 and the area of the walls is $6l^2$. In this vessel (Fig. 1.10) we suppose that there are N molecules moving in all directions at an average speed of u metres per second. If the different velocities of the molecules are resolved into their x, y and z components the average number of molecules moving parallel to one of the axes, say the x-axis, is $N/3$. If one of the three pairs of walls of the cube is considered to be at right angles to this axis, the average number of molecules moving with speed u towards each wall is then $N/6$; an equal number are moving towards the opposite wall. Therefore, since each molecule has to travel a distance l at speed u before striking the wall, the number of impacts on the wall per second is $Nu/6l$. As each molecule strikes the wall and bounces back its momentum changes from $+mu$ to $-mu$, so that the change of momentum per collision is $2mu$. It is this change of momentum which causes the molecules, like hailstones hitting a roof, to exert pressure on the wall of the cube. The total force felt by the wall is given by Newton's second law of motion as the rate of change of momentum at the wall, which in this case is equal to the number of impacts per second multiplied by the momentum change per impact, or $Nmu^2/3l$. The pressure P is then equal to the force divided by the area, which comes to $Nmu^2/3l^3$, or $Nmu^2/3V$. We therefore have the important result:

$$PV = Nmu^2/3 \qquad (1.18)$$

The kinetic energy of a molecule moving at velocity u is $\frac{1}{2}mu^2$, so that for the N molecules in our cube the stored kinetic energy is $\frac{1}{2}Nmu^2$. (Here for simplicity we ignore the difference between the mean value of u and the mean value of u^2.) The square of the quantity PV is seen to be proportional to this stored energy. We now *assert* that the amount of stored energy is proportional to the temperature T, so Eq. 1.18 can be rewritten as

$$PV = \text{constant} \times T \qquad (1.19)$$

which is of the same form as Eq. 1.15. We can make the two equations the same by identifying the constant in Eq. 1.19 as nR. A comparison of Eqs 1.15 and 1.18 shows that the energy stored as kinetic energy of translation in an ideal gas is $3RT/2$ per mole. Each molecule has three independent modes of translational motion, or translational 'degrees of freedom', corresponding to translation parallel to each of the three coordinate axes. Thus the amount of thermal energy stored by a mole of gas is $\frac{1}{2}RT$ per degree of freedom. We shall find a use for this result in Chapter 4.

Exercises

1.1 A mole of electrical charge amounts to 96 487 coulombs. Calculate the charge on an individual electron, given that Avogadro's number is 6.0225×10^{23}.

1.2 The mass of an electron is 1/1840 of the mass of the proton. Calculate the mass of a mole of electrons.

1.3 Using the Planck formula, Eq. 1.3, and the expression $c = \nu\lambda$, calculate the energy of a mole of photons of wavelength 500 nm in a beam of green light.

1.4 Using the Einstein formula, Eq. 1.4, calculate the mass of a mole of photons of wavelength 500 nm.

1.5 Calculate the mass of a mole of dust motes (average mass 0.1 μg).

1.6 Using the de Broglie relationship, Eq. 1.6, calculate the wavelengths of the waves associated with an electron, a water molecule, a dust mote (mass 0.1 μg), and a standard shot-putter's shot (mass 16 lbs = 7.3 kg), all at a particle velocity of 10^{-3} m s^{-1}.

1.7 Consider a particle of mass m moving in one dimension in a region where its potential energy V is zero. Show that a possible wave function for such a particle is

$$\Psi = A \sin \frac{(2\pi m\nu x)}{h} + B \cos \frac{(2\pi m\nu x)}{h}$$

where A and B are constants and ν is the velocity of the particle.

1.8 A particle moving in one dimension only is excluded from the regions $x < 0$ and $x > a$; thus a is the length of a 'one-dimensional box' holding the particle. Show that for this situation the wave function given in exercise 1.7 reduces to

$$\Psi_n = A \sin 2\pi nx/a$$

where n is an integer, and show that the allowed energy levels for the particle in the box are given by

$$E_n = n^2h^2/2ma^2$$

1.9 Calculate the root-mean-square velocity of oxygen molecules in air at 300 K.

1.10 Examine models of crystal lattices, observing the kinds of close-packed lattices of larger ions that are present and the nature of the cavities occupied by smaller ions within the close-packed arrangements.

2 Atomic Structure and the Periodic Table

2.1 Introduction

In this chapter we give more detailed consideration to a topic which was introduced in Section 1.2, namely the interaction between electrons and atomic nuclei. We recall that an atom consists of an extremely small, massive, positively charged nucleus composed of protons and neutrons, surrounded by a cloud of negatively charged electrons. For purposes of reference some of the properties of protons, neutrons and electrons have been summarized in Table 2.1. The behaviour of electrons is of fundamental significance in chemistry; for example, a chemical reaction consists basically of a redistribution of electrons among atoms, the distribution of electron clouds in space plays a major role in determining the shapes of molecules and ions, and chemical bonds are essentially the result of electrons being either shared between two atoms, in what is called a covalent bond, or completely transferred from one atom to another to produce charged species held together by an ionic bond. The chemical properties of an atom are entirely determined by the structure of its electron cloud.

Table 2.1 Some properties of the electron, proton and neutron

Particle	Mass	Charge	Spin	Magnetic moment
Electron	9.109×10^{-31}	-1	$\frac{1}{2}$	1
Proton	1.6725×10^{-27}	$+1$	$\frac{1}{2}$	1
Neutron	1.6748×10^{-27}	0	$\frac{1}{2}$	1

Units: *mass*: kg; *charge*: 1.60210×10^{-19} coulomb; *spin*: $h/2\pi$; *magnetic moment*: $eh/4\pi mc$ (where m is the mass of the particle in question). For m = electron mass, this unit is the Bohr magneton.

We noted in Section 1.3 that an electron in an atom or molecule is described by means of a wave function, usually designated by the Greek letter psi (Ψ), and has energy E, where Ψ and E are given by the Schroedinger equation:

$$H\Psi = E\Psi \qquad (2.1)$$

where H is the Hamiltonian operator for the electron. If Eq. 2.1 is written out in full for a single particle it takes the form of a second-order differential equation in three dimensions. (For a system of several particles the equation contains three coordinates for each particle.) The wave function Ψ for an electron contains within itself all the information there is to know about the electron it describes. This information is not necessarily as complete as one would like, since in general we obtain from Ψ not a single value of a quantity such as the position or momentum of the electron, but rather a probability distribution for the value of the quantity. By a suitable choice of function it may be possible to obtain a probability distribution with a very sharp peak at some particular value of one measurable quantity, but when this is done it always turns out that the value of some other measurable quantity becomes even more uncertain. This is the famous *uncertainty principle*, which was first expressed by Heisenberg in the form

$$\Delta p_x \Delta x \geq h/2\pi \qquad (2.2)$$

The inequality 2.2 states that for any particle the product of the uncertainties, i.e. the ranges of probable error, in simultaneously determined values of momentum in the x direction and position along the x coordinate, is greater than or equal to Planck's constant divided by 2π. The equality in 2.2 applies to a theoretical determination of p_x and x from the wave function; an experimental determination would necessarily give a worse result. Every conceivable experiment to determine p_x and x at the same time fails to do better than is allowed by the uncertainty principle. For example, if the position of a particle were to be determined by allowing it to scatter photons into an optical microscope, its momentum after the position determination would be uncertain because of the recoil produced by photon impact (cf. the Compton effect mentioned in Section 1.3). To reduce this uncertainty it would be necessary to reduce the photon energy, which would mean increasing the wavelength of the radiation being scattered. The position determination would then suffer because of the effect of diffraction, which prevents an object from producing an image smaller than the wavelength of the light which it scatters.

An uncertainty relationship similar to Eq. 2.2 applies to a number of other pairs of measurable quantities, the most important of these probably being energy E and time t.

It was mentioned in Section 1.3 that it is frequently impossible to solve the Schroedinger equation for systems of chemical interest, except by more or less approximate numerical methods. However, for a system which consists of a single particle, or whose motion is mathematically equivalent to the motion of a single particle, it is usually possible to obtain solutions to the Schroedinger equation in closed mathematical form. This means that we obtain wave functions expressed in the form of algebraic functions rather than as tables of numbers. Such a system is the hydrogen atom, which is mathematically equivalent to a single light particle moving in the field of a fixed nucleus. The solutions of the Schroedinger equation for atomic hydrogen are of great importance in chemistry because, apart from their own intrinsic interest, they provide us with our basic qualitative ideas about the shapes and energies of electron clouds in atoms and they serve as a starting point for numerous approximate calculations for more complex systems. As invariably happens with a bound system, where the motion of the particle is restricted to a particular region of space, acceptable wave functions for the atom are obtained only for certain values of the energy. In a bound system we always find discrete energy levels rather than a continuous range of energy. The fixed energy levels are defined by particular sets of values of three numbers n, l and m, called *quantum numbers*, which appear in the formula for the wave function and which are restricted to integral i.e. whole number, values. The next section is devoted to consideration of the forms of the wave functions and the pattern of energy levels which the Schroedinger equation reveals for a hydrogen atom.

2.2 Atomic Orbitals: the Hydrogen Atom

The wave function Ψ which describes the position of an electron in space is called an *orbital*. It is a rather complicated algebraic function, which can be written out most simply in terms of spherical polar coordinates r, θ and ϕ (Fig. 2.1) but can also be expressed in ordinary cartesian coordinates x, y and z. The detailed form of the algebraic function need not concern us now. The square of the wave function at a point gives the actual probability of finding the electron at that point, and can therefore be called the *electron density*. The various acceptable wave functions that can be obtained as solutions of the Schroedinger equation are distinguished by the values assigned to three quantum numbers, n, l and m. The quantum numbers serve both as convenient labels for the orbitals and as parameters in the mathematical expressions for the orbitals. The allow-

able combinations of these numbers which determine the orbitals are fixed in accordance with the following rules:

1 The *principal quantum number n* can take any positive integral value, thus

$$n = 1, 2, 3, 4, \ldots \qquad (2.3)$$

2 For given value of n, the *azimuthal quantum number*, or *angular momentum quantum number*, l can take any of the n values

$$l = 0, 1, 2, 3, \ldots \text{ to } (n - 1) \qquad (2.4)$$

3 For a given value of l, the *magnetic quantum number m* can take any of the $2l + 1$ values

$$m = -l, -(l - 1), \ldots, 0, \ldots (l -), l \qquad (2.5)$$

The energy of an electron in an orbital increases with increasing n; in the hydrogen atom, orbitals with the same n but different l or m all correspond to the same energy level.

With the aid of these rules it is possible to draw up a table of orbitals for the hydrogen atom, as shown in Table 2.2 for values of n from 1 to 4. To distinguish orbitals with different values of l we conventionally use the letters s, p, d, f, g etc., according to whether l is equal to 0, 1, 2, 3, 4 etc. Thus an orbital with $n = 2$ and $l = 1$ is labelled 2p.

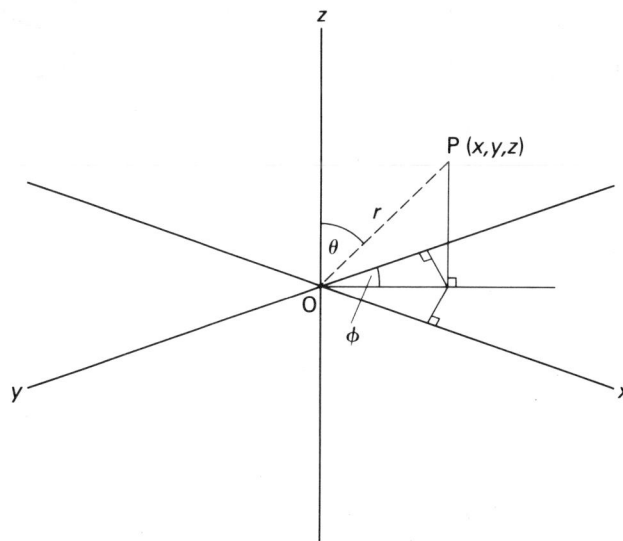

Fig. 2.1 Spherical polar coordinates (r, θ, ϕ) of a point P. In orinary cartesian coordinates P is (x, y, z), where $x = r \sin \theta \cos \phi$, $y = r \sin \theta \sin \phi$, and $z = r \cos \theta$

If we work in terms of coordinates x, y and z having their origin at the nucleus, the three p orbitals which we obtain for any value of n greater than 1 can be labelled p_x, p_y and p_z. The five d orbitals which we obtain when n is greater

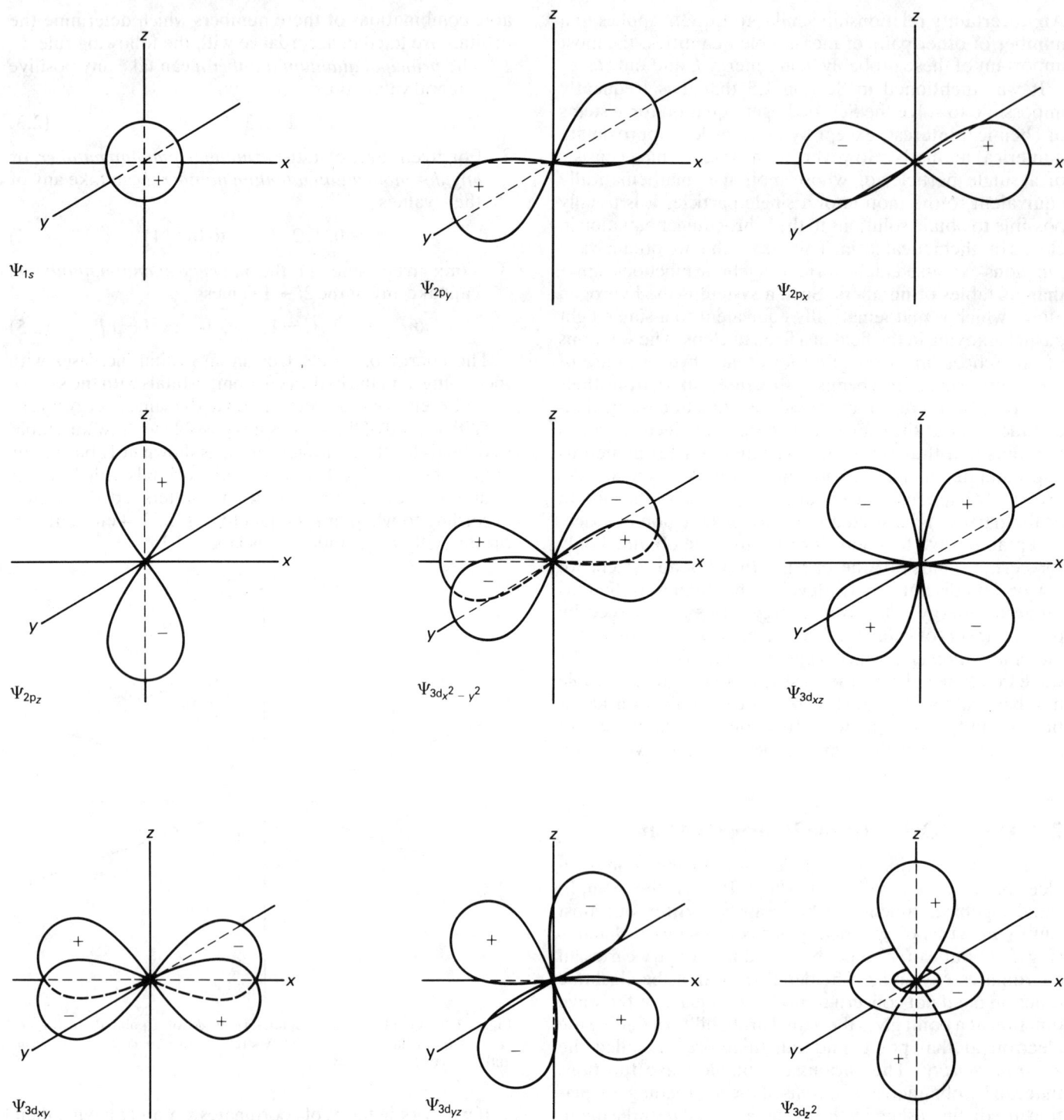

Fig. 2.2(a) Drawings to illustrate shapes of 1s, 2p and 3d orbitals; + and − signs indicate regions of space where Ψ is positive and negative, respectively

than 2 are similarly labelled d_{zy}, d_{yz}, d_{zy}, $d_{x^2-y^2}$ and d_{z^2}. Two-dimensional representations of these orbitals are given in Fig. 2.2a, where the line at the boundary of the orbital corresponds to an arbitrary surface within which there is some fixed probability, say 95%, of finding the electron. Similar diagrams can be drawn for Ψ^2; alternatively one can draw contours of electron density as in Fig. 2.2b. The electron density Ψ^2 is necessarily positive, but the wave function may be positive or negative, as shown by the + and − signs on the lobes in Fig. 2.2a. The surface on which $\Psi = 0$, between regions of positive and negative Ψ, is called a *node*. Orbitals with n greater than 1 have radial nodes, as shown in Fig. 2.2c. The presence of positive and negative regions of Ψ leads to the possibility of interference between different electron waves, as when two functions of opposite sign overlap to produce a region where the resultant electron density Ψ^2 is very small, or two functions of the same sign reinforce one another to produce a region where Ψ^2 is unusually large. Interference phenomena of this sort turn out to be very important in connection with chemical bonding, as we shall see in Chapter 3. The directional properties

of p and d orbitals are also important in relation to chemical bonding, because before electrons can be shared between atoms there must be a physical overlap of the orbitals in which the electrons reside. Thus for p and d orbitals, bonding will tend to be restricted to certain fixed directions in space.

Table 2.2 Quantum numbers and orbitals for the hydrogen atom ($n = 1, 2, 3$ and 4)

n	l	m	Orbital name
1	0	0	1s
2	0	0	2s
	1	$0, \pm 1$	2p
3	0	0	3s
	1	$0, \pm 1$	3p
	2	$0, \pm 1, \pm 2$	3d
4	0	0	4s
	1	$0, \pm 1$	4p
	2	$0, \pm 1, \pm 2$	4d
	3	$0, \pm 1, \pm 2, \pm 3$	4f

The energy levels for the different orbitals in a hydrogen atom are given by the expression

$$E_n = -\frac{2\pi^2 m e^4 Z^2}{n^2 h^2} \tag{2.6}$$

where the charge on the electron is − e, m is the mass of the electron (strictly the *reduced mass* $m_1 m_2/(m_1 + m_2)$ for the system electron plus proton), Ze is the charge on the nucleus, h is Planck's constant, and n is the principal quantum number. An identical expression was obtained by Niels Bohr prior to the advent of the Schroedinger equation, on the basis of a model in which the electron was treated as a classical particle in orbit around the nucleus, with the arbitrary restriction that the angular momentum was limited to values which were integral multiples of $h/2\pi$. The negative energies given by Eq. 2.6 correspond to formation of a *bound* system. in addition there is a whole range of positive energies corresponding to the allowable wave functions of a system in which the electron is not bound to the proton. These positive energy levels form a smooth continuum rather than a set of discrete levels.

The discrete energy levels given by Eq. 2.6 agree precisely with those that are found experimentally by studying the line spectrum of atomic hydrogen. The spectrum can be observed, for example, from an electrical discharge through the gas at low pressure. Transitions occur between levels with different values of n, as shown in Fig. 2.3. The energy difference $E_{n1} - E_{n2}$ is equal to $h\nu$ where $\nu = c/\lambda$ is the frequency of the emitted radiation. Experimentally we find

$$1/\lambda = R_H(1/n_2^2 - 1/n_1^2) \tag{2.7}$$

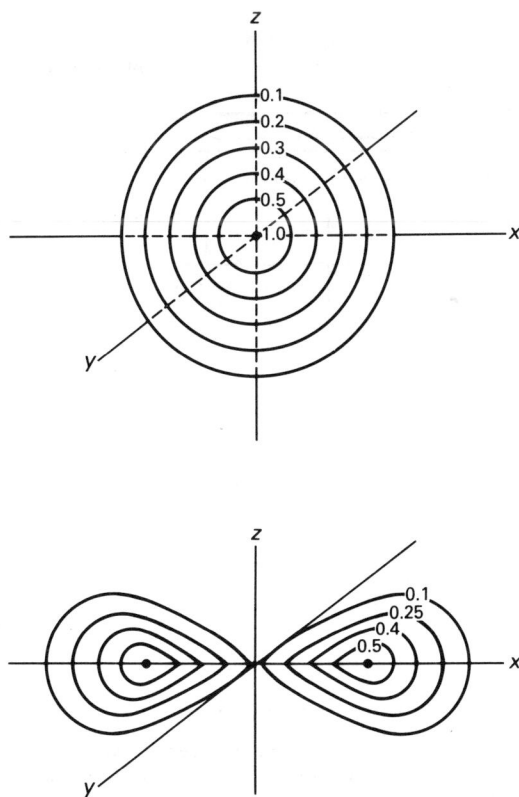

Fig. 2.2(b) Contours of relative electron density for 1s and $2p_x$ orbitals

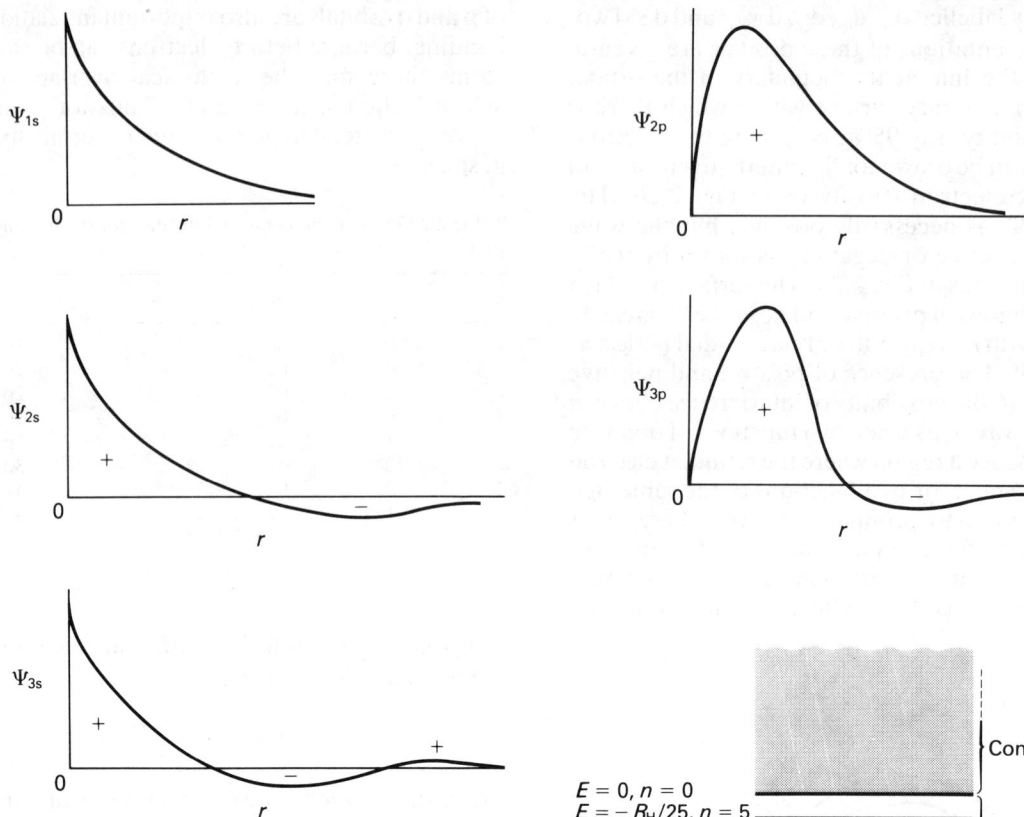

Fig. 2.2(c) Dependence of Ψ on the distance from the nucleus for 1s, 2s, 3s, 2p and 3p orbitals

where R_H is an empirical constant known as the Rydberg constant and n_1 is the larger of the two values of n involved. On the basis of Eq. 2.6 we should expect that R_H should be equal to $2\pi^2 me^4 Z^2/h^3 c$, which works out to $1.097\ 6777 \times 10^6\ m^{-1}$ when we substitute the best current values of the fundamental constants. This agrees very well with the experimental value of R_H; for example, the value obtained by Balmer in 1885 was $1.097\ 21 \times 10^6\ m^{-1}$. Good agreement with current experimental data would not be surprising, because the Rydberg constant is one of the experimental quantities whose precisely determined values are now used in fixing best values for fundamental constants such as e, h, m and c.

A noticeable feature of Eq. 2.6 is that the energy levels do not depend on the value of l, i.e. the energy levels having the same n but different values of l and m *coincide*, and are said to be degenerate. The degeneracy of each level E_n is seen to be equal to n^2, which we express in words by saying that the first level is non-degenerate, for $n = 2$ there is a four-fold degeneracy, for $n = 3$ there is a nine-fold degeneracy, and so on.

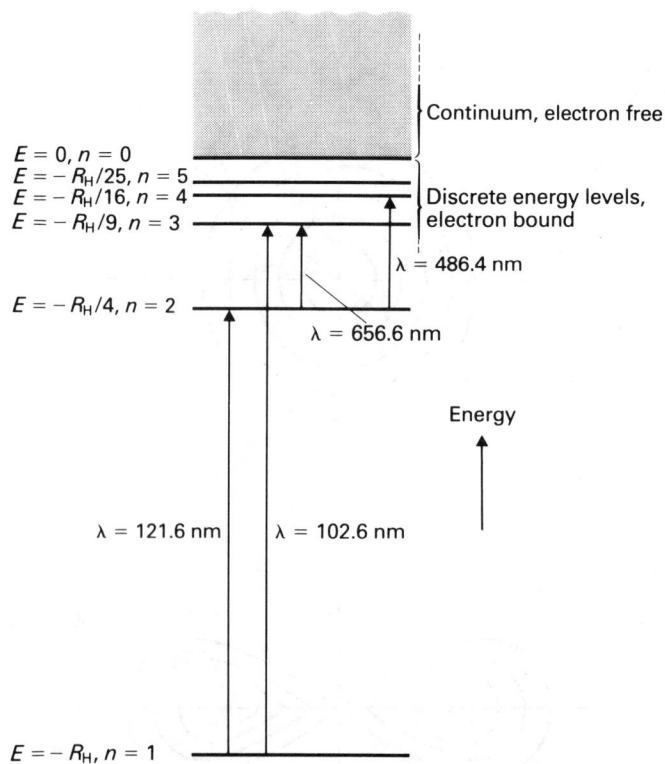

Fig. 2.3 Some energy levels of atomic hydrogen, showing transitions which give rise to characteristic spectral lines. The lines at wavelengths of 121.6 nm and 102.6 nm lie in the extreme ultraviolet and are absorbed by air. The line at 656.6 nm is in the red region of the spectrum, and that at 486.4 nm is in the blue-green

2.3 Atomic Orbitals in Many-Electron Atoms

For atoms more complex than hydrogen the Schroedinger equation has to be solved by approximate methods. Quite accurate results can be obtained by successive-approximation methods, whereby the result obtained by one calculation is used as the starting point for an improved calculation. The solutions so obtained show that the shapes of the orbitals of multi-electron atoms are essentially the same as for hydrogen, with small differences in the relative distances of regions of maximum electron density from the nucleus.

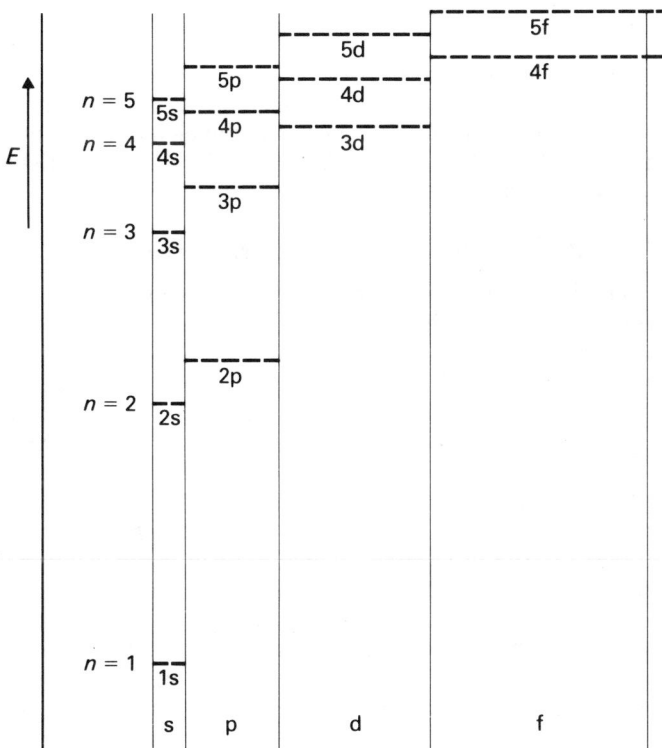

Fig. 2.4 Energy level diagram for a many-electron atom

The main difference between the results for many-electron atoms and those for hydrogen is that the energies of the orbitals now depend upon the quantum number l as well as on n. This means, for example, that the 2s and 2p orbitals no longer have the same energy. The reason is that when there are electrons in both 2s and 2p orbitals, the 2s electrons lie closer to the nucleus and are therefore more firmly held than the 2p electrons. We can say that the 2s electrons penetrate further into the atom than the 2p electrons; the order of decreasing penetration is s > p > d > f. The orbital energy level diagram for a many-electron atom is given in Fig. 2.4 . One of the most important features of this dia-

gram is that the 4s orbitals are of lower energy than the 3d orbitals. This has an important effect on the electron configurations of the atoms of elements with sufficient electrons to fill one but not both of these sub-shells, since electrons will enter the 4s sub-shell in preference to 3d.

Before we can consider the structure of many-electron atoms in detail it proves to be necessary to introduce a fourth quantum number s, associated with *electron spin*. This quantum number is not obtained from the solution of the Schroedinger equation, but is introduced to take account of the experimental result that the electron has an intrinsic magnetic moment (cf. Table 2.1). This magnetic moment is associated with an internal angular momentum which we can regard somewhat loosely as being analogous to the spinning of a macroscopic body, such as a tennis ball, about its own axis. The angular momentum has the value ½ in units of $h/2\pi$; the spin quantum number s (strictly it should be m_s) takes one of the two values $+ ½$ and $- ½$, corresponding to the two possible orientations of this angular momentum vector with respect to some external field. Thus s is related to the spin in the same way that m is related to l. The two possible values of s are often represented in diagrams by arrows or half-arrows pointing in opposite directions, thus: $\uparrow$ and $\downarrow$.

We can now build up the structure of a many-electron atom by placing electrons successively in higher energy orbits until the number of electrons equals the atomic number Z. In doing this we have to observe two rules, namely:

1 *The Pauli Exclusion Principle*   In its simplest form this rule states that no two electrons can have the same set of quantum numbers. This means that each atomic orbital defined by a set of values of n, l and m can hold a maximum of two electrons, and when two electrons are present in the same orbital they must have opposite spin. Thus for a helium atom, which has the electron configuration $(1s)^2$, the quantum numbers for the two electrons are $(n, l, m, s) = (1, 0, 0, + ½)$ and $(1, 0, 0, - ½)$ and there are no other possibilities for $n = 1$.

2 *Hund's Rule*   This rule says that where we have a degenerate set of orbitals (e.g. $2p_x$, $2p_y$, $2p_z$) the orbitals first accept one electron each, all with the same spin, and only after the sub-shell is half-filled will electrons begin to pair up;

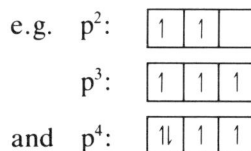

e.g. p^2: $\boxed{\uparrow\,|\,\uparrow\,|\,}$

p^3: $\boxed{\uparrow\,|\,\uparrow\,|\,\uparrow}$

and p^4: $\boxed{\uparrow\downarrow\,|\,\uparrow\,|\,\uparrow}$

An alternative statement of this rule is that for a given electron configuration the lowest energy state is the one with the highest total spin. With these two rules the electronic configuration of any atom in its lowest energy state (the *ground state*) may be arrived at by a simple building-up process, sometimes referred to as the *aufbau* principle. Fig. 2.5 shows the electron configurations of a selected range of commonly encountered elements; Appendix 2 contains a complete list of electron configurations for all of the elements in their ground states. The relatively late filling of 3d and 4f orbitals is a consequence of the higher energies of these orbitals in comparison with inner sub-shells having the next higher value of the principal quantum number.

Within a configuration, a set of electrons which all have the same principal quantum number is called a *shell*; those with the same value of both l and n form a *sub-shell*. An incomplete outer shell of electrons is called a *valence shell*. To reduce the labour of writing out the complete electron configuration for the heavier elements it is a common practice to label the collection of filled shells inside the valence shells according to the inert gas whose configuration they correspond to. Thus for antimony, whose configuration is written in full as:

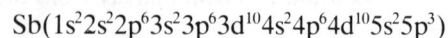

$$Sb(1s^2 2s^2 2p^6 3s^2 3p^6 3d^{10} 4s^2 4p^6 4d^{10} 5s^2 5p^3)$$

we can write $Sb([Kr\ core]4d^{10}5s^2 5p^3)$

As for hydrogen, transitions of electrons between the valence shells and unoccupied outer orbitals give rise to the observed visible and ultraviolet line spectra of the elements. Transitions involving the inner electrons give rise to characteristic X-ray spectra.

2.4 The Periodic Table of the Elements

The list of electron configurations of the elements leads us naturally to consider the Periodic Table, which is a table of elements in order of increasing atomic number, arranged in such a way that chemically related elements, which means in effect elements with similar valence shell configurations, are brought together. Such a table can take various forms; compact block and spiral forms have been advocated by different authors. For many purposes the simplest and most informative version is the so-called long form, shown in Fig. 2.6. In this table the elements are grouped in blocks, called the s, p, d and f blocks, in which the outermost filled or partly filled orbitals are the s, p, d and f orbitals, respectively. The f block includes the 4f series of lanthanide ('rare earth') elements and the 5f series of fourteen actinide elements, of which uranium is the best known member. An alternative classification separates the elements into metals and non-metals, as in Fig. 2.7.

Elements in a *group* lie vertically above and below one another in the table. They have the same outer electron configuration, or valence shell, and therefore can be expected to have rather similar properties. For example, consider the group IV elements, carbon, silicon, germanium, tin and lead, which resemble each other chemically, and also show an interesting steady trend, from non-metallic to metallic properties, with increasing atomic number. An even closer family resemblance is shown by the halogens, fluorine, chlorine, bromine and iodine in group VII, which show the same trend towards increasing metallic character with increasing atomic number. Similar generalizations can be made about other typical groups, such as the alkali metals in group I, the oxygen family in group VI, or the noble

	1s	2s	2p			3s	3p			4s	3d					4p		
He	↑↓																	He
Li	↑↓	↑																Li
Be	↑↓	↑↓																Be
B	↑↓	↑↓	↑															B
C	↑↓	↑↓	↑	↑														C
N	↑↓	↑↓	↑	↑	↑													N
O	↑↓	↑↓	↑↓	↑	↑													O
F	↑↓	↑↓	↑↓	↑↓	↑													F
Ne	↑↓	↑↓	↑↓	↑↓	↑↓													Ne
Na	↑↓	↑↓	↑↓	↑↓	↑↓	↑												Na
Mg	↑↓	↑↓	↑↓	↑↓	↑↓	↑↓												Mg
Al	↑↓	↑↓	↑↓	↑↓	↑↓	↑↓	↑											Al
Si	↑↓	↑↓	↑↓	↑↓	↑↓	↑↓	↑	↑										Si
P	↑↓	↑↓	↑↓	↑↓	↑↓	↑↓	↑	↑	↑									P
S	↑↓	↑↓	↑↓	↑↓	↑↓	↑↓	↑↓	↑	↑									S
Cl	↑↓	↑↓	↑↓	↑↓	↑↓	↑↓	↑↓	↑↓	↑									Cl
Ar	↑↓	↑↓	↑↓	↑↓	↑↓	↑↓	↑↓	↑↓	↑↓									Ar
K	↑↓	↑↓	↑↓	↑↓	↑↓	↑↓	↑↓	↑↓	↑↓	↑								K
Ca	↑↓	↑↓	↑↓	↑↓	↑↓	↑↓	↑↓	↑↓	↑↓	↑↓								Ca
Sc	↑↓	↑↓	↑↓	↑↓	↑↓	↑↓	↑↓	↑↓	↑↓	↑↓	↑							Sc
Ti	↑↓	↑↓	↑↓	↑↓	↑↓	↑↓	↑↓	↑↓	↑↓	↑↓	↑	↑						Ti
V	↑↓	↑↓	↑↓	↑↓	↑↓	↑↓	↑↓	↑↓	↑↓	↑↓	↑	↑	↑					V
Cr	↑↓	↑↓	↑↓	↑↓	↑↓	↑↓	↑↓	↑↓	↑↓	↑	↑	↑	↑	↑	↑			Cr
Mn	↑↓	↑↓	↑↓	↑↓	↑↓	↑↓	↑↓	↑↓	↑↓	↑↓	↑	↑	↑	↑	↑			Mn
Fe	↑↓	↑↓	↑↓	↑↓	↑↓	↑↓	↑↓	↑↓	↑↓	↑↓	↑↓	↑	↑	↑	↑			Fe
Co	↑↓	↑↓	↑↓	↑↓	↑↓	↑↓	↑↓	↑↓	↑↓	↑↓	↑↓	↑↓	↑	↑	↑			Co
Ni	↑↓	↑↓	↑↓	↑↓	↑↓	↑↓	↑↓	↑↓	↑↓	↑↓	↑↓	↑↓	↑↓	↑	↑			Ni
Cu	↑↓	↑↓	↑↓	↑↓	↑↓	↑↓	↑↓	↑↓	↑↓	↑	↑↓	↑↓	↑↓	↑↓	↑↓			Cu
Zn	↑↓	↑↓	↑↓	↑↓	↑↓	↑↓	↑↓	↑↓	↑↓	↑↓	↑↓	↑↓	↑↓	↑↓	↑↓			Zn
Ga	↑↓	↑↓	↑↓	↑↓	↑↓	↑↓	↑↓	↑↓	↑↓	↑↓	↑↓	↑↓	↑↓	↑↓	↑↓	↑		Ga
Ge	↑↓	↑↓	↑↓	↑↓	↑↓	↑↓	↑↓	↑↓	↑↓	↑↓	↑↓	↑↓	↑↓	↑↓	↑↓	↑	↑	Ge

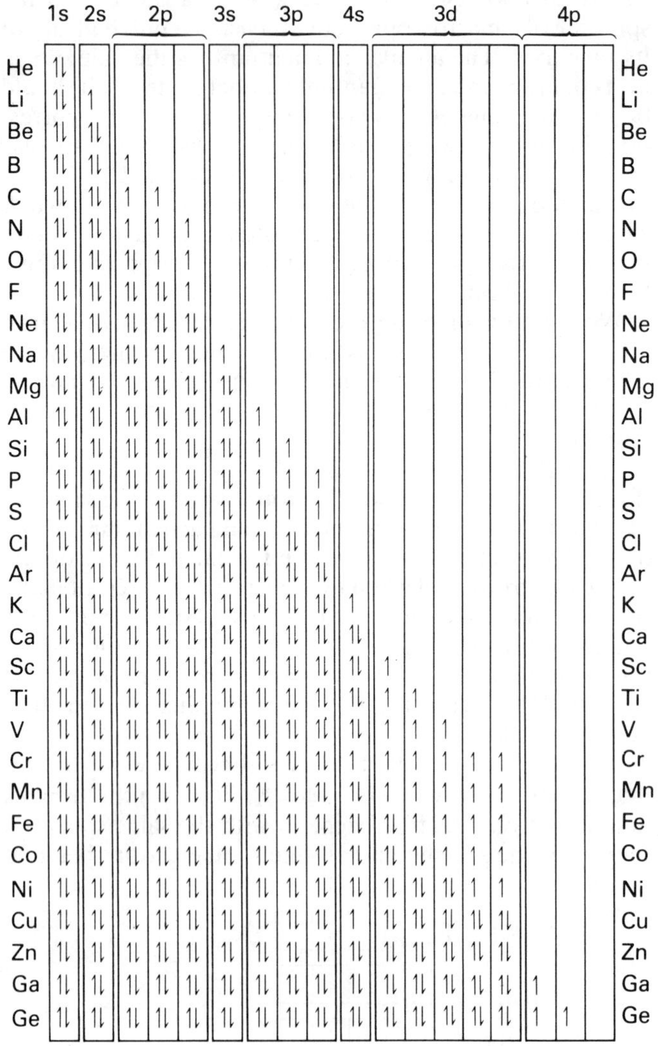

Fig. 2.5 Electron configurations of some elements. Note the extra stability of a half-filled d sub-shell for chromium, and a filled d sub-shell for copper

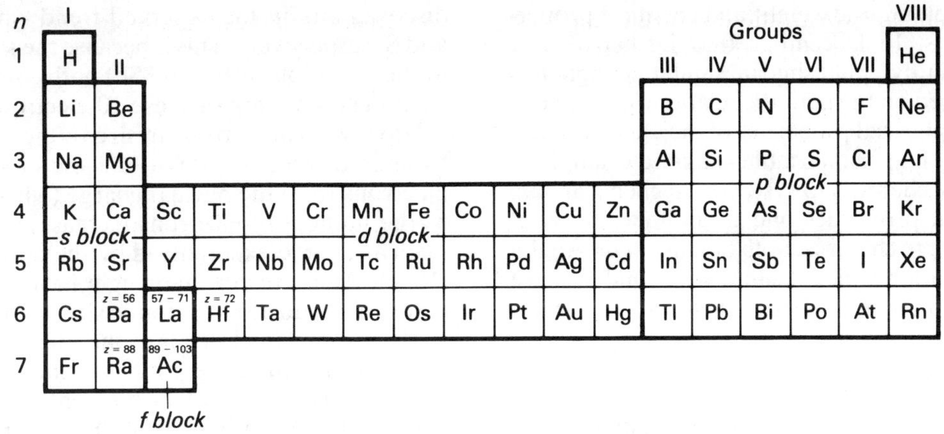

Fig. 2.6 Elements grouped according to the outermost filled or partly filled orbitals

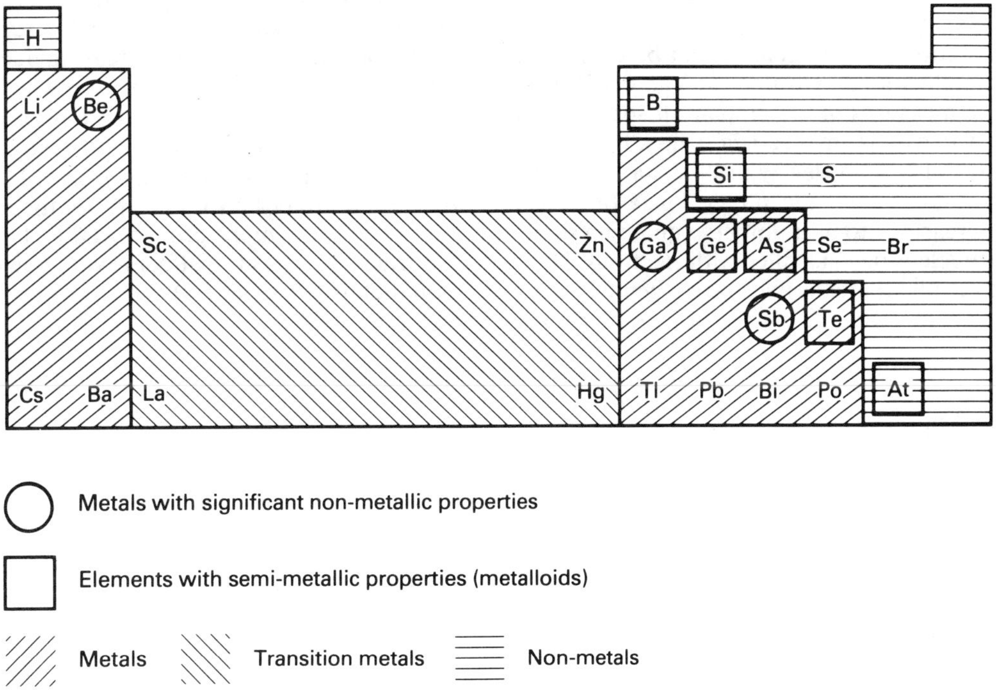

○ Metals with significant non-metallic properties

□ Elements with semi-metallic properties (metalloids)

▧ Metals   ▨ Transition metals   ▤ Non-metals

Fig. 2.7 Metallic or non-metallic character of the elements

gases in group VIII. The increase in atomic number as we move down a group corresponds to the addition of filled shells of electrons inside the valence shell. Thus, on going from C to Si, Z increases by 8 due to adding s^2p^6 and from Si to Ge, Z rises by 18 due to adding $s^2p^6d^{10}$.* From Ge to Sn there is a similar increase of 18, and from Sn to Pb there is a further increase of 32, the additional completed shells being $(5s^25p^65d^{10}4f^{14})$.

Elements in a *period* lie in the same horizontal line of the table; for example, the first short period contains Li, Be, B, C, N, O, F, Ne, each of which contains one more electron than the previous member of the period.

The periodic table was discovered long before anything was known about electron configurations, indeed long before the electron itself was discovered. In March 1869 Dmitri Mendeleeff published a table which he based on the periodic

* The p orbitals are not all equivalent of course: going from C to Si adds $p^2s^2p^4$, and from Si to Ge adds $p^2s^2p^4d^{10}$.

relationship between atomic weight and chemical properties of the elements. In December 1869 Lothar Meyer, working independently, published a similar though less complete table, based on the periodic relationship between atomic weight and physical properties of the elements such as atomic volume. The periodic table is a good example of the efforts made by chemists to systematize chemical data and relate different properties one to another and is of immense importance to chemists, both because of the wealth of chemical information which it summarizes and because of its usefulness as a guide in further chemical research.

2.5 Atomic Properties and the Periodic Table

A variety of atomic properties show interesting trends that can be discussed in terms of the periodicity of the elements. We now consider a selection of these.

(a) Atomic Size

The size of the atoms of an element can be expressed in terms of its metallic, ionic, covalent or Van der Waals' radius. These are measures of the amount of space taken up by an atom when it is involved in a metallic, ionic, or covalent bond, or a Van der Waals' crystal, respectively. As one would expect, the physical size of atoms increases with increasing atomic number as quantum shells further from the nucleus become filled with electrons. Data for two groups, the alkali metals and the oxygen family of group VI, are shown in Fig. 2.8. In each case there is a

discontinuity in the observed trend which occurs after K and S, respectively. This is because the subsequent elements in the two groups (Rb and Se) both contain a filled 3d subshell. The addition of these 10 electrons corresponds to a relatively small increase in size of the atoms because electrons in d orbitals are not efficient at screening the outer electrons from the nuclear charge (which is also increasing as the number of electrons increase). Hence the effective nuclear charge experienced by the outermost electrons of Rb and Se is greater than might have been expected. This leads to a reduction in atomic size as the negative electron shells are pulled closer into the nucleus. The influence of this contraction in the p block elements of the 3rd period (Ga to Kr) and the s block elements of the 4th period (Rb and Sr) has been called the 'middle row anomaly'. The smaller size of the atoms of these elements has an influence on properties such as their ionization energies, reduction potentials, and electronegativities, all of which affect their chemical properties in relation to the regular group trends.

Inefficient screening of increasing nuclear charge by the added electrons is also the reason why there is a contraction of size as we travel across any of the periods of the periodic table; for example, on going from Li to F the covalent radius falls steadily from 123 to 64 picometres (0.123 nm to 0.064 nm). The greater the number of elements in a period the more pronounced is the final effect of the contraction. Thus the effect is very marked for the transition metals (10 members) and is greatest for the lanthanides (14 members).

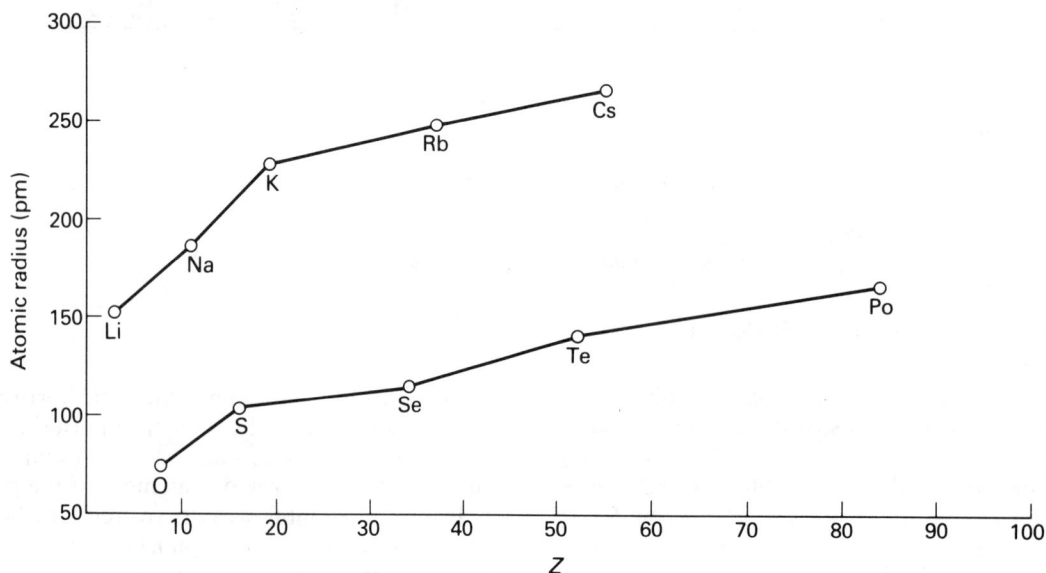

Fig. 2.8 Atomic radius versus atomic number for alkali metals (metallic radius) and oxygen group (covalent radius)

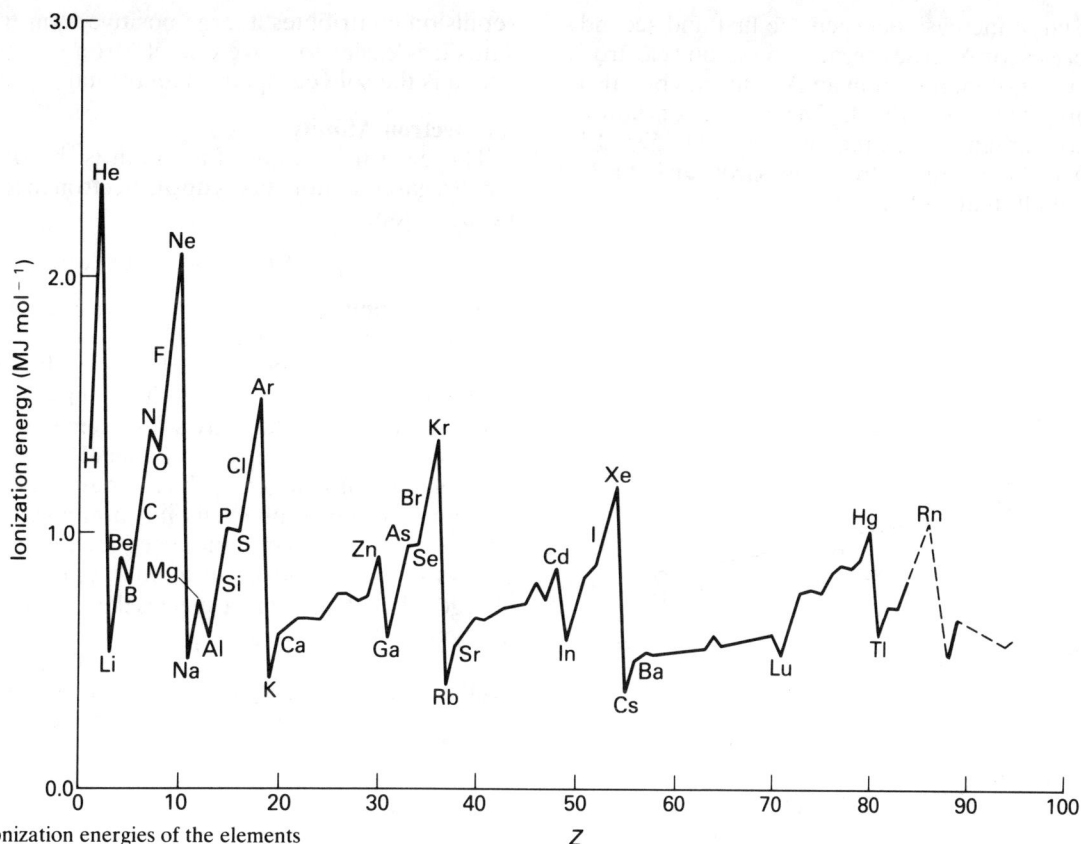

Fig. 2.9 First ionization energies of the elements

(b) Ionization Energy

The ionization energy (formerly 'ionization potential') of an element is the energy which must be absorbed to remove an outer electron from a gaseous atom or from a gaseous ion:

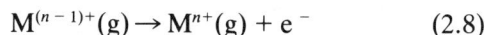

$$M^{(n-1)+}(g) \rightarrow M^{n+}(g) + e^- \qquad (2.8)$$

If $n = 1$ the energy is called the first ionization energy, if $n = 2$ the second ionization energy, and so on. The ease of removal of an electron depends on the strength of the attraction between electron and nucleus. Therefore the largest atoms have the smallest ionization energies, because their outermost electrons are furthest removed from the influence of the nucleus. This means that, with a few exceptions that are to be discussed below, ionization energies increase as we move from left to right across a period, or upwards in a group. This is seen in Fig. 2.9, which shows that the alkali metals have lowest first ionization energies while the noble gases, with their filled (ns^2np^6) sub-shells, have the highest first ionization energies. Trends in ionization energies down a group are influenced by the middle row anomaly as shown in Fig. 2.10; the first ionization energies of Rb and Se are clearly greater than expected.

Second, third, etc. ionization energies are as easy to define but the corresponding ions are not always observable outside a discharge tube. This is because the energy required to remove an electron from an ion becomes greater as the positive charge increases on the ion. An example is given in Table 2.3. The total energy for a particular ionization is the sum of the energies for that and all previous ionizations; for example, to bring about the process

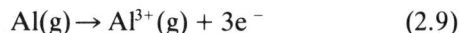

$$Al(g) \rightarrow Al^{3+}(g) + 3e^- \qquad (2.9)$$

the energy required is 5138 kJ mol^{-1} which is the sum of the first, second and third ionization energies.

Table 2.3 Ionization energies for aluminium

Species	Electron Configuration	Energy to produce species (kJ mol^{-1})
Al	$(1s)^2(2s)^2(2p)^6(3s)^2(3p)^1$	0
Al$^+$	$(1s)^2(2s)^2(2p)^6(3s)^2$	577
Al^{2+}	$(1s)^2(2s)^2(2p)^6(3s)^1$	1 816
Al^{3+}	$(1s)^2(3s)^2(2p)^6$	2 745
Al^{4+}	$(1s)^2(2s)^2(2p)^5$	11 575
Al^{5+}	$(1s)^2(2s)^2(2p)^4$	14 840

The rather large increase between the first and second ionization energies for Al arises because the second electron is being removed not merely from an Al^+ species but also from the more stable 3s orbital. The fourth ionization energy for aluminium is extremely large (11 575 kJ mol^{-1}), because the fourth electron comes from an entirely new quantum shell, that with $n = 2$.

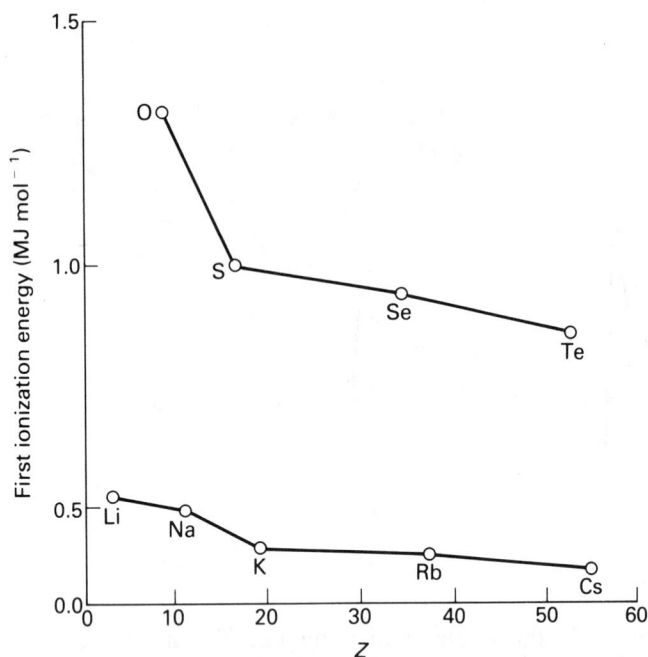

Fig. 2.10 1st ionization energies for alkali metals and elements of the oxygen group

The increase in the first ionization energy with atomic number (Fig. 2.9) is not uniform. In particular, the values for boron and oxygen are noticeably lower than one might have expected. Boron's value is less than that for beryllium because the electron that has to be removed is the single 2p electron, which is less firmly held than the pair of 2s electrons in beryllium. The value for oxygen is lower than for nitrogen because the electron that is removed is one of the paired electrons in a 2p orbital:

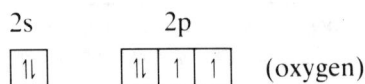

2s   2p

| ⇅ | | ⇅ | ↑ | ↑ | (oxygen)

How this works can be understood if we recall that it is implied by Hund's rule that a pair of electrons in the same orbital have higher energy than if they were unpaired in different orbitals. This is because being in the same orbital keeps them close together in space, so that electron–electron

repulsion contributes a large positive term to the energy. Thus it is easier to move one electron of a pair than one which is the sole occupant of an orbital.

(c) Electron Affinity

The electron affinity of an atom is the *energy released* when a gaseous atom takes up an electron to form a gaseous negative ion.

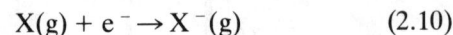

$$X(g) + e^- \rightarrow X^-(g) \qquad (2.10)$$

Electron affinity increases as we move towards the top right hand corner of the periodic table, provided we do not move far enough to be among the noble gases. It is more difficult to determine electron affinity than ionization energy, and reliable values exist for only a few elements, notably the p block elements of groups V, VI and VII. Electron affinities are greatest for halogens, which require only one electron to complete the p sub-shell and form a stable halide ion. The addition of two or more electrons to a gaseous ion X^- always involves absorption of energy, because a negative charge is being added to a species that is already negatively charged.

Table 2.4 Pauling electronegativities of s and p block elements

H 2.1						
Li 1.0	Be 1.5	B 2.0	C 2.5	N 3.0	O 3.5	F 4.0
Na 0.9	Mg 1.2	Al 1.5	Si 1.8	P 2.1	S 2.5	Cl 3.0
K 0.8	Ca 1.0	Ga 1.6	Ge 1.8	As 2.0	Se 2.4	Br 2.8
Rb 0.8	Sr 1.0	In 1.7	Sn 1.8	Sb 1.9	Te 2.1	I 2.5
Cs 0.7	Ba 0.9	Tl 1.8	Pb 1.9	Bi 1.9	Po 2.0	At 2.2

(d) Electronegativity

The electronegativity of an element is a measure of its ability to attract electrons to itself in a compound or in a bond. It is a property which depends very much on the particular bond, i.e. the electronegativity of an element can vary with its environment and oxidation state. The accepted electronegativity values of 4.0 for fluorine and 3.0 for chlorine indicate simply that in the bonds X—F and X—Cl to some element X, the fluorine atom always acquires a greater share of the bonding electrons than does the chlorine atom. Electronegativity really cannot be considered much more precisely than this. A number of methods have been used to estimate electronegativities; the best known is that due to Pauling, based on bond strength data (Table 2.4). More recent electronegativity estimates agree in general with Pauling's table.

Electronegativity is an empirical quantity and the values are significant only in relation to one another, so the scale used is quite arbitrary. The electronegativity of the elements increases as we move across the periods and up the groups. Thus it shows the same trends as electron affinity, but the two quantities should not be confused.

(e) Oxidation States

The chemical properties of an atom vary in a systematic way with the extent to which it has gained or lost electrons, i.e. with its state of combination. For example, a fluorine atom in F_2 is very reactive, whereas the fluorine atom in NaF is quite inert. The state of combination of an element is conveniently described in terms of its *oxidation state*, or *oxidation number*. The terms are virtually interchangeable; one speaks of the $+2$ oxidation state or of the state in which the oxidation number is $+2$. The following rules are used to determine the value of the oxidation number for any element:

(1) The free elements have a zero oxidation number, e.g. S in S_8, Cl in Cl_2.

(2) Monatomic ions have an oxidation number equal to the charge they carry, e.g. calcium in the ion Ca^{2+} is in the $+2$ oxidation state, suphur in the sulphide ion S^{2-} is in the -2 oxidation state.

(3) In their compounds the alkali metals (group I) always exhibit the oxidation state $+1$ and the alkaline earth metals (group II) exhibit only the oxidation state $+2$.

(4) Fluorine always has the oxidation number -1, except in F_2 where it is zero in accordance with rule 1.

(5) The oxidation state of oxygen is always -2 except in peroxides (value -1), superoxides ($-\frac{1}{2}$), and the compound OF_2 ($+2$).

(6) The oxidation state of hydrogen in its compounds is generally $+1$, except in the ionic hydrides (alkali metal and alkaline earth metal hydrides) and in related compounds such as $LiAlH_4$, where it is -1.

(7) For a neutral compound the algebraic sum of the oxidation numbers of the constituent atoms is zero; for an ion the sum is equal to the charge on the ion.

Examples of how these rules can be used to determine oxidation states are given in Table 2.5.

Table 2.5 Examples of the assignment of oxidation states to elements in compounds and ions

Species	Element	Oxidation number	Rule invoked
FNO	F	-1	4
	O	-2	5
	N	$+3$	7
$HClO_4$	O	-2	5
	H	$+1$	6
	Cl	$+7$	7
$Cr_2O_7^{2-}$	O	-2	5
	Cr	$+6$	7
NaO_2	Na	$+1$	3
	O	$-\frac{1}{2}$	7
P_4	P	0	1
$LiAlH_4$	Li	$+1$	3
	H	-1	6
	Al	$+3$	7
$KMnO_4$	K	$+1$	3
	O	-2	5
	Mn	$+7$	7
SO_3^{2-}	O	-2	5
	S	$+4$	7

Transition metals (d block elements) are noted for their ability to exist in a wide range of oxidation states. Trends in the oxidation states of d block elements are discussed in Chapter 12. The s and p block elements show striking *periodic* variations in their oxidation states, as illustrated in Fig. 2.11. Elements of groups I and II have only one oxidation state, which is given by the group number ($+N$); elements in the remaining groups can show other oxidation states in addition to $+N$. The positive oxidation states of non-metals occur when they are bonded to a more electronegative element. For example, chlorine in the oxyanions ClO^-, Cl_2^-, ClO_3^- and ClO_4^- has oxidation number $+1$, $+3$, $+5$ and $+7$, respectively. For elements in groups III to VIII the negative oxidation states ($N-8$) may also occur.

Elements which have more than one oxidation state e.g. sulphur (-2, $+2$, $+4$ and $+6$) and arsenic (-3, $+3$, $+5$) have in their lower positive oxidation states one or more pairs of valence electrons which are not involved directly in bonding. These are called 'inert pairs'. An inert pair becomes more stable as a group is descended; for example, the principal oxidation state of germanium is $+4$ whereas for lead it is $+2$. The term 'inert pair' is a name given to the phenomenon of an element existing in a low oxidation state with an unused pair of electrons, and is not to be confused with an explanation of the phenomenon. The explanation involves a complex compromise between ionization energies, crystal lattice energies, and electron promotion energies.

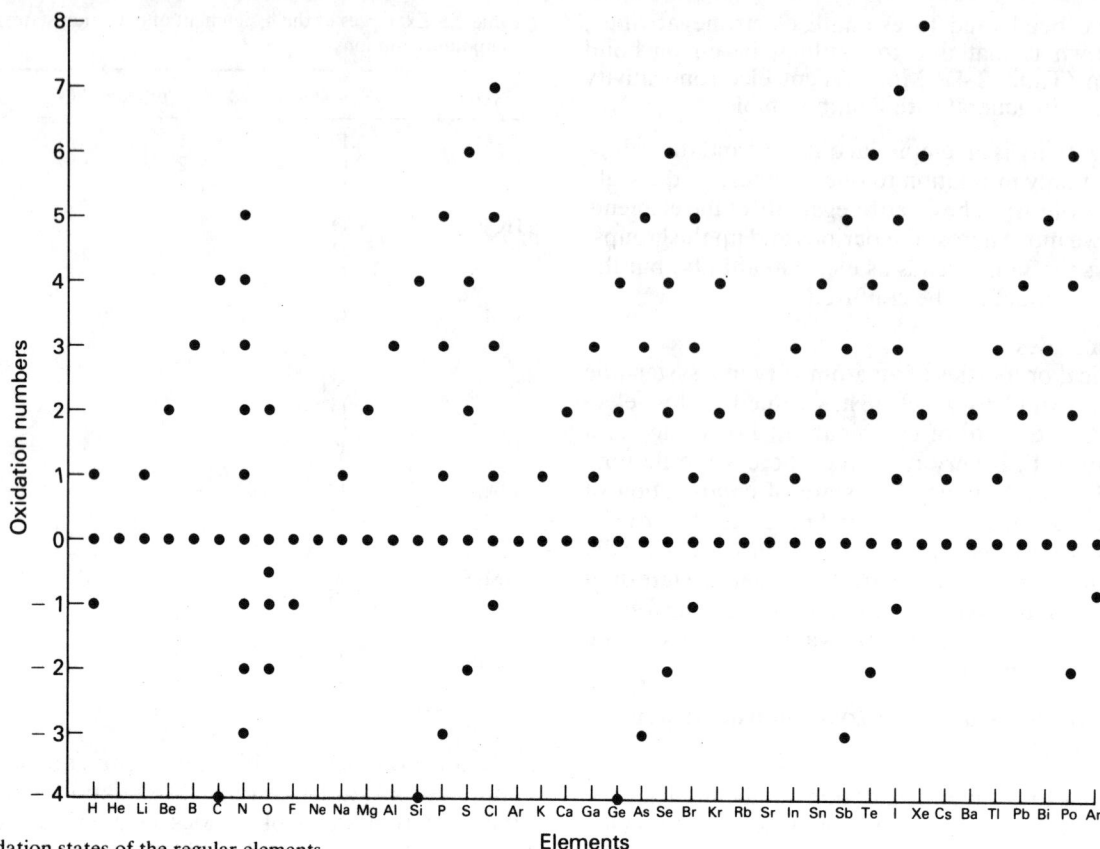

Fig. 2.11 Oxidation states of the regular elements

2.6 Uses of the Periodic Table

When Mendeleeff originally proposed his periodic table a number of gaps existed and he was able to use the table to predict the physical and chemical properties of the missing elements. For example, his predictions for the then unknown element germanium are compared with the element's subsequently observed properties in Table 2.6. There are now no gaps left in the body of the table, but it is still possible to predict, for example, that element 105 would be the first member of the fourth transition metal series. It has been suggested that there might be 'islands of stability' for elements of atomic number near 114 and 126. The periodic table allows us to predict the properties of these elements, and thereby should help in the search for them. The principles employed by Mendeleeff to predict the properties of missing elements are still used by chemists to predict whether or not an element will form a particular type of compound. From the opposite viewpoint, much of the interest excited by the initial investigations of boron and silicon hydrides arose because chemists were curious to see how closely their chemistry resembled that of the hydrides of carbon.

Table 2.6 Mendeleeff's predictions of the chemistry of 'eka-silicon' (germanium) and the experimental results

Predictions for eka-silicon (Es)	*Property*	*Found for germanium* (Ge)
71	Relative atomic mass	73.32
5.5	Specific gravity	5.47 (20°C)
13 cm^3	Atomic volume	13.22 cm^3
Dirty grey	Colour	Grey-white
0.073	Specific heat	0.076
White EsO$_2$	Result of heating in air	White GeO$_2$
Slight	Action of acids	None by HCl
EsO$_2$ + C K$_2$EsF$_6$ + Na	Preparation	GeO$_2$ + C K$_2$GeF$_6$ + Na
Refractory Sp.Gr. 4.7 Mol.Vol. 22 cm^3	Properties of dioxide	Refractory Sp.Gr. 4.703 Mol.Vol. 22.16 cm^3
B.Pt 100°C Sp.Gr. 1.9 Mol.Vol. 113 cm^3	Properties of tetrachloride	B.Pt 86°C Sp.Gr. 1.887 Mol.Vol. 113.35 cm^3

Exercises

2.1 The momentum of an electron moving with a velocity of one metre per second is known with an accuracy of $\pm\,1\%$. Calculate the minimum uncertainty in the position of the electron.

2.2 Calculate the wavelength of the H atom transition corresponding to transfer of an electron from the level with $n = 2$ to that with $n = 1$.

2.3 Calculate the wavelength of the transition of the He$^+$ ion which corresponds to the Lyman-α (121.6 nm) line of atomic hydrogen.

2.4 Bohr's theory of the hydrogen atom was based on the then arbitrary assumption that the orbital angular momentum of the electron has to be an integral multiple of $h/2\pi$. If electrons are regarded as waves of wavelength given by the de Broglie formula $\lambda = h/mv$, what does this assumption imply about the electron waves around a hydrogen atom?

2.5 The rule for adding angular momentum vectors in quantum mechanics is that if two parts of a system are described by quantum numbers l_1 and l_2, the resultant L for the whole system must take one of the values $(l_1 + l_2)$, $(l_1 + l_2 - 1), (l_1 + l_2 - 2), \ldots (l_1 - l_2)$, when $l_1 \ge l_2$. Calculate resultant orbital angular momenta for atoms containing: (a) one s and one p electron; (b) two p electrons; (c) two d electrons; (d) one d and one p electron.

2.6 Determine the oxidation numbers of (a) O in H_2O_2; (b) Cr in K_2CrO_4; (c) Mn in MnO_4^-; (d) I in I_4O_9; (e) Fe in Fe_3O_4.

2.7 Predict the properties of astatine, element 85, by comparison with those of nearby elements in the periodic table. Do the same for francium, element 87.

2.8 Calculate the ionization limit of atomic hydrogen on the basis of Eq. 2.7.

2.9 The wave function for atomic hydrogen is the product of an angular part and a radial part. In the ground state the angular part reduces to a constant, while the radial part is

$$\Psi_{1s} = (\pi\alpha^3)^{-\frac{1}{2}}\, e^{-r/\alpha}$$

where α is the Bohr radius (52.9 pm). What is the most probable distance of the electron from the nucleus?

3 The Structure of Molecules

3.1 Introduction: The Origin of Bonding Forces

In Chapter 2 we considered the behaviour of electrons in atoms; it will be helpful to begin by summarizing what we found there. The solutions of the Schroedinger equation for an atom form a regular pattern of energy levels, each level being associated with its own wave function, or orbital, Ψ. These orbitals are distinguished from one another by the values assigned to the quantum numbers n, l and m, numbers which serve both as labels for the energy levels and as parameters which appear in the mathematical expression for Ψ. To a good approximation the wave function Ψ for each orbital can be written as a function of three space coordinates, either (x, y, z) or (r, θ, ϕ), defined by a set of coordinate axes centred on the nucleus. The value of Ψ^2 at a point gives the relative probability of finding an electron at that point. Each orbital can hold at most two electrons, which are assigned to orbitals in order of increasing orbital energy in accordance with the Pauli exclusion principle and Hund's rule. We now need to consider how these results may be modified when an atom becomes incorporated into a molecule, and how the properties which a chemist habitually associates with a chemical bond can be understood in terms of the behaviour of electrons in a molecular environment.

To solve the Schroedinger equation for a molecule is a much more difficult proposition than solving it for an atom. Nevertheless, extremely accurate results have been obtained for a few molecules and useful approximate solutions are available for several hundred others.* All the theoretical calculations agree that the difference in energy between a tightly-bound molecule and its separate component atoms amounts to only a few tenths of one percent of the total energy of the system of electrons plus nuclei. Therefore, as far as an individual atom is concerned the formation of a chemical bond amounts to only a very small perturbation of the overall situation. In practice we can therefore assume that an atom's inner-shell electrons are essentially unaffected by chemical bonding, and confine our attention to the electrons in the valence shell. For these electrons we can see, in

very simple terms, how bonding forces might arise by considering what happens to the energy of an electron when atomic orbitals on different atoms overlap in such a way that the electron can move from one atom to the other.

Suppose Ψ_A to be an orbital containing just one electron on atom A, and Ψ_B to be a similar half-filled orbital on atom B. If Ψ_A and Ψ_B are chosen to be acceptable (approximate) solutions of the Schroedinger equation for the system of two atoms, then, by the properties of wave equations, so also is $(\Psi_A + \Psi_B)$. The Pauli principle allows this molecular orbital $(\Psi_A + \Psi_B)$ to accommodate both electrons comfortably. Provided the two wave functions overlap, as in Fig. 3.1, each of the two electrons now finds itself in an orbital which extends over both atoms. When an electron is near nucleus B its wave function approximates to Ψ_B. When it is near nucleus A its wave function approximates to Ψ_A.

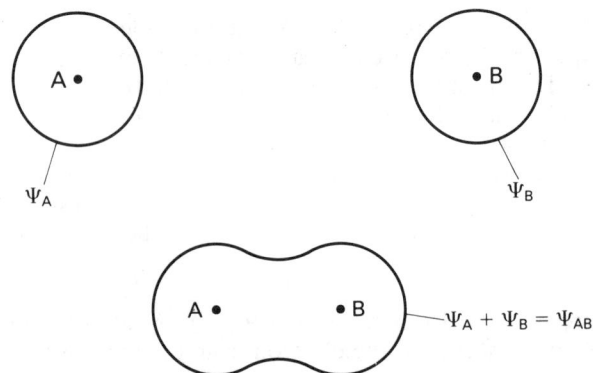

Fig. 3.1 Overlap of atomic orbitals Ψ_A and Ψ_B to form a molecular orbital Ψ_{AB}. (The orbitals are represented by boundary surfaces as in Fig. 2.2a)

Between the nuclei the wave functions reinforce one another, and the electron density $(\Psi_A + \Psi_B)^2$ is greater than the sum of $\Psi_A{}^2$ and $\Psi_B{}^2$ from the separate atoms. The wave function for each electron is now spread over a much larger volume of space, which means that the wavelength of the de Broglie wave associated with the electron can be

---

* In a rare triumph of theory over experiment, the calculations of Kolos and Roothaan for the hydrogen molecule revealed the presence of a numerical error in data derived from extremely precise spectroscopic measurements on H_2.

larger than before. The maximum possible wavelength corresponds roughly to twice the distance between the boundaries of the region in which the electron can move (cf. the vibrating string in Fig. 1.5). The de Broglie wavelength is equal to h/p, where p is the momentum of the electron. Thus a longer wavelength corresponds to lower momentum, which also means lower kinetic energy. This reduction in kinetic energy in turn results in a lower total energy for each electron, and thus a lower energy for the whole system of two atoms. It is the reduction in electronic energy, due essentially to confining the electrons in a bigger box, which is mainly responsible for bonding between the two atoms. The build-up of electron density between the nuclei helps by reducing the effect of the internuclear repulsion.

The behaviour of the energy of the system of two atoms A and B as a function of the internuclear distance is illustrated in Fig. 3.2. Let us consider what happens as the two atoms are slowly brought together. At large internuclear distances the total energy is simply $E_A + E_B$, the energy of the separate atoms. As the atoms come closer together the energy begins to fall, because of the orbitals overlapping as in Fig. 3.1, and the energy reaches its minimum value E_{AB} at the *equilibrium internuclear distance* r_e. At distances shorter than r_e the energy rises steeply as a result of repulsion between the positively charged nuclei. When the internuclear distance is close to r_e and the energy is near its minimum value the two atoms comprise a stable, bound system, because energy has to be provided from outside to bring them up to the level corresponding to separate atoms. The energy which is required to separate the bound atoms, $(E_A + E_B - E_{AB})$, is called the *dissociation energy* D_{AB}, of the A—B bond. Experimental values of bond dissociation energy are discussed in Chapter 4.

3.2 Covalent versus Ionic Bonding: Bond Polarity

The most important properties of a chemical bond are its strength, which is reflected in the value of the dissociation energy D_{AB}, its length, which is fixed by r_e, and its polarity, which will be our major concern in this section. A further important property is the angle between bonds formed when two or more atoms are bound to a single atom. The factors which govern bond angles will be discussed in Section 3.3.

By polarity we mean an uneven distribution of bonding electrons between two atoms A and B, leading to an excess of negative charge on one of the atoms and an excess of positive charge on the other. We found in Chapter 2 that the strength of the attraction of an atom for electrons in a chemical bond can be described in terms of the electronegativity of the element in question. The value of the electronegativity can be looked up in a table such as Table

2.4. In general non-metallic elements of small atomic radius are found to have large electronegativity, and metallic elements of large atomic radius have small electronegativity. When two atoms of similar electronegativity are bound together the electrons are equally shared in a pure *covalent* bond. An extreme example would be a bond between two identical atoms, as in H_2. When two atoms of very different electronegativity are combined together the electrons of the bond tend to spend virtually all their time close to the more electronegative atom, which therefore behaves as a negative ion, while the less electronegative atom becomes a positive ion. These ions are attracted to one another by their opposite charges, but there is nothing to tie any given ion to one specific ion of opposite charge. Thus in solid sodium chloride there are no NaCl molecules, but only a regularly spaced array of Na^+ and Cl^- ions. Even in the molten salt or in a salt solution there are no NaCl molecules, but only transient encounters between independent Na^+ and Cl^- ions. In the gas phase sodium chloride exists mainly as Na^+Cl^- ion pairs, but even here there is a tendency to form neutral clusters of four or more ions. This is an extreme example of an ionic bond. The interaction of a pair of ions follows a potential energy curve whose form is similar to that in Fig. 3.2.

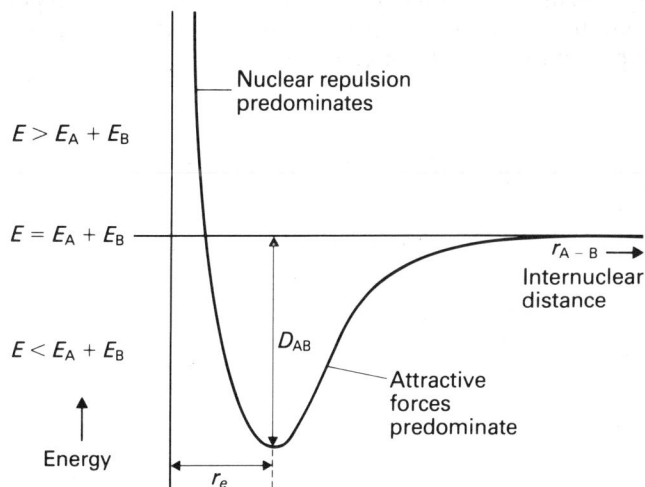

Fig. 3.2 Variations of energy with internuclear distance for molecule AB of Fig. 3.1

Between the extremes of purely covalent and purely ionic bonding there can be found examples of all kinds of intermediate polarity. For an unsymmetrical molecule such as HCl or H_2O the uneven charge distribution results in the molecule having a permanent *dipole moment*. In the case of a diatomic molecule having a charge of $+q$ on the less electronegative atom and $-q$ on the more electronegative atom, with an internuclear distance r, the magnitude of the dipole moment is given by

$$D = rq \qquad (3.1)$$

For a polyatomic molecule such as H_2O or $CHCl_3$ the observed dipole moment is the vector sum of the individual bond dipoles; this is illustrated for H_2O in Fig. 3.3. Values of dipole moment in the literature are mostly tabulated in terms of a unit called the debye, which is equal to 10^{-18} esu cm. The electronic charge is 4.803×10^{-10} esu and 1 Ångstrom unit is 10^{-8} cm, so that the complete transfer of an electron from one atom to another over a distance of 1 Å would give rise to a dipole moment of 4.803 debye. The same moment expressed in SI units would be 1.602×10^{-29} coulomb metre. Molecular dipole moments are fairly easy to obtain experimentally, either from dielectric constant measurements or by microwave spectroscopy. Once the internuclear distance is known the value of the displaced charge q can be calculated from the observed dipole moment. Dipole moment data can therefore provide valuable insight into the manner in which electrons are shared between bonded atoms. In addition the agreement, or lack of it, between experimental and calculated values of dipole moment is frequently used as a test of the quality of an approximate solution to the Schroedinger equation. A representative selection of experimental and predicted values of dipole moment is given in Table 3.1.

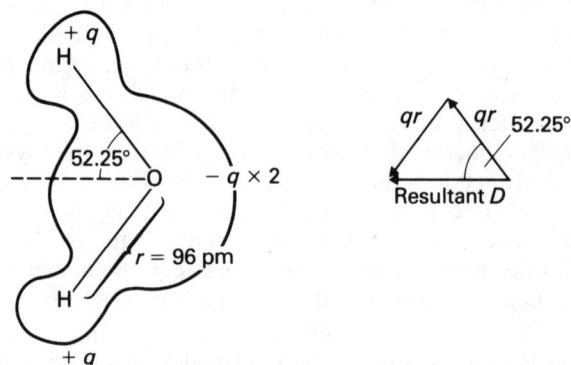

Fig. 3.3 Resultant dipole moment for H_2O (1.846 debyes) as a vector sum of two bond dipoles. [$D = 2qr \cos 52.25°$ gives $q = 5.2 \times 10^{-20}$ coulombs, or about one third of an electronic charge]

Table 3.1 Some experimental and theoretical values of dipole moment. (Observed values are expressed in *(a)* debyes and *(b)* coulomb metres $\times 10^{-30}$.)

Molecule	Calculated D (a)	Observed D (a)	(b)
$CH_2 = CH—CH_3$	0.36	0.364	1.21
HF	1.85	1.8195	6.069
CHF_3	1.66	1.645	5.49
H_2O	2.10	1.846	6.16
NH_3	1.97	1.468	4.90
HCN	2.48	2.986	9.96
CH_3OH	1.94	1.69	5.64
CH_3COCH_3	2.90	2.90	9.67
CH_3NH_2	1.86	1.326	4.42
CH_3CN	3.05	3.92	13.1
CH_3NO_2	4.38	3.46	11.5
HNO_3	2.24	2.16	7.20
HNO_2	2.27	1.85	5.00

Data from J. A. Pople and D. L. Beveridge, *Approximate Molecular Orbital Theory*, McGraw-Hill, New York, 1970.

[Theoretical values obtained by the semi-empirical procedure known as CNDO/II.]

3.3 A More Detailed Consideration of Covalent Bonding

In Section 3.1 we saw how the overlap of two orbitals from different atoms could lead to the formation of a bonding orbital, in which electrons could be placed so as to be shared between a pair of atoms. We shall now consider the outcome of this kind of orbital overlap in more detail, and then apply the resulting concepts to a variety of small molecules.

Consider a system of two atoms A and B, as in Fig. 3.4a, with wave function Ψ_A and Ψ_B that can overlap to form a composite wave function Ψ_{AB}. Here Ψ_A and Ψ_B are solutions of the Schroedinger equation for the separate atoms, and the corresponding orbital energies are E_A and E_B. Even when the atoms A and B are close together, an electron in the vicinity of nucleus A will be affected by the field of this nucleus to a much greater extent than by the field of nucleus B. Therefore, in the region of space near nucleus A, and especially on the side of A away from B, the function Ψ_{AB} must bear a strong resemblance to Ψ_A. Similarly, in the vicinity of nucleus B the function Ψ_{AB} must resemble Ψ_B. Thus an obvious choice for an approximation to Ψ_{AB} is one of the simple linear combinations ($\Psi_A + \Psi_B$) or ($\Psi_A - \Psi_B$). This is a particularly plausible choice because it is a property of wave equations, known as the superposition principle, that if two functions Ψ_A and Ψ_B are acceptable solutions to the equation, so also is any linear combination ($a\Psi_A + b\Psi_B$), where a and b are two arbitrary constants. (For example, consider the superposition of different frequencies in a sound wave.)

The two possibilities we have found for Ψ_{AB} are

$$\Psi_{AB} = \Psi_A + \Psi_B \qquad (3.2)$$

which is shown in Fig. 3.4b, and

$$\Psi_{AB}^* = \Psi_A - \Psi_B \qquad (3.3)$$

which appears in Fig. 3.4c. Potential energy curves corresponding to Ψ_{AB} and Ψ_{AB}^* are shown in Fig. 3.5. The most significant feature of the curve for Ψ_{AB} in Fig. 3.4b is the presence of a region where Ψ has a large value between the nuclei, corresponding to a large electron density in this region. Thus a pair of electrons placed in Ψ_{AB} is effectively shared between the atoms and in addition the high electron density between the nuclei effectively screens the nuclei from one another and so cuts down their mutual repulsion.

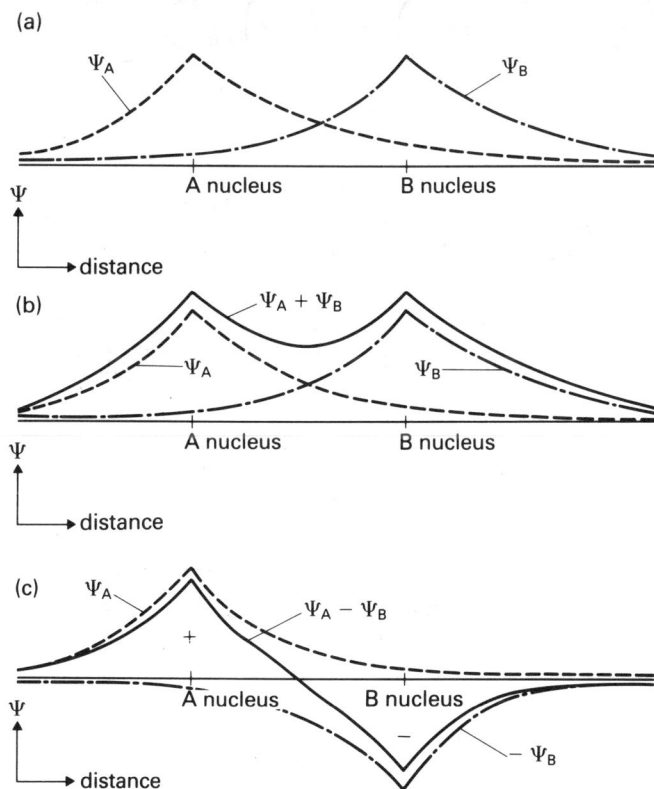

(a)

(b)

(c)

Fig. 3.4 (a) Wave functions Ψ_A and Ψ_B along AB line, with nuclei in position where overlap can occur. (b) Overlap to form bonding orbital $\Psi_A + \Psi_B$. (c) Overlap to form antibonding orbital $\Psi_A - \Psi_B$. [The wave functions are plotted as a function of distance from the nucleus, as in Fig. 2.2c]

This then is the *bonding* orbital that we met in Section 3.1, whose attractive potential energy curve is the lower of the two curves in Fig. 3.5. In contrast, the curve for Ψ_{AB}^* in Fig. 3.4c shows the presence between the nuclei of a *node* i.e. a

two-dimensional surface on which both the wave function and the electron density are zero. The resulting electron density between the nuclei is actually less than it would be if either atom were present by itself. Thus the nuclei must repel each other strongly, and corresponding to this *antibonding* orbital we have the repulsive potential energy curve which is shown as the upper curve in Fig. 3.5.

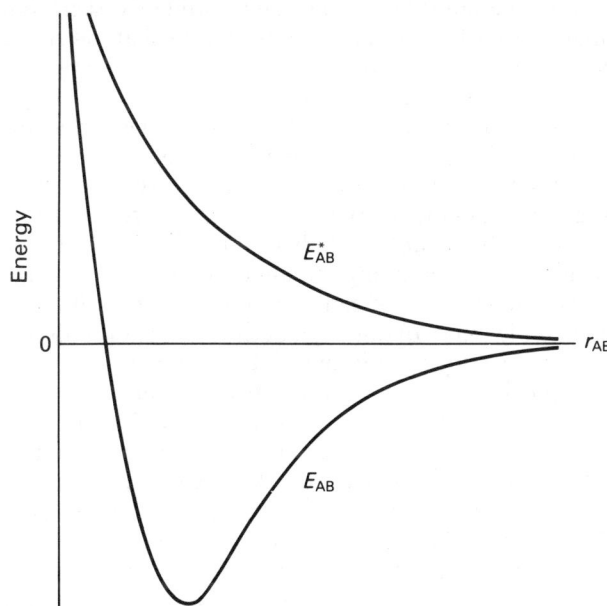

Fig. 3.5 Potential energy curves for antibonding orbital Ψ_{AB}^* and bonding orbital Ψ_{AB}. The zero of energy is taken as the energy of the widely separated atoms. The region where E_{AB} is negative corresponds to formation of a bound AB molecule. E_{AB}^* is always positive, i.e. for this curve the system is always unstable with respect to the system containing separate A and B atoms

We can now conclude that if bonding orbitals in a molecule are occupied by electrons they will contribute to the stability of the molecule, while if antibonding orbitals are occupied this will tend to make the molecule unstable. The relative strengths of the two effects are such that if both bonding and antibonding orbitals are occupied to the same extent the antibonding orbitals must win, i.e. the molecule will be unstable. Thus stability in a molecule always results from electrons occupying an excess of bonding over antibonding orbitals. Bonding orbitals are characterized by a build-up of electron density between nuclei, due to the overlap of wave functions with the same positive or negative sign; antibonding orbitals are characterized by the overlap of wave functions having opposite sign, which results in formation of a node in the wave function and a region of zero electron density between the nuclei. (As an extension of these ideas, we may note that in a polyatomic molecule it is possible for a molecular orbital to be formed which has a node right at the position of one of the nuclei. As far as that particular nucleus is concerned the molecular orbital is then neither bonding nor antibonding, but simply non-bonding.)

Some of the different ways in which two atomic orbitals can overlap to form a molecular orbital are illustrated in Fig. 3.6. The notation for molecular orbitals is seen to be similar to that for atomic orbitals, except that lower case greek letters σ, π, δ, are used in place of the lower case roman letters s, p, d. . . . Thus an orbital which is symmetrical about the line joining the two nuclei is a σ (sigma) orbital, which is usually written σ^* if it is antibonding. The overlap of two s orbitals to form σ and σ^* orbitals is shown in Fig. 3.6a, and the overlap of an s and a p orbital to form σ and σ^* orbitals in Fig. 3.6b. Two p orbitals which have their lobes directed along the internuclear axis, which we have labelled the z-axis, can also form σ and σ^* orbitals, as shown in Fig. 3.6c. In this case, because the p_z orbitals have their positive lobes pointing in the same direction, the bonding orbital has to be $(p_{zA} - p_{zB})$ and the antibonding orbital $(p_{zA} + p_{zB})$. A molecular orbital which has one nodal plane that includes both nuclei is called a π (pi) orbital. Pi orbitals can be formed by overlap of two p_x or two p_y orbitals from different atoms, in the manner which is shown in Fig. 3.6d for both bonding and antibonding examples. (Note that pi orbitals are antisymmetric with respect to reflection in the nodal plane). Continuing in this way, we could similarly produce delta (δ) and phi (ϕ) molecular orbitals, analogous to atomic d and f orbitals, but it turns out these are seldom encountered in practice. With the aid of a judicious admixture of ionic bonding, and a modicum of dipole–dipole interactions, virtually the whole of chemistry can be understood in terms of sigma and pi bonding.

In the next section we shall consider some specific examples which illustrate this statement.

3.4 Bonding in Some Representative Molecules

(a) Hydrogen

Here the two 1s orbitals overlap to form form σ_{1s} and σ_{1s}^* orbitals, of which the bonding σ_{1s} orbital has the lower energy. Two electrons need to be accommodated, and both can go into the bonding orbital, leaving the antibonding orbital empty. The result is a stable molecule held together by a single bond which comprises the pair of electrons in the shared bonding orbital. It is interesting to consider what happens if, instead of putting both electrons in the same orbital, we put one in σ_{1s} and the other in σ_{1s}^*. We then obtain a configuration whose lowest energy state is given by Hund's rule as the state in which the two electrons have

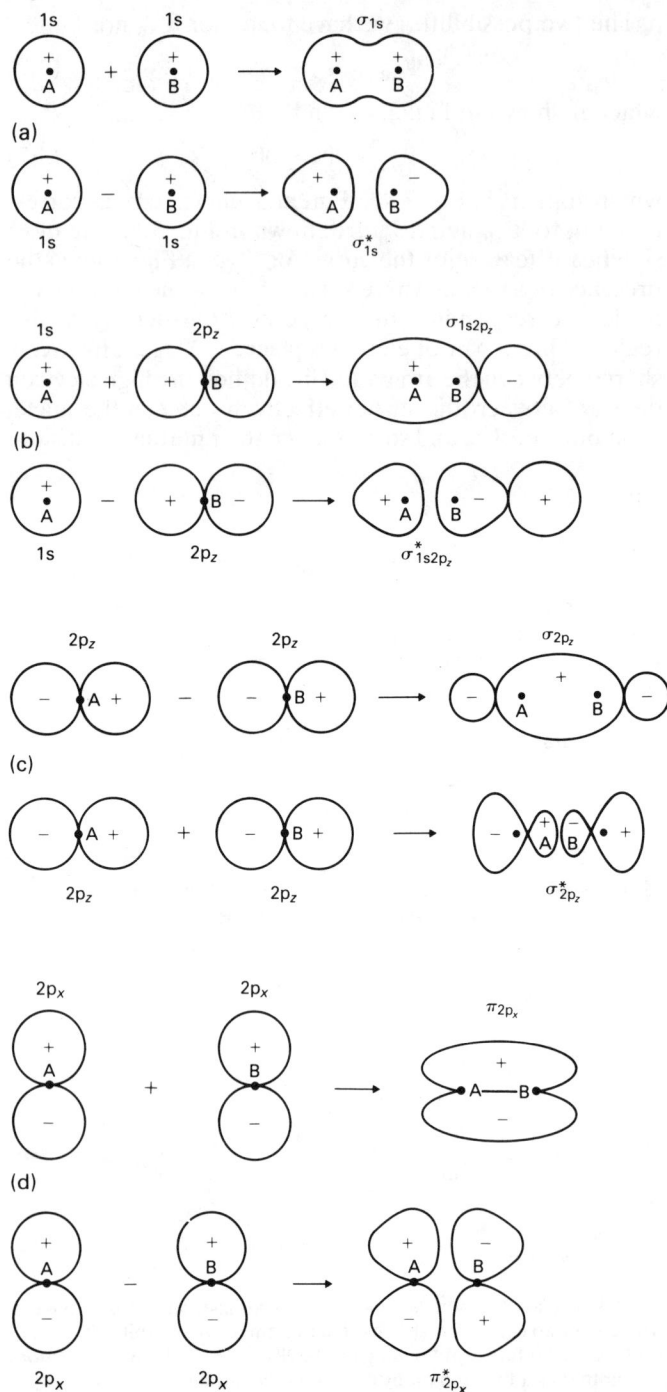

Fig. 3.6 Formation of sigma and pi molecular orbitals: (a) overlap of two 1s atomic orbitals to form σ_{1s} and σ_{1s}^* molecular orbitals; (b) overlap of 1s and $2p_z$ orbitals to form σ_{1s2p_z} and $\sigma_{1s2p_z}^*$; (c) overlap of two $2p_z$ orbitals to form σ_{2p_z} and $\sigma_{2p_z}^*$; (d) overlap of two $2p_x$ orbitals for form π_{2p_x} and $\pi_{2p_x}^*$ molecular orbitals. [Wave functions represented as in Fig. 2.2b]

parallel spins. The total electron spin for the molecule is then 1, and the molecule is said to be in a *triplet state*. To see why this state should be called a triplet, we note that if a molecule has all its electrons paired the resultant total spin is zero, and we have just one spin state which is referred to as a singlet. If a molecule or radical has one unpaired electron the magnitude of its total spin is ½ and there are two possible spin states, namely, + ½ and − ½. Because there are two spin states the overall state is called a doublet. With total spin equal to 1 there are three spin states 1, 0 and − 1, so the system is a triplet. With total spin 1½ there are four states, 1½, ½, − ½ and − 1½, forming a quartet. In general, just as there are $2l + 1$ values of the magnetic quantum number m corresponding to a particular value of l in an atom, so there are $2S + 1$ spin states corresponding to a given total spin S. Returning to the case of H_2 in its lowest triplet state, we can conclude that the effect of the electron in the σ_{1s} orbital will be outweighed by the effect of the electron in the σ_{1s}^* orbital and the molecule will not be stable in this state.

(b) Helium

The helium atom has two 1s electrons, so that in its ground state an He_2 molecule must have a pair of electrons in each of the σ_{1s} and σ_{1s}^* orbitals. Thus the He_2 molecule is necessarily unstable in its ground state. To obtain a stable He_2 entity we have either to remove one of the antibonding electrons, thus forming an He_2^+ ion (which is a stable species under the conditions of an electrical discharge), or we can excite one of the antibonding electrons to a higher bonding orbital, e.g. σ_{2s}, as in the known bound excited states of He_2.

(c) Lithium

The lithium atom has the configuration $1s^2 2s$. Ignoring the filled 1s shell, we can allow the 2s orbitals to overlap and form σ_{2s} and σ_{2s}^* molecular orbitals. Both valence electrons can now go into the σ_{2s} orbital to form a bound Li_2 molecule analogous to H_2.

(d) Nitrogen

The nitrogen atom has the configuration $1s^2 2s^2 2p^3$. Here it is of interest to consider all of the orbitals in the outer shell. To help decide which atomic orbitals will overlap with which we note that inner-shell electrons with relatively large negative energies are not affected very much by molecule formation, i.e. their wave functions have very little tendency to mix with the wave functions of outer-shell electrons. On the other hand the outer-shell electrons do have their energies altered considerably through mixing of their wave functions with orbitals from the other atom. This can only happen through mixing with orbitals of similar energy, since we have just stated that inner-shell orbitals are not involved. Hence, as a guide, we can say that in

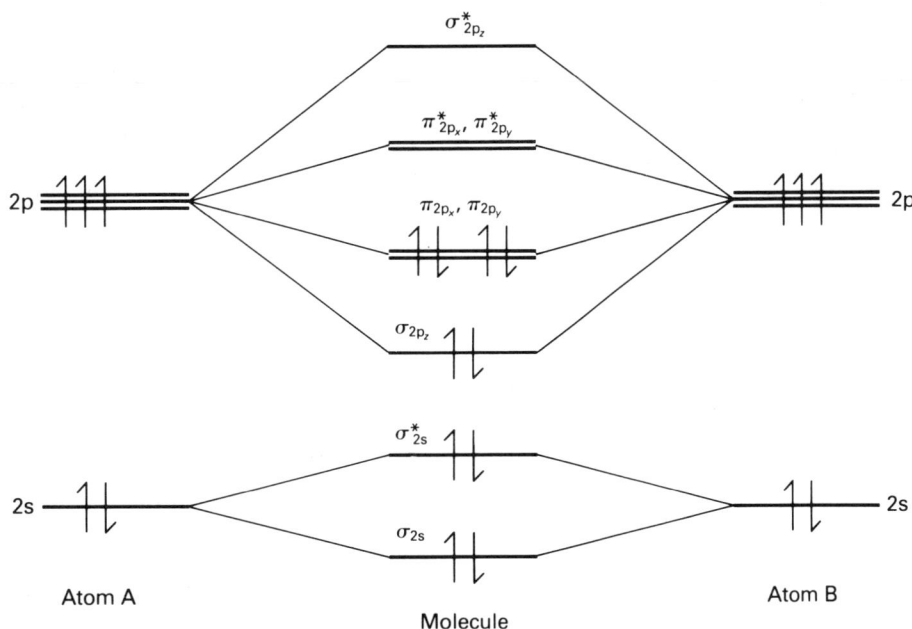

Fig. 3.7 Orbitals involved in bonding in N_2, with N_2 valence electrons shown. For other diatomic molecules formed from the first short period elements B, C, N, O and F, the order of molecular orbitals is the same but the number of electrons in the orbitals differ.

molecule formation atomic orbitals tend to overlap with orbitals of similar energy. On this basis, allowing the orbitals to overlap in pairs, we arrive at the pattern of molecular orbitals for N_2 shown in Fig. 3.7. The order of molecular orbital energies shown is that which leads to best agreement with experimental observations, namely, σ_{2s}, σ_{2s}^*, σ_{2p_z}, $\pi_{2p_x} = \pi_{2p_y}$, $\pi_{2p_x}^* = \pi_{2p_y}^*$, $\sigma_{2p_z}^*$. The ten outer-shell electrons are now able to be accommodated very neatly by placing four in the 2s sub-shell, which according to this model has little or no effect on bonding, two in the σ_{2p_z} bonding orbital, and four in the degenerate pair of π_{2p_z} orbitals. There are seen to be a total of six bonding electrons, and the result is what we normally regard as an N≡N triple bond.

(e) Fluorine

The flourine atom has the configuration $1s^2 2s^2 2p^5$. Thus in the F_2 molecule all the orbitals of Fig. 3.7 are doubly occupied except $\sigma_{2p_z}^*$, and we can disregard the pi orbitals except to note that they will have a small antibonding effect. The net bonding arises from the two electrons in the σ_{2p_z} orbital, which constitute the F—F single bond.

(f) Oxygen

The configuration of the oxygen atom is $1s^2 2s^2 2p^4$. Thus the O_2 molecule contains two more electrons than N_2, and the lowest level at which the two extra electrons can be accommodated is in the degenerate pair of π_{2p}^* orbitals (Fig. 3.8). According to Hund's rule these electrons have to be accommodated singly, one in each of the degenerate orbitals, with parallel spins; therefore the total spin of the molecule in the ground state is 1, i.e. the ground state is a triplet. Because the magnetic moment associated with electron spin is not cancelled out in the ground state of O_2 the molecule is paramagnetic, unlike the vast majority of substances, which are diamagnetic. (In the middle of the last century Michael Faraday used to demonstrate this fact, for which the explanation was then unknown, by showing that soap bubbles filled with pure oxygen were attracted to the poles of a magnet, whereas bubbles filled with other common gases were repelled.) The excess of bonding over antibonding electrons in O_2 is four, so we have an O=O double bond.

(g) Hydrogen Fluoride

The valence shell of hydrogen consists solely of the 1s orbital, so the bonding in HF must involve only one fluorine orbital, and this must be the singly occupied $2p_z$ orbital which was used for bonding in F_2. Therefore we form σ and σ^* orbitals by overlap of $1s_H$ and $2p_{zF}$, and place the two electrons in the bonding orbital. Because of the large difference between the electronegativities of H and F it is to be expected that the shared bonding electrons will actually

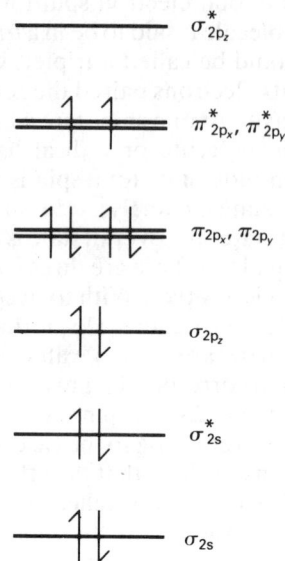

Fig. 3.8 Molecular orbitals and electron configuration for O_2

spend most of their time in the vicinity of the fluorine atom. This can be allowed for in terms of the simple theory we are using by writing the bonding orbital as a general linear combination of the constituent atomic orbitals, thus:

$$\sigma_{H-F} = a(1s_H) + b(2p_{zF}) \qquad (3.4)$$

and specifying that the constant b is much larger than the constant a. (In practice the values of a and b are chosen so as to minimize the energy of the σ_{H-F} orbital, and in order to minimize this energy it is found necessary to have b much larger than a.) Clearly the possibility of choosing optimum values of numerical constants in expressions such as Eq. 3.4 provides a very useful degree of flexibility in the theory.

(h) Water and Ammonia: a Simple Picture

By analogy with what we have just done for HF, the structures of H_2O and NH_3, shown in Fig. 3.9, can be arrived at by allowing σ bonds to be formed by overlap of hydrogen 1s orbitals with singly occupied 2p orbitals of O and N. (By Hund's rule the nitrogen atom has a quartet ground state with three singly occupied 2p orbitals, while the oxygen atom has a triplet ground state, with two singly occupied p orbitals.) The fact that the three p orbitals point in three directions at right angles to one another accounts nicely for the bent and pyramidal geometries of H_2O and NH_3, respectively. The experimentally measured bond angles are actually somewhat larger than 90° (105° for water and 107° for ammonia) but we can tentatively account for this discrepancy by noting that there must be repulsive forces between the hydrogen atoms, which will all carry excess positive charge as a consequence of being bound to the very electronegative O and N atoms. An alternative and generally superior way of describing the bonding in these molecules will become apparent when we consider the bonding in methane (CH_4).

difference that in this case linear combinations are to be formed of orbitals which are all on the same atom.

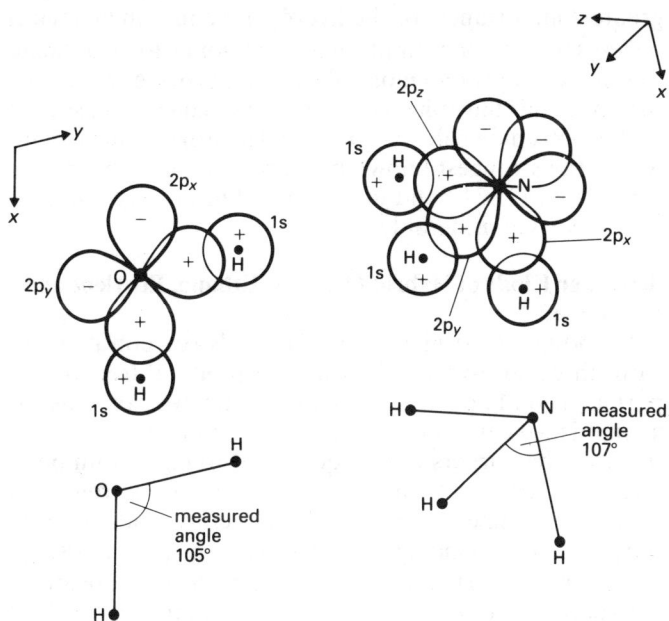

Fig. 3.9 Preliminary interpretation of structures of H_2O and NH_3 in terms of bonding by p orbitals

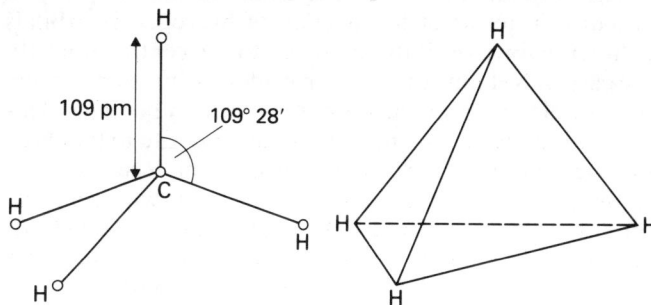

Fig. 3.10 Structure of methane, CH_4, and its relationship to the regular tetrahedron

(i) Methane: the Use of Hybrid Orbitals

The configuration of a carbon atom is $1s^2 2s^2 2p^2$, so that carbon might be expected to be divalent, forming two bonds by sharing its two unpaired 2p electrons. In fact, as is well known, carbon is tetravalent in most of its compounds, of which the prime example could well be methane, CH_4. Methane is a rather stable molecule having four equivalent hydrogen atoms arranged in the tetrahedral configuration shown in Fig. 3.10, with bond angles of 109° 28′. How can we explain this? If one of the 2s electrons of the carbon atom were assumed to be excited, or 'promoted', into the vacant 2p orbital (a process which requires energy) the resulting atom would have four unpaired electrons and so would be tetravalent. Four hydrogen atoms might then be attached, three joined via the three 2p orbitals and one via the single 2s orbital. Three of the H atoms would then be fixed in space because of the directional character of the p orbitals, while the fourth would be relatively free to wander because s orbitals have spherical symmetry. Clearly this will not do. Once again we have to seek a solution on the basis of linear combinations of atomic orbitals, with the

The four orbitals 2s, $2p_x$, $2p_y$ and $2p_z$ are independent solutions of the Schroedinger equation for the carbon atom. From these four orbital functions we can construct four other independent functions which are also acceptable solutions by taking linear combinations of the form

$$\Psi = a\Psi_{2s} + b\Psi_{2px} + c\Psi_{2py} + d\Psi_{2pz} \qquad (3.5)$$

Provided the total set of four functions can still be resolved into one 2s and three different 2p orbitals, the result is merely an alternative way of describing the same atom. However, by a suitable choice of the numerical coefficients a, b, c and d it is possible to obtain four *equivalent* orbitals, which are called sp^3 hybrid orbitals, and which are found to be directed towards the corners of a regular tetrahedron. The shapes of these orbitals are such that they overlap very efficiently with orbitals from other atoms (Fig. 3.11). As a result, the energy gained from bonding is more than sufficient to make up for the energy which was used initially to promote an electron from the 2s to the vacant 2p orbital. Thus the bonding in methane is to be understood in terms of overlap of hydrogen 1s orbitals with carbon sp^3 hybrid orbitals directed towards the corners of a regular tetrahedron.

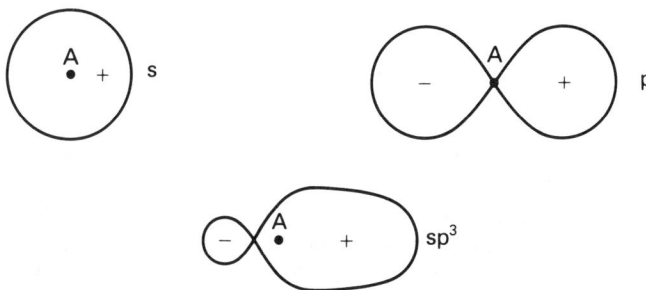

Fig. 3.11 Comparison of shapes of s, p, and hybrid sp^3 orbitals

(j) Reconsideration of Water and Ammonia; the Hydrogen Bond

The structures of water and ammonia are also easy to visualize in terms of the overlap of hydrogen 1s orbitals with sp^3 hybrid orbitals belonging to the central atom. In the case of NH_3 one of the sp^3 hybrids is occupied by a 'lone pair' of electrons which are not involved in bonding. This lone pair constitutes a negative pole, opposite to the three hydrogen atoms at the positive end of the molecular dipole. In the case of H_2O there are two lone pairs, whose mutual repulsion tends to increase the angle between the hybrids on their side of the O atom, with the result that the angle between the hydrogen atoms is reduced slightly from the tetrahedral angle of 109° 28'. This explanation of the bond angle is known as the 'lone pair repulsion theory'.

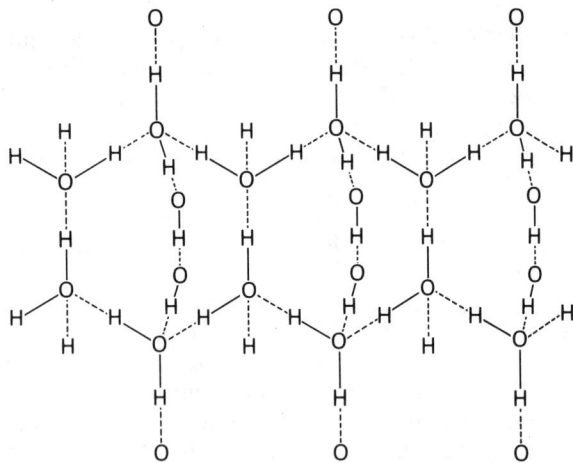

Fig. 3.12 A portion of the structure of ice, showing the tetrahedral arrangement of H atoms around each O atom; - - - - represents a hydrogen bond (see text)

We now have two ways of looking at the bonding in H_2O and NH_3, one in terms of pure p orbitals and the other in terms of sp^3 hybrids on the central atom. The use of sp^3 hybrids in the case of H_2O results in the two lone pairs of electrons being equivalent to one another and directed at approximately tetrahedral angles, whereas with pure p orbitals this is not the case. The crystal structure of ice, shown in Fig. 3.12, therefore supports the hybrid orbital picture, because each oxygen atom is found to be joined tetrahedrally to four other oxygen atoms by electrostatic attraction between the hydrogen atoms, with their excess positive charges, and the oxygen atom lone pairs. This kind of interaction, in which two electronegative atoms (normally limited to F, O, or N) are held together by the electrostatic attraction of one atom for a hydrogen atom bound to the other, is called a *hydrogen bond*. The hydrogen bond is

responsible for the remarkably open structure of ice, and plays a major part in deciding the structures of many other crystals. It is also the cause of the anomalous properties of associated liquids, i.e. liquids whose molecules are attracted to one another by hydrogen bonds, which are much stronger than ordinary Van der Waals' forces. However the area of greatest importance for the hydrogen bond is in biological systems where, for example, it is the major factor in deciding the shapes of protein molecules, and it provides the multitude of connecting rungs between the two component strands of the DNA molecule. An individual hydrogen bond is quite weak, the strongest known being only about one tenth as strong as a C—H bond in methane, but in large numbers they have a considerable effect.

(k) Other Kinds of Hybrid Orbitals: Ethane, Ethylene and Acetylene*

In addition to forming four sp^3 hybrids a carbon atom can form three sp^2 hybrids, leaving one p orbital free to take part in pi bonding, or two sp hybrids, with two p orbitals left available for other purposes. The sp^2 hybrids are directed towards the corners of an equilateral triangle, with bond angles of 120°, while the sp hybrids point away from each other at an angle of 180°. The best overlap, and therefore most effective bonding, is obtained with sp^3 hybrids, sp^2 hybrids being next best and sp the least effective. For atoms with partially occupied d orbitals many other types of hybrid are possible. The most important cases are dsp^2, a planar arrangement in which the four equivalent orbitals are directed towards the corners of a square, and d^2sp^3, a three-dimensional arrangement in which the six equivalent orbitals are directed towards the corners of a regular octahedron. The geometrical characteristics of the more common types of hybrid orbital are summarized in Fig. 3.13.

In *Ethane*, C_2H_6, the bond angles are all close to 109° 28' and the molecule is basically a methane molecule with one hydrogen atom replaced by a CH_3 group (Fig. 3.14a). The two CH_3 groups are fairly free to rotate about their connecting bond, with only a small amount of resistance as the hydrogen atoms on different carbons rotate past one another.

Ethylene, C_2H_4, has the planar configuration shown in Fig. 3.14b with all bond angles close to 120°. Chemists normally think of ethylene as two CH_2 groups connected by a rigid double bond, as is shown by the existence of stable, chemically distinguishable, *cis* and *trans* isomers of substituted ethylenes of the type CHX=CHY (Fig. 3.14c). The presence of bond angles of 120° implies that the molecule is held together by a framework of sigma bonds formed from sp^2 hybrids. With this arrangement the remaining p orbitals, which we arbitrarily choose to be $2p_y$, can overlap as shown in Fig. 3.14b to form a pi bond. Thus the double bond is formed by superposition of a sigma and a pi bond.

---

* In this section we call ethylene and acetylene by their traditional names. Their systematic names are ethene and ethyne. (See Chapter 13.)

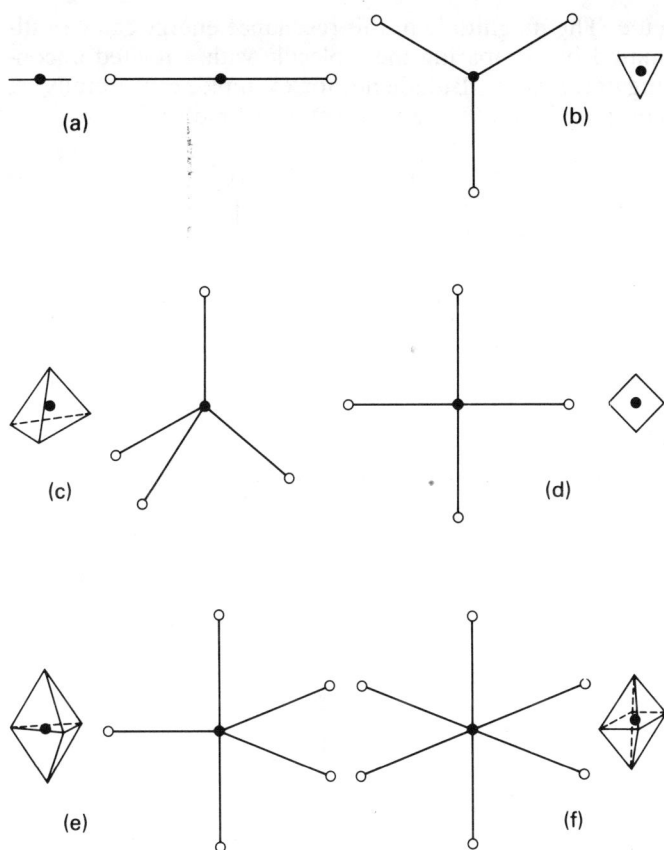

Fig. 3.13 Geometrical characteristics of hybrid orbitals: (a) linear sp, as in CO_2; (b) triangular sp^2, as in BF_3; (c) tetrahedral sp^3, as in CH_4; (d) square planar dsp^2, as in $Ni(CN)_4^{2-}$; (d) trigonal bipyramidal, $sp^2 + pd$, as in PCl_5; (f) octahedral d^2sp^3, as in SF_6

The accessibility of the pi electrons of the double bond to attack by electron-hungry reagents makes ethylene, and olefins in general (molecules with carbon–carbon double bonds), much more reactive than ethane and the rest of the paraffin series (molecules with only single carbon–carbon bonds).

Acetylene, C_2H_2, has the linear structure shown in Fig. 3.14d, with a *triple* bond joining the two carbon atoms. This structure can be understood by assuming the linear framework of the molecule to be based on carbon sp hybrid orbitals, with overlap of the $2p_x$ and $2p_y$ orbitals to produce π_x and π_y molecular orbitals (Fig. 3.14e). Just as in the case of N_2, with which acetylene is isoelectronic (they have the same number of electrons), the triple bond is formed by superposition of one sigma and two pi bonds. Because of the presence of the triple bond, acetylene is even more reactive than ethylene. Acetylene is actually unstable with respect to decomposition into its component elements, and

at high pressure is liable to decompose explosively. (For this reason the cylinders of acetylene used with cutting and welding torches contain acetylene dissolved in acetone, CH_3COCH_3, and not simply the compressed gas.)

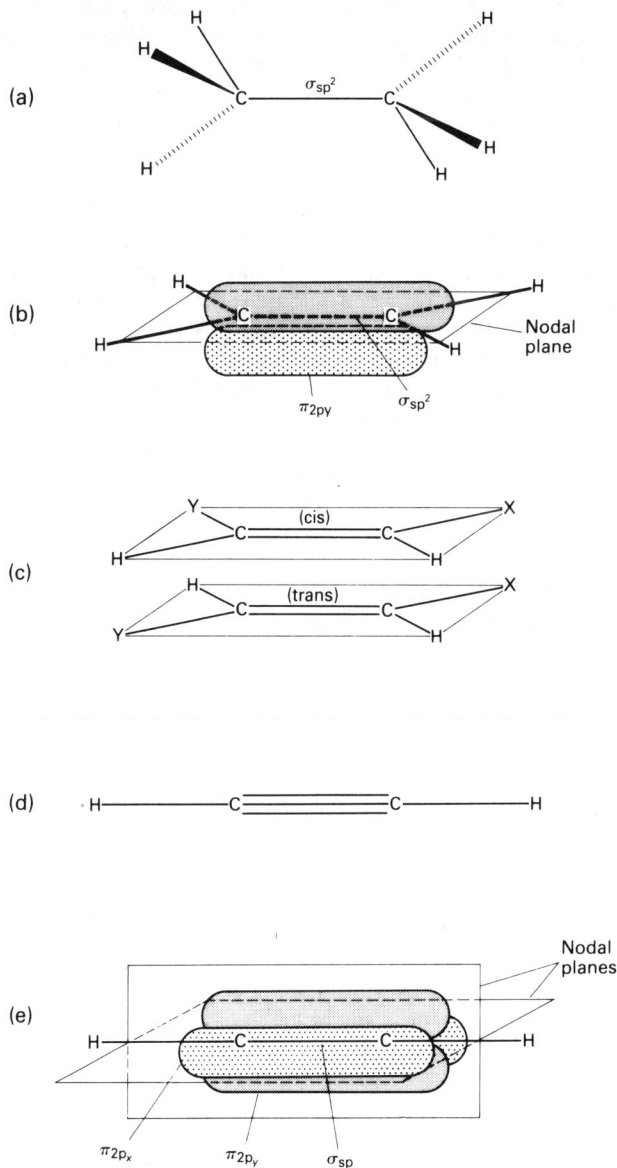

Fig. 3.14 Bonding in ethane, ethylene and acetylene. (a) Ethane, C_2H_6, with skeleton formed from sp^3 hybrid orbitals. (b) Ethylene, C_2H_4, with sp^2 hybrid skeleton and π_{2p_y} orbital of double bond. (c) Cis (same side) and trans (opposite side) isomers of ethylene derivative CHXCHY. (d) Linear configuration of acetylene, C_2H_2. (e) Acetylene, showing sp hybrid skeleton with π_{2p_x} and π_{2p_y} orbitals of triple bond

3.5 Conjugated and Aromatic Hydrocarbons: Delocalized Electrons

An interesting situation arises when a molecule contains a series of alternate single and double bonds along a chain or around a ring. Such a series of alternating bonds is called a system of *conjugated double bonds*. Two examples of conjugated molecules, namely butadiene and benzene, are shown in Fig. 3.15. The chemical and physical properties of these and related substances reveal two important consequences of this type of arrangement. The first is that electrons associated with the pi bonds of the conjugated system are relatively free to move around the molecule throughout the length of the conjugated system: the pi electrons are said to be *delocalized*. The second is that the presence of a conjugated double bond system makes the whole molecule much more stable than it would be if the single and double bonds were not arranged alternately. Benzene represents an extreme case, in which the pi electrons are completely delocalized about the planar ring, and the whole molecule is both remarkably stable and very much less reactive than a small molecule with three double bonds has any right to be. Chemical reactions of benzene generally leave the ring system unscathed—most reactions merely involve the substitution of one group for another in one of the positions occupied by hydrogen atoms in the parent molecule. Molecules of this kind are said to be *aromatic*.

There are two ways of approaching the problem of providing a theoretical description of a conjugated system. The first, known as the *valence bond method*, is to write the wave function for the molecule as a linear combination of the wave functions for all the possible covalent structures which can be written for the molecule without altering the positions of any atoms. Structures which would be included in such a wave function for butadiene and benzene are shown in Fig. 3.15b, and a possible wave function for butadiene is given below.

$$\Psi = a\Psi_1 + b\Psi_2 + c\Psi_3 + b\Psi_4 + a\Psi_5 \qquad (3.6)$$

Clearly the structures will not all be of equal importance in the final wave function; the best values of the coefficients in the linear expansion of Eq. 3.6 are therefore chosen to be those which result in the lowest total energy for the molecule. The resulting molecular structure is termed a *resonance hybrid* of the individual structures; the double-ended arrows between structures in Fig. 3.15b are a conventional means of indicating that the structures belong to a single resonance hybrid. The possibility of moving electrons about in the molecule without destroying the basic framework is what we understand by electron delocalization in the valence bond model. The reduction in energy which results from delocalization is termed the *resonance energy* of the mole-

cule. The magnitude of this resonance energy can be estimated by comparing the molecule with a related unconjugated system. Butadiene, for example, can usefully be compared with two separate ethylene molecules.

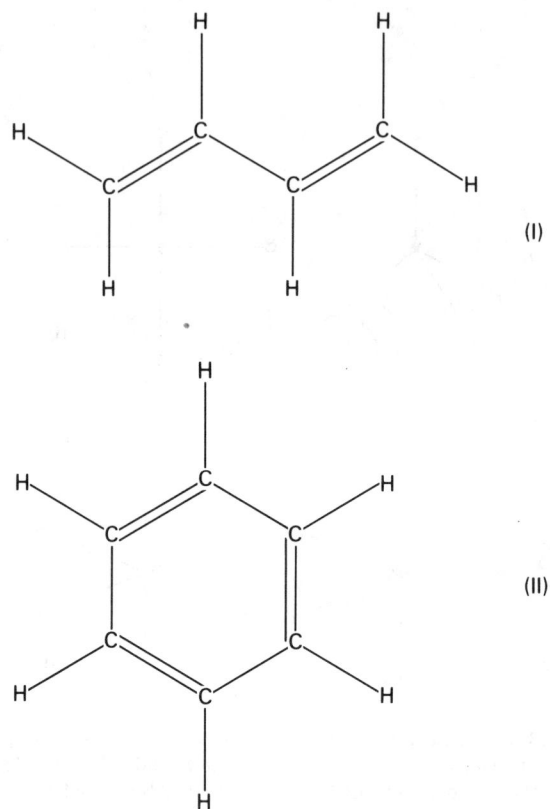

Fig. 3.15 (a) Molecules with conjugated double bonds: (I) Butadiene; (II) Benzene

The other approach to conjugated and aromatic systems is by way of the *molecular orbital theory* that we have already used a great deal in this chapter. With this approach the molecular framework in butadiene or benzene is assumed to be based on carbon sp^2 hybrid orbitals, as illustrated in Fig. 3.15c, with the pi bond system being formed by overlap of adjacent $2p_y$ orbitals as in ethylene. The special properties of conjugated molecules then arise from the fact that overlap of $2p_y$ orbitals occurs along the whole length of the conjugated system, and the molecular orbitals which describe the pi system must therefore include contributions from all atoms in the system.

For butadiene there are four $2p_y$ orbitals which combine to form four molecular orbitals, two of which are found to be bonding orbitals and two antibonding. The four electrons of the pi system must go into the two lowest molecular

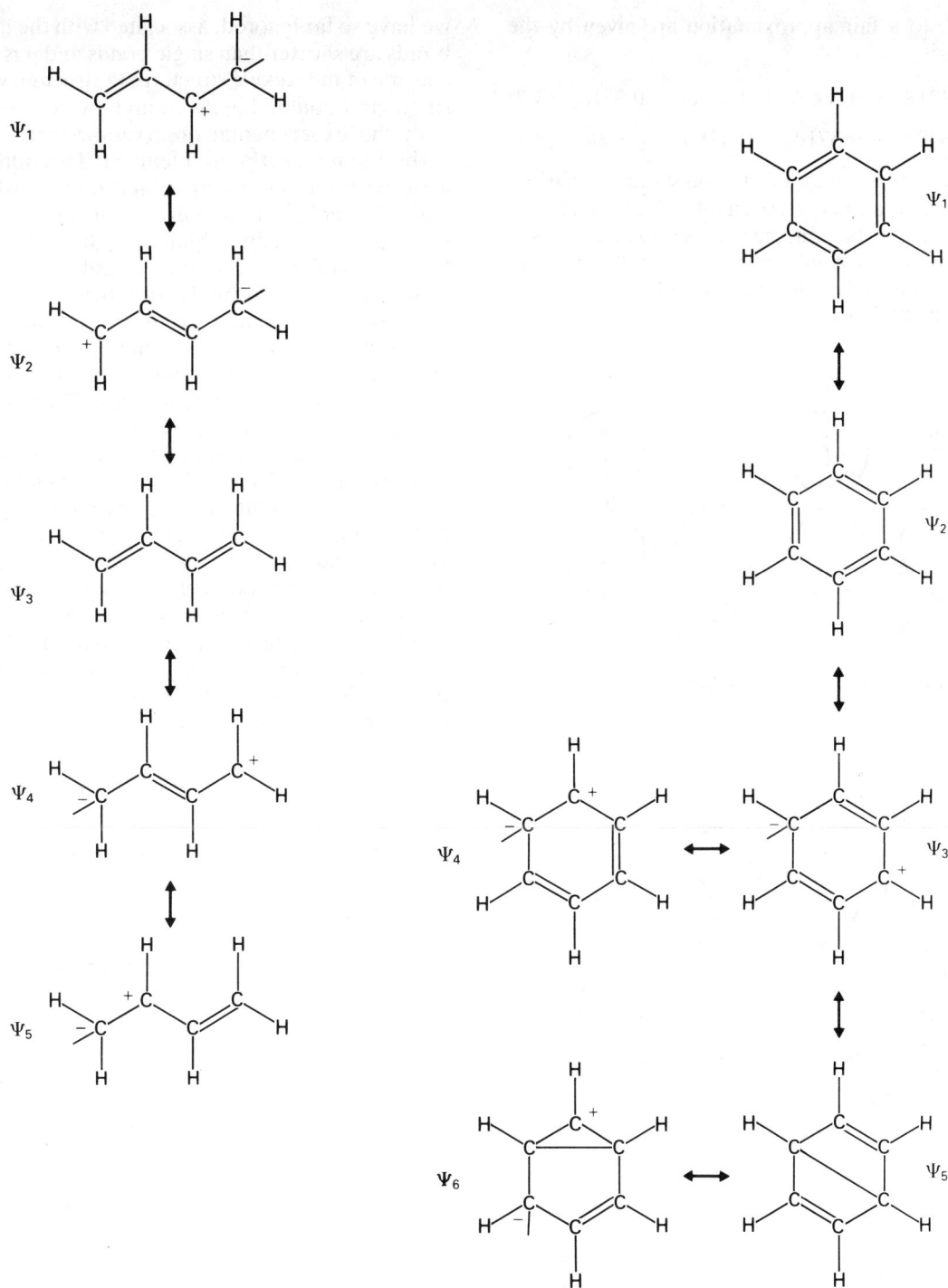

Fig. 3.15 (b) Structures which contribute to valence bond wave functions for butadiene and benzene. Note the presence of localized positive and negative charges on carbon atoms of butadiene structures 1, 2, 4 and 5

orbitals, which to a fair approximation are given by the equations

$$\Psi_1 = 0.371\phi_1 + 0.600\phi_2 + 0.600\phi_3 + 0.371\phi_4 \quad (3.7)$$

and $\quad \Psi_2 = 0.600\phi_1 + 0.371\phi_2 - 0.371\phi_3 - 0.600\phi_4 \quad (3.8)$

where we have written the $2p_y$ orbitals on successive carbon atoms as ϕ_1, ϕ_2, ϕ_3 and ϕ_4 respectively. The lowest molecular orbital is seen to be bonding between all three pairs of adjacent atoms; the second lowest is bonding between atoms 1 and 2, and between atoms 3 and 4, but is antibonding between atoms 2 and 3.

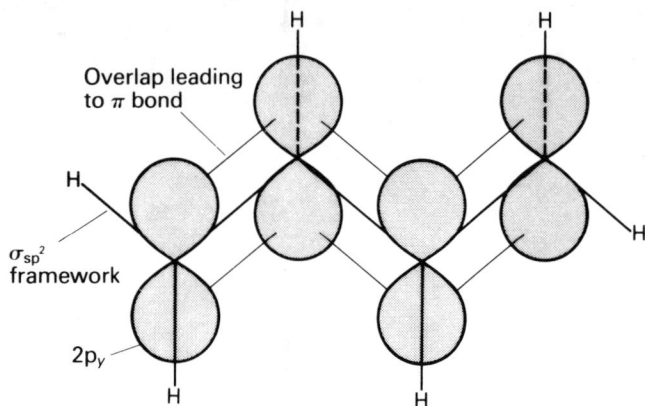

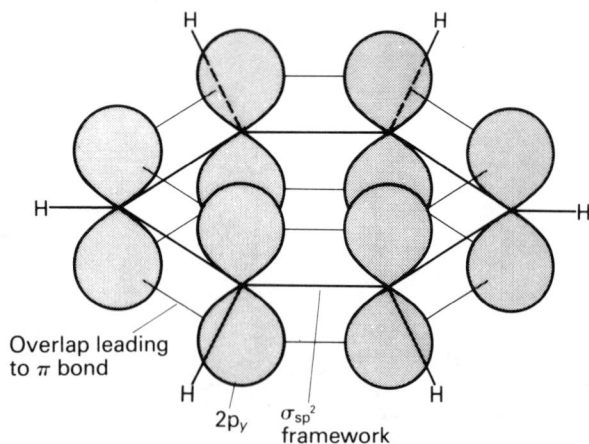

we have so far ignored, associated with the fact that double bonds are shorter than single bonds and it is hard to reconcile one of the Kekulé structures (a structure with alternating single and double bonds around the ring as in Fig. 3.15a) with the experimental observation that all the carbon–carbon bond lengths are identical. This difficulty does not arise with the molecular orbital theory, where we begin with the molecular framework in the form of a regular hexagon. The six $2p_y$ orbitals at right angles to the plane of the ring overlap to form six molecular orbitals which comprise two single orbitals and two pairs of degenerate orbitals. Three of the orbitals are bonding and three antibonding, with the non-degenerate orbitals lying highest and lowest as shown in Fig. 3.16. The six pi electrons occupy the three bonding orbitals. The very high stability and large resonance energy of benzene is associated with the rigid, planar structure and with the fact that complete delocalization of the electrons around the ring can be achieved without introducing any ionic structures, such as were required for butadiene. Several other examples of aromatic molecules are shown in Fig. 3.17, including the interesting species borazole (V) and azulene (VI), which is an isomer of naphthalene (II), both having the formula $C_{10}H_8$. Azulene has a dipole moment of 0.796 debye, with the five-membered ring negative. In terms of valence bond theory this shows that ionic structures play a significant part in the overall wave function.

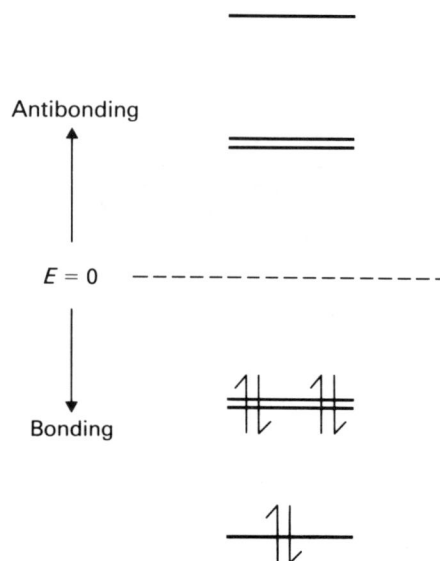

Fig. 3.15 (c) Overlap of $2p_y$ orbitals leading to formation of extended π-orbital systems in butadiene and benzene

For benzene the 120° bond angle of the sp^2 hybrid orbitals proves to be ideal for constructing a six-membered ring. In terms of the valence bond picture there is a difficulty, which

Fig. 3.16 Energy levels associated with π orbitals of benzene

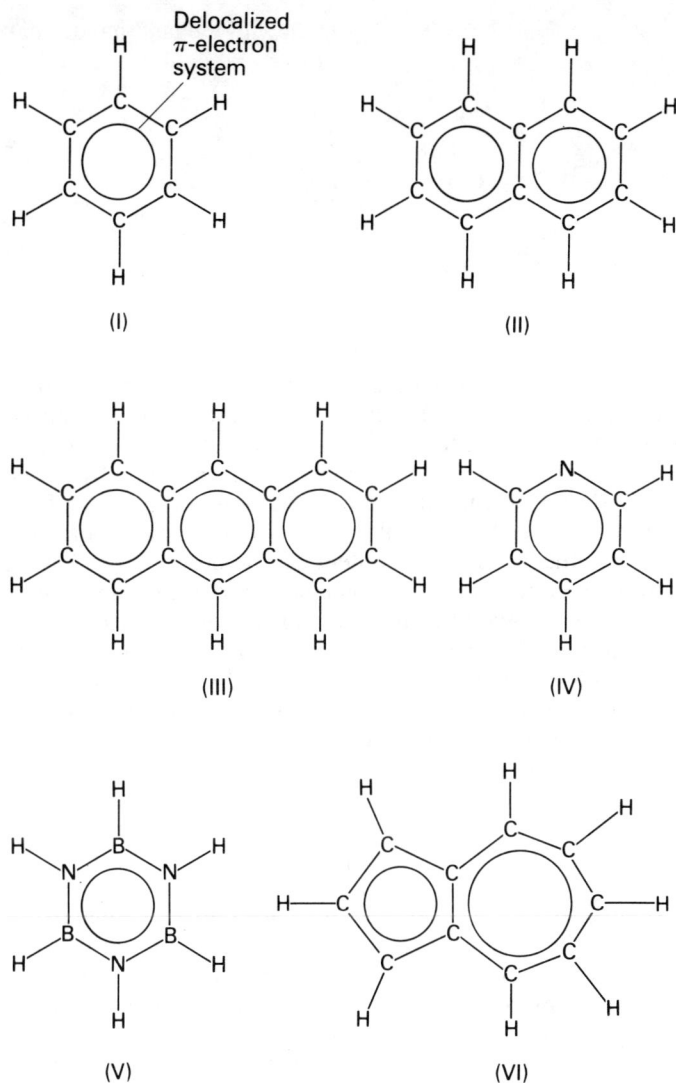

Fig. 3.17 Some aromatic molecules: (I) Benzene; (II) Naphthalene; (III) Anthracene; (IV) Pyridine; (V) Borazole; (VI) Azulene

3.3 The stable gaseous ion CH_5^+ appears to violate the rule that only four bonds may be formed to a carbon atom. Describe its structure in terms of sp^2 hybridization of the carbon atom.

3.4 Describe and discuss electronic structures of (a) LiH; (b) CO; (c) NF.

Exercises

3.1 The bond length in HF is 91.7 pm. On the basis of the experimental dipole moment of 1.85 debye, calculate the effective positive charge on the H atom, as a fraction of the electron charge.

3.2 On the basis of the expression for the energy levels of a particle in a one-dimensional box, derived in exercise 1.8, account for the change in energy of a valence electron when it enters a molecular orbital (a) between two adjacent atoms, and (b) in an extended system of conjugated single and double bonds.

4 Matter and Energy

4.1 Some Basic Concepts

In the three previous chapters we paid considerable attention to the detailed structure of matter and the properties of atoms and molecules at the sub-microscopic level. We now alter our viewpoint for a time in order to discuss the interaction of matter with energy on a macroscopic scale. The concepts and laws that will emerge from this discussion are part of *thermodynamics*, a subject which is to be classed with quantum mechanics as one of the main foundations of modern chemistry. Like quantum mechanics, thermodynamics is built on a framework of generalizations based on experimental observations. Unlike quantum mechanics, thermodynamics does not require us to learn to think in terms of concepts which are far removed from everyday life. There is a link between quantum theory and thermodynamics, in the form of statistical mechanics, a field of study that is concerned with the statistical behaviour of very large numbers of molecules, but the validity of thermodynamics as a description of nature is not dependent upon the existence of this link. As is true for any branch of science, thermodynamics must stand or fall by the accuracy of its predictions, which in this case are predictions about the behaviour of macroscopic samples of matter.

The information that we get from thermodynamics is concerned with the states of chemical systems. By a *system* we mean some portion of the universe that we can mentally separate from the rest for the purpose of study and/or contemplation. A system may be completely isolated from its surroundings, or it may be able to exchange matter or energy with the surroundings. A system can be homogeneous, i.e. of uniform composition and without internal boundaries, or heterogeneous, which covers all the other possibilities. The concept of a system is extremely flexible, being capable of accommodating a bubble of gas, a living organism, or a planet.

The *state of a system* is described by specifying the values of measurable quantities which are associated with the system. Typical examples of quantities which could be used to specify the state of a system are temperature, pressure, volume, mass, surface area and the concentration of a chemical species. The important point about a quantity which is to be used in this way is that its value should be characteristic of the state of the system as it actually exists, and should not be affected by the previous history of the system. Thus, for example, if our system is a pipe containing water, the state of the system can be described by saying how much water is in the pipe now, but not by saying how much has flowed through the pipe at some previous time. Quantities that can be used to specify the state of a system are called state variables, or *state functions*. The quantities in the above list are therefore all state functions, which we represent by the symbols T, P, V, M, A and c_i (where i identifies the substance whose concentration is c_i). It is often useful to classify state functions as being either intensive or extensive, where an intensive variable or function, such as T, P or c_i, is one whose value is independent of the size of the system, and an extensive variable or function, such as V, M or A, is one whose value is proportional to the size of the system. It is important to note that, because the value of a state function is independent of the history of the system, any change in the value of a state function depends only on the system's initial and final states, and not at all on the intermediate states through which the change occurred.

The most basic concepts of thermodynamics are the twin concepts of heat and temperature. Consider thermal energy which, according to the kinetic–molecular theory of matter, is present in all substances as the kinetic energy of random motion of the atoms of which the substance is composed. The average kinetic energy of the atoms is reflected in the temperature of the substance, as indicated by some convenient form of thermometer. Thus in Section 1.4 we asserted that for a perfect gas the temperature and the internal kinetic energy are exactly proportional to one another. Whatever the precise form of the relationship between temperature and internal energy, the quantity that we actually measure is the temperature, and our temperature scales are constructed in such a way that a body at high temperature holds more energy than the same body at low temperature. When two bodies of different temperature are placed in contact we find experimentally that the hotter one cools and the cooler one becomes hotter, a result which we interpret as indicating a flow of heat from the hotter to the cooler body. This statement about the direction of heat flow is sometimes regarded as one of the laws of thermodynamics; being more fundamental than the law which is universally accepted as the First Law, it has to be called the *zeroth law*. Heat, then, is energy in transit.

The idea that temperature and internal kinetic energy are directly related to one another implies the existence of an absolute zero of temperature, a point on the temperature scale at which the kinetic energy has fallen to its lowest possible value. This lower limit occurs at 0 K, or −273.15°C. (Note that in the SI System of units we say zero kelvin not zero degrees kelvin.) The kelvin and celsius scales of temperature are so constructed that their basic units, the kelvin and the degree celsius, are identical at all temperatures.

Heat, like any other form of energy, is measured in joules, which is very convenient in practice because of the ease with which heat can be supplied to a system through an electrical heating coil. The wattage of the coil is equal to the number of joules supplied per second. Heat was formerly distinguished from other forms of energy by being measured in calories, where 1 calorie (equal to 4.184 J) is very close to the amount of heat required to raise the temperature of a gram of liquid water by 1°C.

All forms of energy, including thermal energy, can be made to do work, in the course of which the energy may be changed from one form to another but not destroyed. Thus heat and work are both just energy in the process of conversion from one form to another. Historically, the development of thermodynamics arose from the need to calculate the amount of work which could be obtained from various forms of engine whose operation depended on the flow of heat energy into and out of a system. The relationship which was found to exist between heat, work and the internal energy of a system is embodied in the first law of thermodynamics which we consider in the next section.

4.2 Heat and Work: the First Law of Thermodynamics

The first law of thermodynamics simply says that the internal energy of a system is a state function, and provides a formula which can be used to relate changes in internal energy to the amount of heat supplied to a system and the amount of work done by a system. It is a law of nature, by which we understand a generalization based on experimental observations rather than something which is to be proved or deduced on the basis of more fundamental laws. We accept the law as true because no exception to it has

ever been observed, despite the many determined efforts that have been made. A formal statement of the first law is as follows:

> There exists a function U, called internal energy, which is a state function for any system, and is such that when the state of the system alters the change in U can be calculated from the formula
>
> $$dU = q - w \qquad (4.1)$$
>
> where q is the heat supplied to the system and w is the work done by the system.

Some other properties of the function U which are sometimes included in the statement of the law are that U is an extensive function, and that U is mathematically well-behaved, by which we mean that it is finite, single-valued and continuous. It is important to note that Eq. 4.1 contains dU on the left hand side, i.e. an infinitesimal change in U, so that any measurable change in U has to be calculated from the integral of dU between limits which correspond to the initial and final states of the system. On the other hand, the right hand side of Eq. 4.1 contains the quantities q and w which are not infinitesimal and may in fact be very large. It follows that for Eq. 4.1 to be true q and w must be almost equal to one another whenever U changes by an infinitesimal amount. The heat supplied and work done clearly are not state functions, since their values depend on the detailed history of the system, but the *difference* between q and w is equal to the change in the state function U. Some authors prefer to define w as the work done on the system, rather than work done by the system, in which case the minus sign in Eq. 4.1 becomes a plus sign. This is merely a matter of personal preference; the essential point is that dU is always given by the difference between heat supplied to the system and work done by the system. Provided we are aware of which sign convention is being used for external work the two forms are equally satisfactory. Eq. 4.1 can be regarded as an expression of the law of conservation of energy, since it says that work can be obtained from a system only at the expense of the heat supplied to the system and the system's internal energy.

There are two main forms of work which are encountered in chemical systems, namely electrical work and pressure–volume work. The first occurs when a current I flows through an electrical resistance R so that a potential drop $E = IR$ occurs across the resistance. The power dissipation in the resistance is then EI watts, and the electrical work done in time t is EIt joules. This kind of work may be important in systems where reactions occur between ions. Pressure–volume work is done when a system changes its volume in the presence of an externally applied pressure. It is especially important in systems that contain gases. The work done during an infinitesimal volume change is given by

$$dw = P \, dV \qquad (4.2)$$

where dV is the change in volume and P the external pressure. This result can be obtained by considering Fig. 4.1, which shows a volume V of some compressible material contained in a cylinder by a piston of cross-sectional area A, through which there is exerted an external pressure P. If the piston moves a distance dl the change in volume is given by

$$dV = A \, dl \qquad (4.3)$$

Pressure is defined as force per unit area, so the force on the piston is equal to PA. The work dw is given by the product of force and distance moved, which in this case comes to $PA \, dl$. Hence, replacing $A \, dl$ by dV, we arrive at Eq. 4.2.

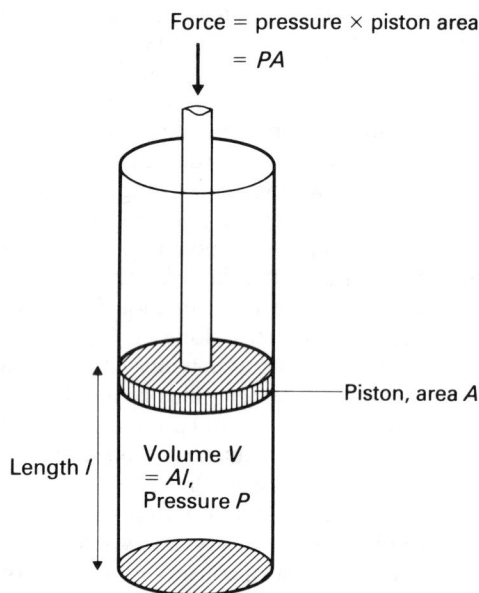

Force = pressure × piston area
$$= PA$$

Piston, area A

Length l

Volume V
$= Al$,
Pressure P

Fig. 4.1 Model used for calculation of pressure–volume work

Addition of heat to a system normally results in an increase in the system's temperature (an exception occurs when the heat is all used to bring about some process which occurs at a constant temperature, such as the vaporization of a liquid at its boiling point). The quantity of heat required to bring about a temperature change of 1 K, or 1°C, is called the heat capacity of the system and is given the symbol C.

The definition of heat capacity can be expressed as

$$dq = C \, dT \qquad (4.4)$$

Using Eq. 4.2 and 4.4 we can now rewrite Eq. 4.1 in the form

$$dU = C \, dT - P \, dV \qquad (4.5)$$

where we are considering infinitesimal values of q and w and it is assumed that the only work that can be done is pressure–volume work. At constant volume ($dV = 0$) this reduces to

$$dU = C_v \, dT \qquad (4.6)$$

Here we have given the heat capacity a subscript v to indicate that the result applies to one particular set of conditions, i.e. to show that C_v represents *the heat capacity at constant volume*. Eq. 4.6 is important because it gives us a relationship between internal energy and temperature, and because it shows that if we measure a heat change at constant volume we are actually measuring a change in the internal energy U.

For many purposes it is convenient to work with the quantity H, called enthalpy (or sometimes heat content), which is defined by

$$H = U + PV \qquad (4.7)$$

From the form of this definition we conclude that H, like U, is an extensive state function for any system. To find an expression for an infinitesimal change in H we differentiate both sides of Eq. 4.7 and obtain

$$dH = dU + P \, dV + V \, dP \qquad (4.8)$$

We now substitute for dU using Eq. 4.5, which gives

$$dH = C \, dT + V \, dP \qquad (4.9)$$

At constant pressure ($dP = 0$) this reduces to

$$dH = C_p dT \qquad (4.10)$$

where the subscript p is used to distinguish *the heat capacity at constant pressure*. Eq. 4.10 is important because it provides a relationship between enthalpy and temperature, and because it shows that when we measure a heat change at constant pressure we are actually measuring a change in the enthalpy H. A finite change in a quantity such as H or U is conventionally written ΔH or ΔU (pronounced delta H or delta U). In practice most experiments are performed at a constant pressure of one atmosphere, with the result that changes in enthalpy are measured rather than changes in internal energy. The difference between ΔH and ΔU is not usually significant when we are dealing with processes that involve only solids or liquids, but is likely to be very important when gases are involved.

Example C_p for liquid water is given as 4.2 J g^{-1} K^{-1}. Calculate the enthalpy change when a kilogram of water cools from 90°C to 10°C.

Since C_p is given as a constant, Eq. 4.10 can be integrated to

$$\Delta H = C_p \Delta T$$

where

C_p is $4.2 \times 1000 \text{ J}^{-1} \text{ kg}^{-1} \text{ K}^{-1}$ and $\Delta T = -80$ K.

Hence $\qquad \Delta H = -336$ kJ.

4.3 Applications of the First Law to Chemical Reactions

(a) Enthalpies of Reaction

Many chemical reactions are accompanied by the evolution of heat, and are said to be *exothermic*. A smaller number occur with the absorption of heat and are termed *endothermic*, while a very few reactions involve no heat change and are termed *thermoneutral*. If a reaction gives out heat this must represent energy lost from the system, so that the enthalpy change ΔH is negative. Similarly, a positive value of ΔH corresponds to an endothermic reaction.

When a reaction can occur in several stages the overall enthalpy change must equal the sum of the enthalpy changes for the individual steps. This statement, which is usually called *Hess's law*, can be regarded either as an expression of the law of conservation of energy, or as a consequence of the fact that H is a state function. Hess's law is most useful in practice when it can be applied to a *thermodynamic cycle* such as the one shown in Fig. 4.2. Here we have an example of a process which can occur directly, with an unknown enthalpy change ΔH, and also by a roundabout path made up of several steps for which the enthalpy changes ΔH_1, ΔH_2, etc. are known. The algebraic sum of the enthalpy changes around the closed cycle is necessarily zero, because taking the system around the cycle corresponds to a process for which the initial and final states are identical.

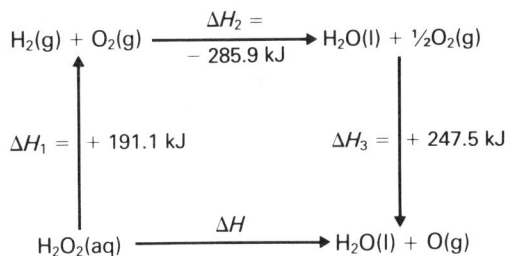

Fig. 4.2 Use of a thermodynamic cycle to calculate an unknown ΔH value. $\Delta H = \Delta H_1 + \Delta H_2 + \Delta H_3$

Hence $\quad (\Delta H_1 + \Delta H_2 + \Delta H_3 + \ldots) - \Delta H = 0 \quad$ (4.11)

and the value of ΔH can be calculated. For the example of Fig. 4.2 we find ΔH to be $+152.7$ kJ mol^{-1}. An important point to notice about the reactions in Fig. 4.2 is that the state of each reactant and product is specified—aqueous (i.e. dissolved in water) for H_2O_2, liquid for H_2O, and gaseous for the others. This is done because the enthalpy of a substance depends on its physical state. In some cases it is even necessary to specify a particular crystalline form; for example, carbon may be diamond or graphite, phosphorus may be red or white, tin may be grey or white, and sulphur may be rhombic or monoclinic.

The number of different chemical substances is very large, but the number of possible chemical reactions is even larger. Therefore, instead of tabulating enthalpies of reaction for individual reactions it is more convenient to list characteristic enthalpies of formation for chemical substances, and to calculate enthalpies of reaction from the enthalpies of formation of reactants and products with the aid of Hess's law. The *standard enthalpy of formation* of a substance is defined as the enthalpy change for formation of a mole of the substance from its constituent elements in their standard states. The value of the standard enthalpy of formation depends on the temperature, which must therefore be specified. The standard state of an element is the pure element in its most stable form at the specified temperature and under a pressure of one atmosphere. Thus the standard states for carbon, oxygen, sulphur and mercury at 298 K are graphite, oxygen gas at 1 atmosphere pressure, rhombic sulphur and liquid mercury, respectively. For substances other than the elements in their most stable forms the standard state is specified as the pure solid, liquid, or gas at a pressure of one atmosphere; if the substance is in solution the standard state is represented by an arbitrary fixed concentration, which is usually chosen to be the concentration corresponding to one mole per litre of an ideal solution. (We shall have more to say about ideal solutions in the next chapter.)

The standard enthalpy of formation is represented by the symbol $H_f^{\ominus}(T)$, where the superscript indicates that standard states are involved and the value of T in kelvins has to be stated.[*] A collection of values of $H_f^{\ominus}(298)$ is given in Table 4.1. A few examples of elements in their standard states are included in the table as a reminder that their enthalpy of formation is zero by definition. For elements in non-standard states, such as gaseous bromine or atomic hydrogen, the value of $H_f^{\ominus}(298)$ is usually positive. Most stable compounds have negative enthalpies of formation but some, notably the oxides of nitrogen and the unsaturated hydrocarbons, have positive values.

[*] The symbol $^{\ominus}$ is a plimsoll mark; $H_f^{\ominus}(T)$ can be read: *H*-plimsoll-f-*T*.

Table 4.1 Standard enthalpies of formation in kilojoules per mole at 298 K

Substance	$H_f^{\ominus}(298)$	Substance	$H_f^{\ominus}(298)$
$H_2(g)$	0.0	$Br_2(g)$	30.71
$H(g)$	217.9	$I_2(s)$	0.0
$H_2O(l)$	− 285.9	$I_2(g)$	62.26
$H_2O(g)$	− 241.8	$Cl(g)$	121.4
$H_2O_2(aq)$	− 191.1	$Br(g)$	111.8
$C(g)$	718.4	$I(g)$	106.6
C (diamond)	1.896	$HCl(g)$	− 92.30
C (graphite)	0.0	$HBr(g)$	− 36.23
$CO(g)$	− 110.5	$HI(g)$	25.44
$CO_2(g)$	− 393.5	$H_2S(g)$	− 20.17
$CH_4(g)$	− 74.85	S (rhombic)	0.0
$C_2H_6(g)$	− 84.68	S (monoclinic)	0.2971
$C_2H_4(g)$	52.30	$SO_2(g)$	− 296.9
$C_2H_2(g)$	226.7	$SO_3(g)$	− 395.2
$C_6H_6(l)$	49.04	Ca(s)	0.0
$C_6H_6(g)$	82.93	CaO(s)	− 635.6
$N(g)$	472.7	$CaCO_3$ (calcite)	− 1207
$N_2(g)$	0.0	$Hg(l)$	0.0
$NO(g)$	90.37	$Hg(g)$	60.84
$NO_2(g)$	33.85	$Hg_2Cl_2(s)$	− 264.9
$N_2O(g)$	81.55	NaCl(s)	− 411.0
$NH_3(g)$	− 46.19	$Na_2CO_3(s)$	− 1131
$O_2(g)$	0.0	KCl(s)	− 435.9
$O(g)$	247.5	$KNO_3(s)$	− 492.7
$O_3(g)$	142.3	AgCl(s)	− 127.0

To calculate the enthalpy change for any reaction we simply set up a thermodynamic cycle of the form shown in Fig. 4.3a and then apply Hess's law. For the example in Fig. 4.3b the enthalpy of reaction at 298 K works out to be − 113.04 kJ mol^{-1}.

(a)

(b)

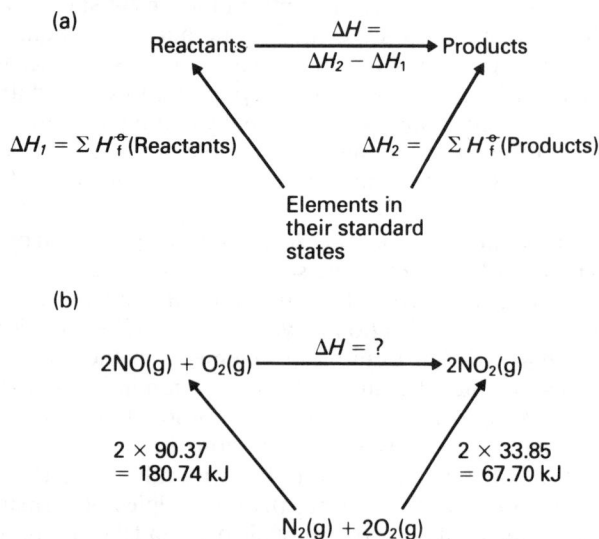

Fig. 4.3 Calculation of enthalpy of reaction from tabulated enthalpies of formation of reactants and products. (a) The basic principle; (b) a specific example

(b) Bond Energies and Dissociation Energies

The dissociation energy of a bond is defined as the enthalpy change involved in breaking the bond in question, when both the parent molecule and the resulting fragments are in the gas phase. Thus the dissociation energy of the H—H bond in hydrogen is twice the enthalpy of formation of atomic hydrogen, and is found from Table 4.1 to be 435.8 kJ mol^{-1} at 298 K. Spectroscopists normally work in terms of dissociation energies at absolute zero, quantities which cannot be obtained by ordinary measurements of heat changes but which are directly accessible to spectroscopic measurement. Bond dissociation energies for diatomic molecules containing two different atoms are readily calculated from data in Table 4.1 by using Hess's law, as in the following example.

Example Calculate the dissociation energy of HCl at 298 K.

Using Hess's law, as in Fig. 4.3 we find the enthalpy change for the reaction

$$HCl(g) \rightarrow H(g) + Cl(g)$$

to be $217.9 + 121.4 - (- 92.3) = 431.6$ kJ mol$^{-1} = D_{H—Cl}$ at 298 K.

The dissociation energy is not usually the same for successive bonds of the same type in a molecule. For example, in H_2O the dissociation energy of the first O—H bond, where the fragments are H and OH, is 502 kJ mol^{-1}, whereas the energy required to dissociate the OH radical is only 423 kJ mol^{-1}. Similarly, the dissociation energies of successive C—H bonds in methane, CH_4, are all different from one another:

$$D_{CH_3—H} = 431 \qquad D_{CH—H} = 523$$
$$D_{CH_2—H} = 364 \qquad D_{C—H} = 339$$

(in kJ mol^{-1} at 298 K).

Despite these differences it turns out to be useful in practice to define a quantity called the *bond energy*, which is the average dissociation energy for bonds of a given type, not merely in a given molecule but in a wide range of molecules. Thus the bond energy for a C—H bond is to be obtained by averaging the dissociation energies of the C—H bonds in CH_4, C_2H_6, C_3H_8, and so on. The resulting bond energies are useful for estimating the total dissociation energy and hence the heat of formation of a molecule for which no experimental data are available. A selection of bond energy values is given in Table 4.2. The accuracy of heats of formation or reaction calculated from average bond energies is typically about ± 2%, unless the molecule is one which is stabilized by resonance in the sense of valence bond theory, as discussed in Chapter 3. The effect of resonance is illustrated by the following example.

Example   Calculate the energy of dissociation of benzene (C_6H_6) into its constituent atoms at 298 K. Comment on the difference between this value and the dissociation energy of a molecule having the Kekulé structure:

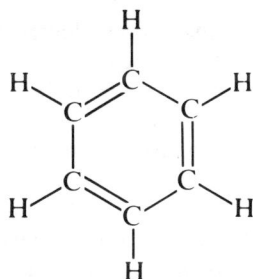

We first use Table 4.1. The required dissociation energy is $6 \times H_f^{\ominus}(C) + 6 \times H_f^{\ominus}(H) - H_f^{\ominus}$ (benzene) $= 6 \times 718.4 + 6 \times 217.9 - 82.93 = 5535$ kJ mol^{-1}.

From Table 4.2 the total bond energy of the Kekulé structure should be $(6 \times 413 + 3 \times 614 + 3 \times 348)$, or 5364 kJ mol^{-1}. The measured dissociation energy of benzene is significantly greater than this and the difference can be attributed to resonance stabilization, as disscussed in Section 3.5.

Table 4.2 Average bond energies (in kJ mol^{-1} at 298 K)

C—H	413	C—O	358	O—O	146
C—C	348	C=O	745	O—H	463
C=C	614	C—N	305	N—H	391
C≡C	839	C=N	615	N—N	163
C—F	489	C≡N	891	N=N	418
C—Cl	348	C—S	272	N—O	201
C—Br	285	C=S	536	N=O	607
C—I	218	Si—H	318	S—H	367
C—Si	285	Si—O	451	C—Si	285

[From Table 10, 'Average bond enthalpies at 25°C', of G. H. Aylward and T. J. V. Findlay, *SI Chemical Data*, Second Edition, p. 101, John Wiley and Sons, 1974.]

The following example illustrates the use of several different kinds of thermodynamic data in a thermodynamic cycle to calculate the value of a quantity which is difficult to measure directly.

Example   Calculate the lattice energy V_1 of a sodium chloride crystal on the basis of the following data:

Enthalpy of formation of NaCl(s)	$= H_f^{\ominus} =$	-411
Ionization energy of Na(g)	$= I$ $=$	$+495$
Electron affinity of Cl(g)	$= A$ $=$	-355
Dissociation energy of Cl_2(g)	$= D$ $=$	$+242$
Enthalpy of sublimation of Na(s)	$= H_s$ $=$	$+108$

(all in kJ mol^{-1}).

The thermodynamic cycle that is involved here is known as the *Born–Haber cycle*. It takes the form:

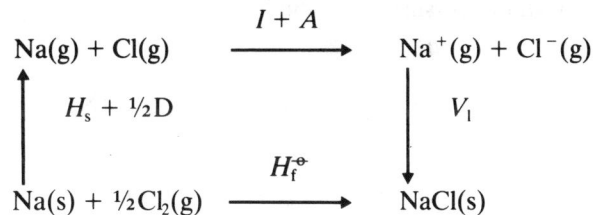

Hence $H_f^{\ominus} = H_s + \frac{1}{2}D + I + A + V_1$

or $V_1 = -411 - 108 - 121 - 495 + 355$
$= -780$ kJ mol^{-1}

Lattice energies estimated using the Born–Haber cycle are used a great deal in discussions of the relative stabilities of crystal lattices and the solubilities of crystalline materials in ionizing solvents. Substances with large negative lattice energies are generally insoluble, because a large amount of energy is required to break down the lattice and allow the ions to go their separate ways in solution.

(c) Dependence of Enthalpy Changes on Temperature

Enthalpies of formation are normally tabulated for a temperature of 298 K, which is not immediately helpful if we happen to be interested in a reaction that occurs at some other temperature. To calculate the dependence of enthalpy change on temperature we again make use of a thermodynamic cycle, as shown in Fig. 4.4. Here we have a hypothetical reaction in which a molecules of substance A react with b molecules of substance B and c molecules of substance C, etc., to produce l molecules of substance L plus m molecules of substance M, and n molecules of substance N, etc. The enthalpy of reaction $\Delta H(T_1)$ at temperature T_1 is presumed known and we wish to calculate the value of

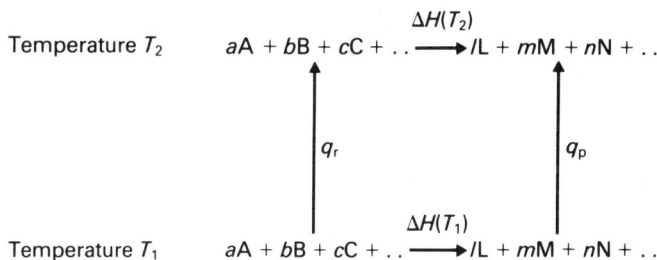

Fig. 4.4 Thermodynamic cycle used to calculate the effect of temperature on enthalpy of reaction

$\Delta H(T_2)$. The quantities q_r and q_p are the enthalpies required to heat the reactants and products, respectively, from T_1 to T_2. Applying Hess's law, we obtain

$$\Delta H(T_2) = \Delta H(T_1) + q_p - q_r \qquad (4.12)$$

To evaluate q_p and q_r we use Eq. 4.10, since we are working at constant pressure, and obtain

$$q_p - q_r = \int_{T_1}^{T_2} [\{lC_p(L) + mC_p(M) + nCp(N)\} \\ - \{aC_p(A) + bC_p(B) + cCp(C)\}]dT \qquad (4.13)$$

The expression inside square brackets represents the difference in heat capacity between products and reactants, and for simplicity we shall write it as ΔC_p. Eq. 4.12 then becomes

$$\Delta H(T_2) = \Delta H(T_1) + \int_{T_1}^{T_2} \Delta C_p dT \qquad (4.14)$$

a result which is usually known as the *Kirchhoff equation*. It applies not only to a chemical reaction, but also to any process having an initial and final state for which the heat capacities are different. *Unless* the heat capacities of the initial and final states differ in the temperature range from T_1 to T_2 the value of ΔH does not vary with temperature. In many applications the dependence of ΔC_p on temperature is small enough to be neglected, so that Eq. 4.14 reduces to

$$\Delta H(T_2) = \Delta H(T_1) + \Delta C_p(T_2 - T_1) \qquad (4.15)$$

where ΔC_p is a constant. Eq. 4.15 is most likely to be applicable when the temperature range is small.

Example   The latent heat of vaporization of mercury is 59.0 kJ mol^{-1} at the boiling point of 357°C. The heat capacities C_p of liquid and gaseous mercury are 28.0 and 21.0 J K^{-1} mol^{-1}, respectively. Calculate the latent heat of vaporization of mercury at 0°C.
The process is

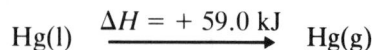

$$Hg(l) \xrightarrow{\Delta H = +59.0 \text{ kJ}} Hg(g)$$

We use Eq. 4.15 with $\Delta C_p = -7.0$ J K^{-1} mol^{-1} and $T_2 - T_1 = -357$ K.

Hence $\Delta H(0°C) = +59.0 + (7 \times 357/1000)$
$$= +61.5 \text{ kJ mol}^{-1}$$

where the factor of 1000 is needed to convert from joules to kilojoules.

If great accuracy is required, or if the temperature range is large, it is necessary to take account of the dependence of ΔC_p on temperature and use Eq. 4.14. Heat capacities for different substances vary with temperature in the manner shown in Fig. 4.5. The value of C_p is most conveniently expressed as a power series in T, thus:

$$C_p = a + bT + cT^2 \qquad (4.16)$$

where values of a, b and c are selected to fit the experimental data for C_p over a specified range of temperature. Some values of a, b and c for gases in the temperature range 300–1500 K are listed in Table 4.3. Sometimes one finds heat capacity values expressed in the alternative form

$$C_p = d + eT \times f/T^2 \qquad (4.17)$$

where d, e and f are chosen to fit the experimental data; this form of expression has no particular advantage over Eq. 4.16. The next two examples show how Eqs 4.14, 4.15 and 4.16 are used in practice.

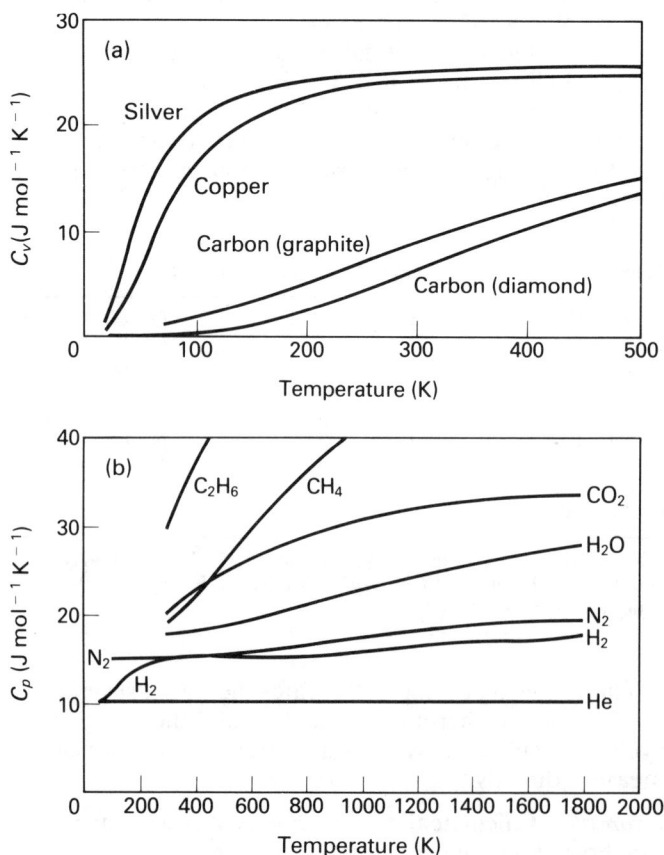

Fig. 4.5 Examples of the dependence of heat capacity on temperature: (a) solids; (b) gases. [Adapted from F. Daniels and R. A. Alberty, *Physical Chemistry*, Wiley, New York, 1966]

Example 1 Calculate the enthalpy change ΔH for the gas phase reaction $N_2(g) + 3H_2(g) \rightarrow 2NH_3(g)$ at 298 K and at 500 K, using the data in Tables 4.1 and 4.3.

Table 4.3 Molar heat capacities of gases at constant pressure (J K^{-1} mol^{-1}). $C_p = a + bT + cT^2$, over the temperature range 300–1500 K.

Gas	a	$10^3 b$	$10^7 c$
H_2	29.07	− 0.8326	20.12
N_2	26.98	5.910	− 3.376
O_2	25.50	13.61	− 42.56
Cl_2	31.70	10.14	− 40.38
CO	26.54	7.683	− 11.72
CO_2	26.76	42.65	− 147.8
HCl	28.17	1.810	15.47
HBr	27.52	3.995	6.615
H_2O	30.07	9.930	8.719
NH_3	25.89	33.00	− 30.46
CH_4	14.32	74.66	− 174.3
C_2H_6	5.753	175.1	− 578.5

The temperature range is quite small so we can use Eq. 4.15. From the listed heat of formation of NH_3 we find immediately that ΔH at 298 K is − 92.38 kJ. To calculate the heat change at 500 K using Eq. 4.15 we should first evaluate the heat capacities in the middle of the temperature range, at 400 K. The results are:

$$N_2 \qquad 29.29$$
$$H_2 \qquad 29.06$$
$$NH_3 \qquad 38.60$$

so that ΔC_p is $(2 \times 38.60 - 3 \times 29.06 - 29.29) = - 39.27$ J K^{-1}.

Hence, from Eq. 4.15, ΔH at 500 K is $- 92.38 - (39.27 \times 202/1000) = - 100.3$ kJ, where the factor of 1/1000 is necessary to convert the second term from joules to kilojoules. It is noticeable that the percentage change in ΔH over the 200 K temperature range is quite small. This would be true of most chemical reactions.

Example 2 Calculate ΔH for the reaction of Example 1 at a temperature of 1500 K.

In this case we have to use Eq. 4.14. When the integration is performed for the case where C_p is given by Eq. 4.16, Eq. 4.14 becomes

$$\Delta H(T_2) = \Delta H(T_1) + \int_{T_1}^{T_2} [\Delta aT + \Delta bT^2/2 + \Delta cT^2/3] \quad (4.18)$$

where Δa, Δb and Δc are the differences between the parameters a, b and c in the expressions for heat capacities of reactants and products. From Table 4.3 we obtain:

$$\Delta a = - 62.41 \; (= 2 \times 25.89 - 26.98 - 3 \times 29.07)$$
$$\Delta b = + 57.59 \times 10^{-3}$$
and $\Delta c = - 117.9 \times 10^{-7}$.

Hence, after substituting 298 for T_1 and 1500 for T_2, and working our way through some fairly lengthy arithmetic, we find ΔH at 1500 K to be − 118.3 kJ. For comparison, the ΔH value obtained by using Eq. 4.15 with ΔC_p evaluated at 900 K is − 111.2 kJ, which represents a substantial error. (At 500 K, where the range of integration is small, the result given by Eq. 4.14 is − 100.73 kJ, which is only slightly different from the figure of − 100.3 kJ that we obtained using Eq. 4.15.)

In general Eq. 4.15 is much easier to use than Eq. 4.14, and introduces far fewer opportunities for arithmetical error in hand calculations. However, if a digital computer is available the evaluation of any number of integrals like that in Eq. 4.14 becomes an entirely painless procedure.

4.4 Applications of the First Law to Gases

(a) Energy and Heat Capacity of an Ideal Gas

An ideal or perfect gas is one that obeys the equation of state

$$PV = nRT \quad (4.19)$$

where P is the pressure, V the volume, n the number of moles, R the ideal gas constant, and T the temperature. For P in newtons per square metre (N m^{-2} = Pa = pascals), V in cubic metres, n in moles, and T in kelvins, R has the value 8.3143 J mol^{-1} K^{-1}. If n is expressed in kilogram moles R is 1000 times larger. For P in atmospheres (1 atmosphere = 1.013 25 × 10^5 Pa), V in litres (litre is a permissible abbreviation for cubic decimetre), n in moles, and T in kelvins, R has the value 0.082 06 l atm mol^{-1} K^{-1}.

The model of an ideal gas that we considered in Section 1.4 consisted of a very large number of structureless particles of negligible size, whose kinetic energy of motion between the walls of their enclosure was the only form of internal energy available to the gas. For a monatomic gas such as helium or argon this would be a good approximation to reality. For this model we obtained the result

$$PV = N m \overline{c^2}/3 \quad (4.20)$$

where N is the number of molecules in the enclosure of volume V, m is their mass, and $\overline{c^2}$ is their mean square velocity. The average kinetic energy per molecule is $\frac{1}{2}m\overline{c^2}$, so that the total kinetic energy for N molecules is given by

$$U = N m \overline{c^2}/2 = 3PV/2$$

i.e. $$U = 3nRT/2 \quad (4.21)$$

According to the model we have chosen for an ideal gas the molecules are so small that pushing them closer together by decreasing V cannot cause them to interact with one another, and so cannot affect the internal energy. Thus *for an ideal gas U depends only on the temperature*, and is independent of P or V. Since H is equal to $U + PV$, and neither U nor PV is dependent on pressure or volume at constant temperature (cf. Boyle's Law, $PV =$ constant), it follows as well that *for an ideal gas H depends only on the temperature*, and is independent of P or V. Consequently the equations

$$dU = C_v dT \qquad (4.6)$$

and

$$dH = C_p dT \qquad (4.10)$$

which apply to any system under conditions of constant volume and constant pressure, respectively, apply to an ideal gas whether or not the pressure or volume are held constant. (Of course the equalities $dU = dq$ and $dH = dq$ are true only at constant volume and constant pressure, respectively.)

If both sides of Eq. 4.21 are differentiated with respect to T we obtain

$$C_v = 3R/2 \qquad (4.22)$$

for one mole of an ideal monatomic gas.

If we add PV to the left hand side and RT to the right hand side of Eq. 4.21 we obtain

$$H = 5RT/2 \qquad (4.23)$$

and differentiating both sides with respect to T now gives

$$C_p = 5R/2 \qquad (4.24)$$

for one mole of an ideal monatomic gas. Because H is equal to $U + PV$, it must always be true that for one mole of ideal gas

$$C_p - C_v = R \qquad (4.25)$$

whether or not the internal energy is given by Eq. 4.21.

Experimental values of heat capacity for monatomic gases (helium, neon, argon, for example) are in excellent agreement with Eqs 4.22 and 4.24, but this agreement is not preserved for gases which contain two or more atoms, even though they may obey Eq. 4.19 very closely. The reason is that polyatomic gases can store energy in the form of rotation and vibration, as well as in the form of translation. The effects of vibration are relatively unimportant, because vibrational energy levels are normally so widely spaced that most molecules remain in their lowest vibrational level at ordinary temperatures and the possibility of storing internal energy as vibration is not properly exploited. The effects of rotation, on the other hand, are always important. In Sec-

tion 1.4 it was mentioned that the internal energy of $3RT/2$ for an ideal monatomic gas amounted to $\frac{1}{2}RT$ per degree of freedom, where a degree of freedom is defined as an independent mode of storing energy. With translation there are three degrees of freedom, corresponding to the independent movement of the molecule in three different directions at right angles to one another. A diatomic or linear polyatomic molecule has in addition two degrees of rotational freedom, corresponding to rotation about two independent axes at right angles to the molecular axis. A non-linear polyatomic molecule has three degrees of rotational freedom. It is reasonable to expect that the same amount of energy will be stored in each degree of freedom that a molecule possesses (a proposition which is known as the equipartition theorem) so that for a linear molecule we expect to have U equal to $5RT/2$ and for a non-linear molecule U should be equal to $3RT$. If there are internal vibrations for which the frequency ν, and therefore the energy level spacing $h\nu$, is low enough for them to store significant amounts of internal energy, U will be larger still. Most large molecules do in fact have several low frequency modes of vibration. The increase of heat capacity with temperature that is observed for all polyatomic gases (cf. Eq. 4.16) results from the increasing use of vibrational degrees of freedom. The total number of vibrational modes is $3N - 5$ for a linear molecule and $3N - 6$ for a non-linear molecule, where N is the number of atoms in the molecule.

The ratio of heat capacities for a gas, C_p/C_v, is used sufficiently often to be given its own symbol, γ (gamma). On the basis of the discussion in the previous paragraph we can conclude that for a monatomic gas γ should be equal to 1.67, for a diatomic or linear polyatomic gas γ should be 1.40, for a non-linear gas with no low-lying vibrational levels γ should be 1.33, and for a non-linear polyatomic gas with accessible vibrational levels γ should be less than 1.33. A number of experimental values of γ are given in Table 4.4. They are seen to be in good agreement with the values predicted by the theory. Hydrogen is an interesting case, in that γ is significantly higher than predicted because the *rotational* energy levels of the molecule are rather widely spaced. It can be shown, using quantum theory, that the spacing of a molecule's rotational energy levels is inversely proportional to its moment of inertia and, of all molecules, hydrogen has the lowest moment of inertia.

Example Calculate the enthalpy change in one mole of an ideal diatomic gas which is heated and compressed from 0°C at one atmosphere pressure to 100°C at ten atmospheres.

For an ideal gas the enthalpy is independent of pressure so we can confine our attention to the temperature change. We assume that the vibrational frequency of the molecule is high, so that C_p can be taken as having a constant value of $7R/2$. For a temperature change of $100°C = 100$ K the value of ΔH is therefore $700R/2$, or 2911 J.

Table 4.4 Comparison of predicted and observed values of $\gamma = C_p/C_v$ at 298 K

Gas	Predicted γ	Observed γ
He	1.67	1.67
Ar	1.67	1.67
Hg	1.67	1.67
H_2	1.40	1.41
N_2	1.40	1.40
O_2	1.40	1.40
Cl_2	1.40	1.34
N_2O	1.40	1.28
CO_2	1.40	1.29
NH_3	1.33	1.31
CH_4	1.33	1.31
SO_2	1.33	1.29
C_2H_6	< 1.33	1.19
C_6H_6	< 1.33	1.11

(b) Heat and Work of Ideal Gas Expansion: Isothermal and Adiabatic Processes

The work done when a gas expands can be calculated by integrating both sides of the equation

$$\mathrm{d}w = P\,\mathrm{d}V \qquad (4.2)$$

between limits corresponding to the initial and final states of the system. Because w is not a state function the value of the integral depends on the path that is followed, i.e. on the nature of the intermediate states. In principle there are an infinite number of possible paths by which a gas expansion can take place, corresponding to an infinite number of ways of selecting the intermediate states. Each path will have its own characteristic values of q and w. There are, however, two particular paths which can be regarded as especially important, namely the isothermal and adiabatic paths, and these we now consider in detail.

As an example of an isothermal process we shall consider the expansion of a sample of ideal gas from a volume of 1 cubic metre at a pressure of 10^5 pascals (roughly one atmosphere) to a volume of 2 cubic metres at 5×10^4 pascals. The pressures and volumes have been chosen to satisfy Boyle's law, so we can be certain that this is an *isothermal* process, with T constant. One way of carrying out the expansion would be to reduce the external pressure on the gas suddenly from 10^5 to 5×10^4 Pa, and then allow the gas to expand against this constant pressure of 5×10^4 Pa. For expansion against a constant pressure P, Eq. 4.2 can be integrated to yield

$$w = P(V_2 - V_1) = P\Delta V \qquad (4.26)$$

and for our example the work done is therefore $5 \times 10^4 \times (2 - 1) = 5 \times 10^4$ J. Here we have to remember that a pascal (Pa) is a newton per square metre (N m^{-2}) and that a Joule (J) is a newton metre (N m), so the product of pascals and cubic metres (m^3) is joules (N m$^{-2} \times$ m^3 = N m = J). The integral in this case is represented by the shaded rectangle in Fig. 4.6.

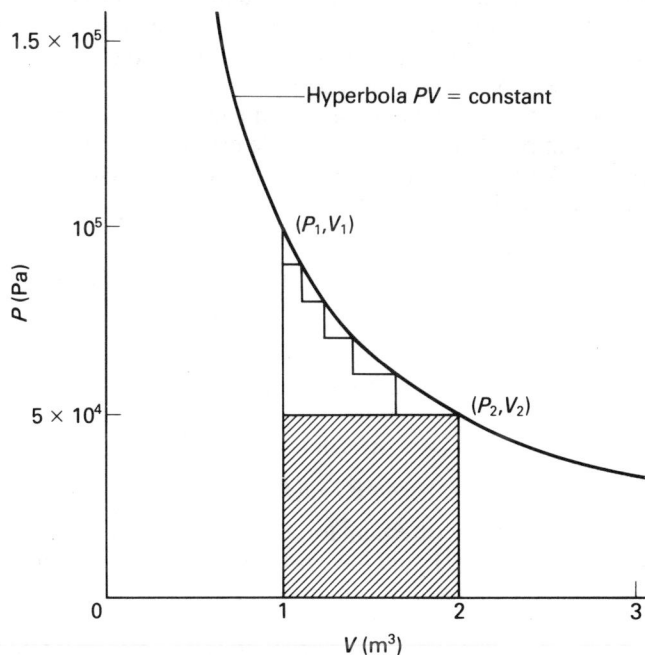

Fig. 4.6 Evaluation of work done during isothermal expansion of an ideal gas. For explanation see text

We could reach the same final state in the expansion by reducing the external pressure by only a very small amount, and allowing the gas to expand to a state of equilibrium at the new pressure before again reducing the external pressure. This might be done, for example, by weighting a piston with a bag of sand and removing the sand one grain at a time. The variation of pressure with volume during the course of the expansion would then be represented by a staircase-shaped graph, similar to the one in Fig. 4.6, which has been drawn for an expansion in just five separate steps. The work done in the expansion is equal to the total area under the staircase, and is clearly greater than that for the one-step expansion. In order to make the expansion take place it is always necessary to reduce the external pressure to less than the pressure of the gas; the expansion of the gas is then a *spontaneous* process. Each small drop in pressure is followed by a spontaneous increase in volume.

No matter how many steps there are in the staircase corresponding to a spontaneous process, the area under the graph, which is the work done by the expanding gas, is always less than the area under the smooth curve which represents the relationship PV = constant. In the limiting case, where the steps in the expansion are vanishingly small, the process is no longer spontaneous, but instead is *reversible*. The term reversible is used in a very restricted sense in thermodynamics to describe a process which occurs in such a way that the system is always in a state of equilibrium while the process is occurring. The concept of a reversible process turns out to be extremely useful, even though it may be impossible to achieve a truly reversible process in practice. In the reversible expansion of an ideal gas Eq. 4.19 is obeyed throughout the process. Hence we can integrate Eq. 4.2 as follows:

$$w = \int_{V_1}^{V_2} P \, dV = \int_{V_1}^{V_2} \frac{nRT \, dV}{V}$$

which gives, with n and T constant,

$$w = nRT \ln(V_2/V_1) \qquad (4.27)$$

or
$$w = nRT \ln(P_1/P_2) \qquad (4.28)$$

where we have written ln for the logarithm to base e, ($\ln x = 2.303 \log_{10} x$), and have used Boyle's law to go from Eq. 4.27 to Eq. 4.28. For the example we are considering the value of w for a reversible expansion is given by $nRT \times 2.303 \times \log_{10} 2$, where, from the ideal gas law, $nRT = 10^5 \times 1$. Therefore $w = 10^5 \times 2.303 \times 0.3010 = 8.343 \times 10^4$ J. This is the area under the curve in Fig. 4.6, as measured between the limits V_1 and V_2.

We now have formulas which allow us to calculate the work done in a spontaneous isothermal expansion and in a reversible isothermal expansion of an ideal gas. We may note also that in any isothermal process ΔU is zero for an ideal gas, so the first law requires that

$$w = q \qquad (4.29)$$

This means that the work done has to be balanced by an equal flow of heat into the gas, or the process will not be isothermal. Eq. 4.26 shows incidentally that when a gas expands freely into a vacuum w must be zero, because the external pressure P is zero. Hence in an isothermal free expansion ΔU, q and w are all zero. The same applies if the final pressure is not zero, as during free expansion into a fixed volume, provided no work is done against an external pressure.

In an isothermal process, heat is allowed to flow in and out of the system under study to the extent which is required to keep the temperature constant. Experimentally this could be achieved by immersing the system in a thermostat, or 'heat bath', at the desired temperature. At the opposite extreme, we can consider a system which is thermally insulated from its surroundings so that no heat flow can occur, i.e. q is zero. This situation is approached very closely in a well-designed calorimeter, and also in cases where the process of interest occurs too rapidly for heat flow to be significant, as in the rapid compressions and rarefactions which accompany the passage of a sound wave through a gas. A process for which q is zero is termed *adiabatic*. In an adiabatic expansion of a gas any work that is done must be done at the expense of the internal energy, since the first law requires

$$\Delta U = - w \qquad (4.30)$$

If w is positive, corresponding to work done by the system, it follows that the temperature of the gas must fall. The magnitude of the temperature change is to be calculated from the equation

$$dU = C_v \, dT \qquad (4.6)$$

To illustrate the procedure to be followed we consider an adiabatic expansion similar to our previous isothermal expansion, from 10^5 Pa in a volume of 1 m^3 to a pressure of 5×10^4 Pa in a volume which has still to be determined. The final volume cannot be obtained simply from Boyle's law because the temperature does not remain constant. To fix the conditions more precisely, we suppose the initial temperature to be 300 K, and we take C_v for the gas to be $3R/2$ per mole, i.e. the gas is presumed to be monatomic.

For the extreme case of a spontaneous adiabatic expansion we allow the gas to expand to a final volume V and final temperature T against the constant pressure of 5×10^4 Pa. The work done is then given by

$$w = P\Delta V \qquad (4.26)$$

where P is 5×10^4 Pa and ΔV is equal to $(V - 1)$m^3. Since C_v is a constant ($= 3nR/2$) we have also that

$$\Delta U = C_v \Delta T \qquad (4.31)$$

and Eq. 4.30 becomes

$$3nR(T - 300)/2 = - 5 \times 10^4 (V - 1)$$

The ideal gas law gives us two further relationships, namely

$$10^5 \times 1 = nR \times 300$$

and
$$5 \times 10^4 V = nRT$$

Hence, by eliminating first nR and then T, we obtain $V = 1.60$ m^3 and $T = 240$ K. The work done is $5 \times 10^4 \times 0.60 = 3 \times 10^4$ J.

To find out what happens in a reversible adiabatic expansion we have to integrate both sides of the equation

$$C_v dT = -P\, dV \qquad (4.32)$$

along the path defined by $PV = nRT$. Hence

$$\int_{T_1}^{T_2} \frac{C_v dT}{T} = \int_{T_1}^{V_2} \frac{-nR\,dV}{V} \qquad (4.33)$$

which becomes, using $C_p - C_v = nR$ and $C_p/C_v = \gamma$,

$$\ln(T_2/T_1) = (\gamma - 1)\ln(V_1/V_1) \qquad (4.34)$$

This can be written in more memorable form as

$$TV^{\gamma - 1} = \text{a constant} \qquad (4.35)$$

Eq. 4.35 enables us to relate the final temperature and volume to the initial temperature and volume in a reversible adiabatic expansion. Two analogous relationships can be obtained by using the ideal gas law to eliminate either T or V from Eq. 4.35. These are

$$PV^\gamma = \text{a constant} \qquad (4.36)$$

and

$$T^\gamma P^{(1-\gamma)} = \text{a constant} \qquad (4.37)$$

(Note that the constants that appear in the last three equations are all different from one another.)

For the example we are considering we are given $P_1 = 10^5$Pa, $V_1 = 1$ m³, $T_1 = 300$ K and $P_2 = 5 \times 10^4$ K. We also know that $\gamma = 5/3$. Eq. 4.37 therefore becomes

$$300^{1.667}/(10^5)^{0.667} = T_2^{1.667}/(5 \times 10^4)^{0.667}$$

from which T_2 can be evaluated with the aid of logarithm tables, or preferably a calculator, as 227.4 K. To calculate the change in internal energy, and hence the work done, we need to know how many moles of gas there are. The ideal gas law gives $10^5 = nR \times 300$, so that $n = 10^3/3R$. Hence $C_v = 3nR/2 = 500$ J K⁻¹, and Eq. 4.31 gives ΔU as -3.63×10^4 J. The work done is therefore $+3.63 \times 10^4$ J.

For a free adiabatic expansion, i.e. an adiabatic expansion into a vacuum or into a fixed volume, the work done against external pressure is zero. Hence, since q is zero, ΔU is also zero and the temperature of the gas does not alter.

Example Calculate the enthalpy changes which occur during all of the gas expansions that we have considered in this section.

Because the enthalpy of an ideal gas depends only on the temperature, ΔH must be zero for all the isothermal processes. For the adiabatic processes we note that $\Delta H = \Delta U \times C_p/C_v$, on account of Eqs 4.6 and 4.10. Therefore, for the spontaneous expansion ΔH is $-1.667 \times 3 \times 10^4$ J $= -5 \times 10^4$ J, and for the reversible expansion ΔH is $-1.667 \times 3.63 \times 10^4 = -6.05 \times 10^4$ J.

The different cases that arise in isothermal and adiabatic gas expansions are summarized in Table 4.5.

Table 4.5 Values of q, w and ΔU and ΔH during isothermal and adiabatic expansions of an ideal gas

Process	Isothermal path	Adiabatic path
Free expansion into vacuum or into a fixed volume	q, w, ΔU, ΔH, ΔT all zero	q, w, ΔU, ΔH, ΔT all zero
Expansion against a constant pressure P. (Spontaneous expansion)	$q = w = P\,\Delta V$ ΔU, ΔH, ΔT all zero	$w = -\Delta U = P\,\Delta V$ $\Delta T = \Delta U/C_v$ $\Delta H = C_p\Delta T$ $q = 0$
Reversible expansion, $PV = nRT$ at all times	$q = w = nRT \ln (V_2/V_1)$ $= nRT \ln (P_1/P_2)$ ΔU, ΔH, ΔT all zero	Calculate final P, V, T using Eq. 4.35, 4.36 or 4.37 as appropriate. $\Delta U = C_v\Delta T$ $w = -\Delta U$ $\Delta H = C_p\Delta T$ $q = 0$

Exercises

4.1 1 mole of an ideal gas is allowed to expand from 2 litres to 4 litres against a constant pressure of 1 atmosphere (101.3 kN m⁻²). At the same time 250 J of heat is supplied from an external source. Calculate ΔU and ΔT for the gas, given that C_v for the gas is 12.5 J K⁻¹ mol⁻¹.

4.2 A bar of metal weighing 10 kg is heated rapidly from 20°C to 250°C. The heat capacity of the metal is 1.1 J g⁻¹. The bar expands as a consequence of the heating and raises a mass of 10 tonnes (10⁴ kg) through a distance of one centimetre. What are q, w and ΔU for the bar of metal. (Gravitational acceleration g = 9.81 ms⁻².)

4.3 Calculate the change in internal energy of an electric battery which delivers a current of one milliampere through a resistance of five kilo-ohms for a period of ten minutes.

4.4 For every joule that is delivered by the battery of exercise 4.3 to its external circuit, 0.5 joules is dissipated as heat in the battery's internal resistance. The heat capacity of the battery is 50 J K⁻¹. Calculate the temperature rise in the battery after ten minutes' operation, neglecting heat losses.

4.5 Calculate the internal energy change and the temperature rise in a spring which is suddenly compressed a distance of 4 cm by a weight of 5×10^3 kg. The weight of the spring is 700 g and the heat capacity of its metal is 1.1 J K⁻¹ g⁻¹ (g = 9.81 m s⁻²).

4.6 Show that for a process involving the production or consumption of an ideal gas, $\Delta H = \Delta U + \Delta nRT$, where Δn

is the change in the number of moles of ideal gas due to the process. For the reaction $N_2 + 3H_2 \rightarrow 2NH_3$ the enthalpy change is -92.38 kJ at 298 K. What is ΔU at 298 K?

4.7   Benzene C_6H_6 (liquid) is burnt to CO_2 and H_2O in a bomb calorimeter (i.e. at constant volume). Calculate the heat released per mole of benzene at 298 K. (Use Table 4.1 to calculate ΔH, then convert to ΔU using the formula proven in exercise 4.6.)

4.8   Calculate the energy of dissociation of acetylene, C_2H_2, into gaseous atoms at 298 K, using data from Table 4.1, and compare your answer with the value predicted from the bond energies in Table 4.2.

4.9   Calculate the heat of combustion of ammonia, to N_2 and water vapour, at 298 K, first using data from Table 4.1 and then using bond energy data from Table 4.2.

4.10   Calculate enthalpy changes at 298 K for the following gas phase reactions:

(a) $N + NO \quad\rightarrow N_2 + O$
(b) $NO_2 + SO_2 \rightarrow NO + SO_3$
(c) $O + NO_2 \quad\rightarrow O_2 + NO$
(d) $O_3 + NO \quad\rightarrow NO_2 + O_2$
(e) $3C_2H_2 \qquad\rightarrow C_6H_6$
(f) $H_2S + 1\frac{1}{2}O_2 \rightarrow H_2O + SO_2$

4.11   Estimate the maximum temperature of a hydrogen-air flame, with $H_2:N_2:O_2 = 2:4:1$, using $C_p = 35$ J K^{-1} for N_2 and $C_p = 47$ J K^{-1} for H_2O.

4.12   Calculate the enthalpy change for the gas phase reaction $2NH_3 + 1\frac{1}{2}O_2 \rightarrow N_2 + 3H_2O$ at 1500 K.

4.13   Calculate the enthalpy change for the reaction $H_2(g) + Cl_2(g) \rightarrow 2HCl(g)$ at 1000 K.

4.14   Argon (an ideal monatomic gas) is present in trace amounts in the fuel–air mixture supplied to a high-flying jet engine. Calculate the enthalpy change, per mole of argon, between the ambient air at $-30°C$, the combustion chamber at 2000°C and the exhaust at 1200°C.

4.15   Calculate the ΔU, ΔH, q and w for one mole of an ideal monatomic gas at 273 K which expands from 11.2 l to 22.4 l: (a) reversibly and isothermally; (b) by expanding isothermally into an evacuated volume; (c) by expanding isothermally against a constant pressure of one atmosphere (101.3 kPa); (d) by expanding adiabatically into an evacuated volume; (e) by expanding adiabatically against a constant pressure; (f) by expanding adiabatically and reversibly.

5 Systems in Equilibrium

5.1 Introduction: Physical and Chemical Equilibria

Processes which occur in chemical systems can usefully be classified as either physical or chemical in nature. Physical processes typically involve phase transitions, such as the evaporation of a liquid, the dissolving of a solid in a liquid, or the transformation of a solid from one crystalline form to another. Chemical processes are transformations in which one substance is converted to another by the breaking and re-forming of chemical bonds. With either kind of process the change in the state of the system continues until equilibrium has been reached, and the initial departure of the system from equilibrium can be regarded as providing the driving force for the process. For example, evaporation of a liquid continues until the rate at which molecules escape into the vapour from the surface of the liquid is exactly balanced by the rate at which molecules from the vapour collide with the surface and are recaptured. At this point the pressure of the vapour is the equilibrium vapour pressure for that temperature. If the temperature of the liquid is raised, thereby increasing the rate at which molecules can escape, more liquid will evaporate until the number of molecules in the gas phase is again large enough to make the rate of condensation equal to the rate of evaporation. This kind of dynamic situation at the molecular level is charcteristic of every state of equilibrium. A chemical reaction reaches equilibrium when the rate of the forward reaction is equal to the rate of the reaction by which products are converted back to reactants; a solid dissolves in a liquid until the rate of solution is exactly balanced by the rate of crystallization. Even for a chemical reaction that appears to go to completion, such as the reaction of hydrogen with oxygen to form water, the equilibrium concentrations of the reactants are not zero, although they may be below the lower limits for detection by present analytical techniques.

Thermodynamics is a very powerful tool for dealing with physical and chemical equilibria. It can provide detailed information about the equilibrium state of a system and about the magnitudes of changes in the values of state functions when a system moves towards equilibrium. Thermodynamics can also predict whether a hypothetical process will tend to occur spontaneously, on the basis of whether or not the process corresponds to a move in the direction of equilibrium. A process which does correspond to a move towards the equilibrium state is described as thermodynamically allowed, or thermodynamically feasible. However, thermodynamics does not give any information about the rate at which a system will move towards equilibrium, and in practice the rate of attainment of equilibrium varies widely from one system to another. It is not uncommon to find that a system remains indefinitely in a *metastable* state because the processes by which it could reach a state of true thermodynamic equilibrium are extremely slow. Thus, for example, a mixture of hydrogen and chlorine can be kept in the dark at room temperature, even though it is highly unstable with respect to conversion to hydrogen chloride and in the presence of ultraviolet light the conversion may take place explosively. At normal temperatures and pressures, diamonds are unstable with respect to conversion to graphite. On a more mundane level, the process of crystallization which causes window glass, a supercooled liquid, to lose its transparency, is so slow at ordinary temperatures that the glass is more likely to fail for other reasons.

Factors which govern the rate of attainment of equilibrium in chemical systems form the subject of Chapter 7. In the remainder of this chapter we shall examine the consequences of some laws which govern physical and chemical equilibria, and consider some useful generalizations that result from the application of thermodynamics to systems in equilibrium. We first consider equilibria that do not involve the conversion of one substance to another.

5.2 Physical Equilibrium

(i) Raoult's Law and the Ideal Solution

When two liquids are shaken together they may either mix completely and dissolve in one another to form a homogeneous solution, or they may be immiscible, forming two separate layers of different composition. The first type of behaviour is shown by water and alcohol; the second by water and ether. When two liquids are immiscible each has its own tendency to evaporate, and each exerts its own vapour pressure independently of the presence of the other, except in so far as one liquid may dissolve to a small extent in the other (and vice versa) and so modify its properties. The effect of dissolved substances on the properties of liquids in homogeneous solutions is the main concern of Raoult's law.

We consider a binary solution, i.e. one with just two components that we call A and B. If one component is present in a marked excess over the other it can be regarded as the solvent, and the other as the solute, but in general we shall want to cover the whole range of composition, from pure component A to pure component B, so that the roles of solvent and solute will be interchangeable. It is convenient to specify the composition of a solution in terms of the *mole fractions* of the components. If a solution contains n_A moles of component A and n_B moles of component B, the mole fractions of A and B are given by

$$x_A = n_A/(n_A + n_B) \tag{5.1}$$

and

$$x_B = n_B/(n_A + n_B) \tag{5.2}$$

The definition is easily extended to deal with solutions of three or more components. In all cases the sum of the mole fractions of the different components of a mixture is equal to 1.

When a solution containing A and B evaporates the rate at which A and B molecules in the gas phase collide with the surface of the liquid and are recaptured is independent of the composition of the liquid, and depends only on the number of each type of molecule in the vapour. However, of the molecules which are in a position to escape from the surface only the fraction x_A are component A, while the fraction x_B are component B (Fig. 5.1). Therefore the vapour pressure of each component at equilibrium above the solution is likely to be less than it would be for the pure component alone. An *ideal solution* is one in which the escaping tendency of the component molecules is not altered as a result of intermolecular interactions between the two components, so that the partial pressures of A and B in the vapour are simply proportional to their mole fractions in the solution. This is *Raoult's Law*, which is usually written in the form

$$P_A = x_A P_A^0 \tag{5.3}$$

where P_A is the equilibrium pressure of A above a solution in which its mole fraction is x_A, and P_A^0 is the vapour pressure of pure liquid A. An ideal solution is one that obeys Raoult's law. It should be noted that the condition for ideality is not that there should be no interaction between

molecules of A and molecules of B, but that the interaction of A with B should have the same effect on the vapour pressure as the interaction of A with A or of B with B.

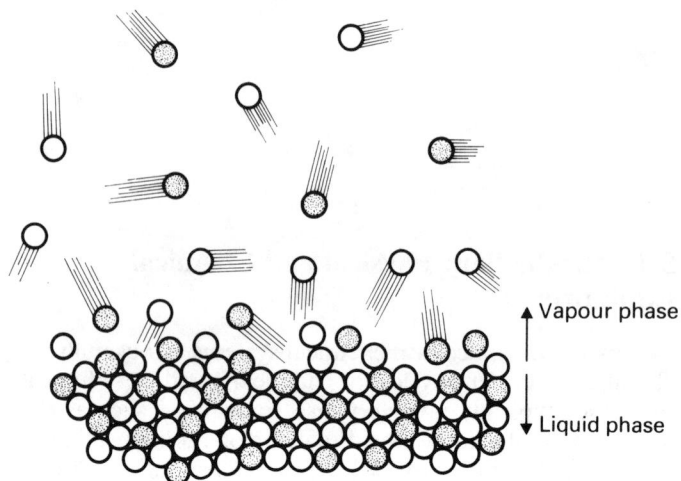

Fig. 5.1 Evaporation at the surface of a two-component solution

Therefore we can expect ideal solutions to be formed by pairs of substances which are chemically very similar to one another. Vapour pressure data for a mixture of bromoethene (CH_2CHBr) and 1-bromopropene ($CH_3CHCHBr$) are shown in Fig. 5.2. The small deviations from Raoult's law for this mixture are within the limits of experimental error.

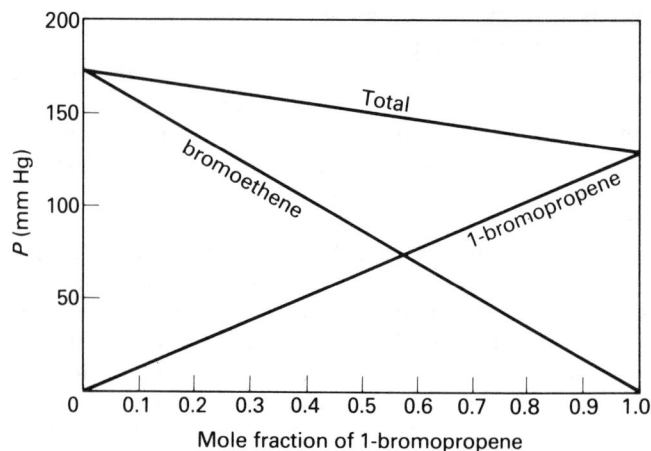

Fig. 5.2 Partial pressures and total vapour pressure of a mixture of bromoethene and 1-bromopropene at 85°C

(ii) Non-ideal Solutions: Henry's Law

Most binary mixtures deviate from the ideal behaviour described by Raoult's law. The deviations may be positive,

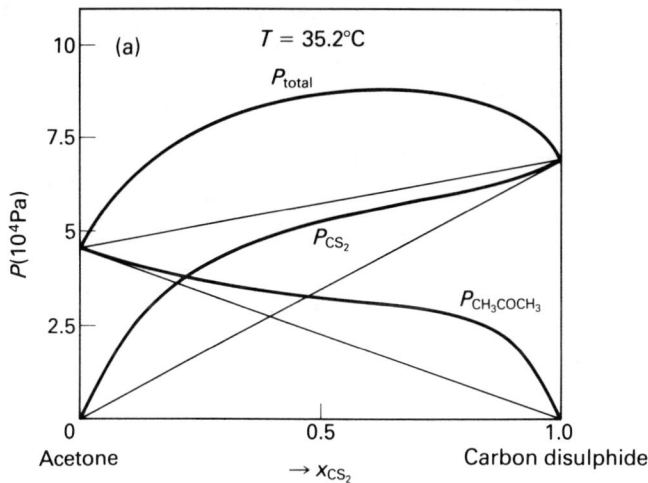

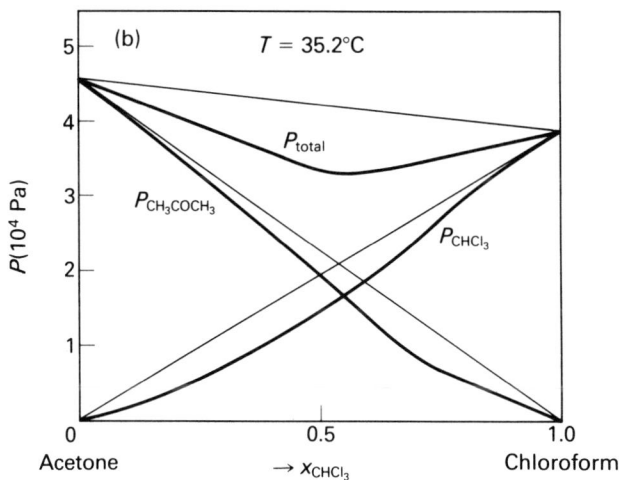

Fig. 5.3 Examples of positive deviations (a) and negative deviations (b) from Raoult's law

This is *Henry's law*. The constant k, which is equal to P_A^0 in the case of an ideal solution, is the Henry's law constant. The Henry's law constant is an empirical quantity, i.e. the value of k has to be found by experiment. We can now say that *in a dilute solution Henry's law is obeyed by the solute*. The validity of Henry's law for a dilute solution does not depend on the possibility of obtaining the pure liquid A, and Henry's law can therefore be applied to a solution of a gas at a temperature either above or below the critical point of the gas. In this situation the law is often expressed in an alternative form as: the solubility of a gas in a liquid is proportional to the partial pressure of the gas above the liquid. The regions in which Raoult's and Henry's laws are applicable to solutions showing marked deviations from ideality are shown in Fig. 5.4.

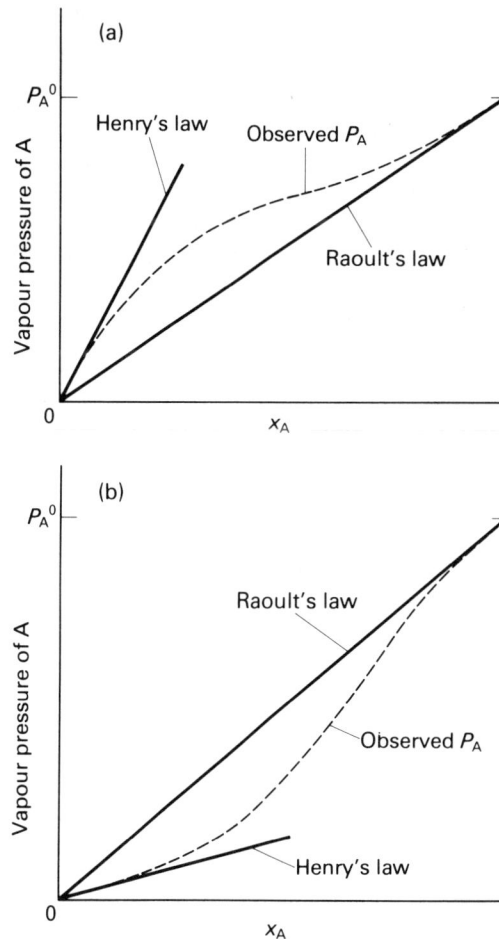

Fig. 5.4 Partial pressure of a substance A as a function of the mole fraction of A in solution, for solutions showing marked positive deviations (a) and negative deviations (b) from Raoult's law. In both cases Raoult's law is obeyed for $x_A > 0.95$ and Henry's law is obeyed for $x_A < 0.05$

as in Fig. 5.3a, or negative, as in Fig. 5.3b, corresponding to an excess of repulsive or attractive interactions, respectively, between the two components. Although the mixtures showing the two types of deviation appear markedly different from one another, there are two important common features in these diagrams. The first is that when the mole fraction of a component approaches unity, the vapour pressure curve for that component approximates very closely to the straight line corresponding to Raoult's law, i.e. *in a dilute solution Raoult's law is obeyed by the solvent*. The second common feature is that when the mole fraction of a component approaches zero, the dependence of vapour pressure on mole fraction is approximately linear, i.e. when x_A is small

$$P_A = k\, x_A \qquad (5.4)$$

The solubility of a gas in a liquid is conveniently defined by

$$x = P s \qquad (5.5)$$

i.e. the solubility s is simply the inverse of the Henry's law constant. Alternatively, the solubility of a gas in a liquid can be expressed in terms of the number of volumes of gas which will dissolve in one volume of liquid at a given temperature. The ratio (volumes of gas)/(volumes of liquid) is known as the Bunsen coefficient. It is left as an exercise for the reader to show, on the basis of Henry's and Boyle's laws, that the value of the Bunsen coefficient is independent of the gas pressure.

(iii) Application of Raoult's Law to Dilute Solutions: Freezing Point Depression and Boiling Point Elevation

We have seen that Raoult's law is always obeyed by the solvent in a dilute solution, even when the deviations from Raoult's law are large for a more concentrated solution. The effect of the solute is to reduce the vapour pressure of the solvent in proportion to the mole fraction of the solute. (From Eq. 5.3, since x_B is equal to $(1 - x_A)$, it follows that $(P_A^0 - P_A)$ is equal to $P_A^0 x_B$). Provided the solute is involatile, this reduction in vapour pressure necessarily leads in turn to an elevation of the boiling point and a reduction of the freezing point of the solution relative to the pure solvent. To see how this comes about we first consider Fig. 5.5a, which represents a *phase diagram* for the pure solvent. The curve AB shows the variation of the vapour pressure of the solid with temperature, which is the same as the variation of the sublimation temperature of the solid with pressure. At values of pressure and temperature which lie on this curve the solid is in equilibrium with its vapour. Below AB the solid evaporates spontaneously; above AB the vapour condenses. The curve BD similarly shows the variation with pressure of the melting point of the solid, and BC shows the temperature dependence of the vapour pressure of the liquid, which is the same thing as the pressure dependence of the boiling point. Point C represents the critical point of the liquid—beyond C there is no longer any distinction between liquid and vapour. The horizontal line across the graph at a pressure of one atmosphere intercepts curve BC at point E, which defines the normal boiling point of the liquid, and curve BD at point F, which defines the normal freezing point. The point B is of some interest, in that it represents a combination of pressure and temperature values such that all three phases, solid, liquid, and vapour, are present together at equilibrium. This point is known as the triple point. The phase diagram provides a useful summary of the properties of a system in which phase changes can occur, and such diagrams are widely used in investigations of the properties of systems of one or more components.

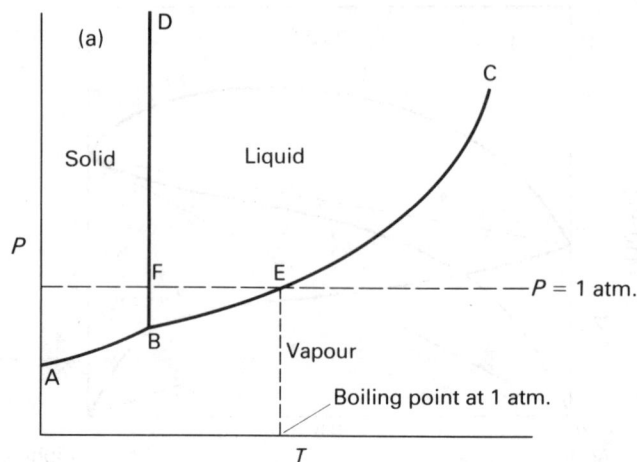

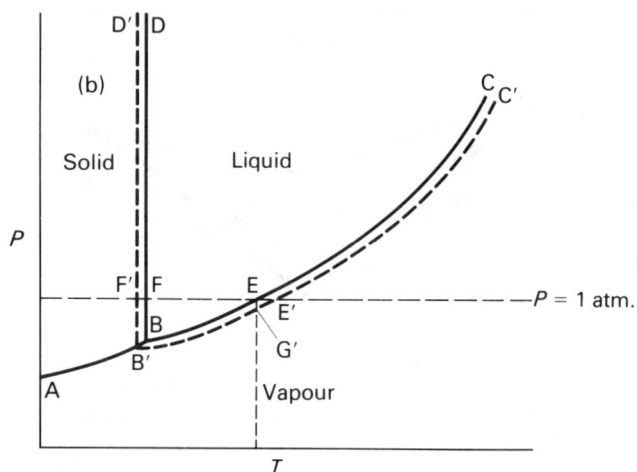

Fig. 5.5 (a) Phase diagram for pure solvent. (b) As in (a), with effect of dissolved solute shown by broken curves

In Fig. 5.5b the effect of a dissolved solute on the vapour pressure of the liquid is shown by the difference between the curves BC and B'C'. The new boiling point E' is defined by the point at which the curve B'C' cuts the line $P = 1$ atmosphere. The vapour pressure reduction at the old boiling point is given by EG'. A new triple point B' is defined by the point at which the new vapour pressure curve cuts the curve AB. The solute is normally insoluble in the solid solvent and therefore does not influence its sublimation pressure, so the curve AB' coincides with the curve AB. For the same reason, the pressure dependence of the melting point of the solid is not significantly affected by the presence of the solute, so the new melting curve B'D' is parallel to BD. A new freezing point F' is defined by the intersection of B'D' with the line $P = 1$ atmosphere. It is apparent that E' corresponds to a higher temperature than E, and F′ corresponds to a lower temperature than F, i.e. the boiling point is elevated and the freezing point depressed.

As the concentration of solute in a dilute solution varies, the size of the vapour pressure depression EG′ varies in accordance with Raoult's law, which means that EG′ is proportional to the mole fraction of solute. Simultaneously, the lengths of the lines EE′, BB′, and FF′ must vary in proportion to EG′, and we can conclude that the elevation of boiling point EE′ and the depression of freezing point FF′ are both proportional to the mole fraction of solute.

This result has been used a great deal in the past to provide a simple means of estimating the relative molecular mass of an unknown substance. The principle of the method is as follows: the depression of freezing point, or elevation of boiling point, is first measured for a solvent containing a known mole fraction of any simple solute. A measured weight of the unknown material is then dissolved in a fixed number of moles of the solvent, and the mole fraction of solute in this solution is determined by comparing the boiling point elevation or freezing point depression with that obtained in the first experiment.

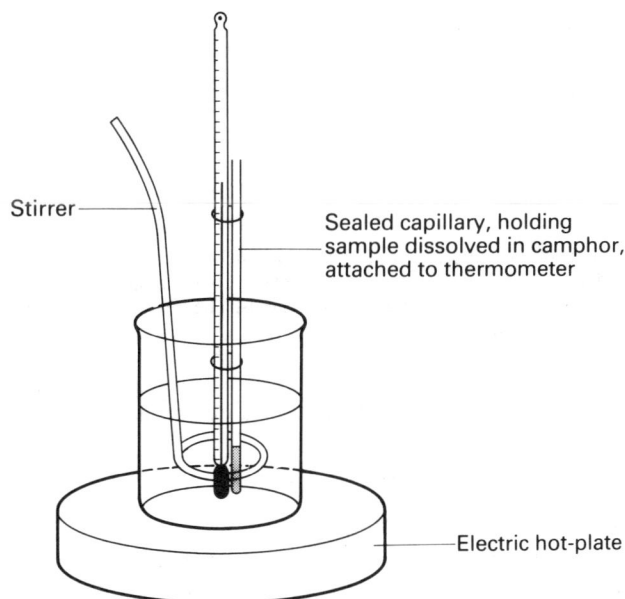

Fig. 5.6 Determination of relative molecular mass ('molecular weight') by measuring the depression of the freezing point of camphor

Since both the mass fraction and mole fraction of the solute are known, the relative molecular mass can be calculated. The molal depression of freezing point (*cryoscopic constant*) or molal elevation of boiling point (*ebullioscopic*

constant) of the solvent is defined as the magnitude of the appropriate temperature change produced when one mole of solute is dissolved in one kilogram of solvent. Some representative values are listed in Table 5.1. The cryoscopic constant for camphor is sufficiently large to allow reasonably accurate relative molecular masses to be determined using an ordinary thermometer graduated in tenths of a degree, as shown in Fig. 5.6. Much more precise relative molecular masses can be obtained with the aid of instruments such as the mass spectrometer shown in Fig. 5.7, but the difference of a factor of 10^4 in cost between the apparatus in Fig. 5.6 and that in Fig. 5.7 has to be borne in mind. At a more elementary level, the melting point is widely used as a convenient criterion of purity for organic compounds, the effect of any minor impurity always being to lower the melting point.

Table 5.1 Values of cryoscopic and ebullioscopic constants (K mole^{-1} kg) for several solvents

Solvent	Cryoscopic constant	Ebullioscopic constant (P = 1 atm.)
Water	1.86	0.512
Acetic acid	3.86	3.07
Benzene	5.12	2.53
Ethanol		1.22
Carbon tetrachloride		5.03
Bromobenzene		6.26
Stearic acid	4.50	
Triphenylmethane	12.45	
Camphor	39.7	

Example When 2.6 g of benzoic acid, C_6H_5COOH, is dissolved in 100 g of benzene the measured freezing point depression is 0.55°C. When 1.0 g of KNO_3 is dissolved in 1000 g of water the observed freezing point depression is 0.036°C. Comment on these observations.

For the benzoic acid solution we find from Table 5.1 that the observed depression corresponds to 0.107 moles of solute in 1000 g of solvent. Hence the relative molecular mass is 26/0.107 = 240. The formula mass of benzoic acid is 122; it follows that in benzene solution it must exist as the dimer $(C_6H_5COOH)_2$.

For the potassium nitrate solution the observed depression corresponds to 0.0194 moles of solute in a kilogram of solvent, i.e. to a relative molecular mass of 1.00/0.0194 = 51.7. The formula mass of KNO_3 is 101; thus the freezing point depression is a factor of two greater than anticipated. This is because KNO_3 gives rise to *two* species in solution, namely K^+ and NO_3^-.

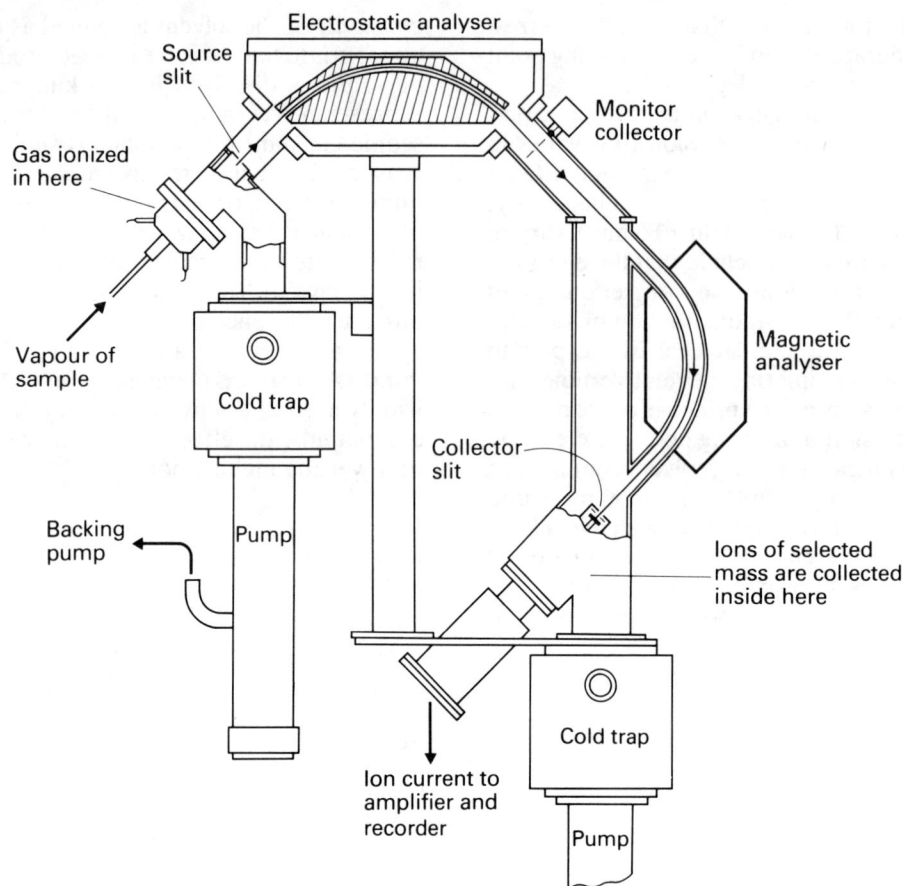

Fig. 5.7 Basic components of a high-resolution mass spectrometer. The ions produced from the sample vapour by bombardment with 75 V electrons are accelerated by a potential of 8 kV. Ions of different masses in the resulting beam are deflected to differing extents by the electric and magnetic fields. Only those in a narrow mass range reach the collector. Ions of a particular mass are selected by varying one of the deflection fields. The heaviest ion observed normally corresponds to the sample molecule minus one electron. Fragment ions are also observed

(iv) The Distribution Coefficient: an Application of Henry's Law. Chromatography

Consider two dilute solutions of a solute in different solvents, placed in separate containers in an evacuated enclosure as in Fig. 5.8a. The solvents are required to be immiscible, so that although their vapours fill the enclosure there is no tendency for either solvent to migrate to the other container. To take a definite example, let us suppose that the solvents are water and chloroform, and the solute is molecular iodine. The solute will also tend to evaporate, and if the vapour pressure of iodine in equilibrium with one solution happens to be greater than that in equilibrium with the other, the iodine will evaporate from the first and dissolve in the second. This process will continue until the equilibrium pressures of iodine vapour are identical for the two solutions. Hence if P is the iodine vapour pressure at equilibrium, we must have

$$P = k_1 x_1 = k_2 x_2 \qquad (5.6)$$

where k_1 and k_2, are Henry's law constants and x_1 and x_2 are mole fractions of solute for solutions 1 and 2. Therefore the mole fractions of solute at equilibrium are related by the *distribution law*

$$x_1/x_2 = K \qquad (5.7)$$

where the distribution coefficient K is given by

$$K = k_2/k_1 \qquad (5.8)$$

The distribution coefficient can also be expressed in terms of the solubility of the solute defined in Eq. 5.5, as

$$K = s_1/s_2 \qquad (5.9)$$

where s_1 and s_2 are the solubilities of the solute in the two solutions.

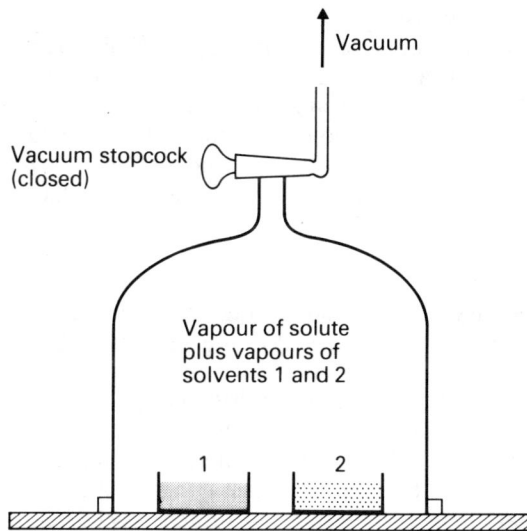

Fig. 5.8 (a) Simple experiment to investigate the migration of solute from one solvent to another

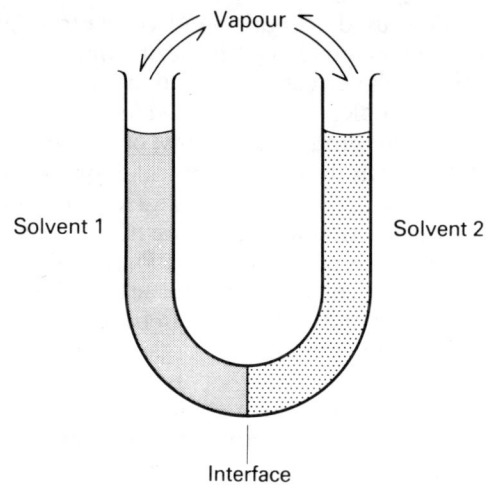

Fig. 5.8 (b) Hypothetical experiment with migration of solute through both the vapour phase and the solvent-solvent interface. The equilibrium distribution is the same as in Fig. 5.8 (a)

The equilibrium ratio of solute concentrations, given by Eq. 5.7, would not be affected if the solutions in equilibrium were brought into contact with one another, as in Fig. 5.8b. Bringing them into contact before equilibrium was reached would simply hasten the attainment of equilibrium, because the iodine would be able to migrate through the interface between the two solutions without the necessity of passing through the gas phase. This arrangement would in fact be an advantage with a solute of low vapour pressure, because the time taken to establish equilibrium by way of the vapour phase alone would be inconveniently long. As a further improvement the equilibrium distribution of solute could be established without any transfer through the gas phase, simply by shaking the solutions together in a separating funnel, as in Fig. 5.8c. The use of a separating funnel is convenient in practice because it permits the denser of the two solutions to be run off into a separate container.

Molecules which are polar tend to dissolve well in polar solvents such as water and poorly in non-polar solvents such as ether, whereas molecules which are non-polar have greatest solubility in non-polar solvents. Hence, on the basis of Eq. 5.9, it is to be expected that with two solvents of different polarity the distribution coefficient will markedly favour the solvent which is most similar to the solute. This is the basis of the technique of liquid–liquid extraction,

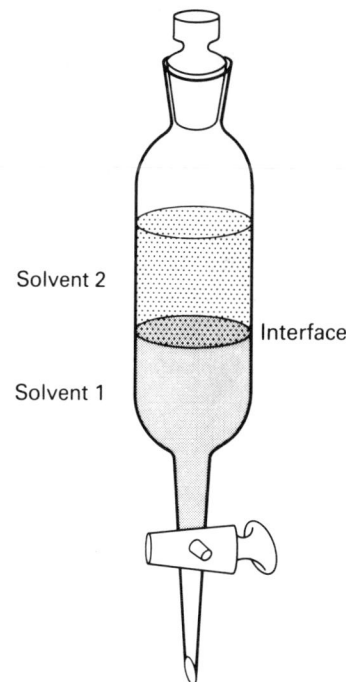

Fig. 5.8 (c) Liquid-liquid extraction. Solute migrates through the interface; the approach to equilibrium is assisted by shaking. The equilibrium distribution is the same as in Fig. 5.8 (b)

which is widely used to purify and concentrate impure materials. The technique is particularly effective for substances whose polarity can be altered at will by chemical means. For example, an amine RNH_2, where R is an unspecified organic group such as ethyl or naphthyl, can be converted to a salt $RNH_3^+X^-$ by addition of an acid H^+X^-. Hence an amine can be separated from non-polar impurities by extraction with an acidic polar solvent, such as aqueous hydrochloric acid. Addition of alkali to the acidic solution then converts the salt back to RNH_2, which is much less polar and can now be separated from polar impurities by extraction with an organic solvent such as ether. For ordinary laboratory separations it is convenient to carry out a liquid–liquid extraction by shaking the liquids together in a separating funnel and then allowing two layers to form. Some highly ingenious automated systems are now available commercially for use in situations where a large number of liquid–liquid extractions have to be performed on a routine basis, as for example in the pathology department of a hospital.

An equation similar to the distribution law, Eq. 5.7, can also be applied to the equilibrium distribution of a substance between a solvent and a solid phase on whose surface the substance is adsorbed. The adsorption process may consist of the sticking of individual molecules on the surface in a layer one molecule thick, a *monolayer*, held in place by Van der Waals' forces, or the absorbed molecules may be retained in cavities within the lattice of the adsorber, as happens with the solids known as 'molecular sieves'. This phenomenon is exploited in the extremely powerful technique of *chromatography*, which is widely used for the separation of individual components in complex mixtures. The purpose of the separation may be to analyse the mixture or it may be to obtain a sizeable sample (milligrams or grams) of one of the components. The basis of chroma-

tography is the distribution equilibrium set up by a solute between a *stationary phase* and a *moving phase*. As Table 5.2 demonstrates, a variety of different kinds of chromatography are possible, depending on the nature of the stationary and moving phases.

In the practice of chromatography, which is to some extent an art, a small sample of the mixture whose components are to be separated is added to the moving phase before it enters a long column of the stationary phase (Fig. 5.9). Once the stationary phase is encountered, the different solutes distribute themselves between the two phases in the proportions that are fixed by their distribution coefficients. At any given time only a fraction, x_1 say, of a given solute is contained in the moving phase, while the rest of the solute is stationary. Hence the average velocity of the solute through the column is equal to the velocity of the moving phase multiplied by x_1. For different solutes there will in general be different values of distribution coefficient, and hence different values of x_1. It follows that the components of a mixture which enter the column together will emerge one after the other, those which cling most tightly to the stationary phase emerging last. Each type of chromatography has its own characteristic methods of monitoring the composition of the different fractions of the mixture as they leave the chromatographic column. A partial listing of these methods is given in the last column of Table 5.2. The main virtue of chromatography is probably its ability to separate substances with almost identical properties, such as structural isomers of hydrocarbons. The ability of thin-layer chromatography, paper chromatography, and gas–liquid chromatography to analyse an extremely minute sample of a mixture is almost equally impressive.

(v) Boiling Point Diagrams and Fractional Distillation

Raoult's and Henry's laws are expressions of the form of

Table 5.2 Important types of two-phase chromatography

Type	Moving phase	Stationary phase	Application	Detection methods
Liquid–solid ('ordinary chromatography')	Organic solvent e.g. ether	Typically powdered alumina	Purification of sizeable samples (100 mg—100 g)	Collect fractions and assay
Paper	Solvent, by capillary attraction	Absorbent paper	Analysis, or purification of ultra-small samples (10 μg—1 mg)	Spray paper with reagents giving colours characteristic of solutes
Thin-layer	Organic solvent	Thin layer of powdered alumina or silica on flat plate	Analysis, or purification of ultra-small samples (10μg—100 mg)	As with paper chromatography
Gas–liquid ('Gas chromatography')	Inert gas: H_2, N_2, He, or Ar	Involatile solvent coated on powdered support, e.g. brick dust	Analysis, or purification of small samples (mg) (10 μg—500 mg)	Thermal conductivity, ionization in flame or by ionizing radiation, mass spectrometry, etc.

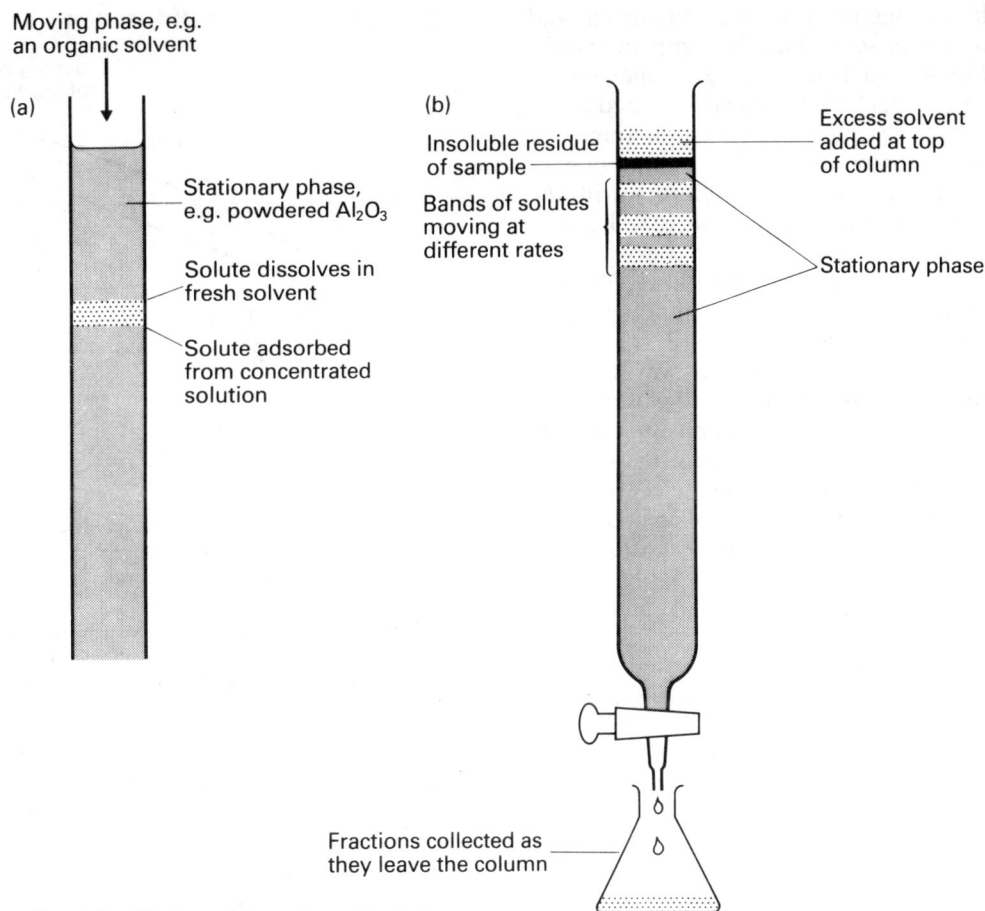

Fig. 5.9 Chromatography: (a) movement of a band of solute through the stationary phase depends on the processes of desorption and readsorption; (b) typical setup for purification by solid-liquid chromatography

vapour pressure diagrams, such as those in Figs. 5.2 and 5.3. The vapour pressure diagrams we have considered so far show the variation of vapour pressure with liquid composition for a solution maintained at a constant temperature. The composition of the vapour generally differs from that of the liquid, the vapour being richer in the more volatile component, as shown by the curve in Fig. 5.10a. An alternative way of representing the properties of the same mixture is to construct a diagram that shows the temperature at which the mixture boils under a fixed external pressure as a function of composition. This is a boiling point diagram. We recall that the boiling point is the temperature at which the total vapour pressure is equal to the external pressure. For a system which obeys Raoult's law the boiling point diagram takes the form shown in Fig. 5.10b. Neither of the lines in the graph is straight because the dependence of vapour pressure on temperature is not

linear. The two curves in the diagram give the composition of the liquid and vapour which are in equilibrium at any given temperature. At the temperature T_1 in Fig. 5.10b the compositions of the liquid and vapour are given by the intersections of the horizontal dotted line with the curves at points x_1 and x_v, respectively. As in Fig. 5.10a, the positions of the curves reflect the fact that the vapour is always richer than the liquid in the more volatile component of the mixture, i.e. in the component having the lower boiling point. At values of T and x_B (for the whole mixture) that lie between the two curves both liquid and vapour are present, their compositions being given by the intersections of the curves with a horizontal line through the point (T, x_B). If the point (T, x_B) lies above the curves only vapour is present; if (T, x_B) lies below the curves only liquid is present.

The fact that the vapour is always richer in the more volatile component of the liquid is the basis of *fractional distillation*, which has long been the most important and widely used technique for separating the components of a liquid mixture. Consider a liquid whose composition is given by $x_B = x_1$ in Fig. 5.11. If this liquid is heated it will boil at a temperature T_1, and the vapour in equilibrium with the liquid will have the composition x_2. If a small amount of this vapour (not enough to upset the equilibrium with the original liquid) is removed and condensed it will give a liquid of composition x_2. This liquid can be boiled at temperature T_2, to give a vapour of composition x_3. A small amount of the vapour of composition x_3 can now be condensed, and boiled in its turn at temperature T_3 to give a vapour of composition x_4. Continuing in this way, the condensate can be made to approach closer and closer to the pure component A. Of course, continuously removing some of the vapour of composition x_2 from the original mixture will not leave its composition unchanged, so that if it is desired to distil a large proportion of the component A out of the initial solution the composition of the liquid left behind must tend towards pure component B. Ideally, at the limit, the distillate would consist of pure A, and pure B would be left behind in the first still pot.

In practice fractional distillation is not carried out as a succession of separate distillations, but in a single operation, the successive stages of equilibration between liquid and vapour being established in a *fractionating column*.

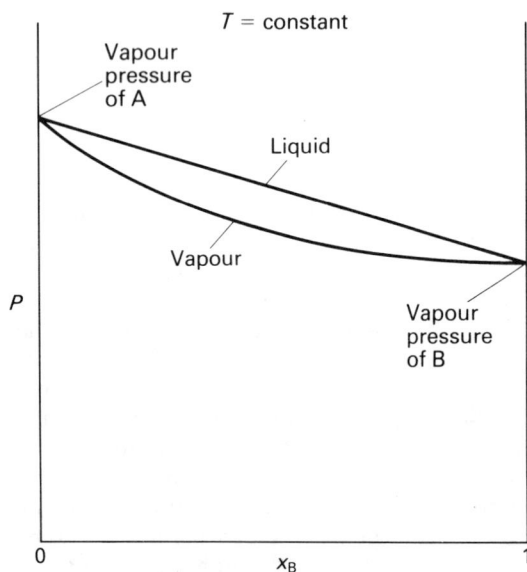

Fig. 5.10 (a) Vapour pressure diagram for a mixture obeying Raoult's Law, showing the total vapour pressure as a function of both liquid and vapour composition

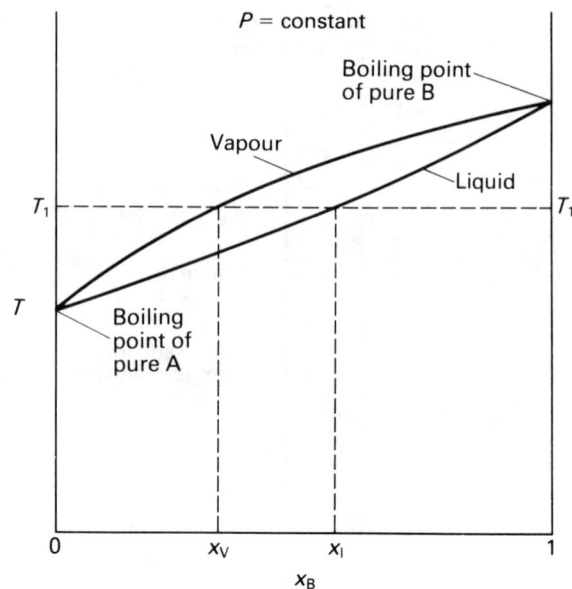

Fig. 5.10 (b) Boiling point diagram for a mixture A + B obeying Raoult's Law. The two curves give the compositions of liquid and vapour which are in equilibrium with one another at temperature T

Fractionating columns take many different forms; the simplest to understand is the bubble-cap column illustrated in Fig. 5.12. This type of column is commonly used in large-scale industrial distillations. The paths of liquid and vapour through the column are shown for plate 1. At each plate the overflow of condensate from the plate above supplies the liquid phase and the distillate from the plate below supplies the vapour phase. The two phases are brought into equilibrium as a result of the vapour being obliged to bubble upwards through the liquid. Ideally a single plate of such a column would be able to produce a separation equivalent to one of the steps in the 'staircase' of Fig. 5.11, for example that from liquid x_1 to liquid x_2. In practice the performance would fall somewhat short of this ideal. The length of fractionating column required to produce a single step in the staircase is known as the *height equivalent to a theoretical plate* (usually abbreviated to HETP) and is a useful figure of merit with which to describe the performance of a fractionating column.

Columns which are used for small-scale separations seldom contain anything that can be identified as a plate, but instead depend on a loose random packing of such materials as short lengths of tubing, glass balls or ceramic beads. Small (5 mm diameter) glass helices are supplied commercially for use as a very efficient column packing in small-scale systems. For distillations involving very small quantities of valuable material the 'hold-up' of the column, i.e.

the quantity of liquid needed to wet the surface of the packing, or to lie in pools on the plates of Fig. 5.12, needs to be minimized. For this kind of application the 'spinning-band' column has been devised. In place of packing, the column contains a central shaft which rotates rapidly so that any liquid falling or condensing into it is flung to the wall of the column, where it flows down as a thin film. Provided the film is thin (0.1 mm or less) the liquid can come to equilibrium sufficiently rapidly with vapour passing up the column.

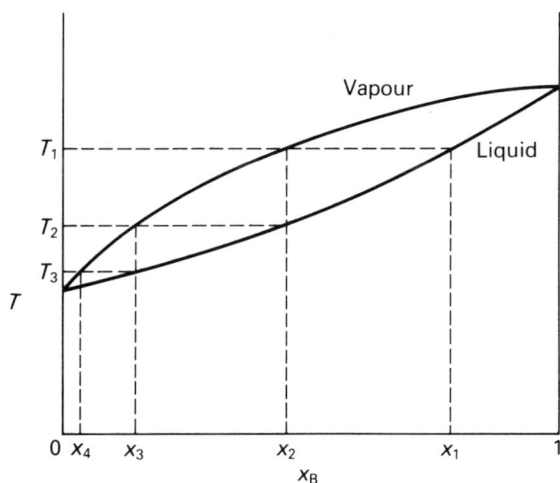

Fig. 5.11 Principle of fractional distillation

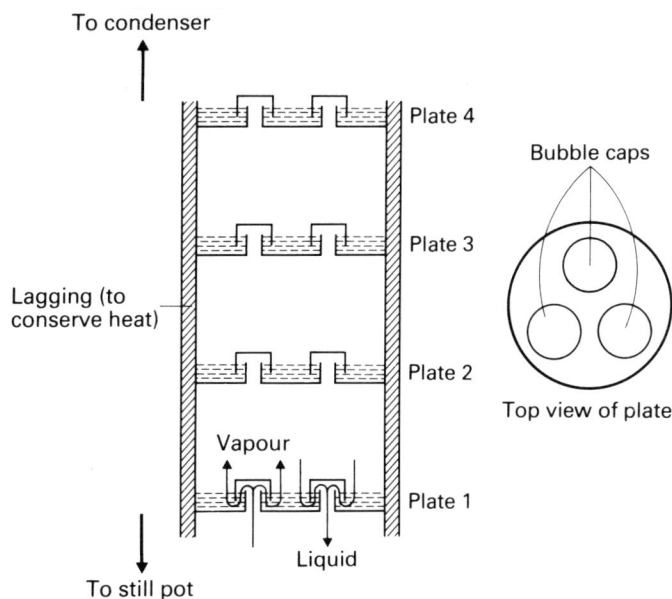

Fig. 5.12 Bubble-cap fractionating column

A typical laboratory set-up for fractional distillation is shown in Fig. 5.13. The basic components are the heat supply, still pot, column, thermometer, condenser and receiver. When a distillation is in progress the two main variables that have to be controlled are the boiling rate, controlled by the rate of supply of heat, and the *reflux ratio*,

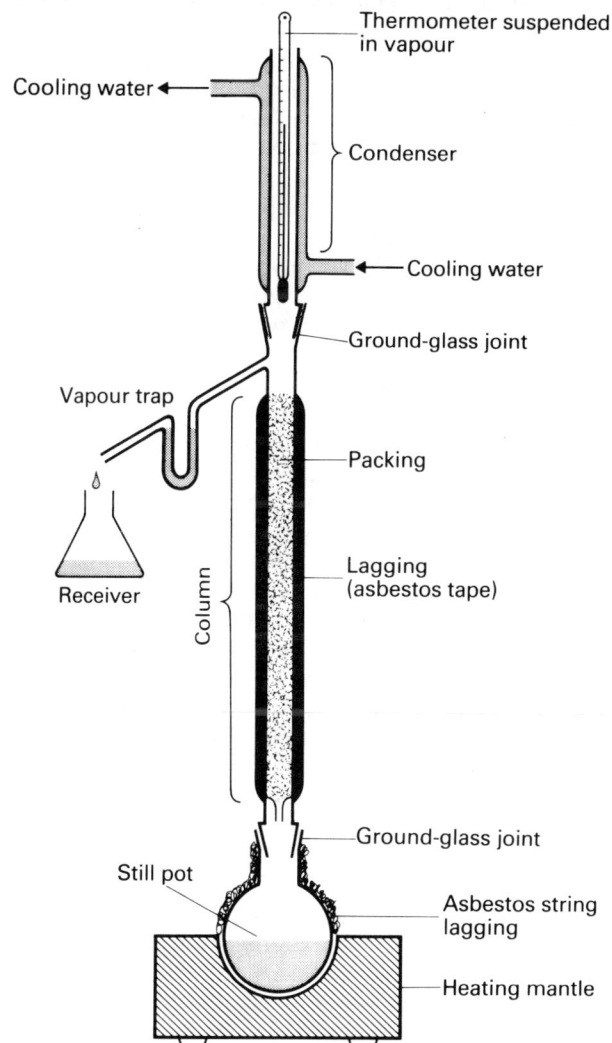

Fig. 5.13 Typical laboratory set-up for fractional distillation

which is the ratio of the amount of condensate which returns to the top of column to the amount which flows through the vapour trap to the receiver. If the boiling rate is too great the column is likely to flood with liquid, with the result that the expanding vapour tends to expel the liquid, the packing and the thermometer out through the hole at the top of the condenser. On the other hand, if the boiling rate is too low the column will dry out and lose efficiency. For optimum efficiency the column should be just on the point of flooding, from which it is apparent that some skill is involved in carrying out the whole operation. The reflux ratio can be fixed by the ratio of the diameter of the side tube to the circumference of the column, or by the intermittent opening of an automatic or manually controlled valve in the side tube leading to the receiver. The reflux ratio should not be too small because efficient fractionation depends on the attainment of an equilibrium distribution of material along the column, and the equilibrium is upset when some material is removed at the top, but if it is too large the distillation takes an excessive amount of time. Typically the optimum value of reflux ratio for a set-up such as that in Fig. 5.13 would be between 10 and 20.

For mixtures which deviate from Raoult's law to a moderate extent the boiling point diagram is basically similar to Fig. 5.10, and the conclusions we have reached so far are still valid. However, where the deviations from Raoult's law are sufficient to result in the existence of a mixture whose vapour pressure is either a maximum or a minimum for the system, as is the case for the examples in Fig. 5.3, the boiling point diagram has one of the characteristic forms shown in Fig. 5.14. In both of the diagrams of Fig. 5.14 there is a constant-boiling mixture, or *azeotrope*, whose composition is that of the minimum or maximum boiling point. Considering Fig. 5.14a, we see that on the left side of the minimum the vapour in equilibrium with a liquid is richer than the liquid in component B, whereas on the right hand side of the minimum the vapour is always richer than the liquid in component A. It follows that precisely at the minimum the vapour cannot be richer than the liquid in either component, i.e. at the minimum the liquid and vapour have the same composition and the minimum boiling mixture distils as if it were a single component. Similar arguments can be used to show that the mixture with maximum boiling point in Fig. 5.14b distils as if it were a single component. The composition of the azeotrope varies with pressure; hence the azeotrope does not correspond to a definite chemical compound of A and B.

The existence of an azeotrope makes it more difficult to separate a mixture of A and B by distillation, because the result of fractionation is always to produce one of the pure components plus the azeotrope. One such system of some practical importance is the ethanol–water system, in which the azeotrope contains 95.6% of ethanol by weight and boils at 78.15°C, 0.15° below the boiling point of pure ethanol. The azeotrope is commonly obtained by distilling the mixture of water and alcohol which results from the fermentation of sugars by yeast organisms. On a laboratory scale the last five percent of water is conveniently removed by distillation from a drying agent such as calcium oxide.

On an industrial scale, absolute alcohol is obtained by adding a calculated amount of benzene and distilling off the water as a three-component benzene–alcohol–water mixture. Water and hydrogen chloride form an azeotrope whose composition is known very accurately as a function of pressure near one atmosphere, so that constant-boiling hydrochloric acid is a useful primary standard in volumetric analysis.

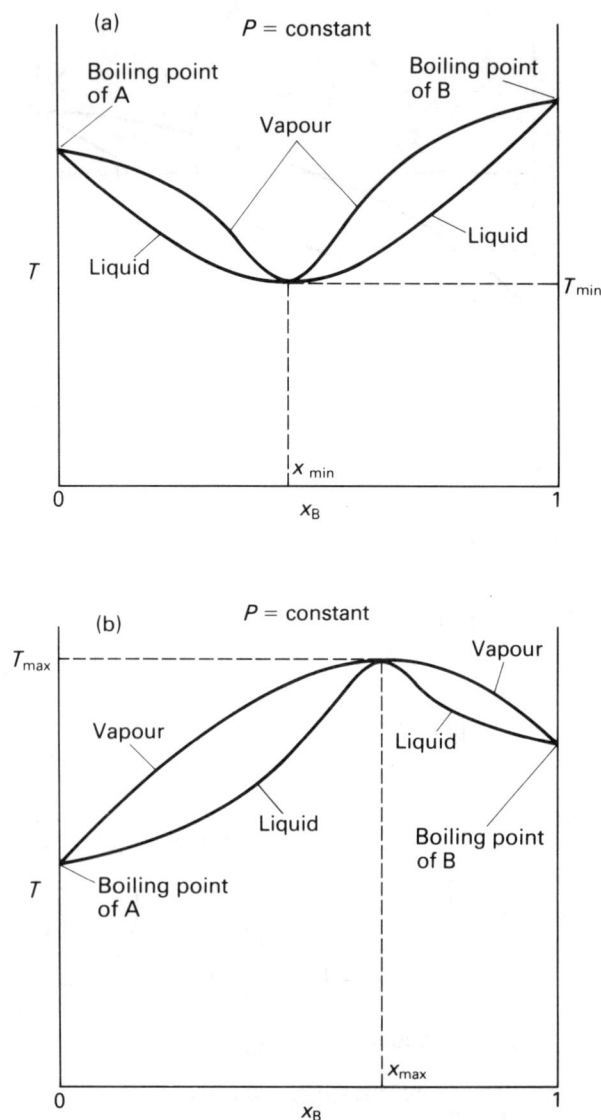

Fig. 5.14 Boiling point diagrams for systems showing marked deviations from Raoult's Law. (a) Positive deviations resulting in the existence of a mixture with a minimum boiling point. (b) Negative deviations giving rise to a mixture with a maximum boiling point

5.3 Chemical Equilibrium

(i) General Considerations

A state of chemical equilibrium is the ultimate outcome of every chemical process. The equilibrium is always a dynamic one, in that the reaction does not stop when equilibrium has been attained, but the rate of the forward reaction diminishes while that of the reverse reaction increases to the point at which the rates of the forward and reverse reactions are equal. At equilibrium the concentrations of reactants and products are related to one another through the *equilibrium constant,* which is given the symbol K. It is probably fair to say that the equilibrium constant is the most important and useful numerical quantity in chemistry. In general K varies with temperature but is independent of the total pressure, volume, or composition of a system. Although it is true that the concentrations of reactants and products are related to one another through the equilibrium constant, the variables to which K refers are not simply concentrations, but are the thermodynamic quantities known as activities, for which we use the symbol a. The activities of substances are directly related to their concentrations, as we shall see.

For the general chemical equilibrium 5.10, in which a molecules of species A react with b molecules of species B and c molecules of species C to produce l molecules of species L, m molecules of species M and n molecules of species N:

$$aA + bB + cC \rightleftharpoons lL + mM + nN \qquad (5.10)$$

the equilibrium constant is given by

$$K = \frac{(a_L)^l (a_M)^m (a_N)^n}{(a_A)^a (a_B)^b (a_C)^c} \qquad (5.11)$$

where, for example, $(a_L)^l$ represents the activity of species L raised to the power l. Some specific examples of chemical equilibria, with their associated equilibrium constants, are:

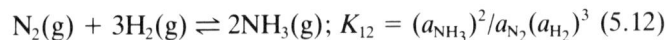

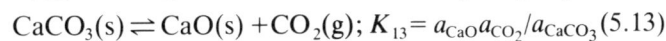

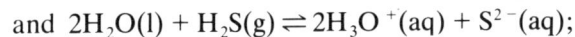

$$N_2(g) + 3H_2(g) \rightleftharpoons 2NH_3(g); \quad K_{12} = (a_{NH_3})^2/a_{N_2}(a_{H_2})^3 \quad (5.12)$$

$$CaCO_3(s) \rightleftharpoons CaO(s) + CO_2(g); \quad K_{13} = a_{CaO} a_{CO_2}/a_{CaCO_3} \quad (5.13)$$

and $2H_2O(l) + H_2S(g) \rightleftharpoons 2H_3O^+(aq) + S^{2-}(aq);$

$$K_{14} = a_{S^{2-}}(a_{H_3O^+})^2/a_{H_2S}(a_{H_2O})^2 \qquad (5.14)$$

We note that as usual it is necessary to specify the physical state of every reactant and product. The examples 5.12–5.14 include species which are solid, liquid, gaseous, and in aqueous solution.

To be able to use the equilibrium constants which appear in Eq. 5.11 to 5.14 we have to know what is meant by the activities of the various reactants and products. The rules which we can use to relate activity to concentration for any species are as follows:

(1) If the species is an ideal gas the activity is equal to the concentration expressed in appropriate units. The most convenient way of stating the activity of a gas is in terms of its partial pressure in atmospheres, but it is equally valid to use mole fraction, or concentration in moles per unit volume. In general the units of concentration must be specified along with the value of the equilibrium constant.

(2) For a species which is a pure solid or a pure liquid, or is the *solvent* in a very dilute solution, the activity is equal to 1.0.

(3) For a neutral, i.e. non-ionic, solute in a dilute solution the activity is simply equal to the concentration expressed in appropriate units, either moles per litre or mole fraction. If the solution is very dilute the activity of the solvent can be set equal to unity; otherwise it is equal to the mole fraction of solvent in the solution. Again the units of concentration need to be specified along with the value of the equilibrium constant. For a solution in which Raoult's law is obeyed the activity can be put equal to the mole fraction for all components over the whole range of composition.

(4) For an ionic species in solution, and for a neutral solute in a concentrated solution that does not obey Raoult's law, the activity is equal to the concentration multiplied by an *activity coefficient,* which we give the symbol f, if concentration is expressed as a mole fraction, and γ (gamma) if concentration is expressed in molar units:

$$a_A = f_A x_A \qquad (5.15)$$

where x_A is the mole fraction of species A, or

$$a_A = \gamma_A [A] \qquad (5.16)$$

where the concentration of A in molar units is symbolized by placing square brackets about the chemical formula for A. On the molar scale, concentration is expressed in moles per litre (or kg moles per cubic metre) of solution. Alternatively one can use moles per kilogram of solvent (the *molal* scale). We shall use only the molar scale, in terms of moles per litre of solution; a 1.0 molar solution contains one mole per litre, and the concentration is written as 1.0 M. The main virtue of the *molal* scale is that the concentration expressed in molal units does not alter when the temperature of the solution is changed, despite the thermal expansion or contraction of the liquid. The rules relating activity to concentration are summarized in Table 5.3.

The effect of including an activity coefficient is always to compensate for any deviations from ideal behaviour that may occur. Thus, for example, if Raoult's law is written in terms of the activity a_i of species i, rather than the mole fraction x_i, the resulting expression (Eq. 5.17) is always obeyed.

$$P_i = a_i P_i^0 \qquad (5.17)$$

Similarly, in very precise work with gases the deviations

from ideal gas behaviour can be taken into account by introducing an activity coefficient, the product of pressure and activity coefficient being known as fugacity. Activity coefficients are useful in practice because they allow the behaviour of any practical system to be described exactly by the laws which govern the behaviour of an ideal system. However, the only situations in which gross errors are likely to result from neglect of the activity coefficient involve ions in solutions which contain large total concentrations of ionic species, or solutions of non-electrolytes (neutral species) in which the deviations from Raoult's law are very large. In the remainder of this book we shall assume, unless otherwise instructed, that activity coefficients can be taken as equal to unity. We shall make no use at all of the concept of fugacity.

Returning to the equilibria 5.12–5.14, we can now apply the rules in Table 5.3 to rewrite the equilibrium constants, as follows:

$$K_{12} = P^2_{NH_3}/P_{N_2}P^3_{H_2} \qquad (5.18)$$

$$K_{13} = P_{CO_2} \qquad (5.19)$$

and $\quad K_{14} = [H_3O^+]^2[S^{2-}]\gamma^2_{H_3O^+}\gamma_{S^{2-}}/P_{H_2S} \qquad (5.20)$

The activity coefficients in Eq. 5.20 can be put equal to 1 if the concentration of ionic species is very low, in which case

$$K_{14} = [H_3O^+]^2[S^{2-}]/P_{H_2S} \qquad (5.21)$$

In Eq. 5.18 we have converted the activities directly to partial pressures in accordance with rule 1. In Eq. 5.19 we have converted the activity of CO_2 to a partial pressure and have put the activities of $CaCO_3$ and CaO equal to unity in accordance with rule 2. In Eq. 5.20 we have used rule 1 for

the H_2S, rule 2 for H_2O, and rule 4 for the ions H_3O^+ and S^{2-}. An example involving rule 3 would be the dissociation of dilute, aqueous acetic acid:

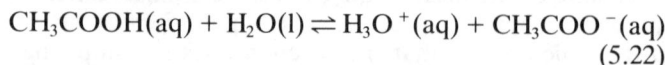

$$CH_3COOH(aq) + H_2O(l) \rightleftharpoons H_3O^+(aq) + CH_3COO^-(aq) \qquad (5.22)$$

for which we should have

$$K = \frac{[H_3O^+][CH_3COO^-]}{[CH_3COOH]} \qquad (5.23)$$

Here the activity of water has been taken as 1 in accordance with rule 2, the activity of acetic acid has been taken as equal to its concentration in accordance with rule 3, and the solution has been assumed to be sufficiently dilute for the activity coefficients, which are required for the H_3O^+ and CH_3COO^- according to rule 4, to be set equal to 1.

In the remainder of this section we shall discuss equilibrium constants for reactions of neutral species. Equilibria involving ions in aqueous solution will be considered in the next section.

(ii) Interconversion of Equilibrium Constants K_p, K_c and K_x

The numerical value of an equilibrium constant depends on the units in which the various activities or concentrations are expressed. For gaseous equilibria there are three possible ways of expressing the equilibrium constant, and we distinguish them by writing K_p for the equilibrium constant in terms of partial pressures, K_c for the equilibrium constant in terms of concentrations, and K_x for the equilibrium constant in terms of mole fractions. The equilibrium constant in terms of activities, defined by Eq. 5.11, can also

Table 5.3 Rules for relating activity to concentration

Rule	Species	Activity is equal to	Formula	Activity coefficient
1	Gas	Partial pressure (atmospheres) or Concentration (moles per litre)	$a_A = P_A$, or $a_A = [A]$	Seldom used
2	Pure solid	1	$a_A = 1$	Not used
2	Pure liquid	1	$a_A = 1$	Not used
3	Non-ionic solute, in dilute solution	Concentration or mole fraction	$a_A = [A]$, or $a_A = x_A$	Usually = 1
3	Solvent in very dilute solution	1, or mole fraction	$a_A = 1$, or $a_A = x_A$	Usually = 1
3	Components of solution obeying Raoult's law	Mole fraction	$a_A = x_A$	Equal to 1
4	Concentrated solution disobeying Raoult's law	Mole fraction times activity coefficient or	$a_A = x_A f_A$	f not equal to 1
4	Ions in solution	Concentration times activity coefficient	$a_A = [A]\gamma_A$	Use Debye–Huckel theory to obtain γ for ions, provided concentration is not too high

be distinguished by writing it as K_a. We shall always work with K_p for gaseous equilibria and K_c, or occasionally K_a, for equilibria in solution. Neverthless it can be useful to know how to convert an equilibrium constant from one set of concentration units to another.

For a gaseous equilibrium of the general form shown in Eq. 5.10 we have

$$K_p = \frac{P_L{}^l P_M{}^m P_N{}^n}{P_A{}^a P_B{}^b P_C{}^c} \qquad (5.24)$$

where, for example, P_A is the partial pressure of species A and a is the number of molecules of A involved in the reaction. The quantity K_p is rigorously independent of total pressure, volume, or composition of the reaction system, but does vary with temperature. Corresponding to the partial pressure P_L of species L there is a mole fraction x_L and a concentration C_L (in moles per litre), given by

$$x_L = P_L/P_{\text{total}} \qquad (5.25)$$

and
$$C_L = [L] = n_L/V$$

$$= P_L/RT \qquad (5.26)$$

Eq. 5.25, in which P_{total} is the total pressure of gas in the reaction system, is simply an expression of Dalton's law of partial pressures. Eq. 5.26 is an application of the ideal gas equation $PV = nRT$ for the case of n_L moles of species L contained in a volume V. If we use these results to substitute for P_L, P_M, P_N, P_A, P_B and P_C in Eq. 5.24 we obtain

$$K_x = \frac{x_L{}^l x_M{}^m x_N{}^n}{x_A{}^a x_B{}^b x_C{}^c}$$

$$= K_p(P_{\text{total}})^{-\Delta n} \qquad (5.27)$$

and
$$K_c = \frac{C_L{}^l C_M{}^m C_N{}^n}{C_A{}^a C_B{}^b C_C{}^c}$$

$$= K_p(RT)^{-\Delta n} \qquad (5.28)$$

where
$$\Delta n = (l + m + n) - (a + b + c) \qquad (5.29)$$

is the change in the number of moles of gas brought about by the reaction, i.e. Δn is the difference between the total number of molecules of products and the total number of molecules of reactants in Eq. 5.10.

If the number of molecules of reactants in the chemical equation is equal to the number of molecules of products, Δn is zero, in which case the numerical value of the equilibrium constant is independent of the choice of units of concentration. This would be true, for example for the equilibrium

$$H_2(g) + I_2(g) \rightleftharpoons 2HI(g) \qquad (5.30)$$

In general, however, Δn is not zero, in which event Eq. 5.27 shows that the equilibrium constant expressed in terms of

mole fractions is *not* independent of variations in the total pressure in the system. An example of a reaction for which Δn is not zero is

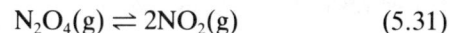

$$N_2O_4(g) \rightleftharpoons 2NO_2(g) \qquad (5.31)$$

For an equilibrium involving gases the equilibrium constants can be converted from one set of concentration units to another by means of Eqs 5.25 and 5.26. For reactions in solution similar conversions can be made between K_c and K_x. The relationship between mole fraction and molar concentration in solution is

$$x_L \rho = C_L M_m \qquad (5.32)$$

where ρ is the density of the solution in grams per litre, and M_m is the mean relative molecular mass of the solution. These quantities are defined by

$$C_L = n_L/V \qquad (5.33)$$

$$x_L = n_L/(\textstyle\sum_i n_i) \qquad (5.34)$$

$$M_m = \textstyle\sum_i x_i M_i \qquad (5.35)$$

and
$$\rho = M_m \textstyle\sum_i n_i/V \qquad (5.36)$$

As before there are n_L moles of species L in a volume V. The sum in the denominator of the right hand side of Eq. 5.34 is the total number of moles of all kinds of molecules in the volume V of solution, n_i being the number of moles of species i that are present. M_i is the relative molecular mass of species number i in the solution, and x_i is the mole fraction of species i. Eq. 5.32 is obtained by combining Eqs 5.33, 5.34 and 5.36. Eq. 5.32 leads to the result

$$K_x = K_c(M_m/\rho)^{-\Delta n} \qquad (5.37)$$

Once again, if Δn is zero the numerical value of the equilibrium constant is independent of the choice of concentration units.

(ii) Calculations using Equilibrium Constants

The most basic type of calculation is to find the equilibrium composition of a reaction system, given the initial composition, the equation for the chemical reaction, and the value of the equilibrium constant in appropriate units. The quickest way to grasp the method of attacking such calculations is to consider some specific problems, as in the following examples

Example 1 The equilibrium constant for the gas phase reaction

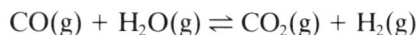

$$CO(g) + H_2O(g) \rightleftharpoons CO_2(g) + H_2(g)$$

is 5.5 at 500°C. Calculate the equilibrium composition of an equimolar mixture of carbon monoxide and water which is heated to 500°C at one atmosphere pressure.

If no reaction occurred the mixture would consist of 0.5 atm. CO and 0.5 atm. H_2O. If the amount of CO which is removed by the reaction is given the symbol x, the partial pressures of the various species are:

$$CO \ : 0.5 - x$$

$$H_2O : 0.5 - x$$

$$H_2 \ \ : x$$

$$CO_2 \ : x$$

Hence
$$K_p = \frac{x^2}{(0.5 - x)^2} = 5.5$$

Taking the square root of both sides of this equation, we find $x = 0.35$ at equilibrium, when the concentrations of CO and H_2O are both 0.15 atm.

Example 2   As in Example 1, but beginning with 2 atm. of CO_2 and 1 atm. of H_2.

Here, if x is the amount of CO formed, the concentrations are:

$$CO : x$$

$$H_2O: x$$

$$H_2 \ : 1 - x$$

$$CO_2: 2 - x$$

so the equilibrium condition is

$$\frac{(2 - x)(1 - x)}{x^2} = 5.5$$

or
$$4.5x^2 + 3x - 2 = 0$$

This is a more usual sort of equation to be solved than that which arose in Example 1, where the occurrence of a perfect square in the expression for the equilibrium constant was rather fortuitous. The values of x which satisfy the quadratic equation are $- 1.08$ and $+ 0.41$. The negative value of x can be discarded as unrealistic, so that the equilibrium pressures are 0.41 atm. for CO and H_2O, 0.59 atm. for H_2, and 1.59 atm. for CO_2.

Example 3   The equilibrium constant K_p for the reaction

$$N_2O_4(g) \rightleftharpoons 2NO_2(g)$$

is 0.141 at 25°C. What is the percentage dissociation of N_2O_4 at equilibrium under a total pressure of one atmosphere.

For the *fraction* (not percentage) of N_2O_4 dissociated we use the symbol α. Let the pressure of N_2O_4 which would be present if there were no dissociation be P. Then the equilibrium partial pressures of N_2O_4 and NO_2 are:

$$N_2O_4 : P(1 - \alpha)$$

$$NO_2 \ : 2P\alpha.$$

The total pressure must be one atmosphere, so we have

$$P(1 + \alpha) = 1, \text{ or } P = 1/(1 + \alpha).$$

The equilibrium condition is

$$\frac{(2P\alpha)^2}{P(1 - \alpha)} = 0.141$$

which reduces to

$$\frac{4\alpha^2}{1 - \alpha^2} = 0.141$$

when we use the relationship $P = 1/(1 + \alpha)$. Hence $\alpha = (0.141/4.141)^{1/2} = 0.185$. The dinitrogen tetroxide is seen to be 18.5% dissociated.

The common feature of these examples is that the system is first visualized as being in a state in which no reaction has occurred. One of the reactants is then assumed to be removed to the extent of x concentration units, or a fraction α, and the stoichiometry of the chemical reaction is used to fix the concentrations of other reactants and products in terms of x or α. The equilibrium constant formula then provides an equation which can be solved to obtain x or α. The reverse process, that of calculating K given x, is quite straightforward, as the following example shows.

Example 4   In a study of the equilibrium

$$2SO_3(g) \rightleftharpoons 2SO_2(g) + O_2(g)$$

Bodenstein and Pohl obtained the following mole fraction data for the equilibrium mixture at 1000 K and one atmosphere pressure:

$$x_{SO_2} = 0.309, \ x_{SO_3} = 0.338, \ x_{O_2} = 0.353.$$

Calculate K_p and K_c.

Since the total pressure is 1 atmosphere the mole fractions given are equal to the partial pressures of the gases in atmospheres. Hence

$$K_p = \frac{P_{SO_2}^2 P_{O_2}}{P_{SO_3}^2} = 0.309^2 \times 0.353/0.338^2 = 0.295 \text{ atm.}$$

To calculate K_c we use Eq. 5.28, $K_c = K_p(RT)^{-\Delta n}$, where in the present example $\Delta n = 1$. In litre atmosphere mole^{-1} Kelvin^{-1} units, R is 0.0821, so K_c becomes $0.295/(0.0821 \times 1000)$, or 3.59×10^{-3} mol 1^{-1}.

More complicated problems arise when two or more equilibrium conditions have to be satisfied simultaneously, as when two chemical reactions occur at once, or when the partial pressure of a gaseous reactant is fixed, through

Raoult's or Henry's law, by the mole fraction of the reactant in solution. Example 5 illustrates such a problem.

Example 5 The equilibrium constant K_p for the reaction

$$C(s) + CO_2(g) \rightleftharpoons 2CO(g)$$

is 1.9 atm. at 1000 K. At the same temperature the equilibrium constant for the dissociation of limestone to quicklime and CO_2

$$CaCO_3(s) \rightleftharpoons CaO(s) + CO_2(g)$$

is 3.9×10^{-2} atm. Calculate the equilibrium pressure of carbon monoxide above a mixture of coke, limestone and quick-lime at 1000 K.

The $CaCO_3$ and CaO are present at unit activity in accordance with rule 2 for relating activity to concentration. The same applies to the solid carbon. The second of the two equilibria is not affected by the presence of the carbon, so the equilibrium pressure of CO_2 is fixed at 3.9×10^{-2} atm. The equilibrium constant for the first reaction then gives

$$P_{CO}^2/P_{CO_2} = 1.9, \text{ or } P_{CO} = 0.272 \text{ atm.}$$

This last example is a special case of the type of problem in which there are two reactions of the type

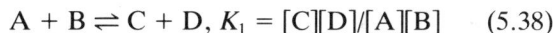

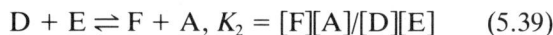

$$A + B \rightleftharpoons C + D, \quad K_1 = [C][D]/[A][B] \quad (5.38)$$

$$D + E \rightleftharpoons F + A, \quad K_2 = [F][A]/[D][E] \quad (5.39)$$

where one or more species which appear as reactants in one case appear as products in the other. The sum of the two reactions is

$$B + E \rightleftharpoons C + F, \quad K_3 = [C][F]/[B][E] \quad (5.40)$$

and we see that $$K_3 = K_1 K_2 \quad (5.41)$$

This is a general result, one which is frequently useful in deriving new equilibrium constants from existing ones by a process akin to the application of Hess's law.

A state of equilibrium is sensitive to changes in the values of state functions such as P, V, or T, in that if the value of the state function alters the position of equilibrium usually alters as well. Thus for the reaction

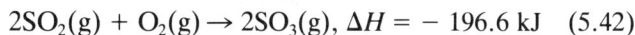

$$2SO_2(g) + O_2(g) \rightarrow 2SO_3(g), \Delta H = -196.6 \text{ kJ} \quad (5.42)$$

one could enquire about the effect on the equilibrium ratio of SO_3 to SO_2 of changing a state function such as P, V, T, or the partial pressure of O_2. These questions can be answered qualitatively with the aid of *Le Chatelier's principle*, which in essence says that if the state of a system in equilibrium is altered by some externally applied perturbation, the position of equilibrium in the system will tend to adjust itself so as to minimize the effect of the perturbation. Thus when

reaction 5.42 is at equilibrium an increase in pressure, or a decrease in volume, will result in the conversion of more SO_2 to SO_3, because this leads to an overall decrease in the number of molecules in the system and so reduces the effect of the pressure or volume change. The forward reaction 5.42 liberates heat, so raising the temperature T would be predicted to cause some SO_3 to be converted to SO_2, with a resulting absorption of heat, i.e. a tendency to counter the temperature increase. Adding O_2 causes more SO_3 to be formed from SO_2 and O_2, a process which uses up some of the added O_2. Le Chatelier's principle is often a useful guide, but should not be used uncritically because it is not always clear or unambiguous how the state of equilibrium can best adjust itself so as to minimize the effect of a perturbation. In general it is preferable to examine the effect of variables such as P, V, or reagent concentration on the formula for the equilibrium constant (if Δn is zero P and V alterations have no effect), and to examine the effect of temperature on the value of K itself. A formula relating the temperature variation of K to ΔH will be derived in Section 5.6.

5.4 Equilibria in Ionic Solutions

(i) The Nature of an Ionic Solution

Ionic solutions are obtained by dissolving electrolytes in ionizing solvents. An *electrolyte* is any substance that dissolves to form an electrically conducting, i.e. ionic, solution. The main characteristic of an ionizing solvent is that it has a large dielectric constant, so that the forces of electrostatic attraction, which tend to make oppositely charged ions stick together and form a crystal lattice, are greatly diminished when solvent molecules are interposed between the ions. Water, liquid ammonia, ethanol, and liquid sulphur dioxide are typical ionizing solvents. A solvent of high dielectric constant is invariably composed of polar molecules; the polar solvent molecules arrange themselves about a dissolved ion so that the ends of the molecules of opposite charge to the ion are directed inwards as in Fig. 5.15. The ion is then said to be *solvated*. The charge of the ion is effectively distributed over the outside of the 'solvation shell', which may contain twelve or more solvent molecules. This distribution of charge stabilizes the solvated ion relative to the ion in the undissolved crystal, so that, provided the crystal's lattice energy is not too large nor the amount of solvent too small, the crystal dissolves completely.

Electrolytes are classified as either strong or weak electrolytes, according to whether they give only ions on dissolving in a solvent such as water, or give mainly neutral solute molecules plus a small proportion (typically 1% or less) of ions. Strong electrolytes are substances such as

NaCl which form ionic crystals, or substances which react with the solvent to form ions, such as hydrogen chloride: HCl reacts completely with water to form H_3O^+ and Cl^-. Weak electrolytes are generally substances such as NH_3, CO_2, H_2S, or acetic acid, CH_3COOH, which react with water to a small extent to form ions. Weak electrolytes are said to be only partially dissociated to ions, strong electrolytes to be completely dissociated. The dissociation of a weak electrolyte, e.g.

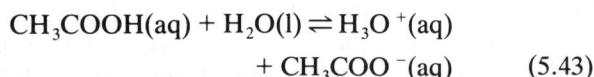

$$CH_3COOH(aq) + H_2O(l) \rightleftharpoons H_3O^+(aq)$$
$$+ CH_3COO^-(aq) \qquad (5.43)$$

is an equilibrium process, and the weakness of the electrolyte is reflected in the fact that the equilibrium constant is small (1.8×10^{-5} mol l^{-1} at 25°C for acetic acid). There also exists a small and relatively unimportant class of substances, known as intermediate electrolytes, for which the dissociation constant is of the order of unity. An example is thallous iodide, TlI. We shall confine our attention to the strong and weak extremes of electrolyte behaviour. The degree of dissociation of an electrolyte in solution can be determined experimentally by measuring the electrical conductivity of the solution and comparing it with the conductivity due to the same electrolyte in a solution which is so dilute that complete dissociation can be assumed. Some very interesting experimental and theoretical problems arise from such measurements, but unfortunately they are outside the scope of this book.

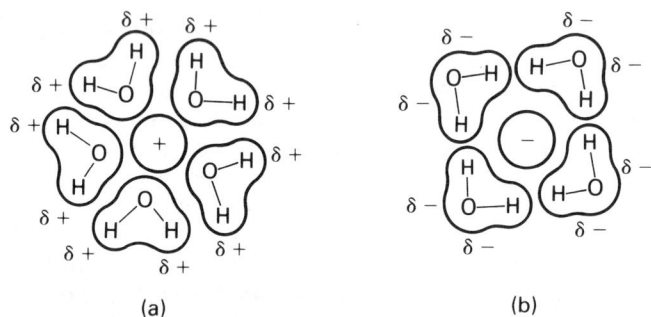

(a)   (b)

Fig. 5.15 Orientation of water molecules in the solvation spheres of a cation (a) and an anion (b)

When a strong electrolyte is in solution each kind of ion interacts with the solvent individually, so that the depression of vapour pressure, elevation of boiling point, and depression of freezing point of the solvent are much greater than they would be if the solute were not dissociated. For example, the depression of the freezing point of water per mole of sodium chloride in solution is almost exactly twice as great as the depression per mole of a non-electrolyte

such as sucrose (cane sugar). Deviations from the ideal behaviour that would be predicted for a solution in which the different ions interacted with the solvent independently of one another are embodied in the activity coefficients of the ions, as described in the next part of this section.

(ii) Activity Coefficients of Ions: The Debye–Huckel Theory

An infinitely dilute ionic solution is one in which the ions produced by dissolving a strong electrolyte in a solvent such as water are so far apart that they can be considered to interact with the solvent quite independently of one another. Each ion is surrounded by its solvation shell, as in Fig. 5.15, and the miniature 'icebergs' that correspond to the separate ions are far enough apart for the electrical forces between them to be negligible. This situation corresponds to an ideal solution, in which the activity of any ion is equal to its concentration, i.e. the activity coefficients of the ions are all unity.

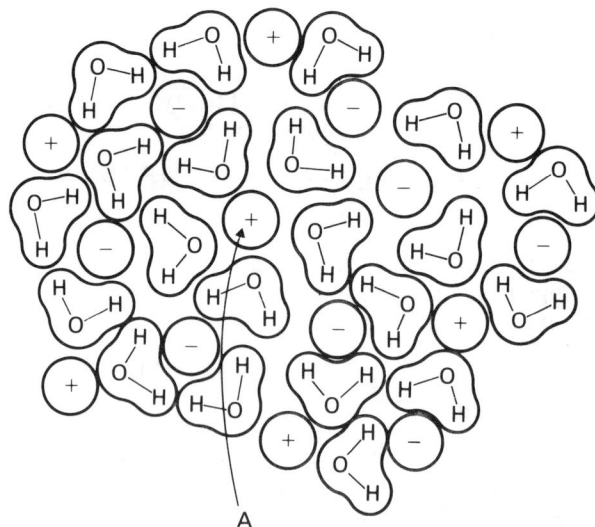

A

Fig. 5.16 Illustration of the ion atmosphere about a cation at point A

In a solution which is not infinitely dilute the electrostatic attraction between oppositely charged ions causes an ordered arrangement to develop, in which an ion of a given charge tends to be surrounded by a cloud of ions of predominantly the opposite charge, as in Fig. 5.16. An oppositely charged ion can even penetrate the solvation shell, to form a more-or-less stable *ion-pair*. The tendency towards forming a perfectly ordered arrangement is, of course, counteracted by the thermal motion of the molecules and ions; otherwise the electrolyte would crystallize. The mobile, partially ordered cloud of oppositely charged species about any given ion is termed that ion's *ionic atmosphere*. The density of the cloud, and therefore the effect of the cloud on the behaviour of the central ion, must increase as the concentration of ions increases. It is also clear that multiply charged ions, such as Ca^{2+} or PO_4^{3-}, must have a greater tendency to accumulate an ionic atmosphere, and be more effective as part of an ionic atmosphere, than singly charged ions such as Na^+ or Cl^-, because they exert stronger electrostatic forces. The presence of the ionic atmosphere around an ion reduces its interaction with the solvent or with other solutes, and so causes the activity coefficient of the ion to be less than unity in moderately concentrated solutions. (In very concentrated solutions the activity coefficient of an ion can be larger than unity, presumably because of the possibility of cooperative interactions between the ions in the ordered arrangement. Such solutions, for which there is at present no satisfactory theory, are outside the scope of our discussion.)

Debye and Huckel derived formulas for the activity coefficient of an ion in solution on the basis of the ion atmosphere model we have just discussed. The activity coefficient was found to be related to the total concentration of ions in the solution through a quantity termed the *ionic strength* of the solution, I, defined by

$$I = \frac{1}{2} \sum_i C_i Z_i^2 \tag{5.44}$$

where C_i is the concentration of ion i in moles per litre, and Z_i is the charge of ion i in units of the charge on a proton. The presence of the square of Z_i in the expression for I takes account of the greater effectiveness of multiply charged ions in forming the ionic atmosphere. The sum on the right hand side of Eq. 5.44 is taken over all the ions in the solution.

Example Calculate the ionic strength of a solution which contains 0.1 mole of $NaNO_3$, 0.1 mole of H_2SO_4, and 0.1 mole of $CrCl_3$ in one litre.

The concentrations and charges of the various ions are tabulated as follows, together with the values of Z_i^2 and $C_i Z_i^2$:

Ion	C_i	Z_i	Z_i^2	$C_i Z_i^2$
Na^+	0.1	1	1	0.1
NO_3^-	0.1	1	1	0.1
H^+	0.2	1	1	0.2
SO_4^{2-}	0.1	-2	4	0.4
Cr^{3+}	0.1	3	9	0.9
Cl^-	0.3	-1	1	0.3
			Total:	2.0

Hence $I = 1.0$. Points to notice are the concentrations of H^+ and Cl^- in column two, the effect of the factor Z_i^2, and the factor of $\frac{1}{2}$ in the definition of I. For a solution of a 1:1 electrolyte such as $NaNO_3$ the ionic strength is seen to be equal to the concentration in moles per litre.

For an ion in a solution which is not too far from infinite dilution (e.g. slightly contaminated distilled water) Debye and Huckel gave a formula known as the Debye–Huckel Limiting Law. This is

$$\log_{10} \gamma = -AZ^2\sqrt{I} \tag{5.45}$$

where γ is the activity coefficient and Z the charge on the ion in question, I is the ionic strength of the solution, and A is a constant which is characteristic of the solvent. For water at 25°C the constant A has the value 0.509. The logarithm of the activity coefficient is negative according to Eq. 5.45, i.e. the activity coefficient is always less than 1.

For a solution which is moderately concentrated (up to ionic strengths of the order of 0.1) there is a second Debye–Huckel formula, namely

$$\log_{10} \gamma = \frac{-AZ^2\sqrt{I}}{1 + B\mathring{a}\sqrt{I}} \tag{5.46}$$

where γ, Z, I and A have the same significance as before, B is a second constant whose value is characteristic of the solvent, and $\mathring{a}$ is equal to the distance of closest approach between the central ion and the components of its ionic atmosphere. The value of $\mathring{a}$ is chosen to give the best agreement with experimental data, i.e. $\mathring{a}$ is an empirical parameter. Various empirical extensions of Eq. 5.46 have been proposed to describe the variation of γ with ionic strength in solutions where I is of the order of 1 or greater. For aqueous solutions near 25°C a useful simplification results from the circumstance that $B\mathring{a}$ is commonly close to 1, so that we can write

$$\log_{10} \gamma = \frac{-AZ^2\sqrt{I}}{1 + \sqrt{I}} \tag{5.47}$$

Eq. 5.47 is probably the most useful form of the Debye–Huckel equation for work with moderately dilute aqueous solutions.

Example Calculate the activity coefficients at 25°C of the following ions in aqueous solutions of ionic strength 0.1, 0.01 and 0.001:

$$Na^+, SO_4^{2-}, Fe^{3+}.$$

The first requirement is to find $\sqrt{I}$ for each of the three solutions; the values obtained are 0.3162, 0.10 and 0.03162. We use Eq. 5.47 with $A = 0.509$. The values of Z^2 are 1, 4 and 9. Hence we obtain the following table.

Ion	Z^2	$\log \gamma$, $I = 0.1$	$\log \gamma$, $I = 0.01$	$\log \gamma$, $I = 0.001$	γ (0.1)	γ (0.01)	γ (0.001)
Na^+	1	− 0.1223	− 0.0463	− 0.0156	0.755	0.899	0.965
SO_4^{2-}	4	− 0.489	− 0.185	− 0.0624	0.324	0.653	0.866
Fe^{3+}	9	− 1.10	− 0.417	− 0.1404	0.079	0.383	0.724

These results and the data in Fig. 5.17 provide an indication of how serious an error would be made by ignoring the activity coefficient for an ion, i.e. assuming activity to be equal to concentration, at the given ionic strengths. For a singly charged ion the error is probably tolerable at $I = 0.01$, and might even be so at $I = 0.1$.

Experimentally the measurement of activity coefficients for individual ions in solution is not practicable; instead one measures a *mean ionic activity coefficient* for an electrolyte, $\gamma \pm$, defined by

$$\gamma \pm = (\gamma_+^{\nu+} . \gamma_-^{\nu-})^{1/\nu} \qquad (5.48)$$

where a mole of the electrolyte gives rise to ν_+ moles of cation, of activity coefficient γ_+, and ν_- moles of anion, of activity coefficient γ_-, and ν is the sum of ν_+ and ν_-. For example, for calcium chloride ν_+ is 1, ν_- is 2 and ν is 3. In Fig. 5.17, experimental values of the mean ionic activity coefficient for KCl and CaCl₂ are compared with the values predicted by Eqs 5.45 and 5.47. Agreement between the predicted and experimental values is better for KCl, but the predictions for CaCl₂ are sufficiently accurate to be useful up to an ionic strength of about 0.5.

(iii) Solubility Product

The process of dissolving solid barium sulphate, an ionic crystal, in water, can be represented by the equation

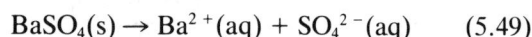

$$BaSO_4(s) \rightarrow Ba^{2+}(aq) + SO_4^{2-}(aq) \qquad (5.49)$$

At equilibrium, when the solution is saturated with $BaSO_4$, we have

$$K_{BaSO_4} = \frac{a_{SO_4^{2-}} . a_{Ba^{2+}}}{a_{BaSO_4}}$$

which immediately reduces to

$$K_{BaSO_4} = a_{SO_4^{2-}} . a_{Ba^{2+}} \qquad (5.50)$$

because the activity of the pure solid barium sulphate is 1.0. The equilibrium constant that we have written K_{BaSO_4} is the *solubility product* of barium sulphate. Such a solubility product governs the concentration of a saturated solution of every strong electrolyte. If the product of the activities on the right hand side of Eq. 5.50 is less than K_{BaSO_4}, the solution will be capable of dissolving more barium sulphate; if the product of activities exceeds K_{BaSO_4}, solid barium sulphate will precipitate. Barium sulphate is generally classed as an insoluble salt, which implies that the ionic strength of a saturated solution will be low unless other electrolytes are present. At low ionic strengths the activities in Eq. 5.50 can be replaced by concentrations without significant error. For a general electrolyte, $M_{\nu_+}N_{\nu_-}$, dissolving to produce ν_+ cations of activity a_M and ν_- anions of activity a_N, the solubility product is given by

$$K_{M_{\nu_+}N_{\nu_-}} = (a_M)^{\nu+}(a_N)^{\nu-} \qquad (5.51)$$

The process corresponding to Eq. 5.49 is

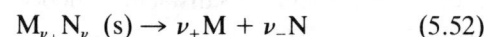

$$M_{\nu_+}N_{\nu_-}(s) \rightarrow \nu_+M + \nu_-N \qquad (5.52)$$

where M and N are positive and negative ions, respectively. It is important to note that the solubility product expression contains the activity of each ion raised to the appro-

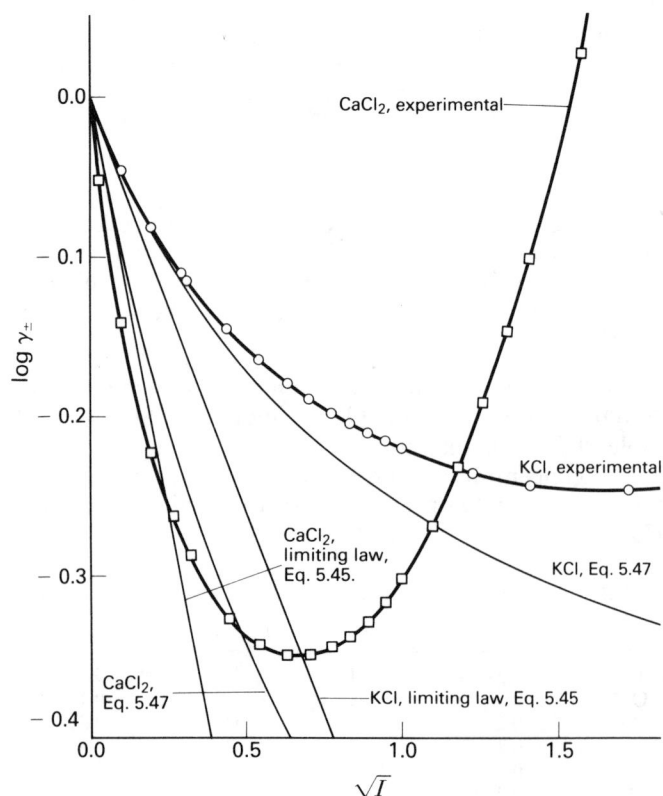

Fig. 5.17 Comparison of experimental values of mean ionic activity coefficient for KCl and CaCl₂ with values predicted by Eqs 5.45 and 5.47

priate power, ν_+ or ν_-. The examples that follow illustrate common ways of applying the solubility product. A number of solubility products of sparingly soluble salts are listed in Table 5.4.

Table 5.4 Solubility products at 25°C in mol l^{-1} units; 2($-$ 10) = 2×10^{-10}. (Values from *SI Chemical Data* by G. H. Aylward and T. J. V. Findlay, Second Edition, John Wiley and Sons, 1974.)

Solid	K	Solid	K
AgBr	5($-$ 13)	Fe(OH)$_2$	8($-$ 16)
AgCN	1($-$ 16)	Fe(OH)$_3$	4($-$ 40)
AgCl	2($-$ 10)	FeS	5($-$ 18)
Ag$_2$CrO$_4$	3($-$ 12)	Hg$_2$Br$_2$	7($-$ 23)
AgI	8($-$ 17)	Hg$_2$Cl$_2$	1($-$ 18)
Ag$_2$S	6($-$ 50)	Hg$_2$I$_2$	5($-$ 29)
Al(OH)$_3$	1($-$ 33)	HgS (black)	1($-$ 52)
BaCO$_3$	5($-$ 9)	HgS (red)	4($-$ 53)
Ba(OH)$_2$	5($-$ 3)	MgCO$_3$	1($-$ 5)
BaSO$_4$	1($-$ 10)	MgNH$_4$PO$_4$	7($-$ 14)
Be(OH)$_2$	1($-$ 21)	Mg(OH)$_2$	1($-$ 11)
CaCO$_3$	5($-$ 9)	Mn(OH)$_2$	2($-$ 13)
CaC$_2$O$_4$	2($-$ 9)	MnS (pink)	3($-$ 10)
CaF$_2$	3($-$ 11)	MnS (green)	3($-$ 13)
Ca(OH)$_2$	4($-$ 6)	Ni(OH)$_2$	6($-$ 18)
CaSO$_4$	2($-$ 5)	PbBr$_2$	9($-$ 6)
Cd(OH)$_2$	4($-$ 15)	PbCO$_3$	6($-$ 14)
CdS	2($-$ 28)	PbCl$_2$	2($-$ 5)
Co(OH)$_2$	6($-$ 15)	PbI$_2$	1($-$ 9)
Co(OH)$_3$	3($-$ 41)	Pb(OH)$_2$	6($-$ 16)
Cr(OH)$_3$	1($-$ 30)	PbS	1($-$ 28)
CuBr	5($-$ 9)	PbSO$_4$	2($-$ 8)
CuCl	2($-$ 7)	Sn(OH)$_2$	8($-$ 29)
CuI	1($-$ 12)	Sr(OH)$_2$	3($-$ 4)
Cu(OH)$_2$	1($-$ 20)	SnS	1($-$ 25)
CuS	6($-$ 36)	Zn(OH)$_2$	3($-$ 17)

Example 1 Calculate the solubilities of silver chloride, mercurous chloride and lead chloride in water at 25°C, ignoring activity coefficients.

The solution process for silver chloride is

$$AgCl(s) \rightarrow Ag^+(aq) + Cl^-(aq)$$

so if the solution contains m moles per litre of AgCl the concentrations of Ag^+ and Cl^- are both equal to m. Hence

$$m^2 = 2 \times 10^{-10} \text{ (from Table 5.4)}$$

or $\quad m = 1.4 \times 10^{-5}$ mol l^{-1}.

For mercurous chloride the process is

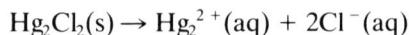

$$Hg_2Cl_2(s) \rightarrow Hg_2^{2+}(aq) + 2Cl^-(aq)$$

so if m moles of mercurous chloride dissolve the concentrations of Hg_2^{2+} and Cl^- are m and $2m$, respectively. Hence

$$4m^3 = 1 \times 10^{-18}$$

or $\quad m = 6.3 \times 10^{-7}$ mol l^{-1}.

For lead chloride the process is

$$PbCl_2(s) \rightarrow Pb^{2+}(aq) + 2Cl^-(aq)$$

so if m moles per litre of PbCl$_2$ dissolve the concentrations of Pb^{2+} and Cl^- are m and $2m$, respectively. Hence

$$4m^3 = 2 \times 10^{-5}$$

or $\quad m = 1.7 \times 10^{-2}$ mol l^{-1}.

(For all of these solutions the ionic strength is low enough to justify ignoring the activity coefficients.)

Example 2 The solubility of strontium hydroxide, Sr(OH)$_2$, in water at 25°C is measured as 9×10^{-2} mol l^{-1}. (a) Calculate the solubility product, ignoring activity coefficients. (b) Determine whether neglect of the activity coefficients is likely to cause a significant error in the calculated value of the solubility product.

For part (a), we note that the process involved is

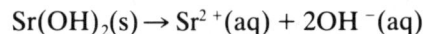

$$Sr(OH)_2(s) \rightarrow Sr^{2+}(aq) + 2OH^-(aq)$$

so that if m moles per litre of strontium hydroxide dissolve, the concentrations of Sr^{2+} and OH^- are m and $2m$, respectively. Hence

$$K = 4m^3 = 2.9 \times 10^{-3}.$$

To answer part (b) we need to know the ionic strength, I. Eq. 5.44 gives, for the specified Sr(OH)$_2$ solution, $I = 0.27$. The value of the mean ionic activity coefficient $\gamma_\pm$ that is given by Eq. 5.47 can be read from Fig. 5.17, using the curve that shows the data predicted for CaCl$_2$. For $I = 0.27$, $\sqrt{I}$ is 0.52, and log $\gamma_\pm$ is -0.34, or $\bar{1}.66$. Hence $\gamma_\pm = 0.46$. From the definition of $\gamma_\pm$, Eq. 5.48, we see that the correct expression for $K_{Sr(OH)_2}$ is

$$K = 4m^3 \gamma_\pm^3 = 2.8 \times 10^{-4}.$$

Thus the neglect of the activity coefficient in this case caused the calculated solubility product to be too low by a factor of ten. This example was purposely chosen to illustrate an extreme case: Sr(OH)$_2$ is actually quite soluble in water (almost 0.1 molar at 25°C) and the doubly charged ion makes a large contribution to the ionic strength. For most of the compounds listed in Table 5.4 the effect of the activity coefficients would be negligible in a saturated solution.

Two other points relating to solubility products can best be understood with the aid of specific examples. The first concerns the *common-ion effect*.

Example 3 Calculate the solubility of silver chloride in 0.01 molar hydrochloric acid at 25°C, and compare it with the value obtained in Example 1. Activity coefficients are to be neglected. (For AgCl the value of $\gamma_\pm$ at $I = 0.01$ is in fact 0.89).

We have $[Ag^+][Cl^-] = 2 \times 10^{-10}$ where $[Ag^+]$ is equal to m, the molarity of AgCl in the solution, and $[Cl^-]$ is $(m + 0.01)$, which is equal to 0.01 to a very good approximation. Hence

$$0.01\, m = 2 \times 10^{-10}$$

or

$$m = 2 \times 10^{-8} \text{ mol l}^{-1}.$$

This is three orders of magnitude less than the solubility of AgCl in pure water. The difference is due to the presence of the 0.01 molar chloride ion, the chloride ion being common to the sparingly soluble salt and the added electrolyte, hydrochloric acid. The common-ion effect always leads to a *reduction* in the equilibrium amount of salt dissolved, and this example shows that the effect of the common ion can be large. The common-ion effect is used industrially to precipitate sodium chloride from brine, additional chloride ions being introduced as gaseous HCl. After the sodium chloride has been filtered off the HCl is recovered by heating the mother liquor.

The next example is concerned with the *electrostatic effect*, also known as the *salt effect*. This is a much smaller effect than the common-ion effect, and is generally less important. We see the effect in operation as follows:

Example 4   With allowance for activity coefficients, calculate the solubility of silver chloride in 0.1 molar potassium nitrate solution and compare it with the value obtained in Example 1.

The first task is to obtain the value of $\gamma_\pm$ for AgCl in a solution of ionic strength 0.1. The value given by Eq. 5.47 must be identical with the prediction for KCl that is plotted in Fig. 5.17. At $I = 0.1$, $\sqrt{I} = 0.316$, and from Fig. 5.17 we find $\log \gamma_\pm = -0.12$, or $\bar{1}.88$. Hence $\gamma_\pm = 0.76$. From the definition of $\gamma_\pm$, Eq. 5.48, we see that

$$m^2\gamma_\pm^2 = K = 2 \times 10^{-10}$$

Hence

$$m = 1.9 \times 10^{-5} \text{ mol l}^{-1}.$$

This is about 50% greater than the value of m calculated in Example 1. The effect of the added electrolyte which has no ion in common with the AgCl is to produce an increase in the quantity of AgCl in the saturated solution. This happens because the increase in ionic strength causes a reduction in the activity coefficient, which must be compensated for by an increase in concentration.

(iv) Equilibria involving Complex Ions

The strength of the interaction between an ion and adjacent neutral solvent or solute molecules, or between the ion and other ions of opposite charge, sometimes becomes so great that it leads to the formation of a stable chemical bond. This most commonly happens with small cations, particularly cations of the transition metals (the d block elements), which become bound to two or more molecules or ions through donation of electrons from the surrounding *ligands* to the central cation. Typical examples of electron donors which can act as ligands are H_2O, CN^-, Cl^-, CO, NO_2^- and NH_3. The ligands occur in a fixed number, commonly 2, 4 or 6, grouped around the central ion in a regular geometrical arrangement, quite different from the essentially haphazard configuration of the solvation shell and the ion atmosphere. Some aspects of the chemistry of such 'coordination complexes' will be described in Chapter 12. For the present we confine our attention to the equilibria which govern their formation. Some typical examples of complex ion equilibria are given in Table 5.5, with values of the corresponding equilibrium constants.

Table 5.5 Some complex ion equilibria

Observation	Reaction	$\log_{10}K$
AgCN precipitate dissolves in excess cyanide	$Ag^+ + 2CN^- \rightleftharpoons Ag(CN)_2^-$	20
AgCl precipitate dissolves in ammonia solution	$Ag^+ + 2NH_3 \rightleftharpoons Ag(NH_3)_2^+$	7
Copper hydroxide dissolves in excess ammonia solution	$Cu^{2+} + 4NH_3 \rightleftharpoons Cu(NH_3)_4^{2+}$	13
Ferric ions removed from solution by addition of cyanide	$Fe^{3+} + 6CN^- \rightleftharpoons Fe(CN)_6^{3-}$	44

The effect of complex formation on the concentration of the parent ion can be very marked, as the following example shows.

Example   Calculate the equilibrium concentrations of silver and ferric ions in 0.01 molar solutions of sodium argentocyanide, $NaAg(CN)_2$, and potassium ferricyanide, $K_3Fe(CN)_6$. Assume that intermediate species such as $Fe(CN)_5^{2-}$ are negligible.

For the concentration of metal ion we write x, so that the concentration of the complex is $(0.01 - x)$, and we assume that this is equal to 0.01 as a first approximation.

For the dissociation of argentocyanide ion the concentration of CN^- is $2x$. Hence

$$K = \frac{0.01}{4x^3} = 10^{20}, \text{ from Table 5.5}$$

This gives

$$x = [Ag^+] = \underline{2.9 \times 10^{-18} \text{ mol l}^{-1}}.$$

For the dissociation of ferricyanide ion the concentration of CN^- is $6x$. Hence

$$K = \frac{0.01}{x(6x)^6} = 10^{44}, \text{ from Table 5.5. This gives}$$

$$x = [Fe^{3+}] = \underline{5.8 \times 10^{-8} \text{ mol } l^{-1}}$$

(a)

(b)

(c)

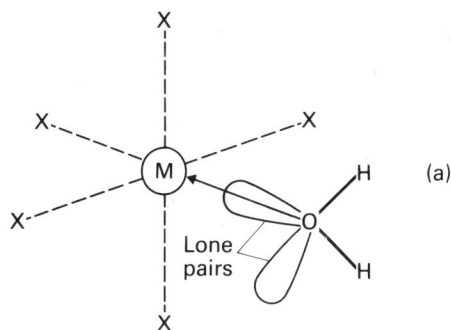

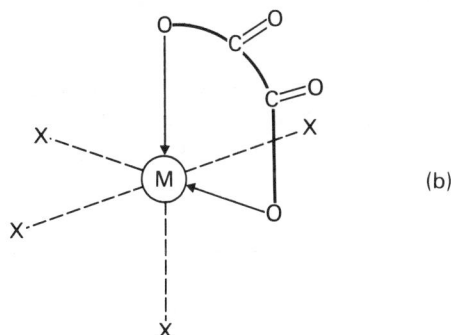

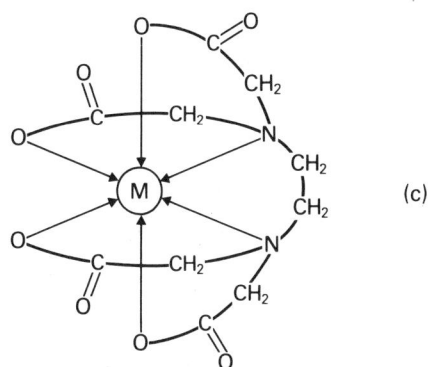

Fig. 5.18 Complex ion formation: (a) bonding of H_2O in an aquo-complex. X is any unspecified ligand. (b) Bonding of the bidentate oxalate ion in an oxalato complex. (c) Bonding of the hexadentate ('six-toothed') ligand in an EDTA complex

The substance ethylenediamine tetraacetic acid, usually referred to as EDTA, is an example of a powerful chelating agent, i.e. of a ligand which will form complexes with a very wide range of metal ions, by a process of wrapping itself around any available ion in the manner that is illustrated in Fig. 5.18. In a strongly alkaline solution EDTA is present as the anion $EDTA^{4-}$, and in this form it has its maximum complexing ability. Equilibrium constants for the reaction

$$M^{m+} + EDTA^{4-} \rightleftharpoons M(EDTA)^{m-4} \qquad (5.53)$$

range from 5×10^8 for Mg^{2+} and 3×10^{10} for Ca^{2+} to 5×10^{21} for Hg^{2+} and 10^{25} for Fe^{3+}. Various methods have been worked out for estimating metal ion concentrations by titrating with a solution of EDTA, substances which form coloured complexes with either the metal or the EDTA being used to indicate the end point. The extent of EDTA complex formation by a metal can be controlled by controlling the alkalinity of the solution, provided the equilibrium constant for reaction 5.53 is not too large. The concentrations of calcium and magnesium ions in tap water are commonly determined in this way.

5.5 Acid–Base Equilibria

(i) The Nature of Acids and Bases

For most purposes an acid can be defined as a substance which gives rise to hydrogen ions in solution. However, it has to be borne in mind that, because of their small size, the hydrogen ions are always solvated, and *free* H^+ does not occur. In aqueous solution the predominant species is the hydronium ion H_3O^+. In other solvents analogous ions are produced; for example, in liquid ammonia an acid is a substance that gives rise to the ammonium ion, NH_4^+. A base is most simply defined as a substance which can react with hydrogen ions, or with the acidic portion of an acid molecule if there are no actual free or solvated hydrogen ions present. Thus NH_3 is clearly a base, because it will react with the acid HCl in the gas phase, in the process forming the salt $NH_4^+Cl^-$. It is necessary to regard water as a base according to this definition, because of the reaction

$$HCl(g) + H_2O(l) \rightarrow H_3O^+(aq) + Cl^-(aq) \quad (5.54)$$

which occurs when HCl dissolves in water. The product ion H_3O^+ in reaction 5.54 is itself an acid, as is shown by its ability to react with the base NH_3:

$$H_3O^+(aq) + NH_3(aq) \rightarrow NH_4^+(aq) + H_2O(l) \quad (5.55)$$

Also, the reverse of reaction 5.54 takes place when hydrochloric acid solution is heated, so the chloride ion must be a base because it reacts with the acid H_3O^+. Similarly, we can conclude that the NH_4^+ ion on the right hand side of Eq. 5.55 must be an acid, and therefore the water molecule must be acting as a base when reaction 5.55 occurs in reverse. Reactions 5.54 and 5.55 are seen to be of the general form

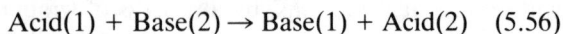

$$Acid(1) + Base(2) \rightarrow Base(1) + Acid(2) \quad (5.56)$$

where Base(1) is the *conjugate base* of Acid(1), and Acid(2) is the *conjugate acid* of Base(2). Conversely, Acid(1) is the conjugate acid of Base(1), and Base(2) is the conjugate base of Acid(2). Substances such as water, which can behave either as an acid or as a base, are termed *amphoteric*. All acid–base reactions are of the form of reaction 5.56. A number of examples of such reactions are listed in Table 5.6.

Table 5.6 Some examples of acid–base reactions

$$Acid(1) + Base(2) \rightarrow Base(1) + Acid(2)$$

Solvent	Acid(1)	Base(2)	Base(1)	Acid(2)
H_2O	HNO_3	H_2O	NO_3^-	H_3O^+
H_2O	H_2SO_4	H_2O	HSO_4^-	H_3O^+
H_2O	HSO_4^-	H_2O	SO_4^{2-}	H_3O^+
H_2O	H_2O	NH_3	OH^-	NH_4^+
H_2O	H_3O^+	OH^-	H_2O	H_2O
NH_3	HCl	NH_3	Cl^-	NH_4^+
NH_3	NH_3	OH^-	NH_2^-	H_2O
NH_3	NH_4^+	NH_2^-	NH_3	NH_3
HF	HF	H_2O	F^-	H_3O^+
H_2SO_4	$H_3SO_4^+$	HSO_4^-	H_2SO_4	H_2SO_4
SO_2	SO^{2+}	SO_3^{2-}	SO_2	SO_2

Our discussion so far has been basically an account of the proton theory of acids and bases, due to Brønsted and Lowry. According to this theory an acid is a substance which can transfer a proton to a base, and a base is a substance which can accept a proton from an acid. Substances which are bases according to this definition invariably possess lone pairs of electrons by which a proton can be attached. It is possible to define an acid as a substance which can accept a lone pair of electrons, and a base as one which can donate a lone pair of electrons. (Hence the inclusion of SO^{2+} and SO_3^{2-} as Acid(1) and Base(2) in Table 5.6.) This definition includes proton acids and their conjugate bases, but is clearly more general. Substances which are acids according to this definition but which do not necessarily contain protons are called *Lewis acids*, after G. N. Lewis who proposed the theory. The definition is quite useful in practice because Lewis acids have much in common with ordinary proton acids, especially in relation to such properties as catalytic activity. Lewis acids which have important applications as catalysts include $FeCl_3$, BF_3 and $AlCl_3$. When a metal ion forms a complex with electron-donating ligands it is behaving as a Lewis acid. The gas phase reaction of BF_3 with NH_3 to form the complex $BF_3.NH_3$ can be regarded as an acid–base reaction, analogous to the reaction of NH_3 with HCl to form the salt NH_4Cl.

(ii) Dissociation of Acids and Bases in Water: pH and pK

In aqueous solution an acid gives rise to the hydronium ion H_3O^+, and a base gives rise to the ion OH^-, the characteristic reactions being

$$HA + H_2O \rightarrow H_3O^+ + A^- \quad (5.57)$$

for an acid HA, and

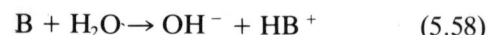

$$B + H_2O \rightarrow OH^- + HB^+ \quad (5.58)$$

for a base B. The equilibrium constants K_a and K_b for these reactions are defined by

$$K_a = \frac{[H_3O^+][A^-]}{[HA]} \quad (5.59)$$

and

$$K_b = \frac{[OH^-][HB^+]}{[B]} \quad (5.60)$$

where we have assumed that activity coefficients may be neglected.

The equilibrium constants in Eqs 5.59 and 5.60 are the acid dissociation constant of HA, and the basic dissociation constant of B, respectively. Similar constants can also be defined for the base A^- and the acid HB^+; thus for A^- the reaction is

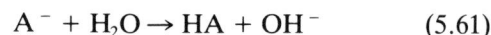

$$A^- + H_2O \rightarrow HA + OH^- \quad (5.61)$$

with

$$K_b = \frac{[HA][OH^-]}{[A^-]} \quad (5.62)$$

and for HB^+ we have

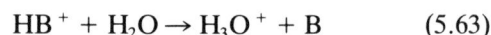

$$HB^+ + H_2O \rightarrow H_3O^+ + B \quad (5.63)$$

with

$$K_a = \frac{[H_3O^+][B]}{[HB^+]} \quad (5.64)$$

Thus for a given conjugate acid–base pair we can define both K_a and K_b. If we multiply corresponding sides of Eqs 5.59 and 5.62 together we obtain, for the acid HA and base A^-, the result

$$K_a K_b = [H_3O^+][OH^-] = K_w \quad (5.65)$$

The quantity K_w is the *ionic product of water,* and is the equilibrium constant of the reaction

$$2H_2O \rightarrow H_3O^+ + OH^- \quad (5.66)$$

The value of K_w in $mol^2 \, l^{-2}$ is usually taken as being exactly 1×10^{-14}, but for precise work it is necessary to take account of its variation with temperature. Values of K_w at 20, 25 and 30°C are 0.681×10^{-14}, 1.008×10^{-14}, and 1.47×10^{-14}, respectively. Eq. 5.65 shows that if a substance is a strong acid, with a large value of K_a, its conjugate base must be weak, with a small value of K_b; similarly, for a strong base with a large value of K_b the conjugate acid must be weak.

From Eq. 5.65 and the value of K_w we can conclude that in pure water at 25°C the concentrations of H_3O^+ and OH^- ions are both equal to 10^{-7} mol l^{-1}. This constitutes a neutral solution. If an acid is introduced the H_3O^+ concentration becomes greater than 10^{-7} and the OH^- concentration simultaneously falls so as to hold the ionic product constant at 10^{-14}. If a base is added the OH^- concentration rises above 10^{-7} molar, and the concentration of H_3O^+ falls. At equilibrium it is always true that

$$a_{H_3O^+} \, a_{OH^-} = K_w \qquad (5.67)$$

In a dilute solution at 25°C this reduces to

$$[H_3O^+][OH^-] = 10^{-14} \qquad (5.68)$$

Quantities which are usually less than 1 and which are liable to vary over many orders of magnitude are conveniently represented by their logarithm to base 10, with sign reversed. Conventionally the prefix 'p' stands for '$-\log_{10}$', and for an equilibrium constant K we thus obtain the quantity pK. For hydrogen ion activity we obtain the quantity pH. Mathematically we can write

$$pK_a = -\log_{10} K_a \qquad (5.69)$$

$$pK_b = -\log_{10} K_b \qquad (5.70)$$

$$pH = -\log_{10} a_{H_3O^+} \sim -\log_{10}[H_3O^+] \qquad (5.71)$$

The pH of a neutral solution is 7 (at 25°C), while acid solutions have pH less than 7 and alkaline solutions pH greater than 7. In terms of pK_a and pK_b we can rewrite Eq. 5.65 as

$$pK_a + pK_b = 14 \qquad (5.72)$$

Substances which behave predominantly as acids in aqueous solution have pK_a's less than 7, while those which normally behave as bases have pK_a's greater than 7. A substance whose pK_a value is exactly 7 has acid and base strengths equal to those of water. Stronger acids force water to act as a base and form H_3O^+, while weaker acids are obliged to accept a proton from water, forming OH^-. The acidic or basic character of a substance is seen to be a function of the solvent, as well as of the nature of the substance itself. Values of pK_a for a number of acids and bases are listed in Table 5.7.

Table 5.7 Values of pK_a at 25°C (concentrations in mol l^{-1})

Acid	Formula	Conjugate base	pK_a
Acetic acid	CH_3COOH	CH_3COO^-	4.76
Ammonium ion	NH_4^+	NH_3	9.24
Bicarbonate ion	HCO_3^-	CO_3^{2-}	10.32
Carbonic acid	$CO_2(aq)$	HCO_3^-	6.38
Chloracetic acid	$CH_2ClCOOH$	CH_2ClCOO^-	2.85
Ethylammonium ion	$C_2H_5NH_3^+$	$C_2H_5NH_2$	10.67
Formic acid	$HCOOH$	$HCOO^-$	3.74
Hydrocyanic acid	HCN	CN^-	9.22
Hydrofluoric acid	HF	F^-	3.17
Hydrogen sulphide	H_2S	HS^-	7.02
Hydrogen sulphide ion	HS^-	S^{2-}	12.89
Methylammonium ion	$CH_3NH_3^+$	CH_3NH_2	10.59
Nitric Acid	HNO_3	NO_3^-	-1.37
Nitrous Acid	HNO_2	NO_2^-	3.14
Phosphoric acid	H_3PO_4	$H_2PO_4^-$	2.13
Dihydrogen phosphate ion	$H_2PO_4^-$	HPO_4^{2-}	7.20
Hydrogen phosphate ion	HPO_4^{2-}	PO_4^{3-}	12.36
Sulphuric acid	H_2SO_4	HSO_4^-	0
Hydrogen sulphate ion	HSO_4^-	SO_4^{2-}	1.99
Sulphurous acid	H_2SO_3	HSO_3^-	1.90
Hydrogen sulphite ion	HSO_3^-	SO_3^{2-}	7.20

(iii) Calculation of pH Variations during Acid–Base Titrations

(a) Strong acid–Strong base. Suppose we gradually add a 0.1 molar solution of a strong base such as sodium hydroxide to 10 ml of 0.1 molar acid. If we have a strong acid such as hydrochloric or nitric acid the initial hydronium ion concentration will be 0.1 molar, corresponding to a pH of 1. When the alkali is added reaction 5.73 occurs. Therefore the addition of any volume less than 10 ml will simply remove an amount of acid equivalent to the amount of alkali added:

$$H_3O^+ + OH^- \rightarrow 2H_2O \qquad (5.73)$$

For example, when 8.18 ml of alkali has been added the amount of acid remaining is the equivalent of 1.82 ml of 0.1 molar solution, which has been diluted to a total volume of 18.2 ml, i.e. the concentration of H_3O^+ is 0.1 molar and the pH is 2. When 9.8 ml of alkali has been added the acid remaining is equivalent to 0.2 ml of 0.1 molar solution in a total volume of 19.8 ml, so the H_3O^+ concentration is approximately 10^{-3} molar, and the pH is 3. When 9.98 ml of alkali has been added the acid remaining is the equivalent of 0.02 ml of 0.1 molar solution in a total volume of 19.98 ml. The H_3O^+ concentration is then 10^{-4} molar, and the pH is 4. When 10.0 ml of alkali has been added the solution is perfectly neutral and the pH is 7. This sharp change in Log (concentration) near the end point is a characteristic feature of titration processes.

Beyond the end point of the titration, the pH of the

solution is governed by the excess of alkali added. To produce a pH change of 1 unit from the end point, corresponding to the OH$^-$ concentration going from 10^{-7} to 10^{-6} molar, requires the addition of only 2.0×10^{-4} ml of 0.1 molar alkali. Successive pH steps involve steadily larger quantities of alkali, a further 2.0×10^{-3} ml being required to go from pH 8 to pH 9, 0.20 ml to go from pH 9 to pH 10, and so on. The variation of pH with volume added for the titration of a strong acid with a strong base is plotted as curves 1 and 2 in Fig. 5.19.

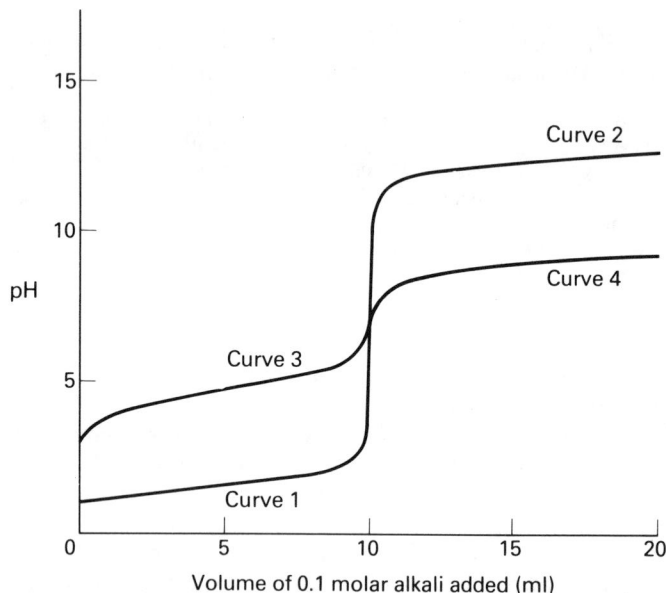

Fig. 5.19 *Titration curves*: Curve 1 + Curve 2, NaOH + HCl; Curve 3 + Curve 2, NaOH + CH$_3$COOH; Curve 1 + Curve 4, NH$_3$ + HCl. Curves calculated for 10 ml of 0.1 molar acid, neglecting activity coefficients

(b) Weak acid–Strong base. Next we consider the case where the strong base NaOH is used to titrate 10 ml of a weak acid, which we suppose to be acetic acid, with pK$_a$ 4.76. Again both solutions are assumed to be 0.1 molar, and the initial volume of acid is taken as 10 ml. The first problem is to calculate the pH of the acid alone. Let α be the fraction of acid dissociated according to the equilibrium

$$CH_3COOH + H_2O \rightleftharpoons CH_3COO^- + H_3O^+ \quad (5.74)$$

To obtain a general formula for this type of situation we suppose the total acetic acid concentration to be C mol l^{-1}. Then at equilibrium the concentrations of H$_3$O$^+$ and acetate ions are αC, and the concentration of undissociated acetic acid is $(1 - \alpha)C$. Hence

$$K_a = \frac{[Ac^-][H_3O^+]}{[HAc]}$$

becomes
$$K_a = \frac{\alpha^2 C}{(1 - \alpha)} \sim \alpha^2 C \quad (5.75)$$

where we have written the formula for acetate ion as Ac$^-$ for brevity, and the degree of dissociation α is commonly much smaller than 1. The pK$_a$ value of 4.76 transforms to $K_a = 1.74 \times 10^{-5}$. The value of α for $C = 0.1$ can be obtained from Eq. 5.75 either by solving a quadratic equation, or by a process of successive approximation, the factor $(1 - \alpha)$ being initially assumed equal to 1. We find $\alpha = 1.31 \times 10^{-2}$, so that the concentration of H$_3$O$^+$ is 1.3×10^{-3} mol l^{-1}, and the pH is 2.88.

When the sodium hydroxide is added it converts part of the acetic acid to the *strong* electrolyte sodium acetate by reaction 5.76, and the concentration of acetate ion is effectively equal to the concentration of sodium acetate, the amount of Ac$^-$ produced by dissociation of HAc being negligible in comparison. The concentration of H$_3$O$^+$ is then fixed by the value of K_a and the ratio of acetate ion to undissociated acetic acid.

$$CH_3COOH + OH^- \rightarrow CH_3COO^- + H_2O \quad (5.76)$$

For example, suppose 1 ml of NaOH has been added. The concentration of HAc becomes $0.1 \times (9/11) = 0.082$, and the concentration of Ac$^-$ becomes $0.1 \times (1/11) = 0.0091$. Hence using

$$[H_3O^+] = K_a[HAc]/[Ac^-] \quad (5.77)$$

we find $[H_3O^+] = 1.5 \times 10^{-4}$ and the pH is 3.81. Continuing in this way until 9.5 ml of NaOH has been added, we obtain the curve labelled curve 3 in Fig. 5.19. An important point to notice is that *when the acid is half neutralized by a strong base the pH is equal to the pK$_a$*, because [HAc] and [Ac$^-$] in Eq. 5.77 are equal to one another. This is a useful general result, which can be expressed in the form

$$pH = pK_a - \log_{10} \left\{ \frac{[Acid]}{[Conjugate\ Base]} \right\} \quad (5.78)$$

To calculate the pH correctly at the end point we have to allow for *hydrolysis* of the salt sodium acetate, i.e. for the reaction of the anion of the weak acid with water, to give OH$^-$ and undissociated acetic acid. We have

$$Ac^- + H_2O \rightarrow HAc + OH^- \quad (5.79)$$

where the equilibrium constant of this reaction, K_b, is related to K_a via Eq. 5.65. This reaction occurs to only a small extent, so the concentration of Ac$^-$ is not affected significantly unless the sodium acetate is extremely dilute, but it has a marked effect on the pH. Because of reaction 5.79, some of the OH$^-$ ions produced by dissociation of water have to be removed according to

$$H_3O^+ + OH^- \rightarrow 2H_2O \qquad (5.73)$$

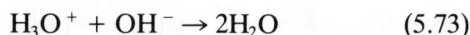

in order that the ionic product of water may remain constant at 10^{-14}. Let the concentration of OH^- resulting from reaction 5.79 be x, and let the concentration of OH^- removed by reaction 5.73 be y. Then at equilibrium the various species in the solution have concentrations as follows:

$$\begin{aligned}
[Ac^-] &= 0.05 - x = 0.05 \text{ approximately} \\
[HAc] &= x \\
[OH^-] &= 10^{-7} + x - y = z + x \\
[H_3O^+] &= 10^{-7} - y = z
\end{aligned}$$

where some simplification results from replacing $(10^{-7} - y)$ by the symbol z. Hence

$$K_a = 1.73 \times 10^{-5} = (0.05 - x)z/x \qquad (5.80)$$

and

$$K_w = 10^{-14} = z(z + x). \qquad (5.81)$$

Making the assumption that x is much less than 0.05 in Eq. 5.80, we find $z = 3.5 \times 10^{-4} x$. Putting this result into Eq. 5.81, we obtain $x = 6.3 \times 10^{-6}$ mol l^{-1} (justifying the initial assumption that x was much less than 0.05) and $z = 2.2 \times 10^{-9}$ mol l^{-1}. This last quantity is the H_3O^+ concentration, so the pH at the end point is 8.66. Curve 3 in Fig. 5.19 meets curves 1 and 2 in the vicinity of the end point. On the right hand side of the end point the pH follows curve 2, because the result of adding excess strong base to a solution of sodium acetate is similar to the result of adding it to a solution of sodium chloride.

(c) Strong acid–Weak base. If a strong acid is titrated by a weak base such as ammonia, the first part of the titration curve follows curve 1 in Fig. 5.19, because at each point of the curve the solution consists of hydrochloric acid mixed with the strong electrolyte ammonium chloride. The NH_3 is converted to NH_4^+ by reaction 5.82:

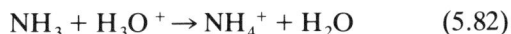

$$NH_3 + H_3O^+ \rightarrow NH_4^+ + H_2O \qquad (5.82)$$

At the end point the solution is acidic because of hydrolysis of the ammonium chloride, i.e. because reaction 5.82 does not go quite to completion when H_3O^+ is not in excess. In a solution of ammonium chloride we have to consider the reverse of reaction 5.82:

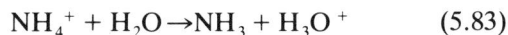

$$NH_4^+ + H_2O \rightarrow NH_3 + H_3O^+ \qquad (5.83)$$

From Table 5.7 the equilibrium constant of reaction 5.83 is found to be 5.76×10^{-10}. Let the concentrations of NH_3 and H_3O^+ produced by reaction 5.83 be x, and the concentrations of H_3O^+ and OH^- removed by reaction 5.73 be y as before. For 0.1 molar solutions of hydrochloric acid and ammonia the concentration of ammonium chloride at the end point is 0.05 molar. Hence the concentration of NH_4^+ at equilibrium is $(0.05 - x)$, and we assume initially

that x is much less than 0.05. The equilibrium constant expressions are therefore

$$K_a = 5.76 \times 10^{-10} = x(10^{-7} + x - y)/0.05$$

and $K_w = 10^{-14} = (10^{-7} - y)(10^{-7} + x - y)$

We now put $z = (10^{-7} - y) = [OH^-]$, and obtain

$$x(z + x) = 2.9 \times 10^{-12}$$

and $z(z + x) = 10^{-14}$.

The last two equations give $x = 290z$, whence $x = 1.7 \times 10^{-6}$ mol l^{-1} (which is much less than 0.05 as we assumed) and $z = 5.9 \times 10^{-9}$ mol $l^{-1} = [OH^-]$. The H_3O^+ concentration is $(z + x)$, which is essentially the same as x, so the pH at the end point works out to be 5.77.

To calculate the pH when NH_3 is in excess we use the formula

$$[H_3O^+] = K_a[NH_4^+]/[NH_3] \qquad (5.84)$$

which can also be written

$$pH = pK_a - \log_{10}\left\{\frac{[\text{Conjugate Acid}]}{[\text{Base}]}\right\} \qquad (5.85)$$

This is seen to be essentially identical with Eq. 5.78. Using this equation we obtain the curve labelled curve 4 in Fig. 5.19. Eq. 5.85 shows that *when a weak base is half neutralized by a strong acid the pH is equal to the pK_a*; when the NH_3 is half neutralized we have $[NH_3] = [NH_4^+]$ in Eq. 5.85.

(d) Weak acid–Weak base. In the titration of acetic acid with ammonia the pH variation would approximate to curve 3 on the left hand side of the equivalence point, and to curve 4 on the right hand side, and the pH at the end point would be close to 7. The total change in $[H_3O^+]$ near the end point of this titration would amount to only three pH units.

Inclusion of activity coefficients for the ions in these calculations would alter the curves to a small extent, because pH is defined in terms of the activity of hydrogen ions rather than the concentration and because the inclusion of activity coefficients would make a difference to the results of using Eq. 5.77. In Eq. 5.84 the activity coefficients of H_3O^+ and NH_4^+ cancel to a good approximation, so that this equation correctly gives the *concentration* of H_3O^+ whether or not the activity coefficients deviate significantly from unity.

(iv) Indicators

One way of following the progress of any of the titrations in Fig. 5.19 would be to monitor the changes in pH with an electrical pH meter. A pH meter is essentially a device for measuring the output voltage of an electrochemical cell whose e.m.f. depends on the concentration of hydrogen ions. We shall consider such devices briefly in the next chapter. In the absence of a pH meter it is customary to detect the end point of a titration with the aid of a chemical indicator, a substance which changes colour when the pH of its solution alters.

Indicators are themselves conjugate acid–base pairs in which the acid form and the basic form are clearly distinguishable by eye. When the pH of the solution is equal to the pK_a of the indicator and the indicator is half neutralized, so that a change of pH by one or two units on either side of the pK_a value will cause the solution to show the colour characteristic of one or other of the two forms. Ideally, therefore, the pK_a value should correspond exactly to the pH at the end point of the titration. In practice, when a strong acid and a strong base are being used the pH change at the end point is so large that virtually any indicator is likely to be satisfactory (cf. Fig. 5.19), but when a weak acid is being titrated with a strong base it is necessary to use an indicator which changes colour on the alkaline side of pH 7, and when a strong acid is being titrated with a weak base it is necessary to use an indicator which changes colour on the acid side. Commonly used indicators are listed with their pK_a values in Table 5.8.

Table 5.8 Some acid–base indicators

Indicator	pK_a	Colour change (low → high pH)	
Thymol blue	1.7	red	→yellow
Methyl orange	3.7	red	→yellow
Bromophenol blue	4.2	yellow	→blue
Methyl red	5.1	pink	→yellow
Bromothymol blue	7.0	yellow	→blue
Cresol red	8.3	yellow	→red
Phenolphthalein	9.6	colourless	→red

(Phenolphthalein is used as a 0.1% solution in 70% ethanol; the rest are used as 0.1% solutions in water.)

(v) Buffer Solutions

A buffer solution is one which will maintain a virtually constant pH despite the addition of appreciable quantities of a strong acid or base. This is a very useful property for a solution to have and it is widely exploited, especially by biochemical systems and by persons studying biochemical systems.

A buffer solution consists essentially of a partially neutralized solution of a weak acid or a weak base. For the case of a weak acid HA, the pH is governed by an equation analogous to Eq. 5.77:

$$[H_3O^+] = K_a [HA]/[A^-] \qquad (5.86)$$

while for a weak base B the corresponding equation is

$$[H_3O^+] = K_a [HB^+]/[B] \qquad (5.87)$$

which is analogous to Eq. 5.84. For both kinds of buffer the pH is given by the *buffer equation*, which we have given as Eqs 5.78 and 5.85. Partially neutralized solutions of weak acids are useful as buffer solutions at pH values below 7; buffers based on weak bases are useful at pH values above 7.

The concentrations of HA and A^-, or of HB^+ and B, can be made very large without altering the pH (apart from the effects associated with activity coefficients) because the pH is governed by the *ratio* of the concentrations of the conjugate acid and base. The pH changes by one unit when this ratio changes by a factor of ten. If, for example, the concentrations of HA and A^- are of the order of 1.0 mol l^{-1}, the addition of 1 ml of 1.0 molar acid to 10 ml of solution changes the ratio of [HA] to $[A^-]$ by only about 10%, and so changes the pH by about 0.05 of a pH unit. In the absence of the buffer the addition of this amount of acid would reduce the pH of water from 7 to 1, a change of 6 units. Buffer solutions are not restricted to the stablilizing of pH; in principle similar equilibria can be devised to control the concentrations of most other ions. The extreme stability of the concentration of calcium ions in blood can be attributed in part to a buffering action which involves the binding of the ions to blood proteins.

5.6 Entropy, Free Energy and Equilibrium

(i) Entropy and the Second Law

So far in our discussion of physical and chemical equilibrium we have implicitly assumed that the nature of the equilibrium state of the system was known, and we have proceeded on the basis that a spontaneous change corresponds to a move in the direction of the equilibrium state. We now seek a systematic way of deciding whether or not a given process is spontaneous, and of predicting where the position of equilibrium lies. The solution to both of these problems is contained in the second law of thermodynamics.

Like the first law, which we considered in Chapter 4, the second law of thermodynamics is to be regarded as a generalization based on experimental observations, rather than as a theoretical construct. The second law also resembles the first in that it amounts to a definition of a state function, the state function in this case being entropy, S. However, in addition to specifying that S exists, the second law goes a step further by specifying that the total change in S during a spontaneous process is always positive. The first law can be paraphrased by saying that the energy of the universe is constant, the second law by saying that the entropy of the universe continually increases. A more complete statement of the law is as follows:

There exists a function S, called the entropy, which is a mathematically well-behaved, extensive state function for any system, and is such that when the state of the system changes

$$dS = dq_{reversible}/T. \tag{5.88}$$

For a *spontaneous* process in an isolated system dS is positive. If the system is not isolated $(dS_{system} + dS_{surroundings}) = dS_{total}$ must be positive.

It is not easy to form an accurate mental picture of what constitutes entropy, but as a fair approximation one can think of entropy as a measure of the degree of randomness, or disorder, in a system. Such disorder might be present in the form of the random arrangement of molecules in a liquid, as opposed to the highly ordered state of the same molecules in a crystal, or as the intimate mixing of the component gases of air, as opposed to the more ordered situation where the component gases are neatly separated from one another. Slightly more difficult to visualize is the randomness associated with a distribution of a population of electrons, atoms, or molecules over a range of available energy levels; this too is an important form of entropy. Entropy differs from energy in one very significant way, namely that it is possible to imagine a perfectly ordered crystalline state for a system, in which state the entropy of the system is zero. For energy there is no such absolute zero, and all the energies which we use to characterize particular substances, such as the standard enthalpies of formation, represent energy differences measured with respect to some standard state. The existence of a state of zero entropy is embodied in the *third law* of thermodynamics, which we shall consider later in this section.

The definition of dS involves the *reversible* heat change, $dq_{reversible}$ (or dq_{rev} for short). Because q is not a state function the differential dq has no fixed meaning unless a path for the change of state is specified. Here the path is specified as a reversible one in the thermodynamic sense, i.e. one in which the system and its surroundings are always in equilibrium (cf. the discussion on p. 52). The actual process by which the change of state occurs need not be reversible, because the change in the value of the state function S depends only on the nature of the initial and final states. However, when we want to calculate the change in S by using Eq. 5.88 we have to visualize the process as occurring by a reversible path, and obtain ΔS as the integral of dq/T along the reversible path. We shall consider some examples of how to do this shortly. An immediate consequence of the last part of our statement of the second law is that in a system at equilibrium the value of S must be a maximum. If it were not a maximum there would still be a spontaneous process by which S could increase. The mathematical condition for S to be a maximum is that dS is zero for any infinitesimal change of state, i.e. *the condition for equilibrium is*

$$dS = 0 \tag{5.89}$$

The flow of heat between two parts of a system at different temperature clearly corresponds to a spontaneous process, because dq is negative for the region at the higher temperature, T_2 say, and positive for the lower temperature T_1, so that dS for the transfer of dq from T_2 to T_1 is equal to $dq(1/T_1 - 1/T_2)$, which is positive. At equilibrium $T_1 = T_2$ and dS is zero as required. The heating of a system from T_1 to T_2 is a reversible process, so the entropy change can be calculated simply by using

$$dq = C_v\, dT \tag{5.90}$$

or

$$dq = C_p\, dT \tag{5.91}$$

Hence at constant pressure:

$$\Delta S = \int_{T_1}^{T_2} \frac{C_p\, dT}{T} \tag{5.92}$$

which simplifies to

$$\Delta S = C_p \ln(T_2/T_1) \tag{5.93}$$

if C_p can be regarded as a constant. At constant volume we replace C_p by C_v in these formulas.

Another type of process which can easily be carried out reversibly is the melting of a solid at the temperature at which solid and liquid are in equilibrium, i.e. at the normal freezing point, or the vaporization of a liquid or solid at a pressure equal to the equilibrium vapour pressure. Each of these processes involves a latent heat, ΔH_f, ΔH_v, or ΔH_{sub} (for fusion, vaporization, or sublimation), which corresponds to q_{rev}, and each process occurs at constant temperature. Therefore, for example, the entropy of vaporization of a liquid is given by

$$\Delta S_v = \Delta H_v/T \tag{5.94}$$

with analogous expressions for entropy of fusion and entropy of sublimation.

Example   Calculate the entropy change in one mole of superheated water which vaporizes at 110°C and one atmosphere pressure. (The heat capacities C_p for liquid water and steam are 75 and 34 J K^{-1} mol^{-1}, respectively, and the latent heat of vaporization of water at 100°C is 40.5 kJ mol^{-1}.) The vaporization of water at 110°C is clearly a spontaneous process, not a reversible one, so we have to find a reversible path for calculating ΔS. The path we choose to follow is shown in Fig. 5.20. We have, since S is a state function,

$$\Delta S = \Delta S_1 + \Delta S_2 + \Delta S_3 \tag{5.95}$$

where $\Delta S_1 = C_{p(\text{water})} \ln(273/373)$

$$\Delta S_2 = \Delta H_v/273$$

and $\Delta S_3 = C_{p(\text{steam})} \ln(373/273)$

so that Eq. 5.95 becomes

$$\Delta S = 41 \times 2.303 \times \log(273/373) + 40.5 \times 10^3/273$$

$$= -12.9 + 148.5 = \underline{135.6 \text{ J K}^{-1} \text{ mol}^{-1}}.$$

Entropy changes for other spontaneous processes can be calculated by making use of a thermodynamic cycle which incorporates a reversible path, just as we have done in this example.

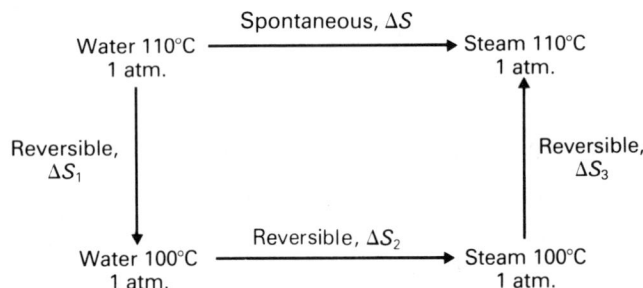

Water 110°C 1 atm. $\xrightarrow{\text{Spontaneous, } \Delta S}$ Steam 110°C 1 atm.

Reversible, ΔS_1 (downward)   Reversible, ΔS_3 (upward)

Water 100°C 1 atm. $\xrightarrow{\text{Reversible, } \Delta S_2}$ Steam 100°C 1 atm.

Fig. 5.20 Cycle used to calculate entropy change for the irreversible vaporization of water at 110°C. $\Delta S = \Delta S_1 + \Delta S_2 + \Delta S_3$

(ii) Entropies of Expansion and Mixing of Ideal Gases

If an ideal gas expands and at the same time undergoes a change in temperature the entropy change for the whole process can be considered as made up of two parts. The first part is a reversible heating or cooling to the final temperature at constant volume, and the second is an isothermal expansion or compression from the initial to the final volume. For the first part the entropy change is simply the integral of $(C_v/T)\mathrm{d}T$ between the initial and final temperature, i.e. the constant volume expression analogous to Eq. 5.92. We now consider how to calculate ΔS for the second part.

For an ideal gas the change in internal energy, ΔU, is zero for an isothermal expansion or compression. Hence, from the first law,

$$\mathrm{d}q_{\text{rev}} = P\mathrm{d}V \tag{5.96}$$

and $$\mathrm{d}S = \frac{P}{T}\,\mathrm{d}V = nR\,\frac{\mathrm{d}V}{V} \tag{5.97}$$

This can be integrated, as in the calculation of reversible work on page 52, to

$$\Delta S = nR \ln(V_2/V_1) \tag{5.98}$$

which is the entropy change for an isothermal expansion, *whether or not the expansion occurs by a reversible path.*

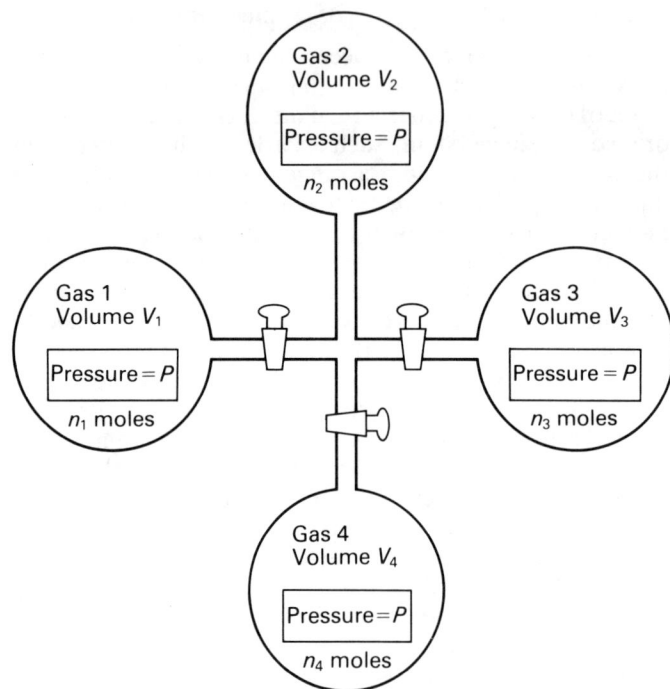

Fig. 5.21 Calculation of entropy of mixing: each gas expands into the total volume $V = V_1 + V_2 + V_3 + V_4$

To calculate the change of entropy associated with the mixing of gases we consider a system set up initially as shown in Fig. 5.21, with the separate components of the mixture all at the same initial pressure. For each individual component the entropy change results from the gas, initially confined in its own flask, being given the freedom of the total volume V, which is equal to $(V_1 + V_2 + V_3 + V_4)$ in Fig. 5.21. Then, from Eq. 5.98, the entropy change for gas j when the stopcocks are opened is given by

$$\Delta S_j = n_j R \ln(V/V_j) \tag{5.99}$$

The total number of moles of gas present is given by

$$n = \sum_j n_j \tag{5.100}$$

and the total entropy change for the mixture is

$$\Delta S = \sum_j \Delta S_j = \sum_j n_j R \ln (V/V_j) \qquad (5.101)$$

The mole fraction of component j, x_j, is given by

$$x_j = V_j/V = n_j/n \qquad (5.102)$$

Hence the *entropy of mixing per mole of mixture* is given by

$$\Delta \bar{S} = - R \left(\sum_j x_j \ln x_j \right) \qquad (5.103)$$

where the bar over the S is a conventional means of showing that this is a molar quantity.

(iii) The Third Law: Absolute Entropies

In order to be able to calculate the entropy change for any chemical process we require a table of characteristic entropy values for individual substances. One way of proceeding would be to use entropies of formation of substances from their component elements in standard states, as we did for enthalpy in Chapter 4. For enthalpy this course was dictated by the absence of an absolute zero of energy with respect to which the enthalpy of any substance could be specified. For entropy, however, there is such an absolute zero, the nature of which is specified by the Third Law of Thermodynamics. A concise statement of the third law is as follows:

> For all pure, crystalline substances the equilibrium value of the entropy tends to zero as the absolute temperature tends to zero.

By virtue of this law the absolute entropy of any substance at 298 K can be evaluated; thus, for a gas:

$$S = \int_0^{298} \frac{C_v}{T} \, dT + \Delta S_f + \Delta S_v \qquad (5.104)$$

The experimental problems associated with measuring heat capacities become particularly formidable as the temperature approaches zero, but precise extrapolation to $T = 0$ is possible by using the theoretical result, due to Debye, that at very low temperatures C_v is proportional to T^3. Our statement of the third law referred to the equilibrium value of the entropy; for a few substances, notably water, nitric oxide, and nitrous oxide, some disorder is frozen into the crystal structure at relatively high temperatures and there is insufficient mobility in the solid at low temperatures for this

disorder to be removed. These substances are said to possess 'residual entropy' at $T = 0$. The effect of the residual entropy is to cause the third law entropy obtained from Eq. 5.104 to be too low, by a predictable amount, in comparison with the value found from a thermodynamic cycle to which the residual entropy does not contribute. This is illustrated for nitric oxide in Fig. 5.22. A selection of values of absolute entropy at 298 K are given in Table 5.9.

Table 5.9 Absolute entropies (J K^{-1} mol^{-1}) at 298 K

Substance	S	Substance	S
Ag(s)	42.7	Hg(l)	77.4
Br_2(l)	152.3	Hg(g)	174.9
Br_2(g)	245.4	K(s)	63.6
Br(g)	174.9	N_2(g)	191.5
Br$^-$(aq)	80.7	N(g)	153.2
$CaCO_3$ (calcite)	92.9	NH_3(g)	192.5
CaO(s)	39.7	N_2O(g)	220.0
C (graphite)	5.7	NO(g)	210.6
C (diamond)	2.4	N_2O_4(g)	304.3
C(g)	158.0	NO_2(g)	240.5
CH_4(g)	186.2	Na(s)	51.0
C_2H_6(g)	229.5	O_2(g)	205.0
C_2H_4(g)	219.5	O(g)	161.0
C_2H_2(g)	200.8	O_3(g)	237.7
C_6H_6(l)	172.8	P (white)	41
CO(g)	197.9	P (red)	23
CO_2(g)	213.6	P (black)	23
Cl_2(g)	223.0	P_4(g)	280
Cl(g)	165.1	S (rhombic)	31.9
Cu(s)	33.3	S (monoclinic)	33
Fe(s)	27.1	SO_2(g)	248.5
H_2(g)	130.6	SO_3(g)	256.2
H(g)	114.6	H_2SO_4(l)	157
H$^+$(aq)	0.0	SO_4^{2-}(aq)	17.2
H_2O(l)	70.0	SO_3^{2-}(aq)	-29
H_2O(g)	188.7	H_2S(g)	205.6

The entropy of formation of an aqueous electrolyte can be measured, and hence the absolute entropy. For individual ions in solution the entropy is defined relative to the entropy of H$^+$(aq), the entropy of aqueous HCl, for example, being all assigned to the chloride ion. Hence it is possible for the tabulated entropy of an ion in solution to be negative, as in the case for the sulphite ion in Table 5.9.

(iv) A General Criterion for Spontaneous Change: the Gibbs Free Energy

The sign of the entropy change for a process is a useful criterion of spontaneity in an isolated system. If the entropy change is positive the process will tend to occur, otherwise it will not. However, processes such as the freezing of a

½N₂(g) + ½O₂(g) $\xrightarrow{\Delta S_2}$ NO(g) at 298 K

ΔS_1

$\Delta S_3 = \int_0^{298} \dfrac{C_v}{T}\, dT$

$+ \Delta S_f + \Delta S_v$

| N—O—N—O—N—O—O—N |
| O—N—N—O—N—O—O—N |
| N—O—N—O—O—N—N—O |

N₂ and O₂ crystals,
S = 0 at 0 K

Nitric oxide crystal
with residual disorder,
S > 0 at 0 K

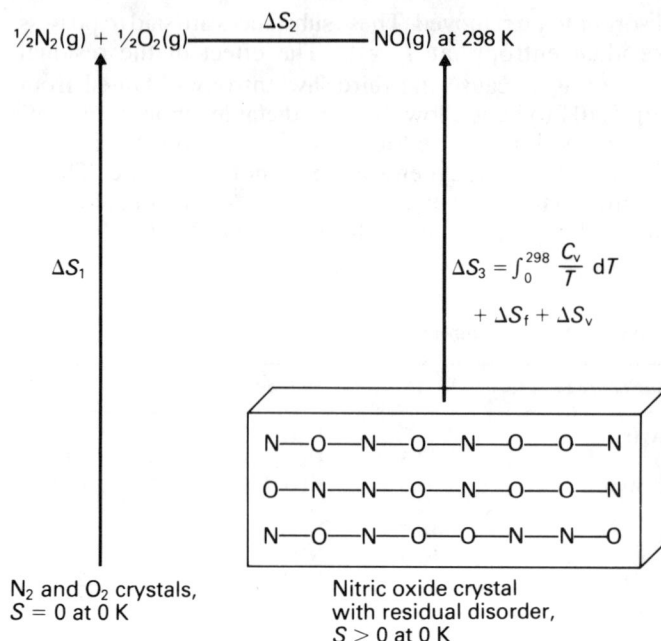

Fig. 5.22 Effect of residual entropy of NO crystal on third law entropy at 298 K. Absolute entropy of NO at 298 K is $\Delta S_1 + \Delta S_2$, which is greater than ΔS_3 by an amount equal to the residual entropy

liquid or the condensing of a vapour do occur spontaneously, even though the entropy of the freezing liquid or condensing vapour actually decreases. This is because the negative entropy change in the system under consideration is counterbalanced by a larger positive entropy change in the surroundings. For example, consider liquid water in equilibrium with ice at 0°C, with the surroundings at a temperature T. If the water freezes the latent heat ΔH_f is transferred to the surroundings, so that

$$\Delta S_{surroundings} = \Delta H_f / T \qquad (5.105)$$

while

$$\Delta S_{water} = -\Delta H_f / 273 \qquad (5.106)$$

If T is less than 273 K the net entropy change for the water plus the surroundings will be positive, and the process will occur spontaneously. If T is greater than 273 K the net entropy change for freezing will be negative, and melting will occur instead. If T is equal to 273 K the system is in equilibrium and neither melting nor freezing will occur. For a system which is not isolated from its surroundings it is the *total* entropy change that has to be considered. If the heat transfer is reversible, this is given by

$$\Delta S_{total} = \frac{-q}{T_{surroundings}} + \Delta S_{system} \qquad (5.107)$$

At constant pressure $-q$ is $-\Delta H$ for the process under consideration. The quantity ΔG, given by

$$\Delta G = \Delta H - T\Delta S \qquad (5.108)$$

is therefore equal to $-T\Delta S_{total}$ for a process which occurs at constant pressure and with the system and its surroundings at the same temperature T. Hence the sign of ΔG is a general criterion for spontaneity for any process in a system which is at constant pressure and in thermal equilibrium with its surroundings. For a spontaneous process ΔG is negative; at equilibrium ΔG is zero.

The state function G, defined by

$$G = H - TS \qquad (5.109)$$

is called the *Gibbs Free Energy*. G is a state function because it is defined in terms of quantities which are themselves state functions. It is an extensive function, *i.e.* one which depends on the size of the system, because H and S are extensive. For most purposes in chemistry the Gibbs Free Energy is the most important and useful of the state functions. An analogous constant volume function can be defined, namely

$$A = U - TS \qquad (5.110)$$

where A is known as the Helmholtz Free Energy.

We have seen how to calculate changes in H and S; corresponding changes in G can therefore be calculated with the aid of Eq. 5.108. For example, for an isothermal expansion of an ideal gas, ΔH is zero and ΔS is given by Eq. 5.98. Hence

$$\Delta G = -nRT \ln (V_2/V_1) \qquad (5.111)$$

and

$$\Delta G = -nRT \ln(P_1/P_2) \qquad (5.112)$$

where the second equality follows from Boyle's law.

Standard Gibbs Free Energies of Formation are tabulated for a temperature of 298 K in the same manner as standard enthalpies of formation. For any substance $G_f^{\ominus}$ is the Gibbs Free Energy change for formation of the substance from its component elements in their standard states. Some values are listed in Table 5.10.

The values of ΔH and ΔS both vary with temperature in a manner that depends on ΔC_p (cf. the discussion of the Kirchhoff equation on p. 49) and so can be regarded as constant over a small temperature range. Hence, by differentiating Eq. 5.108, we find that at constant pressure

$$\frac{d(\Delta G)}{dT} = -\Delta S \qquad (5.113)$$

Eq. 5.113 allows the value of ΔG to be calculated for a temperature other than 298 K, on the basis of the tabulated data for 298 K.

Table 5.10 Standard Gibbs Free Energies of formation $\bar{G}_f^{\ominus}$ at 298 K (kJ mol^{-1})

Substance	$\bar{G}_f^{\ominus}(298)$	Substance	$\bar{G}_f^{\ominus}(298)$
Ag(s)	0.0	Hg(l)	0.0
Br$_2$(l)	0.0	Hg(g)	31.8
Br$_2$(g)	3.1	K(s)	0.0
Br(g)	82.4	N$_2$(g)	0.0
Br$^-$(aq)	-102.8	N(g)	455.6
CaCO$_3$(s)	-112.8	NH$_3$(g)	-16.6
CaO(s)	-604.2	N$_2$O(g)	103.6
C (graphite)	0.0	NO(g)	86.7
C (diamond)	2.8	NO$_2$(g)	51.84
C(g)	672.9	N$_2$O$_4$(g)	98.28
CH$_4$(g)	-50.8	Na(s)	0.0
C$_2$H$_6$(g)	-32.9	O$_2$(g)	0.0
C$_2$H$_4$(g)	68.1	O(g)	230.1
C$_2$H$_2$(g)	209.2	O$_3$(g)	163.4
C$_6$H$_6$(l)	124.5	P (white)	0.0
CO(g)	-137.3	P (red)	-12
CO$_2$(g)	-394.4	P (black)	-33
Cl$_2$(g)	0.0	P$_4$(g)	24
Cl(g)	105.4	S (rhombic)	0.0
Cu(s)	0.0	SO$_2$(g)	-300.4
Fe(s)	0.0	SO$_3$(g)	-370.4
H$_2$(g)	0.0	H$_2$SO$_4$(l)	-690
H(g)	203.3	SO$_4^{2-}$(aq)	-742.0
H$_2$O(l)	-237.2	SO$_3^{2-}$(aq)	-487
H$_2$O(g)	-228.6	H$_2$S(g)	-33.0

(v) Gibbs Free Energy and Maximum Work

A comparison of Eqs 5.111 and 5.112 with the expressions for the work done in the reversible isothermal expansion of an ideal gas, Eqs 4.27 and 4.28, shows that they are the same except for the minus signs in the expressions for ΔG. The reversible work is also the maximum work which can be obtained by this particular process; therefore we can write

$$\Delta G = -w_{max} \qquad (5.114)$$

Eq. 5.114 is in fact an important general result, true for any process, although we shall not prove its generality here. We shall have a use for this result in Chapter 6 in our discussion of electrochemical cells.

If we rewrite Eq. 5.108 in the form

$$\Delta H = \Delta G + T\Delta S \qquad (5.115)$$

we see that ΔH for any process is composed of two parts, a part ΔG which represents the maximum work obtainable from the process and a part $T\Delta S$ which is not available for doing external work.

(vi) Gibbs Free Energy and the Equilibrium Constant

Consider the generalized chemical reaction

$$aA + bB + cC \rightarrow lL + mM + nN \qquad (5.116)$$

in which a molecules of species A react with b molecules of species B and c molecules of species C to produce l molecules of species L plus m molecules of species M and n molecules of species N. If $\bar{G}_i$ is the Gibbs Free Energy of a mole of species i, the free energy change for reaction 5.116 is given by

$$\Delta G = (l\bar{G}_L + m\bar{G}_M + n\bar{G}_N) - (a\bar{G}_A + b\bar{G}_B + c\bar{G}_C) \qquad (5.117)$$

The quantities $\bar{G}_i$ are not *standard* Gibbs Free Energies of formation because the species A, B, C, L, M, N are in general not in their standard states. If the species are all gaseous the standard states correspond to a partial pressure of one atmosphere, and the Gibbs Free Energy at any other partial pressure can be related to the standard value by using Eq. 5.112. If we write

$$\Delta G = \bar{G} - \bar{G}^{\ominus} \qquad (5.118)$$

put $n = 1$, and write $P_2 = P$ and $P_1 = 1$ atmosphere, Eq. 5.112 becomes

$$\bar{G} = \bar{G}^{\ominus} + RT \ln P \qquad (5.119)$$

We can use this equation to substitute for $\bar{G}_i$ in Eq. 5.117, and so obtain

$$\Delta G = \Delta G^{\ominus} + RT \ln \left\{ \frac{P_L^l P_M^m P_N^n}{P_A^a P_B^b P_C^c} \right\} \qquad (5.120)$$

where $\quad \Delta G^{\ominus} = (l\bar{G}_L^{\ominus} + m\bar{G}_M^{\ominus} + n\bar{G}_N^{\ominus}) -$
$$(a\bar{G}_A^{\ominus} + b\bar{G}_B^{\ominus} + c\bar{G}_C^{\ominus}) \qquad (5.121)$$

is the standard Gibbs Free Energy change for the reaction, i.e. the Gibbs Free Energy change when reactants and products are all in their standard states at a partial pressure of one atmosphere.

When the reaction is at equilibrium, ΔG is zero, and the log term in Eq. 5.120 becomes equal to the logarithm of K_p, the equilibrium constant expressed in terms of partial pressures. Hence we obtain the extremely important result

$$\Delta G^{\ominus} = -RT \ln K_p \qquad (5.122)$$

Since $\Delta G^{\ominus}$ is a constant, this amounts to a thermodynamic proof of the constancy of K_p.

If we use Eq. 5.17 to relate activity in solution to partial pressure in the vapour, and choose the standard state as that corresponding to unit activity, we can replace P by a in Eq. 5.119 and obtain the *general* formula

$$\Delta G^{\ominus} = -RT \ln K_a \qquad (5.123)$$

which holds whatever units of concentration are used for the activity, provided the standard state is specified as the state in which the activity a has the value 1.

Example   Using data from Table 5.10, calculate the equilibrium constant K_p for the reaction

$$CO_2(g) + H_2(g) \rightleftharpoons CO(g) + H_2O(l) \quad \text{at 298 K.}$$

From the table we find $\Delta G^{\ominus}$ to be $(-237.2 - 137.3 + 394.4 + 0.0) = +19.9$ kJ.

Hence $\log_{10} K_p = \dfrac{-1.99 \times 10^4}{2.303 \times 8.314 \times 298} = 3.49$

and $\underline{K_p = 3.1 \times 10^3 \text{ atm}^{-1}}$

The units atm^{-1} result from our choice of 1 atm pressure as the standard state for a gas.

Eq. 5.120 can be combined with Eq. 5.122 to give

$$-\Delta G = RT \ln K_p - RT \ln \left\{ \frac{P_L^{\,l} P_M^{\,m} P_N^{\,n}}{P_A^{\,a} P_B^{\,b} P_C^{\,c}} \right\} \quad (5.124)$$

which shows that the driving force for the reaction, $-\Delta G$, depends on the ratio of K_p to the term inside the curly brackets. This reaffirms our earlier statement that the driving force for a spontaneous change is provided by the departure of a system from equilibrium.

(vii) Temperature Dependence of the Equilibrium Constant

To derive an expression for the variation of K_a with temperature we begin with the equation

$$\Delta G^{\ominus} = -RT \ln K_a \quad (5.123)$$

For the case where reactants and products are all in their standard states Eq. 5.108 becomes

$$\Delta G^{\ominus} = \Delta H^{\ominus} - T \Delta S^{\ominus} \quad (5.125)$$

Hence, substituting for $\Delta G^{\ominus}$ in Eq. 5.123, we obtain

$$\ln K_a = -\Delta H^{\ominus}/RT + \Delta S^{\ominus}/R \quad (5.126)$$

This result shows that a graph of $\ln K_a$ versus $1/T$ is a straight line of slope $-\Delta H^{\ominus}/R$, and intercept $\Delta S^{\ominus}/R$ at $1/T = 0$. The line is straight because $\Delta H^{\ominus}$ and $\Delta S^{\ominus}$ are almost independent of T, their temperature variation being governed by the small quantity ΔC_p. If the logarithm to base ten of K_a is plotted, the slope and intercept are smaller by a factor of 2.303. The form of the graph is illustrated in Fig. 5.23.

This deduction from Eq. 5.126 is an extremely important and useful result, which applies to any process that can be represented by an equilibrium constant. A number of such processes are listed in Table 5.11. To determine the standard enthalpy change for such a process it is only necessary to know the value of the equilibrium constant at two different temperatures. The equilibrium constant itself gives the value of $\Delta G^{\ominus}$ at one or other of the temperatures; once

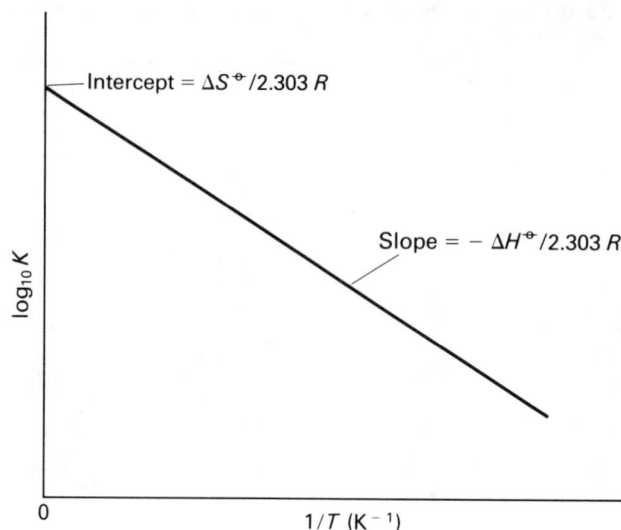

Fig. 5.23 Temperature dependence of the equilibrium constant: a graph of $\log K$ against $1/T$ is a straight line of slope $-\Delta H^{\ominus}/2.303\ R$ and intercept $\Delta S^{\ominus}/2.303R$

Table 5.11 Standard Enthalpy changes which can be determined from the slope of a graph of $\log K$ against $1/T$

Process	K given by:	Slope gives Enthalpy of:
Chemical reaction	Equilibrium constant	Reaction
Complex formation	Stability constant	Formation
Dissolving salt	Solubility product	Solution
Distribution of solute between solvents	Distribution coefficient	Transfer from one solvent to another
Evaporation of liquid	Vapour pressure	Vaporization
Evaporation of solid	Vapour pressure	Sublimation
Gas dissolving in liquid	Solubility (reciprocal) of Henry's Law constant)	Solution
Solid dissolving in liquid (without dissociation)	Solubility	Solution
Dissociation of weak electrolyte	Dissociation constant	Dissociation

$\Delta G^{\ominus}$ and $\Delta H^{\ominus}$ are known the value of $\Delta S^{\ominus}$ follows from Eq 5.125.

Example 1   The equilibrium constant K_p for the reaction

$$SO_2(g) + \tfrac{1}{2}O_2(g) \rightarrow SO_3(g)$$

has the value 7.6×10^7 atm$^{-\frac{1}{2}}$ at 400 K, and 2.1×10^5 atm$^{-\frac{1}{2}}$ at 500 K. Calculate $\Delta H^{\ominus}$ and $\Delta S^{\ominus}$ for this reaction.

The slope of the plot of $\log_{10} K_p$ versus $1/T$ is given by

$$\frac{\log (7.6 \times 10^7) - \log (2.1 \times 10^5)}{(1/400) - (1/500)} = \frac{-\Delta H^{\ominus}}{2.303 \times 8.314}$$

Hence $\Delta H^{\ominus} = -\dfrac{2.56 \times 2.303 \times 8.314}{0.0005} = \underline{-98.3 \text{ kJ.}}$

At 500 K $\Delta G^{\ominus} = -RT \ln K = -8.314 \times 500 \times 2.303 \times 5.32 = \underline{-50.9 \text{ kJ.}}$

Hence $\Delta S^{\ominus} = -(\Delta G^{\ominus} - \Delta H^{\ominus})/T = -47.4 \times 10^3/500 = -94.8 \text{ J K}^{-1}.$

Example 2   Using data from Tables 4.1 and 5.10, estimate the value of the equilibrium constant at 1100°C for the reaction

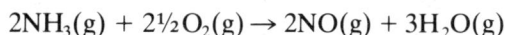

$$2NH_3(g) + 2\tfrac{1}{2}O_2(g) \rightarrow 2NO(g) + 3H_2O(g)$$

Table 4.1 gives $\Delta H^{\ominus}$ as $(2 \times 90.37 + 3 \times -241.8 - 2 \times -46.19 - 2.5 \times 0.0) = -452.3$ kJ at 298 K.

Table 5.10 gives $\Delta G^{\ominus}$ as $(2 \times 86.7 + 3 \times -228.6 - 2 \times -16.6 - 2.5 \times 0.0) = -479.2$ kJ at 298 K.

Hence $\Delta S^{\ominus} = -(-479.2 + 452.3) \times 10^3/298 = 26.9 \times 10^3/298 = +90.4$ J K^{-1} at 298 K. (Table 5.9 gives 89.8 J K^{-1}.)

Using Eq. 5.125 we now find $\Delta G^{\ominus}$ at 1373 K to be $-452.3 - 90.4 \times 1373/10^3 = -576.4$ kJ.

Hence $\log K_p = \dfrac{576.4 \times 10^3}{2.303 \times 8.314 \times 1373} = 21.9,$

and $K_p = 8 \times 10^{21}.$

This value is expected to be only approximate because of the assumption that $\Delta H^{\ominus}$ and $\Delta S^{\ominus}$ are independent of temperature over the range 298–1373 K. An alternative way of obtaining the answer to this type of problem is to use the formula

$$\ln(K_1/K_2) = \frac{-\Delta H^{\ominus}}{R(1/T_1 - 1/T_2)}$$

where K_1 either is given or can be calculated from $\Delta G^{\ominus}$ at 298 K, and T_2, $T_1 = 298$, and $\Delta H^{\ominus}$ are all known.

(viii) Summary

In this section we have covered a lot of territory, and it will be helpful to review the main landmarks along the way. We first introduced the second law of thermodynamics and the associated concept of entropy. In a spontaneous process for any system the total entropy of the system and its surroundings always increases. Equations were derived for calculating entropy changes during several kinds of process. The third law of thermodynamics, which essentially states that the entropy of a pure crystalline material tends to zero as the temperature tends to zero, opened the way to the evaluation of absolute entropies. The Gibbs Free Energy was then introduced as a state function which could be used to decide whether a process would occur spontaneously in any system at constant pressure and temperature, without reference to the surroundings. The criterion for a spontaneous process is that ΔG for the system should be negative. The decrease in Gibbs Free Energy during a process was identified with the maximum work obtainable from the process, and the standard Gibbs Free Energy change was shown to be equal to $-RT \ln K_a$ for any process. It was then shown that a graph of $\ln K_a$ against $1/T$ is linear, with slope $-\Delta H^{\ominus}/R$ and intercept $\Delta S^{\ominus}/R$.

Near the beginning of Section 5.3 it was asserted that the equilibrium constant is the most important and useful numerical quantity in chemistry. In part (iv) of this section it was said that for most purposes in chemistry the Gibbs Free Energy is the most important and useful state function. We can now see that much of the importance of the Gibbs Free Energy must result from its close relationship to the equilibrium constant.

Exercises

5.1   A molar solution of acetic acid in water boils at 100.51°C. Comment on the significance of this observation in relation to the nature of acetic acid in aqueous solution.

5.2   At 20°C the vapour pressures of methanol and ethanol are 11.83 kPa, $(11.83 \times 10^3$ newtons per square metre) and 5.93 kPa, respectively. A mixture of methanol and ethanol obeys Raoult's law quite closely. Calculate the total vapour pressure of a mixture for which the mole fraction of methanol is 0.40 in the liquid, and calculate the mole fraction of methanol in the vapour above the mixture.

5.3   Naphthalene and water do not dissolve in one another to any significant extent. The boiling point of a mixture of naphthalene and water, distilled at a total pressure of 0.964 atmospheres, is 98°C. The vapour pressure of water at 98°C is 0.930 atmospheres. Calculate the mole fraction of naphthalene in the distillate. (This process is 'steam distillation'.)

5.4   The Henry's law constants k for CO_2 in water and benzene are 1.42×10^5 and 9.71×10^3 kPa, respectively, where $P_{vapour} = k \, x_{liquid}$. (a) Calculate the amount of CO_2 dissolved in water and in benzene which are in equilibrium with air, at a pressure of 101.3 kPa, containing 0.33% of CO_2 by volume. (b) Calculate the distribution coefficient for CO_2 between water and benzene.

5.5 Construct the boiling point diagram for the system water + acetic acid at one atmosphere pressure, using the following data:

x_{H_2O} (liquid)	0.0	0.100	0.300	0.500	0.700	1.0
Boiling Point/°C	118.1	113.8	107.5	104.4	102.1	100.0
x_{H_2O} (vapour)	0.0	0.167	0.425	0.626	0.815	1.0

Can these substances be separated completely by fractional distillation? Estimate the number of theoretical plates required to produce a small sample of distillate containing less than ten mole percent water from an equimolar mixture.

5.6 Write down expressions for the equilibrium constants of the following reactions, using Eq. 5.11 and Table 5.3:

(a) $C(s) + CO_2(g) \rightleftharpoons 2CO(g)$
(b) $2S_2O(g) \rightleftharpoons 3S(s) + SO_2(g)$
(c) $I^-(aq) + I_2(aq) \rightleftharpoons I_3^-(aq)$
(d) $Cl_2(g) + 2Ag(s) \rightleftharpoons 2AgCl(s)$
(e) $2HNO_2(aq) \rightleftharpoons NO(g) + NO_2(g) + H_2O(l)$
(f) $2H_2O(l) \rightleftharpoons H_3O^+(aq) + OH^-(aq)$

5.7 At 1483 K and 10 atmospheres pressure the equilibrium composition of sodium vapour is 71.3% of $Na(g)$ and 28.7% of $Na_2(g)$. Calculate the equilibrium constant of the dimerization reaction $2Na(g) \rightleftharpoons Na_2(g)$.

5.8 The equilibrium constant for formation of ethyl acetate and water from anhydrous ethanol and acetic acid has the value 4.0 at 298 K. Calculate the mole fraction of acetic acid in the equilibrium mixture formed by adding one mole of acetic acid to four moles of ethanol.

5.9 The equilibrium constant for formation of ammonia from nitrogen and hydrogen is 1.6×10^{-4} atm^{-2} at 400°C. Calculate the pressure at which ammonia gas is 90% dissociated at this temperature.

5.10 For the ammonia decomposition problem in exercise 5.9, calculate the total pressure for 90% dissociation of NH_3, beginning with an equimolar mixture of ammonia and nitrogen.

5.11 For the reation $Br_2(l) + Cl_2(g) \rightleftharpoons 2BrCl(g)$ at 25°C, $K_p = 2.032$ atmospheres. At this temperature the vapour pressure of liquid bromine is 0.281 atmospheres. Assuming ideal gas behaviour where necessary, calculate the amount of $BrCl(g)$ that is present at 1 atmosphere total pressure in an equimolar mixture of bromine and chlorine.

5.12 Calculate the ionic strength of the following solutions: (a) 0.1 M $ZnCl_2$; (b) 0.2 M KNO_3; (c) 0.1 M $NaHSO_4$ (assume the HSO_4^- ion to be undissociated); (d) 0.001 M $NaHSO_4$ (assume the HSO_4^- ion to be completely dissociated).

5.13 Using the Debye–Huckel law in the form of Eq. 5.47, with $A = 0.509$, calculate the activity coefficients of the individual ions and the mean ionic activity coefficient in each of the solutions of exercise 5.12.

5.14 Solubility products are given in Table 5.4 for both a pink and a green form of manganous sulphide, MnS. Predict what would happen if an aqueous suspension of pink MnS were mixed with an aqueous suspension of green MnS.

5.15 Calculate the solubility of $PbBr_2$ in (a) pure water; (b) 0.05 molar $Pb(NO_3)_2$ solution; (c) 0.15 molar HNO_3.

5.16 For the reactions

$$Cu(NH_3)_{n-1}^{2+} + NH_3 \rightleftharpoons Cu(NH_3)_n^{2+}$$

the equilibrium constants are as given below:

n	1	2	3	4	5
$Log_{10} K_n$	4.3	3.6	3.0	2.2	− 0.5

Calculate the ratio of $Cu(NH_3)_5^{2+}$ to Cu^{2+} in a solution made by dissolving a small crystal of $CuSO_4$ in a large excess of 0.1 M ammonia.

5.17 A weak acid HA is half neutralized by NaOH. The ionic strength of the solution is 0.1 M. Calculate the difference, due to the activity coefficients of the ions A^- and H_3O^+, between the pH of the solution and the pK_a of the acid. (Put the Debye–Huckel constant A equal to 0.509.)

5.18 Calculate the volume of 1.0 molar hydrochloric acid that must be added to 10 ml of 1.0 molar ammonia solution to make a buffer solution of pH = 8.5 exactly.

5.19 Thymol blue is used to indicate the end point during the titration of 10 ml of a 0.1 molar strong acid with a 0.1 molar solution of a weak base. The indicator is in 0.05 molar solution and the quantity added to the acid is 0.5 ml. Estimate the size of the titration error due to the presence of the acidic indicator.

5.20 Calculate the entropies of vaporization at the boiling point for the substances listed in the following table:

Substance	Enthalpy of vaporization	Boiling point (K)
Br_2	31 kJ mol^{-1}	331
Cd	106	1038
Hg	62	630
H_2O	41	373
$(C_2H_5)_2O$ (diethyl ether)	27	308
C_6H_6 (benzene)	34	353
CCl_4	32	350

Comment on any similarities you notice.

5.21 Calculate the entropy change in a mole of supercooled liquid water which freezes at − 20°C. The heat of fusion is 6 kJ mol^{-1} at the normal freezing point; the heat capacities of ice and liquid water are 40 and 80 J mol^{-1} K^{-1}, respectively.

5.22 Calculate the entropy change in the surroundings when a mole of water freezes as in exercise 5.21.

5.23   Calculate the entropy change when 1.5 moles of ideal gas expands from 33.6 litres to 67.2 litres at 0°C.

5.24   Calculate the entropy of mixing of a mole of dry air, containing 78.08% of N_2, 20.96% of O_2, 0.93% of Ar and 0.34% of CO_2.

5.25   Calculate the entropy change for the process

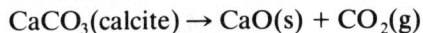

$$CaCO_3(calcite) \rightarrow CaO(s) + CO_2(g)$$

at 298 K and one atmosphere pressure.

5.26   Calculate the entropy change for the reaction

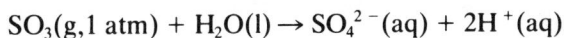

$$SO_3(g, 1 \text{ atm}) + H_2O(l) \rightarrow SO_4^{2-}(aq) + 2H^+(aq)$$

where the ions are present at unit activity in aqueous solution.

5.27   Calculate the Gibbs Free Energy changes for the processes in exercises 5.25 and 5.26.

5.28   Calculate the equilibrium constants for the processes in exercises 5.25 and 5.26.

5.29   Calculate the equilibrium constant for the reaction $N_2O_4 \rightleftharpoons 2NO_2$ at 298 K, using data from Table 5.10.

5.30   Estimate the value of the equilibrium constant for calcite dissociation (exercise 5.25) at 1000 K. (Assume ΔH is independent of temperature.)

6 Electrochemistry

6.1 Introduction

Every chemical phenomenon is a consequence of the electrical forces between electrons and atomic nuclei, and of subtle changes in these forces when electrons are transferred or shared between atoms. Electrochemistry is the study of processes in which electrons are transferred from one reactant to another by a path that lies *outside* the chemical system (Fig. 6.1). Measurements of current and voltage in the external conductor can give precise information about the transfer of electrons from one atom to another, and so provide a probe into the intimate details of electron transfer reactions.

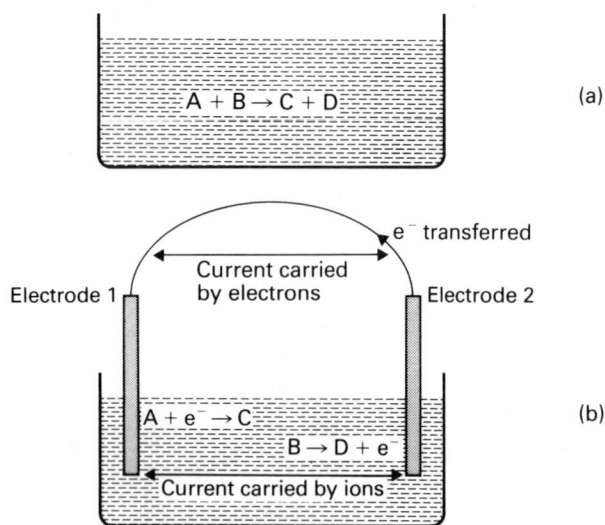

Fig. 6.1 (a) Ordinary chemical system; A and B react in a homogeneous solution to form C and D. (b) Electrochemical system; A is reduced to C at the surface of electrode 1, B oxidized to D at the surface of electrode 2, and an electron is transferred through the external circuit

There are two basic kinds of electrochemical experiment. In the first (Fig. 6.2a) an external voltage source is used to drive a current through a solution of electrolyte.

At the cathode (negative electrode)* electrons enter the solution and a reduction reaction occurs; at the anode (positive electrode) an oxidation reaction occurs and electrons leave the solution. The sum of the anode and cathode reactions is a decomposition process for some component of the electrolyte solution. The driving force for this decomposition process is provided by the external voltage source. In the second type of experiment (Fig. 6.2b) chemical reactions at the electrodes force electrons to flow through the external circuit, reduction occurring at the anode and oxidation occurring at the cathode. The chemical system then comprises an *electrochemical cell*. The driving force of the overall cell reaction is termed its *electromotive force* (e.m.f.). The cell e.m.f. can be measured in the external circuit, and the current of electrons which are driven by this e.m.f. can be made to do external work. Quantities which can be determined by measuring cell e.m.f.'s include equilibrium constants, solubility products, values of ΔG, ΔH and ΔS, activities of ions in solution, and the relative strengths of oxidizing and reducing agents. Such measurements, which will be the main topic of this chapter, can be gathered under the heading of *electrode thermodynamics*.

Electrochemical measurements made with an external voltage source are classified under three main headings, namely *electrolytic conduction*, *electrolysis* and *electrode kinetics*. Careful measurements of electrolytic conduction, mostly made during the last century, provide the basis for our present ideas about the nature of electrolyte solutions. Conduction by an electrolyte solution depends on the movement of the ions in the solution towards one or other of the two electrodes, under the influence of the electric field due to the external voltage (Fig. 6.3a). Negative ions (*anions*) move towards the anode; positive ions (*cations*) move towards the cathode. Measurements of the conductance of electrolyte solutions have been used for the determination of such quantities as ionic mobilities in solution, dissociation constants, and solubility products. Conductivity measurements are a useful tool in studying the structures of salts of complex ions; for example, the salt $[Co(NH_3)_4Cl_2]Cl \cdot H_2O$ gives rise to two ions in solution, $[Co(NH_3)_4Cl_2]^+$ and Cl^-,

---

* Alternatively, the cathode can be defined as the electrode at which electrons appear to enter the cell. For an *electrochemical cell* this interchanges the labels anode and cathode as they are used here.

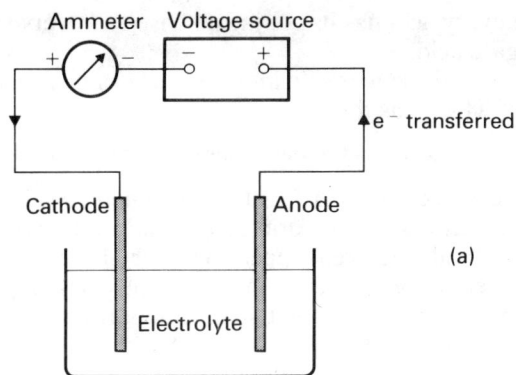

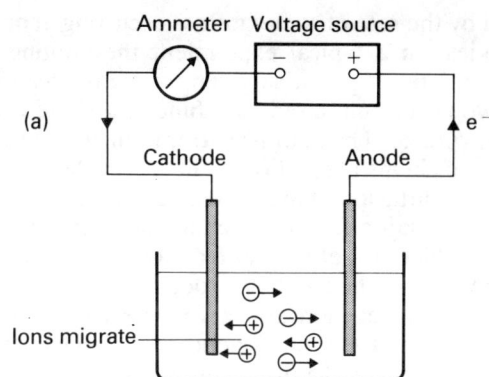

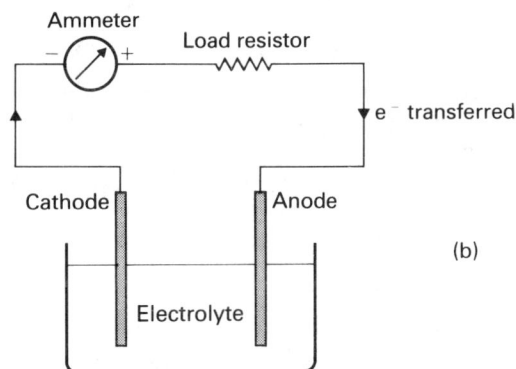

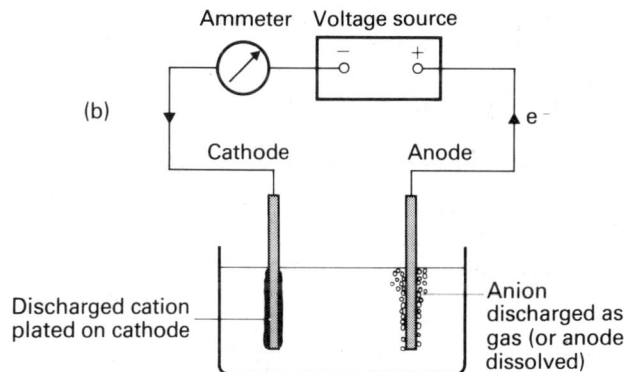

Fig. 6.2 The two basic types of electrochemical experiment. (a) An external voltage source drives electrons through the solution, producing oxidation at the anode and reduction at the cathode. (b) A reaction in the electrochemical cell drives electrons through the external circuit, oxidation occurring at the cathode and reduction at the anode. In both experiments the current flowing is indicated by the ammeter

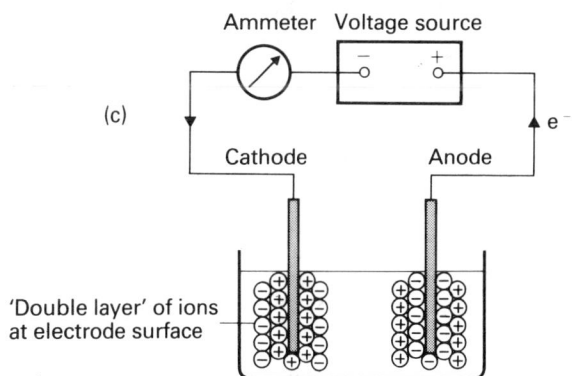

Fig. 6.3 (a) Electrolytic conduction. Current limited by number and mobility of ions in solution (microscopic view). (b) Electrolysis. To obtain electrolysis products at cathode and/or anode (macroscopic view). (c) Electrode kinetics. Current limited by processes at surface of electrodes (microscopic view)

whereas the salt of identical atomic composition, $[Co(NH_3)_4(H_2O)Cl]Cl_2$ gives the three ions $[Co(NH_3)_4(H_2O)Cl]^{2+}$ plus two Cl^-. A solution of the latter will have a much lower electrical resistance, i.e. greater conductance (conductance = 1/resistance), than a solution of the former containing the same weight of electrolyte. Conductance measurements are made under conditions in which processes occurring at the electrodes are not significant, the emphasis being on the migration of ions through the solution.

Electrolysis (Fig. 6.3b) is used in the extraction and purification of elements, notably aluminium, copper and chlorine, and in the industrial production of sodium hydroxide from brine. Electroplating of steel with a corrosion-resistant metal such as nickel or chromium is widely practised for both cosmetic and utilitarian reasons. The problems encountered in this field are almost entirely practical ones, such as how to obtain the maximum yield of aluminium or chlorine per kilowatt-hour of electrical energy, or how to obtain the smoothest and most adherent coating of chromium.

Electrode kinetics (Fig. 6.3c) is concerned with the situation in which the current flow through the external circuit is limited by the rate of some process occurring at one of the electrodes. In a typical experiment the limiting current density (amperes per square centimetre of electrode surface) would be measured as a function of the externally applied voltage. The limiting current can be fixed either by the rate of chemical reaction at the electrode surface or by the rate of diffusion of ions to the surface. Electrode kinetics is the major factor affecting the usefulness of fuel cells, in which a fuel and an oxidant (e.g. hydrogen and oxygen) react at different electrodes in such a way that the energy of the reaction appears as electrical work in an external circuit. Electrode kinetics is also relevant to corrosion: unfavourable electrode kinetics is responsible for the corrosion resistance of chromium and aluminium. The conduction of impulses along nerve fibres has been shown to involve an electrochemical process, so it is probably only a slight exaggeration to say that the speed of thought is governed by electrode kinetics.

6.2 Oxidation–Reduction Reactions

Electron-transfer reactions are usually referred to as oxidation–reduction reactions, or 'redox' reactions. In a redox reaction electrons are transferred from one reactant, the reducing agent, to another, the oxidizing agent. In the process the reducing agent is itself oxidized, and the oxidizing agent reduced. The products which are formed are themselves oxidizing and reducing agents (Eq. 6.1). There is in fact a strong analogy with the situation for acids and bases,

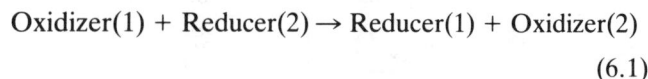

$$\text{Oxidizer}(1) + \text{Reducer}(2) \rightarrow \text{Reducer}(1) + \text{Oxidizer}(2)$$
$$(6.1)$$

where every acid has its conjugate base and every base its conjugate acid.

Reaction 6.1 can be split into two parts, or *half-reactions*, each of which has the form

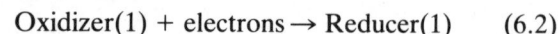

$$\text{Oxidizer}(1) + \text{electrons} \rightarrow \text{Reducer}(1) \qquad (6.2)$$

The overall equation, 6.1, is then obtained by subtracting one half-reaction of the form of 6.2 from another, with the stronger oxidizing agent appearing on the left hand side of the final equation, and the number of molecules of species 1 and 2 being adjusted so that no electrons are left over. For example, consider the reaction of H_2S with SO_2. The reduction processes corresponding to Eq. 6.2 are:

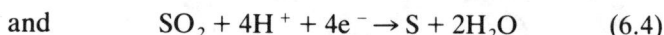

$$S + 2H^+ + 2e^- \rightarrow H_2S \qquad (6.3)$$

and
$$SO_2 + 4H^+ + 4e^- \rightarrow S + 2H_2O \qquad (6.4)$$

We find by experiment that the half-reaction, 6.4, has a greater driving force than reaction 6.3, so 6.3 has to go in reverse. In order that there will be no electrons left over it is necessary to subtract twice Eq. 6.3 from Eq. 6.4, and when this is done it is found that the H^+ ions also cancel. The equation analogous to 6.1 is therefore

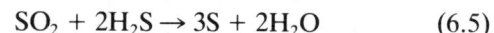

$$SO_2 + 2H_2S \rightarrow 3S + 2H_2O \qquad (6.5)$$

The driving force of a process of the form of 6.2 is expressed by the *standard reduction potential*, $E^{\ominus}$, for the particular half-reaction, and is measured in volts. The electrochemical significance of this statement will become apparent in the next section. A number of standard reduction potentials are listed in Table 6.1. *The more positive the value of the standard reduction potential, the stronger the oxidizing*

Table 6.1 Some standard reduction potentials, $E^{\ominus}$, at 298 K.

Half-reaction		$E^{\ominus}$ (Volts)	Half-reaction		$E^{\ominus}$ (Volts)
$Li^+ + e^-$	$\rightarrow Li(s)$	-3.04	$\frac{1}{2}F_2(g) + e^-$	$\rightarrow F^-$	$+2.87$
$K^+ + e^-$	$\rightarrow K(s)$	-2.92	$MnO_2^- + 4H^+ + 3e^-$	$\rightarrow MnO_2(s) + 2H_2O$	$+1.70$
$Ca^{2+} + 2e^-$	$\rightarrow Ca(s)$	-2.87	$MnO_4^- + 8H^+ + 5e^-$	$\rightarrow Mn^{2+} + 4H_2O$	$+1.51$
$Na^+ + e^-$	$\rightarrow Na(s)$	-2.71	$ClO_3^- + 6H^+ + 5e^-$	$\rightarrow \frac{1}{2}Cl_2(g) + 3H_2O$	$+1.47$
$Mg^{2+} + 2e^-$	$\rightarrow Mg(s)$	-2.36	$PbO_2(s) + 4H^+ + 2e^-$	$\rightarrow Pb^{2+} + 2H_2O$	$+1.46$
$Al^{3+} + e^-$	$\rightarrow Al(s)$	-1.66	$\frac{1}{2}Cl_2(g) + e^-$	$\rightarrow Cl^-$	$+1.36$
$Mn^{2+} + 2e^-$	$\rightarrow Mn(s)$	-1.18	$\frac{1}{2}Cr_2O_7^{2-} + 7H^+ + 3e^-$	$\rightarrow Cr^{3+} + 3\frac{1}{2}H_2O$	$+1.33$
$Zn^{2+} + 2e^-$	$\rightarrow Zn(s)$	-0.76	$MnO_2(s) + 4H^+ + 3e^-$	$\rightarrow Mn^{2+} + 2H_2O$	$+1.23$
$Cr^{3+} + 3e^-$	$\rightarrow Cr(s)$	-0.74	$ClO_4^- + 2H^+ + 2e^-$	$\rightarrow ClO_3^- + H_2O$	$+1.19$
$S(s) + 2e^-$	$\rightarrow S^{2-}$	-0.48	$\frac{1}{2}Br_2(l) + e^-$	$\rightarrow Br^-$	$+1.07$
$Fe^{2+} + 2e^-$	$\rightarrow Fe(s)$	-0.44	$NO_3^- + 4H^+ + 3e^-$	$\rightarrow NO(g) + 2H_2O$	$+0.96$
$Cd^{2+} + 2e^-$	$\rightarrow Cd(s)$	-0.40	$Hg^{2+} + 2e^-$	$\rightarrow Hg(l)$	$+0.85$
$PbSO_4(s) + 2e^-$	$\rightarrow Pb(s) + SO_4^{2-}$	-0.36	$Ag^+ + e^-$	$\rightarrow Ag(s)$	$+0.80$
$Co^{2+} + 2e^-$	$\rightarrow Co(s)$	-0.28	$NO_3^- + 2H^+ + e^-$	$\rightarrow \frac{1}{2}N_2O_4(g) + H_2O$	$+0.80$
$Ni^{2+} + 2e^-$	$\rightarrow Ni(s)$	-0.25	$Hg_2^{2+} + 2e^-$	$\rightarrow 2Hg(l)$	$+0.79$
$Sn^{2+} + 2e^-$	$\rightarrow Sn(s)$	-0.14	$\frac{1}{2}I_2(s) + e^-$	$\rightarrow I^-$	$+0.54$
$Pn^{2+} + 2e^-$	$\rightarrow Pb(s)$	-0.13	$Cu^+ + e^-$	$\rightarrow Cu(s)$	$+0.52$
$CO_2(g) + 2H^+ + 2e^-$	$\rightarrow CO(g) + H_2O$	-0.12	$Cu^{2+} + 2e^-$	$\rightarrow Cu(s)$	$+0.34$
$Fe^{3+} + 3e^-$	$\rightarrow Fe(s)$	-0.02	$\frac{1}{2}Hg_2Cl_2(s) + e^-$	$\rightarrow Hg(l) + Cl^-$	$+0.27$
$H^+ + e^-$	$\rightarrow \frac{1}{2}H_2(g)$	0.0	$AgCl(s) + e^-$	$\rightarrow Ag(s) + Cl^-$	$+0.22$

agent, i.e. the greater the driving force behind the reduction process. Thus when two half-reactions are combined to give a redox reaction of the form of Eq. 6.1, the half-reaction with the less positive reduction potential will be forced to go in reverse. The standard reduction potential for the process

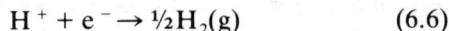

$$H^+ + e^- \rightarrow \frac{1}{2}H_2(g) \qquad (6.6)$$

is *defined* as zero; this means that the reduction potential always expresses the strength of an oxidizing agent relative to the oxidizing power of H^+. The pair of reagents Oxidizer(1)/Reducer(1) of Eq. 6.2 comprise an oxidation–reduction *couple*. If the value of $E^\oplus$ for a redox couple is positive the couple comprises a more powerful oxidant than the H^+/H_2 couple. If $E^\oplus$ is negative the couple is a more powerful reducing agent than H^+/H_2.

Example Obtain chemical equations for the reactions between the following pairs of oxidation–reduction couples:

(a) NO_3^-/NO and Cu^{2+}/Cu
(b) $Cr_2O_7^{2-}/Cr^{3+}$ and I_2/I^-
(c) Fe^{2+}/Fe and H^+/H_2

For example (a), Table 6.1 shows that the NO_3^-/NO couple has a more positive $E^\oplus$ value than Cu^{2+}/Cu, so the Cu^{2+}/Cu half-reaction must go in reverse. The nitrate reduction involves three electrons, and the copper oxidation two; therefore the nitrate half-reaction is multiplied by two and the copper half-reaction by three. The result is:

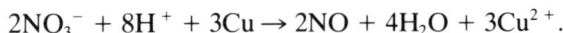

$$2NO_3^- + 8H^+ + 3Cu \rightarrow 2NO + 4H_2O + 3Cu^{2+}.$$

In example (b) the dichromate half-reaction of Table 6.1 is best multiplied by two to get rid of the non-integral number of water molecules. *Six* electrons are now involved, so we must subtract six times the iodine half-reaction, which has the smaller positive $E^\oplus$ value, to obtain:

$$Cr_2O_7^{2-} + 14H^+ + 6I^- \rightarrow 2Cr^{3+} + 7H_2O + 3I_2.$$

For example (c) we take twice the hydrogen half-reaction and subtract from this the ferrous half-reaction, which has the less positive $E^\oplus$ value, to obtain:

$$Fe + 2H^+ \rightarrow Fe^{2+} + H_2.$$

6.3 The Electrochemical Cell

(i) Electrode Reactions

An electrode immersed in a conducting solution acquires a potential which is controlled by redox reactions that take place at the surface of the electrode. Suppose, for example, that an electrode in the form of a piece of copper wire is dipped into a solution of a cupric salt, as in Fig. 6.4a. A dynamic equilibrium will be established between the metal

and the solution. At the surface of the metal copper ions are discharged according to the process

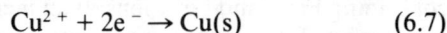

$$Cu^{2+} + 2e^- \rightarrow Cu(s) \qquad (6.7)$$

and electrons will therefore tend to be removed from the metal, giving the wire an excess of positive charge. At the same time the reverse of reaction 6.7 will also occur, copper metal dissolving and giving up electrons to the metal so as to neutralize some or all of the excess positive charge. At equilibrium the copper electrode will possess an electrical potential which is sufficient to prevent any further net gain or loss of electrons, i.e. the potential of the wire will be just sufficient to counterbalance the driving force of the half-reaction 6.7.

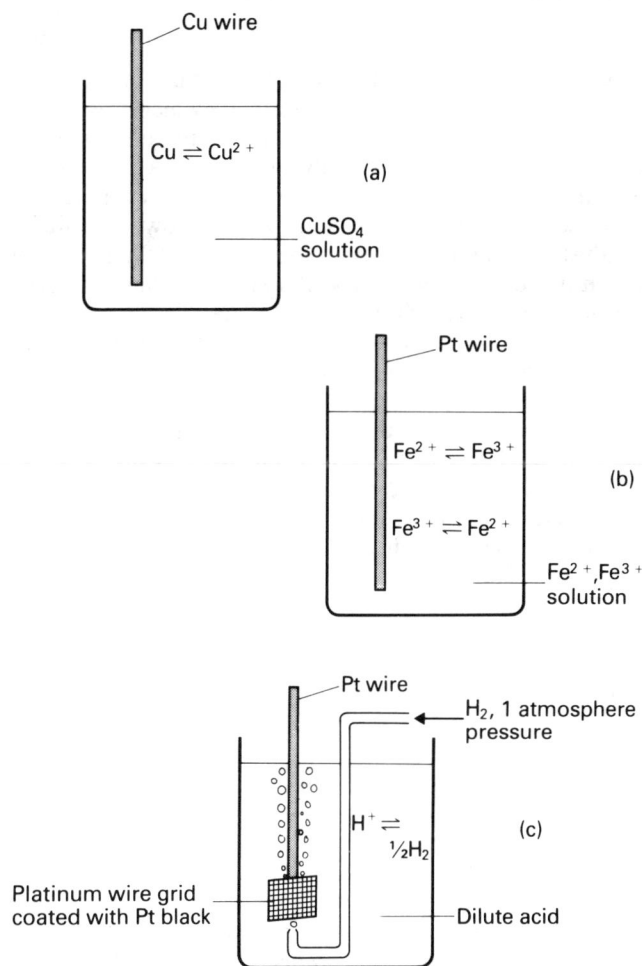

Fig. 6.4 Electrode systems which acquire a potential characteristic of the driving force of a particular half-reaction: (a) $Cu^{2+} + 2e^- \rightarrow Cu(s)$, (b) $Fe^{3+} + e^- \rightarrow Fe^{2+}$; (c) $H^+ + e^- \rightarrow \frac{1}{2}H_2(g)$. In each case the potential is developed between the metal electrode and the solution

It is not necessary for the metal of the electrode to be directly involved in the redox reaction: consider a solution containing Fe^{3+} and Fe^{2+} ions. If an inert conductor, such as a piece of platinum wire, is immersed in the solution (Fig. 6.4b) ferric ions will tend to acquire electrons from the electrode and form ferrous ions, according to

$$Fe^{3+} + e^- \rightarrow Fe^{2+} \qquad (6.8)$$

At the same time ferrous ions will tend to give up electrons to the platinum electrode, reaction 6.8 then occurring in reverse. Thus the inert electrode will acquire an equilibrium potential which is sufficient to prevent any net gain or loss of electrons by reaction 6.8.

Table 6.1 contains several examples of redox reactions in which gases take part. The half-reaction

$$H^+ + e^- \rightarrow \tfrac{1}{2}H_2 \qquad (6.6)$$

would be expected to give rise to a characteristic equilibrium potential on an inert electrode which was dipped into an acid solution saturated with hydrogen gas. However, in the absence of a catalyst it would take a long time for such an electrode to reach its equilibrium potential, because reaction 6.6 is normally very slow. A suitable catalyst for reaction 6.6 exists in the form of finely-divided platinum, or *platinum black*. An electrode having a large surface area coated with platinum black (Fig. 6.4c) rapidly acquires a potential which is sufficient to prevent further gain or loss of electrons by reaction 6.6.

In each of these examples the metal electrode acquires a potential, relative to the solution, which is sufficient to prevent further gain or loss of electrons from or to the solution. The absolute value of this potential cannot be measured, because to measure a potential difference we require an electrical circuit like that in Fig. 6.2b, with *two* electrodes dipping into the solution. Therefore all such potentials are determined relative to that of a standard electrode system. The standard electrode system is that of Fig. 6.4c with the hydrogen ions (actually H_3O^+) present at unit activity. For this system $E^{\ominus}$ is defined to be zero.

(ii) The Cell Reaction

An electrochemical cell can be formed in a variety of ways. The simplest way is to dip two different electrodes into a common electrolyte solution, as in Fig. 6.5a. If the electrode systems are incompatible two solutions can be connected electrically by using some form of *salt bridge*, such as the agar bridge of Fig. 6.5b. An agar bridge consists of a conducting jelly, made by dissolving a powder derived from sea-weed in hot KCl solution and leaving it to cool and solidify inside a U-shaped piece of glass or plastic tubing. Such a bridge is convenient to use because it can be lifted out of one experiment, rinsed, and put to work in another.

A salt bridge can also be used without the jelly, as in Fig. 6.5c, or the bridge can be omitted altogether, as in Fig. 6.5d. If the copper sulphate solution in Fig. 6.5d were to come into contact with the zinc cathode, the Cu^{2+} ions would be reduced by the zinc and a layer of metallic copper would be deposited on the cathode. This would ultimately put the cell out of action; hence the porous pot.

The arrangement of Fig. 6.5d would generally be unsatisfactory for experiments in which the e.m.f. of the cell was being measured because of the likelihood of a large *junction potential* developing between the two electrolyte solutions. A junction potential arises as the result of diffusion of ions from one solution into the other. If one of the ions of an electrolyte diffuses more rapidly than the other, excess charge associated with the faster moving ion (usually the cation) will accumulate on the far side of the boundary. The equilibrium value of the junction potential (typically a few millivolts or tens of millivolts) is such that the electric field gradient due to the potential is just sufficient to counterbalance the tendency of the ions to diffuse, with the result that no further diffusion of charge can occur. A salt bridge is made with an electrolyte such as potassium chloride, or ammonium nitrate, in which the two ions have almost identical rates of diffusion, so that the junction potential is minimized. In the following discussion the junction potential, if any, will be assumed to be negligible.

An electrochemical cell is conveniently represented as

$$\text{electrode(1)/electrolyte(1)//electrolyte(2)/electrode(2)} \qquad (6.9)$$

where the right hand electrode, electrode(2) in the cell 6.9, is conventionally chosen to be the positive electrode of the cell, and the activities of the electrolytes are specified. Some writers prefer to use vertical lines, or semi-colons, in place of the slashes which separate the different sections of the cell in 6.9. A slash is inserted wherever there is a change of phase, e.g. from a solid electrode to a solution, or from one solution to another. If different components of the cell are present in the same phase they are separated by commas. Two adjacent slashes imply the presence of a salt bridge. The cell of Fig. 6.5a is written simply as

$$Pb/PbSO_4(s)/H_2SO_4(1.0M)/PbO_2/Pb \qquad (6.10)$$

because a single electrolyte solution does for both electrodes. A cell of the type shown in Fig. 6.5b is

$$Pt/H_2(1 \text{ atm.})/HCl(1.0M)/\text{agar bridge}/CuSO_4(1.0M)/Cu \qquad (6.11)$$

while the cells in Figs. 6.5c and 6.5d are written as

$$Zn/ZnCl_2(0.1M)/KCl(0.1M)/CdCl_2(0.1M)/Cd \qquad (6.12)$$

and

$$Zn(Hg)/ZnSO_4(2M)/\text{porous pot}/CuSO_4(2M)/Cu \qquad (6.13)$$

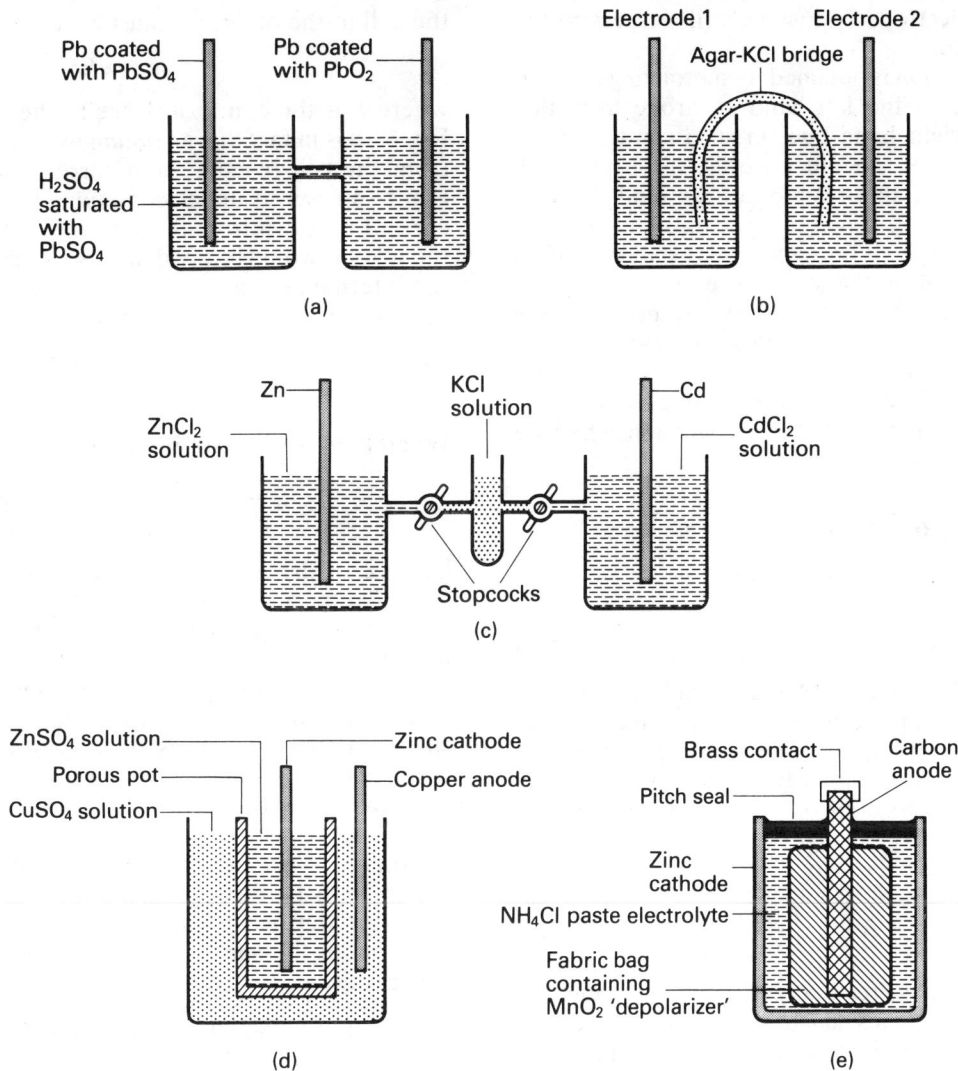

Fig. 6.5 Some electrochemical cells. (a) Storage cell (lead accumulator); the electrode reactions can both be reversed by driving a current throught the cell from an external source. (b) and (c) Forms suitable for e.m.f. measurements in the laboratory. In (c) the stopcocks are kept closed except when a measurement is about to be made. (d) The Daniell cell. The copper sulphate solution is kept away from the zinc cathode by the porous pot. The zinc is treated with mercury to make a zinc amalgam, so H $^+$ will not be discharged. (e) Dry cell form of the Le Clanché cell: the common 'battery'

The observed e.m.f. of a cell is equal to the e.m.f. of the right hand electrode minus the e.m.f. of the left hand electrode:

$$E_{cell} = E_2 - E_1 \qquad (6.14)$$

If all of the species involved in the electrode reactions are present at unit activity, the observed e.m.f. is the *standard e.m.f.*, $E^{\ominus}$, of the cell. The state of unit activity, we recall, corresponds to the pure substance at one atmosphere pressure in the case of a solid or a liquid (for example, lead or

mercury), the ideal gas at one atmosphere pressure in the case of a gas, such as hydrogen, and an activity of 1 mole per litre for a solute such as H $^+$. The standard e.m.f. of a cell containing a hydrogen electrode is, by definition, equal to the standard reduction potential of the other electrode in the cell. Thus, for example, the e.m.f. of the cell 6.11 is equal to the standard reduction potential for the half-reaction

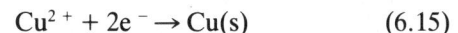

$$Cu^{2+} + 2e^- \rightarrow Cu(s) \qquad (6.15)$$

and the copper electrode is positive with respect to the hydrogen electrode.

The total *cell reaction* is obtained by subtracting the half-reaction occurring at the left hand electrode from that occurring at the right hand one, in such a way that no electrons are left over. In other words, the overall cell reaction is simply a redox reaction of the kind that was discussed in Section 6.2. When the reactants and products of the cell reaction are all present in their standard states the e.m.f. of the cell is the standard e.m.f., $E^{\ominus}$. If the activities of the reactants and products are not unity the e.m.f. is different from $E^{\ominus}$, and in the situation where the cell reaction is at equilibrium, the observed e.m.f. is zero. The dependence of the e.m.f. on the activities of reactants and products is described by the Nernst equation, which we consider in the next section.

6.4 The Nernst Equation

(i) A Derivation of the Equation

Consider the general cell reaction

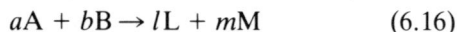

$$aA + bB \rightarrow lL + mM \qquad (6.16)$$

in which a molecules of species A react with b molecules of species B to produce l molecules of species L and m molecules of species M, and suppose that as a result n electrons are transferred through the external circuit. For example, if the cell reaction were reaction 6.5, the value of n would be 4. The Gibbs Free Energy change ΔG for the cell reaction is given by Eq. 5.120 as:

$$\Delta G = \Delta G^{\ominus} + RT \ln \left\{ \frac{(a_L)^l (a_M)^m}{(a_A)^a (a_B)^b} \right\} \qquad (6.17)$$

where $\Delta G^{\ominus}$ is the Gibbs Free Energy change when reactants and products are all in their standard states, and, for example, $(a_L)^l$ stands for the activity of species L raised to the power l.

In Section 5.6 (Eq. 5.114) we stated as a general result that the decrease in Gibbs Free Energy during any process, $-\Delta G$, is equal to the maximum amount of external work that can be obtained from the process. In this case, therefore, Eq. 6.17 gives the maximum work that can be obtained by transferring n electrons through the external circuit. If we consider reaction 6.16 in terms of moles of reactants and products rather than molecules we have to think of the work done by transferring n *moles* of electrons through the external circuit. The quantity of charge equivalent to a mole of electrons is known as the faraday, written F, and amounts to 96 500 coulombs. The work done when one coulomb of charge is transferred between two points whose potential differs by one volt is equal to one joule. Therefore, when nF coulombs are transferred from one pole of

the cell to the other, we must have

$$w = nEF \qquad (6.18)$$

where E is the e.m.f. of the cell. The value of w given by Eq. 6.18 is in fact the *maximum work* that can be obtained from the cell reaction, and corresponds to the value of e.m.f. measured under conditions where no appreciable current is drawn from the cell. If any current were drawn the work w would be divided between the external load and the internal resistance of the cell, as discussed in the next section. Since w is the maximum work, we must have

$$\Delta G = -nEF \qquad (6.19)$$

and

$$\Delta G^{\ominus} = -nE^{\ominus}F \qquad (6.20)$$

When these results are inserted into Eq. 6.17 we obtain

$$E = E^{\ominus} - \frac{RT}{nF} \ln \left\{ \frac{(a_L)^l (a_M)^m}{(a_A)^a (a_B)^b} \right\} \qquad (6.21)$$

Eq. 6.21 is the celebrated *Nernst equation*. It is helpful to note that the activities of reactants and products appear in the logarithm in the same way that they appear in the equilibrium constant. Since $\Delta G^{\ominus}$ is equal to $-RT \ln K_a$, it follows that the observed e.m.f. E is zero when the quantity inside curly brackets is equal to K_a, the equilibrium constant expressed in terms of activities. The activities of reactants and products can be related to their concentrations by way of the rules summarized in Table 5.3.

Example   Write out the Nernst equation for each of the cells 6.10, 6.11, 6.12 and 6.13.

The first step is to work out the equations for the cell reactions by the same procedure as was used in Section 6.2. The results are:

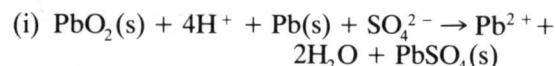

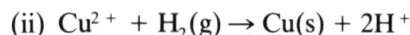

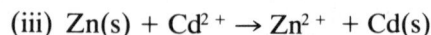

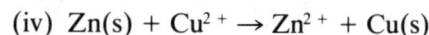

(i) $PbO_2(s) + 4H^+ + Pb(s) + SO_4^{2-} \rightarrow Pb^{2+} +$
$\qquad\qquad\qquad\qquad 2H_2O + PbSO_4(s)$

(ii) $Cu^{2+} + H_2(g) \rightarrow Cu(s) + 2H^+$

(iii) $Zn(s) + Cd^{2+} \rightarrow Zn^{2+} + Cd(s)$

(iv) $Zn(s) + Cu^{2+} \rightarrow Zn^{2+} + Cu(s)$

where H_2O is present as the solvent and the ions are all in aquecus solution.

In reaction (i) we can take the activity as 1 for the solids PbO_2, Pb, and $PbSO_4$, and the solvent H_2O, and these substances may therefore be omitted from the logarithmic term in the Nernst equation. Hence this equation becomes

$$E = E^{\ominus} - \frac{RT}{2F} \ln\{(a_{Pb^{2+}})/(a_{SO_4^{2-}})(a_{H^+})^4\}$$

where $E^{\ominus} = E^{\ominus}_{PbO_2/Pb^{2+}} - E^{\ominus}_{PbSO_4/Pb} = 1.82V.$

In reaction (ii) the activities of copper metal and hydrogen are equal to 1, so the Nernst equation becomes

$$E = E^{\ominus} - \frac{RT}{2F} \ln\{(a_{H^+})^2/(a_{Cu^{2+}})\}$$

where

$$E^{\ominus} = E^{\ominus}_{Cu^{2+}/Cu} - E^{\ominus}_{H^+/H_2} = E^{\ominus}_{Cu^{2+}/Cu} = 0.34V$$

If the activities of hydrogen and cupric ions are also equal to 1, the cell e.m.f. reduces to the standard reduction potential $E^{\ominus}_{Cu^{2+}/Cu}$.

Reactions (iii) and (iv) differ only in the nature of the metal ion that is being displaced from solution by zinc. For the metals themselves the activities are unity. Hence for (iii)

$$E = E^{\ominus} - \frac{RT}{2F} \ln\{a_{Zn^{2+}}/a_{Cd^{2+}}\}$$

where $\quad E^{\ominus} = E^{\ominus}_{Cd^{2+}/Cd} - E^{\ominus}_{Zn^{2+}/Zn} = 0.36V$,

and for (iv)

$$E = E^{\ominus} - \frac{RT}{2F} \ln\{a_{Zn^{2+}}/a_{Cu^{2+}}\}$$

where $\quad E^{\ominus} = E^{\ominus}_{Cu^{2+}/Cu} - E^{\ominus}_{Zn^{2+}/Zn} = 1.11V$.

(ii) The Nernst Equation for a Half-cell

For the cell $Pt/H_2(1 \text{ atm})/HCl(1.0M)/agar$ bridge/$CuSO_4(1.0M)/Cu$ which we considered as part (ii) of the last example, the Nernst equation reduces to

$$E = E^{\ominus}_{Cu^{2+}/Cu} - \frac{RT}{2F} \ln\{a_{Cu}/a_{Cu^{2+}}\} \quad (6.22)$$

in the situation where the hydrogen electrode is a standard hydrogen electrode, i.e. when a_{H_2} and a_{H^+} are both equal to 1. Similarly, for any half-reaction of the type

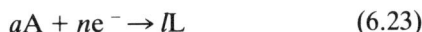

$$aA + ne^- \rightarrow lL \quad (6.23)$$

we can visualize the process as occurring in a cell in which the other electrode is a standard hydrogen electrode, and write *the Nernst equation for the half-cell* in the form

$$E_{A/L} = E^{\ominus}_{A/L} - \frac{RT}{nF} \ln\{(a_L)^l/(a_A)^a\} \quad (6.24)$$

The complete Nernst equation for any cell, Eq. 6.21, is then obtained by subtracting the Nernst equation for the negative half-cell from that for the positive half-cell, as in the following example.

Example For the cell:
$Mg/MgCl_2(0.01M)/NH_4NO_3(0.1M)/MnSO_4(0.01M)$,
$H_2SO_4(0.05M), KMnO_4(0.01M)/Pt$

(a) Write down the Nernst equations for the half-cell reactions and for the overall cell reaction; (b) calculate the e.m.f. for the specified activities of reactants and products.

(a) The ammonium nitrate is present only as a salt bridge. From Table 6.1, the half-reactions are
(i) $Mg^{2+} + 2e^- \rightarrow Mg(s)$, $(E^{\ominus} = -2.36 \text{ V})$ and
(ii) $MnO_4^- + 8H^+ + 5e^- \rightarrow Mn^{2+} + 4H_2O$, $(E^{\ominus} = +1.51V)$.

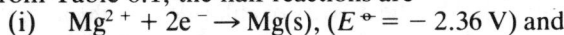

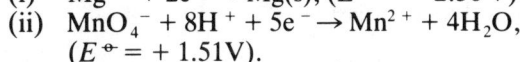

Hence the overall cell reaction is
(iii) $2MnO_4^- + 5Mg(s) + 16H^+ \rightarrow 2Mn^{2+} + 5Mg^{2+} + 8H_2O$, $(E^{\ominus} = +3.87V)$.

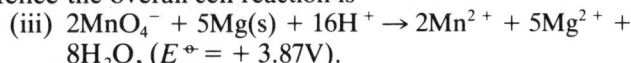

The Nernst equations for reactions (i) and (ii) are

(iv) $E_{Mg^{2+}/Mg} = E^{\ominus}_{Mg^{2+}/Mg} + \frac{RT}{2F} \ln\{a_{Mg^{2+}}\}$, where we

have put the activity of solid magnesium equal to 1, and

(v) $E_{MnO_4^-/Mn^{2+}} = E^{\ominus}_{MnO_4^-/Mn^{2+}} - \frac{RT}{5F} \ln\{(a_{Mn^{2+}})/(a_{MnO_4^-})(a_{H^+})^8\}$

The Nernst equation for reaction (iii) can also be written down directly, as

(vi) $E = E^{\ominus} - \frac{RT}{10F} \ln\{(a_{Mn^{2+}})^2(a_{Mg^{2+}})^5/(a_{MnO_4^-})^2 (a_{H^+})^{16}\}$

Whatever the numbers of electrons involved in the half-reactions, $E^{\ominus}$ for the cell is simply the difference between the $E^{\ominus}$ values for the half-cells. Hence, in equation (vi), $E^{\ominus}$ is 3.87V. Equation (vi) can also be obtained by subtracting (iv) from (v) and combining the logarithmic terms into a single term.

(b) To evaluate the cell e.m.f. E we now substitute the activities of the various ions into equation (vi), together with the value of 3.87V for $E^{\ominus}$. It is usually more convenient to work in terms of logarithms to base ten, rather than to base e, in which case the quantity by which the log term is multiplied is $(2.303RT/nF)$. The factor $(2.303RT/F)$ is equal to 59 millivolts at $T = 298$ K; this is a useful number to remember. The factor inside curly braces in (vi) amounts to 10^6 when the various activities are inserted, so the e.m.f. of the cell works out to be

$$E = 3.87 - 5.9 \times 10^{-3} \times 6 = \underline{3.83V}.$$

(iii) Some Special Electrode Systems

(a) The silver–silver chloride electrode. A silver electrode immersed in a solution containing silver ions rapidly assumes a potential which is governed by the activity of Ag^+ in the solution. The half-reaction is

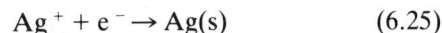

$$Ag^+ + e^- \rightarrow Ag(s) \quad (6.25)$$

and the corresponding Nernst equation is

$$E_{Ag^+/Ag} = E^{\ominus}_{Ag^+/Ag} + \frac{RT}{F} \ln\{a_{Ag^+}\} \qquad (6.26)$$

where we have used the fact that the activity of silver metal is equal to 1. If an excess of a solution containing chloride ions is added to the electrolyte the silver will almost all be precipitated as solid AgCl, the amount left in solution being governed by the activity of chloride ions through the solubility product expression

$$K_{AgCl} = a_{Ag^+} \cdot a_{Cl^-} \qquad (6.27)$$

This equation can be used to substitute for a_{Ag^+} in Eq. 6.26, and obtain

$$E_{AgCl/Ag} = E^{\ominus}_{Ag^+/Ag} + \frac{RT}{F} \ln K_{AgCl} - \frac{RT}{F} \ln\{a_{Cl^-}\} \qquad (6.28)$$

The half-cell e.m.f. is written as $E_{AgCl/Ag}$ rather than as $E_{Ag^+/Ag}$ to show that we are now dealing with a different electrode system, namely the silver–silver chloride electrode. As long as solid AgCl is present the activity of silver ions is given by Eq. 6.27, and the e.m.f. is given by Eq. 6.28 which shows that the observed e.m.f. is governed by the activity of chloride ions in the solution. The following important points can now be made in connection with this electrode system:

(1) An electrode which responds to the activity of an *anion* such as Cl^- can be constructed by taking the sparingly soluble salt of the anion with a suitable metal, and immersing a piece of the metal in a solution saturated with the insoluble salt. For example, a lead–lead sulphate electrode will have its e.m.f. controlled by the activity of sulphate ions in solution, a silver–silver iodide electrode will respond to the activity of iodide ions, and so on.

(2) If the e.m.f. of the silver–silver chloride electrode is measured in a solution which contains a known activity of chloride ions, the term $(RT/F) \ln K_{AgCl}$ can be evaluated and the value of K_{AgCl} determined. Thus this kind of electrode system can be used to measure the solubility product of a sparingly soluble salt. Most tabulated solubility product values for such substances were actually determined in this way.

(3) When electrons pass through the external circuit of a cell containing a silver–silver chloride electrode, the number of silver ions in solution remains controlled by the solubility product of AgCl and the large excess activity of chloride ions. Therefore any silver ions which are lost by reaction 6.25 must be replaced by the process:

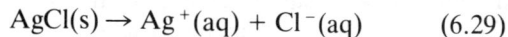

$$AgCl(s) \rightarrow Ag^+(aq) + Cl^-(aq) \qquad (6.29)$$

and the actual half-cell reaction becomes

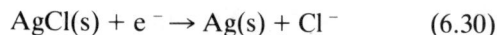

$$AgCl(s) + e^- \rightarrow Ag(s) + Cl^- \qquad (6.30)$$

which is the sum of Eqs 6.25 and 6.29. For reaction 6.30 the Nernst equation is

$$E_{AgCl/Ag} = E^{\ominus}_{AgCl/Ag} - \frac{RT}{F} \ln\{a_{Cl^-}\} \qquad (6.31)$$

Comparison of Eqs 6.28 and 6.31 now reveals that

$$E^{\ominus}_{AgCl/Ag} = E^{\ominus}_{Ag^+/Ag} + \frac{RT}{F} \ln K_{AgCl} \qquad (6.32)$$

An equation analogous to Eq. 6.32 can be written for every electrode system which is based on a metal and its sparingly soluble salt.

(b) The mercury–calomel electrode. This, like the silver–silver chloride electrode, is an important example of an electrode system based on the sparingly soluble salt of a metal. In this case the salt is mercurous chloride, a substance which has the trivial name 'calomel'. In the absence of chloride ions the electrode reaction is

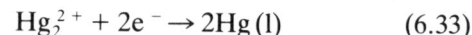

$$Hg_2^{2+} + 2e^- \rightarrow 2Hg(l) \qquad (6.33)$$

and the corresponding Nernst equation is

$$E_{Hg_2^{2+}/Hg} = E^{\ominus}_{Hg_2^{2+}/Hg} + \frac{RT}{2F} \ln\{a_{Hg_2^{2+}}\} \quad (6.34)$$

When chloride ions are present in excess the activity of mercurous ions is governed by the solubility product expression

$$K_{Hg_2Cl_2} = (a_{Hg_2^{2+}}) \cdot (a_{Cl^-})^2 \qquad (6.35)$$

If Eq. 6.35 is used to substitute for the mercurous ion activity in Eq. 6.34 the result is

$$E_{Hg/Hg_2Cl_2} = E^{\ominus}_{Hg/Hg_2Cl_2} - \frac{RT}{F} \ln\{a_{Cl^-}\} \qquad (6.36)$$

where the $E^{\ominus}$ value for the mercury–calomel electrode is given by

$$E^{\ominus}_{Hg/Hg_2Cl_2} = E^{\ominus}_{Hg_2^{2+}/Hg} + \frac{RT}{2F} \ln K_{Hg_2Cl_2} \quad (6.37)$$

analogous to Eq. 6.32.

The silver–silver chloride and mercury–calomel electrodes are commonly used as reference electrodes because their e.m.f.'s are stable and precisely known. In the 'saturated calomel electrode' the activity of chloride ion is fixed by the solubility product of KCl, the solution being saturated with both calomel and potassium chloride, and the e.m.f. at 298 K is thereby fixed at 0.2415 V. A 'normal calomel electrode' contains KCl at a concentration of 1.0 M and has an e.m.f. of 0.2802 V.

(c) *The glass electrode.* The glass electrode (Fig. 6.6) depends on the development of a potential difference across a thin glass membrane as a result of diffusion of hydrogen ions through the membrane, in much the same way that a junction potential develops when ions diffuse across the boundary between two electrolyte solutions. Diffusion continues until the potential drop developed across the membrane is sufficient to counterbalance the concentration gradient and prevent further diffusion. The glass electrode is an ion-selective electrode, i.e. it responds to one particular type of ion in a mixed electrolyte solution. In its usual form it responds to hydrogen ions, and in this form is the basis of most commercial pH meters. Membranes have been produced which are permeable to other ions, notably sodium and calcium, and selective electrodes which operate on the same principle are now available for a growing range of ions, including the fluoride ion. (Silver fluoride is too soluble for a silver–silver fluoride electrode to be useful at low fluoride concentrations.)

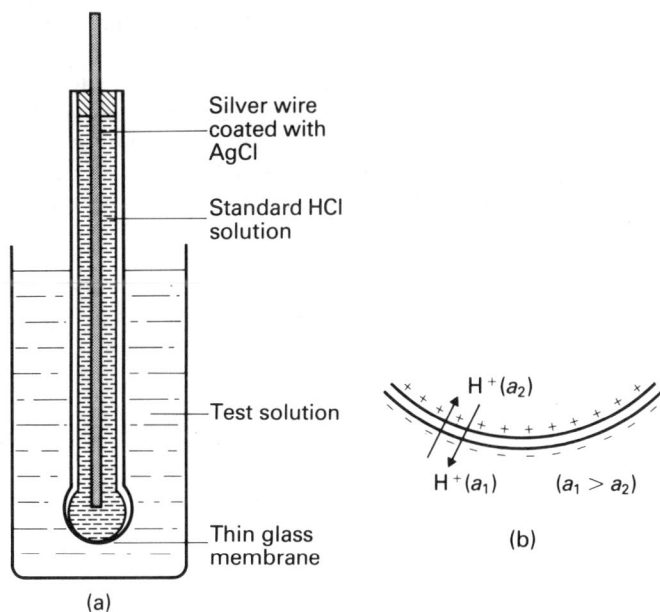

(a)

(b)

Fig. 6.6 (a) Construction of a glass electrode sensitive to hydrogen ions. (b) Diffusion across the membrane continues until the membrane's polarity is sufficient to counterbalance the concentration gradient

For an electrode made of a glass that is permeable only to hydrogen ions the half-reaction is

$$H^+(a_1) \rightarrow H^+(a_2) \qquad (6.38)$$

where a_1 and a_2 are the hydrogen ion activities on the two

sides of the membrane. Hence the Nernst equation for the electrode is

$$E_{\text{glass}} = E_{\text{glass}}^{\ominus} - \frac{RT}{F} \ln\{a_2/a_1\} \qquad (6.39)$$

The term $E_{\text{glass}}^{\ominus}$ must vanish because there is no e.m.f. when the two activities are equal. If the activity a_2 inside the electrode is taken as unity, Eq. 6.39 therefore becomes

$$E_{\text{glass}} = -\frac{2.303RT}{F} \text{ pH} \qquad (6.40)$$

where we have used the definition of pH as $-\log_{10}\{a_{H^+}\}$. In practice the e.m.f. of the glass electrode is usually measured relative to a calomel electrode containing KCl, contact between the KCl solution and the solution under test being made through a porous glass disc. The solution inside the glass electrode is usually hydrochloric acid, and contact with this solution is conveniently made with a silver wire coated with solid silver chloride, i.e. a silver–silver chloride electrode. The e.m.f.'s of the silver–silver chloride and mercury–calomel electrodes are subtracted from the observed e.m.f. and the resulting voltage is converted to a reading on a meter graduated in pH units. Since $2.303RT/F$ is 59 mV at 298 K, Eq. 6.40 shows that the output of the glass electrode changes by 59 mV per unit of pH change at this temperature.

6.5 The Measurement of Cell e.m.f.

In the last section we noted that the e.m.f. described by the Nernst equation for a cell is the e.m.f. measured under conditions where no work is done inside the cell itself. Every electrochemical cell has some internal resistance, so the requirement of zero internal work means in effect that no current must be drawn from the cell during the e.m.f. measurement. An ordinary voltmeter (Fig. 6.7) is actually a current meter in series with a high resistance, and a voltage measurement consists of determining the current which the voltage will drive through the high resistance. Such a device can be used to determine a cell e.m.f., but an extrapolation procedure must be used to obtain the value of the output voltage of the cell in the limit of zero current flow, and the precision of the final value obtained is poor. Precise measurements of cell e.m.f. can be made under conditions where no current is drawn from the cell by using a *potentiometer* (Fig. 6.8).

A potentiometer is said to be balanced when the potential of the external voltage source is exactly matched by the potential drop due to a steady current flowing down a portion of the instrument's uniform slide-wire. To find the position of balance, the moveable contact is slid along the slide-wire until no current flows through the sensitive galvanometer. The distance along the slide-wire then corresponds to the measured voltage. The slide-wire is calibrated, or 'standardized', by using a Weston standard cadmium cell (e.m.f. 1.0183 V) as the external voltage source. If the contact is moved slightly too far along the wire, a small current will be driven backwards through the cell; if the contact is not slid quite far enough the cell will deliver a small current. The balance point corresponds to a position of equilibrium, and the measurement is therefore reversible in the thermodynamic sense that was discussed in Section 4.4(b). The reversible e.m.f. is also the maximum e.m.f., and corresponds to ΔG in the manner specified by Eqs 6.18 and 6.19.

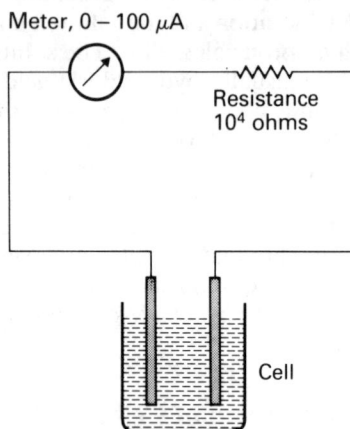

Fig. 6.7 Measurement of cell e.m.f. with a voltmeter. An e.m.f. of one volt will give a current of 100 μA through the resistance of 10^4 ohms, if the internal resistance of the cell is negligible

If the cell whose e.m.f. is to be determined has a very large internal resistance, 10^7 ohms or more, a simple potentiometer set up as in Fig. 6.8 will not be capable of determining the e.m.f. with satisfactory precision. This is because the out-of-balance current which is detected by the galvanometer has to flow through the cell's resistance, and to measure an e.m.f. with a precision of 1 mV in a circuit containing a resistance of ten megohms one would require a galvanometer sensitive to 0.1 nA (10^{-10} amperes). This is beyond the capability of most moving-coil galvanometers, yet for many purposes a precision of 1mV would be barely acceptable. The internal resistance of a glass electrode commonly exceeds 10^7 Ω, so some other means of measuring the e.m.f. must be employed. An electronic circuit which will measure a voltage without drawing appreciable current, or measure current in a circuit containing a very large resistance, is termed an electrometer. More or less expensive electrometers are available which can detect a current down to 10^{-17} amperes (about 60 electrons per second) in

high resistance circuits. For measurements with glass electrodes the sensitivity can safely be many orders of magnitude less than this. To extend the range of usefulness of a potentiometer in the laboratory one could simply replace the galvanometer of Fig. 6.8 by an electrometer. In commercial pH meters no potentiometer balance occurs; instead the voltage output of the electrometer circuit appears directly on a meter calibrated in pH units.

In some practical situations the activity of hydrogen ions either is not well-defined, or varies from one part of a system to another because the system is not truly at equilibrium. Examples of this type of system would be a slurry of soil in water, or the contents of an animal's stomach. Despite the absence of a single, definite value of hydrogen ion activity, the concept of pH can still be useful in such situations, and the results of a pH measurement may convey important information about properties of the system —the fertility of the soil, or the health of the animal. Since the usual definition of pH as $-\log_{10}(a_{H^+})$ no longer applies in such systems, we fall back on what is known as an *operational definition* of pH, namely: pH is the quantity that is measured with a pH meter. The pH value is then simply the reading on the pH meter (which is assumed to be in working order). This procedure does not really evade the issue, because in principle one could always examine the kinetics of the processes which control the e.m.f. of a glass electrode in the system under examination, and so determine the origin of the pH reading in terms of hydrogen ions from different parts of the system.

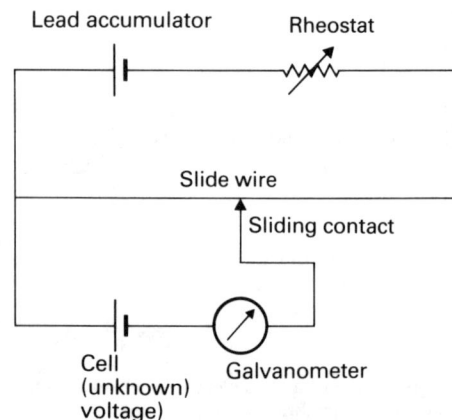

Fig. 6.8 A simple potentiometer. The rheostat is used to adjust the slide-wire current during standardization

Exercises

6.1 Obtain chemical equations and $E^{\ominus}$ values for reactions involving the following redox couples:
- (a) Cu^{2+}/Cu and Pb^{2+}/Pb
- (b) Cl_2/Cl^- and I_2/I^-
- (c) $MnO_2(s)/Mn^{2+}$ and Cl_2/Cl^-
- (d) MnO_4^-/Mn^{2+} and Cl_2/Cl^-

6.2 Give the conventional representations of cells which could be used to measure E values for the following reactions:
- (a) $Fe(s) + Pb^{2+} \rightarrow Fe^{2+} + Pb(s)$
- (b) $2Fe^{3+} + 2I^- \rightarrow 2Fe^{2+} + I_2(s)$
- (c) $2Ag(s) + Hg_2Cl_2(s) \rightarrow 2AgCl(s) + 2Hg(l)$
- (d) $2H^+ + Zn(s) \rightarrow Zn^{2+} + H_2(g)$

6.3 Write down the Nernst equation for each of the cells in exercise 6.2.

6.4 Work out the Nernst equations for the following half-cells:
- (a) $Pt/KMnO_4(m_1), H_2SO_4(m_2)/MnO_2(s)$
- (b) $Pb/PbCl_2(s)/HCl(m)$
- (c) $NH_3(P)/Cu/[Cu(NH_3)_4]SO_4(m_1), CuSO_4(m_2)$
- (d) $Pt/Quinhydrone(s)/HCl(m_1)$ where

quinhydrone is an organic compound which breaks down to give equal amounts of quinone and hydroquinone in solution, and quinone (Q) and hydroquinone (H_2Q) are interconvertible according to

$$Q + 2H^+ + 2e^- \rightarrow H_2Q, E^{\ominus}_{Q/H_2Q} = 0.699 \text{ V}.$$

Note: Hydroquinone is oxidized by air at pH > 7 so the quinhydrone electrode is not useful in alkaline solution.

6.5 Borrow a copy of the instruction manual for a laboratory pH meter and study its mode of operation in relation to the description given in Section 6.5.

6.6 Sketch a graph of the e.m.f. of a hydrogen electrode as a function of added alkali during the titrations of Fig. 5.19.

7 The Rates of Chemical Reactions

7.1 Introduction

For every process that occurs in nature there are two fundamental questions that can be asked. The first is: 'How far does it go?' and the second: 'How fast does it go?' We have already seen that the answer to the first question is provided by thermodynamics in terms of the value of the equilibrium constant, the entropy change, or the free energy change for the process in question. However, many chemical processes which are favoured by thermodynamics take place so slowly that for all practical purposes they can be neglected. Familiar examples include the conversion of diamond to graphite at ordinary temperatures, and the conversion of a mixture of hydrogen and oxygen to water in the absence of heat or ultraviolet radiation. If we are to apply our knowledge of chemistry to the solution of practical problems it is clear that we must know how to find the answer to *both* of the fundamental questions. Thus the experimental and theoretical study of the rates of chemical processes has considerable practical importance in addition to its intrinsic scientific interest. The name of this field of study is *Chemical Kinetics.*

Chemical reactions may be very simple or extremely complex, but in general every reaction can be regarded as taking place by way of a series of individual steps which are called *elementary reactions.* For example, the apparently simple process:

$$H_2 + Br_2 \rightarrow 2HBr \qquad (7.1)$$

occurs by way of the elementary reactions:

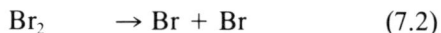

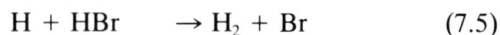

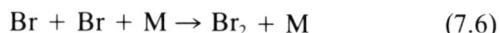

$$Br_2 \rightarrow Br + Br \qquad (7.2)$$

$$Br + H_2 \rightarrow HBr + H \qquad (7.3)$$

$$H + Br_2 \rightarrow HBr + Br \qquad (7.4)$$

$$H + HBr \rightarrow H_2 + Br \qquad (7.5)$$

$$Br + Br + M \rightarrow Br_2 + M \qquad (7.6)$$

where M stands for any stable molecule in the reaction system, which in this case means H_2, Br_2, or HBr. The series of elementary reactions which combine to produce the overall reaction is termed the *reaction mechanism.* Thus the steps 7.2 to 7.6 comprise the mechanism of reaction 7.1. One of the three main aims of chemical kinetics is the elucidation of reaction mechanisms, i.e. the experimental determination of the nature of the elementary reactions which are required to account for the complex reactions that we observe. The way that experimental data about reaction rates are used to obtain information about the reaction mechanisms is illustrated for some representative reactions in Section 7.4.

A second major objective of chemical kinetics is the understanding of the relationship between the structure of a molecule and the rates of its chemical reactions. This problem is still far from a complete solution, but some aspects of the problem, notably the effect of structure on the rates of certain reactions of organic molecules, are quite well understood. The relationship of structure to reactivity will be a prime concern in the later chapters devoted to organic chemistry.

The third main objective of chemical kinetics is simply the accumulation of more and more data about the rates and mechanisms of chemical reactions and the dependence of reaction rates on such variables as temperature, pressure, solvent composition, and the presence of catalysts. It is obvious that this kind of data is useful in practice when it is required to optimize the conditions for a particular reaction. In addition, with a sufficiently large body of data it becomes possible to observe trends and patterns of behaviour which provide both an inspiration and a proving ground for theories of chemical kinetics.

7.2 Experimental Rate Laws

Consider a reaction by which some species A is converted to products. The rate of the reaction can be expressed either as the rate at which the concentration of a specified product P increases, or as the rate at which the concentration of the reactant A decreases, whichever is more convenient. In mathematical form the expression for the reaction rate is

$$\text{Rate} = \frac{d[P]}{dt} = \frac{-d[A]}{dt} \qquad (7.7)$$

where the rates of removal of reactant and formation of product are linked through the chemical equation for the overall reaction.

Experimentally it is found that the rate of reaction depends on the *concentrations* (not the activities) of the different chemical species in the reaction system. For all elementary reactions, and for many complex reactions, the dependence takes the simple form

$$\text{Rate} = k[A]^a[B]^b[C]^c \ldots \qquad (7.8)$$

i.e. it is the product of a numerical constant k and the concentrations of the chemical species A, B, C, etc raised to the powers $a, b, c, \ldots$ respectively. The quantity k is called the *rate constant,* or sometimes the *specific rate,* of the reaction. It is equal to the rate of reaction when all the concentrations are unity. The value of the rate constant k varies with temperature but is independent of the concentrations of the species A, B, C etc. The quantity a, which is the power to which the concentration of species A is raised in the rate equation, is known as the *order of reaction with respect to A.* The order of reaction with respect to any species is usually an integer, and is seldom greater than two or less than zero. The sum of the orders with respect to the different chemical species, $(a + b + c + \ldots)$ is the overall *order of the reaction.* This too is usually a small whole number. Some typical examples of reactions with their rate equations are given in Table 7.1. A reaction for which the overall order is 0, 1, 2, or 3 is spoken of as a zero-order, first-order, second-order, or third-order reaction. One can also say, for example, that a reaction is first-order with respect to a particular reactant.

The usual procedure in investigating the kinetics of a reaction is to determine a number of values of the concentration of a reactant or product at different times after the reactants are mixed together. The resulting data can be plotted as in Fig. 7.1. The rate equation for the reaction is then an expression for the gradient of one or other of the curves of Fig. 7.1, which means that the curve itself can be

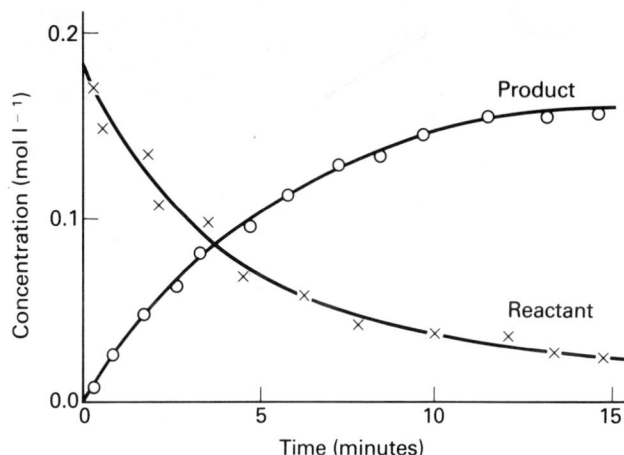

Fig. 7.1 Typical forms of experimental data obtained in a study of the rate of a chemical reaction

obtained by integrating the rate equation. For reactions of zero-, first-, or second-order this integration is quite straightforward, and for a third-order reaction it is merely tedious rather than difficult. The results of the integration for the more important cases are summarized in Table 7.2.

Table 7.2 Integrated rate equations. $[A]_0, [B]_0$ = concentrations of A and B when $t = 0$

Reaction	Order	Result of integration
A → products	0	$[A] = [A]_0 - kt$
A → products	1	$[A] = [A]_0\, e^{-kt}$ or $\ln([A]/[A]_0) = -kt$
A + A → products	2	$1/[A] - 1/[A]_0 = kt$
A + B → products	2	$\ln\{[A]_0[B]/[B]_0[A]\} = ([B]_0 - [A]_0)\, kt$
2A + B → products	3	$(2[B]_0 - [A]_0)^2 kt = \dfrac{(2[B]_0 - [A]_0)([A]_0 - [A])}{[A]_0[A]} + \ln\left\{\dfrac{[B]_0[A]}{[B][A]_0}\right\} \equiv f(A, B)$

One of the most important uses of the equations in the last column of Table 7.2 is in the determination of reaction orders. For each type of reaction there is a function of the

Table 7.1 Some examples of reactions with their rate equations

Reaction	Rate equation	Overall order
$2NH_3 \to N_2 + 3H_2$ (on a hot filament)	$d[N_2]/dt = k$	0
$N_2O_5 \to 2NO_2 + \frac{1}{2}O_2$	$d[N_2O_5]/dt = -k[N_2O_5]$	1
$ClO^- + Br^- \to BrO^- + Cl^-$	$d[ClO^-]/dt = -k[ClO^-][Br^-]$	2
$CH_3COCH_3 + I_2 \to CH_3COCH_2I + HI$	$d[I_2]/dt = -k[CH_3COCH_3][H^+]$	2
$2NO + O_2 \to 2NO_2$	$d[O_2]/dt = -k[NO]^2[O_2]$	3

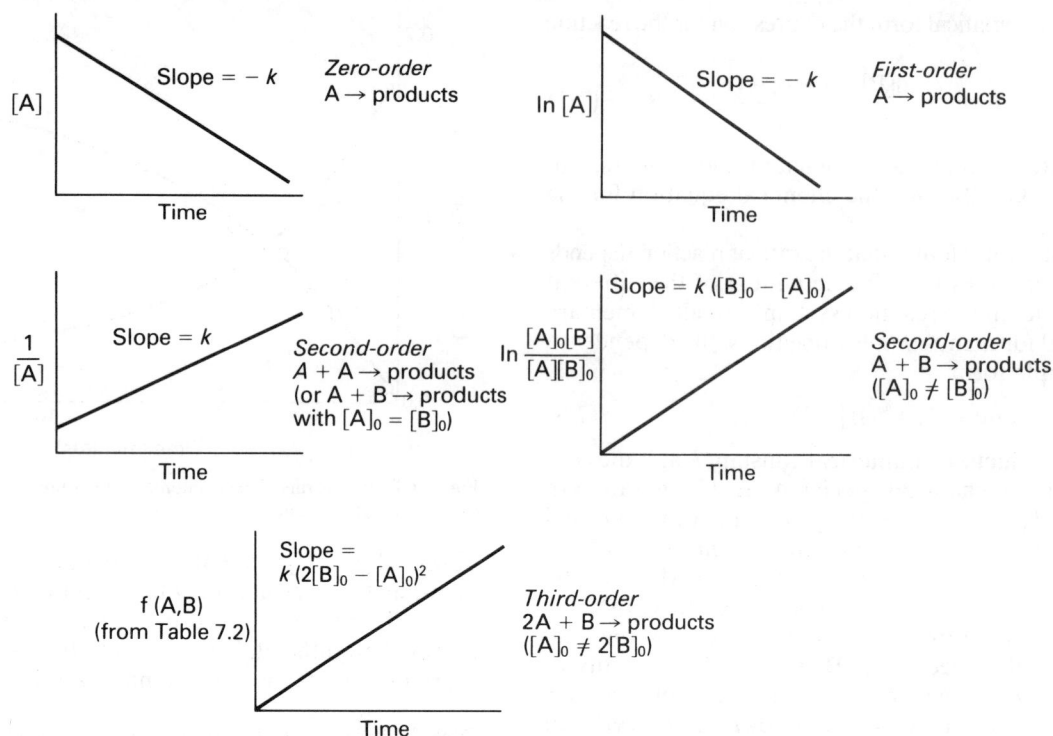

Fig. 7.2 Functions which must be plotted against time to give linear plots for different reaction orders

reactant concentrations which gives a straight line when it is plotted against time. This is shown in Fig. 7.2 for the cases listed in Table 7.2. Given experimental data such as that in Fig. 7.1, it is a fairly simple matter to plot the various functions shown in Fig. 7.2 and so determine whether the order of reaction is 0, 1, 2, or 3. The linear plot which reveals the reaction order also gives the value of the rate constant k, in terms of the slope of the straight line. Once the order of reaction has been found, the rate constant can be calculated either from the slope of the linear plot or by substituting values of concentration and time into the appropriate integrated rate equation of Table 7.2. There are several other ways of determining reaction orders and rate constants, but the graphical method based on the integrated rate equation is probably the most important and direct method.

Table 7.3 Units in which rate constants are expressed, when concentrations are given in moles per litre and time is given in seconds

Order of reaction	Units of k	
0	mole litre^{-1} second^{-1}	(mol l^{-1} s^{-1})
1	second^{-1}	(s^{-1})
2	mole^{-1} litre second^{-1}	(mol^{-1} l s^{-1})
3	mole^{-2} litre^2 second^{-1}	(mol^{-2} l^2 s^{-1})

The units in which the rate constants are expressed depend on the units of concentration and time that are used. For concentration in moles per litre and time in seconds the units are as given in Table 7.3. It is interesting to note that for a first-order reaction the dimensions of k are simply time^{-1}, so that $1/k$ must be a time which is characteristic of the reaction. It can be shown that in fact $1/k$ is the *mean lifetime*, usually written τ, of the species A before it decomposes to products. The quantity $(\ln 2)/k = 0.6931/k$, which is the time required for the concentration of A to decrease to half the initial value $[A]_0$, is known as the *half-life* of the species A and is usually written $t_{1/2}$. Radioactive nuclei decay according to first-order kinetics, the rate of decay of a radioactive sample being proportional to the quantity of radioactive material that is present at any instant. The stability, or lack of it, of a radioactive isotope is customarily specified by quoting the value of the isotope's half-life. It is important to realize that both the mean lifetime and the half-life are independent of the amount of material initially present in the case of a first-order process. For reactions which are not first-order the half-life does depend on the amount of material initially present, and the observed form of this dependence is sometimes used to establish the reaction order.

7.3 Elementary Reactions

Reactions 7.2–7.6 are examples of elementary reactions, whose cumulative effect is expressed in the chemical equation for the overall reaction, Eq. 7.1. For a reaction such as 7.1, which is not an elementary reaction, the chemical equation expresses the observed stoichiometry of the reaction on a macroscopic scale, but gives no information about events which occur on the scale of atoms and molecules. For an elementary reaction, on the other hand, the chemical equation corresponds exactly to what occurs at the atomic and molecular level. Eq. 7.2, for example, states that bromine atoms are generated as a result of a bromine molecule acquiring sufficient energy to split the Br—Br bond. (In practice this energy is acquired through collisions with other gas molecules.) Eq. 7.3 says that when a bromine atom collides with a hydrogen molecule a molecule of HBr can be formed and a hydrogen atom liberated. Eq. 7.6 says that when two bromine atoms collide in the presence of a third body M a bromine molecule can be formed; the third body is required to remove some of the energy which is released by molecule formation. (If this energy were not removed the newly formed bromine molecule would possess sufficient energy to decompose, and would immediately do so.) The elementary reactions 7.2–7.6 do not necessarily occur with high efficiency, in fact it is commonly necessary for the reactant molecules to collide many thousands or even millions of times before the elementary reaction takes place. The efficiency of the collision process is reflected in the value of the rate constant for the elementary reaction. Some very fast reactions are known which occur on essentially every collision between the reactants, but the majority of elementary reactions are much slower than this.

For a reaction which is *not* an elementary reaction the experimental rate equation may bear little or no relation to the overall stoichiometry. One example of this kind is the reaction of acetone with iodine, which is the fourth entry in Table 7.1. The rate of reaction is found by experiment to be proportional to the concentrations of acetone and hydrogen ions, and is independent of the iodine concentration. The hydrogen–bromine reaction 7.1 has the complicated rate law

$$\frac{d[HBr]}{dt} = \frac{k[H_2][Br_2]^{1/2}}{k' + [HBr]/[Br_2]} \qquad (7.9)$$

where k and k' are constants. This rate law, which was found experimentally by Bodenstein and Lind in 1906, can be deduced on the basis of the mechanism 7.2–7.6, as we shall see in the next section.

For an elementary reaction the rate law always corresponds exactly to the chemical equation. Thus reaction 7.2 is first-order, reactions 7.3–7.5 are second-order, and reaction

7.6 is third-order, the rate laws being as shown in Table 7.4. The number of reactant molecules which take part in a reaction is termed the *molecularity* of the reaction. A reaction such as 7.2, in which only one molecule is involved as a reactant, is termed a *unimolecular* reaction. Reactions 7.3–7.5, which involve two reactant molecules, are termed bimolecular. Similarly, reaction 7.6 is *termolecular*. Elementary reactions always have an overall reaction order equal to their molecularity. The molecularity of an elementary reaction is never greater than 3 or less than 1, because of the extreme improbability of a collision involving more than three molecules simultaneously, and because there has to be at least one molecule to take part in the reaction.

Table 7.4 Rate laws for reactions 7.2–7.6

Reaction	Rate law
$Br_2 \rightarrow Br + Br$	$-d[Br_2]/dt = k_{7.2}[Br_2]$
$Br + H_2 \rightarrow HBr + H$	$-d[Br]/dt = k_{7.3}[Br][H_2]$
$H + Br_2 \rightarrow HBr + Br$	$-d[H]/dt = k_{7.4}[H][Br_2]$
$H + HBr \rightarrow H_2 + Br$	$-d[H]/dt = k_{7.5}[H][HBr]$
$Br + Br + M \rightarrow Br_2 + M$	$-d[Br]/dt = 2k_{7.6}[Br]^2[M]$

Once a mechanism has been deduced which is able to account for the observed rate law for a complex reaction, as reactions 7.2–7.6 can account for the rate law 7.9, the next step is to determine the rates of the elementary reactions involved in order to test the ability of the mechanism to predict the actual rate of the overall process. The determination of the rates of elementary reactions is one of the main preoccupations of current work in chemical kinetics, and many rate constant values are listed in the chemical literature. The most obvious value of such data resides in the fact that a particular elementary reaction can be involved in any number of complex reactions, and the rate of an elementary reaction is independent of the system in which it occurs. To give just one example, the rate constant for the fast reaction of an oxygen atom with a molecule of ozone

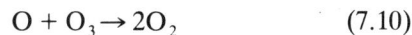

$$O + O_3 \rightarrow 2O_2 \qquad (7.10)$$

has been determined in the laboratory, and the value so obtained has been used in the interpretation of the thermal decomposition of ozone, and in accounting for the chemical behaviour of ozone in photochemical smog, in the earth's stratosphere, and in the atmosphere of Mars.

7.4 Kinetics and Mechanism

In this section we consider a few examples of how reaction mechanisms for complex reactions can be deduced from experimental data. There are two basic conceptual tools which are very useful in making such deductions: the first is the idea of the rate-determining step, and the second is the steady-state hypothesis.

The *rate-determining* step is the elementary reaction whose speed limits the overall reaction rate. In effect it is a bottleneck through which the reactants all have to squeeze on their way to become products. Usually, if a single rate-determining step exists, it is the slowest of the elementary reactions which lead to formation of products; alternatively, a relatively fast process may be rate-determining because it is in competition with an even faster process which operates to reverse the overall reaction. Often the nature of the rate-determining step can be deduced from the form of the experimental rate law. For example, for the reactions

$$N_2O_5 \rightarrow 2NO_2 + \tfrac{1}{2}O_2 \qquad (7.11)$$

and $$CH_3COCH_3 + I_2 \rightarrow CH_3COCH_2I + HI \qquad (7.12)$$

the rate laws which are given in Table 7.1 suggest that for reaction 7.11 the rate-determining step is the unimolecular decomposition of N_2O_5:

$$N_2O_5 \rightarrow products \qquad (7.13)$$

while for reaction 7.12 the rate-determining step may be

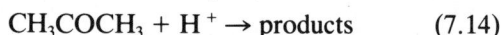

$$CH_3COCH_3 + H^+ \rightarrow products \qquad (7.14)$$

For the hydrogen–bromine reaction, 7.1, the rate law given in Eq. 7.9 suggests that the rate-determining step probably involves the reaction of H_2 with a species whose concentration is proportional to the square root of the Br_2 concentration, and the best choice for such a reaction would appear to be reaction 7.3. The presence of the HBr concentration in the denominator of Eq. 7.9 shows that the product HBr acts as an *inhibitor*, i.e. the opposite of a catalyst, for the reaction. Reaction 7.5 in the proposed mechanism shows one way in which HBr could have this effect, by bringing about the reverse of the reaction that we have identified as the rate-determining step.

The *steady-state hypothesis* is an assumption which is made about the concentrations of highly reactive intermediates in complex reactions. Examples of such intermediates would be the hydrogen and bromine atoms which are postulated as taking part in reaction 7.1. Such highly reactive species are present in the reaction system only at extremely low concentration, because the reactions removing them are invariably very fast relative to the reactions by which they are produced. As soon as a molecule of the intermediate is formed it becomes liable to removal by some fast process. Thus the net rate of production of the intermediate, given by the difference between its rates of formation and removal, is extremely small. The steady-state hypothesis is the assumption that *the rate of change of concentration of a reactive intermediate is zero*. When this is a valid assumption we shall find that it yields an expression for the concentration of the reactive intermediate in terms of the concentrations of stable molecules and the values of various rate constants. The assumption is only likely to be invalid if the reactivity of the intermediate species is not very great, so that its steady-state concentration is large. In this situation the reaction of interest may be over before the intermediate actually attains its steady-state concentration.

Before the steady-state hypothesis can be used with a particular reaction it is necessary to know the identities of the reactive intermediates that are likely to be present, i.e. it is necessary to know something about the general chemistry of the reactants and products. For reaction 7.11, the decomposition of N_2O_5, there are two useful pieces of supplementary information available. The first is that the spectrum of the radical NO_3 has been observed in studies of the absorption of light by decomposing nitrogen pentoxide. The second is that the decomposition of N_2O_5 is catalysed by nitric oxide, i.e. addition of NO speeds up the decomposition. This evidence suggests that NO_3 is a prime candidate for inclusion as a reactive intermediate in the mechanism of N_2O_5 decomposition, and that reaction with NO may be one of the steps by which NO_3 is removed. The most obvious place to introduce NO_3 would appear to be Eq. 7.13, which becomes

$$N_2O_5 \rightarrow NO_2 + NO_3 \qquad (7.15)$$

The reaction of NO with NO_3 is expected to be

$$NO + NO_3 \rightarrow 2NO_2 \qquad (7.16)$$

We still do not have a complete mechanism for the N_2O_5 decomposition, and at this point it is necessary for one's creative instincts to come into play in order to find the minimum set of reactions which will account for all the experimental observations. The mechanism which was proposed by Ogg in 1947 includes the two additional steps

$$NO_2 + NO_3 \rightarrow N_2O_5 \qquad (7.17)$$

and $$NO_2 + NO_3 \rightarrow NO_2 + NO + O_2 \qquad (7.18)$$

where reaction 7.18 is postulated to be much slower than 7.17. At the time it was quite a novel idea that a single pair of reactants could give rise to two different sets of products, as postulated for NO_2 and NO_3 in the last two reactions.

Applying the steady-state hypothesis to NO_3 and NO, we now obtain

$$\frac{d[NO_3]}{dt} = k_{7.15}[N_2O_5] - (k_{7.17} + k_{7.18})[NO_2][NO_3] -$$

$$k_{7.16}[NO][NO_3] = 0 \qquad (7.19)$$

and $\quad \dfrac{d[NO]}{dt} = k_{7.18}[NO_2][NO_3] - k_{7.16}[NO][NO_3] = 0$ (7.20)

Eq. 7.20 now gives the steady-state concentration of NO as

$$[NO] = k_{7.18}[NO_2]/k_{7.16} \qquad (7.21)$$

When this expression for the NO concentration is substituted into Eq. 7.19 the steady-state NO$_3$ concentration becomes

$$[NO_3] = k_{7.15}[N_2O_5]/(k_{7.17} + 2k_{7.18})[NO_2] \quad (7.22)$$

Once the steady-state concentrations of the reactive intermediates are known it is generally a simple matter to calculate an expression for the rate of overall reaction. For the N_2O_5 decomposition, the total rate of removal of N_2O_5 is equal to *twice* the rate of reaction 7.18. This is because reaction 7.18 removes one molecule of NO$_3$ itself, and produces a molecule of NO which immediately removes a second NO$_3$ molecule by reaction 7.16. Every NO$_3$ molecule that is removed in this way corresponds to the irreversible loss of one N_2O_5 molecule by reaction 7.15, since the regeneration of N_2O_5 by reaction 7.17 is no longer possible. Hence

$$- d\,[N_2O_5]/dt = 2k_{7.18}[NO_2][NO_3]$$
$$= 2k_{7.15}k_{7.18}[N_2O_5]/(k_{7.17} + 2k_{7.18}) \quad (7.23)$$

Because reaction 7.17 is postulated to be much faster than reaction 7.18 this reduces to

$$- d[N_2O_5]/dt = 2k_{7.15}k_{7.18}[N_2O_5]/k_{7.17} \qquad (7.24)$$

which is of the same form as the rate equation given in Table 7.1. Additional experimental evidence, relating the temperature dependence of the overall reaction rate to the temperature dependence of the elementary reactions 7.15, 7.18 and 7.17, has confirmed the accuracy of Eq. 7.24.

For the *reaction of iodine with acetone* additional evidence about the mechanism is available in the form of the information that ketones such as acetone exist as an equilibrium mixture of two molecules, one of which is known as the *keto* form and the other as the *enol* form of the ketone. The enol form, which at equilibrium is present to the extent of a small fraction of one percent of the total amount of acetone, contains a carbon–carbon double bond, which makes it very reactive towards halogens such as bromine or iodine. In this connection it is interesting to note that the rate of the acid-catalysed reaction of acetone with iodine is identical with the rate of the analogous reaction with bromine. Therefore the rate-determining step involves the formation of a species which reacts rapidly with halogens. The interconversion of the keto and enol forms is known to be catalysed

by both acids and alkalis, and the same is true of the iodination of acetone.

We now have, in addition to the probably rate-determining step 7.14 for the acid-catalysed reaction, a prime candidate for reactive intermediate in the enol form of the acetone molecule. The steps involved in enol formation are as follows:

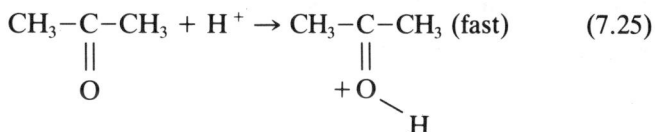

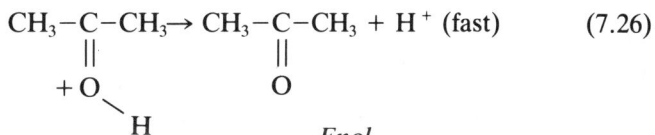

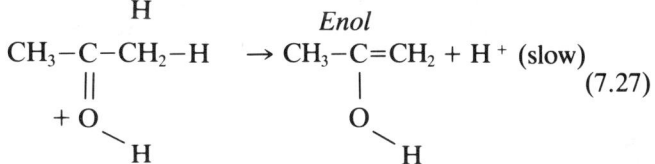

Here $+$ represents a formal unit positive charge on the O atom.

When no halogen is present the reverse of reaction 7.27 occurs to maintain the concentration of the enol form at its low, equilibrium value. The overall reaction leading to formation of enol can be written

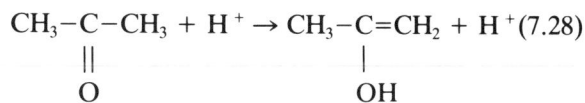

which shows that the H$^+$ ion is a catalyst, taking part in the reaction but not being consumed. Eq. 7.28 can now be identified with the rate-determining step 7.14, but it is apparent, since reaction 7.28 is itself complex, that the true rate-determining step is reaction 7.27. Because the reactions producing and removing it are very fast, the protonated acetone produced by reaction 7.25 is *in equilibrium* with acetone and H$^+$. In the presence of iodine the enol is removed by

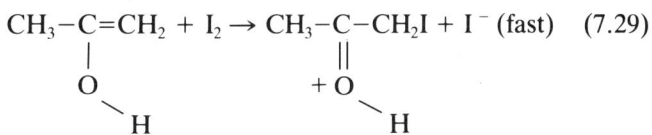

which is immediately followed by

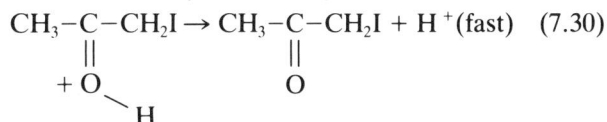

Two further points can be noted here. The first is that the gain or loss of a proton from a lone pair of electrons on oxygen or nitrogen, as in Eqs 7.25, 7.26 and 7.30, is invariably very fast in an ionizing solvent. The second point is that the final step of the overall reaction, 7.30, produces a hydrogen ion, so the reaction is in fact *autocatalytic*.

The steady-state equation which governs the concentration of the enol is

$$\frac{d[Enol]}{dt} = k_{7.27}[CH_3C(OH)^+CH_3] - k_{7.29}[Enol][I_2] = 0 \tag{7.31}$$

where the equilibrium concentration of protonated acetone is given by

$$[CH_3C(OH)^+CH_3] = k_{7.25}[CH_3COCH_3][H^+]/k_{7.26} \tag{7.32}$$

Inserting this into Eq. 7.31, we obtain the expression for the steady-state concentration of enol in the presence of I_2 as

$$[Enol] = k_{7.25}k_{7.27}[CH_3COCH_3][H^+]/k_{7.26}k_{7.29}[I_2] \tag{7.33}$$

The overall rate of the reaction is equal to the rate of removal of I_2, which according to Eq. 7.29 is

$$-d[I_2]/dt = k_{7.29}[Enol][I_2] \tag{7.34}$$

When the steady-state value of the enol concentration is inserted into Eq. 7.34 we obtain the final rate equation

$$-d[I_2]/dt = k_{7.25}k_{7.27}[CH_3COCH_3][H^+]/k_{7.26} \tag{7.35}$$

If we substitute the equilibrium constant K for formation of the protonated acetone in place of the ratio $k_{7.25}/k_{7.26}$ this simplifies to

$$-d[I_2]/dt = k_{7.27}K[CH_3COCH_3][H^+] \tag{7.36}$$

The last two equations are of the same form as the rate law in Table 7.1.

When acetone and acid are present in a large excess over I_2 the concentrations of acetone and acid remain essentially constant with time so that the reaction gives the appearance of being zero-order: the rate of consumption of iodine is constant throughout the reaction. Some typical results of a student experiment designed to test this point are shown in Fig. 7.3. Whenever one of the reactants which appears in the rate equation is present in such a large excess that its concentration remains virtually constant during the course of the reaction, the reaction order appears to be lower than its true value. We then say, for example, that if one of the reactants in a second-order reaction is present in large excess, that the reaction is pseudo first-order. Under the conditions of Fig. 7.3 the iodine–acetone reaction is a pseudo zero-order reaction. It is usually helpful to reduce the apparent order of reaction in this way, because it simplifies the problem of interpreting the kinetic data.

For the *hydrogen–bromine reaction* the mechanism 7.2–7.6 was proposed independently by several workers thirteen years after the rate law given in Eq. 7.9 had been discovered by Bodenstein and Lind. This was among the first reactions for which such a mechanism was proposed, and in this case the evidence for the presence of hydrogen and bromine atoms as intermediates was initially provided by the agreement between the experimental and theoretical rate laws. H and Br were not observed directly.

The steady-state equation for bromine atoms is

$$d[Br]/dt = 2k_{7.2}[Br_2] - k_{7.3}[Br][H_2] + k_{7.4}[H][Br_2] + k_{7.5}[H][HBr] - 2k_{7.6}[Br]^2[M] = 0 \tag{7.37}$$

and for hydrogen atoms

$$d[H]/dt = k_{7.3}[Br][H_2] - k_{7.4}[H][Br_2] - k_{7.5}[H][HBr] = 0 \tag{7.38}$$

When Eq. 7.38 is inserted into Eq. 7.37 the steady-state expression for bromine atoms simplifies to

$$[Br] = (k_{7.2}[Br_2]/k_{7.6}[M])^{1/2} \tag{7.39}$$

If this expression is used to substitute for $[Br]$ in Eq. 7.38 the steady-state concentration of H atoms is found to be

$$[H] = \frac{k_{7.3}[H_2](k_{7.2}[Br_2]/k_{7.6}[M])^{1/2}}{k_{7.4}[Br_2] + k_{7.5}[Br]} \tag{7.40}$$

The rate of formation of HBr is given by

$$d[HBr]/dt = k_{7.3}[Br][H_2] + k_{7.4}[H][Br_2] - k_{7.5}[H][HBr] \tag{7.41}$$

We can use Eq. 7.38 to simplify this to

$$d[HBr]/dt = 2k_{7.4}[H][Br_2] \tag{7.42}$$

and when the steady-state concentration of H atoms is inserted this becomes

$$d[HBr]/dt = k[H_2][Br_2]^{1/2}/(k' + [HBr]/[Br_2]) \tag{7.9}$$

where

$$k = 2k_{7.3}(k_{7.2}/k_{7.6}[M])^{1/2}(k_{7.4}/k_{7.5}) \tag{7.43}$$

and

$$k' = k_{7.4}/k_{7.5} \tag{7.44}$$

An interesting feature of this mechanism is that if reaction 7.3 were very fast, the steady-state hydrogen atom concentration given by Eq. 7.40 would become very large, and it would not be valid to use the steady-state assumption for H atoms. For bromine, reaction 7.3 is in fact quite slow at temperatures of a few hundred degrees Celsius and the most probable fate of a bromine atom is to undergo termolecular recombination according to reaction 7.6. With chlorine, on the other hand, the reaction analogous to 7.3;

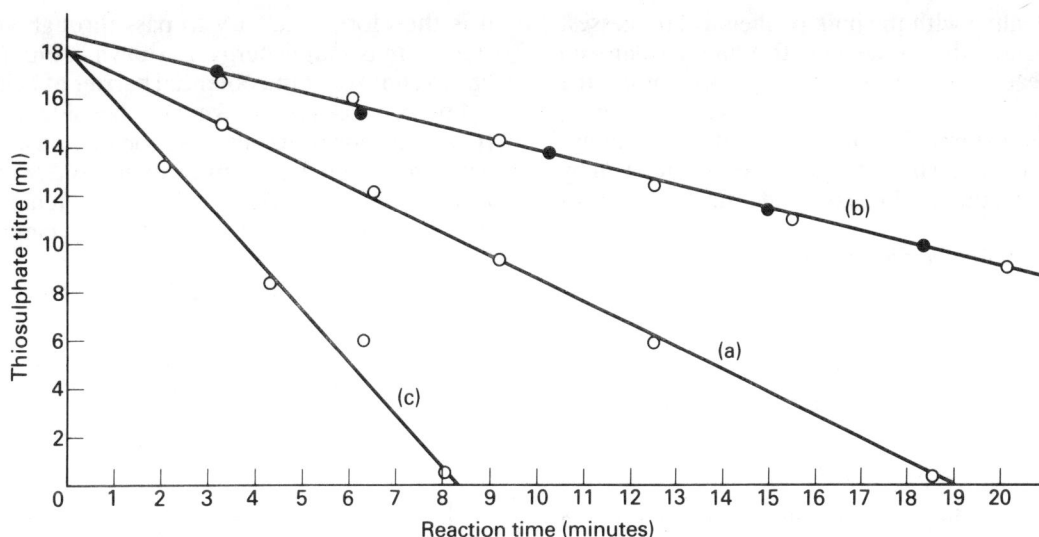

Fig. 7.3 Results of a student's experiments on the reaction of iodine with acetone, with acetone and hydrogen ions present in large excess. (a) First experiments at 25°C. (b) Two experiments at 25°C, one with the acetone concentration halved, the other with the acid concentration halved. (c) As in (a) but with the temperature raised to 40°C. (Reaction stopped by adding excess sodium acetate. I_2 concentrations measured by titration with thiosuphate solution.)

$$Cl + H_2 \rightarrow HCl + H \qquad (7.45)$$

is very fast at moderately elevated temperatures, and together with the reaction analogous to 7.4:

$$H + Cl_2 \rightarrow HCl + Cl \qquad (7.46)$$

it forms a fast, repetitive cycle, or *chain reaction,* which can be repeated many thousands of times before the halogen atom is finally lost by the termolecular recombination process analogous to 7.6:

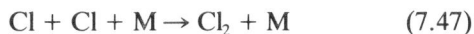

$$Cl + Cl + M \rightarrow Cl_2 + M \qquad (7.47)$$

The heat released in this chain reaction tends to raise the temperature of the system, which produces an acceleration of reaction 7.45, and can ultimately result in a 'thermal' explosion.

Chain reactions are of great practical importance in flames and explosions, and also in the polymerization reactions by which molecules such as ethylene*, C_2H_4, are joined in large numbers to form the highly useful, macromolecular substances known as plastics. The reaction of hydrogen with oxygen includes in its mechanism the step

$$O + H_2 \rightarrow OH + H \qquad (7.48)$$

by which the reactive intermediate atomic oxygen is converted to two reactive intermediates, a hydroxyl radical and a hydrogen atom. Such a reaction, which increases the total population of reactive intermediates, is known as a branching reaction, and the overall process is termed a

branched chain reaction. Reactions such as 7.46, in which the concentration of intermediates stays constant, are termed propagation reactions. The presence of branching reactions allows the radical and atom concentrations to build up very rapidly in the hydrogen–oxygen system, to the point at which an explosion can occur without the necessity of a large preliminary heat release. This type of explosion is known as an isothermal explosion.

7.5 Theories of Chemical Kinetics

Theories of chemical kinetics are aimed at understanding and predicting the rates of elementary reactions. Unimolecular reactions, and to a lesser degree termolecular reactions, can be understood as the outcome of successive bimolecular reactions; therefore in this section we shall concentrate our attention on bimolecular elementary reactions. The effect on a reaction rate of transferring the reactants from the gas phase to a solvent medium is always interesting, but in general the effect is not large unless the solvent is sufficiently viscous to interfere with the ability of the reactant molecules to approach one another. Therefore we shall further restrict our consideration to bimolecular elementary reactions between gaseous molecules.

There are two main theories of chemical kinetics, the first being known as the collision theory and the second as either the transition-state theory or the activated complex theory. The collision theory is the less sophisticated of the two, but despite its simplicity it provides concepts which

---

* Systematic name: ethene. The resulting polymer is polythene.

are useful for dealing with the bulk of chemical processes. The transition-state theory deals with unimolecular and termolecular reactions in a way that is superior to the collision theory approach, but for the case we are considering, that of a simple bimolecular reaction in the gas phase, it yields a result which is equivalent to that given by the simple collision theory. Let us consider the two theories in turn.

The collision theory begins by asserting that the rate of the reaction

$$A + B \rightarrow products \qquad (7.49)$$

is proportional to Z_{AB}, the frequency of collisions between the species A and B. From the kinetic theory of gases Z_{AB} is given by

$$Z_{AB} = n_A n_B \sigma_{AB} (8RT/\pi\mu)^{1/2} \qquad (7.50)$$

where Z_{AB} is the number of A–B collisions in unit volume per unit time, n_A and n_B are the numbers of A and B molecules, respectively, in unit volume, and σ_{AB} is the *collision cross-section* for the molecules A and B, a quantity which can be obtained from measurements of the transport properties of a mixture of A and B. If A and B have radii r_A and r_B, the cross-section σ_{AB} is $\pi(r_A + r_B)^2$. The symbols R, T and π have their usual significance, and μ is the 'reduced mass' of A and B, defined by

$$1/\mu = 1/M_A + 1/M_B \qquad (7.51)$$

where M_A and M_B are the relative molecular masses of A and B, respectively. If the reaction occurred on every collision its rate would be equal to Z_{AB}, so that the quantity Z_{AB} effectively sets an upper limit to the rate of a bimolecular reaction, such that $k_{max} \sim 10^{11} \, mol^{-1} \, s^{-1}$.

One of the factors which may prevent a reaction from occurring on every collision is the need for the molecules to collide with a particular orientation in order that certain chemical bonds may be broken and new ones formed. For example, the reaction of H with HBr, given by Eq. 7.5, may require the hydrogen atom to collide with the hydrogen end of the HBr molecule so that the H—H bond can be formed while the H—Br bond is being broken. To allow for this effect the collision frequency Z_{AB} is multiplied by a 'probability factor' P, which typically has a value between 1 and 1/1000. The probability factor is essentially the probability that the molecules will collide with the correct orientation for reaction to occur.

The second factor that limits the rate of reaction is the need for the colliding molecules to possess sufficient energy for the rearrangement of chemical bonds to be possible. The reactants and products each correspond to states of minimum energy in the system of atoms of which their molecules are composed. To get from reactants to products

it is therefore necessary to pass through an intermediate state of maximum energy, as shown in Fig. 7.4. The system has to climb over the potential barrier of height ΔH^* in Fig. 7.4 before it can roll down the other side to form products. It can be shown that the fraction of collisions which occur with energy greater than or equal to ΔH^* is given by the *Boltzmann factor* $\exp(-\Delta H^*/RT)$. Hence the final collision theory expression for the rate of the bimolecular reaction is

$$Rate = PZ_{AB} e^{-\Delta H^*/RT} \qquad (7.52)$$

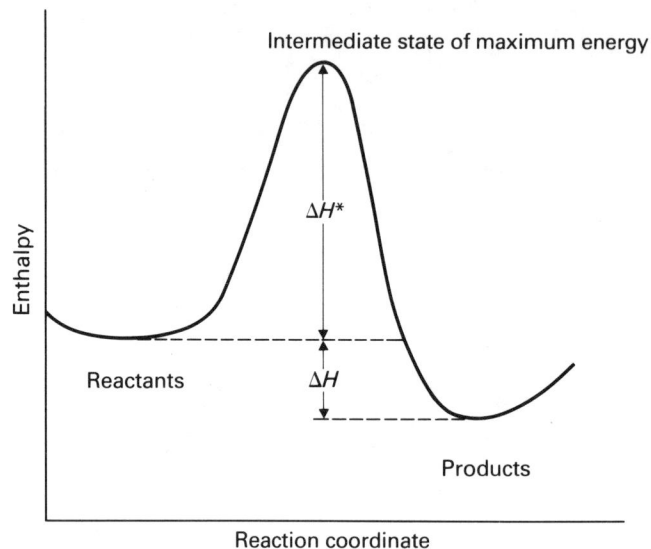

Fig. 7.4 Energy diagram for a chemical reaction. The reaction coordinate is some convenient parameter for measuring the progress of the reaction, e.g. the H—Br distance in reaction 7.5. ΔH is the enthalpy of reaction and ΔH^* is the enthalpy of activation

The quantity ΔH^* is known as the *enthalpy of activation*. In most instances there is no significant difference between ΔH^* and ΔU^*, the energy of activation.

The transition-state theory focusses attention on the intermediate state of maximum energy, and identifies this state as the *transition-state*, or *activated complex*. The transition-state, which we shall write as AB*, is assumed to be *in equilibrium* with the reactants A and B.

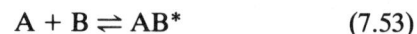

$$A + B \rightleftharpoons AB^* \qquad (7.53)$$

with an equilibrium constant K^*:

$$n_{AB^*} = K^* n_A n_B \qquad (7.54)$$

The activated complex is assumed to dissociate according to

$$AB^* \rightarrow products \qquad (7.55)$$

In advanced treatments of the theory it is shown that the the rate constant of reaction 7.55 is given by

$$k_{7.55} = \kappa \frac{kT}{h} \qquad (7.56)$$

where k is Boltzmann's constant (the gas constant R divided by Avogadro's number N_0), T is the temperature and h is Planck's constant. The 'transmission coefficient' κ allows for the possibility that there may be more than one reaction channel available, i.e. more than one set of possible decomposition products for the activated complex when the potential barrier has been crossed. In a real molecular system the energy does not depend on a single coordinate, in the manner shown in Fig. 7.4, but is a complicated function of the coordinates of all of the atoms in the molecules A and B, and the route over the potential barrier corresponds to a difficult passage across a mountain range in more than three dimensions. The calculation from first principles of the height and shape of the potential barrier for a bimolecular reaction is a very formidable problem, one which has not so far been solved except in a few simple cases.

The overall rate of the reaction is equal to the rate of reaction 7.55, where the concentration of AB* is given by Eq. 7.54. Hence

$$\text{Rate} = \kappa \frac{kT}{h} K^* n_A n_B \qquad (7.57)$$

The equilibrium constant K^* is related to a standard Gibbs Free Energy of activation ΔG^* by the usual expression

$$\Delta G^* = -RT \ln K^* \qquad (7.58)$$

We now use the definition of Gibbs Free Energy to write

$$\Delta G^* = \Delta H^* - T\Delta S^* \qquad (7.59)$$

so that K^* can be written as

$$K^* = e^{\Delta S^*/R} e^{-\Delta H^*/RT} \qquad (7.60)$$

Putting this into Eq. 7.57, we obtain as our final transition-state expression for the reaction rate

$$\text{Rate} = \kappa \frac{kT}{h} e^{\Delta S^*/R} e^{-\Delta H^*/RT n_A n_B} \qquad (7.61)$$

The quantity ΔS^* is the *entropy of activation,* which can be related to the relative probability of the states of the separate reactants and the transition-state. Clearly the factor involving ΔS^* has much in common with the probability factor P of collision theory.

Comparing the meanings of the terms in Eq. 7.52 and 7.61, we find that the terms involving the enthalpy of activa-

tion are identical, and that the probability factor P can be identified reasonably well with the product of κ and $\exp(\Delta S^*/R)$. The remaining factor $(kT/h)n_A n_B$ must therefore play approximately the same role as Z_{AB}. In fact it can be shown that a fully detailed treatment of the transition-state theory for a bimolecular reaction between two atoms A and B gives a result which is identical with the collision theory expression.

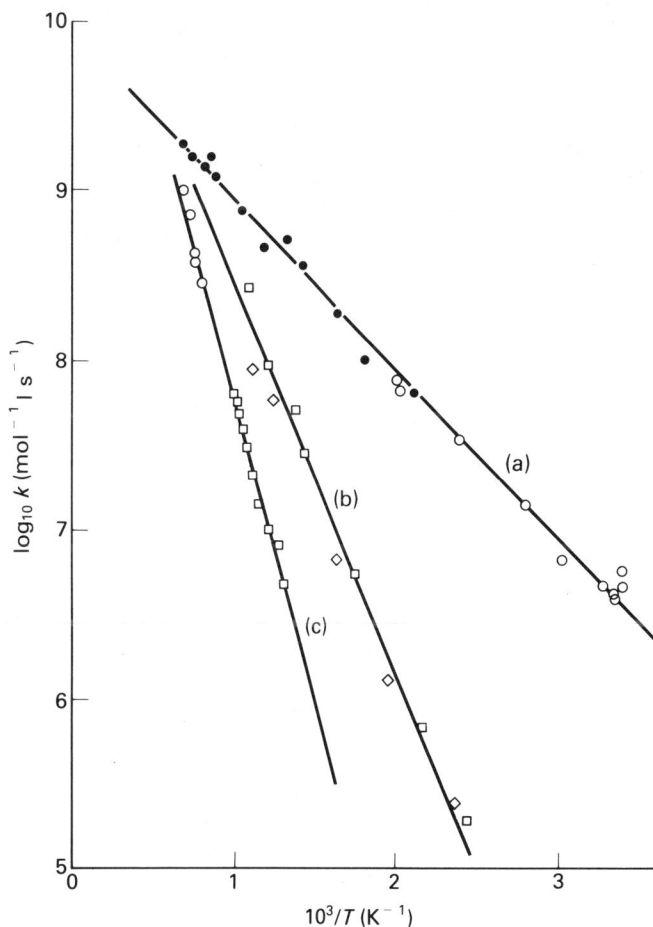

Fig. 7.5 Variation of rate constants for bimolecular reactions with temperature. (a) $H_2 + OH \rightarrow H_2O + H$, (b) $H_2 + O \rightarrow OH + H$, (c) $H + O_2 \rightarrow OH + O$. For each reaction, data from two separate recent papers are plotted

Eqs 7.52 and 7.61 agree in predicting that when the natural logarithm of a rate constant is plotted against $1/T$ the result is a straight line of slope $-\Delta H^*/R$. Some representative results which verify the correctness of this prediction are given in Fig. 7.5. This is the method by which activation energies are determined experimentally. The result

$$k = Ae^{-\Delta H^*/RT} \qquad (7.62)$$

is known as the Arrhenius equation and was first discovered experimentally in 1889. It holds provided ΔH^* is large enough for the temperature dependence of the A factor to be negligible.

When some structural feature of a reactant molecule is altered, by changing the nature of a substituent in an organic molecule, for example, this will affect both ΔS^* and ΔH^*. The two effects can be separated, because one changes the overall reaction rate without affecting its temperature dependence, whereas the other changes both the temperature dependence and the overall rate. There is now a large body of data in the literature about the effects of substituents on the rates of reaction of organic molecules, data which is used day by day to guide the synthesis of new substances.

Exercises

7.1   The decomposition of N_2O into N_2 and O_2 is a first-order process. At 858 K the half-life is 75.09 hours. Calculate
  (a) the rate constant
  (b) the total gas pressure after 75.09 hours at 858 K, the initial pressure being one atmosphere.

7.2   The reaction

$$CH_3CO_2CH_3 + H_2O \rightarrow CH_3CO_2H + CH_3OH$$

(hydrolysis of methyl ethanoate) is first-order with respect to both $CH_3CO_2CH_3$ and H^+ (the reaction is acid catalysed).
  (a) Write down the rate equation;
  (b) calculate the half-life of methyl ethanoate in a 1.0 molar solution of hydrochloric acid, given that the rate constant is $1.3 \times 10^{-4} \, 1 \, \text{mol}^{-1} \text{s}^{-1}$.

7.3   The gas–phase dimerization of butadiene

$$2 \text{ Butadiene} \rightarrow \text{Dimer}$$

was studied at 600 K, giving the following data.

t/minutes	p/kPa
0	84.26
20.78	74.26
49.50	66.41
77.57	61.97
103.58	59.01

Find: (a) the order of the reaction; (b) the rate constant in units of kilopascals and minutes.

7.4   The inversion of sucrose in aqueous solution (hydrolysis to form a mixture of sugars with opposite optical rotation, according to

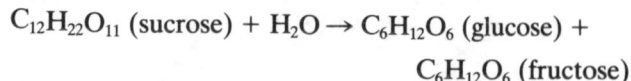

$$C_{12}H_{22}O_{11} \text{ (sucrose)} + H_2O \rightarrow C_6H_{12}O_6 \text{ (glucose)} + C_6H_{12}O_6 \text{ (fructose)}$$

gave the following data:

time/minutes	[*sucrose*]/mole 1^{-1}
0	1.0023
30	0.9022
60	0.8077
90	0.7253

  (a) Determine the reaction order and rate constant.
  (b) Suggest a reason why $[H_2O]$ does not appear in the rate equation.

7.5   The reaction of 1-bromopropane with sodium thiosulphate $(C_3H_7Br + S_2O_3^{2-} \rightarrow C_3H_7S_2O_3^- + Br^-)$ is second-order. Use the data below to determine the rate constant.

t/seconds	0	1110	2010	3192	∞
$[S_2O_3^{2-}]$/mole 1^{-1}:	0.0966	0.0904	0.0863	0.0819	0.0571

7.6   Devise a mechanism for the acidic hydrolysis of methyl acetate (systematic name: methyl ethanoate),

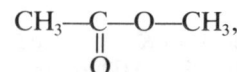

$$CH_3 \!-\! \underset{\underset{O}{\|}}{C} \!-\! O \!-\! CH_3,$$

which will account for the observed second-order kinetics (see exercise 7.2). The products are methanol, CH_3OH, and acetic acid (ethanoic acid), CH_3COOH.

7.7   The mechanism of decomposition of ozone is believed to be

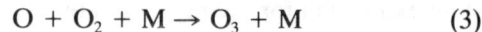

$$\begin{array}{lll} O_3 & \rightarrow O_2 + O & (1) \\ O + O_3 & \rightarrow 2O_2 & (2) \\ O + O_2 + M \rightarrow O_3 + M & & (3) \end{array}$$

  (a) Work out the steady-state concentration of oxygen atoms in terms of k_1, k_2, k_3, $[O_2]$, $[O_3]$ and $[M]$.
  (b) Work out an expression for the rate of decomposition of ozone.

7.8   The experimental rate constant for decomposition of N_2O_5 has the values $7.8 \times 10^{-7} \, \text{s}^{-1}$ at 0°C and $5.0 \times 10^{-3} \, \text{s}^{-1}$ at 65°C. Calculate the activation energy. (Can you relate this activation energy to those of reactions 7.15, 7.17 and 7.18?)

7.9 Calculate the Arrhenius factor A for the N_2O_5 decomposition (exercise 7.8) and compare it with kT/h.

7.10 For the reactions (1) $H_2 + I_2 \rightarrow 2HI$ and (2) $HI + HI \rightarrow H_2 + I_2$ the listed Arrhenius expressions for the rate constants are

$$k_1 = 3.3 \times 10^9 \, T^{1/2} \exp(- 19\,600/T)$$

and $\qquad k_2 = 2.0 \times 10^9 \, T^{1/2} \exp(- 21\,400/T)$

(a) Account for the presence of the $T^{1/2}$ factors (cf. Eq. 7.50).
(b) Calculate the two activation energies.
(c) Calculate the equilibrium constant for $H_2 + I_2 \rightleftharpoons 2HI$ at 298 K.
(d) Calculate the value of ΔH for the reaction in part (c).

8 Extraction of the Elements

8.1 Occurrence of the Elements in Nature

Extraction of the elements from their naturally occurring ores is one of the world's major industrial activities. The form in which an element occurs in nature is dependent on its chemical properties, and therefore is related to its position in the periodic table. Common natural sources of the elements are summarized in Fig. 8.1, in a manner which emphasizes periodic relationships. Some general comments can be made as follows:

1 The occurrence of insoluble salts of the alkaline earth metal ions is a consequence of their dipositive charge, which gives rise to high lattice energies and stable crystal lattices. In contrast, salts of the alkali metal ions have much lower lattice energies and are relatively abundant in solution in the oceans.

2 The majority of elements occur naturally as oxides or sulphides. Again, high lattice energies ensure the insolubility of these compounds. Sulphides are particularly important for the p block elements and for the more electronegative of the d block transition metals. Many elements in these blocks occur as both oxides and sulphides.

3 The least reactive elements, notably the inert gases and the platinum metals, exist as free elements in nature. The more unreactive metals also occur as compounds from which the free metal can be obtained very easily; for example, by heating. For this reason many of these elements have been known since earliest times.

8.2 Extraction of Metals

Most of the elements of the periodic table are metals. A smaller number are definitely non-metals, and a few are of intermediate character. For our present purpose semi-metallic elements such as silicon and antimony, which are usually best regarded as non-metals, can be included with the metals. The process of extraction of a metal generally involves three steps, namely concentration of the ore, reduction of the ore to the metal and refining of the metal. Considering these in order, we find that there are two principal methods of concentration. In the first a finely ground ore is separated from unwanted material by flotation, whereby agitation with a suitably chosen liquid (commonly water or oil) results in separation because of the ore preferentially sticking to the surface of air bubbles. The second method is leaching, which is the removal of the metal-containing fraction from an ore body by washing with acid or alkali or with a solvent.

Table 8.1 Extraction of some elements

Element	$E^{\ominus}(n)*$	Main source	Extraction
K	$-2.92\,(+1)$	KCl, KCl.MgCl$_2$	
Na	$-2.71\,(+1)$	NaCl	
Ca	$-2.87\,(+2)$	CaSO$_4$, CaCO$_3$ Ca$_3$(PO$_4$)$_2$ CaCl$_2$—Solvay process	Electrolysis of fused salts.
Mg	$-2.36\,(+2)$	Mg salts	
Al	$-1.66\,(+3)$	Al$_2$O$_3$	Electrolysis of Al$_2$O$_3$ in Na$_3$AlF$_6$
Mn	$-1.18\,(+2)$	MnO$_2$	Reduction with Al (Thermite process)
Cr	$-0.74\,(+3)$	FeO.Cr$_2$O$_3$	
Zn	$-0.76\,(+2)$	ZnS	
Fe	$-0.44\,(+2)$	Fe$_2$O$_3$, Fe$_3$O$_4$	Chemical reduction with carbon (or hydrogen). First convert sulphides to oxides by roasting.
Co	$-0.28\,(+2)$	CoAsS, Co$_3$S$_4$	
Ni	$-0.25\,(+2)$	Sulphides	
Sn	$-0.14\,(+2)$	SnO$_2$	
Pb	$-0.13\,(+2)$	PbS	
Cu	$+0.34\,(+2)$	Metal, sulphide	
Hg	$+0.85\,(+2)$	HgS	
Ag	$+0.80\,(+1)$	Metal, Ag$_2$S, AgCl	Cyanide extraction
Au	$+1.7$	Metal, tellurides	
Cl	$+1.36\,(-1)$	NaCl	Electrolysis
F	$+2.87$	KHF$_2$	

* $E^{\ominus}$ for reaction $M^{n+} + ne^- \rightarrow M$ (n given in parenthesis).

The preferred method of reduction of an ore to the free metal is decided mainly by the relative ease of achieving the reaction

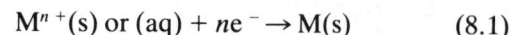

$$M^{n+}(s) \text{ or } (aq) + ne^- \rightarrow M(s) \qquad (8.1)$$

Some methods used for reduction of ores of the more common elements are listed in Table 8.1 and summarized in Fig. 8.2. The methods fall into four main groups:

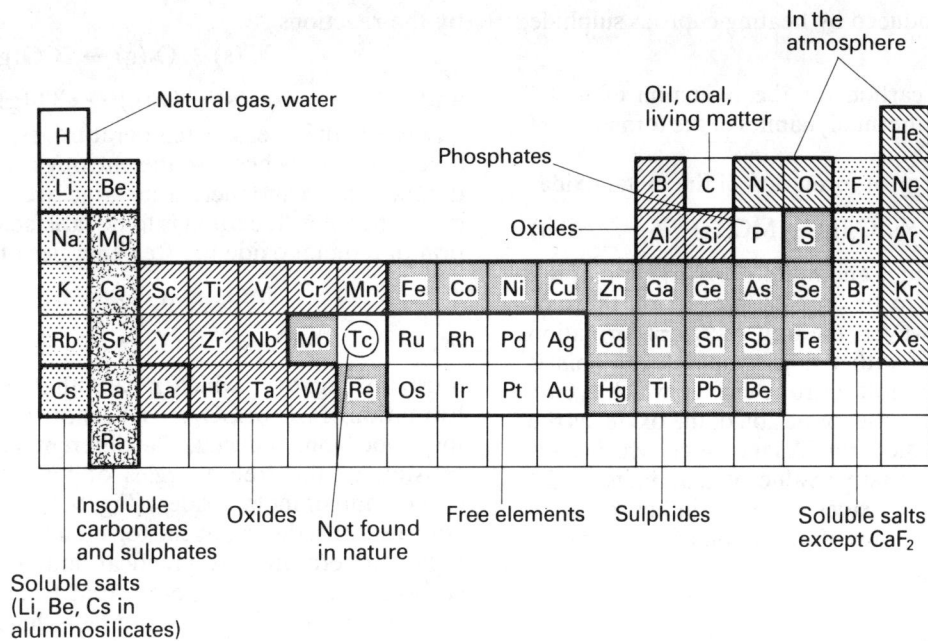

Fig. 8.1 Occurrence of the elements

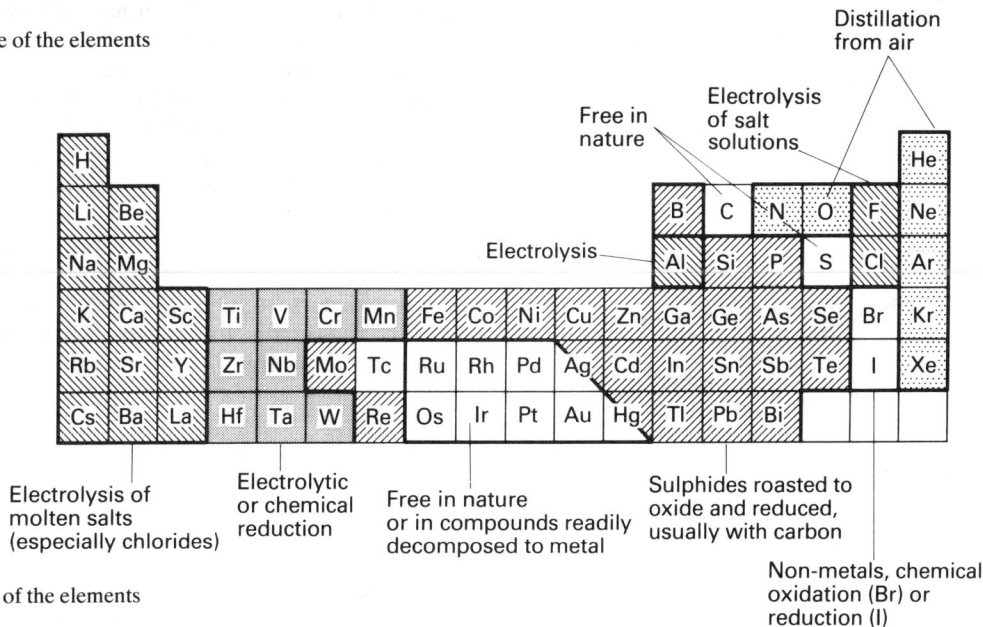

Fig. 8.2 Extraction of the elements

(a) Thermal decomposition of an ore, without the need to use a chemical reducing agent.
(b) Reduction with carbon or hydrogen.
(c) Reduction by a reactive metal (e.g. aluminium or magnesium).
(d) Electrolytic reduction of a molten salt.
Let us now consider each of these in more detail.

(a) Thermal decomposition to the metal is feasible only for metals with positive reduction potentials, and is therefore of limited application. Mercury can be obtained in this way:

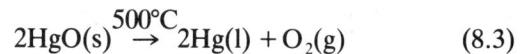

$$2HgS(s) + 3O_2(g) \rightarrow 2SO_2(g) + 2HgO(s) \quad (8.2)$$

$$2HgO(s) \xrightarrow{500°C} 2Hg(l) + O_2(g) \quad (8.3)$$

and copper can be produced by heating cuprous sulphide (Cu_2S).

(b) The wide use of carbon for the reduction of metal oxides is related to the thermodynamics of the formation of the oxides CO and CO_2.

The Gibbs free energy of formation of the metal oxide

$$M(s) + \tfrac{1}{2}nO_2(g) \rightarrow MO_n(s) \qquad (8.4)$$

is negative for most oxides and therefore extraction of the metal is not expected to be an easy process on thermodynamic grounds. However, the free energy of formation becomes less negative with rising temperature. This is because of the unfavourable entropy change which occurs as oxygen is consumed in the formation of the oxide; as the temperature increases the term $T\Delta S$ increases, which leads to a reduction in the negative value of ΔG. Plots of $G_f^{\ominus}$ versus T for a number of oxides are given in Fig. 8.3.

The negative free energies of formation of CO and CO_2

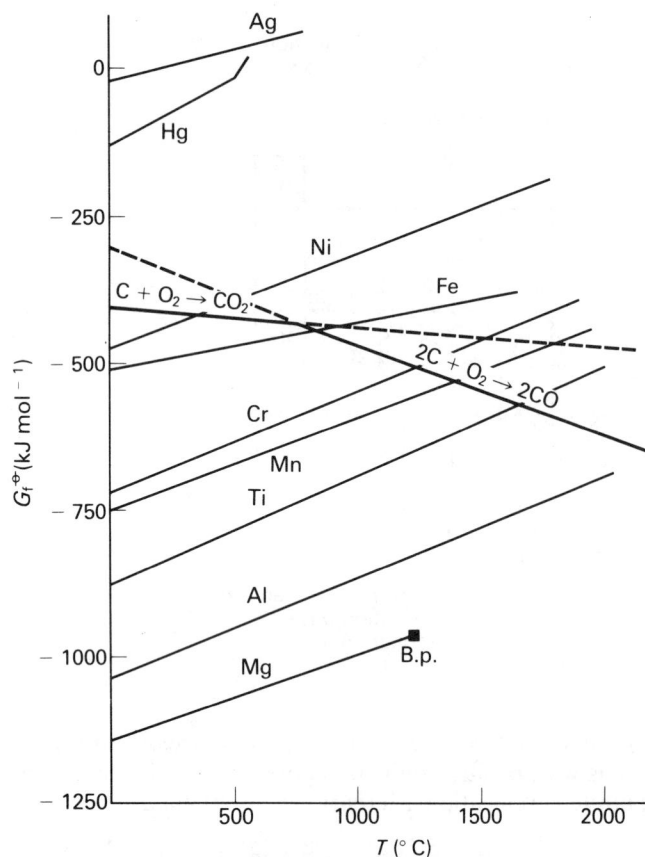

by the reactions

$$2C(s) + O_2(g) \rightarrow 2CO(g) \qquad (8.5)$$

and

$$C(s) + O_2(g) \rightarrow CO_2(g) \qquad (8.6)$$

increase with increasing temperature, especially for the first reaction. This is because the entropy change is favourable in reaction 8.5 and there is no large overall entropy change in reaction 8.6. If carbon is to be satisfactory for the reduction of a metal oxide the free energies of the reactions

$$MO_n + nC \rightarrow M + nCO \qquad (8.7)$$

or

$$MO_n + (n/2)\,C \rightarrow M + (n/2)CO_2 \qquad (8.8)$$

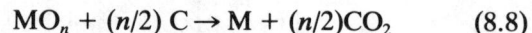

must be negative, which means that the free energy of formation of CO or CO_2 must be more negative than that of the oxide being reduced. The different temperature characteristics of the free energies of formation of oxides of carbon and of metal oxides (Fig. 8.3) cause carbon to be a very useful reducing agent for metal oxides, especially at high temperatures. A practical limitation on the use of carbon is that the temperature at which the carbon line crosses the metal line must not be so high that the metal is vaporized, nor must the temperature be so high that it cannot be achieved at a reasonable cost, as would be the case for magnesium or aluminium.

The most important industry which is based on reduction of an oxide by carbon is the steel industry, where crude ore is reduced by carbon in a blast furnace or rotary kiln. Limestone is added to remove silicon-containing impurities by forming a calcium silicate slag. The reactions taking place in the furnace are complex, but as the iron ore moves down the furnace and as the temperature rises, progressive reduction occurs from Fe_2O_3 to Fe_3O_4 (200°C), then to FeO (300°C), and finally to molten iron (1300–1500°C). The overall reactions are:

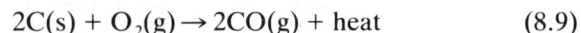

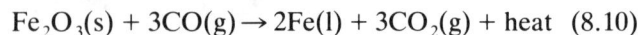

$$2C(s) + O_2(g) \rightarrow 2CO(g) + heat \qquad (8.9)$$

$$Fe_2O_3(s) + 3CO(g) \rightarrow 2Fe(l) + 3CO_2(g) + heat \quad (8.10)$$

$$CaCO_3 \rightarrow CaO(s) + CO_2(g) \qquad (8.11)$$

$$CaO(s) + SiO_2(s) \rightarrow CaSiO_3(l) \qquad (8.12)$$

The molten iron and slag ($CaSiO_3$) are removed at the bottom of the furnace, and the crude iron then has its carbon content reduced in furnaces which allow free access of oxygen in order to remove the carbon as CO_2. The final product is usually alloyed with other materials, such as chromium, titanium, vanadium, or manganese, to produce steels which are suited to particular purposes.

(c) A metal will reduce a metal oxide if the free energy change of the reaction

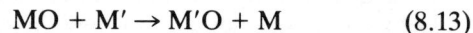

$$MO + M' \rightarrow M'O + M \qquad (8.13)$$

Fig. 8.3 $G_f^{\ominus}$ versus T for metal oxides, CO and CO_2 (from D. J. C. Ives, *Principles of Extraction of Metals* R.I.C 1966)

is negative. This means that any metal will reduce the oxide of another metal if the latter lies above it in the ΔG versus T plot (Fig. 8.3). Both aluminium and magnesium are good reducing agents in this respect, and aluminium in particular is often used in the thermite (or Goldschmidt) process for obtaining free metals from their oxides. Chromium, for example, is produced by the reaction:

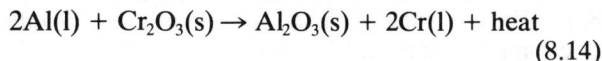

$$2Al(l) + Cr_2O_3(s) \rightarrow Al_2O_3(s) + 2Cr(l) + \text{heat} \tag{8.14}$$

(Both aluminium and chromium are shown as liquid because of the highly exothermic nature of the reaction.)

Titanium is produced by first converting the oxide rutile, TiO_2, to liquid $TiCl_4$, which can be purified by distillation, and then reducing the chloride with magnesium at 800°C in an atmosphere of argon. Excess magnesium and $MgCl_2$ may be removed by volatilization at around 1000°C.

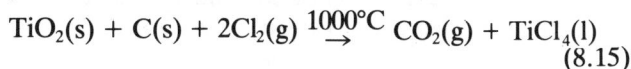

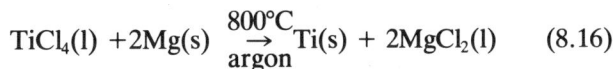

$$TiO_2(s) + C(s) + 2Cl_2(g) \xrightarrow{1000°C} CO_2(g) + TiCl_4(l) \tag{8.15}$$

$$TiCl_4(l) + 2Mg(s) \xrightarrow[\text{argon}]{800°C} Ti(s) + 2MgCl_2(l) \tag{8.16}$$

(d) Metals with large negative reduction potentials cannot be obtained from their ores by the chemical means described above and it is necessary to use more expensive electrolytic reduction techniques, for which a cheap and plentiful supply of electricity is required. It is invariably necessary to work with molten salts because in aqueous solution hydrogen is produced preferentially at the cathode. For example, electrolysis of aqueous sodium chloride gives:

$$\textit{at the anode: } 2Cl^- \rightarrow Cl_2 + 2e^- \tag{8.17}$$

$$\textit{at the cathode: } 2H_2O + 2e^- \rightarrow 2OH^- + H_2, \tag{8.18}$$

a reaction sequence which is used in the preparation of chlorine and caustic soda. Electrolysis of molten sodium chloride gives:

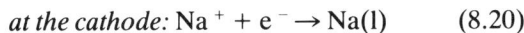

$$\textit{at the anode: } 2Cl^- \rightarrow Cl_2(g) + 2e^- \tag{8.19}$$

$$\textit{at the cathode: } Na^+ + e^- \rightarrow Na(l) \tag{8.20}$$

The temperature necessary to produce molten NaCl is lowered by the addition of anhydrous calcium chloride.

The production of aluminium requires electrolytic reduction of the ore bauxite $Al_2O_3.xH_2O$. If it is to be worthwhile economically the process requires a cheap supply of electricity. Iron and silicon impurities must be removed from the ore before electrolysis. Addition of caustic soda allows the removal of iron:

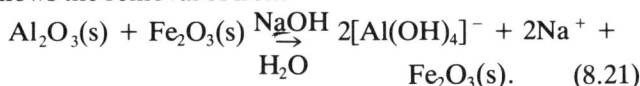

$$Al_2O_3(s) + Fe_2O_3(s) \xrightarrow[H_2O]{NaOH} 2[Al(OH)_4]^- + 2Na^+ + Fe_2O_3(s). \tag{8.21}$$

Addition of carbon dioxide then precipitates $Al(OH)_3$, leaving any silicates in solution. The hydroxide decomposes on heating to give purified Al_2O_3. The Al_2O_3 is then dissolved in a mixture of cryolite (Na_3AlF_6), CaF_2 and NaF, which gives a molten solution for electrolysis at around 1000°C. Liquid aluminium is produced at the carbon cathode which lines the electrolysis vessel. At the anode a variety of products can be produced, such as oxygen, fluorine, and carbon compounds produced by the attack of oxygen and fluorine on the carbon anodes.

The final refining of a metal can usually be achieved in a variety of ways; we shall briefly consider three of the main methods.

Electrolytic purification is achieved by making the crude metal the anode and using a strip of pure metal as the cathode in an electrolysis cell containing a solution of a salt of the metal. Electrolysis transfers metal from anode to cathode, the impurities either remaining in solution or settling out as a sludge. This method is restricted to metals whose cations are discharged in preference to hydrogen. Either the metal must lie below hydrogen in the electrochemical series (e.g. copper, mercury, silver) or the over-voltage* for discharge of hydrogen at the surface of the metal must be high (e.g. chromium). Virtually all copper is purified in this way.

In the second method the crude metal is converted to a volatile compound which can be purified by distillation and then decomposed. The Mond method for the extraction of nickel makes use of this technique, the volatile compound

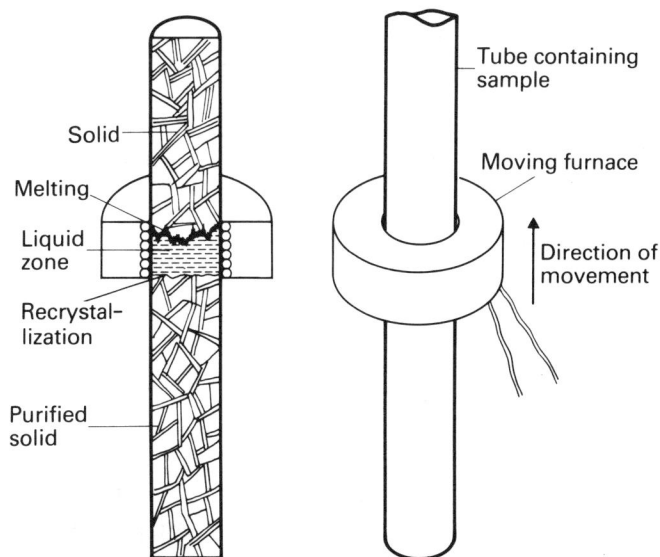

Fig. 8.4 Zone-refining

---

* The over-voltage, or over-potential, is the additional voltage, over and above the reversible e.m.f. of Chapter 6, which must be applied to an electrode to make a significant current flow

being nickel carbonyl, $Ni(CO)_4$.

Zone-refining, or zone-melting, is the technique used to make very pure silicon and germanium semiconductors for use in transistors and other solid-state electronic devices. A cylinder of the crude metal is melted by a narrow band heater (Fig. 8.4), and the heater is moved slowly along the cylinder. The element crystallizes at the cool side of the band and impurities tend to remain in the moving liquid phase. When the heater reaches the end of the rod it is moved quickly back to the beginning and the process is repeated. After several such passes a very high level of purity is achieved. In the semiconductor industry impurity levels of 1 part in 10^8 are routinely specified.

8.3 Extraction of Non-Metals

Less can be said in a systematic way concerning the extraction of non-metals. Reference to Fig. 8.2 shows that reduction of sulphides is used for the heavier members of groups V and VI, distillation of air for the inert gases, oxygen and nitrogen, electrolysis for fluorine and chlorine, chemical oxidation for bromine, and reduction for iodine. Details regarding the extraction of halogens are given in Table 8.2. Carbon and sulphur occur naturally and merely require purification. Phosphorus is obtained by reduction of the oxide with carbon. In general, highly reactive non-metals are obtained by methods similar to those used for highly reactive metals.

Exercises

8.1 Discuss the chemical reasons for the occurrence of the elements in nature according to the details given in Fig. 8.1.

8.2 List the principal methods of extracting metals from their ores in order of increasing chemical difficulty.

8.3 Why has carbon such a wide application in the reduction of metal oxides to give the free metal?

8.4 Write out the chemical equations describing the production of Fe in a blast furnace.

8.5 Why are Mg and Al so powerful as reducing agents? Indicate one use of each of the elements in the extraction of metals.

8.6 Write out, using chemical equations, the chemical steps in the production of aluminium from bauxite.

8.7 Explain how zone-refining purifies a metal.

8.8 Write out the ion–electron half-reactions that are involved in the extraction of Br_2 and I_2 (Table 8.2) and verify the overall reactions given in the Table.

Table 8.2 Extraction of the halogens

Halogen	Source	Method
F_2	CaF_2	$CaF_2 + H_2SO_4 \rightarrow CaSO_4 + 2HF$ KF + HF (1:2 mixture) is electrolysed $2F^- - 2e^- \rightarrow F_2$ (carbon anode) $2H^+ + 2e^- \rightarrow H_2$ (steel cathode)
Cl_2	NaCl	(1) Electrolysis of brine $\quad 2Cl^- - 2e^- \rightarrow Cl_2$ (carbon anode) and either $\quad Na^+ + e^- \rightarrow Na/Hg$ (mercury cathode) or $\quad 2H_2O + 2e^- \rightarrow 2OH^- + H_2$ (steel cathode) (2) Electrolysis of fused NaCl (see p. 119)
Br_2	NaBr	$2Br^- + Cl_2 \rightarrow Br_2 + 2Cl^-$ absorb Br_2 into Na_2CO_3 $3Br_2 + 3CO_3^{2-} \rightarrow 5Br^- + BrO_3^- + 3CO_2$ $5Br^- + BrO_3^- + 6H^+ \rightarrow 3Br_2 + 3H_2O$
I_2	$Ca(IO_3)_2$	$2IO_3^- + 6HSO_3^- \rightarrow 2I^- + 6SO_4^{2-} + 6H^+$ add to fresh mother liquor $5I^- + IO_3^- + 6H^+ \rightarrow 3I_2 + 3H_2O$

9 The Chemistry of Hydrogen

9.1 Introduction

The correct place to insert hydrogen in the periodic table is not at all obvious. In terms of the electron configuration hydrogen is a member of the 's block' elements; however, it is certainly not typical of these elements. It might also be considered as a pseudo-halogen (i.e. an element that is one electron short of having a filled valence shell), since the species H^- is known (though it is of lower thermodynamic stability than the halide ions), and because hydrogen forms a wide range of covalent compounds with elements that also form covalent halides. In fact the small size of the hydrogen atom, and its single electron, (configuration $1s^1$), make hydrogen unique among elements, and it is therefore natural to consider it on its own. Some of the properties of hydrogen are listed in Table 9.1.

Table 9.1 Properties of hydrogen

Ionization energy	1312 kJ mol^{-1} (13.59 eV)
Electron affinity	-73 kJ mol^{-1}
Electronegativity	2.1
Covalent radius	28–37.5 pm*
Ionic radius H^-	137–154 pm**
Ionic radius H^+	approx. 1.5×10^{-3} pm
Van der Waals radius	120 pm
Bond length (H_2)	74.9 pm
Bond energy (H_2)	436 kJ mol^{-1}

* The covalent radius varies with the type of molecule, e.g. 37.5 pm from H_2 bond length, 32 pm from group IV hydrides, 28 pm from hydrogen halides.

** The calculated radius of free H^- is 208 pm; the radius of H^- varies with the cation associated with it.

In considering the chemistry of hydrogen we must take account of the following classes of compounds:
1 The ionic or salt-like hydrides of the alkali and alkaline earth metals, in which the anion H^- exists.
2 Covalent hydrides formed with the p block elements.
3 Compounds which give rise to the solvated cation H^+, (e.g. H_3O^+) when dissolved in a polar solvent. This class includes covalent hydrides and oxyacids.
4 Transition metal hydrides, in which the hydrogen occupies some of the cavities in the close-packed structure of the metal.
5 Complexes such as MH_4^- (M = B, or Al), and transition metal hydrido-complexes.
6 Compounds in which special bonding situations arise, e.g. hydrogen bonded compounds, or hydrogen bridge compounds (boron hydrides).

The occurrence of these classes of compounds for elements in different sections of the periodic table is summarized in Fig. 9.1.

9.2 Ionic or Salt-like Hydrides

The hydrides of the alkali metals and of calcium, strontium and barium are prepared by direct combination of the elements within the temperature range 300–700°C. Evidence for the ionic nature of the hydrides and for the presence of H^- comes from electrical conductivity of the molten salt and the observation that hydrogen is liberated at the anode during the electrolysis.

Solid ionic hydrides are stabilized by their high lattice energies, which counterbalance the endothermic formation of $H^-(g)$. The comparison of H^- with F^-, which is smaller in size, is instructive:

ΔH for process:	H	F
$\frac{1}{2}X_2(g) \rightarrow X(g)$	$+218$	$+79$ (kJ mol^{-1})
$X(g) + e^- \rightarrow X^-(g)$	-73	-333
$\frac{1}{2}X_2(g) + e^- \rightarrow X^-(g)$	$+145$	-254
Lattice energy of NaX:	-806	-904 (kJ mol^{-1})

Both the stronger X—X bond and lower electron affinity for hydrogen contribute to the positive (i.e. endothermic) enthalpy of formation of the gaseous H^- ion; therefore coexistence with H^- in the solid state occurs only for the most electropositive elements. In aqueous solution the H^- ion, which is a stronger base than the hydroxide ion, removes a proton from water in the reaction

$$H^- + H_2O \rightarrow H_2 + OH^- \qquad (9.1)$$

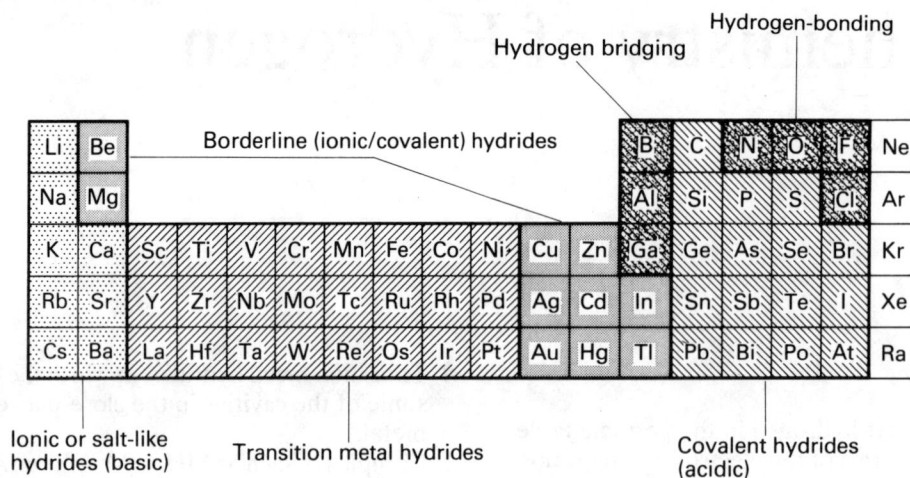

Fig. 9.1 Classification of hydrides

The calculated radius of the free H^- ion is 208 pm (approximately twice the radius of the isoelectronic helium atom) which is greater than the value observed in hydride compounds. The observed radius of H^- varies with the nature of the associated cation:

Cation:	Li^+	Na^+	K^+	Rb^+	Cs^+	Ca^{2+}
Radius of H^-(pm):	137	146	152	154	152	138

Because the H^- ion has two electrons under the influence of a single proton the 1s orbital is rather diffuse and large, and its size is easily modified by the positive electric field of the cation. The greater the field (i.e. the smaller the cation) the more effective becomes the nuclear charge of the hydrogen, and the smaller the orbital size.

The ionic hydrides are useful drying agents in the laboratory, and are powerful reducing agents. For example, the following reductions can be achieved:

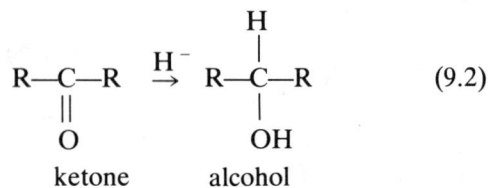

$$R{-}\underset{\underset{O}{\|}}{C}{-}R \xrightarrow{H^-} R{-}\underset{\underset{OH}{|}}{\overset{\overset{H}{|}}{C}}{-}R \quad (9.2)$$

$$\text{ketone} \qquad\qquad \text{alcohol}$$

$$SO_4^{2-} \xrightarrow{H^-} S^{2-} \quad (9.3)$$

$$NO_3^- \xrightarrow{H^-} NH_3 \quad (9.4)$$

9.3 Covalent Hydrides

There are three general methods of preparation of covalent hydrides;

(1) Direct combination of the elements

$$\text{e.g. } H_2 + Cl_2 \xrightarrow{h\nu} 2HCl \quad (9.5)$$

(2) Hydrolysis of binary compounds such as phosphides, borides etc.

$$\text{e.g. } Ca_3P_2 \xrightarrow{H_2O} PH_3 \quad (9.6)$$

(3) Reaction of a halide or oxide with an ionic hydride or hydrido-complex,

$$\text{e.g. } SiCl_4 + 4LiH \rightarrow SiH_4 + 4LiCl \quad (9.7)$$

or $4AsCl_3 + 3LiAlH_4 \rightarrow 4AsH_3 + \underbrace{3LiCl + 3AlCl_3}_{\downarrow} \quad (9.8)$

$$3LiAlCl_4$$

The covalent hydrides are formed between hydrogen and all 'p block' elements except the inert gases. The bonding can be considered as covalent, becoming more polar for the elements towards the top right of the periodic table as the electronegativity difference increases (Table 9.2). A similar trend is observed for bond strength (except for CH_4), which is to be expected because bond strength is influenced by the relative electronegativities of the bonded atoms. (This is the basis of the Pauling electronegativity scale.) Size is also important, as the smaller atoms give more effective overlap between their bonding orbital and the hydrogen 1s orbital. Both an increase in thermal stability and a decrease in reducing power occur as a group is ascended, which can also be related to trends in bond strength and polarity.

The covalent hydrides display trends in shape and bond angles which are a function of both the number of lone pairs of electrons on the central atom and the size of the central atom (Table 9.3). The major change in bond angle (groups

V and VI) which occurs between the first row elements and the rest is directly related to the change in the central atom radius.

Table 9.2 Covalent hydrides: bond strength, polarity, thermal stability and reducing power

increasing bond strength and bond polarity →

	CH$_4$	NH$_3$	H$_2$O	HF
	414	389	464	565
	SiH$_4$	PH$_3$	H$_2$S	HCl
	318	322	386	431
	GeH$_4$	AsH$_3$	H$_2$Se	HBr
	285	297	305	364
	SnH$_4$	SbH$_3$	H$_2$Te	HI
	251	255	238	299

increasing thermal stability ↑ increasing reducing power ↓ increasing bond strength and bond polarity ↑

bond strength in kJ mol^{-1}

There is also a group of hydrides known as borderline hydrides (Fig. 9.1) which are intermediate between the ionic and covalent types. They tend to have polymeric structures, with hydrogen bridges.

Table 9.3 Covalent hydrides: bond angle and bond lengths

	CH$_4$	NH$_3$	OH$_2$	HF
M—H (pm)	109.3	101.2	95.7	91.7
HMH (deg)	109.5	106.7	104.5	—
	SiH$_4$	PH$_3$	SH$_2$	HCl
M—H (pm)	148	141.9	132.8	127
HMH (deg)	109.5	93.5	92.2	—
	GeH$_4$	AsH$_3$	SeH$_2$	HBr
M—H (pm)	152.7	152	146	141
HMH (deg)	109.5	91.8	91.0	—
	SnH$_4$	SbH$_3$	TeH$_2$	HI
M—H (pm)	170	171	170	161
HMH (deg)	109.5	91.3	89.5	—

increasing size of central atom ↓

← increasing size of central atom

9.4 The Solvated Proton

Covalent hydrides to the right of the 'p block' elements in the periodic table dissociate in water to give the proton H$^+$, which exists only as a solvated species, such as H$_3$O$^+$ or H$_5$O$_2^+$.

The enthalpy change for the reaction

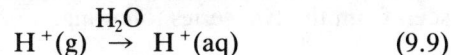

$$H^+(g) \xrightarrow{H_2O} H^+(aq) \qquad (9.9)$$

is estimated to be -1075 kJ mol^{-1}, and it is this high value which makes it energetically possible for strong covalent bonds to hydrogen to be broken when an acid dissolves in water.

The pK_a's of some covalent hydrides are listed in Table 9.4. It will be noticed that two trends are apparent, the first being increasing acidity as we move across the periodic table (note that bond strength increases in the same direction) and the second being increasing acidity as we move down the table (in the direction corresponding to decreasing bond strength). The opposing trends highlight two aspects of the acidity of the hydrides MH$_n$, namely, the increasing electron affinity of M and hydration energy of M$^-$ as we move across the table produce increasing acidity, and decreasing M—H bond strength as we move down the table also produces increasing acidity.

Table 9.4 Acidity of binary hydrides

	CH$_4$	NH$_3$	H$_2$O	HF
pK_a	58	35	16	3
		PH$_3$	H$_2$S	HCl
pK_a		27	7	-7
			H$_2$Se	HBr
pK_a			4	-9
			H$_2$Te	HI
pK_a			3	-10

increasing bond strength ↑ increasing acidity ↓

→ increasing acidity and increasing bond strength

(Adapted from Kneen, Rogers, and Simpson, *Chemistry—Facts, Patterns and Principles*, Addison-Wesley, 1973.)

The ionization of the hydrogen halides may be represented by the thermodynamic cycle shown in Fig. 9.2. Figures for the various free energy changes are given in Table 9.5. The weak acidity of HF can be seen to be due to: (a) positive ΔG_v, probably due to hydrogen bonding between HF and H$_2$O (see below); (b) high positive free energy of bond breaking, ΔG_d; (c) the low electron capture of fluorine (ΔG_a) as compared, for example, with chlorine. Even though the sum of the negative hydration energies ($\Delta G_{hH^+} + \Delta G_{hX^-}$) for HF is large it does not compensate for the positive free energy terms. However, the high negative value for free energy of hydration ΔG_{hF^-} is probably a reason why the fluoride is the strongest acid in the series CH$_4$, NH$_3$, H$_2$O and HF. The hydrated proton is also present in solutions of oxyacids. Again the acidity varies in a systematic manner, as can be seen from the two series following.

Again the acidity varies in a systematic manner, as can be seen from the two series following.

Acid	First pK_a	Acid	First pK_a
Si(OH)$_4$	10.0	Cl(OH)	7.2
OP(OH)$_3$	2.1	OCl(OH)	2.0
O$_2$S(OH)$_2$	-3	O$_2$Cl(OH)	-1
O$_3$Cl(OH)	-10	O$_3$Cl(OH)	-10

In both series the acidity increases as a result of electron withdrawal from the central atom by the increasing number of oxygen atoms bonded to it. As electron withdrawal from the central atom increases more charge is withdrawn from the OH group, which increases the positive character of the oxygen and leads to an easier cleavage of the O—H bond. The same approach, in terms of electron withdrawal from the OH group, can be applied to the relative acidities of series of organic acids.

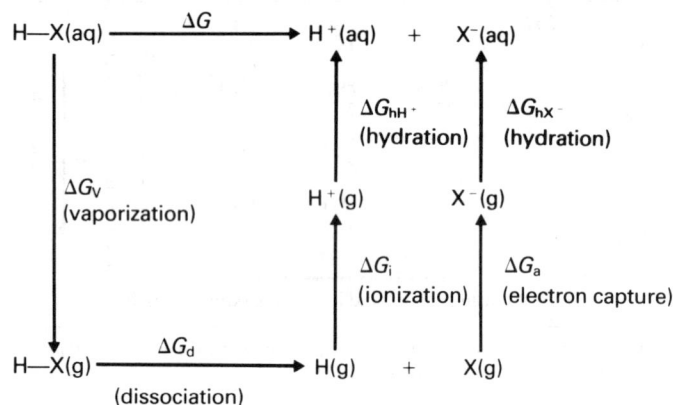

Fig. 9.2 Thermodynamic cycle of ionization of HX in water

Table 9.5 Free energy changes for hydrogen halides

	HF	HCl	HBr	HI
ΔG_v	23.8*	-4.2	-4.2	-4.2
ΔG_d	534.7	403.8	338.9	272
ΔG_i	1319	1319	1319	1319
ΔG_a	-347.3	-366.5	-345.2	-315.1
$\Delta G_{hH}{}^+$ } + $\Delta G_{hX}{}^-$ }	-1512.5	-1392.4	-1362.7	-1328.9
ΔG	17.7	-40.8	-54.2	-57
$\dfrac{\Delta G}{5.69} = pK_a$	3.1	-7.2	-9.5	-10.2

* In kJ mol^{-1}

9.5 Transition Metal Hydrides

Direct union of the transition metals ('d block elements') with hydrogen gas leads to formation of transition metal hydrides. Except for palladium hydride, their formation becomes more difficult as we move towards the bottom right of the periodic table. Non-stoichiometric compounds such as TiH$_{1.8}$, NbH$_{0.7}$ and PdH$_{0.6}$, as well as stoichiometric compounds such as TiH$_2$ and CrH are formed. The hydrides maintain many of the characteristic properties of the parent metal, and may be better regarded as alloys, or solid solutions.

In a number of cases the structures relate to the parent metal structure, but in others there are distinct changes. As yet no agreement has been reached as to the nature of the hydrogen species and its bonding in the hydrides. It appears that the hydrides do not consist simply of H atoms occupying cavities in the existing metallic structure (hence the name interstitial hydride is a misnomer) but they contain either H$^-$ or H$^+$, bonded to the metal atoms, mainly in tetrahedral coordination. One physical change which is always observed is an increase in the volume of the unit cell of the metallic structure when the hydrogen content increases. For example, PdH$_{0.03}$ has a unit cell edge of 389 pm, close to that of the pure metal, whereas the value is 402 pm for PdH$_{0.6}$.

9.6 Complex and Transition Metal Hydrido-Complexes

The hydride ion H$^-$ is a strong Lewis base and will form species of the type MH$^-$ with the group III elements boron, aluminium and gallium. These compounds are useful reducing agents in synthetic inorganic and organic chemistry. They may be prepared by reactions such as:

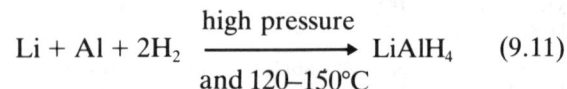

$$4LiH + AlCl_3 \rightarrow LiAlH_4 + 3LiCl \qquad (9.10)$$

$$Li + Al + 2H_2 \xrightarrow[\text{and 120–150°C}]{\text{high pressure}} LiAlH_4 \qquad (9.11)$$

The borohydrides are the least reactive as boron is small and electronically saturated, whereas aluminium and gallium can expand their valence shells. The transition metals also form hydrido-complexes where the hydrogen can be regarded as H$^-$, e.g. [HCo(CO)$_4$], K$_2$[ReH$_9$], [H$_2$Fe(CO)$_4$] and [RhH$_2$Cl(Ph$_3$P)$_2$C$_2$H$_4$]. In the preparation of such compounds the source of hydrogen can be molecular hydrogen or a hydride such as LiAlH$_4$. The complex [HCo(CO)$_4$] is an important intermediate in the catalytic formation of aldehydes from olefins; [RhH$_2$Cl(Ph$_3$P)$_2$C$_2$H$_4$] is an efficient hydrogenation catalyst in which the two hydride ions react with the ethylene to give ethane (Fig. 9.3).

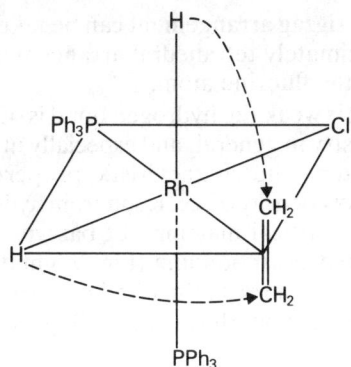

Fig. 9.3 Hydrogenation of C_2H_4 in the complex $[RhH_2Cl(Ph_3P)_2C_2H_4]$

9.7 Hydrogen Bonds and Hydrogen Bridges

Hydrogen bonds can be formed between the electronegative atoms F, N and O (and perhaps also Cl and S). The hydrogen bond may be considered as arising from the attraction between the positive charge on the hydrogen in the polar bond $X—H^{\delta+}$ and the negative charge of a lone pair of electrons on atom Y, to give $X—H\cdots Y$. As a result of this interaction the X and Y atoms approach one another more closely than the distance corresponding to the sum of their Van der Waals' radii.

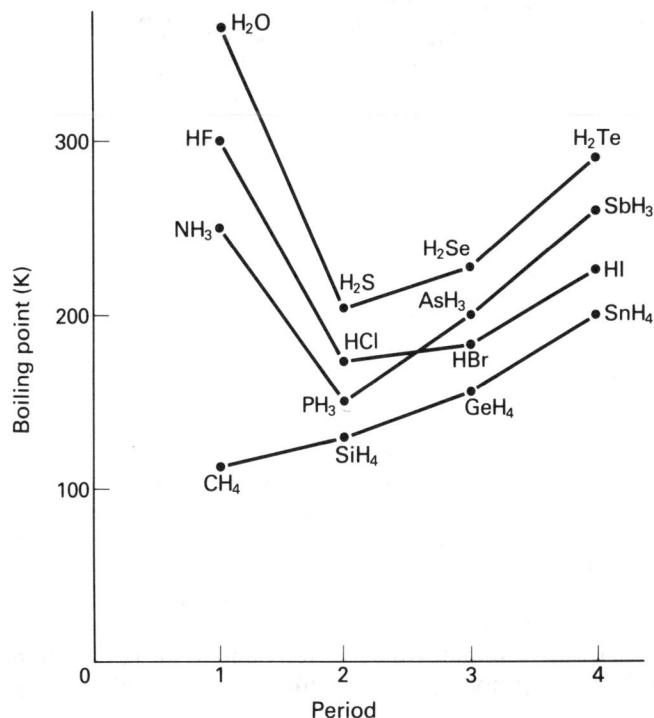

Fig. 9.4 Boiling points of the covalent hydrides

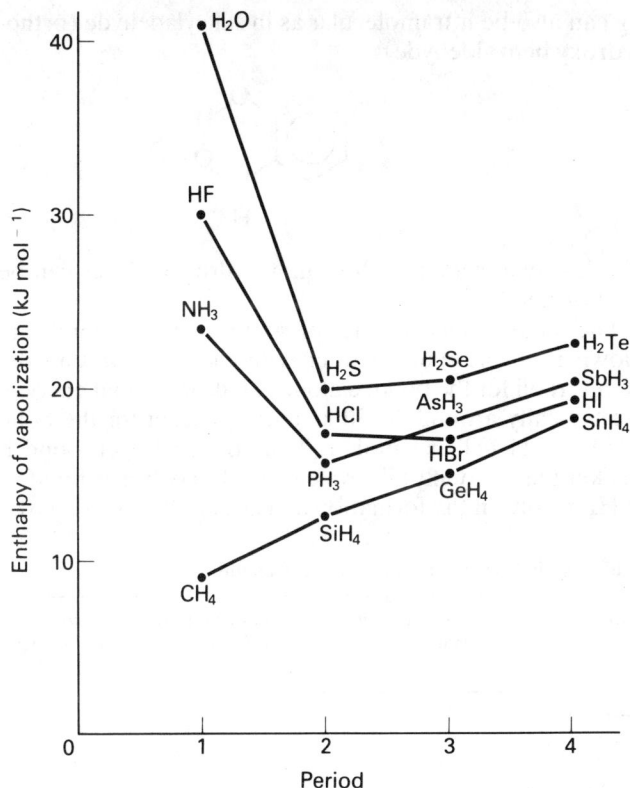

Fig. 9.5 Enthalpies of vaporization of the covalent hydrides

Evidence for this type of bond comes from spectroscopic and diffraction studies, as well as from a consideration of thermodynamic properties, such as the boiling points and enthalpies of vaporization of the covalent hydrides (Figs 9.4 and 9.5). The anomalously high boiling points and enthalpies of vaporization of HF, H_2O and NH_3 result from the need to overcome the intermolecular attraction due to hydrogen bonding. Molecular association due to hydrogen bonding can give rise to species of well-defined composition. For example, in non-polar solvents benzoic acid normally dimerizes, as:

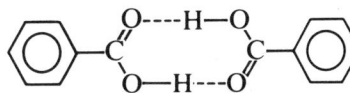

The compound quinhydrone

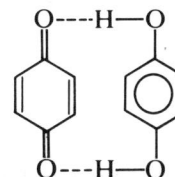

results from hydrogen bonding between a molecule of hydroquinone and a molecule of quinone. Hydrogen bonding can also be intramolecular as in salicyladehyde (ortho-hydroxy benzaldehyde):

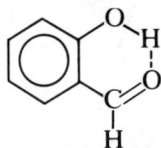

The last two examples show that hydrogen bonds can be bent or linear.

The lengths and energies of some hydrogen bonds are shown in Table 9.6, together with the sums of Van der Waals' radii for the bonded atoms. Hydrogen bond energies are typically around 20–30 kJ mol^{-1} except for the bond H—F····H (113 kJ mol^{-}). Thus the hydrogen bond is weaker than virtually all covalent bonds. Hydrogen bonding in HF results in the formation of a zigzag chain (Fig. 9.6).

Table 9.6 Hydrogen bond lengths and strengths

Bond	X—Y distance (pm)	Sum of Van der Waals' radii (pm)	Hydrogen bond energy (kJ mol^{-1})
F—H—F (symmetrical)	226–232	270	113
F—H····F (unsymmetrical)	255	270	28.0
O—H—O (symmetrical)	240	280	
O—H····O (unsymmetrical)	248–276	280	19–30
N—H····N	294–300	300	25
N—H····F	263	285	
N—H····O	291	290	
N—H····Cl	313	330	

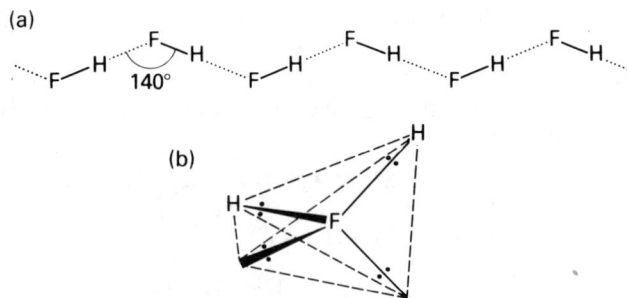

(a)

(b)

Fig. 9.6 Structure of hydrogen fluoride: (a) hydrogen bonded chain; (b) approximately tetrahedral arrangement of one HFH group with fluorine lone pairs

With the linear chain, non-bonded H····H repulsions would be reduced. The preference for a non-linear arrangement is attributed to the hydrogen bond being formed with a fluorine lone pair; the zigzag arrangement can be taken as evidence for an approximately tetrahedral arrangement of the lone pairs around the fluorine atom.

Although it is weak the hydrogen bond is of great importance in chemistry in general, and especially in the chemistry of living matter. The characteristic properties of water, including the low density of ice, result from hydrogen bonding (Fig. 9.7); the critical matching of base pairs in DNA is achieved by hydrogen bonding (Fig. 9.8 and Chapter 17); structures and properties of silk and nylon are the result of extensive hydrogen bonding (Fig. 9.10, page 128).

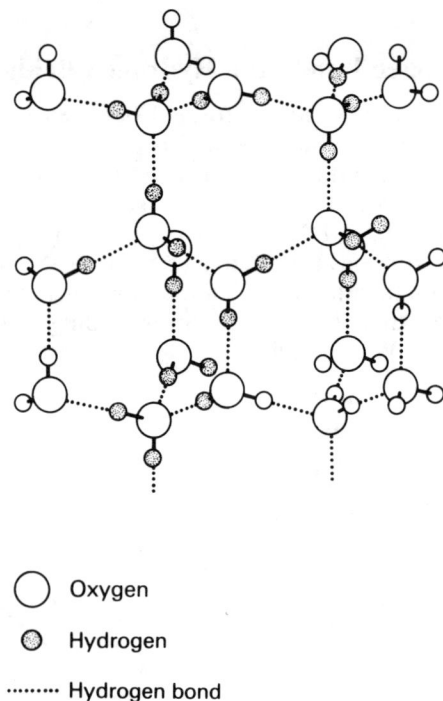

○ Oxygen

◎ Hydrogen

········ Hydrogen bond

Fig. 9.7 The structure of ice

A hydrogen atom may also form a bridge between a pair of atoms where the hydrogen atom and the other two are associated in a three-centre bond. The bonding is covalent, and markedly stronger than the essentially electrostatic hydrogen bond we have just considered. This type of bonding is found particularly in the hydrides of boron, and to a lesser extent in those of aluminium and gallium. The unusually strong, symmetrical hydrogen bond in the [FHF]$^{-}$ ion of KHF_2 is probably best regarded as a covalent bridging bond.

More than twenty boron hydrides are known, as well as

Thymine

Adenine

←280 pm→

←300 pm→

To chain

To chain

Cytosine

Guanine

←290 pm→

←300 pm→

←290 pm→

To chain

To chain

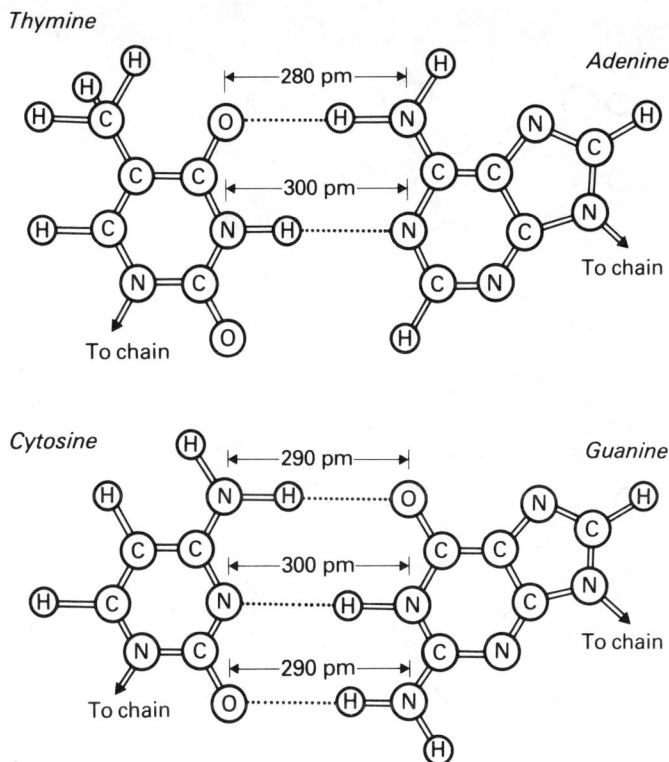

Fig. 9.8 Arrangement of hydrogen bonded base pairs in DNA

numerous derivatives (Table 9.7). The investigation of these compounds by Stock during the period 1912–1932 arose from his curiosity as to whether boron would form a series of hydrides similar to that of carbon. Since boron has one electron less than carbon, the simplest boron hydride, diborane B_2H_6, cannot be structurally identical with ethane C_2H_6 because it is electron deficient compared with C_2H_6. The electron deficiency of the boron hydrides is overcome by the formation of *three-centre bonds*. Three atomic orbitals,

one on each of the three atoms, combine to form a bonding orbital which in the normal way accommodates two electrons (Fig. 9.9). Linear combination of the three atomic orbitals results in formation of the bonding orbital, which is occupied, plus one non-bonding and one anti-bonding orbital. This is a case of the quite widespread phenomenon of multi-centre bonding. The usual two-centre bond formed by a pair of electrons is a special case which, because it normally allows maximum bond strength to be achieved, is the most common type of bonding in chemical compounds.

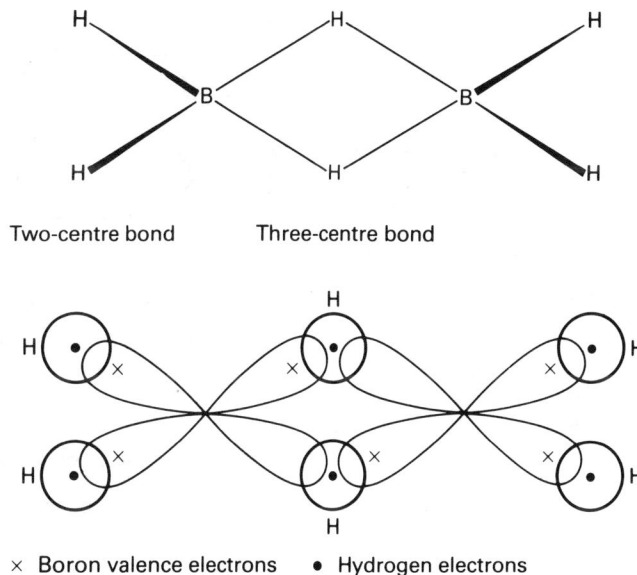

Two-centre bond   Three-centre bond

× Boron valence electrons   • Hydrogen electrons

Fig. 9.9 B_2H_6

9.8 Hydrogen Isotopes

The three principal isotopes of hydrogen are ^1_1H, ^2_1H (or D, deuterium) and ^3_1H (or T, tritium) (Table 9.8). The relatively large mass differences among the isotopes have a marked effect on the physical and to some extent the chemical properties of the compounds in which the isotopes are present (Table 9.9).

Table 9.7 Some boron hydrides and derivatives

Boron hydrides	B_2H_6	B_8H_{18}
	B_4H_{10}	$\alpha, \beta B_{10}H_{18}$
	B_5H_{11}	$B_{10}H_{20}$
	B_8H_{12}	$B_{18}H_{22}$
Boron hydride ions	$[B_6H_6]^{2-}$	$[B_nH_n]^{2-}$ series
	$[B_8H_8]^{2-}$	
	$[B_4H_7]^-$	
	$[B_5H_{10}]^-$	$[B_{10}H_{13}]^-$
	$[B_9H_{14}]^-$	$[B_{11}H_{13}]^{2-}$
Carboranes	$C_2B_4H_6$	
	$C_3B_3H_7$	
	$C_2B_{10}H_{12}$	
	$[C_2B_8H_{10}]^{2-}$	

Table 9.8 Isotopes of hydrogen

Isotope	Name	Symbol	Natural abundance	Stability
^1_1H	hydrogen	H	99.98%	stable
^2_1H	deuterium	D	0.02%	stable
^3_1H	tritium	T	$10^{-17}\%$	β-emitter
				($^3_1H \rightarrow {}^3_2He + {}_{-1}^{0}e$)
				$t_{1/2} = 12.4$ years

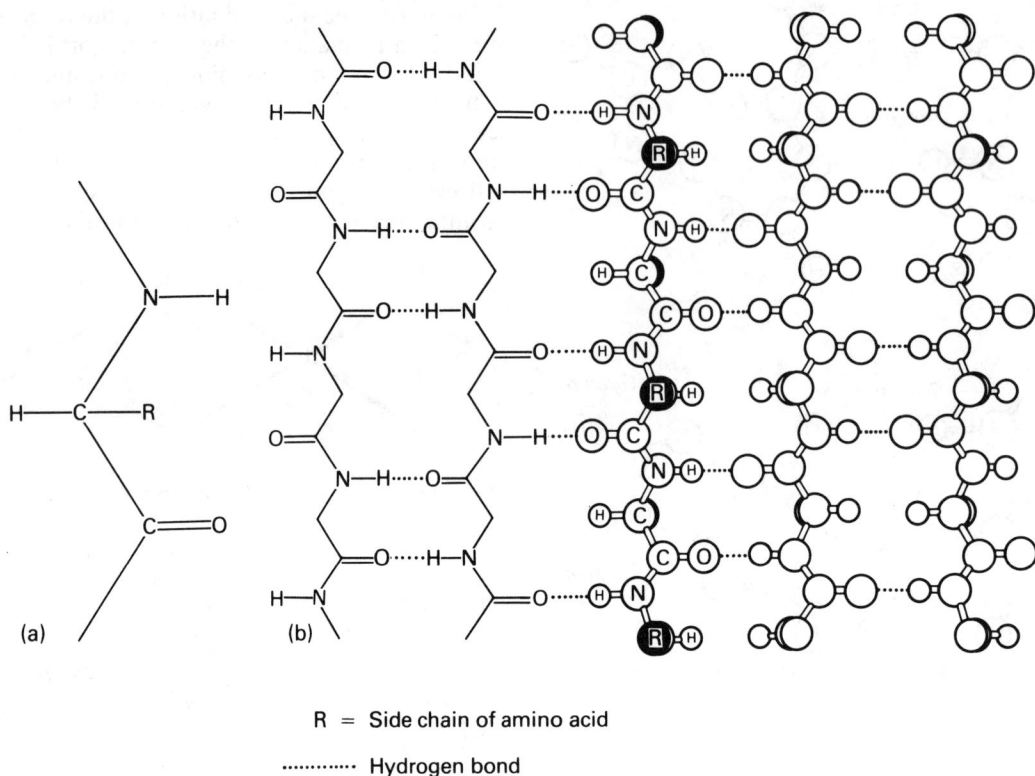

R = Side chain of amino acid

·············· Hydrogen bond

Fig. 9.10 Structure of silk. (a) Amino acid sub-unit in a protein chain. (b) Linking by hydrogen bonds of the protein chains in silk

Table 9.9 Some properties of H_2O, D_2O and T_2O

	H_2O	D_2O	T_2O
Molecular weight	18.015	20.028	22.032
Melting point (°C)	0.00	3.81	—
Temperature of maximum density	3.98	11.23	13.4
Maximum density (gm^{-3})	1.00	1.11	1.22
Boiling point (°C)	100.00	101.42	101.51
ΔH_{fusion} at melting point (kJ mol^{-1})	6.01	6.34	—

An important use of the isotopes of hydrogen is the replacement of H in compounds by either D or T, in order to study compounds and their reactions by tracer techniques. The presence of the isotopic species can be detected in a number of ways; for example, the infrared spectra of deuterated compounds show significant differences in the energies of vibrational transitions which involve the hydrogen atom. The changes in frequency of the O—H stretching vibration in water on deuteration and tritiation are shown in Fig. 9.11. Proton magnetic resonance spectroscopy is one of the most powerful methods we have of studying compounds containing hydrogen. Deuterium, with nuclear spin 1, behaves quite differently from ordinary hydrogen (nuclear spin = ½). The weak β emission from tritium is

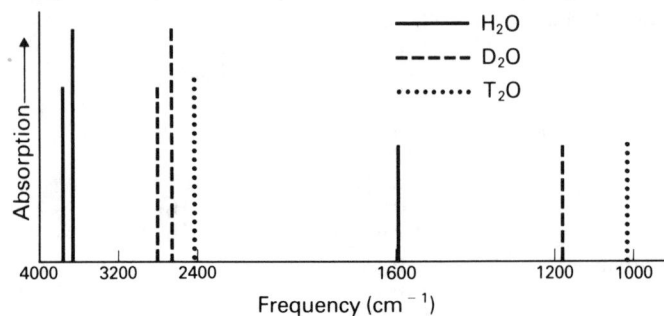

Fig. 9.11 Vibrational frequency of H_2O, D_2O and T_2O

made use of in radioactive tracer studies of chemical reactions involving hydrogen.

Exercises

9.1 Which would you predict to have the higher ionization energy—H or H_2? Why?

9.2 Which hydride in each pair of hydrides would ionize, in aqueous solution, to the greatest extent: H_2S, H_2Se; HBr, HF; H_2S, HCl? Why?

9.3 Discuss why an ionic hydride, such as CaH_2, gives a basic solution in water, while a covalent hydride, such as HCl, gives an acidic solution. Write equations to represent the processes occurring in solution.

9.4 Explain how it is that hydrogen can form both the H^- ion and the H^+ ion.

9.5 Describe why the H^+ ion does not exist by itself in chemical systems.

9.6 Would you expect liquid LiH to conduct electricity? If the answer is yes, write out the electrode reactions.

9.7 Suggest a reason(s) why H_2 is seldom used as a commercial reducing agent for metal oxides.

9.8 Discuss with reasons, how the acid strength varies along the series:

| HF, | HCl, | HBr, | HI; |
| HClO, | $HClO_2$, | $HClO_3$, | $HClO_4$ |

9.9 Explain why the $(HF)_n$ polymer has a zigzag chain structure.

9.10 Account for the fact that ethanol is more soluble in water than is dimethyl ether. Which do you expect to be the more soluble in water—NH_4Cl or $(CH_3)_4NCl$? Why?

9.11 Discuss the influence of H-bonding on the boiling points of liquids in which H-bonding occurs. Illustrate your answer with examples.

9.12 Explain why water is 4°C is denser than ice.

9.13 Give examples where H-bonding is important to the existence of life.

9.14 Explain why binary boron hydrides are not simple mononuclear compounds.

9.15 Write the formula of a boron hydride which is isoelectronic with CH_4. Does such a species exist? What important chemical property might it have?

9.16 Why should substitution of one hydrogen isotope for another have significant effects on the properties of a compound?

10 The s block Elements

10.1 General Properties of the s block Elements

The s block elements consist of the six alkali metals of group I, and the six alkaline earth metals of group II (Fig. 10.1). The last two members of each group, francium and radium, are radioactive, the longest lived isotope of francium having a half-life of 21 minutes. The metals are all highly reactive and exist in nature in their ionic form, the ions being M^+ and M^{2+} for the alkali and alkaline earth metals respectively. The valence shell electron configurations are $(ns)^1$ and $(ns)^2$ for the group I and II metals respectively, where n ranges from 2 to 7. The next inner set of electrons are $(n-1)p^6$ except for lithium and beryllium where they are the $(1s)^2$ electrons.

A number of the properties of the metals and their ions show regular trends with increasing atomic number, as outlined in Fig. 10.2. Data relating to these properties are listed in Tables 10.1 and 10.2. Some of the periodic trends for both groups of metals will now be reviewed.

Increasing:
- m.pt., b.pt.
- hardness
- sublimation energy
- electronegativity
- ionization energy
- hydration energy
- polarizing power

Li	Be
Na	Mg
K	Ca
Rb	Sr
Cs	Ba

Increasing:
- atomic radius
- ionic radius
- electropositive character
- reactivity
- reducing power
- anion stabilization

Fig. 10.2 Trends in some properties of the s block elements

Group Ia Group IIa

Li	Be
Na	Mg
K	Ca
Rb	Sr
Cs	Ba
Fr	Ra

Alkali metals   Alkaline earth metals

Fig. 10.1 s block elements of the periodic table

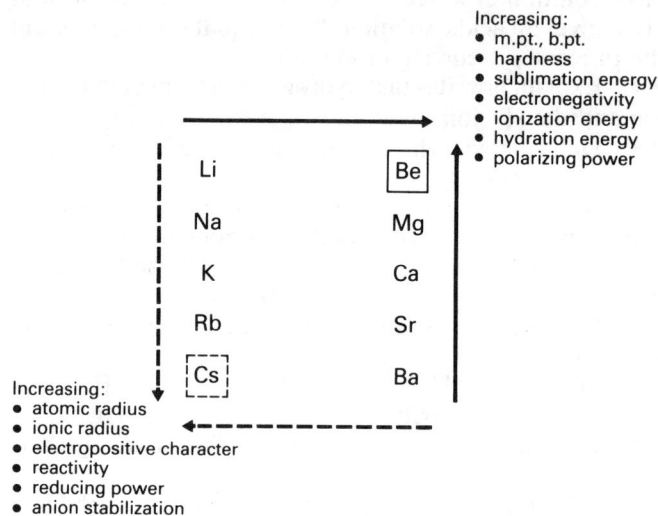

Table 10.1 Properties of the alkali metals

		Li	Na	K	Rb	Cs
Electron configuration $(ns)^1$	$n =$	2	3	4	5	6
Relative atomic mass		6.94	22.99	39.10	85.47	132.90
Metallic radius (pm)		155	190	235	248	267
Ionic radius (pm)		60	95	133	148	169
Crystal form*		bcc	bcc	bcc	bcc	bcc
Ionization energy (kJ mol^{-1})		520	495	418	403	374
Electronegativity		1.15	1.0	0.9	0.9	0.85
Hydration energy						
$M^+(g) \rightarrow M^+(aq)$(kJ mol^{-1})		− 498	− 393	− 310	− 284	− 251
Reduction potential						
Volts		− 3.04	− 2.71	− 2.93	− 2.99	− 3.02
kJ mol^{-1}		181	210	198	205	201
Melting point (°C)		180	97.8	63.2	39	28.5
Boiling point (°C)		1317	881.4	756.5	688	705
Heat of sublimation (kJ mol^{-1})		159	108	90	86	78
Geological abundance (ppm)**		65	28 300	25 900	310	7

    * bcc = body-centred cubic.   ** ppm = grams per tonne.

Table 10.2 Properties of the alkaline earth metals

		Be	Mg	Ca	Sr	Ba	Ra
Electron configuration							
$(ns)^2$	$n =$	2	3	4	5	6	7
Relative atomic mass		9.01	24.31	40.08	87.62	137.34	22.6
Metallic radius (pm)		112	160	197	215	222	
Ionic radius (pm)		31	65	99	113	135	(148)
Crystal form*		hcp	hcp	fcc	hcp	bcc	bcc
Ionization energy							
1st, kJ mol^{-1}		899	738	590	549	503	509
2nd, kJ mol^{-1}		1757	1451	1145	1064	965	979
Electronegativity		1.5	1.25	1.05	1.0	0.95	
Hydration energy							
$M^{2+}(g) \rightarrow M^{2+}(aq)$(kJ mol^{-1})		− 2455	− 1900	− 1565	− 1415	− 1275	
Reduction potential							
Volts		− 1.85	− 2.37	− 2.87	− 2.89	− 2.90	
kJ mol^{-1}		523	439	347	361	369	
Melting point (°C)		1280	649	839	768	727	(700)
Boiling point (°C)		2500	1105	1494	1381	(1849)	(1700)
Heat of sublimation (kJ mol^{-1})		322	150	177	163	176	
Gelogical abundance (ppm)**		6	20 900	36 300	150	430	

    * At room temperature; hcp = hexagonal close-packed, fcc = face-centred cubic = cubic close-packed, bcc = body-centred cubic.   ** ppm = grams per tonne.

(a) Metallic and Ionic Radius

Since the valence electrons are in an s orbital there are significant increases in size for each element as we move down a group and a new quantum shell is started. (The discontinuity in the graph of size versus atomic number, shown in Fig. 10.3, has been discussed in Chapter 2 (p. 22).)

The ionic radius is significantly less than the metallic (or atomic, radius because the ion has had the outer (ns) elec-trons removed. The ionic radius of beryllium is only 31 pm, which is one of the main reasons why its chemistry is not typical of the rest of the metals of group II, and its compounds tend to be more covalent than ionic in character.

The decrease in atomic size on going from group I to group II is due to the contraction effect discussed previously (p. 22). Some increase in size from $(ns)^1$ to $(ns)^2$, due to increased electron repulsion is expected, but clearly this is not the dominating factor.

The importance of atomic size in relation to both the physical and chemical properties of the elements cannot be emphasized too much and this is particularly true for the alkali and alkaline earth metals.

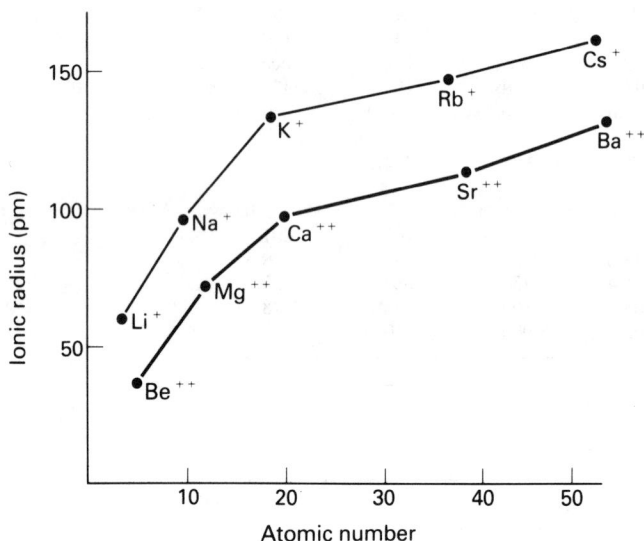

Fig. 10.3 Trends in ionic radius

(b) Polarizing Power

The polarizing power of an ion is dependent on its ratio of charge to radius. The greater the charge and the smaller the radius the greater the polarizing power. The polarizing power of the s block metal ions increases as we move up the groups and as we go from group I to group II. A number of properties can be discussed in terms of the polarizing power of the cations; for example the most polarizing cations have the highest hydration energies (see below).

In the case of lithium and beryllium strong polarization of anions leads to partial covalent bonding (Fig. 10.4). It has been estimated that the bonding in LiI is approximately 50% covalent and in BeI_2 approximately 75% covalent. One effect of the partial covalency of the bonding is that the experimental dipole moments are lower than those calculated for a purely ionic model. The strong polarizing power of lithium and sodium (and to a lesser extent potassium) can destabilize weakly bonded anions such as the polyiodide ions (I_3^-, I_5^-, I_7^-, I_9^-). The polyiodide ions consist of I^- weakly bonded to one or more molecules of iodine. These ions co-exist in the solid state only with large, weakly polarizing cations such as Rb^+, Cs^+ and alkyl ammonium R_4N^+.

The diagonal similarities in the chemistry of pairs of cations such as Li and Mg, Na and Ca, Be and Al or B and Si, are related to similarities in polarizing power; the differences in ionic charge are compensated for by the different ionic radii.

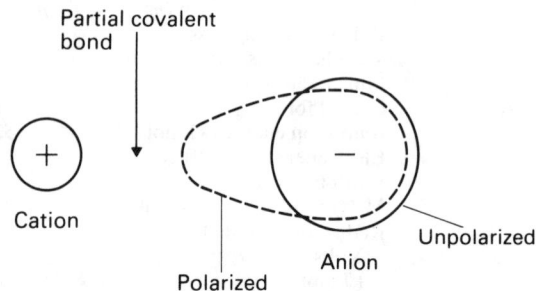

Fig. 10.4 Polarization of an anion by a cation

(c) Crystal Structures

The alkali metals have the body-centred cubic structure, which is not the most efficient form of packing. The metals are soft (easily cut with a knife) with weak metallic bonding (cf. the heats of sublimation and the melting and boiling points in Table 10.1 and 10.2). The harder alkaline earth metals, except for the larger members barium and radium, have much more compact hexagonal and cubic close-packed structures.

(d) Ionization Energy

The ionization energy for the process

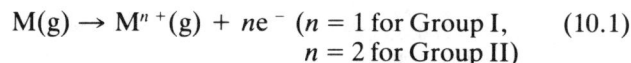

$$M(g) \rightarrow M^{n+}(g) + ne^- \quad (n = 1 \text{ for Group I}, \quad (10.1)$$
$$n = 2 \text{ for Group II})$$

decreases as we move down the groups and as we go from group II to I, as a consequence of the increasing size of the atoms and of the need to remove two electrons from the group II atoms. The ionization energies of the alkali metals are low and it is relatively easy to oxidize the metal to the cation M^+. It is more difficult to obtain the dipositive cation M^{2+} for the alkaline earth metals; indeed the dipositive cation is unstable with respect to $M^+(g)$. Consider magnesium as an example:

$$\Delta H$$
$$Mg(g) \rightarrow Mg^+(g) + e^- \qquad +738 \text{ kJ mol}^{-1} \quad (10.2)$$
$$Mg^{2+}(g) + e^- \rightarrow Mg^+(g) \qquad -1451 \text{ kJ mol}^{-1}, \quad (10.3)$$

hence
$$Mg(g) + Mg^{2+}(g) \rightarrow 2Mg^+(g), \quad -713 \text{ kJ mol}^{-1} \quad (10.4)$$

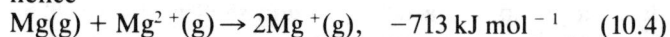

In the solid state or in solution the situation is reversed, and Mg^{2+} becomes the more stable, because of the influence of lattice energies or hydration energies, respectively (see below).

(e) Hydration Energy

The reaction

$$M^{n+}(g) + H_2O \rightarrow M^{n+}(aq) \qquad (10.5)$$

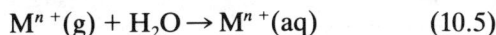

is exothermic, ΔH being largest for the most polarizing of the cations (Tables 10.1 and 10.2). It is this energy which leads to the existence in solution of the dipositive cations of group II metals. If the thermodynamic analysis carried out above is extended to include hydration energies then the reaction

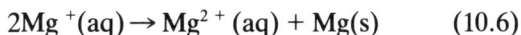

$$2Mg^+(aq) \rightarrow Mg^{2+}(aq) + Mg(s) \qquad (10.6)$$

is exothermic. In order to calculate this enthalpy change it is necessary to have a value for the hydration energy of the Mg^+ cation. For simplicity we will take it as being the same as for Na^+ (this will be an underestimate). Extending the analysis, we have:

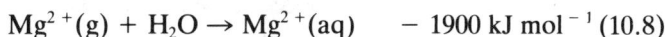

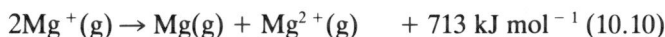

$$2Mg^+(aq) \rightarrow 2Mg^+(g) + H_2O \quad + 786 \text{ kJ mol}^{-1} \ (10.7)$$

$$Mg^{2+}(g) + H_2O \rightarrow Mg^{2+}(aq) \quad - 1900 \text{ kJ mol}^{-1} (10.8)$$

$$Mg(g) \rightarrow Mg(s) \quad - 150 \text{ kJ mol}^{+1} \ (10.9)$$

$$2Mg^+(g) \rightarrow Mg(g) + Mg^{2+}(g) \quad + 713 \text{ kJ mol}^{-1} (10.10)$$

$$2Mg^+(aq) \rightarrow Mg^{2+}(aq) + Mg(s), \ - 551 \text{ kJ mol}^{-1} (10.6)$$

Hydration is clearly an important factor in the stabilization of the $Mg^{2+}(aq)$ cation.

Not only does the hydration energy increase as we move up a group, but so also does the number of water molecules attached to the cation. This alters the effective size of the ions and hence their ionic mobility. Thus lithium, the smallest cation of Group I, is the least mobile because its hydrated size is the largest (Table 10.3).

Table 10.3 Hydration of the alkali metals

Metal	Li$^+$	Na$^+$	K$^+$	Rb$^+$	Cs$^+$
Ionic radius (pm)	60	95	133	148	169
Hydrated radius (pm)	340	276	232	228	228
Relative ionic mobility	1.0	1.3	1.9	2.0	2.0
Hydration number	25.3	16.6	10.5	10	9.9
Hydration energy (kJ mol^{-1})	− 498	− 393	− 310	− 284	− 251

The attraction between water and the Be^{2+} ion is so strong that hydrolysis occurs, with the breaking of an O—H bond in water:

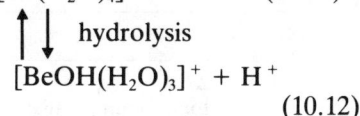

$$Be^{2+} + 4H_2O \rightarrow [Be(H_2O)_4]^{2+} \qquad (10.11)$$

$$\uparrow \downarrow \text{ hydrolysis}$$

$$[BeOH(H_2O)_3]^+ + H^+ \qquad (10.12)$$

In a basic solution the hydroxy anion $[Be(OH)_4]^{2-}$ is obtained.

(f) Electrode Potential

The reduction potential for the reaction

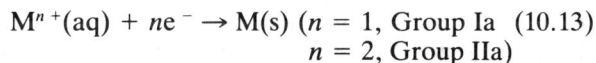

$$M^{n+}(aq) + ne^- \rightarrow M(s) \ (n = 1, \text{ Group Ia} \ (10.13)$$
$$n = 2, \text{ Group IIa})$$

decreases as we move down a group, i.e. it becomes more negative, (except for the anomalous case of lithium) and decreases as we go from group II to I. This means that the reaction as written will proceed less readily for the heavier metals and therefore the reverse reaction

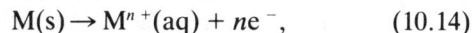

$$M(s) \rightarrow M^{n+}(aq) + ne^-, \qquad (10.14)$$

where the metal is acting as a reducing agent, will occur more readily.

The size of the reduction potential indicates that the alkali metals, and to a lesser extent the alkaline earth metals are very strong reducing agents. They will, for example, reduce water, the reaction occurring with some violence in the case of the group I metals.

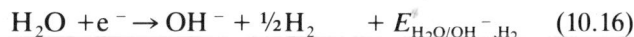

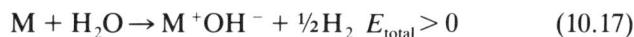

$$M - e^- \rightarrow M^+ \qquad\qquad - E_{M/M^+} \qquad (10.15)$$

$$\underline{H_2O + e^- \rightarrow OH^- + \tfrac{1}{2}H_2 \qquad + E_{H_2O/OH^-,H_2}} \quad (10.16)$$

$$M + H_2O \rightarrow M^+OH^- + \tfrac{1}{2}H_2 \ E_{total} > 0 \qquad (10.17)$$

Lithium does not conform with the general trend within the series, and the reason can be seen if we carry out the analysis which follows. The enthalpy changes (entropy changes can be shown to be less significant in this connection) which contribute to the differences between the reduction potentials of the alkali metals are shown in the Born–Haber cycle:

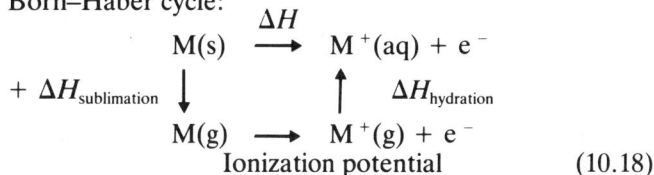

$$\begin{array}{ccc} & \Delta H & \\ M(s) & \longrightarrow & M^+(aq) + e^- \\ +\Delta H_{sublimation} \downarrow & & \uparrow \Delta H_{hydration} \\ M(g) & \longrightarrow & M^+(g) + e^- \\ & \text{Ionization potential} & (10.18) \end{array}$$

The data for the various enthalpy terms in the cycle are listed in Table 10.4 for the alkali metals and for beryllium and magnesium. The figures show that the reaction $M(s) \rightarrow M^+(aq) + e^-$ is least endothermic for lithium, i.e. that lithium metal is the strongest reducing agent. This may be attributed to the high hydration energy of the small lithium

Table 10.4 Enthalpy data for the reaction $M(s) \rightarrow M^+(aq) + e^-$ (in kJ mol^{-1})

	Li	Na	K	Rb	Cs	Be	Mg
$\Delta H_{sublimation}$	159	108	90	86	78	322	150
Ionization potential	520	495	418	403	374	2656	2189
$\Delta H_{hydration}$	-498	-393	-310	-284	-251	-2455	-1900
ΔH	181	210	198	205	201	523	429

ion. The high ionization potential and high negative reduction potential for lithium may appear to contradict one another, but this is not so, because the ionization potential relates to the gas phase and the electrode potential refers to an aqueous solution, the difference being the hydration energy of the ion. The above analysis represents an over-simplification in that enthalpy rather than free energy data were considered, i.e. entropy changes were neglected. This is why the trends in $E^{\ominus}$ and ΔH are not in accord for the other alkali metals.

Beryllium is not anomalous in the same way that lithium is. The data in Table 10.4 indicate that while the hydration energy for the Be^{2+} ion is high it is not as important relative to the total ionization energy and the sublimation energy of beryllium metal as it is for lithium. In view of the high ionization energies for the group II metals it is in fact surprising that they are such good reducing agents. The explanation is provided by the high hydration energies of their dipositive ions.

The ammonium cation NH_4^+ is often considered together with the alkali metals, but it differs in one significant respect, namely that it is often commonly associated with anions through hydrogen bonding. It is also, of course, subject to hydrolysis, reacting with water to form NH_3 and H_3O^+, and with OH^- to form NH_3 and H_2O.

10.2 The Chemistry of the s block Elements

We have seen already that the s block metals, especially the alkali metals, are very reactive. Reactivity generally increases as we move down a group. This is because the important chemical reaction for the metals is:

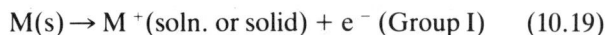

$$M(s) \rightarrow M^+(\text{soln. or solid}) + e^- \text{ (Group I)} \quad (10.19)$$

and

$$M(s) \rightarrow M^{2+}(\text{soln. or solid}) + 2e^- \text{ (Group II)} \quad (10.20)$$

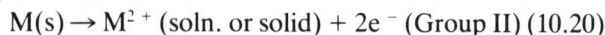

As described above, these reactions tend to become more favourable as we move down the groups, the exceptions being lithium and beryllium.

(a) Alkali Metals

Typical reactions of the alkali metals are given in Table 10.5. The reactions shown do not necessarily provide the best means of obtaining the specified reaction products; the nitrides M_3N (M = Na, K, Rb and Cs) are an example.

Table 10.5 Reactions of the alkali metals

Reaction	Comments
$M + \frac{1}{2}H_2 \rightarrow M^+H^-$	Reaction at high temperatures, products ionic as shown.
$M + \frac{1}{2}X_2 \rightarrow M^+X^-$ (X = halogen)	Reaction occurs readily; poly-halides may also form for heavier metals e.g. $K^+ICl_4^-$.
$2Li + \frac{1}{2}O_2 \rightarrow Li_2O$	Other metals form Na_2O_2, K_2O_2, KO_2, RbO_2 and CsO_2 (see text).
$2M + S \rightarrow M_2S$	(Also Se and Te), reaction proceeds readily, can also obtain polysulphides.
$3M + \frac{1}{2}N_2 \rightarrow 3M^+N^{3-}$	Reaction occurs most readily for lithium; indirect methods used for other metals.
$3M + P \rightarrow M_3P$	(Also As, Sb, Bi); heat required.
$2M + 2C \rightarrow M_2C_2$	Most readily for lithium. Reaction best achieved using C_2H_2 instead of C.
$M + H_2O \rightarrow M^+OH^- + \frac{1}{2}H_2$	Lithium reacts most slowly.
$M + NH_3(l) \rightarrow [M(NH_3)_n]^+ + e^-(NH_3)$ $[M(NH_3)_n]^+ \rightarrow M^+NH_2^- + \frac{1}{2}H_2$	Blue solution of electrons in liquid ammonia. Second reaction needs to be catalysed.

The reaction of the alkali metals with oxygen highlights interesting differences in the reducing powers of the metals, and shows how the reducing power is influenced by the lattice energies of the products. The reactions are:

Reaction	Product	Metal	Oxidation state of oxygen	
$2M + \frac{1}{2}O_2 \rightarrow M_2O$	Oxide $2M^+O^{2-}$	Li	-2	(10.21)
$2M + O_2 \rightarrow M_2O_2$	Peroxide $2M^+O_2^{2-}$	Na	-1	(10.22)
$M + O_2 \rightarrow MO_2$	Superoxide, $M^+O_2^-$	K, Rb, Cs	$-\frac{1}{2}$	(10.23)

The strongest reducing agent, lithium, reduces oxygen to O^{2-} (oxide) whereas sodium produces O_2^{2-} (peroxide) and the heavier metals produce O_2^- (superoxide). The courses of the reactions are influenced by the lattice energies: lithium oxide has a lattice energy of 2885 kJ mol^{-1}, whereas for Na$_2$O it is 2515 kJ mol^{-1}. The higher lattice energy of Li$_2$O makes it possible for the endothermic reduction of ½O$_2$ to O^{2-} ($+$ 940 kJ mol^{-1}) to be achieved. A second factor is that the larger cations are less polarizing and will stabilize the larger, more easily polarizable anions O$_2^{2-}$ and O$_2^-$, whereas the strong polarizing effect of lithium weakens the O—O bond leading to its disruption and producing O^{2-}. The reaction with oxygen produces the thermodynamically most stable compound under the given reaction conditions; this does not mean, however, that the other compounds do not exist (see Table 10.6).

Table 10.6 Binary oxygen compounds of the alkali metals. Heats of formation (kJ mol^{-1}) of solid compounds at 298 K

	Li	Na	K	Rb	Cs
M$_2$O	$-$ 598	$-$ 414	$-$ 360	$-$ 331	$-$ 318
M$_2$O$_2$	$-$ 272	$-$ 506	$-$ 494	$-$ 427	$-$ 402
MO$_2$	*	$-$ 259	$-$ 285	$-$ 289	$-$ 293

* LiO$_2$ known in dilute solid solutions at low temperatures and in the gas phase at high temperatures.

Solutions of the alkali metals in liquid ammonia are deep blue in colour and contain both solvated M$^+$ ions and solvated electrons. The solutions, which are useful reducing agents, conduct electricity. At high concentrations they became copper-coloured and have typical liquid metallic properties.

Table 10.7 Reactions of the alkaline earth metals

Reaction	Comments
M + H$_2$ → M^{2+}2H$^-$	Reaction does not go for Be; Mg only reacts under pressure of H$_2$.
M + X$_2$ → M^{2+}2X$^-$ (X = halogen)	Readily achieved reaction.
2M + O$_2$ → 2MO	Barium and strontium also give some peroxide MO$_2$.
M + S → MS	Heat needed.
3M + N$_2$ → M$_3$N$_2$	Reactions carried out at red heat.
M + 2C → MC$_2$	Formed at high temperatures, compounds ionic.
M + 2H$_2$O → M(OH)$_2$ + H$_2$	Ca, Sr, Ba will react with cold water; Be and Mg require steam.
M + 2OH$^-$ → MO$_2^{2-}$ + H$_2$	Only occurs for beryllium.
M + 2H$^+$ → M^{2+} + H$_2$	Be is the least reactive, and may become passive (protected by a surface oxide film).
M + NH$_3$(l) → M^{2+}(NH$_3$) + 2e$^-$(NH$_3$)	For Ca, Sr, Ba; in gaseous ammonia nitrides are formed.

(b) Alkaline Earth Metals

The main reactions of the alkaline earth metals closely follow those of the alkali metals (Table 10.7). The main differences are;

(i) amphoteric properties become evident; beryllium for example, dissolves fairly readily in strong alkali but reacts with acid with difficulty;

(ii) the trend of increasing reactivity down a group is more obvious for the group II metals; and

(iii) all of the metals burn in air to give the oxide MO.

The exclusive formation of normal oxides MO by burning the metals in air must be related to the very high lattice energies of the oxides (3800 kJ mol^{-1} for MgO). The peroxides of Mg, Ca, Sr and Ba and superoxides of Ca, Sr and Ba can be obtained by other methods.

10.3 Compounds of the Alkali Metals

(a) Oxides

The oxides of the alkali metals (other than Li$_2$O) may be prepared by reduction of the metal nitrate with free metal.

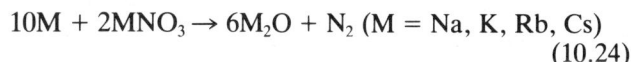

$$10M + 2MNO_3 \rightarrow 6M_2O + N_2 \quad (M = Na, K, Rb, Cs)$$
$$(10.24)$$

The oxides have the antifluorite structure, i.e. cubic close-packing of O^{2-} ions with the metal ions filling all the tetrahedral holes. There are twice as many tetrahedral holes as O^{2-} ions, hence the O^{2-}:M$^+$ coordination is 8:4 (Fig. 10.5). (Fluorite, CaF$_2$, has Ca^{2+} ions in cubic close-packing with F$^-$ ions in the tetrahedral holes).

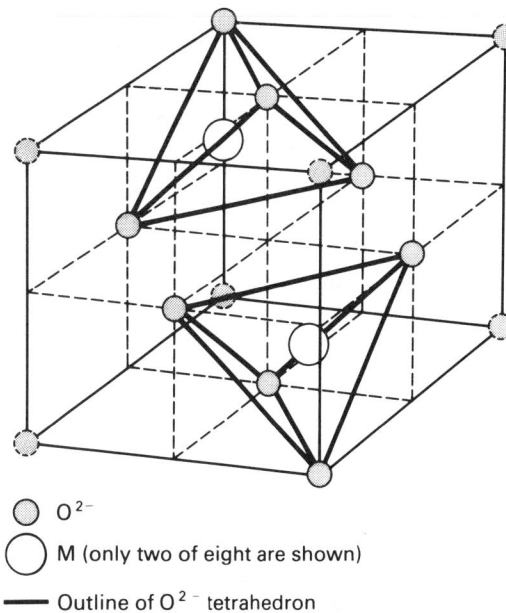

○ O^{2-}

◯ M (only two of eight are shown)

— Outline of O^{2-} tetrahedron

Fig. 10.5 Structure of alkali metal oxides (antifluorite)

The oxides react with water to form hydroxides:

$$O^{2-}(s) + H_2O \rightarrow 2OH^- \qquad (10.25)$$

The oxide ion O^{2-} cannot exist in water, but immediately undergoes hydrolysis. The peroxides react with water to form hydrogen peroxide:

$$O_2^{2-} + 2H_2O \rightarrow H_2O_2 + 2OH^- \qquad (10.26)$$

while the superoxides react to form H_2O_2 and liberate oxygen,

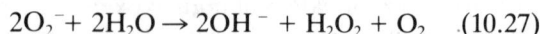

$$2O_2^- + 2H_2O \rightarrow 2OH^- + H_2O_2 + O_2 \qquad (10.27)$$

(b) Halides

The alkali metal halides are an extremely stable group of compounds because of their high lattice energies and the stability of the cations M^+ and anions X^-. The formation of the halides can usefully be considered in terms of the Born–Haber cycle:

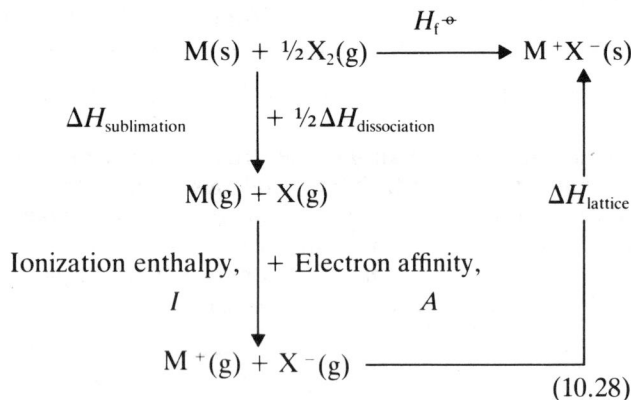

$$
\begin{array}{ccc}
M(s) + \tfrac{1}{2}X_2(g) & \xrightarrow{\;H_f^{\ominus}\;} & M^+X^-(s) \\
\end{array}
$$

$\Delta H_{\text{sublimation}}$ $+ \tfrac{1}{2}\Delta H_{\text{dissociation}}$

$$M(g) + X(g) \qquad\qquad \Delta H_{\text{lattice}}$$

Ionization enthalpy, I $+$ Electron affinity, A

$$M^+(g) + X^-(g) \qquad\qquad\qquad (10.28)$$

This cycle was discussed in the examples on page 47. Data for some representative halides are given in Table 10.8. The figures indicate the relative importance of the various energy terms, and their variation with the nature of the

anion and cation. The lattice energies and heats of formation of the alkali metal halides are shown graphically in Figs. 10.6 and 10.7.

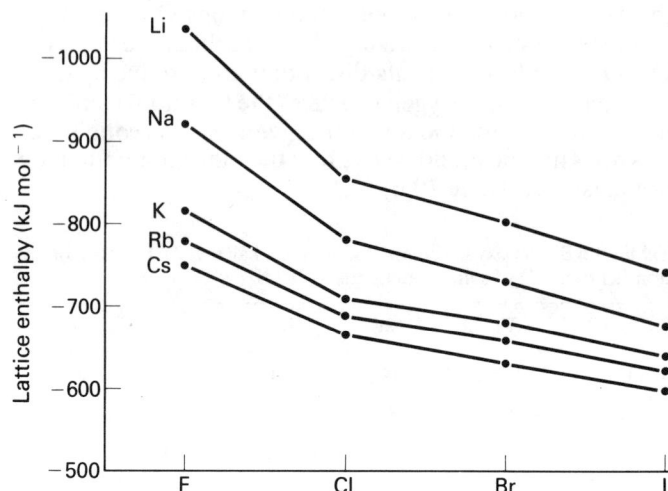

Fig. 10.6 Lattice energies of alkali metal halides

The lattice energies (Fig. 10.6) follow the expected trends, decreasing as we go from fluoride to iodide for a given metal, and from lithium to caesium for a given halide. The value for LiF is significantly higher than for the other halides.

The heats of formation (Table 10.8 and Fig. 10.7) for a given metal, decrease as we go from fluoride to iodide, which is due primarily to a decrease in the term ($\tfrac{1}{2}\Delta H_{\text{diss}} + A$) and a decrease in the lattice energy. For the same halogen two trends are observed:

(a) The magnitude of the heat of formation decreases from Li to Cs for the fluorides, as a result of the dominating influence of the lattice energies for LiF and NaF.

Table 10.8 Data for the Born–Haber cycles of some alkali metal halides (kJ mol^{-1})

	LiF	NaF	NaCl	NaBr	NaI	LiCl	NaCl	KCl	RbCl	CsCl	
ΔH_{sub}	159	108	108	108	108	159	108	90	86	78	} Metal
I	520	495	495	495	495	520	495	418	403	374	}
$\tfrac{1}{2}\Delta H_{\text{diss}}$	79	79	121	97	76	121	121	121	121	121	} Halogen
A	−333	−333	−355	−330	−302	−355	−355	−355	−355	−355	}
ΔH_{latt}	−1035	−920	−780	−730	−675	−853	−787	−710	−688	−665	
$H_f^{\ominus}$	−610	−571	−411	−360	−298	−408	−411	−436	−433	−447	

($H_f^{\ominus} = \Delta H_{\text{sub}} + \tfrac{1}{2}\Delta H_{\text{diss}} + I + A + \Delta H_{\text{latt}}$. In the example on p. 47, ΔH_{diss} is given the symbol D (dissociation energy of the halogen molecule) and $\Delta H_{\text{lattice}}$ is given the symbol U_l.

(b) For the other halogens $-H_f^{\ominus}$ increases from Li to Cs, but not by a great amount. The lattice energies for the chlorides, bromides and iodides do not differ greatly (especially for the K^+, Rb^+ and Cs^+ salts, Fig. 10.7), therefore the decrease in the sublimation and ionization energies of the metal atoms dominates the overall situation, with the result that the caesium halides (Cl, Br and I) are the most stable.

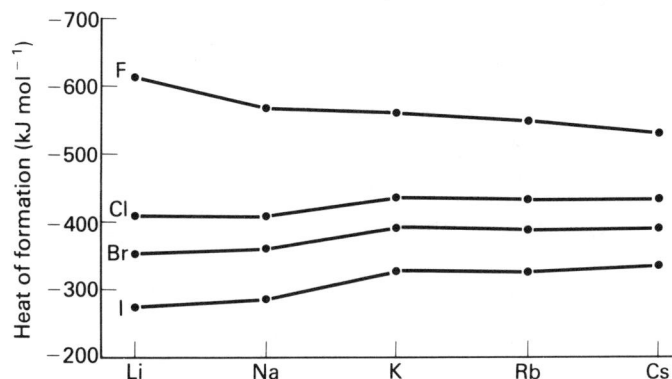

Fig. 10.7 Heats of formation of crystalline alkali metal halides

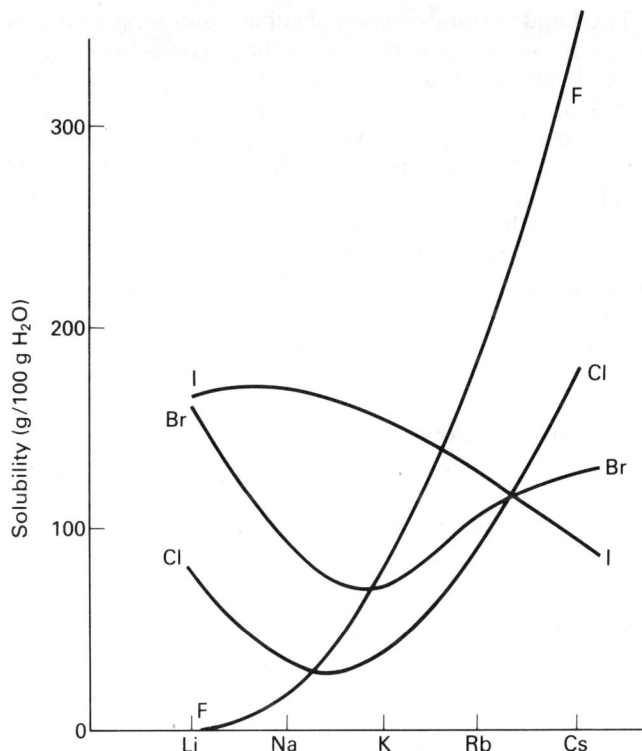

Fig. 10.8 Solubilities of alkali metal halides (25°C)

The alkali metal halides are soluble in water, the solubility depending on the free energy changes involved in breaking down the lattice and forming hydrated ions. Trends in solubility (Fig. 10.8) depend on the interplay of these effects as they vary with cation or anion. The reader may wish to calculate $-\Delta H_{soln.} = (\Delta H_{lattice} - \Delta H_{hydration})$ for a number of halides and decide for himself which factors are responsible for the various trends (Fig. 10.8). Since $\Delta H_{lattice}$ and $\Delta H_{hydration}$ are of comparable magnitude for the alkali metal halides, it turns out that ΔH_{soln} is generally small (Table 10.9). Hence the solubilities of the halides do not change greatly with temperature, in contrast with, for example, KNO_3 whose solubility increases rapidly with temperature (Fig. 10.9).

Table 10.9 Heats of solution for some alkali metal halides (kJ mol^{-1})

	$-\Delta H_{lattice}$	$-\Delta H_{hydration}$	ΔH_{soln}
NaI	686	698	-12
KF	817	770	$+47$
KCl	699	695	$+4$
KI	615	615	0
RbI	596	589	$+7$

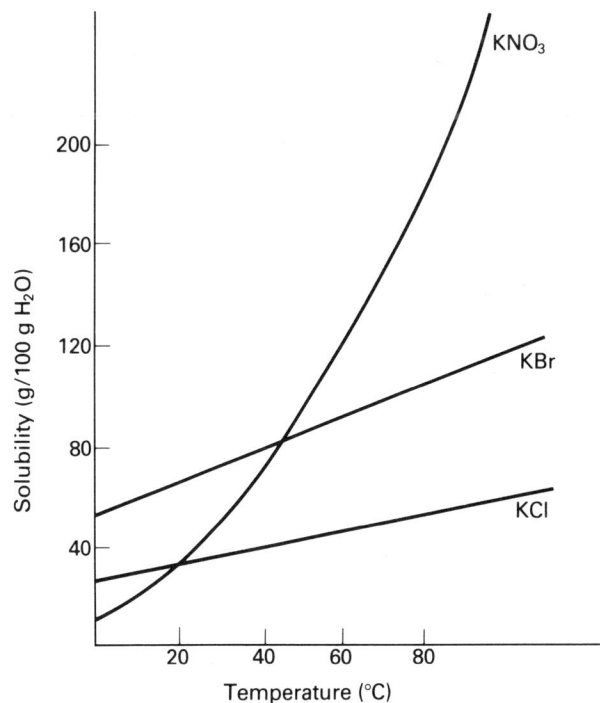

Fig. 10.9 Variation of solubility with temperature

The halide crystals consist of cubic close-packed lattices of halide ions with the metal ions filling cavities in the close-packed lattice. (Except for lithium, the metal ions are sufficiently large to prevent the halide ions being in contact with one another.) The two basic structural types are the NaCl and CsCl structures (Table 10.10, Figs 10.10 and 10.11). In the former the metal ion occupies octahedral cavities, with six-fold coordination of halide ions; in the latter the metal ion is surrounded by eight halide ions (a body-centred cubic structure). The alkali metal halides form polyhalide salts with the halogens (see page 132) and with other halides such as $CaCl_2$ and $MgCl_2$. The latter type are best considered as double salts or solid solutions of one halide in another. The halides can also be used to form halogeno-complexes, e.g.

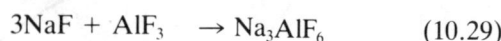

$$3NaF + AlF_3 \rightarrow Na_3AlF_6 \qquad (10.29)$$

$$2KCl + PdCl_2 \rightarrow K_2PdCl_4 \qquad (10.30)$$

$$2RbBr + SnBr_4 \rightarrow Rb_2SnBr_6 \qquad (10.31)$$

Table 10.10 Structures of alkali metal halides

| Structure | Compounds | Coordination number | |
		Anion	Cation
NaCl	Li, Na, K, Rb halides (F, Cl, Br and I) and CsF	6	6
CsCl	CsCl, CsBr, CsI	8	8

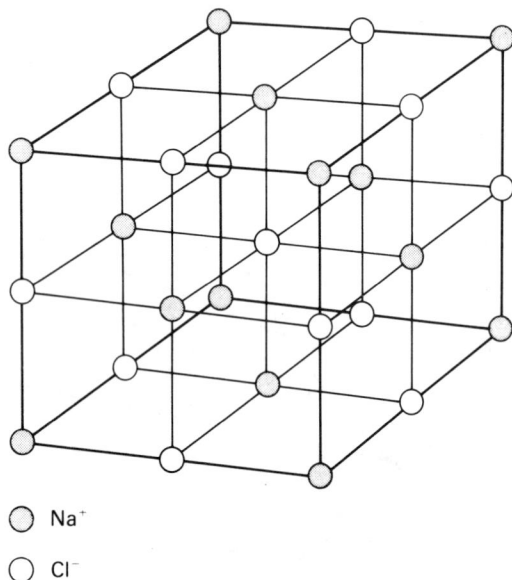

Cs$^+$

Cl$^-$

Fig. 10.11 CsCl structure

Na$^+$

Cl$^-$

Fig. 10.10 NaCl structure

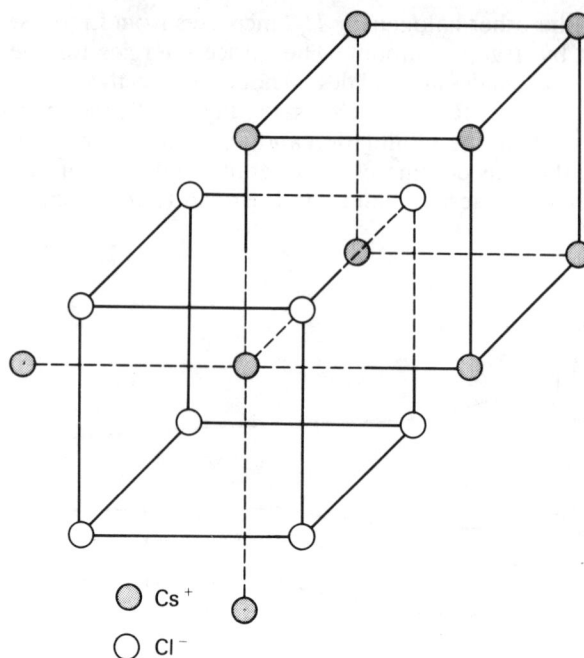

(c) Hydroxides

Alkali metal oxides react with water to give hydroxides. These are the most basic of hydroxides and their basicity increases from lithium to caesium, since the dissociation reaction:

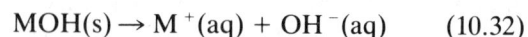

$$MOH(s) \rightarrow M^+(aq) + OH^-(aq) \qquad (10.32)$$

occurs most readily for the large cations. Reaction of the hydroxides with acids produces salts; reaction even occurs with CO_2 to give carbonates and bicarbonates.

Both NaOH and KOH are important industrial chemicals. Sodium hydroxide is used in soap manufacture, and both are used in other industrial processes such as the production of rayon and of paper, and the purification of bauxite for aluminium production. Sodium hydroxide is prepared commercially by the electrolysis of aqueous sodium chloride. The cell reactions are:

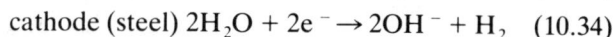

anode (carbon) $2Cl^- \rightarrow Cl_2 + 2e^-$ (10.33)

cathode (steel) $2H_2O + 2e^- \rightarrow 2OH^- + H_2$ (10.34)

If a mercury cathode is used the reaction at the electrode is:

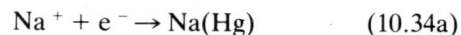

$$Na^+ + e^- \rightarrow Na(Hg) \qquad (10.34a)$$

The sodium forms sodium amalgam which is made to react with water in a separate vessel to give NaOH and H_2 gas.

The chlorine and hydrogen are collected and may be combined to give HCl, or used in other ways. The solid hydroxides have the sodium chloride structure.

(d) Oxyanion Salts

A large range of oxyanion salts are known for the alkali metals. They are readily prepared by the addition of a metal hydroxide to the appropriate acid. The salt Na_2CO_3, 'washing soda', is an important commercial chemical which is prepared by the Solvay process according to the reactions:

$$CaCO_3 \xrightarrow{heat} CaO + CO_2 \qquad (10.35)$$

$$CO_2 + NH_4OH \rightarrow NH_4HCO_3 \qquad (10.36)$$

$$NH_4HCO_3 + NaCl \rightarrow NaHCO_3 + NH_4Cl \qquad (10.37)$$

$$2NH_4Cl + CaO + H_2O \rightarrow 2NH_3 + 2H_2O + CaCl_2 \qquad (10.38)$$

$$2NaHCO_3 \xrightarrow{heat} Na_2CO_3 + H_2O + CO_2 \qquad (10.39)$$

Recently, extensive mining of $Na_2CO_3NaHCO_3$ ('trona'), which gives entirely Na_2CO_3 on heating, has provided some competition for the Solvay process. A third process, namely the addition of CO_2 to caustic soda, is also being used.

The effect of heat on some oxyanion salts is summarized in Table 10.11. Generally lithium salts decompose most readily because of the strong polarizing power of the Li^+ ion.

The solubility of the oxyanion salts depends to some extent on the relative sizes of the cations and anions. If the anion is small (OH^-), trends in lattice energy dominate and solubilities increase down the group. If the anions are larger (e.g. SO_4^{2-}) the lattice energy is determined mainly by the anion (see later p. 142) and therefore is reasonably constant for a series of metals. Hence trends in the hydration energy dominate and solubilities increase up the group.

(e) Complexes and Organo-metallic Compounds

The alkali metals, particularly lithium, form compounds termed 'coordination complexes', in which a central atom is surrounded by a symmetrical arrangement of four or more atoms belonging to 'ligand' molecules. For example, the reaction of sodium hydroxide and the diketone pentane-2,4-dione gives a neutral complex in which electron pairs are shared between the sodium atom and four oxygen atoms, two of which belong to water molecules:

Sodium pentane-2,4-dione complex   (10.41)

Lithium, having the greatest tendency to form covalent bonds, forms complexes most readily. Direct covalent bonds to carbon atoms of organic compounds can also be formed, e.g. by reaction with an alkyl halide:

$$2M + RX \rightarrow MR + MX \qquad (10.42)$$

as in

$$2Li + C_2H_5Cl \rightarrow LiC_2H_5 + LiCl \qquad (10.42a)$$

These organo-metallic compounds are highly reactive and are used as sources of organic groups in a variety of reactions. The butyl lithiums are used extensively as polymerization catalysts.

10.4 Compounds of the Alkaline Earth Metals

(a) Oxides

The oxides of the alkaline earth metals can be prepared by burning the metals in oxygen:

$$2M(s) + O_2(g) \rightarrow 2MO(s) \qquad (10.43)$$

Table 10.11 Effect of heat on alkali metal oxyanion salts (°C)

		CO_3^{2-}	HCO_3^-	NO_3^-		SO_4^{2-} and OH^-
Li^+	800°		Give CO_3^{2-}	$Li_2O + NO_2 + O_2$		
Na^+	950°	Decompose	CO_2 and	380°	Give	Stable
K^+	1000°	to give	H_2O below	400°	NO_2^-	to
Rb^+	900°	the oxide	200°C		and O_2	heat
Cs^+	620°			600°		

The compounds are refractory (high-melting) and MgO and CaO are used as lining materials for furnaces, especially if acidic impurities are to be removed. The high thermal stability of the oxides is a consequence of their high lattice energies and heats of formation.

The oxides all have the NaCl structure, except for BeO which has the wurtzite (ZnS) structure. The latter structure has 4:4 anion:cation coordination, and consists of a hexagonal close-packed lattice of anions with half of the available tetrahedral cavities occupied by cations.

The oxides are converted to hydroxides by reaction with water:

$$MO + H_2O \rightarrow M(OH)_2 \qquad (10.44)$$

(b) Halides

The dihalides are obtained by burning the metal in the appropriate halogen. The important halide calcium fluoride, CaF_2, occurs naturally as fluorspar and is the main mineral source of fluorine. Calcium fluoride is also the principal source of fluoride ions in water and in many places is added to drinking water (1 to 1.5 ppm) in order to reduce the incidence of dental caries.

The dihalides are thermodynamically stable compounds, as the data in Table 10.12 indicate. Their heats of formation increase on going from Be to Ba for a given halogen, and for a given metal increase from iodine to fluorine. The trends correlate well with trends in lattice energy. The repulsive forces between the non-bonding electrons on the halogen atoms may also contribute to the trends in the heats of formation. For a given halogen the repulsion will be least for the biggest metal atoms and for a given metal they will be least for the smallest halogens.

Table 10.12 Melting points (°C) and heats of formation (kJ mol^{-1}) for alkaline earth halides

	F	Cl	Br	I	
Be	522°	405°	488°	510°	M.Pt
	− 1014	− 502	− 332	− 165	$H_f^{\ominus}$ (298)
Mg	1263°	708°	711°	700°(d)	M.Pt
	− 1124	− 640	− 519	− 360	$H_f^{\ominus}$ (298)
Ca	1418°	772°	730°	740°	M.Pt
	− 1215	− 795	− 675	− 535	$H_f^{\ominus}$ (298)
Sr	1400°	873°	643°	507°	M.Pt
	− 1215	− 828	− 716	− 567	$H_f^{\ominus}$ (298)
Ba	1320°	963°	847°	711°	M.Pt
	− 1200	− 862	− 755	− 602	$H_f^{\ominus}$ (298)

It is instructive to consider the relative thermodynamic stabilities of the halides MgCl, $MgCl_2$ and $MgCl_3$. The energy terms which vary in calculation of $H_f^{\ominus}$ from a Born–Haber cycle are the ionization energy of magnesium, the lattice energy, and the term involving ($\frac{1}{2}\Delta H_{diss} + A$) for chlorine (this term varies because of the number of chlorine atoms required for each compound). The values of these energies and $H_f^{\ominus}$ are plotted in Fig. 10.12. It is clear that the high energy of ionization to obtain Mg^{3+} prevents the formation of $MgCl_3(s)$. MgCl(s) is stable ($\Delta H = - 168$ kJ mol^{-1}) relative to dissociation into the constituent atoms, but it is unstable with respect to disproportionation into $MgCl_2(s)$ and Mg(s) ($\Delta H = - 306$ kJ mol^{-1}):

$$2MgCl(s) \rightarrow MgCl_2(s) + Mg(s) \qquad (10.45)$$

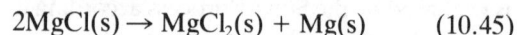

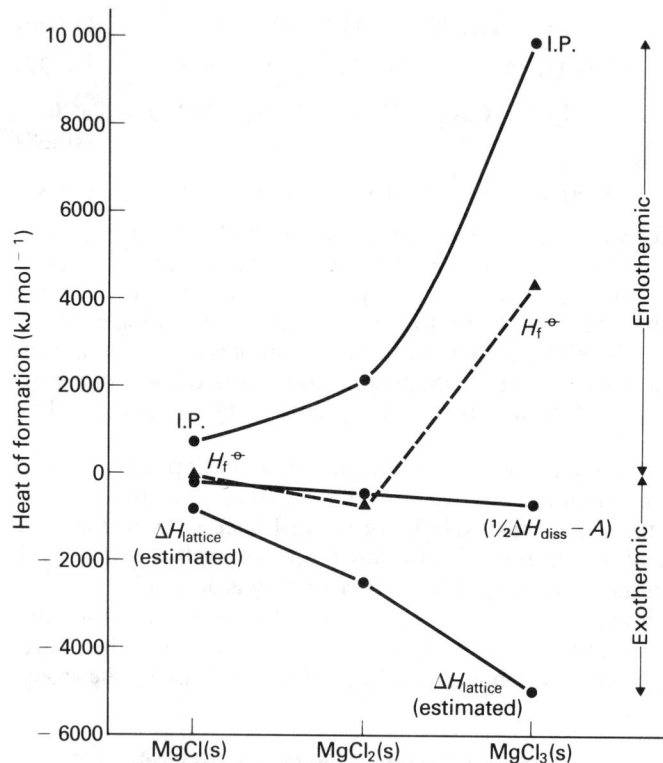

Fig. 10.12 Energy terms involved in the heats of formation of MgCl, $MgCl_2$ and $MgCl_3$

There is evidence for covalency in the bonding of the solid halides, and for beryllium the halides may be considered predominantly covalent (cf. the melting points in Table 10.12). The tendency to covalency results in the structures of the halides differing from the usual ionic structural types. Thus, for example, calcium chloride and bromide have a distorted rutile (TiO_2)-type structure with coordination numbers of 6 and 3, while magnesium bromide and iodide have the CdI_2 layer lattice structure. (The halogen atoms form a hexagonal close-packed lattice, with each metal atom bonded to halogens in one layer of the lattice only.) The halides BeF_2, $BeCl_2$ and $BeBr_2$ are polymeric and have chain structures (Fig. 10.13) with each beryllium tetrahedrally coordinated. The coordination number of beryllium drops as the compound is heated and vaporized to give Be_2Cl_4 and finally $BeCl_2$ (Fig. 10.13).

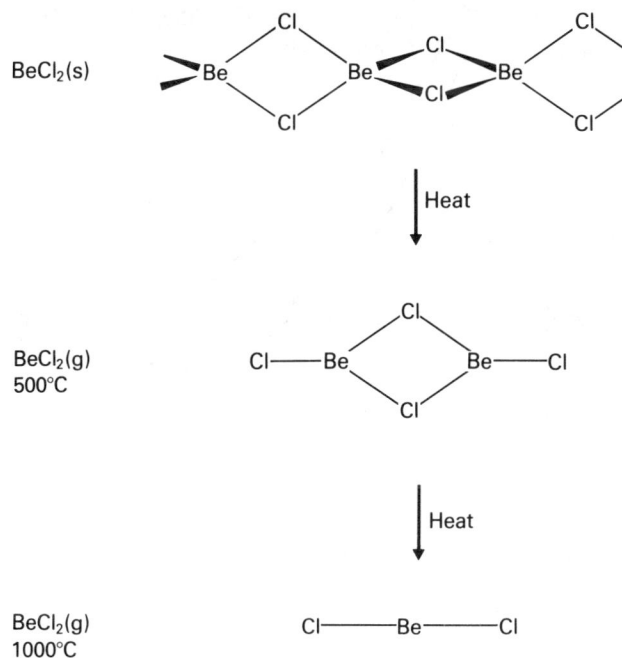

$BeCl_2(s)$

| Heat

$BeCl_2(g)$
500°C

| Heat

$BeCl_2(g)$
1000°C

Fig. 10.13 Structure of $BeCl_2$

The gaseous halides are monomeric covalent molecules, but surprisingly the structures are not all linear as predicted by the electron pair repulsion theory (Table 10.13). The bond angles for the bent molecules lie within the range of 100–160°. The incorporation of d orbitals in the bonding could account for the bent structures. The anhydrous halides dissolve in water and can be recovered as hydrates, such as $BaCl_2.2H_2O$ and $MgCl_2.6H_2O$, in which the water is associated with the cation by ion–dipole interactions.

Table 10.13 Structures of gaseous alkaline earth metal halides

	F	Cl	Br	I
Be	l	l	l	l
Mg	l	l	l	l
Ca	b	l	l	l
Sr	b	b	l	l
Ba	b	b	b	b

l = linear   b = bent

(c) Hydroxides

The alkaline earth metal oxides react with water to form the hydroxides $M(OH)_2$. These are weaker bases than the hydroxides of the alkali metals, because of their lower solubility and reduced ionic character. The layer lattice structures of both $Mg(OH)_2$ and $Ca(OH)_2$ give evidence of an increased tendency to covalent bonding.

Certain hydroxides can be precipitated from a solution of a metal salt by introducing OH^- ions, a process which depends on the concentration of OH^- ions being sufficiently large in relation to the solubility product of the hydroxide (Table 10.14). Solutions of ammonia produce sufficient hydroxide ions to exceed the solubility product of $Mg(OH)_2$, but if ammonium chloride is added no precipitate forms because the hydroxide ion concentration is reduced, the equilibrium

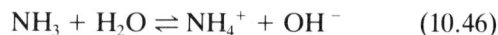

$$NH_3 + H_2O \rightleftharpoons NH_4^+ + OH^- \qquad (10.46)$$

being displaced to the left by the increase in the ammonium ion concentration.

Table 10.14 Alkaline earth hydroxides

	Be	Mg	Ca	Sr	Ba
NH_3/H_2O solution	$Be(OH)_2$	$Mg(OH)_2$	No precipitate	No precipitate	No precipitate
NH_3/H_2O + NH_4Cl solution	$Be(OH)_2$	No precipitate	No precipitate	No precipitate	No precipitate
NaOH	$Be(OH)_2$ $Be(OH)_4^{2-}$ in excess	$Mg(OH)_2$	$Ca(OH)_2$	$Sr(OH)_2$ Slight precipitate	No precipitate
Solubility product (mole l^{-1} units)	1.6×10^{-26}	8.9×10^{-12}	1.3×10^{-6}	3.2×10^{-4}	5×10^{-3}

The solubility of the hydroxides increases as we move down the group (Table 10.14). Since the OH$^-$ and M^{2+} ions are comparable in size the lattice energy of the hydroxides depends on the cation size and so decreases as we move down the group. However, the hydration energy of the cation also decreases on moving down the group, and it happens fortuitously that the lattice energy dominates, giving rise to the increasing solubility of the heavier hydroxides.

(d) Oxyanion Salts

The alkaline earth metals form a wide range of oxyanion salts, but it is significant that solid bicarbonates cannot be produced, presumably because of instability resulting from the strong polarizing power of the dipositive cations.

The effect of heat on some oxyanion salts is summarized in Table 10.15; it is clear that thermal stability is lower than for the corresponding alkali metal compounds. The carbonates decompose according to the reaction

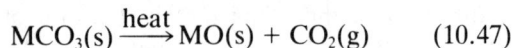

$$MCO_3(s) \xrightarrow{\text{heat}} MO(s) + CO_2(g) \qquad (10.47)$$

the ease of decomposition increasing from barium to beryllium. The heat of the reaction (Fig. 10.14) is given by the difference between the lattice energy of the oxide (ΔH_{latt} (MO)) and the sum of the lattice energy of the carbonates (ΔH_{latt} (MCO$_3$)) and the dissociation energy (ΔH_{diss}) for the reaction

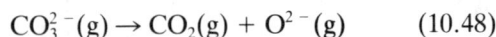

$$CO_3^{2-}(g) \rightarrow CO_2(g) + O^{2-}(g) \qquad (10.48)$$

Table 10.15 Effect of heat on alkaline-earth oxyanion salts

	Carbonates	Sulphates	Nitrates	Hydroxides
Be		Gives BeO + SO$_3$ 650°C		
Mg	All give MO	Stable	All give metal oxide	All give metal oxide
Ca	+ CO$_2$	Stable	+ NO$_2$ and O$_2$	+ H$_2$O
Sr		Gives MO + SO$_3$		
Ba		$T \sim 1580°C$		

The reason for the decrease in thermal stability of the carbonates on going from Ba to Be is that the lattice energy of the oxide increases more rapidly from Ba to Be than it does for the carbonates, which is due to the difference in the size of the anions. The lattice energy is inversely proportional to the sum of ionic radii ($r_+ + r_-$). For the carbonates $r_- \gg r_+$; hence the lattice energy does not change

greatly with a change in the radius of the cation (r_+). For the oxides r_+ and r_- are comparable, so the lattice energy is significantly influenced by the increase in the cation radius.

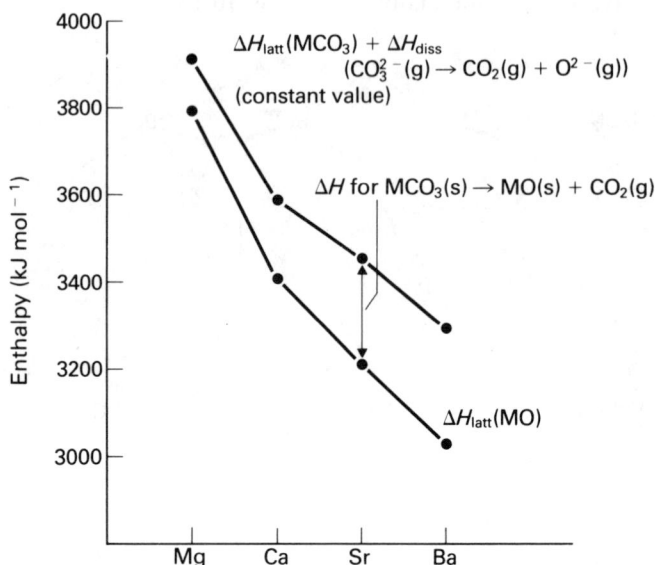

Fig. 10.14 Thermodynamic quantites for the reaction $MCO_3(s) \rightarrow MO(s) + CO_2(g)$

A similar type of reasoning may be used to explain why the solubilities of the carbonates, sulphates and chromates increase as we move up the group (Table 10.16). The rise in hydration energy from Ba to Be is more rapid than the increase in lattice energy, so the hydration energy dominates.

Table 10.16 Solubility products (mol^3 litre^{-3}) for alkaline earth oxyanion salts

	SO$_4^{2-}$	CO$_3^{2-}$	CrO$_4^{2-}$
Mg	10	1×10^{-5}	—
Ca	2.4×10^{-5}	4.8×10^{-9}	7.1×10^{-1}
Sr	2.8×10^{-7}	9.4×10^{-10}	3.6×10^{-5}
Ba	9.9×10^{-11}	4.9×10^{-9}	2×10^{-10}

Beryllium forms complex oxyanion compounds such as [Be$_4$O (acetate)$_6$] and [Be$_4$O(NO$_3$)$_6$] (Fig. 10.15). They consist of a tetrahedral arrangement of beryllium atoms with an oxygen at the centre, and each pair of beryllium atoms bridged by the acetate or nitrate ligands. This type of structure also occurs for zinc and copper.

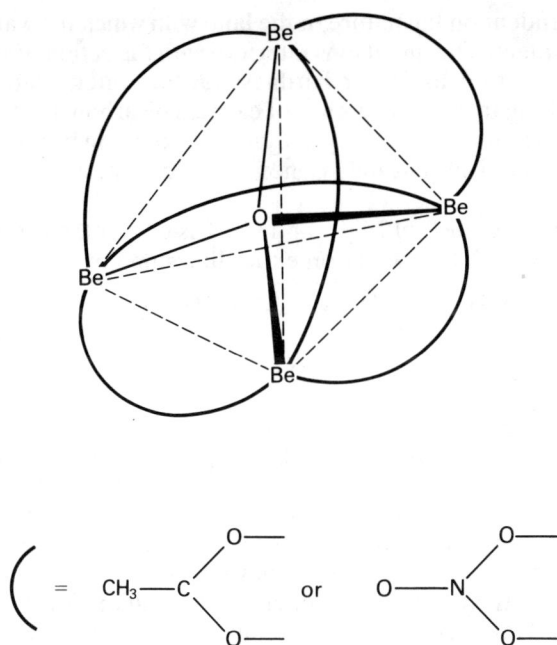

Fig. 10.15 Shape of [Be₄O (acetate)₆] and [Be₄O (NO₃)₆]

(e) Complexes

The group II metals form a wider range of complexes than the alkali metals, the beryllium complexes [Be₄O (ligand)₆] mentioned above being examples. Chlorophyll, a magnesium complex, is of outstanding importance. It contains a planar arrangement of four nitrogen atoms, part of a porphyrin ring, around the metal (Fig. 10.16). The chlorophylls are the green coloured pigments in plants; they absorb visible radiation at 400 nm (violet) and 650 nm (red). The energy so absorbed is transferred to active sites within the chloroplast which contains the chlorophyll. At the active sites the energy is used in the photosynthesis of simple sugars from CO_2 and water.

Ethylenediaminetetra-acetic acid, EDTA (10.49)

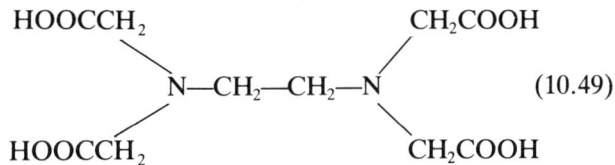

(10.49)

is an important reagent in analytical chemistry. It forms complexes with a wide range of metal ions, including Mg^{2+} and Ca^{2+}, and is used to estimate metal ion concentrations by titration with a suitable indicator. The ligand has six electron-donor atoms for linking to metal atoms,

and is called a chelating ligand because of the claw-like manner in which its six donor atoms are able to grasp a single metal atom (Fig. 10.17). One important use of EDTA is in the estimation of hardness of water, i.e. the measurement of the amount of Mg^{2+} and Ca^{2+} ions in solution.

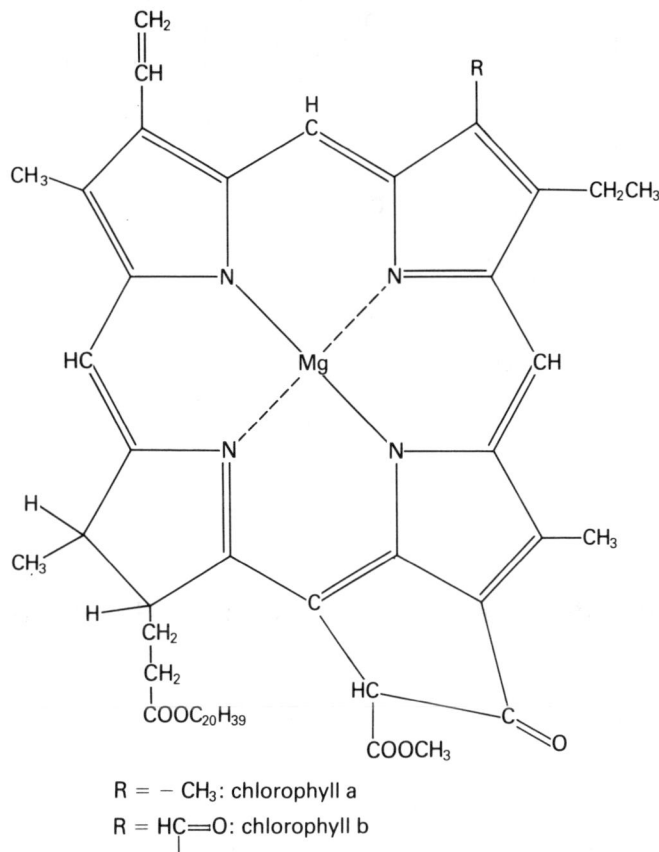

R = – CH₃: chlorophyll a

R = HC=O: chlorophyll b

Fig. 10.16 Chlorophylls

(f) Organo-metallic Compounds

By far the most important organo-metallic compounds of the group II metals are the Grignard reagents R—Mg—X. The reagents are normally prepared and used in ether solution (Eq. 10.50). The nature of the organic group R is governed by the initial choice of organic halide.

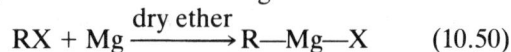

$$RX + Mg \xrightarrow{\text{dry ether}} R—Mg—X \qquad (10.50)$$

The compounds, whose structures are not well understood, are probably polymers with halogen bridges. Their practical importance results from their ease of preparation and their ability to transfer a group R— to another molecule, such as an organic molecule with an electron-deficient site, or to another metal in the preparation of other organo-metallic compounds.

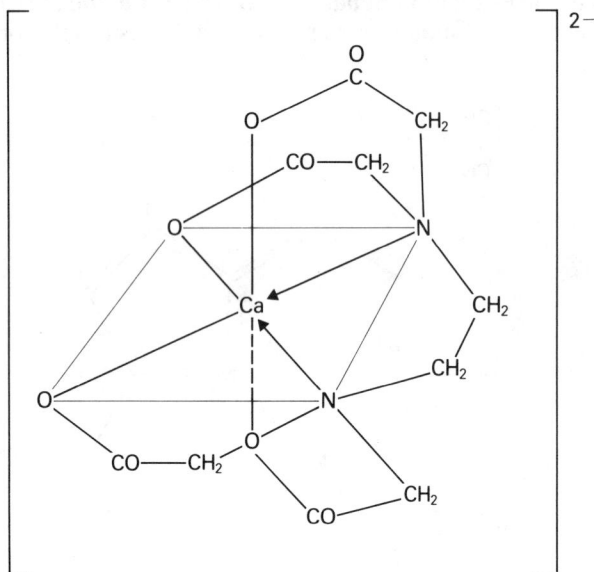

Fig. 10.17 EDTA complex of calcium

10.5 Practical Aspects of Group Ia and IIa Metals and their Compounds

Some of the common and major uses of the metals and their compounds are listed in Table 10.17. The production of NaOH, of Na_2CO_3, and of cement are three of the major chemical industries in the world.

Sodium and potassium are both essential to life. Of the two, sodium is the more important to animal life while potassium is the more important to plant life. Magnesium and calcium are also essential to life; as noted before, magnesium is contained in the chlorophylls which are vital to photosynthesis in plants. Both metals are involved in phosphate transport in animals and calcium, as calcium phosphate, is a major constituent of bone. Beryllium and barium and their compounds are generally toxic to man, but it is of interest that the insolubility of $BaSO_4$ allows it to be taken internally in order to enable the shape of the gut to be determined with X-rays. A major hazard associated with nuclear fission is that one of the products, radioactive ^{90}Sr (half-life 28.1 years), is very similar in its chemistry to calcium. If ingested it becomes incorporated into bone structure where it can do great harm.

The alkaline earth metal cations Ca^{2+} and Mg^{2+} are commonly present in fresh water, their concentrations being dependent on the nature of the land with which the water is in contact. The metal ions interfere with the action of soaps and are said to confer hardness on the water. Hardness resulting from the presence of calcium bicarbonate in solution is termed 'temporary hardness', since it can be removed by boiling. Calcium and magnesium halides cause 'permanent hardness'.

The presence of Mg^{2+} and Ca^{2+} is due to the action of dissolved CO_2 in water on either limestone or dolomite.

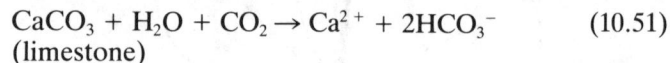

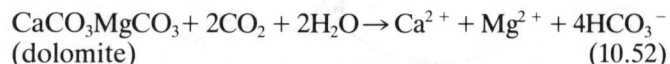

$$CaCO_3 + H_2O + CO_2 \rightarrow Ca^{2+} + 2HCO_3^- \qquad (10.51)$$
(limestone)

$$CaCO_3MgCO_3 + 2CO_2 + 2H_2O \rightarrow Ca^{2+} + Mg^{2+} + 4HCO_3^-$$
(dolomite) $\qquad (10.52)$

The processes are reversed by evaporation, or heating. In the formation of limestone caves the carbonate is first dissolved, and then is reprecipitated as the water evaporates, to produce the various types of limestone cave formations.

Hardness in water produces solid deposits of $CaCO_3$ in hot water systems and boilers and results in wastage of soap. Hardness may be overcome in a variety of ways, as outlined in Table 10.18. The removal of hardness is expensive and in recent times, with the increase in the use of phosphates and synthetic detergents, has led to water pollution problems.

Table 10.17 Commercial and industrial uses of group I and II metals and their compounds

Na	80% used in preparation of Et_4Pb and Me_4Pb (anti-knock additives for petroleum); beginning to be used as wire, plastic coated, for electrical transmission lines.
Na, K	Heat exchangers in nuclear reactors.
Na, Ca, Mg	Thermal reduction of oxides for preparation of metals (e.g. Ti).
Rb, Cs	In photocells.
LiH	Hydrogen generator (reaction with water).
LiOH $\atop$ Li_2CO_3 $\Big\}$	Sources of lithium for making other compounds.
LiButyl	Polymerization catalyst.
NaCl	Common salt, source of Na, preparation of Cl_2.
NaOH $\atop$ KOH $\Big\}$	Wide use in organic and inorganic chemistry and industrial uses.
Na_2CO_3	Use in glass, paper and pulp and soap production.
K compounds	Use in fertilizers e.g. KNO_3.
Mg	In light structural alloys, Al/Zn/Mn/Mg; in preparation of Grignard reagents.
Ca, Sr, Ba	Oxygen getters in steel manufacture.
$CaCO_3$	Use in cement production—Portland cement has approximate composition 60–67% CaO, 17–25% SiO_2, 3–8% Al_2O_3, 6% Fe_2O_3.
$Ca(OH)_2$	Preparation of bleaching agent 'chloride of lime', $Ca(OH)_2 + Cl_2 \rightarrow Ca(OCl)\,Cl + H_2O$.
$CaSO_4 \cdot \frac{1}{2}H_2O$	Plaster of paris.

Table 10.18 Methods of removing hardness from water

Method	Reaction	Comments
Excess soap	$Ca^{2+} + 2C_{18}H_{35}O_2^- \rightarrow Ca(C_{18}H_{35}O_2)_2(s)$	Wasteful in places where water is very hard.
Boiling	$Ca^{2+} + 2HCO_3^- \rightarrow CaCO_3(s) + H_2O + CO_2$	Deposits occur which may be difficult to handle.
Na_2CO_3 (washing soda)	$Ca^{2+} + CO_3^{2-} \rightarrow CaCO_3(s)$	For permanent hardness only.
$Ca(OH)_2$ (slaked lime)	$HCO_3^- + OH^- \rightarrow CO_3^{2-} + H_2O$ $CO_3^{2-} + Ca^{2+} \rightarrow CaCO_3(s)$	For temporary hardness only.
Na_2CO_3 $\Big\}$ lime $+$ $Ca(OH)_2$ $\Big\}$ soda	$HCO_3^- + OH^- \rightarrow CO_3^{2-} + H_2O$ $Ca^{2+} + CO_3^{2-} \rightarrow CaCO_3(s)$ $\Big\}$ $Mg^{2+} + 2OH^- \rightarrow Mg(OH)_2(s)$	Useful for both temporary and permanent hardness.
Phosphate e.g. $Na_3(PO_3)_3$	$2(PO_3)_3^{3-} + 3M^{2+} \rightarrow M_3[(PO_3)_3]_2$ and $PO_4^{3-} + H_2O \rightarrow HPO_4^{2-} + OH^-$ $OH^- + grease \rightarrow soap$	Metal is complexed and now harmless with respect to soap, but the use of phosphate leads to pollution problems.
Synthetic detergents e.g. $ROSO_3^-Na^+$	$2ROSO_3^- + Ca^{2+} \rightarrow (ROSO_3)_2Ca$ not precipitated	Calcium tied up but not precipited. In sodium lauryl sulphate, $R = CH_3(CH_2)_{11}$. If R is linear the chain is biodegradable; if R is branched it is not biodegradable.

Two of the most important methods of softening water are the lime-soda process (which has the advantage of removing both Ca^{2+} and Mg^{2+}, by replacing them with Na^+, and also removing the bicarbonate ion) and by the use of ion-exchange techniques. The latter method is efficient and clean in the sense that no precipitate has to be removed. The ion-exchange can be either a cation-exchange resin, Resin—$SO_3^-M^+$ or an anion-exchange resin, Resin—$NR_3^+X^-$. The resins are insoluble and consist of a polymeric hydrocarbon framework with either anionic or cationic groups. Before the use of organic resins, clay minerals called zeolites were used in a similar manner as ion-exchange water softeners. If both anion- and cation-exchange is performed, deionized water may be produced:

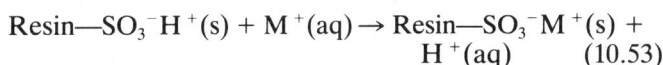

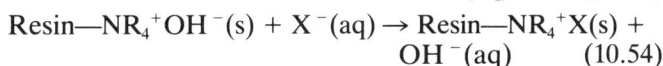

$$Resin—SO_3^-H^+(s) + M^+(aq) \rightarrow Resin—SO_3^-M^+(s) + H^+(aq) \quad (10.53)$$

$$Resin—NR_4^+OH^-(s) + X^-(aq) \rightarrow Resin—NR_4^+X(s) + OH^-(aq) \quad (10.54)$$

When its capacity to hold M^+ or X^- is exhausted the exchanger may be regenerated by treating with a strong solution of the appropriate ions: Na^+ for a resin which is to act as shown at the bottom of Table 10.18, and H^+ and OH^- for resins which are to act as in Eqs 10.53 and 10.54, respectively.

The use of inorganic phosphates and synthetic detergents (which may also contain added phosphates) gives rise to two pollution problems. One is the large amount of phosphate that turns up in lakes and encourages eutrophication, i.e. the runaway growth of plant life which ultimately converts the lake into a swamp. The other problem is that certain detergents have branched hydrocarbon chains and are not biodegradable. These cause foaming in water systems. Water pollution is an increasing problem in our society. To prevent and remove the present pollution requires the combined endeavours of chemistry, biochemistry and biology. The chemistry and biochemistry of a variety of chemicals in water and their effects on living matter is now an important and necessary study. The need for fast, accurate analyses of water and other samples for trace impurities resulting from pollution has led to many new developments in analytical chemistry. A number of important sources of pollution in rivers, lakes and the sea are listed in Table 10.19.

Table 10.19 Water pollution

Source of chemicals involved	Comments
Sewage, mainly organic compounds, phosphates	Sewage demands oxygen for its breakdown and this often reduces the amount of dissolved O_2 to levels dangerous to fish. Also if $[O_2]$ is low complete destruction of dangerous organic chemicals will not occur.
Fertilizers, NO_3^-, PO_4^{3-}	Run-off from land. Can promote eutrophication of lakes.
Crop spraying, pesticides e.g. DDT	Often slow to degrade to harmless materials. Concentrated by animals high in food chains, e.g. fish-eating birds, people, other predators.
Industrial wastes. A great variety, including mercury, cyanides, lead, acids, bases, oils, dyes, and waste solvents, especially chlorinated solvents.	Many methods are now available to remove pollutants before they are discharged into water supplies.
Silt, finely divided clays	From earthworks and erosion.
Radioactive wastes from nuclear reactors. A variety of elements.	Long lifetime of radioactive nuclei makes safe disposal extremely difficult.
Heat, cooling water from power plants.	Raised temperature and decreased O_2 content of water affect local ecology. Not invariably harmful.

Exercises

10.1　Discuss and give reasons for the trends in the following properties of (a) the alkali metals and (b) the alkaline earth metals: ionic radius, ionization energies and hydration energies.

10.2　Why does the I_3^- ion exist in association with Cs^+ but not with Na^+?

10.3　Explain why the electrode potential of lithium does not follow the normal group trend.

10.4　Discuss the difference in reaction of different alkali metals with oxygen; write the appropriate equations.

10.5　Discuss the trends in the heats of formation of the alkali metal halides, and the alkaline earth metal halides (Tables 10.8 and 10.12). Indicate the dominate energy term in each case.

10.6　Compare the chemistry of lithium and beryllium.

10.7　Why is LiF less soluble than CsI and why is LiI more soluble than CsI?

10.8　Predict the following properties of francium; ionic radius, ionization potential (1st), oxidation potential, melting point and density.

10.9　Write balanced equations for the following reactions: $K + H_2O$, preparation of NaOH from NaCl, preparation of Rb_2O.

10.10　Why is $NaHCO_3$ less basic than Na_2CO_3?

10.11　Why are potassium and caesium used in photo-electric cells, in preference to sodium and lithium?

10.12　Give the chemical reactions which occur in making Na_2CO_3 by the Solvay process.

10.13　Starting with NaCl indicate how you would prepare: Na, $NaNH_2$, $NaNO_3$ and Na_2O_2.

10.14　Plot the melting points of NaCl, KCl, RbCl and CsCl against their formula weights. Extrapolate to find the value for LiCl. Does this agree with the experimental value? Explain any difference.

10.15　Explain why the group II metals are harder than the group I metals.

10.16　Explain why plots of the first ionization potential of (a) the alkali metals and (b) the alkaline earth metals against atomic number are not linear. Do similar plots of ionic radius have the same appearance?

10.17　Why is the first ionization potential of beryllium greater than that of lithium, while the second ionization potential of beryllium is less than that of lithium?

10.18　What is the basis of the diagonal relationship between elements of the periodic table? For example, why are some features of the chemistry of Li and Mg similar?

10.19　Explain the trend in the solubilities of the alkaline earth metal hydroxides and sulphates.

10.20　Why is it not possible to produce the Mg^{3+} ion. Give an analysis based on enthalpy data.

10.21　If the decomposition of the alkaline earth metal sulphates proceeds according to the reaction:

$$MSO_4(s) \rightarrow MO(s) + SO_3(g)$$

state with reasons, which sulphate you expect to be most stable and which the least stable.

10.22　Why is the Be^{2+} ion the only group II metal ion that undergoes hydrolysis?

10.23　Write equations for the reaction of water with: Ca, $MgSO_4$, $CaCl_2$, MgO, CaH_2.

10.24　Why does $CaCO_3$ gradually dissolve in water when CO_2 is added? Indicate how this process manifests itself in nature.

10.25　Discuss the factors that cause the hardness of water, and describe with equations three ways of removing permanent hardness.

10.26　How would you make permanent hard water in the laboratory?

10.27　Discuss the properties of ethylenediaminetetra-acetic acid (EDTA) which make it useful in analytical studies of Ca^{2+} and Mg^{2+}

11 The p block Elements

11.1 General Properties of the Elements

The location of the p block elements in the Periodic Table is shown in Fig. 11.1. Except for a few members to the left and bottom of the block the chemistry of these elements is the chemistry of the non-metals. The valence shell electron configuration is ns^2np^x where x varies from 1 to 6, and n from 2 to 6. There are thirty-one p block elements, six of which are the inert gases, and three of which are radioactive (Fig. 11.2).

Despite the wide range of elements involved, it is possible to make a number of useful generalizations concerning trends in the properties of the p block elements. Some of the observed trends are indicated in Fig. 11.2, where the arrows point to increases in the properties written beside them. The trends in size (covalent radius) follow the normal increase down and decrease across the table, and the variation of the first ionization potential closely follows the variation in atomic size. The trends in reactivity and oxidizing power parallel the change in electronegativity; fluorine is the most reactive element of all and has the strongest oxidizing power.

It is convenient to consider first trends in the properties of the elements across the first and second short periods, and then compare the first short period elements with the other periods. Some data are listed in Tables 11.1 and 11.2 (the s block elements are included for completeness). The main structural trend for the elements and representative compounds is from ionic solids early in the periods, through covalent and often polymeric solids around group IV, to liquid and gaseous covalent compounds later in the periods.

The valence orbitals of the first short period elements, 2s and 2p, can hold at most a total of four electron pairs. The elements beryllium and boron (also lithium) have less than the maximum possible number in such compounds as $BeCl_2(g)$ and BF_3. They tend to achieve electronic saturation, i.e. fully occupied outer electron shells, in a variety of ways, namely:

(a) Polymerization, e.g. $BeCl_2$ is a chain polymer in the solid state.
(b) Multiple bonding, e.g. BF_3, in which a lone pair on each fluorine atom may be used in a $B \leftarrow F$ dative π bond involving the empty p orbital on the boron atom.
(c) Formation of complexes in which electrons are received from a donor molecule, e.g. $F_3B \leftarrow NH_3$.
(d) Loss of all valence electrons to form a cation, e.g. Na^+ and Ca^{2+}.

Polymerization generally results in formation of high-melting solids such as B_2O_3 and Al_2O_3. Polymerization also occurs in compounds of group IV elements; thus extended arrays of single bonds around carbon or silicon occur in compounds such as organic polymers and silicates, as well as in the elements themselves. Beyond carbon the elements

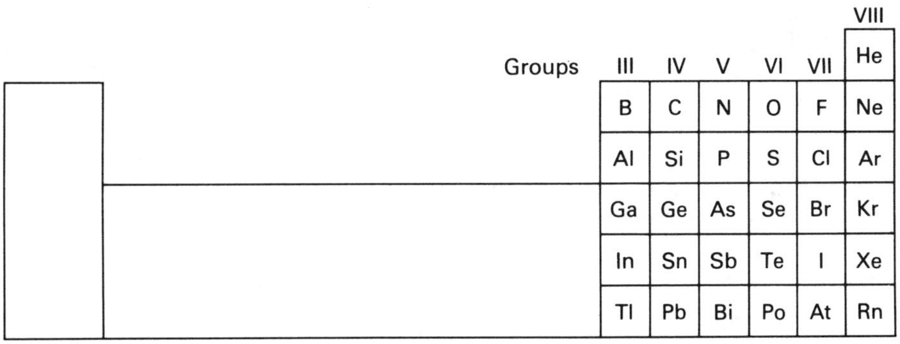

Groups	III	IV	V	VI	VII	VIII
						He
	B	C	N	O	F	Ne
	Al	Si	P	S	Cl	Ar
	Ga	Ge	As	Se	Br	Kr
	In	Sn	Sb	Te	I	Xe
	Tl	Pb	Bi	Po	At	Rn

Fig. 11.1 The 'p block' elements

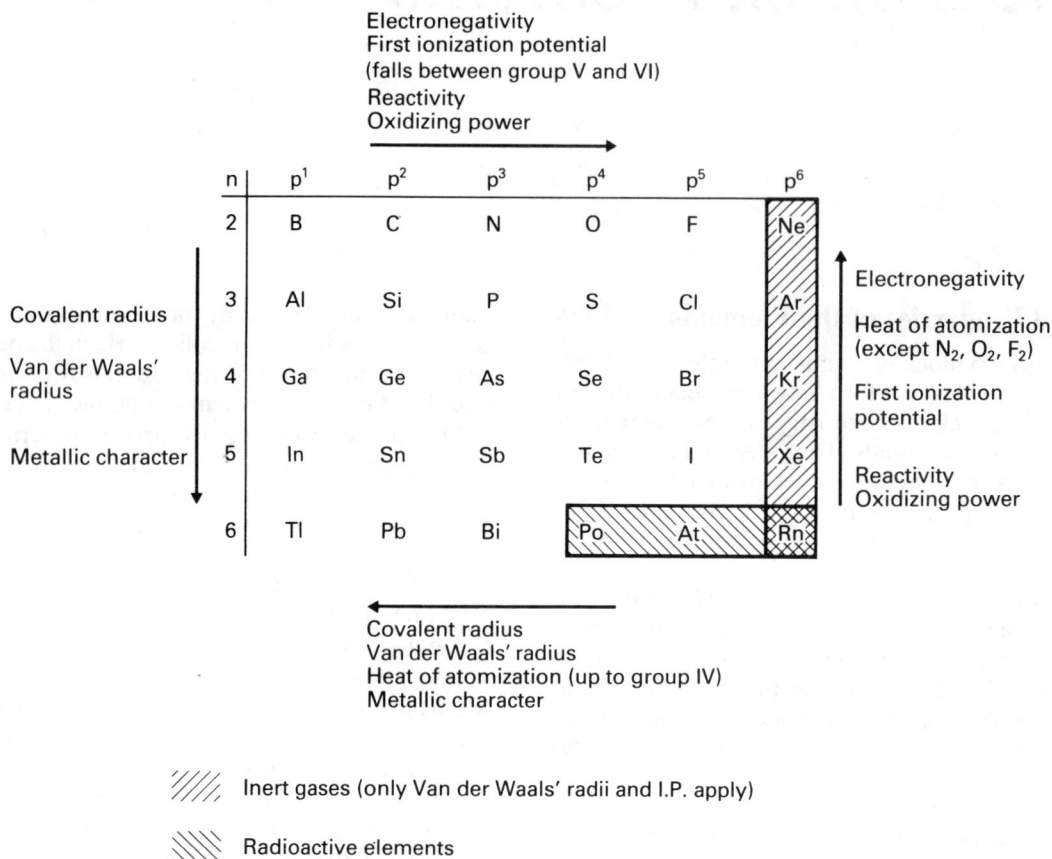

Fig. 11.2 Trends in some properties of the p block elements

are electronically saturated, i.e. they have sufficient or more than sufficient valence electrons for their bonding needs. Hence we find the free elements and their compounds generally exist as simple covalent molecules. Compounds formed by elements to the left of carbon are Lewis acids, because of their ability to accept further electrons into the 2p sub-shell, while elements to the right form compounds which are Lewis bases. The formation of a complex by BF_3 and NH_3, Eq. 11.1, can therefore be regarded as an acid–base reaction, analogous to the reaction of NH_3 with HCl to form ammonium chloride.

$$BF_3(g) + NH_3(g) \rightarrow F_3B \cdot NH_3(s) \qquad (11.1)$$

$$NH_3(g) + HCl(g) \rightarrow NH_4Cl(s) \qquad (11.2)$$

The bonding in the elemental states of Li, Be, B and C

ranges from metallic to covalent and polymeric. For the next two elements (N and O) multiple bonding occurs in their elemental state, $N\equiv N$ and $O=O$. Extended polymeric structures based on frameworks of single bonds for elemental N and O would lead to severe lone pair repulsions and weaken the bonding (Fig. 11.3a). The same lone pair repulsions are responsible for the weakness of the single bonds N—N, O—O and F—F relative to P—P, S—S and Cl—Cl (Tables 11.1 and 11.2). For the latter groups of elements the lone pair repulsions are less because the atoms are bigger. The larger atomic size also provides an explanation for the lack of multiple bonding in the elemental forms of phosphorus and sulphur. Overlap to form π orbitals is reduced because of the larger atoms and consequent greater bond lengths (Fig. 11.3b, page 152).

Table 11.1 First short period elements (including s block elements)

	Li	Be	B	C	N	O	F	Ne
Structure [bcc = body-centred cubic hcp = hexagonal close-packed]	Metal bcc soft	Metal hcp hard	Non-metal interlinked icosahedra	Non-metal (diamond or graphite)	Non-metal gas $N\equiv N$	Non-metal gas $O=O$ paramagnetic	Non-metal gas F—F	Non-metal gas Ne
Melting point (°C)	180	1280	2000	3530 (diamond)	− 210	− 219	− 219.6	− 248.6
Boiling point (°C)	1317	2500	2550	4200	− 195.8	− 183	− 187.9	− 246.0
Atomic radius (pm)	155	112	98	91	70	66	64	—
Covalent radius (pm) (single bond)	(123)	(89)	82	77	70	66	64	—
Ionic radius (pm)	60	31	20	15(+ 4) 260(− 4)	11(+ 5) 171(− 3)	9(+ 6) 140(− 2)	7(+ 7) 136(− 1)	—
Electronegativity	1.15	1.5	2.0	2.5	3.05	3.5	4.1	—
First ionization potential (kJ mol^{-1})	520	899	800	1086	1043	1314	1681	2081
Heat of atomization (kJ mol^{-1})	159	322	562	716	473	250	79.1	0.0
M—M single bond strength (kJ mol^{-1})	109 (Li$_2$)	unknown	301 (B$_2$F$_4$)	347 (alkanes)	159 (N$_2$H$_4$)	142 (H$_2$O$_2$)	158 (F$_2$)	—
Oxide structure	Li$_2$O ionic solid, anti-fluorite structure	BeO ionic/ covalent solid, wurtzite structure	B$_2$O$_3$ solid covalent polymeric	CO$_2$ (also CO) gas covalent O=C=O	various gases covalent	O$_3$ gas covalent	F$_2$O gas covalent	None
Acid–base character	base	amphoteric	amphoteric	acidic (CO$_2$)	acidic	neutral	neutral	
Chloride	LiCl ionic solid, NaCl structure	BeCl$_2$ solid, chain polymer	BCl$_3$ gas, b.pt 12.5°C	CCl$_4$ liquid, tetrahedral b. pt 76.7°C	NCl$_3$ liquid b.pt 2°C, pyramidal	OCl$_2$ gas b.pt 3.8°C, bent	FCl gas b. pt − 100°C	None

Certain elements show a marked tendency to form covalent bonds with themselves in either finite or infinite arrays. The elements that show this property, termed catenation (chain formation), are listed in Table 11.3; the ones which have the property most strongly developed are within the box. Details of the forms which catenation takes for elements and compounds are given in Tables 11.4 and 11.5, respectively. The factors that appear most relevant to catenation are atomization energy, single bond strength, and atomic size. A high heat of atomization is an indication that extended bonding networks exist in the solid element. Such networks will usually consist of single bonds (graphite is the only exception, and even here the average bond order is only slightly greater than 1). Hence strong single bonds will favour catenation. Strength of single bond formation is correlated with small atomic size, because atoms need to be close

together for effective orbital overlap to occur. Hence one would expect catenation to be most important for the elements at the top of the periodic groups, as is mostly observed. Note, however, the anomalous nature of oxygen and nitrogen in this respect, because of the lone pair repulsions discussed above and because of their readiness to form multiple bonds.

Table 11.3 Elements which display catenation

B	C			
	Si	P	S	
	Ge	As	Se	
	Sn	Sb	Te	I
	Pb	Bi		

Table 11.2 Second short period elements (including s block elements)

	Na	Mg	Al	Si	P	S	Cl	Ar
Structure [bcc = body-centred cubic hcp = hexagonal close-packed fcc = face-centred cubic (cubic close-packed)]	metal (soft) bcc	metal (hard) hcp	metal (hard) fcc	polymeric tetrahedral configuration	P_4 pyramids in solid	S_8 rings in solid	Cl_2 gas	Ar gas
Melting point (°C)	98	651	660	1410	44	119	− 101	− 189
Boiling point (°C)	892	1107	2467	2355	280	445	− 34.5	− 186
Atomic radius (pm)	190	160	143	132	128	104	99	—
Covalent radius (pm)	(156)	136	125	117	110	104	99	—
Ionic radius (pm)	95	65	50	41(+ 4) 271(− 4)	34(+ 5) 212(− 3)	29(+ 6) 184(− 2)	26(+ 7) 181(− 1)	—
Electronegativity	1.0	1.25	1.45	1.74	2.05	2.45	2.85	—
First ionization potential (kJ mol^{-1})	495	738	577	787	1060	1000	1255	1520
Heat of atomization (kJ mol^{-1})	108	150	326	456	315	280	122	0.0
M—M single bond-strength (kJ mol^{-1})	72 (Na_2)	unknown	155 (Al_2)	226 (Si_2H_6)	209 (P_4)	264 (S_8)	244 (Cl_2)	—
Oxide	Na_2O ionic solid,	MgO ionic solid,	Al_2O_3 ionic solid,	SiO_2 covalent polymeric solid	P_4O_6 P_4O_{10} solids	SO_2 SO_3 gases	Cl_2O gas b.pt 9.7°	None
Structure	anti-fluorite structure	NaCl structure	ccp oxygen, Al in 2/3 of the octahedral cavities		based on tetrahedral P_4 grouping	SO_2 angular SO_3 planar	bent	
Acid–base character	base	base	amphoteric	acid	acid	acid	acid	
Chloride	NaCl ionic solid	$MgCl_2$ ionic solid	$AlCl_3$ ionic solid	$SiCl_4$ covalent liquid	PCl_3 PCl_5 PCl_3, covalent liquid PCl_5, ionic solid (PCl_4^+ PCl_6^-)	SCl_2 covalent liquid	Cl_2 covalent gas	None

11.2 Some Important Classes of Compounds

Three of the most important classes of compounds of the p block elements are the oxides, the oxyanions and the halides. We consider these in turn.

(a) Oxides

Oxygen forms one or more compounds with every element except helium, neon, argon and krypton. The neutral, binary compounds of oxygen with other elements, called oxides, can generally be obtained as the end product of reaction between an element and oxygen. In many cases the oxide is the form in which the element occurs naturally. Oxides can usefully be classified according to their acid–base properties and their structure and type of bonding. In terms of their acid–base properties there are four types of oxides, namely acidic, basic, amphoteric and neutral. The acidity, or otherwise, is determined by the nature of their reaction with water, or in the cases of insoluble and amphoteric oxides, their reactions with acids and bases.

Table 11.4 Structures of p block elements

IIIb	IVb	Vb	VIb	VIIb
Boron, icosahedron B_{12}	Carbon, diamond, graphite	Nitrogen $N\equiv N$	Oxygen $O=O$	Fluorine F—F
Aluminium, metal ccp	Silicon, diamond structure	Phosphorous white P$\diagup$$\mid$$\diagdown$P P red chains black corrugated sheets	Sulphur rings S_8 chains $-S\diagup S\diagup S-$ 	Chlorine Cl—Cl
	Germanium diamond structure semi-metallic	Arsenic yellow As_4 black corrugated sheets (metallic)	Selenium rings Se_8 chains (semi-metallic) sheets (metallic)	Bromine Br—Br
	Tin, α diamond structure β each atom 6—coordinated (metallic)	Antimony yellow Sb_4 white corrugated sheets (metallic)	Tellurium chains (semi-metallic)	Iodine I—I

Table 11.5 Examples of compounds showing catenation

IIIb	IVb	Vb	VIb	VIIb
BORON Boron hydrides (fragments of B_{12} structure)	CARBON Whole range of organic chemicals			
	SILICON Silicon hydrides Si_nH_{2n+2} $n = 1 \rightarrow 8$ chlorosilanes	PHOSPHORUS Phosphorus sulphides P_4S_3, P_4S_5, P_4S_7 containing 3, 2 and 1 P—P bonds, respectively	SULPHUR Polycations e.g. S_8^{2+}, S_{16}^{2+}, and anions S_6^{2-}	
	GERMANIUM Germanium hydrides Ge_nH_{2n+2} $n = 1 \rightarrow 5$	ARSENIC Cations As_4^{2+} organo-arsenic compounds $(RAs)_n$	SELENIUM Polycations Se_4^{2+}, Se_8^{2+} and anions	
	TIN Organo-tin compounds $[(C_6H_5)_3Sn]_4Sn$ and cyclic polytins $[(C_6H_4)_3Sn]_6$	ANTIMONY Cation Sb_4^{2+}	TELLURIUM Polycations Te_4^{2+}, Te_6^{2+}	IODINE Polyiodides I_n^-, $n = 3,5,7,9$
	LEAD As for Tin	BISMUTH Isolated examples exist		

Basic oxides react with water as in the following examples:

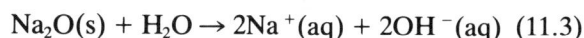

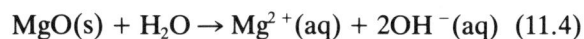

$$Na_2O(s) + H_2O \rightarrow 2Na^+(aq) + 2OH^-(aq) \quad (11.3)$$

$$MgO(s) + H_2O \rightarrow Mg^{2+}(aq) + 2OH^-(aq) \quad (11.4)$$

If a basic oxide is insoluble it reacts with an acid to form a salt:

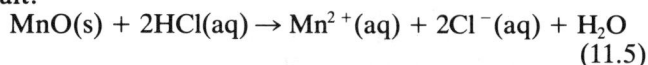

$$MnO(s) + 2HCl(aq) \rightarrow Mn^{2+}(aq) + 2Cl^-(aq) + H_2O \quad (11.5)$$

The common reaction for basic oxides, which contain the oxide ion O^{2-}, is:

$$O^{2-} + H_2O \rightleftharpoons 2OH^- \quad (11.6)$$

The equilibrium constant for this reaction is greater than 10^{22}, i.e. the O^{2-} ion cannot exist at significant concentrations in aqueous solution. The oxide ion is a stable constituent of solid ionic oxides because of the high lattice energies of such materials (e.g. the lattice energy of MgO is 3800 kJ mol^{-1}) but in water, although the hydration energy of O^{2-} is expected to be reasonably large (1000–1200 kJ mol^{-1}),

Strong lone pair – lone pair repulsions

(a)

N   140 pm   N

Strong interaction

P   220 pm   P

Very weak interaction

(b)

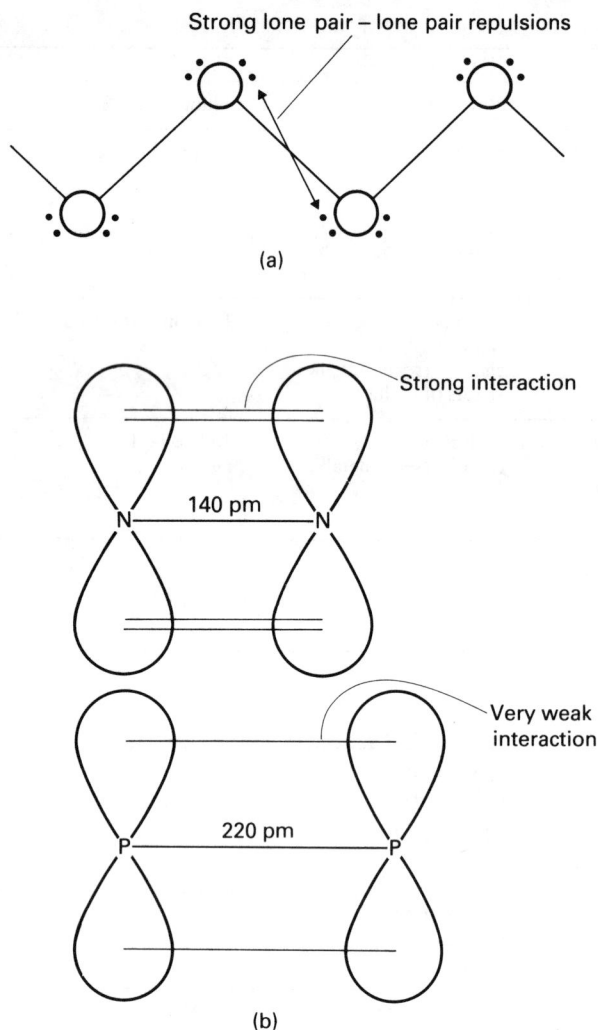

Fig. 11.3 (a) Bonding in hypothetical O_n. (b) Comparison of possibilities for π bonding in N_2 and P_2

the strong polarizing power of the ion is sufficient to cause cleavage of an O—H bond of water. The oxides of most non-metals, and also those of some metals, are *acidic*, i.e. they react with water to give the hydronium ion, H_3O^+ or $H^+(aq)$. Thus:

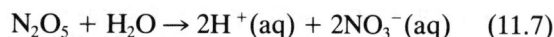

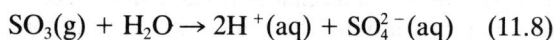

$$N_2O_5 + H_2O \rightarrow 2H^+(aq) + 2NO_3^-(aq) \quad (11.7)$$

$$SO_3(g) + H_2O \rightarrow 2H^+(aq) + SO_4^{2-}(aq) \quad (11.8)$$

An insoluble acidic oxide will react with a strong base to form an oxyanion salt, as:

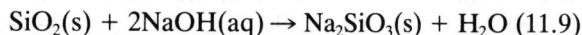

$$SiO_2(s) + 2NaOH(aq) \rightarrow Na_2SiO_3(s) + H_2O \quad (11.9)$$

Amphoteric oxides display both acidic and basic properties, i.e. they form salts by reaction with both acids and alkalis. For example, zinc oxide:

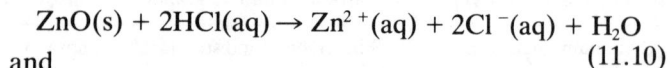

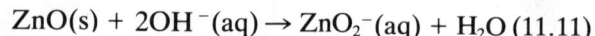

$$ZnO(s) + 2HCl(aq) \rightarrow Zn^{2+}(aq) + 2Cl^-(aq) + H_2O$$

and   (11.10)

$$ZnO(s) + 2OH^-(aq) \rightarrow ZnO_2^-(aq) + H_2O \quad (11.11)$$

A few oxides, notably CO, N_2O and NO, are *neutral*, in that they do not react with water even when, as in the case of N_2O, they dissolve fairly readily. Nor do they react with acids or bases. The oxide of fluorine (better described as the flouride of oxygen), OF_2, is neutral in an interesting way. It does not form an oxyanion when it reacts with water; instead an oxidation–reduction reaction occurs and the acid HF is produced, as shown in Eq. 11.12.

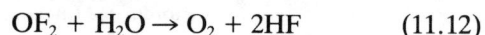

$$OF_2 + H_2O \rightarrow O_2 + 2HF \quad (11.12)$$

Trends in the acid–base character of the oxides of s and p block elements are shown in Fig. 11.4. Similar trends are found for the transition metals (d block elements). For the transition metals, which are noted for their variable oxidation numbers, the oxidation state of the metal strongly influences the acid–base character of the oxide. Generally, when the metals are in high oxidation states the oxides are acidic, while in low oxidation states they are basic:

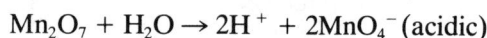

$$Mn_2O_7 + H_2O \rightarrow 2H^+ + 2MnO_4^- \text{ (acidic)}$$

Increasing acidity →

↓ Increasing basicity

Li	Be	B	C	N	O	F
Na	Mg	Al	Si	P	S	Cl
K	Ca	Ga	Ge	As	Se	Br
Rb	Sr	In	Sn	Sb	Te	I
Cs	Ba	Tl	Pb	Bi		

Basic oxides   Amphoteric oxides   Acidic oxides (except CO, NO, N_2O, F_2O)

(Partial shading indicates one oxide amphoteric while the other is basic or acidic)

Fig. 11.4 Acid and basic properties of the s and p block elements

$$MnO(s) + 2H^+ \rightarrow Mn^{2+}(aq) + H_2O \text{ (basic)} \quad (11.14)$$

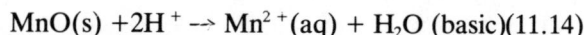

Intermediate oxidation states are amphoteric:

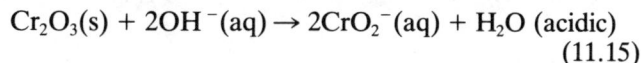

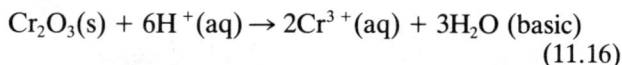

$$Cr_2O_3(s) + 2OH^-(aq) \rightarrow 2CrO_2^-(aq) + H_2O \text{ (acidic)} \quad (11.15)$$

$$Cr_2O_3(s) + 6H^+(aq) \rightarrow 2Cr^{3+}(aq) + 3H_2O \text{ (basic)} \quad (11.16)$$

When oxides are classified according to their bonding and structural types it is found that basic oxides are ionic in character and contain the O^{2-} ion, whereas acidic oxides are covalent or at most only partly ionic. The classification of oxides according to their bonding highlights a large group of oxides which are polymeric and covalent (Fig. 11.5). The subset of polymeric oxides is almost identical with the subset of amphoteric oxides, but differences occur, especially on the non-metal side. Polymeric structures can be further subdivided into three-dimensional networks, layer lattices, and chain lattices, making a total of five distinct structural types. These are:

(1) Infinite three-dimensional ionic lattices.
(2) Infinite three-dimensional lattices with mainly covalent bonding.
(3) Layer lattices, in two dimensions.
(4) Chain lattices, polymeric in one dimension.
(5) Discrete molecular species.

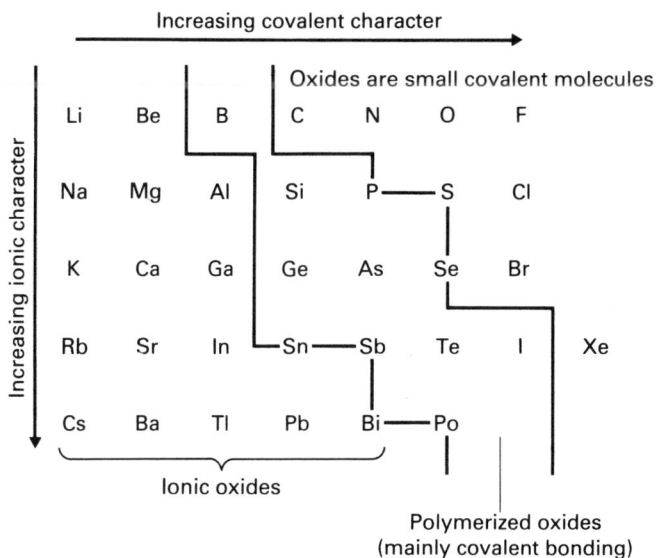

Fig. 11.5 Bonding in oxides of s and p block elements

The change from structures of type 1 through to 5 correlates with increasing covalency in the bonding. Two examples which illustrate this point are shown in Table 11.6.

Table 11.6 Correlation of structures and properties for MO_3 and MO_2 oxides

CrO_3	chain lattice	increasing acidity
MoO_3	layer lattice	increasing covalent bonding
WO_3	3-D lattice	
SO_2	molecular	increasing acidity
SeO_2	chain lattice	increasing covalent bonding
TeO_2	3-D lattice	

For the regular (i.e. s and p block) elements the most common structural type is the infinite three-dimensional ionic or partly ionic lattice. The type of packing within the lattice depends on ionic size, bonding and stoichiometry.

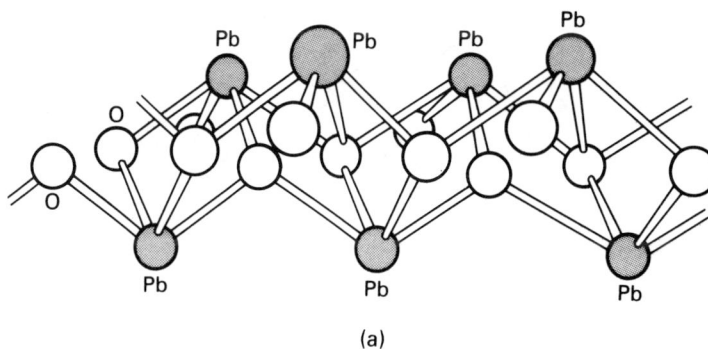

(a)

Fig. 11.6a Structure of PbO

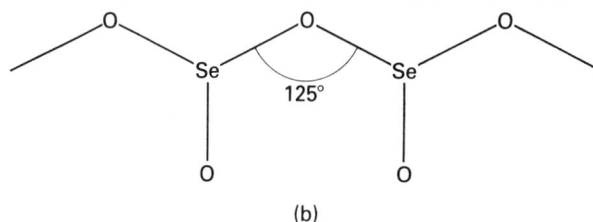

(b)

Fig. 11.6b Structure of SeO_2

Thus lead monoxide (PbO) is an example of layer lattice structure, based on PbO_4 pyramids (Fig. 11.6a), while SeO_2 (Fig. 11.6b) is an oxide with a chain lattice. Sulphur trioxide in its β-modification also has a chain lattice, while the γ-form is a ring structure $(SO_3)_3$ (Figs. 11.6c and d). The oxides of phosphorus (Figs. 11.7a and b) are examples of oxides which display some degree of polymerization but are still in the form of discrete molecular units. The physical properties of the oxides are very markedly dependent on their crystal structure. Thus ionic solids and infinite three-dimensional lattices are generally hard and high-melting

(c)

(d)

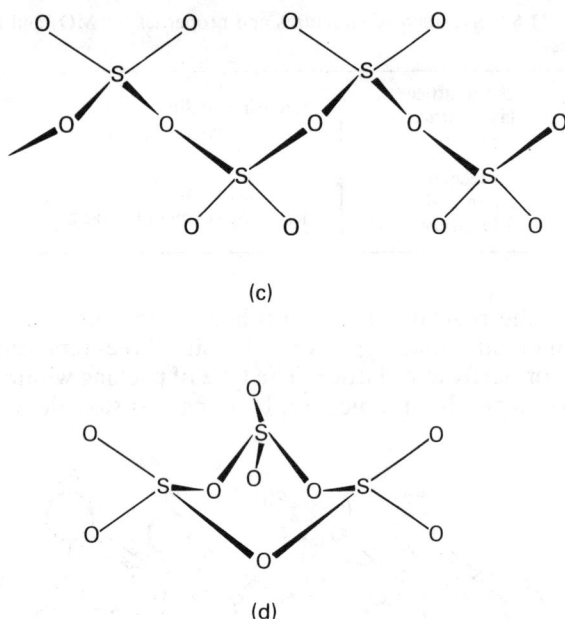

Fig. 11.6c and d Structures of SO_3

P$_4$O$_6$ P$_4$O$_{10}$

(a)   (b)

Fig. 11.7a and b Structures of phosphorus oxides

Table 11.7 Structures of the covalent oxides

Linear	CO_2, N_2O (CO, NO, O_2)
Bent	H_2O, NO_2, O_3, OF_2, S_2O, SO_2, ClO_2, Cl_2O, BrO_2, Br_2O
Trigonal planar	SO_3(g), N_2O_4 (direct N—N bond), N_2O_5 (bridging oxygen)
Pyramidal	XeO_3, I_2O_5 (bridging oxygen)
Tetrahedral	XeO_4, Cl_2O_7 (bridging oxygen)

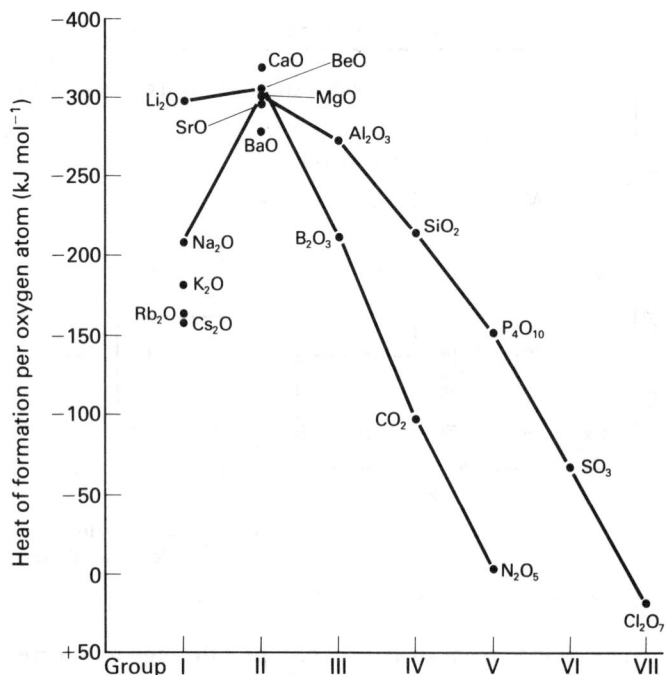

(e.g. quartz, SiO_2), layer lattices tend to flake readily (graphite is the classic example), and materials with chain lattices are soft and may be fibrous like asbestos (a hydrated magnesium silicate). Oxides which do not form infinite lattices exist in the solid state as weak molecular (Van der Waals') crystals, and are soft with low melting points. The shapes of molecular oxides (Table 11.7) can be predicted with some success on the basis of the electron-pair repulsion theory, taking into account the multiple bonding that often occurs between oxygen and the other elements. The importance of multiple bonding can be assessed by comparing corresponding bond lengths.

The heats of formation of some oxides per oxygen atom, $H_f^{\ominus}/n$, are plotted as a function of valence shell configuration in Fig. 11.8. The high (negative) values for the elements of groups I, II and III reflect their lower ionization energies and greater lattice energies compared with the covalent oxides of the non-metals. The rise in $-H_f^{\ominus}/n$ between groups I and II probably results from more efficient packing in the alkaline earth oxides. Trends within groups I and II are governed by competition between the lattice energy of the oxide and the ionization energy of the element bonded to oxygen. In group I the lattice energy trends dominate the picture.

The preparation and properties of some important oxides of the p block elements are summarized in Table 11.8.

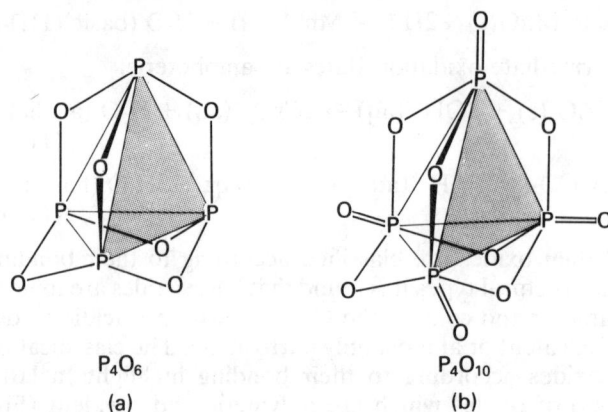

Fig. 11.8 Heats of formation of oxides, per oxygen atom

Table 11.8 Oxides of p block elements

(a) Group III oxides

Oxide	Preparation	Comments
B_2O_3	$2B(OH)_3 \xrightarrow[100°C]{heat} B_2O_3 + 3H_2O$ (boric acid)	Polymeric compound—basic unit is a planar trigonal BO_3 moiety. Acidic.
Al_2O_3	Strong heat on nitrate, hydroxide or sulphate.	8 polymorphic forms; most important is α-Al_2O_3, *corundum*, hcp lattice of O^{2-} with Al^{3+} in 2/3 of octahedral holes. Cr^{3+} impurity gives ruby. Fe^{2+}, Fe^{3+} or Ti^{4+} give blue sapphire. $H_f^{\ominus} = -1676$ kJ mol^{-1}, hence Al is useful reducing agent for other oxides (cf. chapter 8). Amphoteric.

(b) Group IV oxides

Oxide	Preparation	Comments
CO	$C(g) + \frac{1}{2}O_2 \rightarrow CO$ (limited supply of oxygen) $HCOOH \xrightarrow{H_2SO_4} CO + H_2O$ Commercially produced in form of producer gas and water gas.	Strong $C\equiv O$ bond (1071 kJ mol^{-1}), iso-electronic with NO^+, N_2 and CN^-. Forms transition metal carbonyls readily. Toxic because competes with oxygen for haemoglobin in the blood. Physical properties resemble N_2.
CO_2	$C(s) + O_2 \rightarrow CO_2$ (plentiful supply of oxygen) Carbonates + acid $\rightarrow CO_2$ $CaCO_3 \xrightarrow[800°C]{heat} CaO + CO_2$ (an industrial source)	Linear molecule $O{=}C{=}O$. Gives some H_2CO_3 in water; weak acid, $K_1 = 4.3 \times 10^{-7}$ and $K_2 = 4.8 \times 10^{-11}$. (Solubility of CO_2 in water is low.) Solid CO_2 sublimes at $-78.5°C$.
SiO_2	$SiCl_4 + 2H_2O \rightarrow SiO_2 + 4H^+ + 4Cl^-$	Several forms of SiO_2 are known. Structure is 3-D polymer. Acidic oxide, reacts with alkalis, HF and fluorine; otherwise relatively unreactive.
GeO_2	$Ge + O_2 \rightarrow GeO_2$ $GeCl_4 + 2H_2O \rightarrow GeO_2 + 4H^+ + 4Cl^-$	Structure similar to SiO_2; compound not very reactive, but will react with HF.
SnO	$Sn(OH)_2 \xrightarrow[100°C]{heat} SnO + H_2O$	Amphoteric oxide. Layer lattice, similar to PbO (Fig. 11.6a).
SnO_2	$3Sn + 4HNO_3 \rightarrow 3SnO_2 + 4NO + 2H_2O$	Occurs naturally as cassiterite. SnO_2 unreactive except to hot acids and alkalis (amphoteric).
PbO	$Pb + \frac{1}{2}O_2 \rightarrow PbO$ $Pb(acetate)_2 + NH_4OH \rightarrow$ very pure PbO	Two forms exist, yellow and red; can be interconverted by heating. Example of layer lattice structure (Fig. 11.6a).
PbO_2	$PbO + NaClO_3$ or $NaNO_3 \xrightarrow{fusion} PbO_2$ $Pb(acetate)_4 \xrightarrow{hydrolysis}$ very pure PbO_2	Main use in lead-acid storage cell. $PbSO_4(s) + 2e^- \rightarrow Pb(s) + SO_4^{2-}$, $E^{\ominus} = -0.36V$. $PbO_2(s) + SO_4^{2-} + 4H^+ + 2e^- \rightarrow PbSO_4(s) + 2H_2O$, $E^{\ominus} = 1.68V$. $E^{\ominus}_{cell} = 2.04V$. PbO_2 is a strong oxidizing agent, e.g. releases Cl_2 from HCl
Pb_3O_4	PbO or $PbCO_3 \xrightarrow[in\ air]{heat} Pb_3O_4$	$2PbO.PbO_2$ ('red lead').

(c) Group V oxides

Oxide	Preparation	Comments
N_2O	$NH_4NO_3(s) \xrightarrow[170°C]{melt} N_2O(g) + 2H_2O(g)$	Neutral, rather inert; mild anaesthetic
NO	$3Cu + 8HNO_3 \rightarrow 3Cu(NO_3)_2 + 4H_2O + 2NO$ (dilute nitric acid). $6NaNO_2 + 3H_2SO_4 \rightarrow 4NO + 2H_2O + 3Na_2SO_4 + 2HNO_3$ (dilute sulphuric acid). Industrial preparation $4NH_3 + 5O_2 \xrightarrow[catalyst]{Rh-Pt} 4NO + 6H_2O$	Neutral, reactive odd-electron compound; forms NO^+ readily. Readily oxidized, $2NO + O_2 \rightarrow 2NO_2$. Dimeric in solid and liquid state. Forms nitrosyl complexes with transition metals.
NO_2 $\updownarrow$ N_2O_4	$Cu + 4HNO_3 \rightarrow 2NO_2 + Cu(NO_3)_2 + 2H_2O$ (concentrated nitric acid)	Acidic, reactive odd-electron compound. Bent molecule. Forms NO_2^+ (linear) and NO_2^- (more bent).

Table 11.8 continued   *(c) Group V oxides (continued)*

Oxide	Preparation	Comments
NO$_2$ ⇅ N$_2$O$_4$	2Pb (NO$_3$)$_2$ → 4NO$_2$ + 2PbO + O$_2$	N$_2$O$_4$ exists in several forms in solid state. Most stable form is planar with an unusually long N—N bond. NO$_2$ brown, N$_2$O$_4$ colourless. solid − 11.2°C 100% N$_2$O$_4$ liquid 21.15°C { 99.9% N$_2$O$_4$ / 0.1% NO$_2$ } gas (1 atm) 21.15°C { 84.1% N$_2$O$_4$ / 15.9% NO$_2$ } gas 135°C { 1% N$_2$O$_4$ / 99% NO$_2$ } gas 600°C decomposes to NO + O$_2$
N$_2$O$_3$	4HNO$_3$ + As$_4$O$_6$ + 4H$_2$O → 4H$_3$AsO$_4$ + 2N$_2$O$_3$ 4NO + O$_2$ → 2N$_2$O$_3$ 2NO + N$_2$O$_4$ → 2N$_2$O$_3$	Blue solid; decomposes above − 20°C to NO and NO$_2$. Planar molecule with a very long N—N bond. With alkali gives NO$_2^-$ ion (nitrite), as does equimolar mixture of NO + NO$_2$.
N$_2$O$_5$	4HNO$_3$ + P$_4$O$_{10}$ → 2N$_2$O$_5$ + 4HPO$_3$ 2NO$_2$ + O$_3$ → N$_2$O$_5$ + O$_2$	Structure, planar O$_2$N—O—NO$_2$ in gas, NO$_2^+$.NO$_3^-$ in solid. Reacts with water to give HNO$_3$.
P$_4$O$_6$	P$_4$ + 3O$_2$ → P$_4$O$_6$ (limited supply of O$_2$)	Structure given in Fig. 11.7a. Reacts with water to give H$_3$PO$_3$. Reacts with halogen X$_2$ to give POX$_3$.
P$_4$O$_{10}$	P$_4$ + 5O$_2$ → P$_4$O$_{10}$	Structure in Fig. 11.7b. Very powerful drying agent, reacts with water to give H$_3$PO$_4$ (intermediate products are HPO$_3$ and H$_4$P$_2$O$_7$).
As$_4$O$_6$	As$_4$ + 3O$_2$ → As$_4$O$_6$ (limited supply of O$_2$)	Amphoteric oxide, slightly soluble in water. Same structure as P$_4$O$_6$ in gaseous state.
As$_2$O$_5$	As$_4$O$_6$ + HNO$_3$ → H$_3$AsO$_4$H$_2$O $\xrightarrow[170°C]{heat}$ As$_2$O$_5$	Structure unknown; acts as an oxidizing agent, in water gives H$_3$AsO$_4$.
Sb$_4$O$_6$	Sb$_4$ + 3O$_2$ → Sb$_4$O$_6$	Same structure as P$_4$O$_6$; amphoteric.
Sb$_2$O$_5$	Sb + HNO$_3$ $\xrightarrow{heat}$ Sb$_2$O$_5$ SbCl$_5$ + OH$^-$ $\xrightarrow{heat}$ Sb$_2$O$_5$	Structure unknown.
Bi$_2$O$_3$	Heat on Bi(OH)$_3$ or Bi(NO$_3$)$_3$	Soluble in acids, but not in alkali.
Bi$_2$O$_5$	Bi$_2$O$_3$ + strong oxidising agent → Bi$_2$O$_5$	Loses O$_2$ thermally.

(d) Group VI oxides

Oxide	Preparation	Comments
O$_3$	3O$_2$ $\xrightarrow[\text{electrical discharge}]{\text{silent}}$ 2O$_3$	Triangular (117°). Neutral, reactive. Powerful oxidizing agent.
SO$_2$	S + O$_2$ $\xrightarrow{burn}$ SO$_2$ Sulphides + O$_2$ $\xrightarrow{roast}$ SO$_2$	Triangular (119°). Useful non-aqueous solvent for ionic compounds: liquid range − 10° to − 75.5°C at 1 atmosphere. Used in manufacture of sulphuric acid.
SO$_3$	2SO$_2$ + O$_2$ $\xrightarrow{500°}$ 2SO$_3$ catalyst (surface) Pt, V$_2$O$_5$ or Fe$_2$O$_3$.	$\Delta G^{\ominus} = -70$ kJ; reaction thermodynamically favoured but is slow, hence raised temperature and catalysts. In water gives H$_2$SO$_4$. Powerful oxidizing agent; converts P$_4$ to P$_4$O$_{10}$ and HBr to Br$_2$. Two of several known structures shown in Figs 11.6c and d. In gas phase SO$_3$ is planar and symmetrical.
SeO$_2$	Se + O$_2$ $\xrightarrow{heat}$ SeO$_2$	Polymeric chain structure: see Fig. 11.6b. Weakly basic.
TeO$_2$	Te + O$_2$ $\xrightarrow{heat}$ TeO$_2$ Te + HNO$_3$ $\xrightarrow{190°C}$ TeO$_2$.	More basic than SeO$_2$. 3-D infinite lattice structure.

Table 11.8 continued.

(e) Group VII oxides		
Oxide	Preparation	Comments
OF_2	$F_2 + NaOH \rightarrow OF_2$ (2% solution NaOH)	Explosive; powerful oxidizing agent. Reacts with water to form O_2 and HF.
O_2F_2	$O_2 + F_2 \xrightarrow[\text{electric discharge}]{\text{high voltage}} O_2F_2$	Decomposes above $-50°C$. Powerful oxidizing and fluorinating agent.
Cl_2O	$2Cl_2 + 2HgO \rightarrow HgCl_2 . HgO + Cl_2O$	Yellow gas, explosive and powerful oxidizing agent. Acidic: in water gives HClO.
ClO_2	$2NaClO_3 + H_2SO_4 \rightarrow 2ClO_2 + 2NaHSO_4,$ or various commercial processes.	Explosive; powerful oxidizing agent; odd-electron compound. Weakly acidic. Used as a bleaching agent, e.g. for wood pulp.
Cl_2O_6	$2ClO_2 + O_3 \rightarrow Cl_2O_6 + \frac{1}{2}O_2$	Decomposes at its melting point. Acidic.
Cl_2O_7	$2HClO_4 + P_2O_5 \rightarrow Cl_2O_7$	Most stable of chlorine oxides. Acidic: in water gives back $HClO_4$.
Br_2O	$2HgO + 2Br_2$ in $CCl_4 \rightarrow Br_2O + HgBr_2 . HgO$	Above $-40°C$ gives $Br_2 + O_2$; good oxidizing agent.
BrO_2	$Br_2 + 2LiO_3 \xrightarrow[-78°C]{\text{fluorocarbon}} 2BrO_2 + Li_2O_2$	Also unstable above $-40°C$.
I_2O_5	$2HIO_3 \xrightarrow{200°C} I_2O_5 + H_2O$	Acidic; strong oxidizing agent; reacts with CO to give $I_2 + CO_2$: used to analyse for CO by titrating the iodine liberated.

(f) Group VIII oxides		
Oxide	Preparation	Comments
XeO_3	$3XeF_4 + 6H_2O \rightarrow XeO_3 + 2Xe + 3/2\ O_2 + 12HF$ $XeF_6 + 3H_2O \rightarrow XeO_3 + 6HF$	Dangerously explosive solid. In basic solutions gives $HXeO_4^-$, which disproportionates to give $XeO_6^{4-} + Xe + O_2$. Structure: triangular pyramid.
XeO_4	$Ba_2XeO_6 \xrightarrow{H_2SO_4} XeO_4 + BaSO_4 + H_2O$	Unstable, explosive gas. Tetrahedral structure.

Several oxides of p block elements are associated with air pollution. Gaseous emissions from oil- or coal-fired power plants, various kinds of industry, and internal combustion engines typically contain CO, CO_2, hydrocarbons, NO, NO_2 and SO_2. The action of sunlight on air containing NO_2 leads to formation of ozone as the first step in producing photochemical smog. Even in the absence of man-made pollution air normally contains traces of these pollutant gases, but the amounts are small and local regions of high concentration rarely occur. The annual amounts of different pollutants introduced into the atmosphere, on a world-wide basis, are given in Table 11.9. It is apparent that the amount introduced from natural sources is greater than the man-made pollution for all except SO_2, CO and CO_2. However, the high local concentrations of hydrocarbons and nitrogen oxides that can arise in cities make these materials significant and dangerous pollutants (Table 11.10). The residence times of pollutants in the atmosphere are limited mainly by washing out in rainfall and by bacterial action. Ordinary soil has a surprisingly large capacity to absorb traces of SO_2 and CO from air.

Concern and interest in pollution has led to the development of many new methods, and the revival of some old methods, of chemical analysis, especially those useful for trace quantities. The removal of pollutants from emitted gases is generally expensive; hence a compromise must be arrived at between the acceptable level of pollution and the cost of the final product. Once the initial cost of the removal equipment is met the former waste materials can often be collected and converted to useful products, but this gain is seldom sufficient to offset more than a fraction of the added running costs.

(b) Oxyanions
At first glance the oxyacids and oxyanions of the p block elements may appear to the student to comprise a haphazard list of complicated formulae, but fuller study reveals a number of systematic trends. The products of reactions of p block element oxides with water are commonly hydroxy compounds in which, because of the high electronegativity of the central atom (M) and the strong M—O bond, the O—H bond is readily cleaved by water. Hence the species

Table 11.9 Sources and quantitites of gaseous pollutants

Compound	Major pollution source	Natural source	Annual amount pollution	Annual amount natural sources	Residence time in the atmosphere
SO_2	Coal and oil combustion.	Volcanoes.	149×10^6 tonnes	—	4 days
H_2S	Chemical plants. Sewage.	Volcanoes. Biological action.	3×10^6	100×10^6	2 days
CO	Combustion.	Forest fires.	200×10^6	11×10^6	3 years
N-oxides	Combustion.	Biological action.	54×10^6	500×10^6	5 days
NH_3	Waste treatment.	Biological action.	4×10^6	6000×10^6	2 days
Hydrocarbons	Combustion. Chemical processes.	Biological action.	89×10^6	490×10^6	16 years
CO_2	Combustion.	Biological action.	$13\,000 \times 10^6$	$10 \times 10^6 \times 10^6$	4 years
CF_2Cl_2, CCl_3F	Refrigeration. Aerosol sprays.	None.	10×10^6	None	Over 20 years.

Table 11.10 Effects of pollutant gases on living matter and materials

Pollutant	Man	Animals	Vegetation	Materials
H_2S	400–700 ppm lethal ½–1 hour exposure	Lethal.	Marking on leaves, growing tips killed.	Discolouration of paintwork.
SO_2	5–10 ppm severe distress in 10 minutes. Lower concentrations cause bronchial irritation.	0.2 ppm affects central nervous system in 10 s.	Bleached spots on leaves.	Sulphuric acid aerosol causes corrosion of metals. Paint is discoloured and softened by SO_2 and H_2SO_4. Leather, paper, textiles weakened.
Nitrogen oxides	500 ppm lethal after 48 hours. Causes lung cancer, pneumonia.	20–100 ppm severe distress.	Bleached spots on leaves.	Fading of dyes, and as for SO_2 because of acidic nature of oxides.
O_3	1.5–2.0 ppm severe distress in 2 hours; impairs lung function. Products of reaction with hydrocarbons and NO cause eye irritation (photochemical smog).		Bleaching and growth suppression (0.03 ppm in 4 hours).	Cracking and weakening of rubber.
CO	100 ppm in 15 hours, severe distress. 4000 ppm, lethal in 1 hour.	Lethal.		

are acids, which dissociate to form H_3O^+ and an oxyanion. Alternatively, reaction of the oxide with alkali gives the oxyanion (cf. Eqs 11.7–11.9). In a few cases an oxy-acid exists without a corresponding oxide or acid anhydride; an example is H_3PO_2, hypophosphorous acid.

Bonds between non-metals and oxygen are relatively strong (Table 11.11), and there is the possibility of forming single or double bonds. Hence there are many ways of arriving at stable oxyanions, and the range of possibilities is greatest when M—O single bonds occur, because the oxygen atom can then span two elements. Thus much of the complexity of oxyanion chemistry must result from the number of options available in relation to condensation or oxygen bridging. The extreme example of complexity arising from this source is provided by the range of structures corresponding to various stages of condensation with the Si—O system. We now consider these compounds in some detail.

Silicates

The chemistry of silicon is largely dominated by the great strength of the Si—O bond (452 kJ) relative to the Si—Si bond (222 kJ). Silicon tends to form extended —Si—O—

Table 11.11 Strength of bonds between oxygen and p block elements (kJ mol^{-1})

B—O	C—O	N—O	O—O	F—O
523 (esters)	360 (organic)	163 (NH_2OH)	142 (H_2O_2)	213 (F_2O)
	$C=O$ 803 (CO_2)	$N=O$ 594 (NOCl)	$O=O$ 498 (O_2)	
	$C\equiv O$ 1073 (CO)	$N\equiv O$ 632 (NO)		
		$N\equiv O$ 1050 (NO^+)		
Si—O 464 ($SiO_2(s)$)	P—O 368 (P_4O_6)	$S=O$ 340 (SO_3)		Cl—O 205 (Cl_2O)
$Si=O$ 640 ($SiO_2(g)$)	$P=O$ 545 (POF_3)	$S=O$ 523 (SO_2)		
Ge—O 360 ($GeO_2(s)$)	As—O 331 (As_4O_6)			Br—O 238 (NOBr)
				I—O 185 (IO radical)
				Xe—O 74 (XeO_3)

Si—O— chains and networks, in contrast to carbon whose catenation depends on the versatility and strength of carbon—carbon bonds. Stable silicon compounds make little or no use of multiple bonds. The basic structural unit of silicates is the tetrahedral grouping SiO_4. By itself this grouping carries four negative charges, $[SiO]^{4-}$. Silica, SiO_2, can be regarded as the acid anhydride of $Si(OH)_4$. Silica is insoluble in water, but when fused with alkali it gives the SiO_3^{2-} species. Various possible condensations of SiO_4 groups are listed below (see also Fig. 11.9).

(1) Discrete SiO_4^{4-} ions exist in the orthosilicates, e.g. Be_2SiO_4 and Mg_2SiO_4.

(2) If two tetrahedral groups condense and share one oxygen atom the species $Si_2O_7^{6-}$, pyrosilicate, is formed, e.g. $Sc_2Si_2O_7$ and $Zn_4(OH)_2Si_2O_7$.

(3) If the number of shared oxygen atoms per silicon atom is increased to two, then two structural types are formed. The first consists of infinite chains, $[(SiO_3)^{2-}]_n$, the pyroxenes. An example is diopside $(Ca,Mg)_n(SiO_3)_{2n}$, where the cations

link the chains by electrostatic attraction. The second structural type consists of finite rings, e.g. three tetrahedral groups per ring $[Si_3O_9]^{6-}$ as in $BaTiSi_3O_9$ and $Ca_2BaSi_3O_9$, or six tetrahedral groups per ring, $[Si_6O_{18}]^{12-}$, as in beryl $Be_3Al_2Si_6O_{18}$ (green beryl is emerald).

(4) The formation of double chains, i.e. condensation of two $[(SiO_3)^{2-}]_n$ chains, gives the amphiboles, in which half the silicon atoms have two shared oxygens per silicon and half have three. The general formulation is $[(Si_4O_{11})^{6-}]_n$. Compounds with this structure are fibrous in nature; some forms of asbestos have double-chain structures.

(5) When three oxygens are shared per silicon atom we have infinite sheet structures based on rings cross-linked in one plane, $[Si_4O_{10}^{6-}]_n$. The best examples of this structure are the micas and many of the clay minerals, which also contain aluminium. Cations occupy the regions between the sheets, and cleavage occurs readily along planes parallel to the sheets. This results in the characteristic flakiness of mica.

(6) When all the oxygen atoms are shared three-dimensional structures are obtained in which there is complete cross-linking via chains of alternate silicon and oxygen atoms. Silica $(SiO_2)_n$ in its various crystalline and glassy forms (e.g. fused quartz) is the chief example of this type. If some silicon is replaced by aluminium to give aluminosilicates the additional positive charge required for electrical neutrality must be provided by a separate cation. In this way materials such as the felspars (e.g. orthoclase $K[AlO_2(SiO_2)_3]$) and the open framework zeolite structures are produced.

Clays are aluminosilicates with the infinite sheet structure, and are usually colloidal in nature. They have plastic properties, but when fired they lose water and undergo further condensation, becoming hard and brittle. Use is made of this property in the production of bricks, pipes and pottery. Kaolinite, $Al_4Si_4O_{10}(OH)_8$, is a clay widely used in making pottery and refractories. Bentonite is used as a bonding agent, for example, in making pellets of metal ore prior to reduction.

Asbestos is a fibrous silicate with either the chain or sheet structures. Chrysotile, $3MgO.2SiO_2.2H_2O$, has the sheet structure. Various forms of asbestos are used in composite fibre materials such as corrugated roofing sheets. Their resistance to chemical attack and good thermal stability is widely exploited in protective clothing. The extremely deleterious long-term effects of inhaled asbestos fibres have only recently become appreciated.

Micas have infinite sheet structures, with corresponding cleavage properties. An example is muscovite: $K_2Al_4[Si_6Al_2O_{20}](OH)_4$. Their high electrical resistance, fair transparency, and good thermal stability lead to their use as electrical insulators and supports for heating elements.

$[SiO_4]^{4-}$

Orthosilicates

$[Si_2O_7]^{6-}$

Pyrosilicates

$[Si_3O_9]^{6-}$

Cyclic silicates

$[SiO_3^{2-}]_n$

Pyroxenes

$[Si_4O_{11}^{6-}]_n$

Amphiboles

$[Si_4O_{10}^{4-}]_n$

Infinite sheet silicates

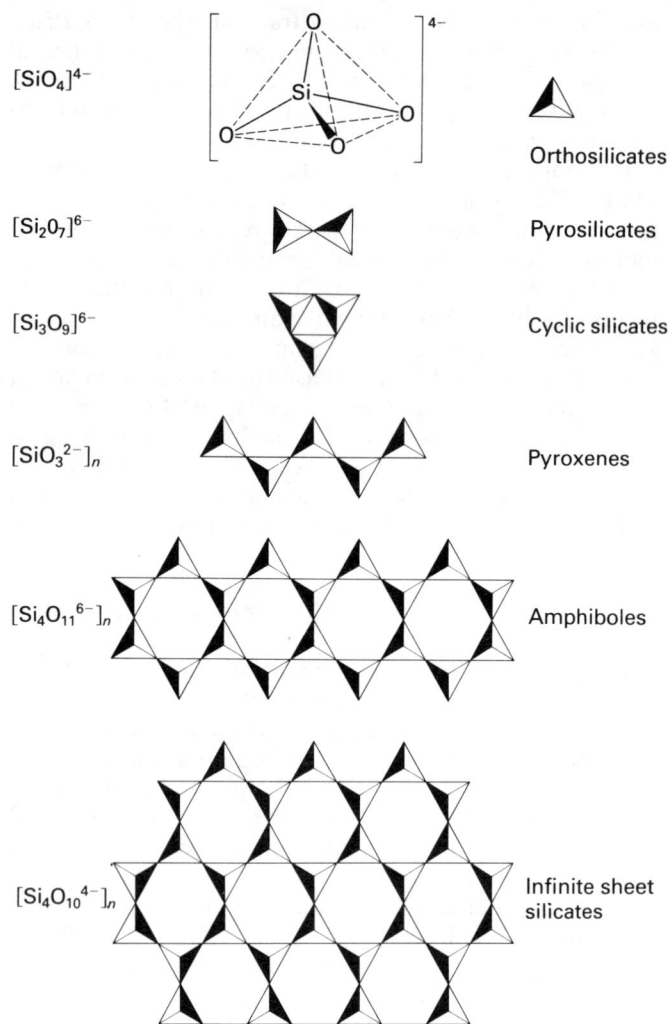

Fig. 11.9 Silicate structures

Zeolites are three-dimensional silicates $[(Al,Si)_nO_{2n}]^{x-}$ which have unusually open structures. The negative charge on the lattice which results from the replacement of Si by Al is counterbalanced by cations which fit loosely in the cavities in the structure. Zeolites can be used as ion-exchangers, whose operation depends on replacing one kind of cation with another in these cavities. Dehydrated zeolites are useful for drying organic solvents and for adsorbing impurities from gases such as N_2 or H_2. They are termed molecular sieves, because the small H_2O molecules can get into the cavities but larger molecules cannot. Molecular sieves are available in a range of pore sizes for different purposes.

Glass is a supercooled silicate. It is obtained by rapidly cooling a melt so that the disorder which characterizes the liquid state is maintained. The randomly oriented and often strained bonds are frozen into position and the structure is termed amorphous, non-crystalline, or just glassy. Glass does not melt sharply, because the most strained bonds break first when it is heated, and it does not cleave in a regular manner because of the lack of the order which characterizes crystalline materials. Ordinary window or bottle glass is principally sodium silicate. The thermal, optical and tensile properties of glass can be modified by the addition of other cations. Adding boron lowers the coefficient of thermal expansion, as in Pyrex; adding lead increases the refractive index, as in 'crystal'.

Silicones

The silicones are a group of organo-silicon-oxygen compounds of considerable industrial importance. They contain an —O—Si—O— backbone with organic groups bonded to the silicon atoms. Their useful properties arise from the thermal strength of the —Si—O—Si—O—Si— system and the water repellance and resistance to chemical attack of the organic groups. Depending on the relative molecular mass and extent of cross-linking, the polymers range from liquids to hard, thermally stable solids. They are prepared according to the scheme:

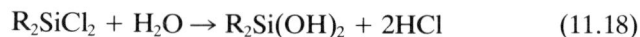

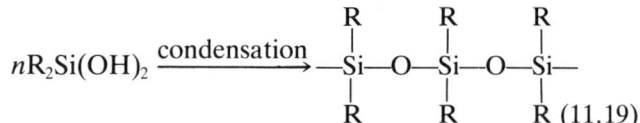

$$RCl + Si \xrightarrow{\text{Cu catalyst}} R_xSiCl_{4-x}(x = 1, 2 \text{ or } 3) \quad (11.17)$$

$$R_2SiCl_2 + H_2O \rightarrow R_2Si(OH)_2 + 2HCl \quad (11.18)$$

$$nR_2Si(OH)_2 \xrightarrow{\text{condensation}} \underset{\underset{R}{|}}{\overset{\overset{R}{|}}{-Si}}-O-\underset{\underset{R}{|}}{\overset{\overset{R}{|}}{Si}}-O-\underset{\underset{R}{|}}{\overset{\overset{R}{|}}{Si}}- \quad (11.19)$$

Nitrogen Oxyanions

The two principal oxyanions of nitrogen are the nitrite, NO_2^- and nitrate NO_3^-. The shapes of the ions and of nitric acid are given in Fig. 11.10. Nitrous acid HNO_2, exists in the gaseous state and in aqueous solution. It is formed by treating nitrite salts with acid or by adding the mixed oxide N_2O_3 to ice-cold water. Solutions of the acid decompose on warming:

$$3HNO_2 \rightarrow HNO_3 + H_2O + 2NO \quad (11.20)$$

The acid is an oxidizing agent:

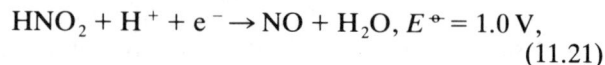

$$HNO_2 + H^+ + e^- \rightarrow NO + H_2O, E^\ominus = 1.0\,V, \quad (11.21)$$

and its redox reaction with iodine may be used for estimating NO_2^-

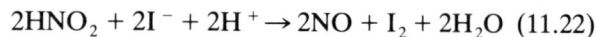

$$2HNO_2 + 2I^- + 2H^+ \rightarrow 2NO + I_2 + 2H_2O \quad (11.22)$$

The acid is also a reducing agent

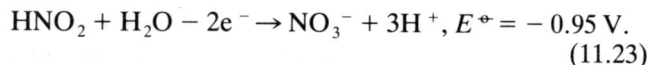

$$HNO_2 + H_2O - 2e^- \rightarrow NO_3^- + 3H^+, E^\ominus = -0.95\,V. \quad (11.23)$$

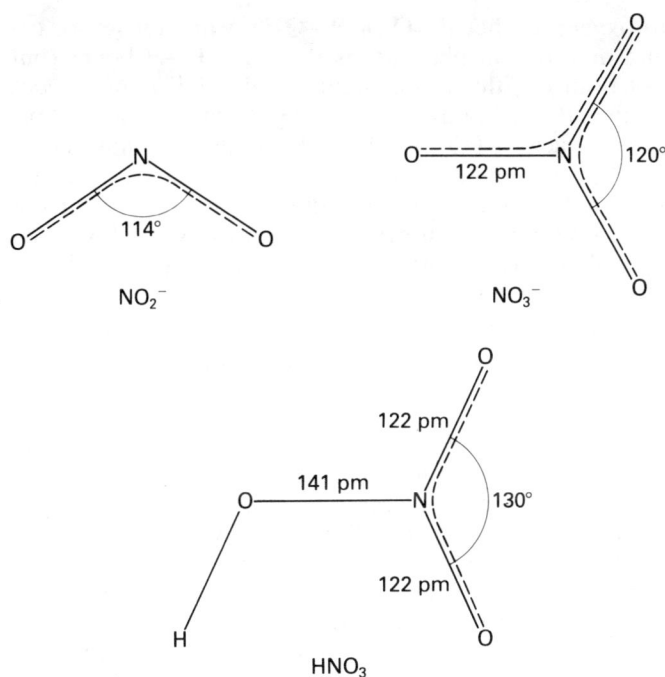

Fig. 11.10 Structures of HNO_3, NO_2^- and NO_3^-

Nitrous acid is used in organic chemistry for the preparation of diazonium salts. The $-NO_2$ group bonds to organic groups and transition metals in two different ways, namely as R—ONO (nitrito compounds) and as R—NO_2 (nitro compounds).

Nitric acid is obtained commercially by the catalytic oxidation of ammonia using a platinum–rhodium catalyst. The catalyst causes the reaction to lead to the formation of NO rather than taking the thermodynamically more favourable path to give nitrogen.

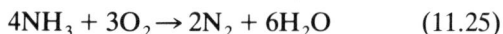

$$4NH_3 + 5O_2 \rightarrow 4NO + 6H_2O \qquad (11.24)$$

$$4NH_3 + 3O_2 \rightarrow 2N_2 + 6H_2O \qquad (11.25)$$

The nitric oxide so produced is oxidized with air or oxygen, to form NO_2:

$$2NO + O_2 \rightarrow 2NO_2 \qquad (11.26)$$

which is dissolved in water in the presence of excess O_2 to give nitric acid.

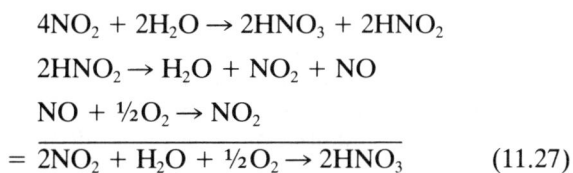

$$4NO_2 + 2H_2O \rightarrow 2HNO_3 + 2HNO_2$$

$$2HNO_2 \rightarrow H_2O + NO_2 + NO$$

$$NO + \tfrac{1}{2}O_2 \rightarrow NO_2$$

$$= \overline{2NO_2 + H_2O + \tfrac{1}{2}O_2 \rightarrow 2HNO_3} \qquad (11.27)$$

A major application of nitric acid is to prepare nitrate salts for use as fertilizers or in explosives. The first explosive to be invented, gunpowder, is a mixture of potassium nitrate, sulphur and charcoal. Nitric acid is a strong oxidizing agent and can be reduced all the way from HNO_3, containing N in the $+5$ state, to NH_3, containing N in the -3 state. The reduction potentials corresponding to various intermediate stages in this process are given in Table 11.12.
(11.13) e 11.12 contains a great deal of information about the chemistry of HNO_3. For example, the relative magnitudes of the $E^{\oplus}$ values show that NO_3^- and NO_2^- are much stronger oxidizing agents in acid than in alkaline solution.

Table 11.12 Reduction potentials for NO_3^- and related species in acidic and basic solutions

Reaction	$E^{\oplus}$ (V)
Acidic solution	
$2NO_3^- + 8H^+ + 2e^- \rightarrow N_2O_4 + 2H_2O$	$+0.90$
$NO_3^- + 3H^+ + 2e^- \rightarrow HNO_2 + H_2O$	$+0.94$
$NO_3^- + 4H^+ + 3e^- \rightarrow NO + 2H_2O$	$+0.96$
$2NO_3^- + 10H^+ + 8e^- \rightarrow N_2O + 5H_2O$	$+1.11$
$HNO_2 + H^+ + e^- \rightarrow H_2O + NO$	$+1.00$
$N_2O + 2H^+ + 2e^- \rightarrow N_2 + H_2O$	$+1.77$
$N_2 + 8H^+ + 6e^- \rightarrow 2NH_4^+$	$+0.27$
Alkaline solution	
$NO_3^- + H_2O + 2e^- \rightarrow NO_2^- + 2OH^-$	$+0.01$
$NO_3^- + 2H_2O + 3e^- \rightarrow NO + 4OH^-$	$+0.15$
$2NO_3^- + 5H_2O + 8e^- \rightarrow N_2O + 10OH^-$	$+0.10$
$NO_2^- + H_2O + e^- \rightarrow NO + 2OH^-$	-0.46
$2NO + H_2O + 2e^- \rightarrow N_2O + 2OH^-$	$+0.76$
$N_2O + H_2O + 2e^- \rightarrow N_2 + 2OH^-$	$+0.94$
$N_2 + 6H_2O + 6e^- \rightarrow 2NH_3 + 6OH^-$	-0.73

The presence of varying numbers of hydrogen ions in the equations given for acid solutions implies that the nature of the products of reduction of HNO_3 must depend markedly on pH. This is borne out by the observation that, for example, copper metal reduces concentrated nitric acid mainly to N_2O_4 (or NO_2), whereas it reduces dilute nitric acid mainly to NO. In acid solution nitrous acid is unstable with respect to disproportionation into NO_3^- and NO, according to

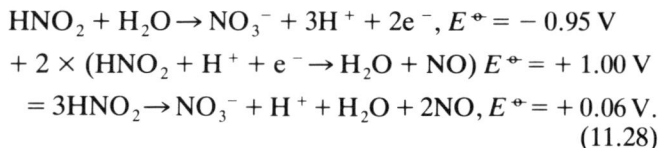

$$HNO_2 + H_2O \rightarrow NO_3^- + 3H^+ + 2e^-, E^{\oplus} = -0.95 \text{ V}$$

$$+ 2 \times (HNO_2 + H^+ + e^- \rightarrow H_2O + NO)\, E^{\oplus} = +1.00 \text{ V}$$

$$= 3HNO_2 \rightarrow NO_3^- + H^+ + H_2O + 2NO, E^{\oplus} = +0.06 \text{ V}. \qquad (11.28)$$

In alkaline solution this does not apply, the $E^{\oplus}$ value for conversion of NO_2^- to NO being negative. The driving force for the process which occurs in acid solution is provided by the excess of $E^{\oplus}$ for reduction of HNO_2 to NO over $E^{\oplus}$

for reduction of NO_3^- to HNO_2. (Recall that a positive $E^{\ominus}$ corresponds to a negative $\Delta G^{\ominus}$.) The information in Table 11.12 can be expressed more succinctly in the form of the reduction potential diagrams of Fig. 11.11. The necessary thermodynamic condition for a species to disproportionate spontaneously, as with HNO_2 in acid soultion, is that the $E^{\ominus}$ value for the process removing the species should be more positive than $E^{\ominus}$ for the process by which the species is formed. Concentrated HNO_3 oxidizes concentrated hydrochloric acid to chlorine, itself being reduced to nitrosyl chloride (Eq. 11.29).

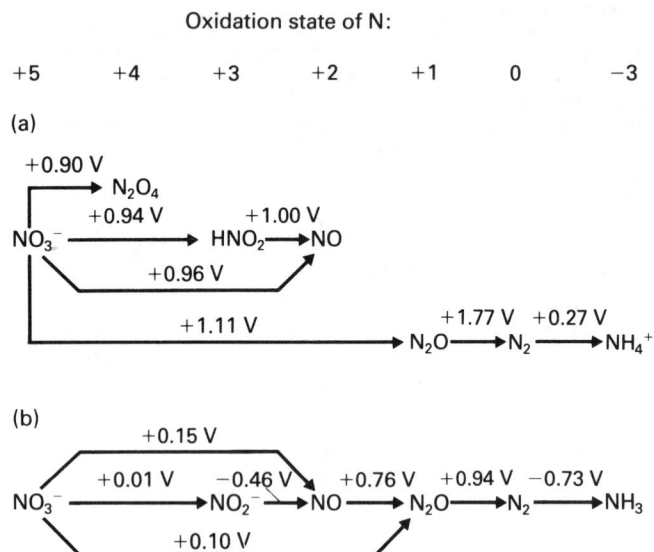

$$HNO_3 + 3HCl \rightarrow 2H_2O + NOCl + Cl_2 \quad (11.29)$$

Oxidation state of N:

+5   +4   +3   +2   +1   0   −3

Fig. 11.11 Reduction potential diagrams for HNO_3 and related species: (a) in acid solution; (b) in alkaline solution

The resulting greenish solution, containing free chlorine, is called aqua regia ('King of Waters') and will dissolve noble metals such as gold and platinum. A mixture of concentrated nitric and sulphuric acids is used as a nitrating agent in organic chemistry, and is used in industry to manufacture nitrocellulose (guncotton) and glyceryl trinitrate (nitroglycerine), the active ingredient of dynamite and gelignite. Kinetic studies have shown that the species responsible for the nitrating action is the nitrosyl cation NO_2^+:

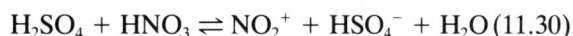

$$H_2SO_4 + HNO_3 \rightleftharpoons NO_2^+ + HSO_4^- + H_2O \quad (11.30)$$

Phosphorus oxyanions

Numerous oxyanions and corresponding oxyacids are known for phosphorus. In all cases the phosphorus is four-coordinated. When the phosphorus is pentavalent it bonds to oxygen (either P=O or P—OH), while for tetra-, tri- and monovalent phosphorus P—P and P—H bonds (but not both together) also occur (Table 11.13). Acids containing P—H bonds are reducing agents e.g. hypophosphorous acid H_3PO_2. The P—H bonds do not confer acidity on the acid; the acidity always arises from the hydroxyl proton. Condensation occurs either by oxygen bridging or by formation of direct P—P bonds. Complex polyphosphorus acids exist, as well as cyclic compounds similar to cyclic silicates.

Orthophosphoric acid may be obtained by the reaction:

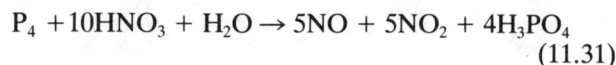

$$P_4 + 10HNO_3 + H_2O \rightarrow 5NO + 5NO_2 + 4H_3PO_4 \quad (11.31)$$

Commercially it is produced by treating calcium phosphate with concentrated sulphuric acid. It is a moderately strong tribasic acid and forms three types of salts. For example, with NaOH it forms NaH_2PO_4, Na_2HPO_4 or Na_3PO_4. The relative acid strengths of H_3PO_4, $H_2PO_4^-$ and HPO_4^{2-} are reflected in the pK_a values: 2.13, 7.20 and 12.36. Phosphates are important fertilizers and are also used as water softeners. Superphosphate, a mixture of soluble calcium phosphate and calcium sulphate, is the most common phosphate fertilizer:

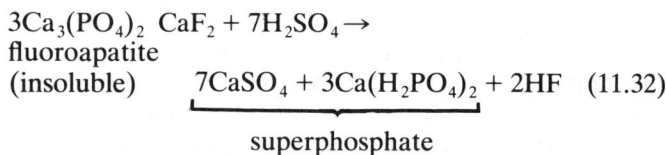

$$3Ca_3(PO_4)_2 \cdot CaF_2 + 7H_2SO_4 \rightarrow$$
fluoroapatite
(insoluble) $\quad \underbrace{7CaSO_4 + 3Ca(H_2PO_4)_2}_{\text{superphosphate}} + 2HF \quad (11.32)$

Animal bones are composed of sub-microscopic calcium phosphate crystals deposited in a framework of the protein collagen. The important molecule adenosine triphosphate, ATP, contains three phosphate groups. It is involved in the Krebs cycle, by which sugars are converted into carbon dioxide and water with the release of energy in animal muscles. ATP molecules store energy at an intermediate stage of the cycle. Hydrolysis of ATP splits off a phosphate group, forming adenosine diphosphate, ADP, and liberating 33 kJ mol^{-1} of energy for transfer to other processes in the cycle.

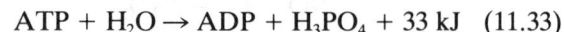

$$ATP + H_2O \rightarrow ADP + H_3PO_4 + 33 \text{ kJ} \quad (11.33)$$

Sulphur oxyanions

A varied assortment of oxyacids of sulphur exists. The sulphur is mostly four-coordinated and bonded to O=, HO—, —S— or sometimes —O—O—. It is noteworthy that phosphorus and sulphur, both of which display catenation in their elementary forms, also show this tendency in their oxyacids. A number of oxyacids of sulphur are listed in Table 11.14.

Table 11.13 Oxyacids of phosphorus

Oxidation state of phosphorus	Acid	Name	Preparation	Structure
+ 1	H_3PO_2	Hypophosphorous	$P_4 + alkali \rightarrow H_2PO_2^- + H_2$	H—P(=O)(OH)—H
+ 3	$(HPO_2)_n$	Metaphosphorous		$\left[P{-}O{-}P(OH){-}O \right]_n$ P ?
	$H_4P_2O_5$	Pyrophosphorous	$PCl_3 + H_3PO_3$	H—P(=O)(OH)—O—P(=O)(OH)—H
	H_3PO_3	Orthophosphorous	P_2O_3 or $PCl_3 + H_2O$	H—O—P(=O)(H)—OH
+ 4	$H_4P_2O_6$	Hypophosphoric	red P + alkali	(HO)(HO)(O=)P—P(=O)(OH)(OH)
+ 5	$(HPO_3)_n$	Metaphosphoric	Heat on orthophosphoric acid	$\left[P{-}O{-}P(=O)(OH){-}O \right]_n$ P
	$H_4P_2O_7$	Pyrophosphoric	Heat on phosphates	(HO)(HO)(O=)P—O—P(=O)(OH)(OH)
	H_3PO_4	Orthophosphoric	$P_2O_5 + H_2O$	HO—P(=O)(OH)—OH

Sulphuric acid is an extremely important industrial chemical. Although it seldom appears in the final product, it is involved at some stage in many chemical-based manufacturing processes. Most sulphuric acid is produced by the 'contact' process, the main step of which is the catalytic oxidation of SO_2 to SO_3.

$$2SO_2 + O_2 \xrightarrow{\text{Pt or } V_2O_5} 2SO_3 (\Delta H = -98 \text{ kJ}) \quad (11.34)$$

The reaction is thermodynamically favoured, but requires heat and a catalyst to make it proceed at a useful rate. To help displace the equilibrium of reaction 11.34 to the right the process is carried out under pressure with an excess of oxygen. The best catalyst, platinum, is readily poisoned by traces of selenium or arsenic, so vanadium pentoxide, which is a less efficient catalyst but is less readily poisoned, is generally used. The conditions required for a rapid reaction

Table 11.14 Some oxyacids of sulphur

Number of sulphur atoms	Acid (S oxid. no.)	Name	Preparation	Structure
1	$H_2SO_3(+4)$	Sulphurous	$SO_2 + H_2O$ or acid on sulphites. Free acid does not exist.	$HO-\underset{\underset{}{}}{\overset{O}{\overset{\|}{S}}}-OH$
	$H_2SO_4(+6)$	Sulphuric	$SO_3 + H_2O$.	$O{=}\underset{\underset{OH}{\|}}{\overset{O}{\overset{\|}{S}}}-OH$
2	$H_2S_2O_3\left(\begin{smallmatrix}+6\\-2\end{smallmatrix}\right)$	Thiosulphuric	Sodium thiosulphate + HCl. Unstable at room temperature.	$HO-\underset{\underset{O}{\|\|}}{\overset{OH}{\overset{\|}{S}}}{=}S$
	$H_2S_2O_4(+4)$	Dithionous	Free acid unknown. Zn reduction of H_2SO_3.	$HO-\overset{O}{\overset{\|\|}{S}}-\overset{O}{\overset{\|\|}{S}}-OH$
	$H_2S_2O_7(+6)$	Pyrosulphuric	$SO_3 + H_2SO_4$.	$HO-\underset{\underset{O}{\|\|}}{\overset{O}{\overset{\|\|}{S}}}-O-\underset{\underset{O}{\|\|}}{\overset{O}{\overset{\|\|}{S}}}-OH$
3 or more	$(H_2S_2O_6)S_n\left(\begin{smallmatrix}+6\\-2\end{smallmatrix}\right)$	Polythionic		$HO-\underset{\underset{O}{\|\|}}{\overset{O}{\overset{\|\|}{S}}}-S_n-\underset{\underset{O}{\|\|}}{\overset{O}{\overset{\|\|}{S}}}-OH$
Peroxoacids	$H_2SO_5(+6)$	**Peroxymonosulphuric**	H_2O_2 + chlorosulphuric acid (HO_3SCl).	$H-O-O-\underset{\underset{O}{\|\|}}{\overset{O}{\overset{\|\|}{S}}}-OH$
	$H_2S_2O_8(+6)$	**Peroxodisulphuric**	anodic oxidation of HSO_4^{-}.	$HO-\underset{\underset{O}{\|\|}}{\overset{O}{\overset{\|\|}{S}}}-O-O-\underset{\underset{O}{\|\|}}{\overset{O}{\overset{\|\|}{S}}}-OH$

do not necessarily give the best yield of SO_3 because, since the reaction is exothermic, an increase in temperature reduces the equilibrium constant. In practice typical conditions for a 98% yield are 500°C at 1 atmosphere pressure, with excess O_2 and using V_2O_5 catalyst. When SO_3 is dissolved in water undesirable acid aerosols are formed. To avoid this problem the SO_3 is normally dissolved in concentrated sulphuric acid, to give oleum or 'fuming sulphuric acid', $H_2S_2O_7$. This acid is then diluted as required. Solid H_2SO_4 is a white crystalline material, melting at 10.4°C. The concentrated aqueous acid has a great affinity for water and is used as a laboratory drying agent, especially for gases which can be dried by being bubbled through the acid.

$$H_2O + H_2SO_4 \rightarrow H_3O^+ + HSO_4^- \qquad (11.35)$$

The acid removes the elements of water from organic compounds such as oxalic acid and carbohydrates:

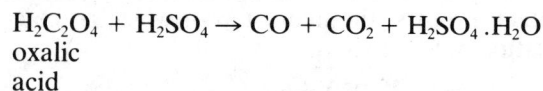

$$H_2C_2O_4 + H_2SO_4 \rightarrow CO + CO_2 + H_2SO_4 \cdot H_2O$$
oxalic
acid

$$(11.36)$$

$$C_{12}H_{22}O_{11} + 11H_2SO_4 \rightarrow 12C + 11H_2SO_4 \cdot H_2O \tag{11.37}$$

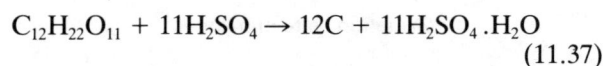

The concentrated acid is a strong oxidizing agent, and resembles nitric acid in having a range of possible half-reactions available (see Table 11.15 and Fig. 11.12). SO_2 is the usual product of reduction, but free sulphur and H_2S can also be produced. Copper metal reduces hot concentrated sulphuric acid to SO_2, in the process forming $CuSO_4$, CuS and S.

Table 11.15 Reduction potentials for H_2SO_4 and related species in acidic and basic solutions

Reaction		$E^\ominus$ (V)
Acid solution		
$SO_4^{2-} + 4H^+ + 2e^-$	$\rightarrow SO_2 + 2H_2O$	$+0.17$
$4SO_2 + 4H^+ + 6e^-$	$\rightarrow S_4O_6^{2-} + 2H_2O$	$+0.51$
$2SO_2 + 2H^+ + 4e^-$	$\rightarrow S_2O_3^{2-} + H_2O$	$+0.40$
$S_4O_6^{2-} + 2e^-$	$\rightarrow 2S_2O_3^{2-}$	$+0.08$
$S_2O_3^{2-} + 6H^+ + 4e^-$	$\rightarrow 2S + 3H_2O$	$+0.50$
$S + 2H^+ + 2e^-$	$\rightarrow H_2S$	$+0.14$
Alkaline solution		
$SO_4^{2-} + H_2O + 2e^-$	$\rightarrow SO_3^{2-} + 2OH^-$	-0.98
$2SO_3^{2-} + 3H_2O + 4e^-$	$\rightarrow S_2O_3^{2-} + 6OH^-$	-0.58
$SO_3^{2-} + 3H_2O + 6e^-$	$\rightarrow S^{2-} + 6OH^-$	-0.59
$S_2O_3^{2-} + 3H_2O + 4e^-$	$\rightarrow 2S + 6OH^-$	-0.74
$S + 2e^-$	$\rightarrow S^{2-}$	-0.51

Oxidation state of S:

+6	+4	Mixed	Mixed	0	−2

(a)

$$SO_4^{-2} \xrightarrow{+0.17\ V} SO_2 \xrightarrow{+0.51\ V} S_4O_6^{-2} \xrightarrow{+0.08\ V} S_2O_3^{-2} \xrightarrow{+0.50\ V} S \xrightarrow{+0.14\ V} H_2S$$

$$+0.40\ V$$

(b)

$$SO_4^{-2} \xrightarrow{-0.98\ V} SO_3^{-2} \xrightarrow{-0.58\ V} S_2O_3^{-2} \xrightarrow{-0.74\ V} S \xrightarrow{-0.51\ V} S^{-2}$$

$$-0.59\ V$$

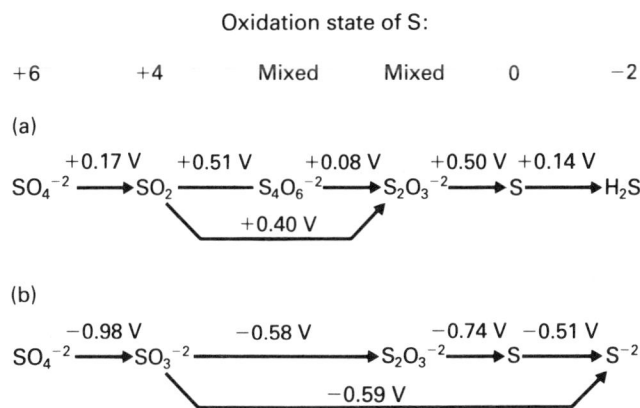

Fig. 11.12 Reduction potential diagrams for H_2SO_4 and related species: (a) in acid solution; (b) in alkaline solution

The $E^\ominus$ values in Table 11.15 show that sulphates are oxidizing agents only in acid solution; in alkaline solution sulphides, thiosulphates (containing the $S_2O_3^{2-}$ ion) and sulphites (containing the SO_3^{2-} ion) are quite strong *reducing* agents. The $E^\ominus$ value for reduction of thiosulphate to sulphur exceeds the $E^\ominus$ value for oxidation of SO_2 to thiosulphate.

Hence in acid solution thiosulphates decompose according to Eq. 11.38, for which the $E^\ominus$ value is $0.50 - 0.40 = +0.10$ V.

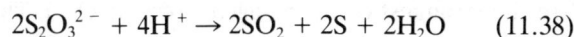

$$2S_2O_3^{2-} + 4H^+ \rightarrow 2SO_2 + 2S + 2H_2O \tag{11.38}$$

The tetrathionate ion $S_4O_6^{2-}$ is readily produced by oxidation of thiosulphate with iodine, a reaction which is the basis of the thiosulphate titration for estimating molecular iodine concentrations.

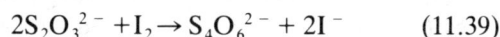

$$2S_2O_3^{2-} + I_2 \rightarrow S_4O_6^{2-} + 2I^- \tag{11.39}$$

Gaseous SO_2 and H_2S reduce and oxidize one another, the product in both cases being finely divided sulphur. This is the basis of the process by which H_2S in Canadian natural gas is converted to elemental sulphur. Partial combustion of the H_2S produces SO_2 (reaction 11.40), which then reacts with the remaining H_2S as in Eq. 11.41.

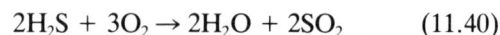

$$2H_2S + 3O_2 \rightarrow 2H_2O + 2SO_2 \tag{11.40}$$

$$2H_2S + SO_2 \rightarrow 2H_2O + 3S \tag{11.41}$$

The peroxy-acids and their salts contain the peroxo grouping —O—O— and are therefore very strong oxidizing agents. For reduction of the peroxydisulphate ion according to Eq. 11.42, $E^\ominus$ is $+2.01$ V.

$$S_2O_8^{2-} + 2e^- \rightarrow 2SO_4^{2-} \tag{11.42}$$

In practice oxidations with peroxysulphuric acids are often slow and they commonly require a catalyst such as Ag^+ ions.

The thiosulphate anion, in which an S atom replaces one O atom of the sulphate ion, is well known, although the parent acid $H_2S_2O_3$ is not stable at room temperature. The anion can be produced simply by boiling sulphur with a solution containing the sulphate ion. Thiosulphates are widely used in photography because of their ability to dissolve silver halides, with formation of the complex anion $[Ag(S_2O_3)_2]^{3-}$. In photography an image composed of black elemental silver is *developed* by gentle controlled reduction of AgBr which has been exposed to light. The image is *fixed* by dissolving away excess AgBr with sodium thiosulphate ('hypo') solution.

Oxyanions of Halogens

The known oxyacids of the halogens are listed in Table 11.16. The only fluorine compound, HOF, is rapidly oxidized in water, but for the others the aqueous solutions are stable. Some of the acids are known only in aqueous solution. Besides being acids, the compounds are also oxidizing agents in both acid and alkaline solutions. Reduction potential diagrams for the various oxyanions are given in Fig. 11.13. The oxidizing powers generally decrease on going from Cl to I, although in some cases the bromo-compounds are the strongest oxidizing agents (e.g. BrO_3^-). The acids and their salts are used extensively in redox reactions e.g.

$$HOCl + 2Fe^{2+} + H^+ \rightarrow 2Fe^{3+} + Cl^- + H_2O \tag{11.43}$$

$$IO_3^- + 6H^+ + 5I^- \rightarrow 3I_2 + 3H_2O \tag{11.44}$$

Reaction 11.44 is used for standardizing thiosulphate solutions (which react with the liberated iodine),KIO_3 being suitable for use as a primary standard. In several places in Fig. 11.13 the condition for disproportionation is fulfilled; thus in alkaline solutions Cl_2, Br_2 and I_2 disproportionate to halide and hypohalite ions, X^- and XO^-. Other reactions by which these species can be interconverted are:

$$3OX^- \rightarrow XO_3^- + 2X^- \tag{11.45}$$
$$\text{(hot solution)}$$

$$3NaClO_2(s) \xrightarrow{heat} 2NaClO_3(s) + NaCl(s) \tag{11.46}$$
and

$$2KXO_3(s) \xrightarrow{300°C} 2KX(s) + 3O_2(g) \ (X = Cl, Br) \tag{11.47}$$

If X = Cl the last reaction gives $KClO_4$ + KCl at temperatures below 300°C.

Table 11.16 Oxyacids of the halogens

Oxidation state	F	Cl	Br	I
+1	FOH	ClOH*†	BrOH*†	IOH*†
+3		HClO₂*†	HBrO₂*†	
+5		HClO₃*†	HBrO₃*†	HIO₃†
+7		HClO₄†	HBrO₄*†	HIO₄
+7				H₅IO₆
+7				H₄I₂O₉†

* Not obtainable in a pure state.   † Known in aqueous solution.

Acid strength increases on going from HOX to HXO₄. The perchlorate ion ClO_4^- forms insoluble salts with a wide range of cations, especially transition metal complex cations, and so can be used to isolate these species from solution. However, there is the problem that when the salts contain organic materials they are sometimes dangerously explosive.

Hypochlorites and hypobromites (generally as their calcium salts) are used as bleaching agents, hypobromite being the stronger of the two. The bleaching action can consist of oxidation of the coloured material, or addition of HOX

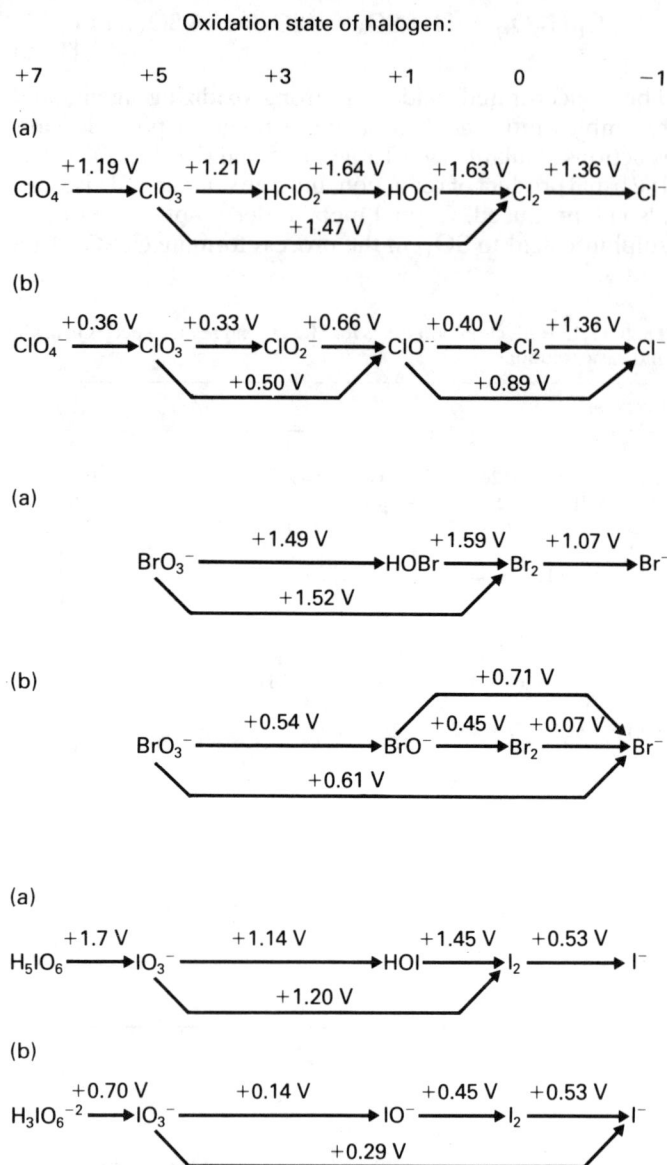

Fig. 11.13 Reduction potential diagrams for halogen oxy-anions: (a) in acid solution; (b) in alkaline solution

across a double bond, or halogenation of a saturated compound. Sodium chlorite, $NaClO_2$, is an excellent bleaching agent for cellulosic materials such as paper and straw.

(c) Halides
The majority of elements form binary halides, especially fluorides and chlorides. The compounds cover the whole range from ionic or semi-ionic polymeric lattices to discrete molecules that form Van der Waals' crystals. Approximate

divisions between the various types are indicated in Fig. 11.14. For a given metal, metal halides in which the metal is in a high oxidation state tend to be covalent and molecular (because of the large polarizing power of a highly charged ion) while lower oxidation state halides are normally ionic or semi-ionic and polymeric. This trend is most obvious for the d block elements, but is also shown, for example, by $PbCl_2$ and $PbCl_4$ or $SnCl_2$ and $SnCl_4$. Trends in the bonding and structure of halides are evident from trends in their boiling and melting points (Fig. 11.15), which give an indication of the strength of intermolecular forces. The boiling and melting points of the molecular compounds increase on going from fluorides to iodides as would be expected because of the mass increase; the reverse situation holds for the ionic and semi-ionic halides because of the influence of the lattice energy. The heats of formation per halogen atom generally show an increase on going from the molecular to the ionic halides (Fig. 11.16). This again reflects the influence of lattice energy. The most stable halides thermodynamically are the fluorides, especially the molecular fluorides of the p block elements. This reflects the stronger bonding to fluorine which results from its extreme electronegativity and small atomic radius. Because of these factors fluorine tends to cause other elements to achieve their maximum oxidation states and coordination numbers, as for example in SF_6 and IF_5.

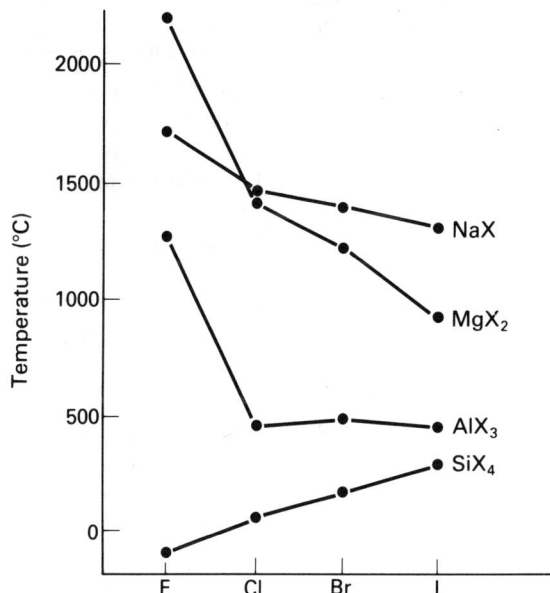

Fig. 11.15 Boiling points of some representative halides

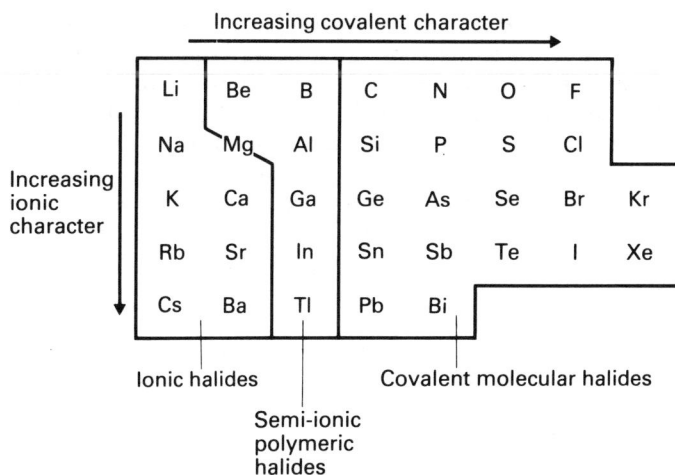

Fig. 11.14 Bonding and structural properties of the halides of the regular elements

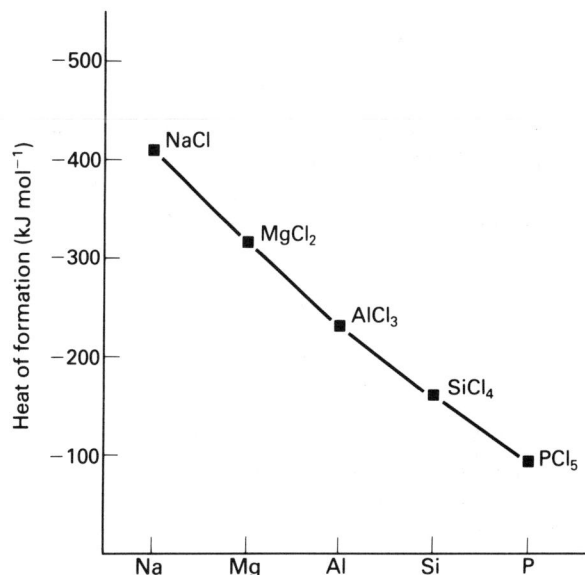

Fig. 11.16 Heats of formations of some chlorides per atom of chlorine

Numerous preparative methods are available for halides; some are listed with examples in Table 11.17. It is often necessary to use anhydrous conditions, since many halides hydrolyse readily.

Hydrolysis occurs readily for the majority of the p block element halides, whereas for the ionic s block element halides the compounds dissolve in water with ionic dissocia-

Table 11.17 Preparative methods for halides

	Method	Example
Dry		
1	Direct reaction of elements.	$Sn + 2Cl_2 \rightarrow SnCl_4$
2	Metal + HX	$Cr + 2HCl \xrightarrow{900°} CrCl_2 + H_2$
3	Halogenation of oxide or sulphide.	$Cr_2O_3 + 3C + 3Cl_2 \rightarrow 2CrCl_3 + 3CO$ $2BeO + CCl_4 \rightarrow 2BeCl_2 + CO_2$
4	Dehydration of hydrated halide.	$CuCl_2 2H_2O \xrightarrow[HCl\ gas]{150°C\ in} CuCl_2 + 2H_2O$
5	Halogen exchange.	$FeCl_3 + BBr_3 \rightarrow FeBr_3 + BCl_3$ $3SiCl_4 + 4SbF_3 \rightarrow 3SiF_4 + 4SbCl_3$
6	Reduction of higher halide.	$3WBr_5 + Al \xrightarrow{240-475°C} 3WBr_4 + AlBr_3$
7	Thermal decomposition of a higher halide.	$AuCl_3 \xrightarrow{160°} AuCl + Cl_2$
Wet		
8	Element + aqueous HX	$Fe + 2HCl(aq) \rightarrow FeCl_2 6H_2O + H_2$
9	Oxide, carbonate or hydroxide + aqueous HX	$Rb_2CO_3 + 2HCl \rightarrow 2RbCl + H_2O + CO_2$
10	Precipitation.	$Ag^+ + Cl^- \rightarrow AgCl$

tion. Hydrolysis is initiated by donation of an oxygen lone pair of electrons to the central atom:

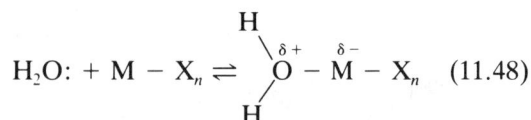

$$H_2O: + M - X_n \rightleftharpoons \overset{H}{\underset{H}{\overset{|}{O}}} \overset{\delta+}{-} \overset{\delta-}{M} - X_n \quad (11.48)$$

This can occur provided the oxygen can get close enough to the central atom M and provided this atom has a vacant orbital to accept the lone pair. The next step is cleavage of the M—X and O—H bonds. Some examples are:

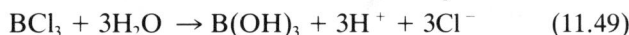

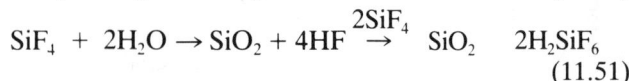

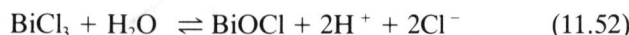

$$BCl_3 + 3H_2O \rightarrow B(OH)_3 + 3H^+ + 3Cl^- \quad (11.49)$$

$$SiCl_4 + 2H_2O \rightarrow SiO_2 + 4H^+ + 4Cl^- \quad (11.50)$$

$$SiF_4 + 2H_2O \rightarrow SiO_2 + 4HF \overset{2SiF_4}{\rightarrow} SiO_2 \quad 2H_2SiF_6 \quad (11.51)$$

$$BiCl_3 + H_2O \rightleftharpoons BiOCl + 2H^+ + 2Cl^- \quad (11.52)$$

The ease of hydrolysis decreases in the order $F > Cl > Br > I$; thus it is related to the size and electronegativity of the halogen. With fluorides the small fluorine atom does not completely shield the central atom from attacking agents, and because of its electronegativity the effective positive charge on the central atom is greatest for fluorides, which results in maximum attraction for the negatively charged oxygen atom of water. Hydrolysis occurs most readily for halides of second and third row elements, because they have accessible vacant orbitals to accept the oxygen lone

pair (e.g. the d orbitals of silicon in $SiCl_4$). Although the reaction

$$CF_4(g) + 2H_2O(g) \rightarrow CO_2(g) + 4HF(g) \quad (11.53)$$

is thermodynamically favoured ($\Delta G = -151$ kJ) it does not occur because the carbon atom cannot expand its valence shell and also is small enough to be well shielded by the surrounding fluorine atoms. Hydrolysis of BCl_3 does occur because the boron atom has a vacant p orbital and the planar geometry of BCl_3 makes it easy for the water molecule to get close to the boron atom. In the case of NCl_3, where there is no vacant orbital but there is a lone pair on the nitrogen atom, hydrolysis is initiated by donation of the nitrogen lone pair to a proton of water:

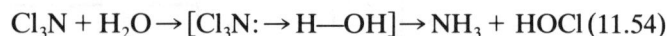

$$Cl_3N + H_2O \rightarrow [Cl_3N: \rightarrow H—OH] \rightarrow NH_3 + HOCl \quad (11.54)$$

Hydrolysis does not occur for NF_3 and OF_2; here the influence of the fluorine is to reduce the donor power of the nitrogen and oxygen. It is worth noting that whereas fluorides are thermodynamically the most stable of the halides with respect to the free elements, they are also the most reactive with water. Clearly one cannot assume that stability of one type guarantees stability in other directions.

Since the halide ions have lone pairs they can act as Lewis bases and form halogeno-complex anions, as for example:

$$SnCl_4 + 2Cl^- \rightarrow SnCl_6^{2-} \quad (11.55)$$

$$CoCl_2 + 2Cl^- \rightarrow CoCl_4^{2-} \quad (11.56)$$

The two principal types of halogeno-complex anions are MX_4^{n-}, either tetrahedral (e.g. $CoCl_4^{2-}$, $FeCl_4^-$, HgI_4^{2-}) or square planar (e.g. $PdCl_4^{2-}$, $AuCl_4^-$), and MX_6^{x-}, octahedral (e.g. AlF_6^{3-}, $SnCl_6^{2-}$). Species of the type MX_6^{2-} exist for a wide range of tetravalent atoms, including the group IV and group VI elements and particularly the second and third row transition metals. The anions MX_6^{2-} and associated cations commonly form a cubic crystal lattice similar to the NaCl structure. Polynuclear halogeno-complexes also exist, especially for the heavy transition metals to the left of the transition metal series. Some examples are, $[Cr_2Cl_9]^{3-}$, $([Mo_6Cl_8]Cl_6)^{2-}$, and $[Re_3Cl_{12}]^{3-}$ (Fig. 11.17). The stability and existence of a complex A_mMX_n depends on whether the energy of the reaction:

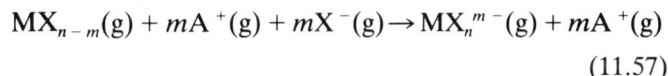

$$MX_{n-m}(g) + mA^+(g) + mX^-(g) \rightarrow MX_n^{m-}(g) + mA^+(g)$$

$$(11.57)$$

compensates for the loss of lattice energy on changing from small compact ions in AX to a large anion MX_n^{m-} in A_mMX_n. The difference between these two lattice energies decreases as A^+ becomes larger, and hence we find that large cations tend to stabilize the complex halogeno anions. For example, the anion $FeCl_6^{3-}$ only exists in the presence of large cations such as $[Co(NH_3)_6]^{3+}$. Salts with large cations, comparable in size to the MX_n^{m-} anion, are often of low solubility compared with a salt where the anion and cation size differ greatly. For this reason the strong oxidizing anion $RhCl_6^{2-}$ is stabilized by precipitation from solution as the caesium salt.

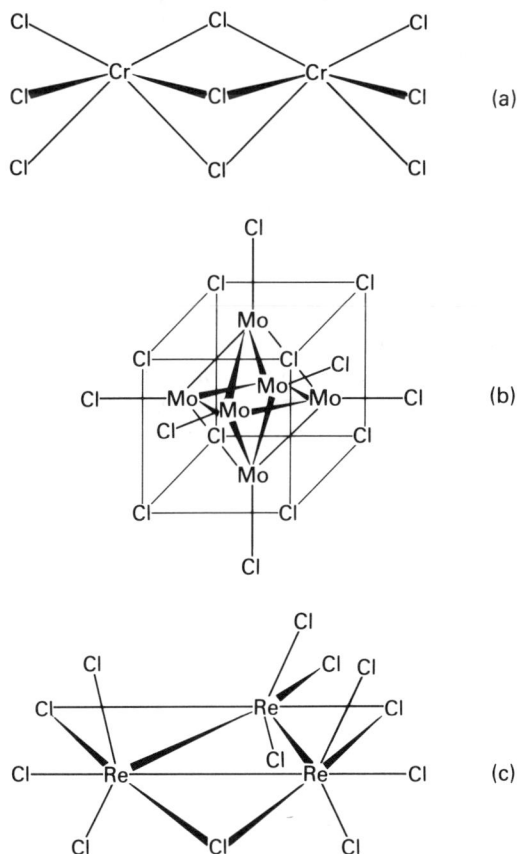

11.3 Boron and Aluminium

The ionic radius of boron (Table 11.18) is so small that the cation B^{3+} does not exist in solution and therefore there is no value listed for the heat of hydration. The Al^{3+} cation does exist and has a heat of hydration nearly double that of Be^{2+}. Whereas boron is a distinct non-metal, aluminium is a typical metal, although some of its compounds (e.g. the oxide and halides) have properties which are borderline between those of metals and non-metals. The elements are not very reactive (Table 11.19), and aluminium is even less reactive than would be suggested by its electrode potential because of a coherent protective coating of oxide which rapidly forms on its surface. Thus aluminium will not dissolve in nitric acid.

(a)

(b)

(c)

Fig. 11.17 Structures of (a) $[Cr_2Cl_9]^{3-}$, (b) $[Mo_6Cl_8]Cl_6$, and (c) $[Re_3Cl_{12}]^{3-}$

Table 11.18 Some properties of boron and aluminium

	B	Al
Electron configuration	$1s^2 2s^2 2p^1$	$KL 3s^2 3p^1$
Ionic radius (pm)	20	50
Covalent radius (pm)	82	125
Heat of atomization (kJ mol^{-1})	562	326
Heat of hydration (kJ mol^{-1})	—	-4630
Electronegativity	2.0	1.45
Ionization potential (kJ mol^{-1}) 1st	800	577
2nd	2427	1816
3rd	3658	2745
Electron affinity (kJ mol^{-1})	15	26
Electrode potential M^{3+}/M (V)	$+0.73$	-1.67
Oxidation state	III	III

Table 11.19 Products of some reactions of boron and aluminium

Reaction	B	Al
Heat in air	Mixture of oxide and nitride.	
Heat in nitrogen	1200°, BN	1600°, AlN
Heat with X = S, Se or Te	B_2X_3	Al_2X_3
Heat with halogen X_2, X = F, Cl, Br, I	BX_3	AlX_3
Concentrated HCl	No reaction	$Al^{3+} + H_2$
Hot concentrated alkali	$H_2BO_3^-$	$Al(OH)_4^-$

(a) Halides

The halides of boron and aluminium are Lewis acids, i.e. they rapidly accept electrons from donor atoms. The acceptor ability increases in the order $BF_3 < BCl_3 < BBr_3 < BI_3$. This is not the order which would be expected on the basis of the halogen electronegativities. In BF_3 the fluorine should tend to increase the effective positive charge on the metal and increase its acceptor ability. In part this is true, but in fact the boron atom satisfies some of its acceptor power by forming a π bond with the fluorine lone pairs. The π bond is so well developed that the vacant p orbital of boron is much less readily available to a donor atom than in BI_3, where the π bond is less important. A less important factor is that the shapes of the halides change on coordination, from planar to approximately tetrahedral, and this requires some 'reorganisation energy' which in the case of AlX_3 compounds is 134 (Cl), 117 (Br), and 79 (I) kJ mol^{-1}. This also favours the iodides as electron acceptors.

The structure of aluminium trichloride changes with temperature (Fig. 11.18). This type of reduction in coordination number with increasing temperature is fairly common. Aluminium forms a number of fluoro-anions, such as the tetrahedral species AlF_4^- as well as species AlF_6^{3-}, AlF_5^{2-} and AlF_4^-, all of which contain an octahedral grouping of fluorine atoms around aluminium, with differing degrees of condensation between octahedra (Fig. 11.19).

(b) Oxy-Compounds

The single bond B—O is strong (523 kJ mol^{-1}) and boron–oxygen multiple bonding does not occur, so the oxyanion chemistry of boron is similar to that of silicon. Boron oxyanions contain planar BO_3 groups and tetrahedral BO_4 groups; examples are metaborate, $(BO_2^-)_3$ (planar) and $[B_4O_5(OH)_4]^{2-}$ (tetrahedral) in borax, $Na_2B_4O_7 10H_2O$.

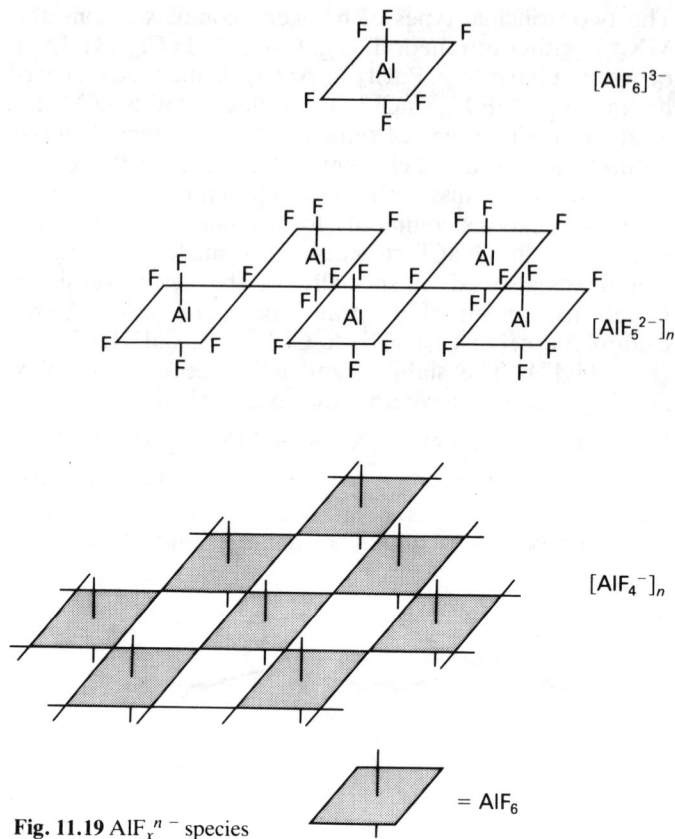

Fig. 11.19 AlF_x^{n-} species

(c) Boron–Nitrogen Compounds

The reaction of boron with nitrogen or BCl_3 with ammonia at 750°C produces boron nitride, BN, which has essentially the same structure as graphite (Figs 11.20a and b), based on extensive covalent bonding within layers of

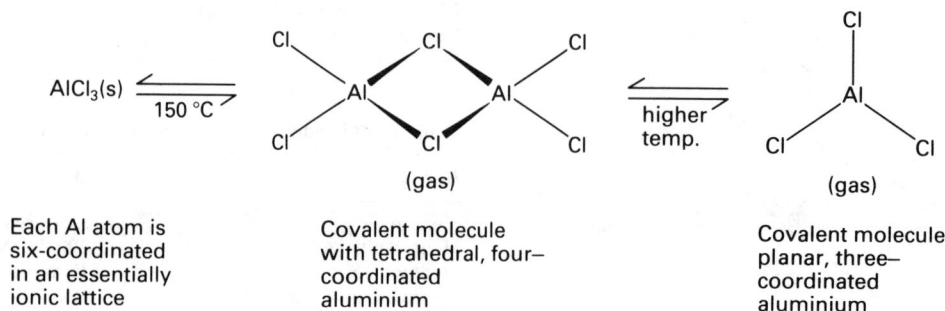

Each Al atom is six-coordinated in an essentially ionic lattice

Covalent molecule with tetrahedral, four–coordinated aluminium

Covalent molecule planar, three–coordinated aluminium

Fig. 11.18 Effect of temperature on $AlCl_3$

(a)

Top layer

Weak bond

Second layer

(b)

Top layer

Weak bond

Second layer

(c)

(d)

etc.

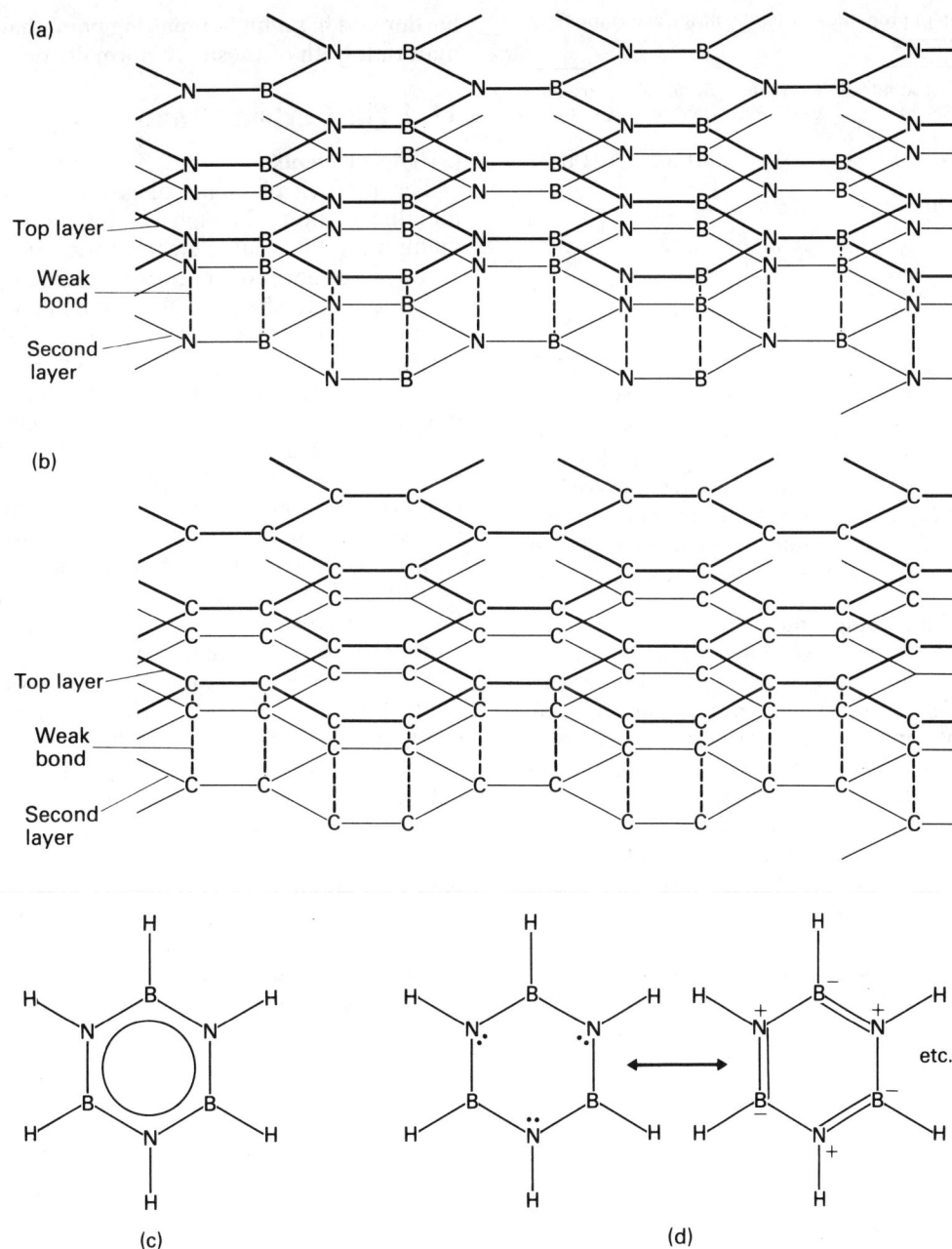

Fig. 11.20 (a) Boron nitride; (b) graphite; (c) borazine; (d) two valence-bond structures for borazine

B_3N_3 hexagons and weak bonding ($\sim 16\,\mathrm{kJ\,mol^{-1}}$) between layers. Another compound, borazine (Fig. 11.20c), can be made by reaction of ammonia with diborane:

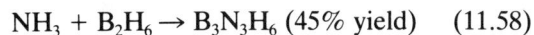

$$NH_3 + B_2H_6 \rightarrow B_3N_3H_6 \ (45\% \ \text{yield}) \quad (11.58)$$

Borazine has some physical properties similar to those of benzene (Table 11.20), and similar delocalization of π electrons occurs around the two six-membered rings (Fig. 11.20d), but the chemical reactions of borazine do not resemble those of benzene.

Table 11.20 Comparison of properties for boron nitride and graphite and for borazine and benzene

	Boron nitride	Graphite	Borazine	Benzene
Bond length within rings (pm)	BN 144.6	CC 145	BN 144	CC 142
M—H bond lengths (pm)	—	—	NH 102 BH 120	CH 108
Interlayer distance (pm)	330	334.5	—	—
Density (g cm^{-3})	2.29	2.255	0.81	0.81
Boiling point (Kelvin)	—	—	328	353

(d) Alums

Double salts of the general formula M(I) M(III) $(SO_4)_2 12H_2O$, where M(III) is Al, Ti, V, Cr, Mn, Fe, Co, Ca, In, Rh, or Ir, are called alums. The trivalent cations are hexahydrates $[M(H_2O)]^{3+}$ in the crystalline state. M(I) is normally K^+ or NH_4^+. The salts form an isomorphous series, and it is possible to crystallize a layer of one alum over a crystal of another member of the series.

(e) Organo-Aluminium Compounds

Compounds of the type R_3Al, where R is an organic radical or a hydrogen atom, are used extensively in reduction reactions in transition metal chemistry. They have an important industrial application as polymerization catalysts for olefins at ordinary temperature and pressure. Catalysts formed between aluminium alkyls and titanium, called Ziegler–Natta catalysts, are obtained by mixing titanium chloride and R_3Al. These catalysts can adsorb ethylene at room temperature and convert it to polyethylene. The polymers so produced are 'stereo-regular' i.e. the successive monomer units are joined on to the growing chain in a sterically regular fashion, in contrast to the irregular result of polymerization in solution. The resulting polymer is harder and has a higher melting point than the non-regular material; both of these are normally desirable properties.

11.4 The Carbon Group

(a) The Elements

The p block elements of group four range from non-metallic carbon, through semi-metallic silicon and germanium, to the weakly electropositive metals tin and lead. Some important properties of the elements are summarized in Table 11.21. The transition from non-metals to metals is manifest in the chemical properties of the elements and their compounds, in the structures of the crystalline elements, and in the trends in atomization energy and melting point between carbon and lead. The elemental state changes from polymeric with strong covalent bonding at the top of the group to metallic bonding at the bottom. The covalent radius of the atom increases very markedly between carbon and silicon (Fig. 11.21); for this reason the chemistry of carbon (as with other first short period elements) is not very typical of the rest of the group. The increasing stability of the divalent oxidation state with atomic number is demonstrated by the change in the oxidizing power of the dioxides (Table 11.22); lead dioxide is a strong oxidizing agent. The figure for carbon dioxide is slightly anomalous, nevertheless CO_2 is a stable, non-oxidizing compound.

The elements carbon and silicon are very plentiful in nature. Carbon occurs in living matter, in fossil fuels and in rocks (3rd most abundant), and silicon in rocks (7th most abundant). On a terrestrial scale silicon is second only to oxygen in abundance (27.72% by weight). Ge, Sn and Pb are less abundant, but tin and lead have been well known for thousands of years because of their useful metallic properties. Germanium and silicon promise to become equally well known because of their use in transistors,

Table 11.21 Properties of the carbon group

	C	Si	Ge	Sn	Pb
Electron configuration	$1s^2 2s^2 2p^2$	$3s^2 3p^2$	$3d^{10} 4s^2 4p^2$	$4d^{10} 5s^2 5p^2$	$4f^{14} 5d^{10} 6s^2 6p^2$
Ionic radius (M^{2+})(pm)	—	—	93	112	120
Covalent radius	77	117	122	140	154
Heat of atomization (kJ mol^{-1})	716	456	376	302	195
Melting point (°C)	3550	1410	937	232	327
Electronegativity	2.5	1.74	2.0	1.7	1.55
Electron affinity (kJ mol^{-1})	123	135	—	—	—
Electrode potential M^{2+}/M (V)	—	—	—	− 0.14	− 0.13
Structure of the elements	Graphite Diamond	Diamond	Diamond	Diamond Metallic	Metallic
Metal/Non-metal properties	Non-metal	Semi-metal	Semi-metal	Weakly electropositive metals	
M—M bond energy (kJ mol^{-1})	356	210–250	180–210	105–150	—
Oxidation states	+ 4	+ 4	+ 4 (+ 2)	+ 4, + 2	(+ 4) + 2

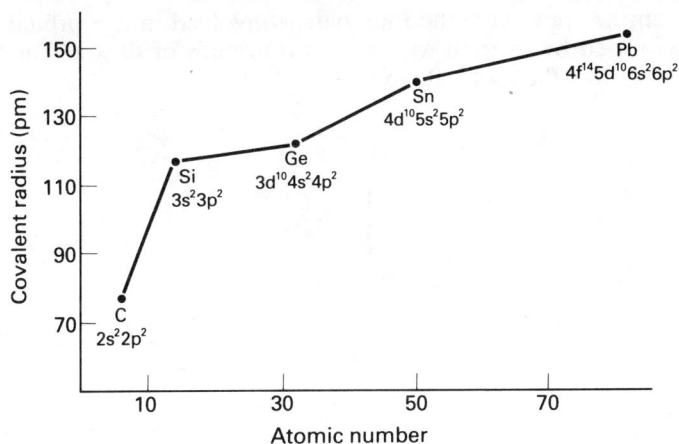

Fig. 11.21 Covalent radius of carbon group elements

at 18°C. At low temperatures the transition leads to crumbling of the metal.

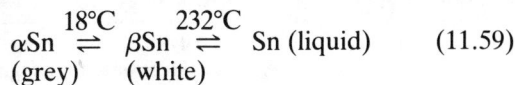

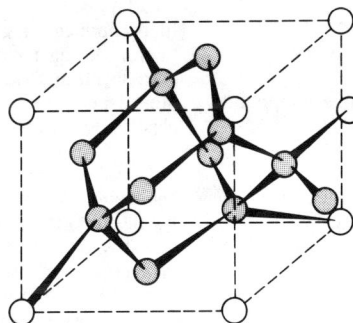

$$\alpha Sn \underset{}{\overset{18°C}{\rightleftharpoons}} \beta Sn \underset{}{\overset{232°C}{\rightleftharpoons}} Sn \text{ (liquid)} \qquad (11.59)$$
$$\text{(grey)} \quad \text{(white)}$$

Diamond   C—C distance: 155.45 pm

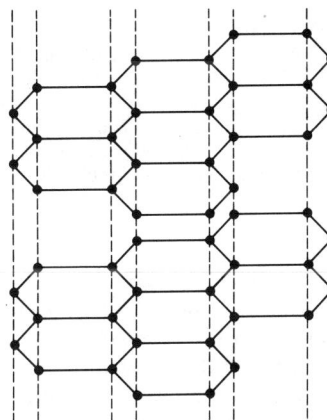

Graphite   C—C distance (in planes): 145.1 pm
C—C distance (between planes): 324.5 pm

Fig. 11.22 Structures of diamond and graphite

integrated circuits and other forms of solid-state electronic device. Carbon exists in three allotropic forms, diamond and two different kinds of graphite. So-called amorphous carbon is microcrystalline graphite. The diamond and graphite structures are compared in Fig. 11.22. The weak forces between the planes in graphite allow them to slide over one another and make graphite a useful dry lubricant. The delocalization of π electrons within the layers confers electrical conductivity along the planes. Graphite is slightly more stable than diamond, $\Delta G^{\ominus}$ for the conversion of diamond to graphite being -2.85 kJ mol^{-1} at 298 K.

Table 11.22 Oxidizing power of + 4 oxidation state for the carbon group

	C	Si	Ge	Sn*	Pb
$E^{\ominus}$ for $MO_2 + 4H^+ +$ 4e$^- \rightarrow M + 2H_2O$	+ 0.38	− 0.86	− 0.15	+ 0.01	+ 1.33

* Reaction for tin is $Sn^{4+} + 4e^- \rightarrow Sn$

However, since the interconversion requires a considerable amount of bond breaking and rearrangement, it is extremely slow. Diamond has a more compact structure than graphite, so that high pressures favour its existence, and at 2000°C and 100 000 atmospheres pressure it is possible to convert graphite to diamond with the aid of transition metal catalysts. The process is used commercially to produce diamonds of industrial quality for use in cutting tools and in heat sinks for solid-state electronic devices. (Diamond is the hardest substance known and also has the highest thermal conductivity.) The allotropes of tin, white metallic tin and a grey form with a diamond structure, interconvert

The elements of group IV and their compounds have a number of important uses, some of which are listed in Table 11.23. The elements as a whole are not very reactive, but reactivity increases as we go down the group. Table 11.24 lists representative reactions of the elements. Silicon is only attacked by one acid, namely HF, and this presumably because of the stability of the final product SiF_6^{2-}:

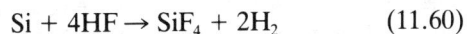

$$Si + 4HF \rightarrow SiF_4 + 2H_2 \qquad (11.60)$$
$$SiF_4 + 2HF \rightarrow H_2SiF_6 \qquad (11.61)$$

Table 11.23 Uses of the carbon group elements

Element	Form	Uses
C	Diamond	Drilling and cutting. Heat sinks for integrated circuits. Jewellery.
	Graphite	Lubricant. 'Lead' pencils. Inert conductor in electrolytic cells.
	Amorphous	Fuel; reducing agent in extraction of metals from oxides.
	Organic compounds	Most aspects of life: examples— plastics, fertilizers, drugs, dyestuffs, fats, oils, soaps, pesticides, proteins, carbohydrates.
Si	Very pure metal	Semiconductors and transistors.
	Silicates	As glass, Portland cement, asbestos, quartz, etc.
	Silicones	Protective coatings, lubricants, structural materials.
Ge	Very pure metal	Semiconductors, transistors.
Sn	Metal	Tin plating. Alloys, e.g. soft solder (50% Sn 50% Pb), bronze (10% Sn 90% Cu), type metal (10% Sn 75% Pb 15% Sb).
	Organic compounds	Fungicides, insecticides.
Pb	Metal	Alloys (see Sn above). Screen for nuclear radiation. Lead accumulators. Protective material, e.g. on roofs, and water pipes.
	Lead compounds	Lead tetraethyl main antiknock agent in petrol; lead oxides used in paints.

Lead is somewhat less reactive than would be expected because of a tendency to form surface coatings. However, lead will react slowly with soft water, to form poisonous $Pb(OH)_2$, which is a good reason for not using lead piping in household water supplies. Lead poisoning may have been a contributing factor in the decline of the Roman Empire, because lead ducting was used in their water systems.

It is of interest that most bonds to silicon are stronger than those to carbon, except the M—H and M—M bonds. It is of course the strength of the C—H and C—C bonds, and the ability of carbon to form multiple bonds with itself and with oxygen and nitrogen, which are responsible for the extent of organic chemistry. One of the interesting differences between carbon and silicon, which is a consequence of silicon having vacant d orbitals for bonding, is in the shape of the two amines $(CH_3)_3N$ and $(SiH_3)_3N$ (Fig. 11.23). The pyramidal shape of trimethylamine results essentially from sp^3 hybridization, with one vertex of the tetrahedron being occupied by the nitrogen lone pair, as in NH_3. In trisilyl-

amine, however, the lone pair is involved in a π orbital formed by overlap with vacant d orbitals of silicon. This requires the molecule to be planar.

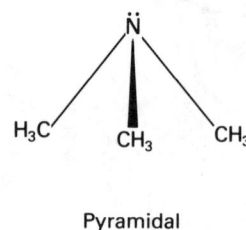

Pyramidal

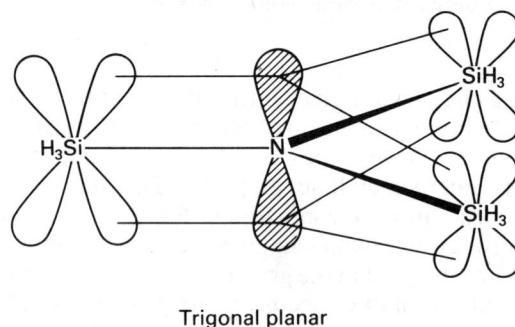

Trigonal planar

Fig. 11.23 Shapes of $(CH_3)_3N$ and $(SiH_3)_3N$

(b) Carbides

Binary compounds of carbon with elements whose electronegativity is less than that of carbon (i.e. less than 2.5) are termed carbides. There are three main types: *salt-like*, formed by groups I and II and aluminium, *interstitial*, formed by most transition metals, and *covalent*. Binary carbon compounds with elements whose electronegativity is greater than 2.5 (e.g. CCl_4, CO), as well as carbon hydrides, are not termed carbides because their properties are in general quite different. Carbides may be prepared by direct union of the elements, by heating an oxide with carbon (the formation of carbides is often a problem when carbon is being used as a reducing agent in the production of metals from the oxides), or by heating a metal in the vapour of a hydrocarbon. The salt-like carbides are of two sorts, the methanides, which contain the C^{4-} ion and give CH_4 on hydrolysis (e.g. Be_2C), and the acetylides, which contain the C_2^{2-} ion and give C_2H_2 on hydrolysis (e.g. CaC_2). Transition metal carbides are also of two types. In interstitial carbides the carbon ions fit into cavities in the metallic structure. They are very hard and brittle compounds, (e.g. tungsten carbide). The second type arises

Table 11.24 Reactions of the carbon group elements

Reagent	Reaction	Comments
Oxygen, or air	$M + O_2 \rightarrow MO_2$	Pb reacts slowly and gives PbO and Pb_3O_4 above 450°.
H_2O (room temp.)	$Pb + H_2O \xrightarrow{O_2} Pb(OH)_2$	(in soft water).
	$Pb + H_2O \xrightarrow{O_2} $ Insoluble $PbSO_4$ and $PbCO_3$	(in hard water).
Steam	$M + 2H_2O \rightarrow MO_2 + 2H_2$	Not Ge.
	$C + H_2O \rightarrow CO + H_2$	
Chlorine	$M + 2Cl_2 \rightarrow MCl_4$	Pb gives $PbCl_2$.
Aqueous alkali	$Si + 2OH^- + H_2O \rightarrow SiO_3^{2-} + 2H_2$	Not C, Ge, Pb. Sn reacts slowly.
Molten alkali		Gives SiO_4^{4-}, GeO_4^{4-}, $Sn(OH)_6^{2-}$, $Pb(OH)_4^{2-}$, no reaction with carbon.
Hot conc. HCl	$Sn + 2H^+ \rightarrow Sn^{2+} + H_2$	Pb slow due to insoluble $PbCl_2$.
Hot conc. H_2SO_4	$C + 2H_2SO_4 \rightarrow CO_2 + 2SO_2 + 2H_2O$	Not for Si; $n = 2$, Pb; $n = 4$,
	$M + H_2SO_4 \rightarrow M^{n+} + SO_2$	Ge or Sn.
Hot conc. HNO_3	$3M + 4HNO_3 \rightarrow 3MO_2 + 4NO + 2H_2O$	With Ge or Sn; Pb gives Pb^{2+}.
Metals	C gives carbides, Si gives silicides, Sn and Sb gives alloys.	

when the cavities in the metallic lattice are not large enough to hold a carbon atom, and in this case chains of carbon atoms exist in the structure (e.g. Fe_3C). The true interstitial carbides are chemically inert, whereas the other type are more reactive and hydrolyse to give a variety of hydrocarbons. Examples of covalent carbides are SiC and B_4C. Carborundum, SiC, is an infinite three-dimensional array of four-coordinated silicon and carbon. It is extremely hard and is widely used as an abrasive.

(c) Oxides of Carbon

Carbon monoxide has a triple bond and is isoelectronic with the triple bonded species, N_2, CN^-, and NO^+ (Table 11.25). CO has a small dipole moment with the carbon atom at the negative end of the dipole. This would not be predicted on the basis of relative electronegativities; theoretical calculations show that the multiple bonding influences the charge distribution. The oxide is a weak Lewis base, and is coordinated by the transition metals to form carbonyl complexes, with the carbon bonded to the metal and the M—C—O group linear. An example is:

$$Ni(s) + 4CO(g) \rightarrow Ni(CO)_4(l) \qquad (11.62)$$

Many carbonyls of the transition metals are known and they find important uses in chemistry. In the purification of nickel the metal is first converted to $Ni(CO)_4$, which can be purified by distillation, after which the application of heat gives back pure nickel metal. The cobalt carbonyl $Co_2(CO)_8$ is used as a catalyst in the conversion of alkenes to aldehydes.

Table 11.25 Comparative properties of CO, N_2, CN^- and NO^+

Oxide	Bond length (pm)	Bond energy (kJ mol^{-1})	Chemical properties
CO	113	1080	Reducing agent, forms transition metal carbonyl complexes.
N_2	110	946	Unreactive, forms transition metal dinitrogen complexes.
CN^-	105	866	Forms a number of ionic salts; also forms transition metal cyanide complexes.
NO^+	106	1050	Formed from reactive molecule NO; occurs as cation $NO^+ClO_4^-$ and in transition metal complexes.

Carbon dioxide, a linear molecule, is acidic and dissolves in water to give hydrated CO_2 plus some carbonic acid. The acid dissociation constants

$$\frac{[H^+][HCO_3^-]}{[H_2CO_3]} = 4.16 \times 10^{-7} \qquad (11.63)$$

and

$$\frac{[H^+][CO_3^{2-}]}{[HCO_3^-]} = 4.84 \times 10^{-11} \qquad (11.64)$$

suggest that carbonic acid is a very weak acid, but this may be misleading because the concentration of undissociated H_2CO_3 is probably much less than that calculated on the basis of the amount of dissolved CO_2. If a correction is made for the amount of hydrated CO_2 in solution the first acid dissociation constant becomes about 2×10^{-4}, a value more in keeping with the structure

$$
\begin{array}{c}
H-O \\
\diagdown \\
C=O \\
\diagup \\
H-O
\end{array}
\qquad (11.65)
$$

The role of CO_2 in photosynthesis is described in Chapter 17. Other group IV oxides are considered in section 11.2.

11.5 The Nitrogen Group

The nitrogen group is dominated by nitrogen itself, whose chemistry is diverse and whose compounds are of great practical importance. The elements show the usual trends within the group (Table 11.26) with the first short period element being, as usual, somewhat atypical. The wide range of oxidation states that exist for nitrogen is a special aspect of its chemistry (Table 11.27). The more important oxidation states appear in Fig. 11.11 which we have discussed in relation to the chemistry of the nitrate ion.

Molecular nitrogen is very stable in both acid and alkaline media; although it is thermodynamically unstable with respect to the formation of NH_4^+, the reaction does not proceed at a reasonable rate except in certain bacterial enzyme systems. For the group V elements other than nitrogen the -3 oxidation state becomes less stable as we move down the group, and at the same time the $+5$ oxidation state becomes less stable relative to $+3$ (Table 11.28). The oxidizing power of the $+5$ state decreases in the order $Bi \gg N > Sb \gg As > P$. (Nitrogen, as nitrate, is seen to be out of order in this sequence).

Further aspects of nitrogen chemistry which distinguish it from the other elements of the same group are as follows:
   (a) the strong $N\equiv N$ bond and absence of catenation.

Table 11.27 Oxidation states of nitrogen in various compounds

Oxidation state	Example
5	N_2O_5, NO_3^-
4	NO_2, N_2O_4
3	NO_2^-, NF_3, NOF
2	NO, N_2F_4
1	N_2O, $N_2O_2^{2-}$
0	N_2
-1	NH_2OH
-2	N_2H_4, $N_2H_5^+$
-3	NH_3, NH_4^+

Table 11.28 Oxidation states $+5$ and $+3$ for the nitrogen group elements

Couple (acid solution)	Reduction potential	$+5$ State	$+3$ State
N(V)/N(III)	$+0.94$	oxidizing	weakly reducing
P(V)/P(III)	-0.28	stable	reducing
As(V)/As(III)	$+0.56$	oxidizing	mild reducing
Sb(V)/Sb(III)	$+0.58$	oxidizing	mild reducing
Bi(V)/Bi(III)	$+1.6$	strongly oxidizing	stable

   (b) The ability of nitrogen to form multiple bonds to oxygen causes its oxides to be small molecules, not solid polymeric substances with single M—O bonds as in the case of the other elements.
   (c) Nitrogen is limited to a maximum coordination number of four. The other elements can expand their valence shell to accommodate a coordination number of six; higher coordination numbers also occur for the bigger elements of the group.
   (d) Nitrogen can take part in hydrogen bonding. The anomalous physical properties of NH_3 by comparison with PH_3 are due to hydrogen bonding.

Table 11.26 Some properties of the nitrogen group elements

Property	N	P	As	Sb	Bi
Electron configuration	$2s^22p^3$	$3s^23p^3$	$3d^{10}4s^24p^3$	$4d^{10}5s^25p^3$	$4f^{14}5d^{10}6s^26p^3$
Covalent radius (pm)	70	110	121	141	148
Ionic radius (pm)	$171(-3)$	$212(-3)$	$222(-3)$	$62(+5)$	$120(+3)$
Heat of atomization (kJ mol^{-1})	473	315	287	262	207
Melting point (°C)	-210	44 (white)	817 (grey)	630 (grey)	271
Electronegativity	3.05	2.05	2.2	1.8	1.65
Oxidation states	$-3, -2, -1,$ $0, 1, 2, 3,$ $4, 5$	$-3, 1, 3, 5$	$-3, 3, 5$	$-3, 3, 5$	$3, 5$

(e) Molecular nitrogen is unusually inert, for both thermodynamic and kinetic reasons. Remarkably, certain bacteria in soil can 'fix' molecular nitrogen, and convert it to ammonia at ordinary temperature and pressure.

Nitrogen is obtained by fractional distillation of liquid air and is generally converted to useful nitrogen compounds such as NH_3, HNO_3, or calcium cyanamide:

$$CaC_2 + N_2 \xrightarrow{1000°} C + CaNCN$$
$$\text{calcium cyanamide} \quad (11.66)$$

$$CaNCN \xrightarrow{3H_2O} Ca(OH)_2 + (NH_2)_2CO$$
$$\text{urea} \quad (11.67)$$

$$(NH_2)_2CO \xrightarrow{2H_2O} (NH_4)_2CO_3$$
$$\text{ammonium carbonate}$$
$$\text{(fertilizer)} \quad (11.68)$$

Phosphorus occurs as phosphate ores, such as apatite $Ca_3(PO_4)_2$ and fluoro-apatite $3Ca_3(PO_4)_2.CaF_2$. When apatite is heated with sand and coke in an electric furnace at 1500°C white phosphorus is formed

$$2Ca_3(PO_4)_2 + 6SiO_2 + 10C \rightarrow 6CaSiO_3 + P_4 + 10CO$$
$$(11.69)$$

The remaining group of elements (As, Sb and Bi) occur as sulphides and are obtained by reduction of the ore.

Phosphorus has a number of allotropes, the most common being white and red phosphorus. White phosphorus contains discrete P_4 molecules with the atoms arranged tetrahedrally (Fig. 11.24). Red phosphorus is a polymer whose structure is obtained by breaking one P—P bond in P_4 and linking the P_4 units together to give P_n. Black phosphorus (also P_n) has a graphite-like structure, but the layers are corrugated rather than planar. Interconversion of the allotropes can be effected as shown in Fig. 11.25. White phosphorus is extremely reactive; for example, finely divided white phosphorus ignites spontaneously in air, and is poisonous. Red and black phosphorus are relatively inert and are non-poisonous.

Nitrogen Fixation

One of the world's most important chemicals is ammonia, which is virtually all produced by the Haber process or its modifications

$$N_2(g) + 3H_2(g) \underset{\text{catalyst}}{\rightleftharpoons} 2NH_3(g) \quad (11.70)$$
$$(\Delta G = -33.4 \text{ kJ})$$

Although the reaction is favourable on thermodynamic grounds it is very slow at room temperature and pressure; therefore optimum conditions are found at high pressure,

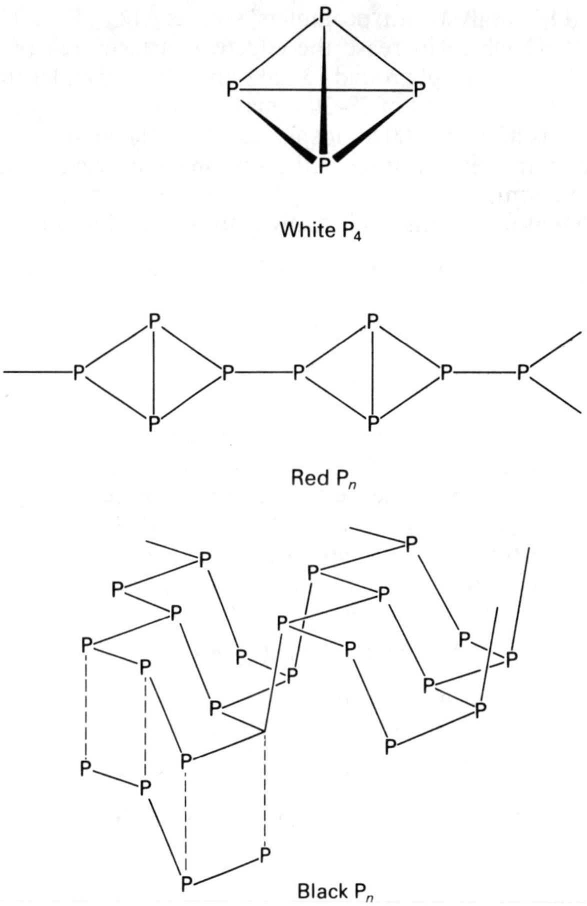

Fig. 11.24 Structures of elemental phosphorus

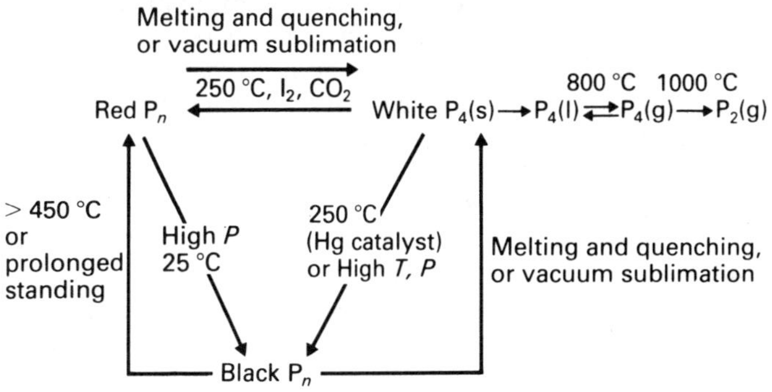

Fig. 11.25 Interconversion of allotropes of phosphorus

which drives the equilibrium to the right, and high temperature, which speeds the reaction up but also drives the equilibrium to the left. Most industrial plants use $P = 300$–325 atmospheres and $T = 530$–560°C, and an iron oxide

(Fe_3O_4) catalyst with 'promoters' such as Al_2O_3, K_2O, CaO or MgO which increase the effective surface area of the catalyst. Yields of around 13–15% are achieved under these conditions; a yield of 28–30% may be obtained using pressures as high as 1000 atmospheres. After the ammonia has been removed the unreacted gases can be recycled through the system.

Ammonia burns in air to give nitrogen and water:

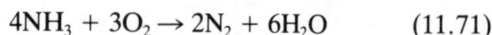

$$4NH_3 + 3O_2 \rightarrow 2N_2 + 6H_2O \qquad (11.71)$$

However, if a platinum catalyst is used nitric oxide is obtained

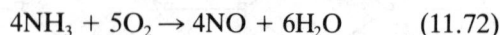

$$4NH_3 + 5O_2 \rightarrow 4NO + 6H_2O \qquad (11.72)$$

This reaction is the basis of the commercial production of nitric acid. Ammonia dissolves in water to give NH_3(aq) (not discrete molecules of 'ammonium hydroxide' NH_4OH) which is partly hydrolysed to NH_4^+ and OH^-. Ammonia also reacts with acids to produce the tetrahedral ammonium ion NH_4^+. A number of ammonium salts volatilize with dissociation at around 300°C, e.g. for ammonium chloride

$$NH_4Cl(s) \rightarrow NH_3(g) + HCl(g) \qquad (11.73)$$

If the anion is an oxidizing agent the cation may be decomposed, e.g. with ammonium dichromate

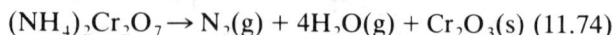

$$(NH_4)_2Cr_2O_7 \rightarrow N_2(g) + 4H_2O(g) + Cr_2O_3(s) \quad (11.74)$$

for which $\Delta H = -315$ kJ mol^{-1}. Once this reaction is initiated the heat evolved ensures its continuation.

The conversion of molecular nitrogen to a form in which it is useable by plants ('nitrogen fixation') is a vital step in the nitrogen cycle (Fig. 11.26). Nitrogen is an essential component of the amino acids which are the building blocks of proteins; hence it is essential to all forms of life on earth. Nitrogen fixation is achieved by certain plants, e.g. the legume family, which have root nodules that contain symbiotic micro-organisms which convert nitrogen to ammonia. As an alternative to using artificial fertilizers one may plant nitrogen-fixing plants in places where nitrogen is required. The Haber process is relatively inefficient and costly; in recent years the possibility has arisen of developing a more efficient process using transition metal complexes which coordinate molecular nitrogen and, when the complex is not too stable, reduce it to ammonia. The metal atoms employed by the nitrogen-fixing bacteria, Mo and Fe, also figure in these complexes. Recent work with complexes of molybdenum shows some promise in providing a means of emulating the chemical accomplishments of the bacteria.

Ammonia has a liquid range extending from -77.8 to -33.4°C. At the boiling point its heat of vaporization is high (1.4 kJ mol^{-1}) because of hydrogen bonding. Liquid ammonia is easily handled in usual laboratory equipment and is a useful non-aqueous solvent with properties fairly similar to those of water. However, its lower dielectric constant (17, versus 82 for water) means that liquid ammonia is a better solvent for organic compounds than water, and a poorer solvent for ionic compounds.

Hydrazine, N_2H_4, may be considered as NH_3 with one H atom replaced by NH_2. It is made commercially by oxidation of ammonia with hypochlorite

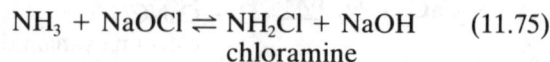

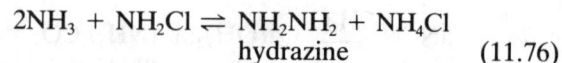

$$NH_3 + NaOCl \rightleftharpoons NH_2Cl + NaOH \qquad (11.75)$$
$$\text{chloramine}$$

$$2NH_3 + NH_2Cl \rightleftharpoons NH_2NH_2 + NH_4Cl \qquad (11.76)$$
$$\text{hydrazine}$$

Hydrazine burns vigorously in oxygen, (Eq. 11.77) and has been used as a rocket fuel. Concentrated nitric acid and hydrazine ignite spontaneously when brought together, which makes them a useful combination in steering rockets that are designed to be fired intermittently.

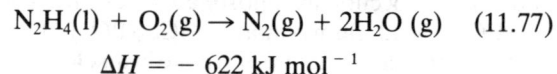

$$N_2H_4(l) + O_2(g) \rightarrow N_2(g) + 2H_2O (g) \quad (11.77)$$
$$\Delta H = -622 \text{ kJ mol}^{-1}$$

Cations $N_2H_5^+$ and $N_2H_6^{2+}$ can be produced in which the N—N bonds are shorter than in hydrazine. As protons are added the lone pairs become bonding pairs, thereby reducing the repulsions between the nitrogen atoms.

The azide ion N_3^- (from the parent hydrazoic acid HN_3) exists in ionic azides which decompose smoothly on heating to give N_2 gas (except for LiN_3 which gives Li_3N). Covalent azides, on the other hand, explode when heated or even when struck. Silver azide, formed from the ion $Ag(NH_3)_2^+$, is a hazardous byproduct of old silvering solutions which have not been discarded after use in making mirrors. The azide ion is linear with both N—N bonds the same length, whereas in the covalent azides two different N—N bond lengths are observed (Fig. 11.27).

All of the nitrogen group elements form trihalides, MX_3. Except for NF_3, the nitrogen halides are the least stable, in fact NCl_3 is notoriously explosive. The halides NBr_3 and NI_3 exist only as ammoniates $NBr_3(NH_3)_6$ and $NI_3(NH_3)$. The trihalides of phosphorus can be oxidized as follows:

$$2PX_3 + O_2 \rightarrow 2POX_3 \qquad (11.78)$$

$$PX_3 + X_2 \rightarrow PX_5 \qquad (11.79)$$

The ease of oxidation decreases as we move down the nitrogen group elements. However, the nitrogen compounds cannot be oxidized at all because of the limited capacity of the nitrogen atom's valence shell. Some reactions of PCl_3 and PCl_5 are given in Fig. 11.28. The stability of the pentahalides decreases down the group and on going from fluoride to bromide. The structures of both PCl_5 and PBr_5 vary with their physical state. The gaseous compounds are

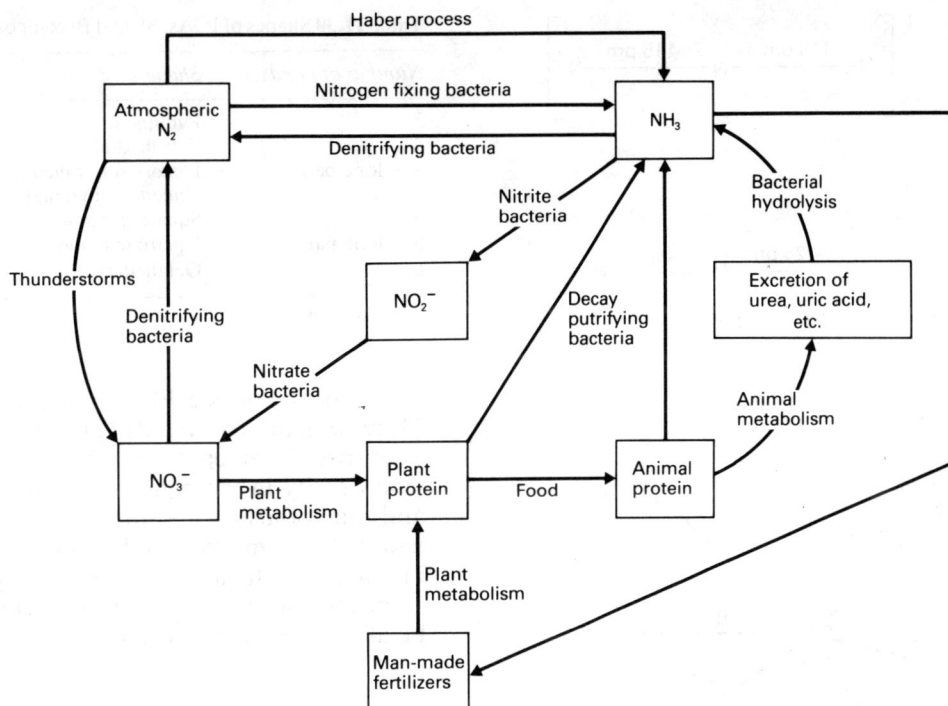

Fig. 11.26 The nitrogen cycle

Table 11.29 Modes of bonding for nitrogen and molecular shape

Type of bonding to N	Electron configuration	Shape	Example
Ionic	N^{3-}	Linear (symmetrical)	Li_3N (nitride)
Covalent, sp^3 hybridization	(H–N⁻–H)	Bent	KNH_2 (amide)
	(H₃N)	Pyramidal Tetrahedral	NH_3 (ammonia) NH_4^+ (ammonium)
Covalent, sp^2 plus π bonds	O=N(–O)(O)	Planar (trigonal)	NO_3^- (nitrate)
	N(O)(O)	Bent	NO_2^- (nitrite)
	N=N	Bent (around N)	RN=NR (diazenes)
	N=O	Bent (around N)	NOCl (nitrosyl chloride)
Covalent, sp plus π bonds	N≡N	—	N_2 nitrogen
	N≡C—C≡N [C≡N]⁻ O=N=O	— — Linear	C_2N_2 cyanogen CN^- cyanide ion NO_2^+ (nitronium)

N_3^-

Azide ion

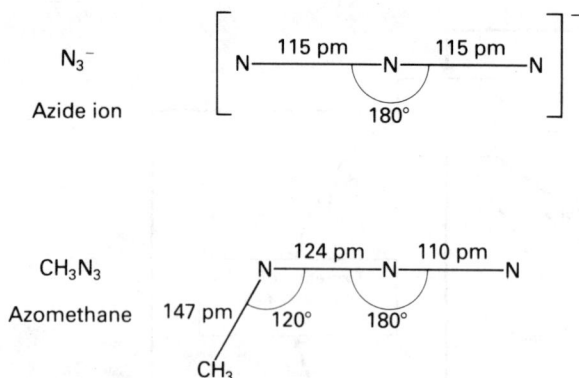

CH$_3$N$_3$

Azomethane

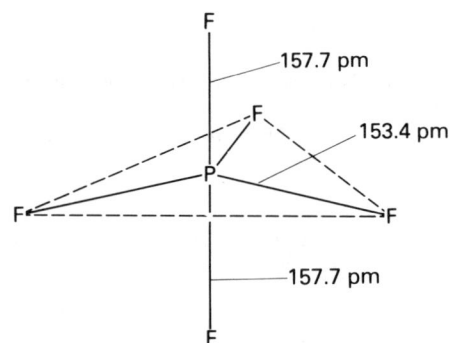

Fig. 11.27 Structures of azides

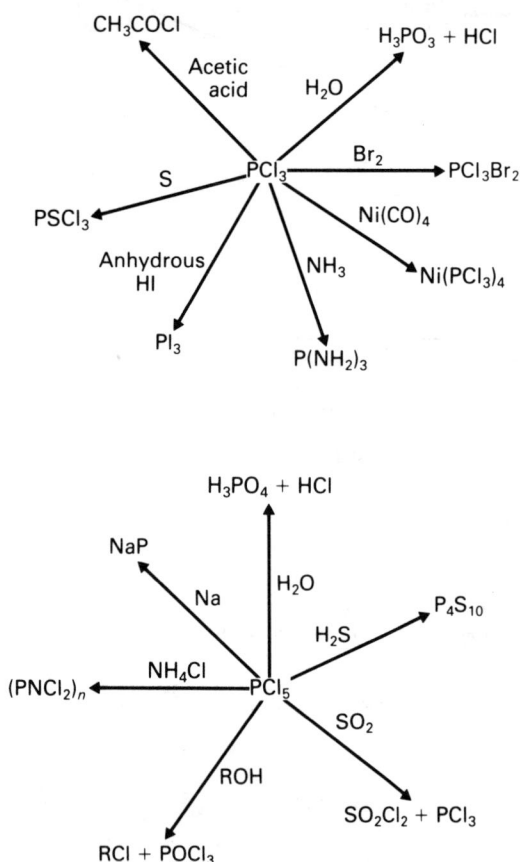

Fig. 11.28 Some reactions of PCl$_3$ and PCl$_5$

molecular and have a trigonal bipyramidal shape, but in the solid they are ionic PCl$_4^+$PCl$_6^-$ and PBr$_4^+$Br$^-$ respectively (PBr$_6^-$ does not exist because of the difficulty of accommodating six bromine atoms around the phosphorus atom).

Table 11.30 Shapes of P, As, Sb and Bi compounds

Number of bonds	Shape	Examples
3	Pyramidal	PH$_3$, AsCl$_3$
4	Tetrahedral	PH$_4^+$, POCl$_3$
4 + lone pair	Distorted tetrahedron	SbCl$_3$.NH$_2$Ph*
5	Trigonal bipyramid	PF$_5$, SbCl$_5$
5	Square pyramid	SbPh$_5$*
5 + lone pair	Square pyramid	[SbCl$_5$]$^{2-}$
6	Octahedral	PCl$_6^-$, SbBr$_6^-$

* Ph = Phenyl, C$_6$H$_5$

The pentafluoride PF$_5$ has the structure given in Fig. 11.29, with the axial bonds 4 pm longer than the equatorial. This arises from repulsion of the axial fluorines by the electron density in the equatorial plane. However, the axial and equatorial fluorine atoms interchange their positions in less than a microsecond, and to certain methods of investigating their environment, e.g. nuclear magnetic resonance of the ^{19}F nucleus, the fluorine atoms appear to be all equivalent to one another.

Fig. 11.29 Structure of PF$_5$

A number of oxyhalides of the nitrogen group elements are known, such as NOX, NXO$_2$ and POX$_3$. The nitrosyl halides have a bent configuration and are strong oxidizing agents. Nitrosyl chloride, NOCl, is one of the active reagents in aqua regia (HNO$_3$ + 3HCl). The P—O bond in the oxyhalides of phosphorus, POX$_3$, is basically a double bond.

The reaction of PCl$_5$ with NH$_4$Cl gives polymeric species called phospho-nitrilic chlorides (PNCl$_2$)$_n$

$$nPCl_5 + nNH_4Cl \rightarrow (PNCl_2)_n + 4nHCl \quad (11.80)$$

When n is 3 the molecule consists of a six-membered planar ring with some electron delocalization around it (Fig. 11.30). Larger rings, as well as chain polymers, are also obtainable. It was hoped at first that useful polymeric

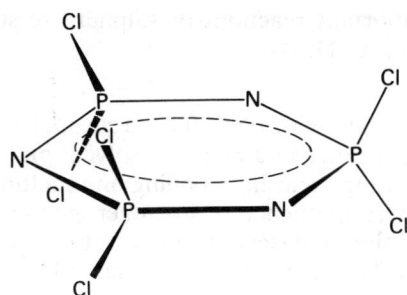

Fig. 11.30 Structure of $(PNCl_2)_3$

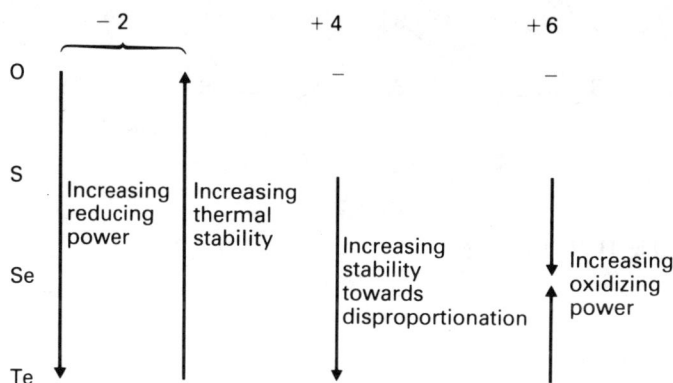

Fig. 11.31 Relative properties of the oxidation states of group VI p block elements

materials would be produced in this way, but unfortunately the compounds are reactive, especially towards hydrolysis which destroys their rubber-like properties. If bulky, unreactive side chains were to replace the halogen atoms the problem might still be overcome.

11.6 The Oxygen Group

Some of the properties of the elements of the oxygen group, O, S, Se and Te, are listed in Table 11.31. The usual trends are observed, such as increasing metallic character down the group, which leads to decreasing acidity of the oxides. The oxidation states of the elements range from $+6$ to -2 in steps of two. The positive oxidation states do not occur for oxygen, and the high positive oxidation states become less stable as we go down the group. Some trends in thermal stability and oxidizing/reducing power of the different oxidation states are represented in Fig. 11.31.

The dinuclear species O_2^+, O_2, O_2^- and O_2^{2-} display regular trends in the properties of the O—O bonds as the number of electrons varies. The additional electrons are added to the π^* antibonding energy levels and therefore reduce the overall bonding between the oxygen atoms. This shows up clearly in the bond length and bond strength, as well as in the vibrational frequency of the O—O bond (Table 11.32).

Table 11.32 Some properties of O_2 species

Property	O_2^+	O_2	O_2^-	O_2^{2-}
Designation	oxygenyl	oxygen	superoxide	peroxide
Bond length (pm)	112	121	133	148
Bond energy (kJ mol^{-1})	623	498	unknown	142
Bond stretching frequency (cm^{-1})	1860	1556	1145	770
Bond order	$2\frac{1}{2}$	2	$1\frac{1}{2}$	1
Valence electrons	11	12	13	14

The allotropy of sulphur is rather complex. The basic unit is the puckered S_8 ring (Fig. 11.32) which packs in different ways in the solid state to give different modifications. The S_8 rings also persist in the liquid and gaseous states, although linear chains and species such as S_4 and S_2 become important below the boiling point. Some of the well-established interconversions between the different allotropes and phases are shown in Fig. 11.33. A wide variety of other allotropes have been reported; not all of them have been substantiated.

Table 11.31 Some properties of the oxygen group elements

Property	O	S	Se	Te
Electron configuration	$2s^2 2p^4$	$3s^2 3p^4$	$3d^{10}4s^2 4p^4$	$4d^{10}5s^2 5p^4$
Ionic radius M^{2-} (pm)	140	184	198	221
Covalent radius (pm)				
double bond	62	94	107	127
single bond	66	104	117	137
Heat of atomization (kJ mol^{-1})	250	280	207	197
Melting point (°C)	-218	119 (monoclinic)	217 (grey)	450
Electronegativity	3.5	2.45	2.5	2.0
Electrode potential M/M^{2-} (V)		-0.51	-0.92	-1.14
Oxidation states	$-2, -1$	$-2, 2, 4, 6$	$-2, 2, 4, 6$	$-2, 2, 4, 6$

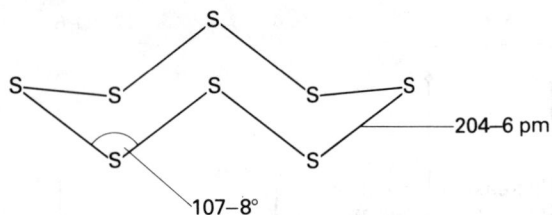

Fig. 11.32 The S_8 ring

107–8°

204–6 pm

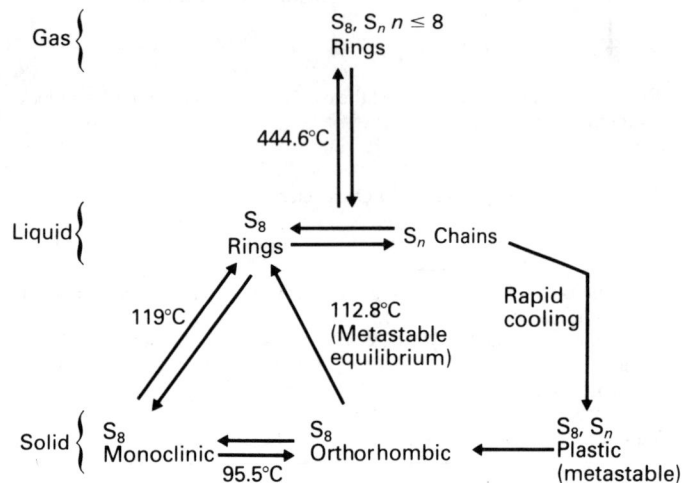

Fig. 11.33 Interconversion of some of the allotropes of sulphur

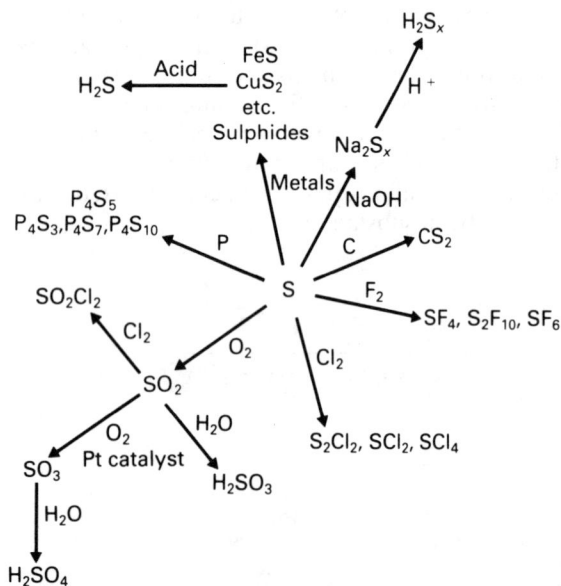

Fig. 11.34 Some reactions of sulphur

Some important reactions of sulphur are shown schematically in Fig. 11.34.

Hydrides

Water, besides being plentiful and cheap, has a very convenient liquid range and excellent solvent properties. The wide liquid range and high boiling and melting points of water are a consequence of hydrogen bonding, which is extensive in the solid state and persists to a marked degree in the liquid. The high dielectric constant 81.7 (18°C) makes it a good solvent for ionic materials and not so good for covalently bonded materials. The force of attraction between two charges e_1 and e_2 a distance r apart is given by

$$F = \frac{-e_1 e_2}{4\pi\epsilon_o D r^2} \qquad (11.81)$$

The force is reduced when D, the dielectric constant, is high. Thus a solvent of high dielectric constant can insert itself between ions in a crystal and promote dissolution of the lattice.

Water can be incorporated in solid compounds in a variety of ways. For example, hydration of cations (e.g. alkali metal ions) or anions (e.g. halide ions) arises from the attraction of the charged species to either the positive or negative end of the water molecule (Fig. 11.35). The energy of this process is the heat of hydration. When the attraction is strong the OH bond is weakened, in extreme cases leading to hydrolysis. The water molecule has a bent structure,

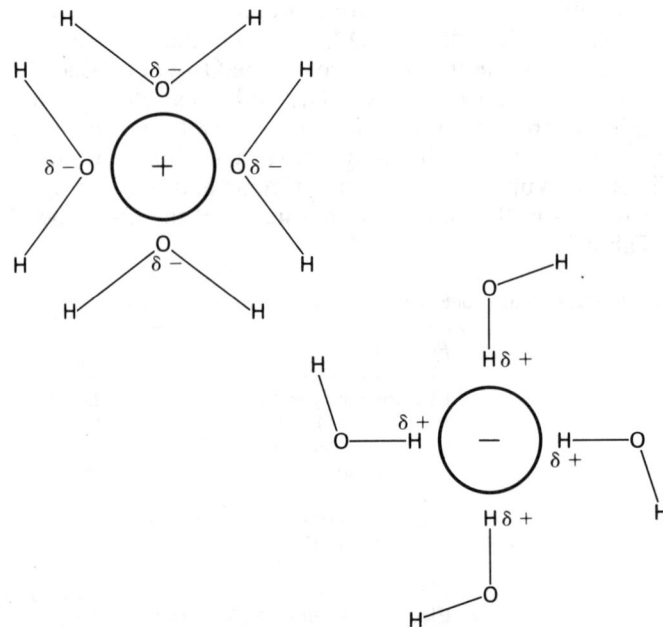

Fig. 11.35 Hydration of cation and anion

with two lone pairs on the central oxygen. In simple hydrated cations the lone pairs of the water are symmetrically disposed with respect to the cation (Fig. 11.36a). In aquo-complexes, on the other hand, one lone pair is donated to the central atom (e.g. a transition metal ion) and the water molecule is not symmetrically placed with respect to the cation (Fig. 11.36b). Sometimes water occupies sites in crystal lattices without being specifically attached to another species. The alums have six water molecules coordinated to the tripositive cation (Al or transition metal) while six others surround the monopositive cation. The lattice water that occupies cavities in zeolites has already been mentioned.

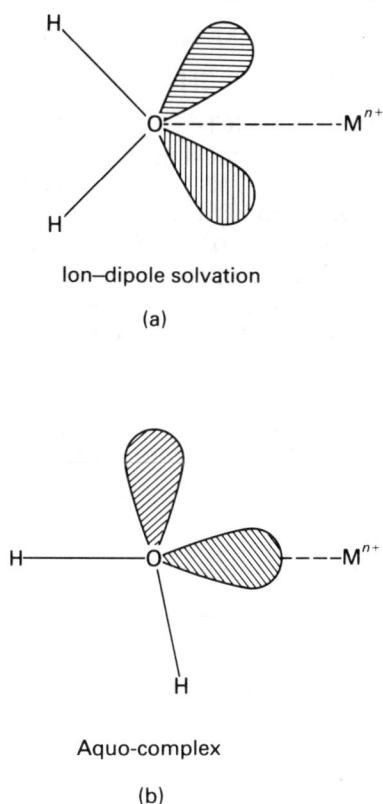

Fig. 11.37 A representation of the structure of $CuSO_4 5H_2O$

Ion–dipole solvation

(a)

Aquo-complex

(b)

Fig. 11.36 Hydrated cations

Another type of anion water occurs, for example, in crystalline sulphates, where the anion (SO_4^{2-}) is hydrogen bonded to one water molecule. Consider $CuSO_4 5H_2O$ (Fig. 11.37). When the salt is heated two water molecules (labelled (a)) can be removed at 30°C, two more (b) are lost at 100°C, while the anion water (c), bonded to the SO_4^{2-} ions, is not removed until 250°C.

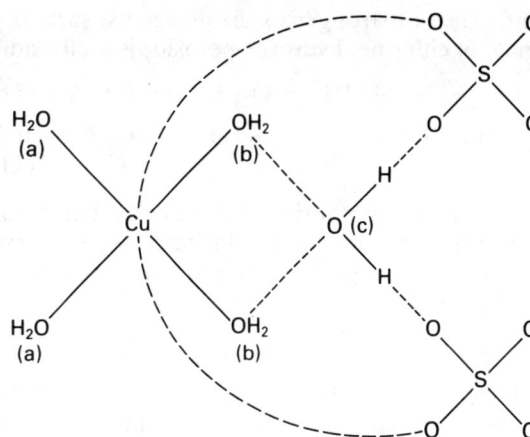

Hydrogen peroxide has the molecular shape shown in Fig. 11.38. The shape arises from a compromise between the various electron pair repulsions around the oxygen atoms. The O—O bond is a single bond (148 pm). Pure liquid peroxide is a very powerful oxidizing agent and explodes before it boils. The liquid is thermodynamically unstable with respect to water and O_2:

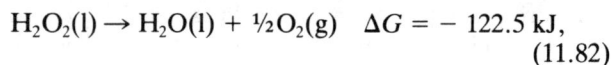

$$H_2O_2(l) \rightarrow H_2O(l) + \tfrac{1}{2}O_2(g) \quad \Delta G = -122.5 \text{ kJ},$$
$$(11.82)$$

but decomposition is slow at room temperature unless a catalyst, such as a transition metal oxide, is present. H_2O_2 is a strong oxidizing agent under both acid and alkaline conditions:

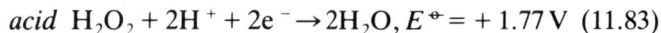

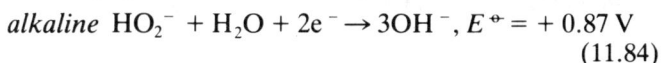

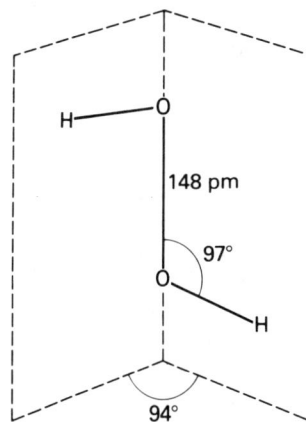

acid $\quad H_2O_2 + 2H^+ + 2e^- \rightarrow 2H_2O, E^{\ominus} = +1.77 \text{ V} \quad (11.83)$

alkaline $\quad HO_2^- + H_2O + 2e^- \rightarrow 3OH^-, E^{\ominus} = +0.87 \text{ V}$
$$(11.84)$$

Fig. 11.38 Structure of H_2O_2

In the presence of stronger oxidizing agents, such as permanganate or chlorine, hydrogen peroxide is itself oxidized:

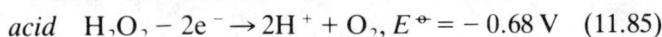

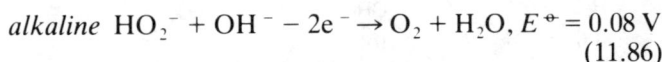

acid $H_2O_2 - 2e^- \rightarrow 2H^+ + O_2, E^{\leftrightarrow} = -0.68\,V$ (11.85)

alkaline $HO_2^- + OH^- - 2e^- \rightarrow O_2 + H_2O, E^{\leftrightarrow} = 0.08\,V$ (11.86)

Note: (The last two potentials shown are oxidation potentials; reduction potentials would refer to the reverse reactions.) Hydrogen peroxide is best as an oxidizing agent in acid solution; however, the reaction is often slow and in need of catalysis. For example, acidic H_2O_2 oxidation of Cr(III) to $Cr_2O_7^{2-}$ can be speeded up by adding Ag^+ ions. As a reducing agent H_2O_2 is strongest in alkaline conditions.

Hydrogen peroxide is produced commercially by the electrolytic formation of the peroxydisulphate ion, $S_2O_8^{2-}$, followed by hydrolysis.

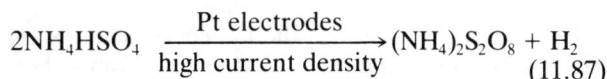

$$2NH_4HSO_4 \xrightarrow[\text{high current density}]{\text{Pt electrodes}} (NH_4)_2S_2O_8 + H_2 \quad (11.87)$$

$$(NH_4)_2S_2O_8 + 2H_2O \rightarrow H_2O_2 + 2NH_4HSO_4 \quad (11.88)$$

Hydrogen peroxide has uses in organic polymerization reactions, as a bleaching agent, and as a source of oxygen in rocket fuels.

Hydrogen sulphide, an evil smelling and highly poisonous gas, is produced in nature by the action of bacteria on sulphates or organic sulphides. Natural gas may contain as much as 10% of H_2S. In the laboratory H_2S is commonly produced by the reaction:

$$FeS + 2HCl \rightarrow FeCl_2 + H_2S \quad (11.89)$$

Like water, the molecule has a bent structure, the HSH angle being 92° 20′. H_2S is a weak dibasic acid:

$$H_2S + H_2O \rightleftharpoons H_3O^+ + SH^-, \quad K = 10^{-17} \quad (11.90)$$

and $$HS^- + H_2O \rightleftharpoons H_3O^+ + S^{2-}, \quad K = 10^{-14} \quad (11.91)$$

Since H_2S is not very soluble in water (10^{-1} molar) there is little free sulphide ion in solution. Hydrogen sulphide is a mild reducing agent and will reduce Fe(III) to Fe(II), SO_2 to sulphur, and NO_3^- to NO_2. Other hydrogen sulphides are known, e.g. H_2S_2, H_2S_6 (sulphanes) which contain chains of sulphur atoms. Salts of these materials are called polysulphides, e.g. K_2S_2, K_2S_5. They contain an S_x^{2-} chain.

Most elements react with sulphur to give binary sulphides e.g. Na_2S, Cu_2S and FeS. Reaction of an alkali with H_2S may also be used. Except for the sulphides of the alkali and alkaline earth metals, sulphides are almost without exception insoluble in water (hence their occurrence in nature). The soluble sulphides hydrolyse in a similar manner to oxides:

$$S^{2-} + H_2O \rightleftharpoons 2OH^- + H_2S \quad (11.92)$$

Group I metal sulphides are the least readily hydrolysed. The sulphides of tin, arsenic and antiomony are amphoteric (Fig. 11.39) and will dissolve in excess sulphide ion to give thioanions; e.g.

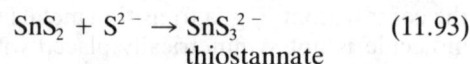

$$SnS_2 + S^{2-} \rightarrow SnS_3^{2-} \quad (11.93)$$
$$\text{thiostannate}$$

Differences in the solubility of sulphides in hydrochloric acid are exploited in qualitative analysis for identifying and separating elements. If a sulphide dissolves and reacts with HCl then H_2S is formed

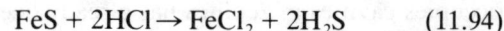

$$FeS + 2HCl \rightarrow FeCl_2 + 2H_2S \quad (11.94)$$

Solubility in HCl depends on the solubility product of the sulphide and the dissociation equilibrium of H_2S:

$$MS(s) \rightleftharpoons M^{2+}(aq) + S^{2-}(aq) \quad (11.95)$$

$$S^{2-} + 2H^+ \rightleftharpoons H_2S \quad (11.96)$$

If the concentration of HCl is increased the second reaction is driven to the right, removing S^{2-} ions from solution, so the sulphide dissolves. This occurs for sulphides for which $[M^{2+}][S^{2-}] = K_{sp}$ is 10^{-20} to 10^{-30}. Sulphides insoluble in dilute HCl have $K_{sp} < 10^{-30}$; if $K_{sp} < 10^{-50}$ (e.g. CuS, Bi_2S_3 and HgS) the sulphide is insoluble in concentrated HCl. Acid-soluble sulphides may be precipitated under alkaline conditions, where the concentration of S^{2-} is increased.

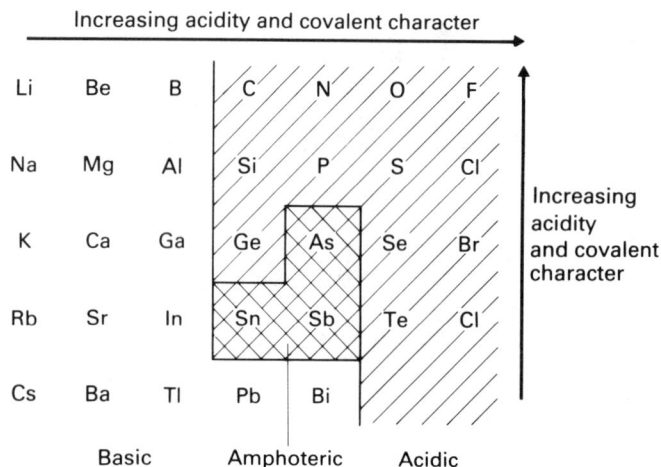

Fig. 11.39 Properties of sulphides of the regular elements

The tetravalent halides such as SF_4, $SeCl_4$, $TeBr_4$ all have lone pairs of electrons. Both the selenium and tellurium compounds can add halide ions to form octahedral halogeno-anions $SeCl_6^{2-}$ and $TeBr_6^{2-}$. The two oxychlorides of sulphur are $SOCl_2$, sulphinyl chloride (or thionyl chloride), and SO_2Cl_2 sulphonyl chloride (or sulphuryl chloride). The former is a useful chlorinating agent, e.g.

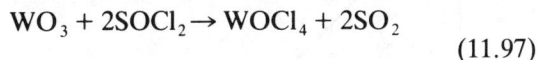

$$WO_3 + 2SOCl_2 \rightarrow WOCl_4 + 2SO_2 \tag{11.97}$$

and has dehydrating properties

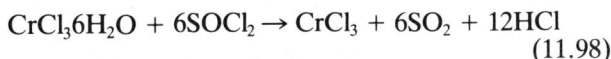

$$CrCl_3 6H_2O + 6SOCl_2 \rightarrow CrCl_3 + 6SO_2 + 12HCl \tag{11.98}$$

However, the products of thionyl chloride reactions are often rather impure.

An interesting compound sulphur nitride, S_4N_4, of which there are many derivatives, can be made by the reaction of sulphur with ammonia in carbon tetrachloride. Its structure (Fig. 11.40a) is cage-like with a weak S....S interaction. The corresponding arsenic sulphide As_4S_4 (Fig. 11.40b) has the same structure but with the atom positions reversed, allowing As—As bonds to form.

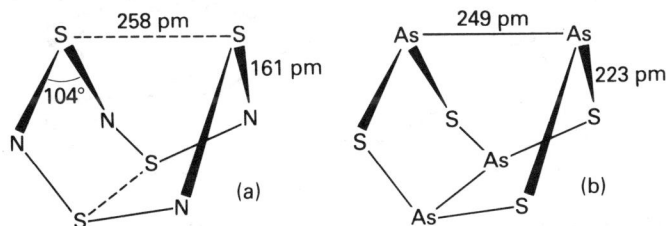

Fig. 11.40 Structures of S_4N_4 (a), and As_4S_4 (b)

11.7 The Halogens

The halogens F, Cl, Br and I, are non-metallic with closely related properties. They all exist as the diatomic species X_2, iodine being a solid, bromine a liquid and chlorine and fluorine being gases at normal temperature and pressure. The electron configuration of the halogens is ns^2np^5, and the predominant features of their chemistry are the reaction

$$X + e^- \rightarrow X^- \tag{11.99}$$

to give the halide ion, or alternatively the formation of single covalent bonds by electron sharing as in HF and $SiCl_4$. In both ways the inert gas valence shell configuration is achieved and the oxidation state of the halogen becomes -1. When predominantly covalent bonding occurs the lone pairs on the halogens can be used to increase the coordination number of the halogen by bridging, as in Al_2Cl_6 (cf. Figs. 11.17–11.19). Except for fluorine the halogens also have the positive oxidation states $+1$, $+3$, $+5$ and $+7$, which vary in steps of two. Generally the oxidizing power of the positive oxidation states increases in the order $I < Cl < Br$. The halogens in positive oxidation states do not occur as free cations (except possibly I^+ in ICl) because the enthalpy change for the reaction

$$\frac{1}{2}X_2 \rightarrow X^+(g) + e^- \tag{11.100}$$

is prohibitive ($+1760$ (F), $+1377$ (Cl), $+1253$ (Br) and $+1113$ (I) kJ mol^{-1}). The removal of one electron from a halogen does not reduce its size greatly and therefore solvation and lattice energies are not sufficient to compensate for the high positive energies given above. The positive oxidation states exist in association with atoms whose electronegativity is comparable with or greater than that of the halogen, i.e. either oxygen, giving oxides and oxyanions, or other halogens, giving interhalogen compounds (see below).

The usual trends occur in the properties of the elements (Table 11.33) except for certain anomalous properties of fluorine, notably its atomization energy, bond energy and electron affinity.

Bond Energies
The low atomization and bond energies of fluorine arise from the strong repulsion between lone pairs on the two

Table 11.33 Some properties of the halogens

Property	F	Cl	Br	I
Electron configuration	$2s^22p^6$	$3s^23p^6$	$3d^{10}4s^24p^6$	$4d^{10}5s^25p^6$
Ionic radius (pm) (X^-)	136	181	195	216
Covalent radius (pm)	64	99	114	133
Heat of atomization (kJ mol^{-1})	79	122	111	106
Bond energy (kJ mol^{-1})	158	244	192	150
Melting point (°C)	-220	-101	-7.3	113
Heat of hydration of X^-(g) (kJ mol^{-1})	460	385	351	305
Electronegativity	4.1	2.85	2.75	2.20
Electron affinity (kJ mol^{-1})	333	348	340	297
Reduction potential (V)	$+2.87$	$+1.36$	$+1.07$	$+0.54$
Oxidation states	-1	$-1, 1, 3, 5, 7$	$-1, 1, 3, 5$	$-1, 1, 3, 5, 7$

atoms in F_2. The repulsions are particularly strong because the fluorine atoms are very small and so need to get close to each other in order to form a bond. A consequence of this weak bond is that fluorine is very much more reactive than the other halogens, and fluorine will most readily form compounds with other elements in high oxidation states.

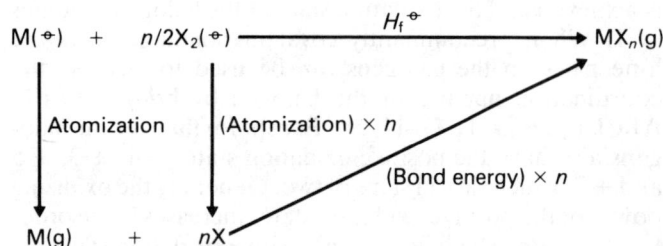

Fig. 11.41 Formation of a gaseous halide MX_n

With some oxidation states only fluorides exist. The thermodymanic cycle for formation of a gaseous covalent halide is given in Fig. 11.41; relevant data for phosphorus chlorides and fluorides are listed in Table 11.34. The relative weakness of the F—F bond and strength of the P—F bond are very apparent. In general, stability with respect to dissociation of halides to the free elements decreases in the order F > Cl > Br > I. The greater bond energies in covalent fluorides (Fig. 11.42) occur because the fluorine atom has a high electronegativity and is small enough to get close to the atom it bonds to, so there is good orbital overlap.

Fig. 11.42 Covalent bond energies of Si, P, S and Cl halides

Table 11.34 Heats of formation of some gaseous phosphorus halides (in kJ mol^{-1})

	PX$_3$		PX$_5$	
	F	Cl	F	Cl
1/4 P$_4$(s) → P(g)	+ 315	+ 315	+ 315	+ 315
3/2 X$_2$ → 3X	+ 237	+ 366		
5/2 X$_2$ → 5X			+ 395	+ 610
3(P—X)	− 1494	− 993		
5(P—X)			− 2490	− 1655
$H_f^{\ominus}$	− 942	− 312	− 1780	− 730

Electron Affinity

At first sight it is surprising that the electron affinity of fluorine is lower than that of chlorine. The reason is again related to the small size of the fluorine atom. The addition of an electron leads to increased interelectronic repulsion

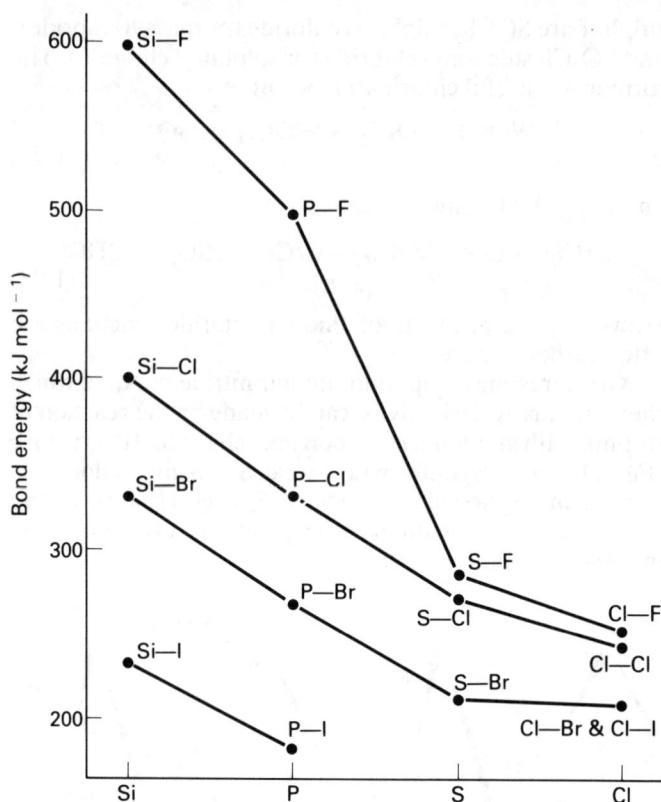

which is greatest for fluorine where the electrons are closest together. Electron affinity relates to the process

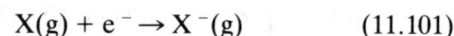

$$X(g) + e^- \rightarrow X^-(g) \qquad (11.101)$$

and must not be confused with electronegativity, which is a property of an atom in a molecule (a measure of its ability to attract more than its share of electrons in a covalent bond).

Electrode Potential

Elemental fluorine is the strongest chemical oxidant known, under all conditions. For example, fluorine will oxidize water to oxygen

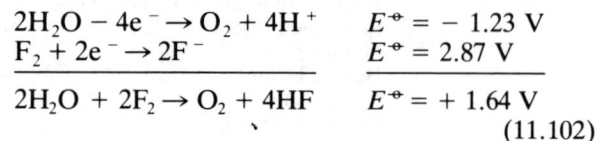

$$\begin{array}{ll} 2H_2O - 4e^- \rightarrow O_2 + 4H^+ & E^{\ominus} = -1.23 \text{ V} \\ F_2 + 2e^- \rightarrow 2F^- & E^{\ominus} = 2.87 \text{ V} \\ \hline 2H_2O + 2F_2 \rightarrow O_2 + 4HF & E^{\ominus} = +1.64 \text{ V} \end{array}$$

$$(11.102)$$

The same oxidation process occurs with chlorine, but HOCl is formed and only slowly decomposes to HCl and oxygen.

$$Cl_2 + OH^- \rightarrow OCl^- + Cl^- + H^+ \quad (11.103)$$

$$2ClO^- \rightarrow 2Cl^- + O_2 \quad (11.104)$$

Even bromine is a strong oxidizing agent. Iodine is relatively weak; atmospheric oxygen is sufficient to oxidize solutions of hydriodic acid, HI, to free iodine.

Preparation of the Halogens

Since the electrode potential for the reaction

$$F^- - e^- \rightarrow \tfrac{1}{2}F_2 \quad (11.105)$$

is so highly negative, the production of fluorine has to be achieved electrolytically and in the absence of other anions which would be preferentially oxidized. Moisture also has to be excluded—otherwise fluorine oxidizes it to oxygen! Anhydrous mixtures of KF and HF (ratio approximately 1:2) are electrolysed at around 80–90°C, (with less HF in the mixture the melting point is higher). The fluorine evolved at the carbon anode and hydrogen evolved at the steel cathode have to be kept separate from one another as the mixture ignites spontaneously. Fluorine can be handled in vessels of teflon (polytetrafluoroethylene), copper, nickel, or monel metal (65% Ni, 32% Cu, 3% Fe, Mn). The last three form protective fluoride coatings.

Other halogens may be obtained in the laboratory by heating NaX with MnO_2:

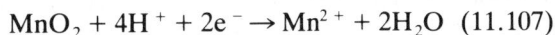

$$2X^- - 2e^- \rightarrow X_2 \quad (11.106)$$

$$MnO_2 + 4H^+ + 2e^- \rightarrow Mn^{2+} + 2H_2O \quad (11.107)$$

which gives

$$2X^- + MnO_2 + 4H^+ \rightarrow Mn^{2+} + X_2 + 2H_2O \quad (11.108)$$

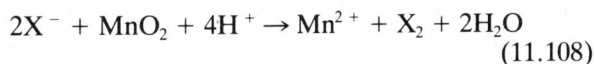

Commercially chlorine is obtained by electrolysis either of fused NaCl or of a concentrated aqueous solution of NaCl. Bromine is obtained from sea water by treatment with chlorine:

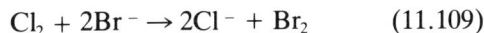

$$Cl_2 + 2Br^- \rightarrow 2Cl^- + Br_2 \quad (11.109)$$

The production of iodine depends on the type of ore used. If it is iodate the reaction is:

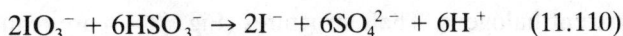

$$2IO_3^- + 6HSO_3^- \rightarrow 2I^- + 6SO_4^{2-} + 6H^+ \quad (11.110)$$

followed by

$$5I^- + IO_3^- + 6H^+ \rightarrow 3I_2 + 3H_2O \quad (11.111)$$

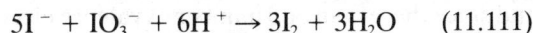

Iodine is also obtained from iodide by treatment with chlorine.

The halogens and their compounds have many uses: for example, halo-organic chemicals range from solvents through disinfectants to insecticides, and plastics; silver halides are used in photography; 1,2-dibromoethane is used to remove lead from internal combustion engines (via the exhaust!).

The reactivity of the halogens is high and decreases in the order $F_2 > Cl_2 > Br_2 > I_2$. They react with most elements, including other halogens, forming halide ions and polyhalide ions. They react readily with unsaturated organic compounds and they are among the most powerful oxidizing agents. Some typical reactions (for Cl_2, Br_2 and I_2) are:

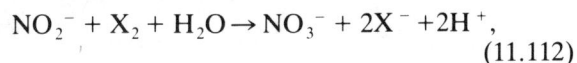

$$NO_2^- + X_2 + H_2O \rightarrow NO_3^- + 2X^- + 2H^+, \quad (11.112)$$

$$2NH_3 + 3X_2 \rightarrow N_2 + 6H^+ + 6X^- \quad (11.113)$$

(the reaction is more complex with iodine),

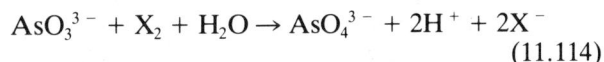

$$AsO_3^{3-} + X_2 + H_2O \rightarrow AsO_4^{3-} + 2H^+ + 2X^- \quad (11.114)$$

(a useful analytical reaction when $X_2 = I_2$),

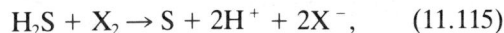

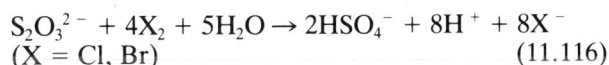

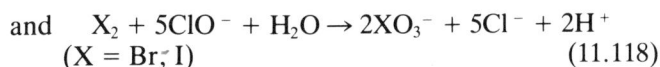

$$H_2S + X_2 \rightarrow S + 2H^+ + 2X^-, \quad (11.115)$$

$$S_2O_3^{2-} + 4X_2 + 5H_2O \rightarrow 2HSO_4^- + 8H^+ + 8X^- \quad (X = Cl, Br) \quad (11.116)$$

$$Sn^{2+} + X_2 \rightarrow Sn^{4+} + 2X^-, \quad (11.117)$$

and $\quad X_2 + 5ClO^- + H_2O \rightarrow 2XO_3^- + 5Cl^- + 2H^+$
$\quad (X = Br, I) \quad (11.118)$

Similar reactions will proceed for fluorine, but are complicated by the fact that fluorine reacts with water.

Interhalogen Species

The compounds or ions formed between the halogens (Table 11.35) are generally very reactive and similar to the

Table 11.35 Some interhalogen compounds and ions and polyhalide ions

AB	AB$_3$	AB$_5$	AB$_7$	ABC$^-$	AB$_2^-$	AB$_4^-$	X$_n^-$	AB$_2^+$	AB$_4^+$
ClF	ClF$_3$			IBrF$^-$	ClBr$_2^-$	ClF$_4^-$	Br$_3^-$	BrF$_2^+$	IF$_4^+$
BrF	BrF$_3$	BrF$_5$		IBrCl$^-$	BrCl$_2^-$	BrF$_4^-$	I$_3^-$	ICl$_2^+$	
		IF$_5$	IF$_7$		ICl$_2^-$	IF$_4^-$	I$_5^-$		
BrCl						ICl$_4^-$	I$_7^-$		
ICl							I$_9^-$		
							I$_8^{2-}$		

parent halogens. Their structures (Fig. 11.43) are readily understood in terms of the electron pair repulsion theory. Since in the compounds AB_n all electrons are paired up, and since there is an odd number of valence electrons in the valence shell of the central atom A, then n is always odd. For the ions, e.g. AB_n^+ and AB_n^-, n is always even.

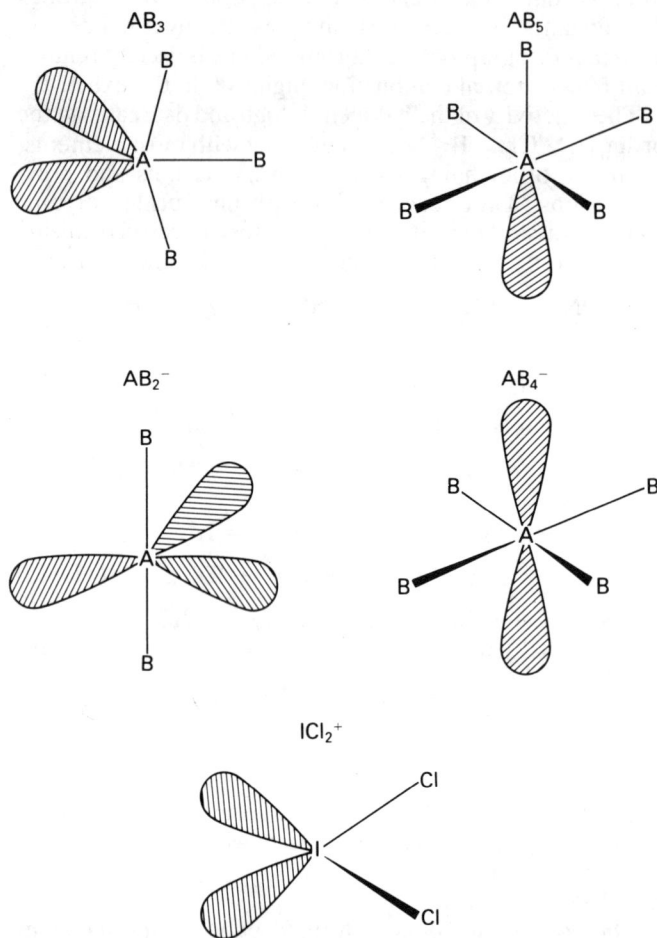

AB₃

AB₅

AB₂⁻

AB₄⁻

ICl₂⁺

Fig. 11.43 Shapes of some interhalogen compounds or ions

Preparation of interhalogens is achieved by direct reaction, e.g.

$$Cl_2 + F_2 \xrightarrow{200°C} 2ClF \qquad (11.119)$$

or by replacement of one halogen by a heavier halogen, e.g.

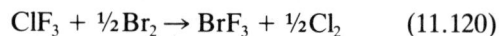

$$ClF_3 + \tfrac{1}{2}Br_2 \rightarrow BrF_3 + \tfrac{1}{2}Cl_2 \qquad (11.120)$$

The physical properties, such as boiling points, of the AB interhalogen compounds are intermediate between those of the parent halogens. Thermal stability may be correlated with electronegativity difference, as may the free energy of formation, (Table 11.36). The most stable compounds have the greatest electronegativity difference, i.e. stability IF > BrF > ClF > ICl > IBr ~ BrCl.

Table 11.36 Thermodynamic stabilities of AB interhalogen compounds

Compound	Electronegativity difference	$G_f^{\ominus}$ (gas) (kJ mol^{-1})
IF	1.9	− 117.5
BrF	1.35	− 73.5
ClF	1.25	− 57.8
ICl	0.65	− 5.9
IBr	0.55	+ 4.1
BrCl	0.10	− 0.8

Silver Halides

Some properties of the silver halides AgF, AgCl, AgBr and AgI are listed in Table 11.37. Their low solubility in water (except for AgF) results from the polarizability of the cation and anion, which favours covalent bonding in the crystal. Use is made of the low solubility in quantitative analysis for the elements Ag, Cl, Br and I. The silver ion differs from the alkali metals in that the single s electron in the valence shell overlies a filled $4d^{10}$ shell, which is highly polarizable. Hence the bonding between the Ag^+ and polarizable anions (especially I^- and Br^-) is largely covalent, and the compounds are insoluble in water. The halides do dissolve in certain reagents, such as NH_3, $S_2O_3^{2-}$ and CN^-, with the formation of silver complex ions.

Whether the halide dissolves, or not, depends on the relative magnitudes of the equilibrium constants for reactions 11.121 and 11.123:

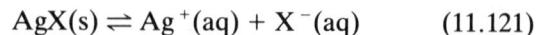

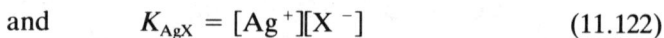

$$AgX(s) \rightleftharpoons Ag^+(aq) + X^-(aq) \qquad (11.121)$$

and

$$K_{AgX} = [Ag^+][X^-] \qquad (11.122)$$

and

$$[AgL_2]^+(aq) \rightleftharpoons Ag^+(aq) + 2L(aq) \qquad (11.123)$$

$$K_d = \frac{[Ag^+][L]^2}{[AgL_2^+]} \qquad (11.124)$$

If sufficient Ag^+ ions arise from the dissociation of the complex for the solubility product of AgX to be exceeded the halide will not dissolve.

The silver halides AgBr and AgI are used in photography as they readily undergo photochemical decomposition with blue or green light.

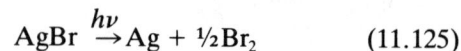

$$AgBr \xrightarrow{h\nu} Ag + \tfrac{1}{2}Br_2 \qquad (11.125)$$

Table 11.37 Properties of the silver halides

	AgF	AgCl	AgBr	AgI
$AgNO_3 + X^-$	No precipitate	White precipitate	Cream precipitate	Yellow precipitate
Solubility of AgX in water	Soluble	Insoluble $K_{sp} = 1.7 \times 10^{-10}$	Insoluble $K_{sp} = 3.3 \times 10^{-13}$	Insoluble $K_{sp} = 8.5 \times 10^{-17}$
Concentrated NH_3	Soluble	Soluble $[Ag(NH_3)_2]^+$ $K = 2.04 \times 10^3$	Soluble	Insoluble
$S_2O_3^{2-}$ solution	Soluble	Soluble $[Ag(S_2O_3)_2]^{3-}$	Soluble	Insoluble
CN^- solution	Soluble	Soluble $[Ag(CN)_2]^-$ $K = 6.3 \times 10^{20}$	Soluble	Soluble
Nitric acid	Soluble	Insoluble	Insoluble	Insoluble

Table 11.38 Some properties of the noble gases

Property	He	Ne	Ar	Kr	Xe
Electron configuration	$1s^2$	$2s^2 2p^6$	$3s^2 3p^6$	$3d^{10} 4s^2 3p^6$	$4d^{10} 5s^2 5p^6$
Van der Waals' radius (pm)		131	174	189	210
Melting point (°C)	-272	-249	-189	-157	-112
Heat of vaporization ($kJ\ mol^{-1}$)	0.08	1.8	6.5	9.0	12.6
Ionization energy ($kJ\ mol^{-1}$)	2372	2081	1520	1350	1170
% by volume in dry air	5.2×10^{-4}	1.8×10^{-3}	0.93	1.1×10^{-4}	8.7×10^{-6}

Light hitting a photographic plate initiates the above reaction, producing a latent image of free silver atoms. The halogen product reacts with the gelatin of the emulsion. During development the free silver atoms serve as nuclei for further reduction of silver ions around them, and the image is accentuated. When the image has been developed to the desired extent, unaffected AgBr is dissolved by a fixing solution containing thiosulphate ions.

11.8 The Noble Gases

Because of the paucity of their chemistry, the discovery of the noble gases was based entirely on physical methods of identification. The unreactive nature of the gases (hence the name sometimes used: 'inert gases') results from their completed octet configurations $ns^2 np^6$. The inert character of the complete octet is expressed in the 'octet rule', which is that elements attempt to achieve by chemical combination a completed shell similar to the nearest noble gas. Simple illustrations are provided by the ready loss of an electron from an alkali metal, or gain of an electron by a halogen, and the sharing of electrons in a covalent bond as in CCl_4. There are many exceptions to the octet rule, but it is still a useful generalization, particularly for the first short period elements.

Some of the properties of the noble gases are listed in Table 11.38. Interactions between the atoms in the solid or liquid state are minimal. The interactions which do occur are typical Van der Waals' forces. These forces increase with the size of the atoms, due to increasing polarizability. The noble gases have uses as inert atmospheres, e.g. argon is used in welding and in incandescent light bulbs. Helium is used for filling balloons. Neon, and to a lesser extent xenon, are used in electric discharge lamps. Over the years chemists have suggested that stable compounds of the noble gases, particularly the heavier members, could exist. The first such compound was prepared by Bartlett in 1962. He reasoned that since PtF_6 (a powerful oxidizing agent) oxidizes O_2 to O_2^+ to produce $O_2^+ PtF_6^-$, and O_2 and Xe have similar first ionization potentials (1180 and 1170 $kJ\ mol^{-1}$ respectively), that $Xe^+ PtF_6^-$ should be produced in a similar reaction. A xenon compound was produced, and since then the fluorides XeF_2, XeF_4 and XeF_6, oxyfluorides, such as $XeOF_4$ and oxides, such as XeO_3 have been isolated. Radon (the heaviest, radioactive member of the series, for which properties are not listed in Table 11.38) and krypton also form fluorides.

The fluorides of xenon are stable white crystalline compounds with negative free energies of formation. By making use of the experimental value of the heat of formation of

XeF$_4$(g) an Xe—F bond energy of 133 kJ can be deduced (Fig. 11.44), which corresponds to a fairly strong bond, similar to that in I$_2$. The structures of XeF$_2$ and XeF$_4$ (Fig. 11.45) are explicable in terms of the electron pair repulsion theory. The hexafluoride has seven electron pairs in the valence shell of Xe, so a distorted structure is to be expected. As yet definitive evidence about the structure is not available but the majority of the evidence points to the distorted structure of Fig. 11.45(c). The existence and compositions of the xenon fluorides are in line with the periodic trends of the elements, as is shown by Table 11.39.

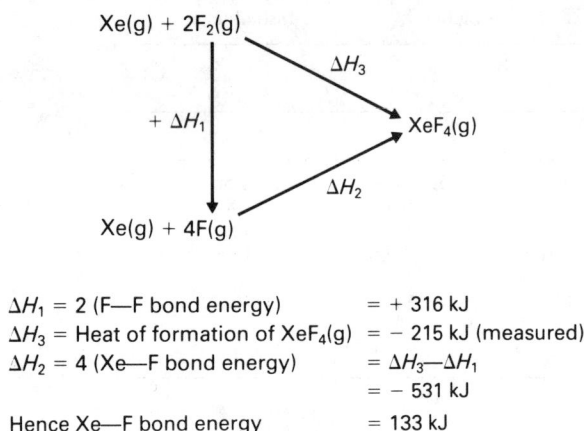

$$\Delta H_1 = 2 \text{ (F—F bond energy)} \quad = + 316 \text{ kJ}$$
$$\Delta H_3 = \text{Heat of formation of XeF}_4\text{(g)} = - 215 \text{ kJ (measured)}$$
$$\Delta H_2 = 4 \text{ (Xe—F bond energy)} \quad = \Delta H_3 - \Delta H_1$$
$$= - 531 \text{ kJ}$$
$$\text{Hence Xe—F bond energy} \quad = 133 \text{ kJ}$$

Fig. 11.44 Xe—F bond energy in XeF$_4$

XeF$_4$ is obtained by heating F$_2$ and Xe (5:1) at 400°C and six atmospheres pressure are in a nickel vessel. The compound, like the other xenon fluorides, is a strong fluorinating agent and oxidizing agent. In water the acidic oxide XeO$_3$ is produced from XeF$_4$. The oxide is dangerously explosive in the solid state. The oxide reacts with base to give HXeO$_4^-$ (xenate) and XeO$_6^{4-}$ (perxenate). The latter species is a very powerful oxidizing agent; for example it oxidizes Mn^{2+} to MnO$_4^-$

Table 11.39 The relationship between xenon fluorides and p block element fluorides

Valence shell	Sn	Sb	Te	I	Xe	Cs
8 electrons (4 pairs)	SnF$_4$	SbF$_3$	TeF$_2$	IF	Xe	Cs$^+$
10 electrons (5 pairs)		SbF$_5$	TeF$_4$	IF$_3$	XeF$_2$	CsF$_{(gas)}$
12 electrons (6 pairs)	SnF$_6^{2-}$	SbF$_6^-$	TeF$_6$	IF$_5$	XeF$_4$	
14 electrons (7 pairs)				IF$_7$	XeF$_6$	

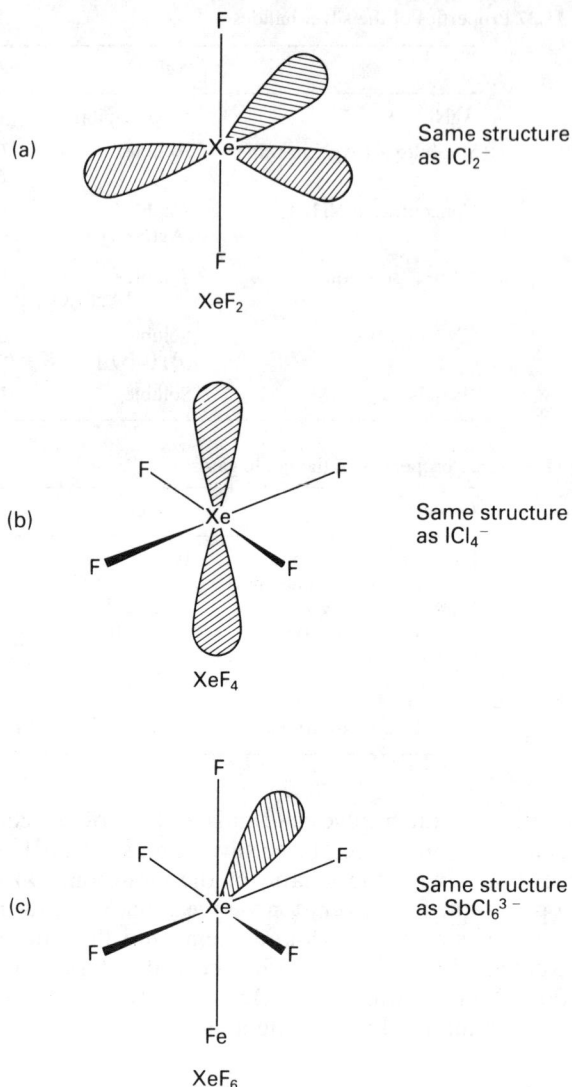

Fig. 11.45 Structures of xenon fluorides

Exercises

11.1 Discuss the trends in the following properties of the elements of the first short period: (a) atomic radius, (b) electronegativity, (c) structures of the oxides.

11.2 Discuss the trends in the same properties for the second short period elements.

11.3 Explain the term catenation and indicate its prevalence for the p block elements.

11.4 Discuss why BX$_3$ and AlX$_3$ (halides) are good electron acceptors (Lewis acids).

11.5 Why is it possible to make significant comparisons between boron nitride and graphite?

11.6 Why is the melting point of boron so high?

11.7 What is the principal difference between the oxygen chemistry of boron and that of carbon and nitrogen?

11.8 Discuss the reasons for the trends in the M—M bond strengths for C, Si, Ge, Sn and Pb (cf. Table 11.21).

11.9 Describe the structural differences in the allotropes of carbon. Do some of the forms bear any relationship to organic compounds? Give examples.

11.10 Discuss the differences in the various types of carbides, and indicate the relationship between the types and the position of the elements in the periodic table.

11.11 Discuss the reasons for the similarity of a number of the properties of the binuclear species CO, N_2, CN^-, NO^+.

11.12 Explain why carbonic acid appears to be a weak acid.

11.13 Why is Sn^{2+} more extensively hydrolysed than Pb^{2+}?

11.14 Explain why CO_2 and CCl_4 have such different properties from the analogous silicon compounds SiO_2 and $SiCl_4$ respectively.

11.15 The temperature of the thermal decomposition of the tetrahydrides MH_4 are 600, 450, 300, 100 and 0°C, for M = C, Si, Ge, Sn and Pb respectively. Account for this trend.

11.16 Beryl, a mineral, $Be_3Al_2Si_6O_{18}$ contains the discrete anion $Si_6O_{18}{}^{12-}$. Suggest a structure for the anion.

11.17 Outline the range of oxidation states that are achieved by nitrogen and give an example for each.

11.18 Discuss the features of dinitrogen (N_2) that make it a dominating influence on the chemistry of nitrogen.

11.19 Describe with explanations, the structures of $PCl_5(g)$, $PCl_5(s)$, $POCl_3$, $NH_2{}^-$, $N_3{}^-$.

11.20 Discuss the chemistry of and give equations for, the industrial preparations of NH_3 and HNO_3.

11.21 Describe how to obtain NH_3 from ammonium salts. Does this apply to all ammonium salts? Give examples.

11.22 Why is NH_3 a useful refrigerating agent?

11.23 Starting with NO how would you prepare N_2O, NO_2, N_2 and NH_3? Give equations.

11.24 What do you understand by the term 'nitrogen fixation'? How is the process achieved in nature and how is it accomplished industrially?

11.25 Why are compounds containing N—N bonds uncommon?

11.26 What factor influences the structure of NH_4F making it different from the structure of the other ammonium halides?

11.27 Discuss the difference between basic and acidic oxides.

11.28 How do the structures of oxides relate to their chemical properties?

11.29 What is the structure of the oxides: CO_2, NO_2, $SO_3(g)$ and XeO_3?

11.30 Write equations for the laboratory preparation of the oxides: Al_2O_3, SiO_2, NO, P_4O_{10} and SO_3.

11.31 Discuss the influence of covalent oxides on atmospheric pollution.

11.32 Discuss the implication of the strong Si—O bond, and the divalency of oxygen on the oxy-chemistry of silicon.

11.33 Explain why the reaction $O^-(g) + e^- \rightarrow O^{2-}(g)$ is strongly endothermic while the reaction $O(g) + e^- \rightarrow O^-(g)$ is exothermic.

11.34 Discuss the reasons for the trends in the bond lengths and bond energies of the dioxygen species: $O_2{}^+$, O_2, $O_2{}^-$, $O_2{}^{2-}$.

11.35 Write the ion-electron equations for H_2O_2 acting as a reducing agent and as an oxidizing agent under both acid and alkaline conditions (four equations).

11.36 When H_2S is bubbled into HNO_3 solution the following products are obtained: S, NO_2, NO, N_2 and $NH_4{}^+$. Write balanced equations (four) to indicate what is happening.

11.37 H_2S can act as a weak acid, dibasic acid and reducing agent, and can be used in qualitative analysis. Give chemical equations in which H_2S is displaying each of these properties independently.

11.38 Discuss the factors that influence the stablity of SF_6.

11.39 Which of the oxides in the following pairs are the more acidic?

$$\begin{array}{ll} CaO, & CO \\ MnO, & Mn_2O_7 \\ N_2O, & N_2O_5 \end{array}$$

11.40 Why are compounds containing S—S bonds more stable than compounds containing O—O or Se—Se bonds?

11.41 Using bond energy data calculate the enthalpy of the reaction $S_8(g) \rightarrow 4S_2(g)$.

11.42 Discuss, and give equations, for the hydrolysis of the halides $SiCl_4$, SiF_4 and NCl_3.

11.43 Why is the electron affinity of F less than that of Cl and why is the F—F bond weaker than the Cl—Cl bond?

11.44 Discuss why HI is a stronger acid than HF.

11.45 Write the chemical equations for the preparation of I_2 from ores containing iodate.

11.46 Determine, and discuss, the shapes of the interhalogen species ClF_3, IF_5, $ICl_2{}^-$, $ICl_4{}^-$ and $BrF_2{}^+$.

11.47 Explain the trend in the volatility of the halogens X_2.

11.48 Analyse the energetics of the electrode reaction $\frac{1}{2}X_2(g) + e^- \rightarrow X^-(aq)$ for F_2, Cl_2 and Br_2 and suggest why the strength of the halogen as an oxidizing agent decreases down the series.

11.49   Why is it not possible to obtain F_2 from the electrolysis of aqueous HF?

11.50   What happens when I_2 is added to a KI solution? What happens when Cl_2 is added to a KI solution?

11.51   What happens when Cl_2 is bubbled into water?

11.52   Starting from Br_2 explain, and give equations for, the preparation of HBr, $KBrO_3$ and PBr_3.

11.53   What is the oxidation state of the halogen in the compounds: $KBrO_3$, $NaClO_2$, Cl_2O and $Ca(IO_4)_2$?

11.54   Explain the trends in the structures of the chlorides of the 1st and 2nd short period elements.

11.55   Describe a chemical test which would enable you to distinguish the components of the following pairs of compounds: H_2O_2, Cl_2: PF_3, PCl_3; CO, NO; $Na_2S_2O_3$ soln., $NaClO_4$ soln., and H_3PO_3 soln., H_2SO_3 soln.

11.56   Outline the experiments and the reasoning behind them, which finally led to the isolation of xenon fluorides.

11.57   Predict the stoichiometry and shapes of the three xenon fluorides using as a basis your knowledge of the shapes and stoichiometries of the fluorides of the p block elements.

12 The d block Elements

12.1 Introduction

The d block elements, or *transition metals* as they are commonly called, account for 30 elements of the periodic table (Table 12.1). They are classified into three groups, or series, according to whether the 3d, 4d or 5d sub-shells are filling with electrons. There are five d orbitals (Fig. 12.1) which can contain a total of 10 electrons; hence each transition series consists of 10 members. We will be concerned mainly with the first row transition metals, namely scandium, titanium, vanadium, chromium, manganese, iron, cobalt, nickel, copper and zinc. The last member, Zn, has a complete d sub-shell and so lacks many of the typical properties of the transition metals, such as variable valence and a tendency to form coloured complex ions.

Table 12.1 The d block elements

Group:	III	IV	V	VI	VII	VIII			I	II
First row:	Sc	Ti	V	Cr	Mn	Fe	Co	Ni	Cu	Zn
Second row:	Y	Zr	Nb	Mo	Tc	Ru	Rh	Pd	Ag	Cd
Third row:	La	Hf	Ta	W	Re	Os	Ir	Pt	Au	Hg

The first row transition metals differ in a number of respects from the second and third row transition metals, which have closely similar chemistry. One reason for this is clear from atomic radii data (Table 12.2). Atomic radii of the second and third row metals in a given group are the same, or very nearly so, whereas the radii of the elements of the first transition series are markedly smaller. Elements of the third transition metal series have almost the same radii as the corresponding elements of the second series because of the lanthanide contraction (cf. p. 22). .

The outer electron configurations of the elements of the first transition metal series and neighbouring elements are given in Table 12.3. The 4s orbital fills before the 3d as a consequence of 'orbital penetration', as discussed in Chapter 2. The apparently anomalous electron configurations of Cr and Cu, $3d^54s^1$ and $3d^{10}4s^1$ respectively, are a consequence

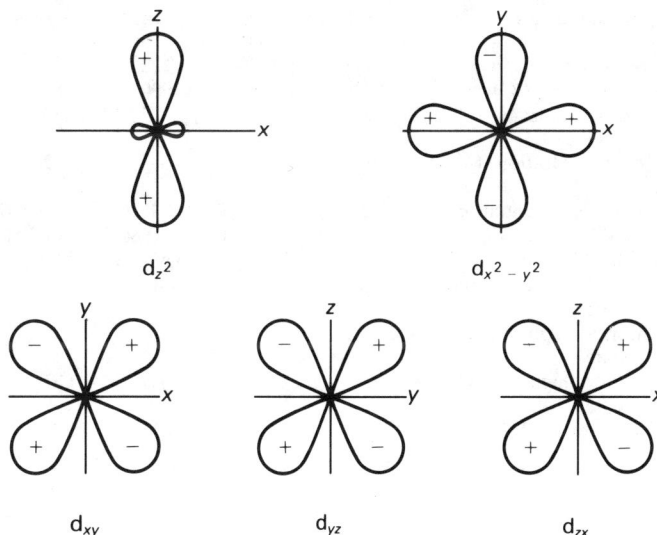

Fig. 12.1 d orbitals

of the high stability associated with either filled or half-filled shells. To help us understand why a half-filled shell is very stable we consider the case of chromium. The configuration $3d^44s^2$ is less stable than $3d^54s^1$ because in the former the paired 4s electrons are permitted to be close to one another in the same orbital, with the result that the inter-electron repulsion energy is high. (A similar effect is the basis of Hund's rule, according to which for a given configuration the state of lowest energy is that of highest total electron spin.) A half-filled shell can have the maximum possible number of unpaired electrons.

Table 12.2 Metallic radii of the transition elements (pm)

Ti	V	Cr	Mn	Fe	Co	Ni	Cu
147	135	130	135	126	125	125	128
Zr	Nb	Mo	Tc	Ru	Rh	Pd	Ag
160	146	139	136	134	134	137	144
Hf	Ta	W	Re	Os	Ir	Pt	Au
160	149	141	137	135	136	139	146

Table 12.3 Outer electronic configuration of some of the 3rd period elements

s block		d block									p block
K $3d^04s^1$	Ca $3d^04s^2$	Sc $3d^14s^2$	Ti $3d^24s^2$	V $3d^34s^2$	Cr $3d^54s^1$	Mn $3d^54s^2$	Fe $3d^64s^2$	Co $3d^74s^2$	Ni $3d^84s^2$	Cu $3d^{10}4s^1$ Zn $3d^{10}4s^2$	Ga $3d^{10}4s^24p^1$

Table 12.4 Some properties of the transition metals

Property	Sc	Ti	V	Cr	Mn	Fe	Co	Ni	Cu	Zn
Metallic radius (pm)	164	147	135	130	135	126	125	125	128	137
Ionic radius (pm)	81(+3)	76(+3) 68(+4)	74(+3) 60(+4)	84(+2) 69(+3)	80(+2) 66(+3)	76(+2) 64(+3)	74(+2) 63(+3)	72(+2)	96(+1) 69(+2)	74(+2)
Covalent radius (pm)	144	132	122	118	117	117	116	115	117	125
Melting point (°C)	1539	1675	1900	1890	1244	1535	1495	1453	1083	419
Heat of atomization (kJ mol^{-1})	390	469	502	397	284	406	439	427	341	130
Heat of hydration (M^{2+})	3915 (Sc^{3+})	—	—	1820	1815	1890	2025	2075	2075	2017
Electrode potential (n) ($M^{n+} + ne^-$)	−2.1(3)	−1.2(3) −1.63(4)	−1.2(2) −0.86(3)	−0.91(2) −0.74(3)	−1.18(2) −0.28(3)	−0.44(2) −0.04(3)	−0.28(2) +0.4(3)	−0.25(2)	+0.52(1) +0.3(2)	−0.76(2)
Colour M^{2+}(aq)		brown	lavender	blue	pink	green	pink	green	blue	colourless
Oxide (highest oxidation state)	Sc_2O_3	TiO_2	V_2O_5	CrO_3	Mn_2O_7	Fe_2O_3	Co_2O_3	Ni_2O_3?	CuO	ZnO
Chloride (highest oxidation state)	$ScCl_3$	$TiCl_4$	VCl_4	$CrCl_4$	$MnCl_3$	$FeCl_3$	$CoCl_2$	$NiCl_2$	$CuCl_2$	$ZnCl_2$
Fluoride (highest oxidation state)	ScF_3	TiF_4	VF_5	CrF_6	MnF_4	FeF_3	CoF_3	NiF_2	CuF_2	ZnF_2
Configuration M^{2+}	—	$3d^3$	$3d^3$	$3d^4$	$3d^5$	$3d^6$	$3d^7$	$3d^8$	$3d^9$	$3d^{10}$
M^{3+}	$3d^0$	$3d^1$	$3d^2$	$3d^3$	$3d^4$	$3d^5$	$3d^6$	$3d^7$	$3d^8$	

Some of the properties of the first row transition metals are listed in Table 12.4. The elements are typical metals with high melting points and reasonably high heats of atomization (or sublimation), suggesting strong metallic bonding. This is due to the large number of valence electrons available for metallic bonding (compared with one electron per alkali metal atom).

The metallic, ionic (for the same charge) and covalent radii decrease as we go across the series with a corresponding increase in the heats of hydration. The electrode potentials ($M^{n+} + ne^- \rightarrow M$) also increase as we go across the series, indicating decreasing electropositive character. Even the most electropositive (Sc and Ti) are much less electropositive than the alkali metals. The ionization energies of the transition metals increase across the series as the atomic size decreases. The 1st and 2nd ionization energies range from 633 to 762 kJ mol^{-1} and from 1235 to 1958 kJ mol^{-1} respectively, values which are not much greater than those for calcium (590 and 1146 kJ mol^{-1}). The significant difference between Ca and the transition metals shows up in the third ionization energy. For calcium it is 4941 kJ mol^{-1} (electron removed from the $3p^6$ sub-shell) whereas

for the first row transition metals it ranges from 2388 to 3556 kJ mol^{-1}. Hence, whereas it is difficult to obtain Ca^{3+} it is relatively easy to obtain the species M^{3+} for many of the transition metals. The increase in successive ionization energies for the transition metals is not very great; hence the metals can display a wide range of oxidation states. Irregularities in the trend for the second ionization energies of Cr and Cu relate to the half-filled and filled d^5 and d^{10} shells respectively (note also the irregularity for the 3rd ionization energy of Mn).

12.2 Properties of Transition Metals

(a) Oxidation States

Transition metals display a wide range of oxidation states, as shown by the crosses for the first row metals in Table 12.5. Oxidation states are designated by Roman numerals I, II etc rather than +1, +2, because ions such as Mn^{7+} or Co^{2+} do not exist in solution by themselves, but only in association with other species. Thus Mn(VII) occurs in MnO_4^- and Co(II) in $[Co(NH_3)_6]^{2+}$. The groups O and NH_3 around the metal atoms in these two complex ions are

Table 12.5 Oxidation states of the first row transition metals

Oxidation state	Ca	Sc	Ti	V	Cr	Mn	Fe	Co	Ni	Cu	Zn	
VII						Ⓧd^0						
VI					Ⓧ	X	Ⓧ					mainly
V				Ⓧ	X	X	X		d^5			oxyanions
IV			Ⓧ	Ⓧ	X	Ⓧ	X	X	X			and fluorides
III		Ⓧ	Ⓧ	Ⓧ	Ⓧ	X	Ⓧ	Ⓧ	X	X		most common
II	Ⓧ		X	X	Ⓧ	Ⓧ	Ⓧ	Ⓧ	Ⓧ	Ⓧ	Ⓧd^{10}	states
I			X	X	X	X	X	X	X	Ⓧ		carbonyl
0		X	X	X	X	X	X	X	X			complexes
− I		X	X	X	X	X	X	X				
− II				X	X	X						
− III					X							

Ⓞ = Most stable oxidation states   − − − = Oxidation states with same d^n electron configuration

called *ligands,* a term which is used frequently in this chapter and elsewhere to stand for any atom, molecule or ion which is one of a group bonded to a central atom.

High oxidation states with oxidation number IV and above are normally associated with ligands such as oxygen and fluorine, as in CrO_4^{2-}, Mn_2O_7 and VF_5. The oxidation states II and III are the most common and occur in association with a wide variety of ligands such as H_2O, NH_3, CN^- and many others. The low oxidation states ($\leq$ (I)) occur with carbon monoxide as a ligand and with unsaturated organic compounds, such as ethylene. Oxidation states may be classified in terms of the d electron configuration of the metal. Metal ions with the same d electron configuration have a number of similar chemical properties. For example, if one is considering the configuration d^8 it spans the elements and oxidation states Cr(−II), Mn(−I), Fe(O), Co(I), Ni(II), Cu(III). (The electronic configuration of metallic iron is $3d^64s^2$, but for zero-valent iron in a complex (e.g. $Fe(CO)_5$) all the valence electrons are forced into the 3d orbitals.)

The most stable, or perhaps more correctly the most common oxidation states are highlighted in Table 12.5. The stability of a particular state is markedly dependent on the ligand associated with the metal. Thus, for example, Mn(IV) is stable as the oxide MnO_2, but unstable in association with carbon monoxide. It is therefore always necessary to specify the ligand when discussing the relative stability of oxidation states of transition metals.

The rise in the maximum oxidation state on going from Sc to Mn is explicable in terms of the increasing number of valence electrons on the metal. After manganese the maximum oxidation number decreases, and it is reasonable to ask why Fe(VIII) does not exist. (Ruthenium and osmium, just below iron, do exist in the VIII oxidation state.) The absence of Fe(VIII) implies that such a species would be so powerful an oxidizing agent that no ligands could

remain un-oxidized in its presence. In order to remove the high formal positive charge on the metal and so reduce its oxidizing power a strongly electron donating ligand would be necessary, or alternatively a large number of ligands coordinated to the metal. However, the maximum possible coordination number of Fe(VIII) is limited by its small size, so this remedy is not available.

In working out the formal oxidation state of a transition metal it is necessary to remember, in addition to the usual rules, that ligands such as NH_3, H_2O and CO are neutral and their coordination does not alter the oxidation state. Some examples of the working out of oxidation states of metal ions in complexes are given in Table 12.6.

Table 12.6 Oxidation states of metal ions in some complexes

Compound or ion	Formal charges on ligands	Formal charge on metal	Metal oxidation state
CrO_2Cl_2	$2O^{2-} + 2Cl^-$	Cr^{6+} to give neutral compound	Cr(VI)
$[Fe(CN)_6]^{3-}$	$6CN^-$	Fe^{3+} to give overall − 3 charge to complex ion	Fe(III)
$[Co(NH_3)_6]^{3+}$	$6(NH_3)^0$	Co^{3+} to give overall + 3 charge to the ion	Co(III)
$Cr_2O_7^{2-}$	$7O^{2-}$	$2Cr^{6+}$ to leave overall − 2 charge on ion	Cr(VI)
$Fe(CO)_5$	$5(CO)^0$	No charge on metal to give neutral complex	Fe(O)
$[Cr(NH_3)_4Cl_2]^+$	$4(NH_3)^0 + 2Cl^-$	Cr^{3+} to leave overall + 1 charge on metal complex ion	Cr(III)

(b) Complex Formation

A covalent bond between C and H in CH_4 is formed by the sharing of the hydrogen 1s electron and a carbon (2s, 2p) electron, both electrons being necessary for bond formation. When ammonia reacts with acids to form the ammonium ion the lone pair on the nitrogen atom is donated to the empty 1s orbital of H^+ to form a *coordinate* or *dative* covalent bond, both electrons in the bond coming from the ammonia. The bond is indistinguishable from the other N—H bonds once the ion is formed. In this process the ammonia is an electron donor (Lewis base) and the proton is an electron acceptor (Lewis acid). In general we can write

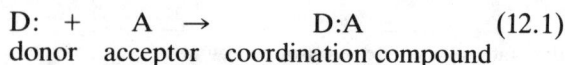

$$\text{D:} \quad + \quad \text{A} \quad \rightarrow \quad \text{D:A} \qquad (12.1)$$
$$\text{donor} \quad \text{acceptor} \quad \text{coordination compound}$$

The product DA is called a coordination compound. There are numerous examples of this sort of reaction, and all transition metal complexes may be considered as being formed in this way. The conditions necessary for formation of a coordination compound are that the metal ion has vacant and accessible orbitals (3d, 4s and 4p for the first transition metal series) and that the ligand has one or more lone pairs of electrons. The number of ligands coordinated is termed the coordination number. It is sometimes not easy to decide which of the available ligands are coordinated to the metal and which are not. The complex $[Cr(NH_3)_6]Cl_3$ has six ammonia ligands coordinated to chromium, while the three chlorine atoms are present as separate ions. However, chloride ions can also act as ligands, and in general it is necessary to decide by experiment which ligands, and how many, are coordinated. Measurements of electrical conductivity are commonly used for this purpose. An example is given in Table 12.7.

Table 12.7 The formulation of cobalt ammonia complexes

Compound	Electrolytic behaviour (type of electrolyte)	Moles of AgCl precipitated per mole of compound	Formulation
$CoCl_36NH_3$	3:1	3	$[Co(NH_3)_6]^{3+}3Cl^-$
$CoCl_35NH_3$	2:1	2	$[Co(NH_3)_5Cl]^{2+}2Cl^-$
$CoCl_34NH_3$	1:1	1	$[Co(NH_3)_4Cl_2]^+Cl^-$
$CoCl_33NH_3$	non-electrolyte	0	$[Co(NH_3)_3Cl_3]$

(c) Ligands

Ions or molecules which contain atoms with lone pairs in their valence shell are potential ligands. Such atoms are listed in Table 12.8, and some of the more common ligands derived from these atoms are given in Table 12.9. The first

Table 12.8 Donor atoms occurring in ligands

			H
Ⓒ	Ⓝ	Ⓞ	Ⓕ
Si	Ⓟ	S	Ⓒⓛ
	As	Se	Br
Sn	Sb	Te	I

○ = Most common donor atoms

group of ligands contains species which bond through one atom only and are termed monodentate. The second group of ligands may bond through two or more donor atoms and are termed polydentate (bidentate, tridentate, etc.) ligands. They are also termed chelating ligands (Greek: Chela = claw). The bonding of oxalate and diethylenetriamine are illustrated in Fig. 12.2.

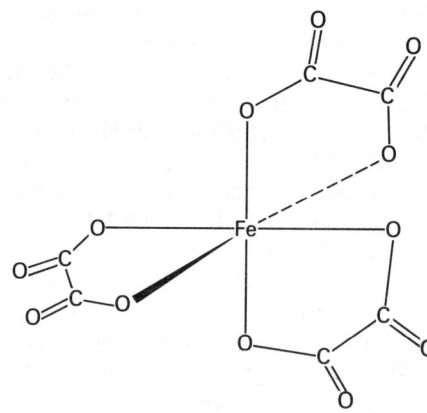

$[Fe(C_2O_4)_3]^{3-}$

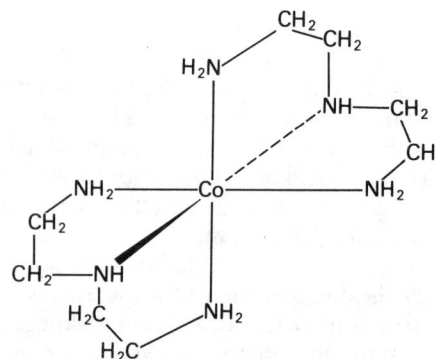

$[Co(diethylenetriamine)_2]^{3+}$

Fig. 12.2 Structures of complexes with polydentate ligands

Table 12.9 Ligands

Ligand	Donor atom	Name
Monodentate		
F^-, Cl^-, Br^-, I^-	X	halo-
NO_2^- ONO^-	N, O	nitro-, nitrito-
CN^-	C	cyano-
OH^-	O	hydroxo-
O^{2-}	O	oxo-
CH_3COO^-	O	acetato-
H_2O	O	aqua
NH_3	N	ammine
CO	C	carbonyl
NO^+	N	nitrosyl
	N	pyridine
R_3P	P	trialkylphosphine
R_2S	S	dialkylsulphide
Polydentate		
NH_2—CH_2—CH_2—NH_2	NN	ethylenediamine
NH_2—CH_2—CH_2—NH—CH_2—CH_2—NH_3	NNN	diethylenetriamine
	NN	dipyridyl
OOC—COO^{2-}	OO	oxalato-
$(OOCCH_2)N(CH_2)_2N(CH_2COO)_2^{4-}$	OONNOO	ethylenediaminetetraacetato-
$NH_2CH_2COO^-$	NO	glycinate

Transition metal complexes require systematic nomenclature to cope with the naming of the wide variety of possible compounds. The rules recommended by IUPAC (International Union of Pure and Applied Chemistry) are given below. For a few complexes common names are used, as for the ferricyanide ion $Fe(CN)_6^{3-}$, or dichromate $Cr_2O_7^{2-}$ but this is permissible only if no ambiguity arises.

Rules of nomenclature for coordination complexes

1 The cation is named first, then the anion, e.g. $K_2(PtCl_4)$ —potassium tetrachloroplatinate(IV); $[Cr(NH_3)_6]$ $(NO_3)_3$—hexamminechromium(III) nitrate.

2 The name of the non-ionic components is given in one word, e.g. $[Co(NH_3)_3(NO_2)_3]$—trinitrotriammine cobalt(III).

3 Neutral ligands are normally named as molecules, anionic ligands end with -o and cationic ligands (which are rare) end with -ium.

$NH_2CH_2CH_2NH_2$	ethylenediamine
$(C_2H_5)_2S$	diethylsulphide
CN^-	cyano
Cl^-	chloro

Two common exceptions are H_2O (aqua) and NH_3 (ammine).

4 The order in which ligands are enumerated is first negative, then neutral, and last positive. Within each group the least complicated comes first, and within the last category the order is alphabetical, e.g. $[Pt(NH_3)_4(NO_2)$ $Cl]SO_4$—chloronitrotetraammine platinum(IV) sulphate.

5 Di, tri, tetra etc., are used as prefixes to indicate the numbers of simple ligands; bis, tris, tetrakis etc., are used for more complex ligands, e.g.

$(Cl)_2$	dichloro-
$(NH_2CH_2NH_2)_3$	tris(ethylenediamine)

6 Anionic complexes end in -ate. $K_3[Fe(CN)_6]$ potassium hexacyanoferrate(III). If the complex is neutral or cationic no ending is used.

7 The oxidation state of the metal is indicated with a Roman numeral.

$Fe(CO)_5$	pentacarbonyl iron (0)
$K_2[CrOCl_5]$	potassium pentachlorooxochromate (V)

8 Bridging ligands are prefixed by the Greek letter μ. $[Pt_2Cl_4((C_2H_5)_2S)_2]$dichloro-$\mu$-dichloro bis(diethylsulphide)platinum(II)

9 If more than one donor atom may bond to the metal, the one that does is indicated by including its symbol in the name, e.g. $(NH_4)_3[Cr(NCS)_6]$ ammonium hexathiocyanato-N-chromate(III).

(d) Bonding in Octahedral Complexes

The most common type of transition metal complex is one with six ligands L around the metal M i.e. ML_6^{+n} ($n = 0, 1. . . .$). The ligands are positioned at the six apices of an octahedron (Fig. 12.3), so that d^2sp^3 hybrid orbitals point directly towards the ligands and can be used in σ bond formation. There are just six of these orbitals, which is the number required to form six bonds. The d orbitals used to form the hybrids are d_{z^2} and $d_{x^2-y^2}$. The three remaining d orbitals (d_{xz}, d_{xy}, d_{yz}, Fig. 12.1) point between the ligands and are non-bonding. Therefore as a consequence of the presence of the ligands the five d orbitals separate into two groups, σ-bonding type and non-σ-bonding type. Two alternative representations of the orbitals in the complex are given in Figs 12.4 and 12.5. To a first approximation the energy of the non-bonding d orbitals is not affected by the presence of the ligands.

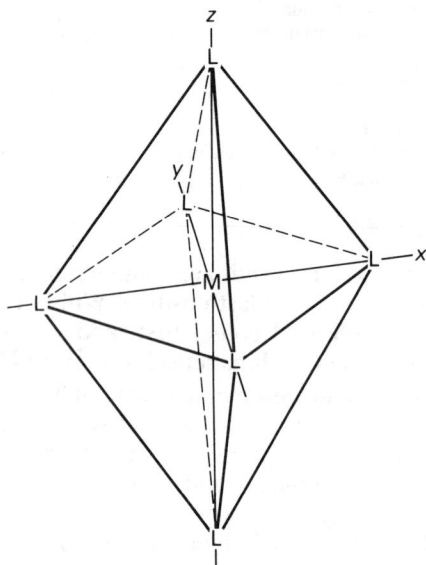

Fig. 12.3 An octahedral complex ML_6

The energy gap between the three non-bonding d orbitals (labelled t_{2g}) and the antibonding component arising from the overlap of the ligand orbitals with the metal orbitals (labelled e_g^*) is customarily given the symbol Δ, and is called the ligand field splitting energy. The labels t_{2g} and e_g are conventional symmetry labels derived from group theory. The bonding electrons, which come entirely from the ligands, fill the six bonding orbitals which are formed by overlap of the ligand orbitals and the d^2sp^3 hybrids. The metal's corresponding antibonding orbitals appear at the top of the diagrams in Figs 12.4 and 12.5. The d electrons fill the t_{2g} and e_g^* orbitals in accordance with the building-up principles outlined in Chapter 2. The number of d electrons

to be accommodated depends on the configuration of the metal ion. With an ion of configuration d^4 (e.g. Cr(II), Mn(III)) the fourth electron has two possible places to go: either it can pair up with one of the other three electrons in a t_{2g} orbital, or it can go into an e_g^* orbital (Fig. 12.6, p. 200). Which one actually occurs depends on the relative magnitudes of the energy separation Δ and the repulsion energy E_R between two electrons which are paired in the same orbital. If $\Delta > E_R$ then the fourth electron goes into a t_{2g} orbital; if $\Delta < E_R$ the fourth electron goes into e_g^*. In the first case there are only two unpaired electrons; in the second there are four. The magnitude of Δ depends on the M—L bond strength, since the stronger the bond the more stable (lower energy) become the bonding orbitals. As the bonding orbitals become lower in energy the antibonding orbitals become higher, thereby increasing the t_{2g}—e_g^* energy gap. The bond strength is influenced by: (a) the charge on the metal, higher charge leading to a stronger bond; (b) the donor power of the ligand, more powerful donors forming stronger bonds; and (c) the polarizability of the ligand, more polarizable ligands giving stronger bonding. When ligands are altered the quantity Δ increases in the order

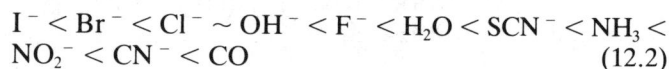

$$I^- < Br^- < Cl^- \sim OH^- < F^- < H_2O < SCN^- < NH_3 < NO_2^- < CN^- < CO \qquad (12.2)$$

This ordered list is known as the *spectrochemical series* of ligands, for reasons we shall see shortly. The non-bonding t_{2g} orbitals can also be involved in bonding by the formation of π bonds with the ligands. This is particularly important for the existence of complexes of the transition metals in low oxidation states, such as carbonyls.

(e) Shapes of Complexes

The two principal coordination numbers for transition metal complexes are four and six. The shapes of four coordinate species are either tetrahedral or square planar (Fig. 12.7) and for six coordination the shape is normally octahedral (Fig. 12.3). Lone electrons on the central metal atom do not influence the shape of the complex as the nitrogen lone pair affects the shape of NH_3. The d electrons can, however, affect the shape of a complex. For example, in an octahedral complex, electrons in the e_g^* orbitals lie directly between the metal and ligand, which leads to repulsion of the ligands to an extent which depends on the magnitude of their negative charge.

(f) Colour

The colour of transition metal compounds, one of their most prominent characteristics, varies with the metal, the type of ligand, and with the shape of the complex (Table 12.10). The colour is due to the absorption of visible radia-

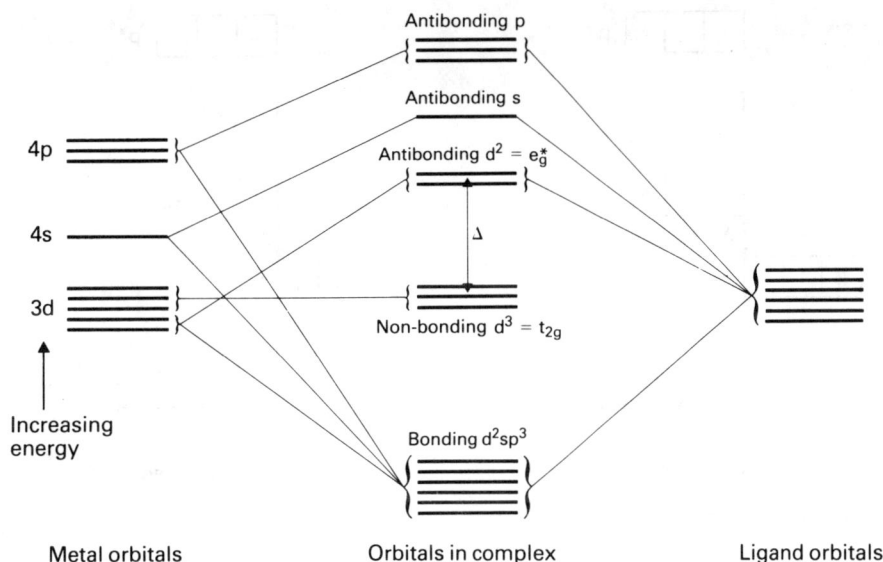

Fig. 12.4 Orbitals in an octahedral transition metal complex. (Note that in the ion the orbitals 3d lie below 4s.)

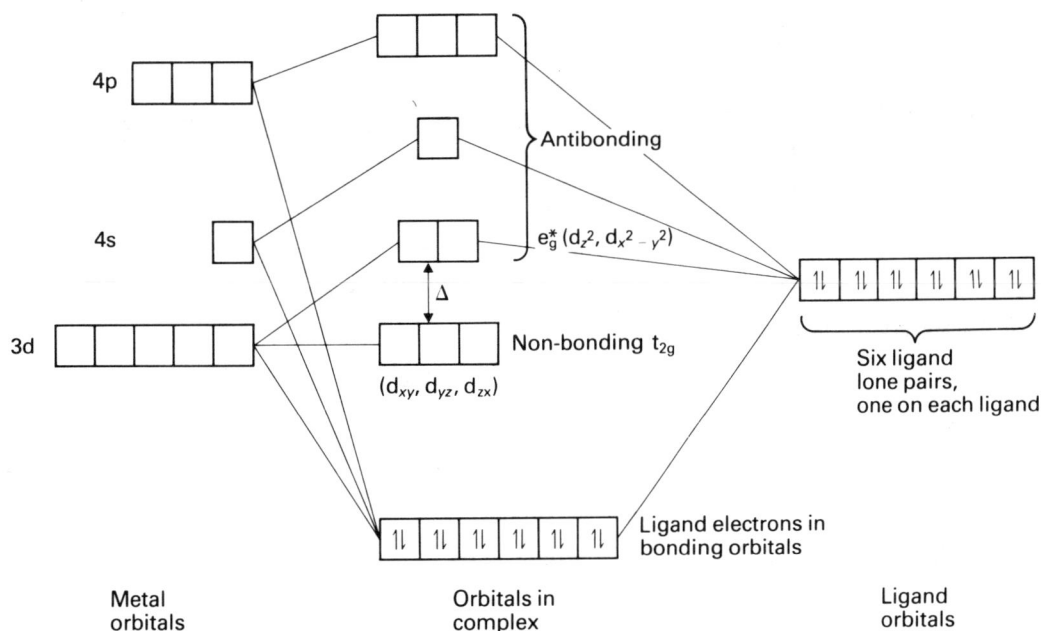

Fig. 12.5 Bonding and antibonding orbitals in an octahedral complex

tion (Fig. 12.8) which excites electrons from one energy level to another. For most transition metal compounds the ligand field splitting energy, Δ, is comparable with the energy $h\nu$ of a quantum of visible radiation. Thus the observed colour is associated with promotion of a metal d electron from one orbital in the ground state to an orbital at higher energy in the excited state.

The simplest example we can consider is titanium(III) in $[Ti(H_2O)_6]^{3+}$. The single d electron undergoes the transition $t_{2g} \rightarrow e_g^*$ when light of wavelength near 500 nm is absorbed by the sample (Figs 12.9 and 12.10). The wavelength of the maximum absorption is altered by changing the ligand because the magnitude of Δ varies with the ligand. Stronger ligands cause the absorption band to shift

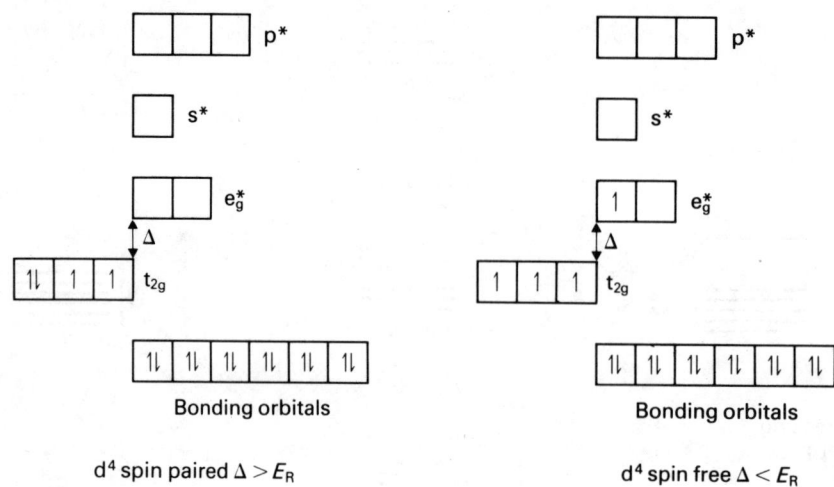

Fig. 12.6 d^4 configuration in an octahedral complex

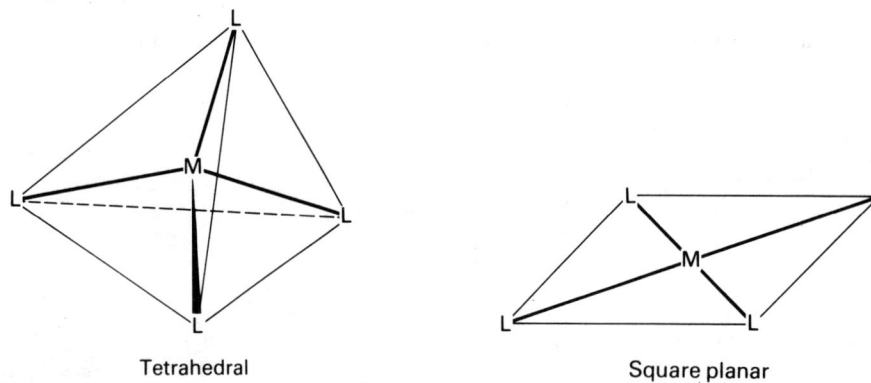

Tetrahedral

Square planar

Fig. 12.7 Tetrahedral and square planar complexes

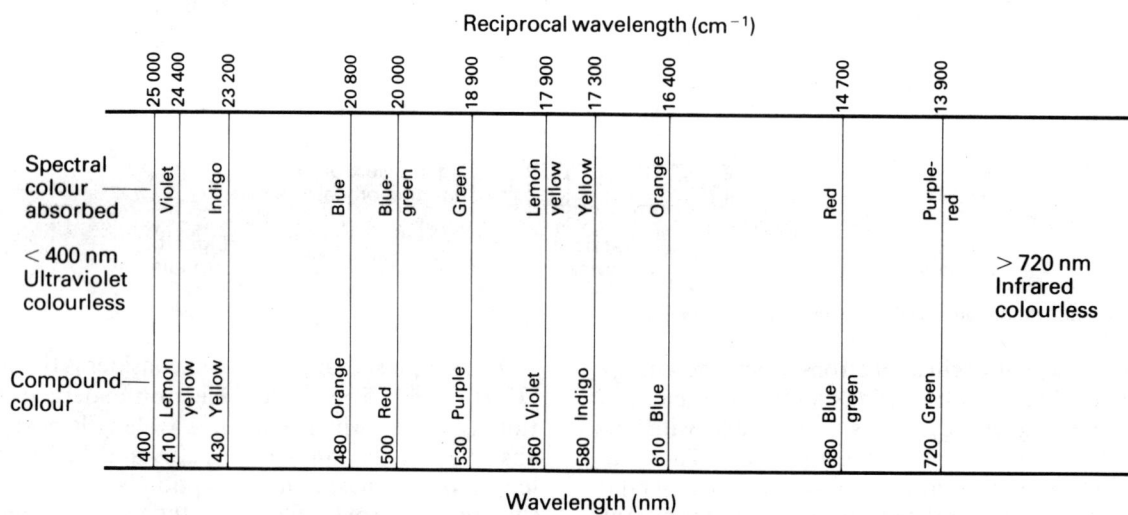

Fig. 12.8 Reciprocal wavelength (cm^{-1}) and wavelength (nm) of light absorbed in relation to the colour of a compound

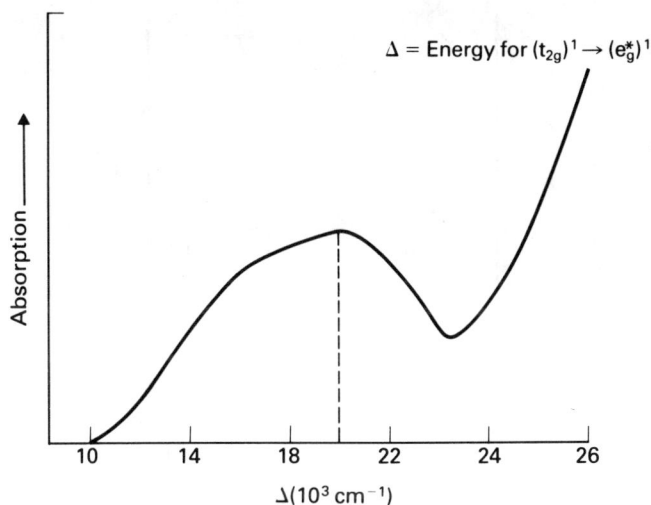

Δ = Energy for $(t_{2g})^1 \rightarrow (e_g^*)^1$

Fig. 12.9 Absorption spectrum of $[Ti(H_2O)_6]^{3+}$

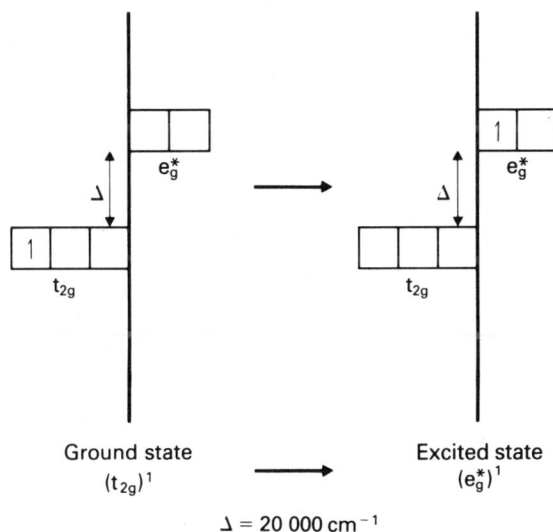

Ground state $(t_{2g})^1$ →   Excited state $(e_g^*)^1$

Δ = 20 000 cm^{-1}

Fig. 12.10 Origin of the absorption spectrum of $[Ti(H_2O)_6]^{3+}$

to higher energy (shorter wavelength). This is the reason for the change in colour of the octahedral complexes of copper listed in Table 12.10, and is the origin of the term 'spectrochemical series' for the list 12.2.

For d electron configurations more complex than d^1 or d^9 more than one ligand field transition can occur. A rigorous quantum mechanical description of the bonding and energy states of the metal complex is required to account for the details of their complex spectra.

The ligand field transitions are not the only ones that can occur in a complex, but are normally the ones that confer

colour on the compounds. Another type of electronic transition, known as a 'charge transfer' transition, occurs when a ligand electron is excited to a metal orbital or vice versa. Absorption bands associated with such transitions most often occur in the ultraviolet, but can extend into the visible region. Charge transfer spectra are normally very intense. The purple colour of the permangante ion (Mn(VII) d^0), which has no manganese d electrons and therefore cannot exhibit a ligand field transition, is due to a charge transfer spectrum involving excitation of an oxygen electron into a metal orbital. The same applies to the highly coloured ions orange $Cr_2O_7^{2-}$, yellow CrO_4^{2-} (both Cr(VI)d^0) and yellow VO_4^{3-} (V(V) d^0).

Table 12.10 Colours of transition metal complexes

Complex ion	Colour	Comments
$[Co(H_2O)_6]^{2+}$	pink	Change of metal ion.
$[Ni(H_2O)_6]^{2+}$	green	
$[Cu(H_2O)_6]^{2+}$	pale blue	Change of ligand coordin-
$[Cu(NH_3)_4(H_2O)_2]^{2+}$	deep blue	ated to the same metal.
$[Cu(en)_2(H_2O)_2]^{2+}$	purple/blue	Change of shape.
$[CuCl_4]^{2-}$ (tetrahedral)	yellow	

(en = enthylenediamine)

(g) Magnetism

Magnetism results from the presence of unpaired electrons in a substance. Compounds placed in a magnetic field are attracted in to it if they contain unpaired electrons and are paramagnetic. Substances with no unpaired electrons are repelled out of a magnetic field. All matter has this property which is called diamagnetism. Paramagnetism is rare for compounds of the regular elements, because stable compounds as a rule have all their electrons paired. Exceptions are O_2, NO, NO_2 and ClO_2. However, paramagnetism is common among the transition metals,* and since there is a direct relationship between the magnitude of the paramagnetism, expressed as a magnetic moment (μ), and the number of unpaired electrons (Table 12.11), the measurement of magnetic moments can give information about such properties as the strengths of ligand fields (bond strengths), oxidation states of metal ions, and even the shapes of complexes.

---

* *Ferromagnetism,* the property of making a much more powerful response to a magnetic field than that corresponding to paramagnetism, occurs when unpaired electron spins are aligned in a crystal so that their magnetic fields reinforce one another. *Antiferromagnetism* occurs when the spins are aligned in such a way that their magnetic effects cancel.

Table 12.11 Magnetic moments and unpaired electrons

Number of unpaired electrons†	Magnetic moment (μ) in bohr magnetons
1	1.73
2	2.83
3	3.87
4	4.90
5	5.92

† For transition metals the maximum possible number of unpaired electrons is 5.

The relationship of magnetic moment to bond strengths and oxidation states can be understood in terms of simple theory with reference to the examples which follow. The two cobalt (IV) d^6 complex ions $[CoF]^{3-}$ and $[Co(NH_3)_6]^{3+}$ have magnetic moments of 4.9 (expressed in bohr magnetrons) and zero, respectively. This shows that for $[CoF_6]^{3-}$, $\Delta < E_R$ and for $[Co(NH_3)_6]^{3+}$, $\Delta > E_R$ (Fig. 12.11).

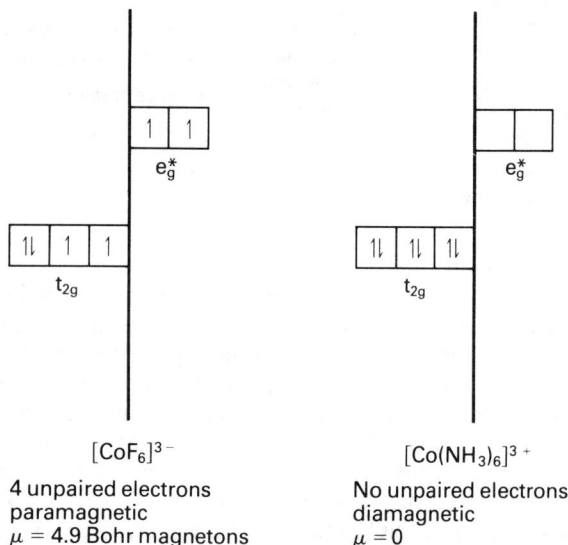

$[CoF_6]^{3-}$

4 unpaired electrons
paramagnetic
$\mu = 4.9$ Bohr magnetons

$[Co(NH_3)_6]^{3+}$

No unpaired electrons
diamagnetic
$\mu = 0$

Fig. 12.11 Magnetic properties of Co(III) complexes

The two cyano-complex ions of iron $[Fe(CN)_6]^{3-}$ and $[Fe(CN)_6]^{4-}$ contain Fe(III) d^5 and Fe(II) d^6 respectively and because the cyano ligand is strong both complexes are spin paired. The magnetic properties show that $[Fe(CN)_6]^{3-}$ has one unpaired electron and $[Fe(CN)_6]^{4-}$ has none (Fig. 12.12) confirming the oxidation states of Fe.

(h) Isomerism

Compounds with the same molecular formula but with differing structural formulas are termed isomers. There are several different types of isomerism; the principal types encountered in transition metal compounds are:

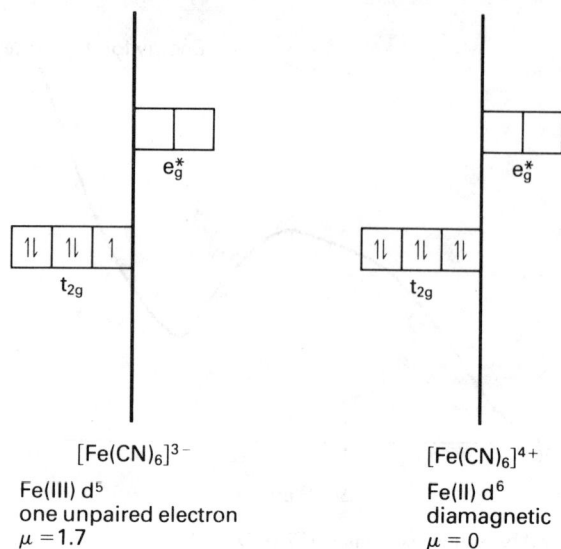

$[Fe(CN)_6]^{3-}$

Fe(III) d^5
one unpaired electron
$\mu = 1.7$

$[Fe(CN)_6]^{4+}$

Fe(II) d^6
diamagnetic
$\mu = 0$

Fig. 12.12 Magnetic properties of Fe(II) and Fe(III) compounds

(i) Geometric isomerism in which the geometrical arrangements of the ligands around the metal ion differ. For example in the square planar complex $[Pt\,Cl_2(NH_3)_2]$ the arrangement of ligands can be either *cis* or *trans*; the same applies in the octahedral complex ion $[Co(NH_3)_4Cl_2]^+$ (Fig. 12.13). Two ligands of the same type are adjacent in the *cis* isomer, and are on opposite sides of the central atom in the *trans* isomer.

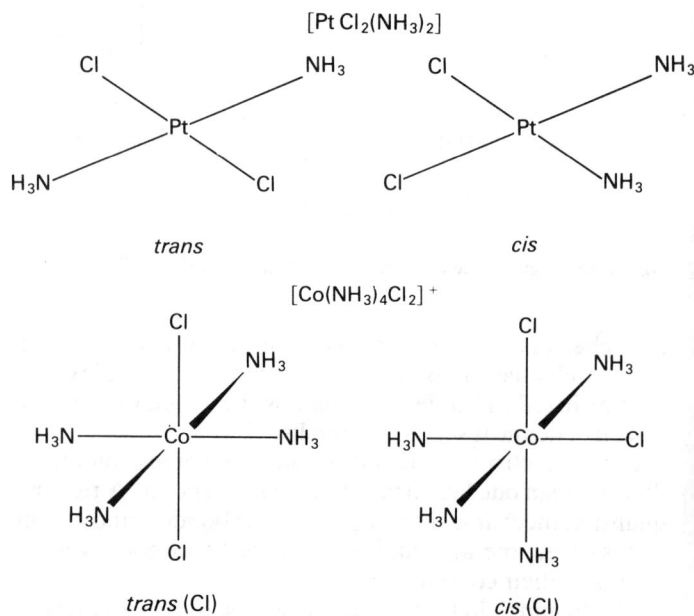

$[Pt\,Cl_2(NH_3)_2]$

trans   *cis*

$[Co(NH_3)_4Cl_2]^+$

trans (Cl)   *cis* (Cl)

Fig. 12.13 Geometric isomerism

(ii) Optical isomerism occurs when a molecule or ion contains no plane or centre of symmetry, i.e. the molecule or ion cannot be superimposed upon its mirror image. The two non-superimposable forms, which are related to one another as mirror images, are termed enantiomers. Substances which show this form of isomerism are optically active, i.e. they can rotate the plane of polarization of polarized light. Some examples of optical isomers are shown in Fig. 12.14. Enantiomers can be distinguished by the direction in which they rotate the plane of polarized light. A fuller discussion of optical isomerism is given in Chapter 14 in connection with the possible arrangements of different groups around a tetrahedral carbon atom.

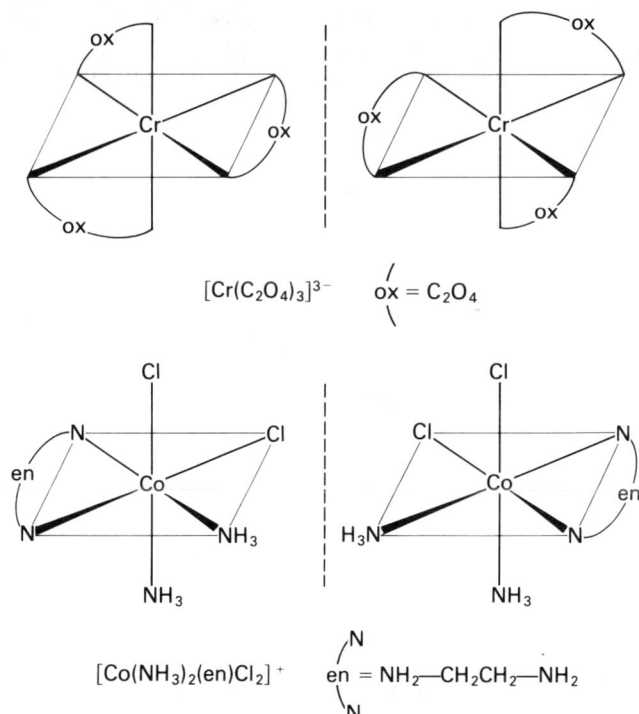

$[Cr(C_2O_4)_3]^{3-}$ $ox = C_2O_4$

$[Co(NH_3)_2(en)Cl_2]^+$ $en = NH_2—CH_2CH_2—NH_2$

Fig. 12.14 Examples of optical isomerism

(iii) Ionization isomers are compounds which have the same molecular formula but give different ions in solution, e.g. $[Co(NH_3)_5Br]^{2+}SO_4^{2-}$ and $[Co(NH_3)_5SO_4]^+Br^-$.

(iv) Hydration isomerism occurs when the number of coordinated water molecules varies, e.g. $[Cr(H_2O)_6]Cl_3$ (violet), $[Cr(H_2O)_5Cl]Cl_2.H_2O$ (light green), and $[Cr(H_2O)_4Cl_2]Cl\ 2H_2O$ (dark green). The usual form of $CrCl_3.6H_2O$ is the dark green isomer.

(v) Linkage isomerism occurs when a ligand may coordinate in more than one manner e.g. $[(NH_3)_5Co—NO_2]^{2+}$ (nitro, $—NO_2$) and $[(NH_3)_5Co—ONO]^{2+}$ (nitrito, $—ONO$).

Another ligand that can bond in two ways is thiocyanate, the alternatives being $—NCS^-$ and $—SCN^-$.

(vi) Coordination isomerism may occur when both the cation and anion of a compound are complex ions, e.g. $[Co(NH_3)_6][Cr(C_2O_4)_3]$ and $[Co(C_2O_4)_3][Cr(NH_3)_6]$. Another example is $[Pt(II)(NH_3)_4][Pt(IV)Cl_6]$ and $[Pt(IV)(NH_4)_4Cl_2][Pt(II)Cl_4]$. In the platinum compound the metal ions are in two oxidation states; it may also be called an intervalence compound. Sometimes such compounds are highly coloured because charge transfer is possible between the two metal ions, as for example in prussian blue (see p. 212).

12.3 Uses of Transition Metals and their Compounds

(a) Structural

The principal use of the majority of the metals is as structural material; this particularly applies to iron which, because of its abundance and ease of working, is one of man's principal structural materials. Alloying of iron with other metals adds to its range of useful properties. For example, the resistance of chromium to corrosion is made use of in producing stainless steel. Alternatively, steel can be protected by electroplating with chromium. Addition of manganese assists in the removal of sulphur and oxygen and also increases the hardness of steel.

A serious disadvantage of iron as a structural material is its moderate reactivity in moist air, combined with the fact that the reaction product, rust, does not protect the metal's surface by adhering to it. Conditions for rusting to occur are the presence of impurities and/or regions of local stress in the iron, and contact with water and oxygen. Rusting is an electrochemical process; at the cathode, which is a point of impurity, the reaction is

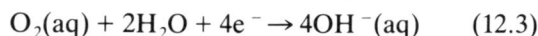

$$O_2(aq) + 2H_2O + 4e^- \rightarrow 4OH^-(aq) \qquad (12.3)$$

At the anode, which is purer iron, the reaction is

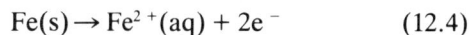

$$Fe(s) \rightarrow Fe^{2+}(aq) + 2e^- \qquad (12.4)$$

Between anode and cathode the reaction forms $Fe(OH)_2$:

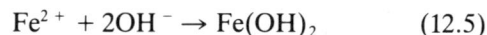

$$Fe^{2+} + 2OH^- \rightarrow Fe(OH)_2 \qquad (12.5)$$

which is rapidly oxidized to yellow-brown $Fe_2O_3.xH_2O$ or rust. Acids assist in the dissolving of iron; therefore pollution of the atmosphere by SO_2, which ultimately forms sulphuric acid, is a cause of corrosion. Alkalis retard rusting, presumably by influencing the cathode reaction. Electrolytes speed up rusting by improving conductivity within the system, which is the reason for more rapid rusting of iron near salt water. Rusting may be reduced or stopped by covering the iron with a protective film such as paint or with a zinc coating. Because of the difference in electrode potential zinc corrodes before iron, and forms a self-protective coating:

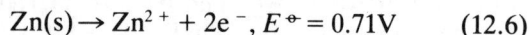

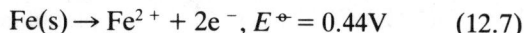

$$Zn(s) \rightarrow Zn^{2+} + 2e^-, E^{\ominus} = 0.71V \qquad (12.6)$$

$$Fe(s) \rightarrow Fe^{2+} + 2e^-, E^{\ominus} = 0.44V \qquad (12.7)$$

(Note that these voltages are oxidation potentials.) Tin coating has also been used, but it is ineffective once the coating has been penetrated because the iron will be oxidized preferentially since;

$$Sn(s) \rightarrow Sn^{2+} + 2e^-, E^{\ominus} = 0.14V. \qquad (12.8)$$

Zinc will actually reverse the corrosion of iron, and magnesium is even better. This process, termed cathodic protection, is used in practice. It is not necessary to completely cover the iron but merely to place blocks of the protective metal Mg or Zn on the iron. These corrode first, e.g.

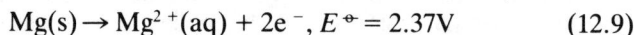

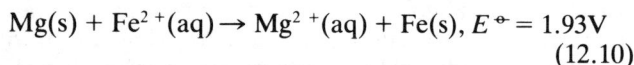

$$Mg(s) \rightarrow Mg^{2+}(aq) + 2e^-, E^{\ominus} = 2.37V \qquad (12.9)$$

$$Mg(s) + Fe^{2+}(aq) \rightarrow Mg^{2+}(aq) + Fe(s), E^{\ominus} = 1.93V \qquad (12.10)$$

To prevent corrosion of steel structures such as floating docks, which are continuously immersed in sea water, it is a common practice to insert a 'sacrificial anode' maintained at a positive potential of about one volt. This ensures that the main steel structure acts as a cathode as in Eq. 12.10, so that corrosion occurs only at the anode (which may itself take the form of an unwanted steel plate).

(b) Catalysis

Compounds of the transition metals find wide use as catalysts for inorganic and organic reactions. Until recently most catalytic reactions were heterogeneous, i.e. the metal or its compound were in a different phase from the reaction mixture. More recently homogeneous catalysts have been developed for a wide variety of organic reactions. For example, the hydroformylation reaction for converting alkenes to aldehydes

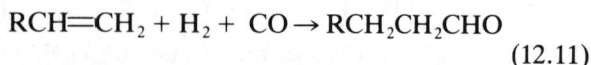

$$RCH{=}CH_2 + H_2 + CO \rightarrow RCH_2CH_2CHO \qquad (12.11)$$

occurs readily with homogeneous catalysts such as $Co_2(CO)_8$ and $RhHCO(Ph_3P)_3$ $(Ph = C_6H_5)$.

(c) Biochemistry

Several transition metal ions are very important in biochemical systems. Thus iron in haemoglobin is essential for oxygen transport in blood, molybdenum and iron are necessary for the fixation of atmospheric nitrogen by bacteria, and cobalt occurs in vitamin B_{12} (p. 219). The biochemistry of these compounds is rather complex, and in practice is often studied using simpler model systems, i.e. relatively uncomplicated transition metal complexes which are presumed to simulate the behaviour of the biochemical systems.

(d) Analysis

Transition metal complexes find wide use in analytical chemistry. The intensity of visible radiation absorbed by a coloured complex is simply related to the concentration of the complex in solution. For example, the intensity of absorption of blue-green light by the bright red ion $[Fe(SCN)]^{2+}$ is used to estimate the amount of iron (III) present in solution. Certain ligands are specific in their ability to form insoluble complexes with particular transition metals, and so enable the concentrations of those metals to be estimated. One example is the method for estimation of nickel based on the complex of nickel with dimethylglyoxime (Fig. 12.15).

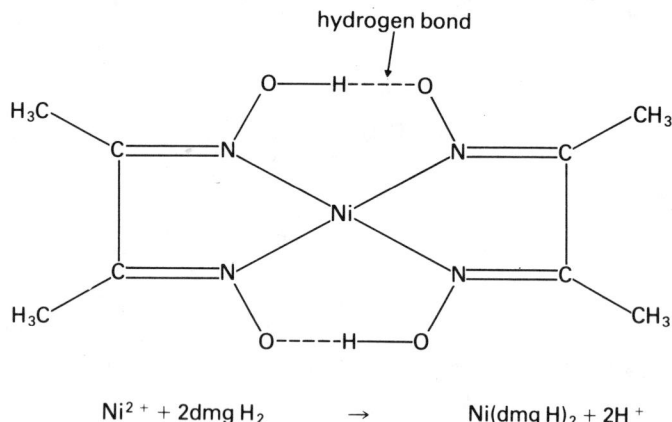

$$Ni^{2+} + 2dmg\,H_2 \quad \rightarrow \quad Ni(dmg\,H)_2 + 2H^+$$

Fig. 12.15 [Ni (dimethylglyoxime)$_2$]

12.4 Chemistry of the Transition Metals

The chemistry of the transition metals can be discussed systematically in terms of compound types, or electron configurations, or periodic groups, or in several other ways. No method is entirely satisfactory. The method used here is to discuss the chemistry of each element in turn, with frequent cross reference to related systems. However, the oxides and halides (cf. Figs 12.16 and 12.17) may be considered as a whole. The transition metal halides and oxides in low valency states have ionic structures, tending to covalent molecular structures with increasing oxidation number, as in VF_5 and CrO_3. The acid/base character of the oxides shows regular trends as indicated in Fig. 12.16. The halides (Fig. 12.17, page 206) show the predominance of fluorides in the high oxidation states. This becomes still

	Sc	Ti	V	Cr	Mn	Fe	Co	Ni	Cu	Zn
I									Cu_2O [b] 169 [am]	
II		TiO [b] 520	VO [b] 420	CrO [b] —	$\underline{MnO}$ [b] 385	FeO [b] 268	$\underline{CoO}$ [b] 239	$\underline{NiO}$ [b] 240	CuO [am] 157	ZnO [am] —
III	Sc_2O_3 1720	Ti_2O_3 [b] 1520	V_2O_3 [b] 1240	$\underline{Cr_2O_3}$ [am] 1140	Mn_2O_3 [am] 960	Fe_2O_3 [b] 820				
IV		$\underline{TiO_2}$ [am] 940	$\underline{VO_2}$ [am] 715	$\underline{CrO_2}$ [am] 595	$\underline{MnO_2}$ [am] 517					
V			$\underline{V_2O_5}$ [a] 1560							
VI				CrO_3 [a] 575						
VII					Mn_2O_7 [a] —					

b– basic   am– amphoteric   a– acidic   ——— – most stable

Fig. 12.16 Standard heats of formation of oxides of the first transition metal series ($kJ\ mol^{-1}$)

clearer for the halides of the second and third transition metal series. The data for oxides and halides show that variable valency (for oxidation states greater than II) mainly occurs to the left of the first transition metal series.

(a) Titanium

Titanium, a reasonably abudant metal, reacts with oxygen, the halogens and steam. As a structural material it is very light and strong, but is relatively expensive to separate from its ores.

The principal oxidation state Ti (IV) occurs as the oxide, TiO_2, and the chloride $TiCl_4$. The oxide TiO_2 occurs naturally in three crystalline modifications, the best known of which is rutile. Titanium dioxide is widely used as a white paint pigment because of its good light-scattering properties. It is obtainable in a pure white form from ilmenite according to the sequence:

$$FeTiO_3 \xrightarrow[H_2SO_4]{\text{digest with}} FeSO_4 + Fe_2(SO_4)_3 + TiOSO_4$$

$$\downarrow \text{add iron (reduce } Fe^{3+})$$

$$FeSO_4 + TiOSO_4$$

cryst. $\downarrow$ $\downarrow$ boil with H_2O, seed with rutile

$$FeSO_4 7H_2O$$

$$TiO_2 \text{ hydrated (rutile form)}$$

$$900\text{–}950°C \downarrow$$

$$TiO_2 \qquad (12.12)$$

The tetrachloride, a colourless liquid that fumes in air, is prepared during the isolation of titanium metal from rutile. The halide is a useful starting material for making other titanium compounds, and is used to make pure rutile by hydolysis:

$$TiCl_4 + 2H_2O \rightarrow TiO_2(\text{hydrated}) + 4HCl \quad (12.13)$$

Trivalent titanium is a mild reducing agent;

$$TiO^{2+} + 2H^+ + e^- \rightarrow Ti^{3+} + H_2O,\ E^{\ominus} = +0.1V$$
Titanyl ion   (12.14)

and will reduce ferric iron to ferrous. Titanium (III) solutions are violet in colour (Ti(IV) is colourless) due to the excitation of the single d electron by visible radiation. The trihalide β-$TiCl_3$ has a weak Ti—Ti bond. This is the only first row transition metal halide that displays metal—metal bonding. Metal—metal bonds are fairly common for the second and third row transition metal on the left of the series. Divalent titanium is a powerful reducing agent, reducing H_2O to H_2, and can only be kept in complete absence of air.

(b) Vanadium

Like titanium, vanadium metal is quite reactive, giving V_2O_5 when heated with oxygen and with the halogens giving VF_5, VCl_4, VBr_3 and VI_3. The highest oxidation state is $+5$, typified by the amphoteric oxide V_2O_5, which with strong bases gives the vanadate ion VO_3^-. If the pH is lowered condensed ions form, such as $V_3O_9^{3-}$ (pH 9–7) and $V_{10}O_{28}^{6-}$ (pH 7–6.5). The formation of polyanions also occurs with titanium and chromium but is most pronounced for molybdenum and tungsten.

	F						Cl						Br						I					
	I	II	III	IV	V	VI	I	II	III	IV	V	VI	I	II	III	IV	V	VI	I	II	III	IV	V	VI
Sc			ScF_3						$ScCl_3$						$ScBr_3$						ScI_3			
Ti			TiF_3	TiF_4				$TiCl_2$	$TiCl_3$	$TiCl_4$				$TiBr_2$	$TiBr_3$	$TiBr_4$				TiI_2	TiI_3	TiI_4		
V	VF_2	VF_3	VF_4	VF_5				VCl_2	VCl_3	VCl_4				VBr_2	VBr_3	VBr_4				VI_2	VI_3			
Cr	CrF_2	CrF_3	CrF_4	CrF_5	CrF_6			$CrCl_2$	$CrCl_3$					$CrBr_2$	$CrBr_3$					CrI_2	CrI_3			
Mn	MnF_2	MnF_3	MnF_4					$MnCl_2$						$MnBr_2$						MnI_2				
Fe	FeF_2	FeF_3						$FeCl_2$	$FeCl_3$					$FeBr_2$	$FeBr_3$					FeI_2				
Co	CoF_2	CoF_3						$CoCl_2$						$CoBr_2$						CoI_2				
Ni	NiF_2							$NiCl_2$						$NiBr_2$						NiI_2				
Cu	CuF_2						$CuCl$	$CuCl_2$					$CuBr$	$CuBr_2$					CuI					
Zn	ZnF_2							$ZnCl_2$						$ZnBr_2$						ZnI_2				

Fig. 12.17 Halides of the first transition metal series

Interconversion of the oxidation states of vanadium V, IV, III and II, is readily achieved, and clearly visible owing to the different colours of the various oxidation states (Table 12.12). The divalent species $[V(H_2O)_6]^{2+}$ is a powerful reducing agent, only stable in the absence of oxygen.

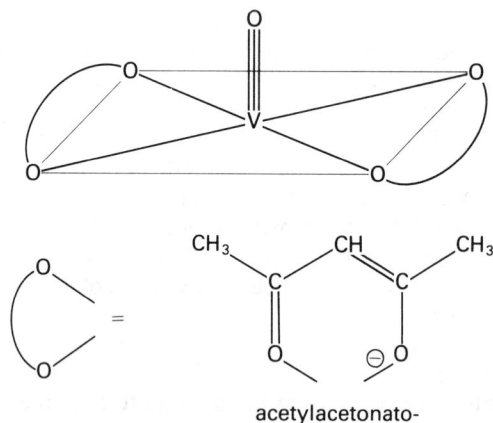

Fig. 12.18 Structure of VO(acetylacetone)$_2$

The tetravalent oxidation state occurs in a number of compounds in which the vanadium is strongly bonded to oxygen as the $[VO]^{2+}$ (vanadyl) species. The $[VO]^{2+}$ group is an oxy-cation which resembles oxy-anions such as SO_4^{2-} or VO_3^- in that it retains its identity throughout a variety of chemical reactions. The oxohalide species $[VOX_4]^{2-}$ are known, and complexes such as $[VO(acetylacetone)_2]$ (Fig. 12.18). The compounds are five coordinate and have a square-based pyramidal structure. Hydrolysis of the tetrachloride VCl_4 gives the oxychloride $VOCl_2$ and the anion $[VOCl_4]^{2-}$.

(c) Chromium

Chromium is not as reactive as titanium or vanadium and has an oxide film that makes it corrosion resistant. Some reactions of the element are given in Fig. 12.19. The principal oxidation states are III and VI, and there is an unstable divalent state.

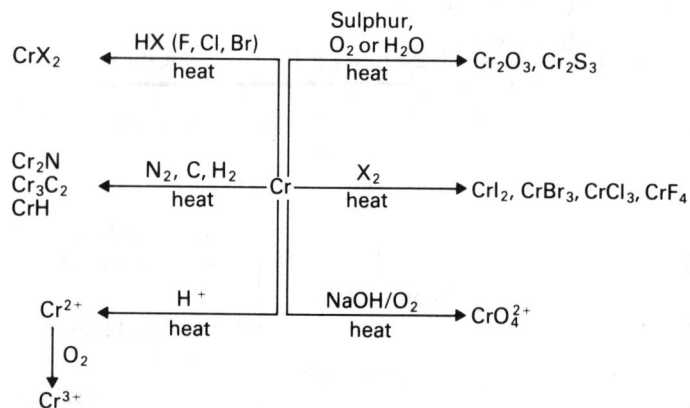

Fig. 12.19 Some reactions of chromium

Table 12.12 Interconversion of the oxidation states of vanadium

Oxidation state	V	IV	III	II
Species	VO_3^- $\xrightarrow{+1.0V}$	VO^{2+} $\xrightarrow{+0.3V}$	$[V(H_2O)_6]^{3+}$ $\xrightarrow{-0.2V}$	$[V(H_2O)_6]^{2+}$
Colour	yellow	blue	green	violet
Reducing agent	$\dfrac{SO_3^{2-}}{\text{or } Fe^{2+}}$ $\longrightarrow$			
		Ti^{3+} $\longrightarrow$		
	Zn/HCl or Zn/Hg + H$^+$ $\longrightarrow$			
Oxidizing agent	$\xleftarrow{\text{heat necessary}}$ KMnO$_4$/H$^+$ $\longrightarrow$			

The oxyanions of chromium (VI), CrO_4^{2-} and $Cr_2O_7^{2-}$ (Fig. 12.20), are important species. The two anions are interconvertible by treatment with acid or alkali.

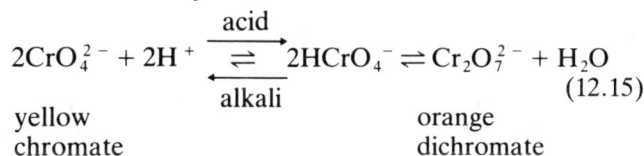

$$2CrO_4^{2-} + 2H^+ \underset{\text{alkali}}{\overset{\text{acid}}{\rightleftharpoons}} 2HCrO_4^- \rightleftharpoons Cr_2O_7^{2-} + H_2O \quad (12.15)$$

yellow   orange
chromate   dichromate

Stable protonated forms of the tetrahedral oxyanions VO_4^{3-}, CrO_4^{2-} and MnO_4^- cannot be formed. The ion $HCrO_4^-$, produced as an intermediate in the chromate–dichromate interconversion is unstable and in acid the condensation

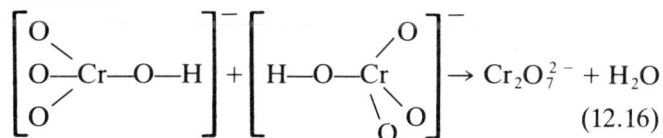

$$\rightarrow Cr_2O_7^{2-} + H_2O \quad (12.16)$$

readily occurs. At low pH the dichromate undergoes further condensation to give ions such as $Cr_3O_{10}^{2-}$. In concentrated acid the oxide CrO_3 is produced. The mixture of CrO_3 and concentrated sulphuric acid, termed chromic acid, is a good cleaning agent for glassware because of its oxidizing ability.

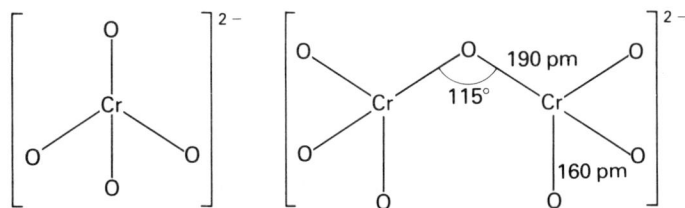

Fig. 12.20 The chromate and dichromate ions

The dichromate ion is a strong oxidizing agent in acid solution:

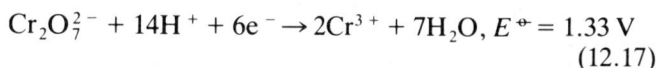

$$Cr_2O_7^{2-} + 14H^+ + 6e^- \rightarrow 2Cr^{3+} + 7H_2O, E^{\oplus} = 1.33 \text{ V} \quad (12.17)$$

and will effect the oxidations:

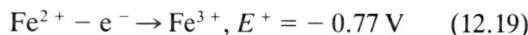

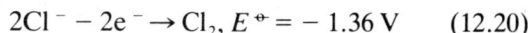

$$2I^- - 2e^- \rightarrow I_2, E^{\oplus} = -0.54 \text{ V} \quad (12.18)$$

$$Fe^{2+} - e^- \rightarrow Fe^{3+}, E^+ = -0.77 \text{ V} \quad (12.19)$$

$$2Cl^- - 2e^- \rightarrow Cl_2, E^{\oplus} = -1.36 \text{ V} \quad (12.20)$$

The oxidation of chloride is slow (with equal activities the equilibrium favours $Cr_2O_7^{2-}$ and Cl^-), so oxidations with dichromate may be carried out in hydrochloric acid solutions. In basic solution (where the chromate ion exists) the oxidizing power is greatly reduced:

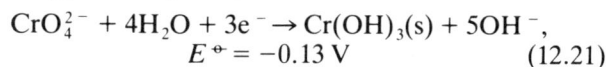

$$CrO_4^{2-} + 4H_2O + 3e^- \rightarrow Cr(OH)_3(s) + 5OH^-,$$
$$E^{\oplus} = -0.13 \text{ V} \quad (12.21)$$

In fact the reverse of reaction (12.21) readily occurs with oxidizing agents such as hydrogen peroxide.

Dichromate salts are more soluble than chromates. Some insoluble chromates, notably Ag_2CrO_4 (brick red), $BaCrO_4$, and $PbCrO_4$ (yellow) are made use of in analytical work. For example, in the presence of a trace of chromate the end point in the titration of Cl^- ions with Ag^+ (in neutral solution) is detected by the precipitation of brick-red Ag_2CrO_4. The concentration of the silver ion at the end point of the titration is obtained from the solubility product of AgCl

$$[Ag^+][Cl^-] = 1.2 \times 10^{-10} \text{ mol}^2 \text{l}^{-2}. \quad (12.22)$$

Since $[Ag^+] = [Cl^-]$ for AgCl we have

$$[Ag^+] = 1.1 \times 10^{-5} \text{ mol}^{-1}\text{l}^{-1} \quad (12.23)$$

If Ag_2CrO_4 is to precipitate exactly at the end point the solubility product relationship

$$[Ag^+]^2[CrO_4^{2-}] = 1.7 \times 10^{-12} \, mol^3 \, l^{-3} \quad (12.24)$$

must be fulfilled. Therefore

$$[CrO_4^{2-}] = \frac{1.7 \times 10^{-12}}{[Ag^+]^2} = \frac{1.7 \times 10^{-12}}{1.2 \times 10^{-10}} = 0.014 \, mol \, l^{-1}$$

Hence the solution is 0.014 mol l^{-1} in chromate ion for exact determination of the end point. However, even if the chromate concentration is 0.003 mol l^{-1} the end point will be detected within 0.02 ml of the true end point using 0.1 mol l^{-1} $AgNO_3$.

An important oxychloride of chromium (VI) is chromyl chloride, CrO_2Cl_2. This is a moderately volatile red liquid, prepared according to the reaction

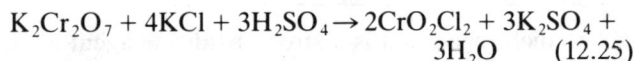

$$K_2Cr_2O_7 + 4KCl + 3H_2SO_4 \rightarrow 2CrO_2Cl_2 + 3K_2SO_4 + \atop 3H_2O \quad (12.25)$$

This reaction can be used as a test for the presence of chloride.

The trivalent oxide Cr_2O_3 is a stable compound with amphoteric properties. In acid $[Cr(H_2O)_6]^{3+}$ is formed, and in alkali $[CrO_3]^{3-}$ or $[Cr(OH)_6]^{3-}$ is produced. The oxide Cr_2O_3 has the corundum structure of Al_2O_3. The Cr^{3+} and Al^{3+} ions have similar radii (64 and 50 pm respectively) which allows isomorphous replacement of the ions to occur. Thus, for example, a complete range of solid solutions of aluminium alum and chromium alum can be obtained without difficulty.

Reduction of acid solutions of Cr(III) by zinc in the presence of dilute acid produces the sky-blue Cr(II) $[Cr(H_2O)_6]^{2+}$ species. Chromium (II) is a good reducing agent and is readily oxidized by atmospheric oxygen. It is so sensitive to oxygen that solutions of chromium (II) can be used to detect small amounts, or to remove small amounts of oxygen from a gas stream. The oxidation state is stabilized to some degree by forming insoluble compounds such as the red chromous acetate dimer $[Cr(acetate)_2 H_2O]_2$ (Fig. 12.21). This type of compound is typical of a number of acetate complexes (notably those of Mo, Re, Rh and Cu) where the ligand spans two metal atoms. The shortness of the Cr—Cr bond in the chromium compound suggests multiple bonding.

Chromium forms a number of peroxy-compounds, for example the blue Cr(VI) compounds CrO_5 (or more correctly $[CrO(O_2)_2]$) and adducts such as $[CrO(O_2)_2$ pyridine] (Fig. 12.22). A red peroxy-complex ion of chromium (V), $[Cr(O_2)_4]^{3-}$ exists which has an identical structure to similar compounds of vanadium, niobium and tantalum. The compound has a tetrahedral grouping of O_2^{2-} ligands around the metal. If the ammonium salt $(NH_4)_3 Cr(O_2)_4$ is warmed to 50°C a chromium (IV) peroxy compound is produced, $(NH_3)_3Cr(O_2)_2$ (Fig. 12.22). The bond

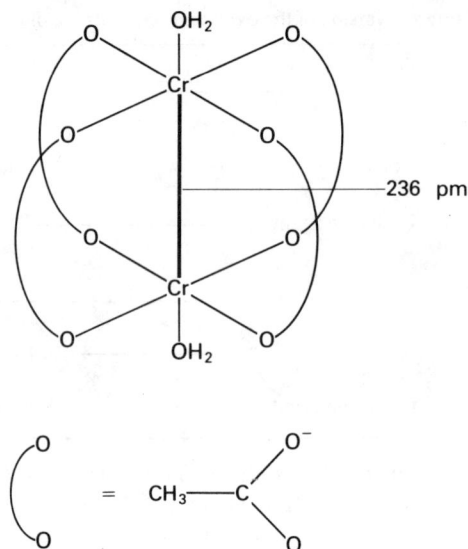

Fig. 12.21 Structure of $[Cr(acetate)_2 H_2O]_2$

length of O—O in these compounds lies in the range 140–143 pm, slightly less than the peroxide bond length (149 pm) in barium peroxide.

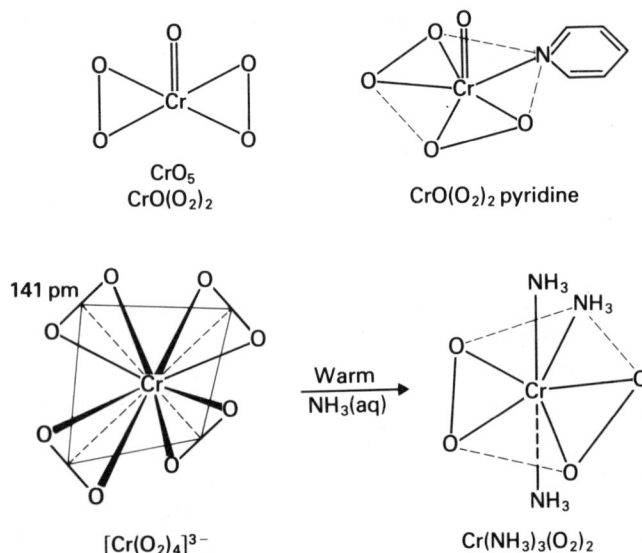

Fig. 12.22 Structures of chromium peroxy-compounds

(d) Manganese

Of all the first row transition metals manganese exhibits the greatest range of oxidation states, namely − III to + VII. The metal is quite reactive and slowly dissolves in water. No protective oxide coating is formed. An import-

Table 12.13 Oxide and oxyanion chemistry of manganese

Oxidation state	Oxide	Cation	Anion	Comments
VII	Mn_2O_7	—	MnO_4^- Permanganate	Oxide is acidic. Anion is stable and a powerful oxidizing agent.
VI	—	—	MnO_4^{2-} Manganate	Stable in alkaline conditions. In acid disproportionates to give: $Mn(VI) \rightarrow Mn(VII) + Mn(IV)$.
V	—	—	MnO_4^{3-}	Unstable oxyanion disproportionates to $Mn(IV) + Mn(VI)$.
IV	MnO_2	Mn^{4+}?	MnO_3^{2-}	Stability of oxide due to insolubility; cation and anion unstable, and cation exists only in complexes.
III	Mn_2O_3	Mn^{3+}	—	Cation exists in complexes; otherwise unstable.
II	MnO	Mn^{2+}	—	Basic oxide, stable cation, also found in many complexes.

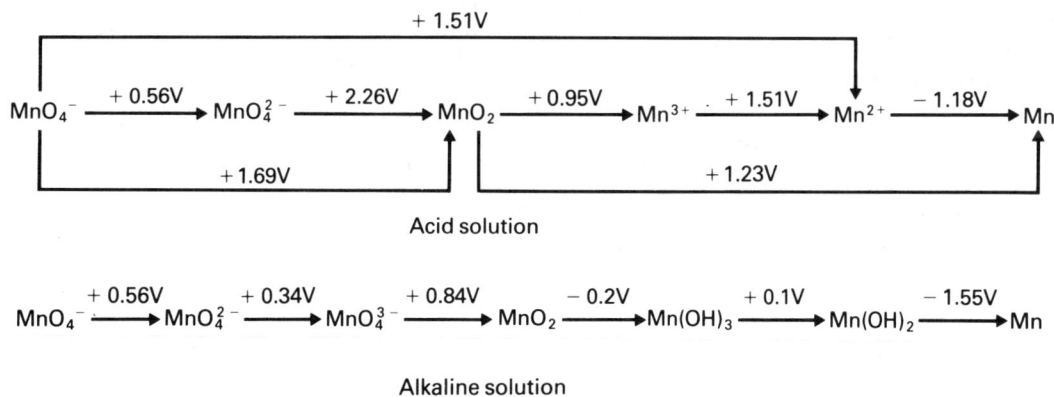

Fig. 12.23 Reduction potentials diagrams for manganese

ant aspect of manganese is its oxide and oxyanion chemistry, especially in relation to redox properties. A summary is given in Table 12.13, and the reduction potentials for the various oxidation states in both acid and alkaline solution are given in Fig. 12.23. In acid the stable state is Mn^{2+}, while in alkali it is $Mn(IV)$, as the oxide MnO_2.

The permangante ion, MnO_4^-, is the most powerful chemical oxidizing agent that can exist in water. It slowly decomposes in solution, decomposition being assisted by light and the presence of impurities.

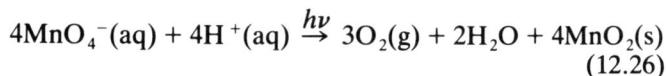

$$4MnO_4^-(aq) + 4H^+(aq) \xrightarrow{h\nu} 3O_2(g) + 2H_2O + 4MnO_2(s)$$
(12.26)

Potassium permanganate is a useful titrimetric redox reagent, which acts as its own indicator. It cannot be used as a

primary standard, but must be standardized against a reducing agent such as oxalic acid. In acid solution the MnO_4^-/Mn^{2+} half-reaction is

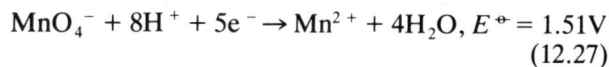

$$MnO_4^- + 8H^+ + 5e^- \rightarrow Mn^{2+} + 4H_2O, E^\ominus = 1.51V$$
(12.27)

and can be used for the following oxidations

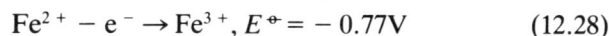

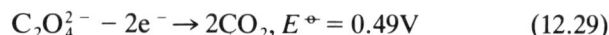

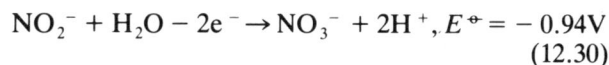

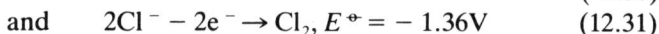

$$Fe^{2+} - e^- \rightarrow Fe^{3+}, E^\ominus = -0.77V$$
(12.28)

$$C_2O_4^{2-} - 2e^- \rightarrow 2CO_2, E^\ominus = 0.49V$$
(12.29)

$$NO_2^- + H_2O - 2e^- \rightarrow NO_3^- + 2H^+, E^\ominus = -0.94V$$
(12.30)

and $$2Cl^- - 2e^- \rightarrow Cl_2, E^\ominus = -1.36V$$
(12.31)

If insufficient acid is added the reduction of permanganate only goes as far as MnO_2:

$$MnO_4^- + 4H^+ + 3e^- \rightarrow MnO_2(s) + 2H_2O, \quad E^{\ominus} = 1.69V \tag{12.32}$$

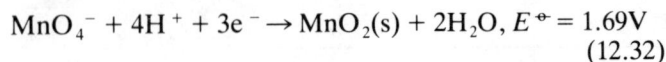

which destroys the quantitative nature of the reaction. The brown MnO_2 precipitate is slow to redissolve when acid is added. In alkaline or neutral solution the permanganate ion is a weaker oxidizing agent, and MnO_2 is the reduction product.

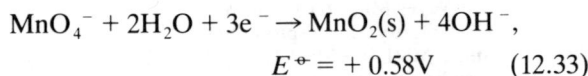

$$MnO_4^- + 2H_2O + 3e^- \rightarrow MnO_2(s) + 4OH^-,$$
$$E^{\ominus} = +0.58V \tag{12.33}$$

The permanganate ion is produced by oxidation of the Mn^{2+} ion with a reagent such as PbO_2 or $S_2O_8^{2-}$ (peroxydisulphate);

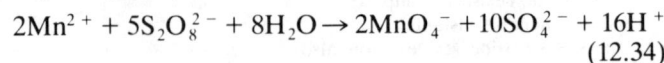

$$2Mn^{2+} + 5S_2O_8^{2-} + 8H_2O \rightarrow 2MnO_4^- + 10SO_4^{2-} + 16H^+ \tag{12.34}$$

This reaction is used in quantitative analysis for manganese. The permanganate produced can be estimated colorimetrically or by a redox titration.

Treatment of permanganate with concentrated sulphuric acid gives the explosive green oil Mn_2O_7. Treatment with alkali gives the green manganate ion $[MnO_4]^{2-}$.

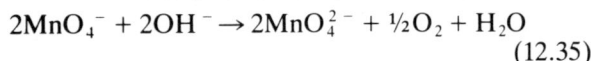

$$2MnO_4^- + 2OH^- \rightarrow 2MnO_4^{2-} + \frac{1}{2}O_2 + H_2O \tag{12.35}$$

Addition of acid to manganate, or allowing an aqueous solution of MnO_4^{2-} to stand, results in a disproportionation reaction represented by the general equation

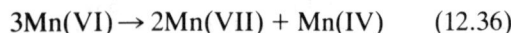

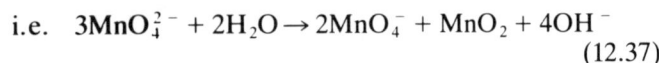

$$3Mn(VI) \rightarrow 2Mn(VII) + Mn(IV) \tag{12.36}$$

i.e. $3MnO_4^{2-} + 2H_2O \rightarrow 2MnO_4^- + MnO_2 + 4OH^- \tag{12.37}$

The tetravalent oxidation state of manganese is not stable except as the oxide MnO_2, where stability is due to its very low solubility in water. The oxide occurs in nature as the ore pyrolusite, which is the starting material for a number of manganese products, such as manganese itself, ferromanganese (Fe/Mn alloy), and $MnSO_4$ for trace element addition to soils. The dioxide is also used as a depolarizing (oxidizing) agent in dry cell batteries, for colouring glass, and as a drier in paint.

The trivalent oxidation state is stabilized in complexes, but in the absence of complexing agents it is readily reduced to the divalent state. Examples of trivalent manganese complex ions are $[Mn(CN)_6]^{3-}$ and $[Mn(C_2O_4)_3]^{3-}$. The latter ion contains trivalent manganese coordinated to the oxalate ion. Normally, in solution, the oxalate ion would reduce Mn(III) to Mn(II). The deep purple solid complex $K_3[Mn(C_2O_4)_3]3H_2O$ exists for only about a day if kept above 0°C; exposure to light speeds up the decomposition. The compound is prepared by a controlled reaction between KMnO₄ and oxalic acid at 0°C or less. The complex $K_3[Mn(C_2O_4)_3]3H_2O$ is identical in composition to oxalato-complexes of Al(III), Cr(III), Fe(III) and Co(III).

The reaction between $KMnO_4$ and $C_2O_4^{2-}$ in acid conditions requires heat to make it fast enough for quantitative work. It is catalysed by the Mn^{2+} ion and there is an induction period at the beginning of the reaction while Mn^{2+} ions are produced from the permanganate. If the reaction is carried out carefully the intermediate purple colour of the Mn(III) oxalato-complex can be observed

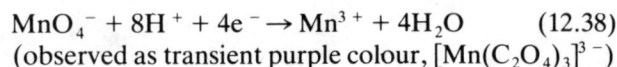

$$MnO_4^- + 8H^+ + 4e^- \rightarrow Mn^{3+} + 4H_2O \tag{12.38}$$
(observed as transient purple colour, $[Mn(C_2O_4)_3]^{3-}$)

$$Mn^{3+} + e^- \rightarrow Mn^{2+} \text{ (catalyst)} \tag{12.39}$$

and $C_2O_4^{2-} - 2e^- \rightarrow 2CO_2 \tag{12.40}$

The d^4 electron arrangement for octahedral complexes of manganese (III) will be one of those given in Fig. 12.6. When the fourth electron is in an e_g^* orbital it points directly towards some of the ligands, with repulsion and distortion of the complex. This is observed for the trifluoride MnF_3 and also for the anhydrous halides of chromium (II) in which the chromium has the d^4 configuration (Fig. 12.24).

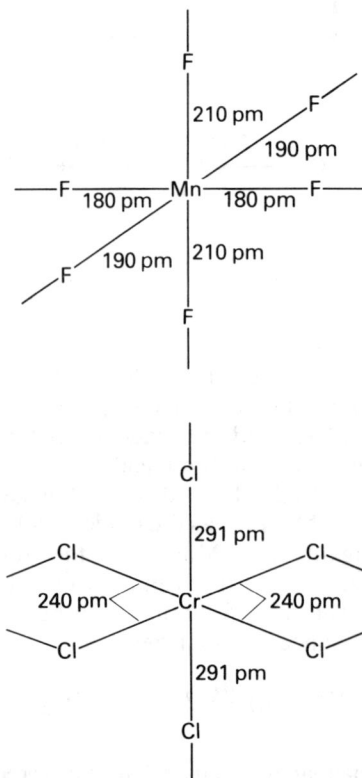

Fig. 12.24 Structures of MnF_3 and $CrCl_2$

The divalent oxidation state is generally less stable than Mn(III) in *spin-paired* complexes. For example, the reaction

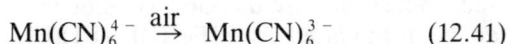

$$Mn(CN)_6^{4-} \xrightarrow{air} Mn(CN)_6^{3-} \qquad (12.41)$$

proceeds readily. However, the Mn^{2+} ion is normally the more stable of the two oxidation states (cf. Fig. 12.19), especially in the form of the aqua cation $[Mn(H_2O)_6]^{2+}$. In this cation the Mn(II) ion has all its d electrons unpaired, an especially stable arrangement because of the half-filled d sub-shell.

(e) Iron

The extraction of iron has been discussed in Chapter 8, and the rusting of iron earlier in this chapter. The two principal oxidation states are Fe(II) and Fe(III). Some reactions of these two oxidation states are listed in Fig. 12.25. Iron exists in higher oxidations states (in association with oxygen), and also in lower oxidation states such as Fe(I) and Fe(O) (in association with organic π-bonded ligands and carbon monoxide). The ferrate ion FeO_4^{2-} is an unstable species, and a more powerful oxidizing agent than permanganate. In acid it rapidly decomposes:

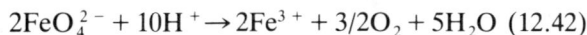

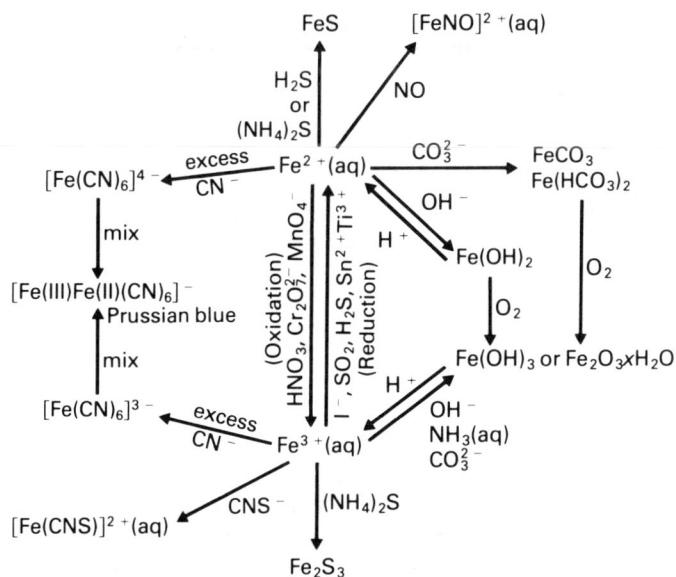

$$2FeO_4^{2-} + 10H^+ \rightarrow 2Fe^{3+} + 3/2O_2 + 5H_2O \quad (12.42)$$

Fig. 12.25 Some reactions of $Fe^{2+}(aq)$ and $Fe^{3+}(aq)$

The properties of the oxides of iron, FeO, Fe_2O_3 and Fe_3O_4 are summarized in Table 12.14. The oxide FeO is generally non-stoichiometric i.e. the ratio of Fe to O is not unity, the iron content being low. To maintain electrical neutrality when some Fe^{2+} is missing, some of the remaining Fe^{2+} ions are replaced with Fe^{3+} ions. The compound can be written as $Fe^{2+}_{1-3X} Fe^{3+}_{2X} O^{2-}$, i.e. X Fe^{2+} are missing from the structure and $2X Fe^{2+}$ of the remaining iron is replaced by $2X Fe^{3+}$. If more Fe^{3+} ions replace Fe^{2+} until there are two Fe^{3+} for every Fe^{2+} ion the oxide Fe_3O_4 ($Fe^{2+}Fe_2^{3+}O_4$) is produced. Non-stoichiometry of solid compounds is not uncommon in oxide chemistry.

Table 12.14 Some properties of iron oxides

Oxide	Structure	Comments
FeO	NaCl structure Fe^{2+} and O^{2-} ions	Prepared by heating FeC_2O_4 at high temperature in vacuo.
α-Fe_2O_3	Has O^{2-} ions in cubic close packing with Fe^{3+} ions randomly arranged in octahedral and tetrahedral sites.	Occurs as ore haematite. Used as a red pigment. A less stable γ-Fe_2O_3 can be prepared.
Fe_3O_4 (Fe(II)2Fe(III) + $4O^{2-}$)	Has ccp O^{2-} ions: Fe^{3+} half in octahedral and half in tetrahedral sites, Fe^{2+} in octahedral sites.	Occurs as ore magnetite. Is strongly ferromagnetic.

The iron sulphide FeS_2, iron pyrites, is a natural ore of iron. It has a deep yellow colour and has the alternative name 'Fool's Gold'. Another form of FeS_2 is marcasite. The compound, which has a distorted NaCl structure, contains the disulphide ion S_2^{2-} and is therefore a compound of Fe(II).

The reduction potential diagrams for iron in both acid and alkaline solution are given in Fig. 12.26. It can be seen that the $+2$ and $+3$ oxidation states of iron are readily interconvertible (cf. Fig. 12.25) and oxidation of Fe(II) to Fe(III) is achieved more readily in alkaline or neutral conditions than in acid. Solid ferrous hydroxide is readily oxidized; the almost white compound prepared by precipitation rapidly darkens, giving probably $Fe(OH)_3$ and finally $Fe_2O_3 3H_2O$. Some of the driving force for this last reaction is provided by the insolubility of $Fe(OH)_3$ by comparison with $Fe(OH)_2$.

(a) $\qquad FeO_4^{2-} \xrightarrow{>1.9V} Fe^{3+} \xrightarrow{+0.77V} Fe^{2+} \xrightarrow{-0.44V} Fe$

(b) $\qquad FeO_4^{2-} \xrightarrow{>0.9V} Fe(OH)_3 \xrightarrow{-0.56V} Fe(OH)_2 \xrightarrow{0.89V} Fe$

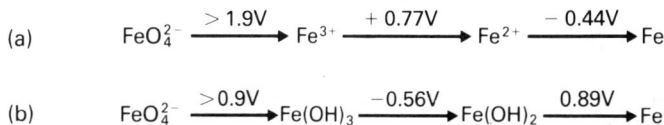

Fig. 12.26 Reduction potential diagrams for iron: (a) acid solution (b) alkaline solution

The relative stabilities of Fe(II) and Fe(III) depend on the nature of the ligands coordinated to the metal, as the following data show:

$$Fe(H_2O)_6^{3+} + e^- \rightarrow Fe(H_2O)_6^{2+}, E^{\ominus} = +0.77V \tag{12.43}$$

$$Fe(CN)_6^{3-} + e^- \rightarrow Fe(CN)_6^{4-}, E^{\ominus} = +0.36V \tag{12.44}$$

$$Fe(o\ phen)_3^{3+} + e^- \rightarrow Fe(o\ phen)_3^{2+}, E^{\ominus} = +1.12V$$
$$(o\ phen = o\ phenanthroline) \tag{12.45}$$

$$Fe(C_2O_4)_3^{3-} + e^- \rightarrow Fe(C_2O_4)_2^{2-} + C_2O_4^{2-},$$
$$E^{\ominus} = +0.02V \tag{12.46}$$

The role of iron in biochemical systems often involves redox reactions, and it is highly probable that the ligands coordinated to the iron in these systems influence the redox potentials in such a way that the required electron transfers occur.

The anhydrous halides of iron have ionic structures in the solid state. In the gaseous state $FeCl_3$ and $FeBr_3$ are dimeric with the Al_2Cl_6 type of structure. Ferric iodide does not exist because ferric ions are reduced by I^- in aqueous solution.

The sulphates of iron are common starting materials for studying the chemistry of iron. Iron (II) is readily available as $FeSO_4 7H_2O$ and as the double salt $Fe(NH_4)_2(SO_4)_2 6H_2O$, and iron(III) as $Fe_2(SO_4)_3 nH_2O$ and the alum $NH_4Fe(SO_4)_2 12H_2O$. Ferrous ammonium sulphate is a convenient form of iron (II), which can be obtained in a good state of purity because it is reasonably stable towards air oxidation and loss of water. Ordinary ferrous sulphate loses water and turns yellow-brown due to air oxidation.

Ferric iron is readily hydrolysed. The cation $Fe(H_2O)_6^{3+}$ only exists in strong acid solutions or in crystalline salts such as the sulphate, perchlorate, and nitrate. In moderately acidic solutions (pH 2 to 3) hydrolysis occurs according to

$$[Fe(H_2O)_6]^{3+} + H_2O \rightleftharpoons [Fe(H_2O)_5OH]^{2+} + H_3O^+ \tag{12.47}$$

and

$$[Fe(H_2O)_5OH]^{2+} + H_2O \rightleftharpoons [Fe(H_2O)_4(OH)_2]^+ + H_3O^+ \tag{12.48}$$

In solutions with pH greater than 3 polymeric bridged species are produced, such as $[Fe(H_2O)_4(OH)_2Fe(H_2O)_4]^{4+}$, and ultimately $Fe_2O_3 nH_2O$ is formed as a brown gelatinous precipitate.

The cyanide complexes of Fe(II) and Fe(III), $K_4[Fe(CN)_6]$ and $K_3[Fe(CN)_6]$, are stable in water, but acid releases HCN. If either of the two cyanides is treated with an aqueous solution of iron in the other oxidation state a deep blue compound called prussian blue, or Turnbull's blue, is produced. The two names arose from the two approaches to obtaining the compound i.e. $Fe(CN)_6^{4-} + Fe^{3+}$ giving prussian blue and $Fe(CN)_6^{3-} + Fe^{2+}$ giving Turnbull's blue. In fact they are the same compound. The compound can be formulated as $K[Fe(II)Fe(III)(CN)_6]$ and has a simple cubic structure (Fig. 12.27). The deep blue colour results from the absorption of visible radiation causing an electron transfer between Fe(II) and Fe(III), interchanging the oxidation states. If the cyanides are mixed with iron in the same oxidation state the compounds $[Fe(III)Fe(III)(CN)_6]$ and $K_2[Fe(II)_2(CN)_6]$ are produced with similar structures to prussian blue. If the potassium ions in prussian blue are replaced with Fe^{2+} or Fe^{3+} an insoluble form of prussian blue is produced, either $Fe^{2+}[Fe(II)Fe(III)(CN)_6]_2$ or $Fe^{3+}[Fe(II)Fe(III)(CN)_6]_3$.

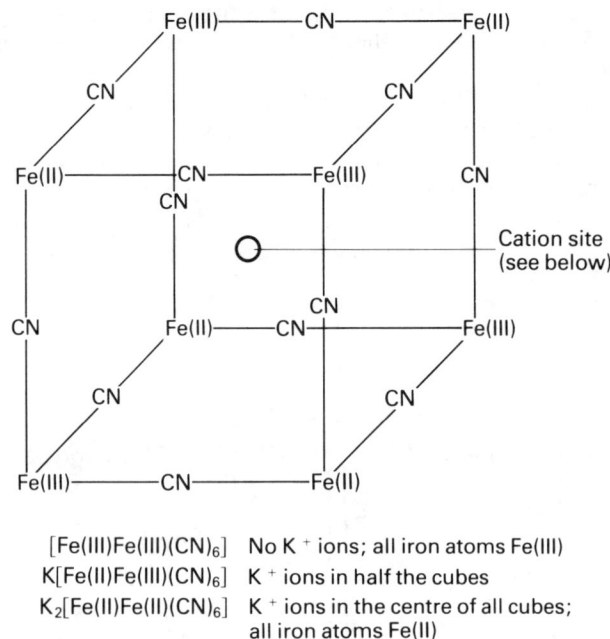

$[Fe(III)Fe(III)(CN)_6]$	No K^+ ions; all iron atoms Fe(III)
$K[Fe(II)Fe(III)(CN)_6]$	K^+ ions in half the cubes
$K_2[Fe(II)Fe(II)(CN)_6]$	K^+ ions in the centre of all cubes; all iron atoms Fe(II)

Fig 12.27 Structure of $[Fe(II)\ Fe(III)\ (CN)_6]^-$ (prussian blue and related compounds)

Iron is an important component of haemoglobin, where it is present in the divalent state and is surrounded by a square planar coordinated porphyrin ring (Fig. 12.28) with a nitrogen base coordinated on one side of the plane and a vacant site on the other. The vacant position is able to coordinate O_2, and the uptake by and release of O_2 from the iron is the basis of oxygen transport by blood. Carbon monoxide is a poison because it coordinates irreversibly with the iron, in preference to oxygen. Iron is also involved in myoglobin, which stores oxygen in muscles (see Chapter 17).

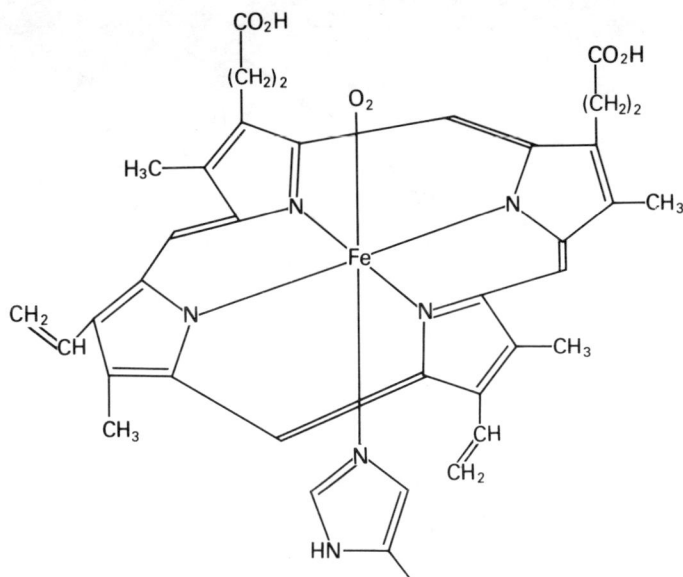

Fig. 12.28 Iron complex in haemoglobin

The unusually shaped organometallic complex of iron in Fig. 12.29, $Fe(C_5H_5)_2$, called *ferrocene*, is of historical importance because its accidental discovery in the early 1950's initiated tremendous development in the organometallic chemistry of the transition metals. The complex consists of an iron atom sandwiched between two planar C_5H_5 rings, each of which has six π electrons in an orbital system similar to that of benzene. All carbon atoms are equidistant from the iron; hence the bonding between the Fe and C_5H_5 involves overlap of the metal d, s and p orbitals with the delocalized π orbitals of the cyclopentadienyl rings.

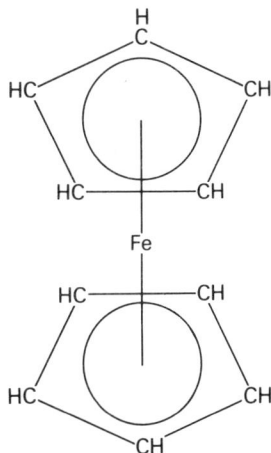

Fig. 12.29 Structure of ferrocene $Fe(C_5H_5)_2$

(f) Cobalt

The two principal oxidation states of cobalt are Co(II) and Co(III). The inter-relationship between the oxidation states is similar to that between Fe(II) and Fe(III), except that Co(III) exists almost entirely in complexes and only a few simple compounds are known. These are compounds such as CoF_3 and $Co_2(SO_4)_3$ which contain non-oxidizable ligands. The aqua complexes of Co(II) are most stable, but ligands such as CN^- and those containing nitrogen donor atoms preferentially stabilize Co(III).

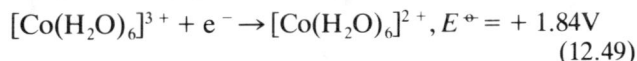

$$[Co(H_2O)_6]^{3+} + e^- \rightarrow [Co(H_2O)_6]^{2+}, E^{\ominus} = +1.84V \quad (12.49)$$

$$[Co(NH_3)_6]^{3+} + e^- \rightarrow [Co(NH_3)_6]^{2+}, E^{\ominus} = +0.1V \quad (12.50)$$

$$[Co(CN)_6]^{3-} + e^- \rightarrow [Co(CN)_6]^{4-}, \quad E^{\ominus} = -0.8V \quad (12.51)$$

The cation $[Co(III)(H_2O)_6]^{3+}$ is a powerful oxidizing agent and will oxidize water to oxygen, whereas $[Co(II)(CN)_6]^{4-}$ is a powerful reducing agent, reducing water to H_2. Cobalt(III) complexes are relatively unreactive because of the filled t_{2g}^6 shell. In all of these complexes $\Delta > E_R$, so that spin-pairing occurs. Some ammonia complexes of Co(III) (cobalt ammines) are listed in Table 12.7. The complex ion $[Co(NO_2)_6]^{3-}$ is useful in practice because it forms a sparingly soluble potassium salt $K_3[Co(NO_2)_6]$ which is the basis of a gravimetric method for estimating K^+ ions.

The colours of cobalt (II) complexes depend on both the coordination number of the cobalt atom and the shape of the complexes. Thus octahedral $[Co(H_2O)_6]^{2+}$ is pink whereas tetrahedral $[CoCl_4]^{2-}$ is deep blue. The colour difference is due to the different magnitude of the d orbital splitting by the two ligand fields. Since Δ, the ligand field splitting, is a function of the strength of bonding to the ligands it is not surprising that it also depends on the number of ligands coordinated. In general, as one would expect

$$\Delta_{octahedral} > \Delta_{tetrahedral}. \quad (12.52)$$

(g) Nickel

With nickel the divalent state is the most important oxidation state, and numerous complexes have been prepared. Nickel (II) compounds have a variety of shapes, namely octahedral (six-coordinate), trigonal biprism or square based pyramid (five-coordinate), and tetrahedral or square planar (four-coordinate). Divalent nickel has the $3d^8$ configuration, which is ideal for the dsp^2 square planar stereochemistry provided all the d electrons are paired up, leaving one vacant d orbital, one s and three p orbitals. If the third p orbital is used five-coordination will occur. Whether four-coordinated nickel forms square planar or tetrahedral complexes depends on the magnitude of Δ, i.e. on the type of ligand. Thus $[Ni(CN)_4]^{2-}$ is square planar, whereas $[NiCl_4]^{2-}$ is tetrahedral (Fig. 12.30).

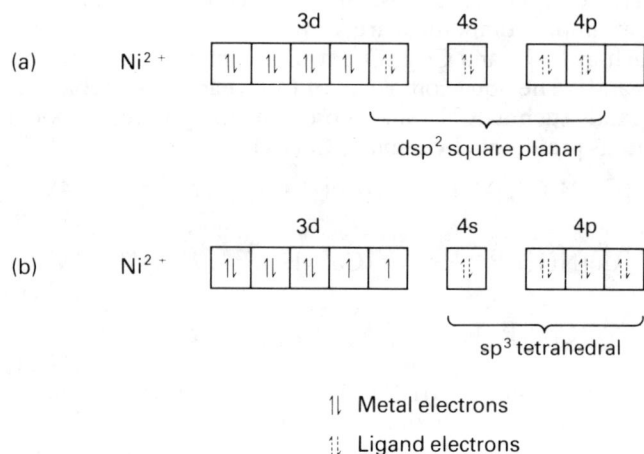

Fig. 12.30 Alternative types of hybridization for four-coordinated nickel (II). (a) Large Δ, promotion energy $3d^2 \rightarrow 4p^2 > E_R$; (b) small Δ, promotion energy $3d^2 \rightarrow 4p^2 < E_R$.

In the extraction of nickel from its ores nickel tetra-carbonyl, $Ni(CO)_4$ is formed (cf. the Mond process, Chapter 8). The carbonyl is a liquid boiling at 42.2°C and is prepared by the direct reaction of Ni with CO. The transition metals form a number of carbonyls (cf. Table 12.15 and Fig. 12.31) in which the metal atom is in the zero oxidation state. The stoichiometry of the compounds and the dimerization in some cases correspond to the metal achieving the inert gas configuration of Kr (Table 12.15) with 18 outer electrons, i.e. $3d^{10}4s^24p^6$.

(h) Copper

Copper is an important metal with a very wide range of uses, the most important of which is probably as an electrical conductor. The main oxidation states are Cu(II) and Cu(I), of which Cu(II) is the more stable. Some reactions of the metal and of Cu(II) are given in Fig. 12.32.

Fig. 12.31 Structure of some transition metal carbonyls

Copper (I) is unstable relative to Cu(II), which in some respects is a little surprising because Cu(I) has the stable d^{10} configuration and Cu(II) is d^9. However, the low charge density of Cu^+ compared with Cu^{2+} (Cu^+ radius 96 pm,

Table 12.15 Transition metal carbonyls

Element	V	Cr	Mn	Fe	Co	Ni
Carbonyl	$V(CO)_6$	$Cr(CO)_6$	$Mn_2(CO)_{10}$	$Fe(CO)_5$	$Co_2(CO)_8$	$Ni(CO)_4$
Configuration d^n	d^5	d^6	d^7	d^8	d^9	d^{10}
CO's per metal atom	6	6	5	5	4	4
CO electrons donated	12	12	10	10	8	8
M—M bond electrons per M atom	0	0	1	0	1	0
Total electrons per M	17	18	18	18	18	18

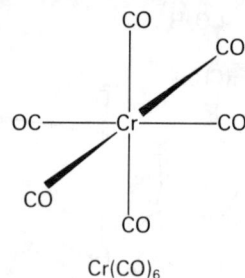

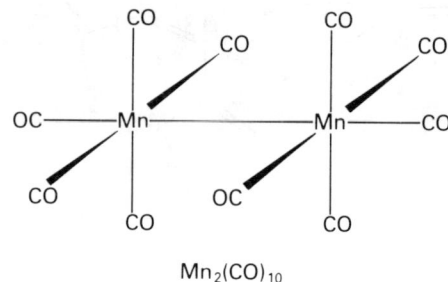

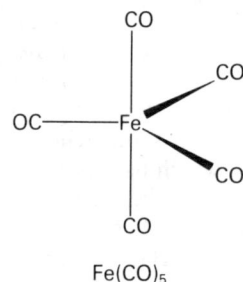

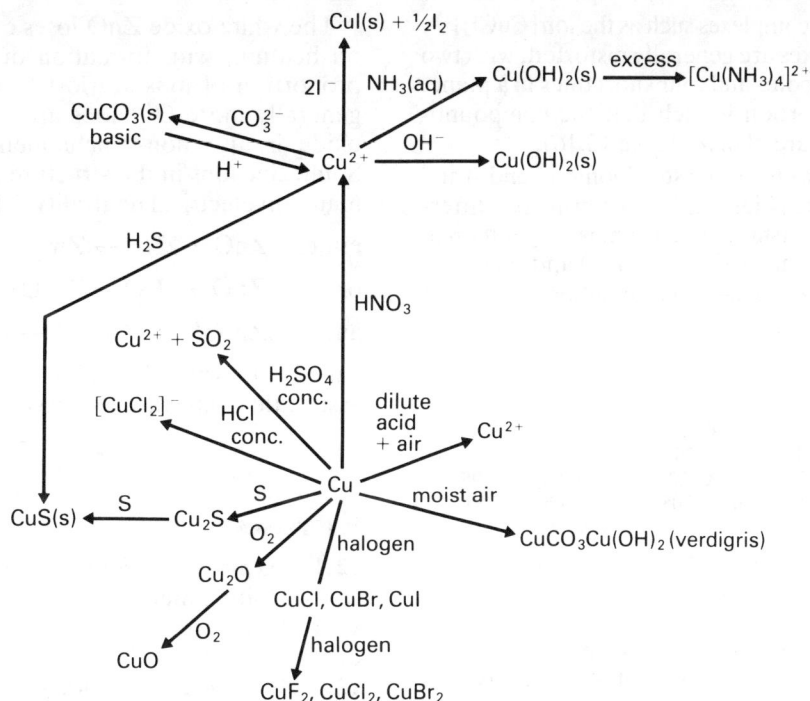

Fig. 12.32 Some reactions of Cu and Cu^{2+}

Cu^{2+} radius 69 pm) means that the hydration energy of Cu^+ is much less than that of Cu^{2+}. Also Cu(I) complexes are just two-coordinate and linear, so that a hydrate would be $[Cu(H_2O)_2]^+$, whereas the Cu(II) hydrate is four-coordinate; therefore the greater bonding favours Cu^{2+} over Cu^+. The following electrode potentials bear this out:

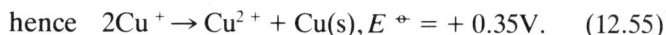

$$Cu^+ + e^- \rightarrow Cu, \qquad E^{\oplus} = +0.52V \qquad (12.53)$$

$$Cu^{2+} + e^- \rightarrow Cu^+, \qquad E^{\oplus} = +0.17V \qquad (12.54)$$

hence $2Cu^+ \rightarrow Cu^{2+} + Cu(s), E^{\oplus} = +0.35V.$ (12.55)

The equilibrium constant for the reaction 12.55 is approximately 10^6. In order to send the reaction to the left the concentration of Cu^+ must be kept small, i.e. Cu^+ ions need to be removed from solution as fast as they are produced. This can be achieved either by forming compounds of low solubility, such as CuI, or by forming copper (I) complexes, e.g. CuCl in excess Cl^- gives $[CuCl_2]^-$. The reaction between Cu^{2+} and I^- ions is as follows:

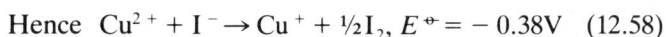

$$Cu^{2+} + e^- \rightarrow Cu^+, \qquad E^{\oplus} = +0.16V \quad (12.56)$$

$$I^- - e^- \rightarrow \tfrac{1}{2}I_2, \qquad E^{\oplus} = -0.54V \quad (12.57)$$

Hence $Cu^{2+} + I^- \rightarrow Cu^+ + \tfrac{1}{2}I_2, E^{\oplus} = -0.38V$ (12.58)

which is not favourable as it stands. If, however, Cu^+ ions

are removed from solution as insoluble CuI

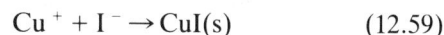

$$Cu^+ + I^- \rightarrow CuI(s) \qquad (12.59)$$

the overall process becomes

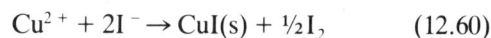

$$Cu^{2+} + 2I^- \rightarrow CuI(s) + \tfrac{1}{2}I_2 \qquad (12.60)$$

for which $E^{\oplus}$ is $+0.34V$. To illustrate the second possibility, if copper metal and a cupric salt are heated together in water no reaction occurs, but if HCl is added reduction of the Cu^{2+} and oxidation of the Cu is achieved, giving the $[CuCl_2]^-$ complex:

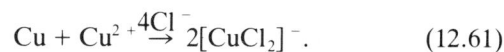

$$Cu + Cu^{2+} \overset{4Cl^-}{\rightarrow} 2[CuCl_2]^-. \qquad (12.61)$$

Once again removal of Cu^+ ions is the reason for the reaction proceeding. If the acid solution of the complex is diluted with water, lowering the Cl^- concentration, insoluble CuCl forms. Similarly, Cu will react with HCl to give Cu(I) in the presence of thiourea, $(NH_2)_2 CS$, because the Cu^+ ions are removed as a stable thiourea complex. In all of these cases removal of Cu^+ ions from the solution gives rise to favourable $E^{\oplus}$ values.

Divalent copper is well known in a wide variety of blue octahedral complexes, such as $[Cu(H_2O)_4]SO_4H_2O$, $[Cu(NH_3)_4]SO_4H_2O$, and $Cu(NO_3)_2 3H_2O$, as well as yellow to green tetrahedral complexes such as the ion $[CuCl_4]^{2-}$. The octahedral complexes are generally distorted, with two long bonds at opposite poles and four short ones in a plane. In some cases the distortion is such that the compounds may be regarded as square planar (Table 12.16).

Copper (II) acetate has the same stoichiometry and structure as the Cr(II) acetate (Fig. 12.21). The principal difference is that the Cu—Cu distance of 264 pm is longer than in the metal (Cu—Cu in metallic Cu is 255 pm) and therefore corresponds to only a weak Cu—Cu interaction.

Table 12.16 Distortion of octahedral copper (II) complexes

Compound	Bond lengths (pm)	
$CuCl_2 2H_2O$	Two Cu—Cl 231	Two Cu—Cl 298
	Two Cu—O 201	
$CuCl_2$	Four Cu—Cl 230	Two Cu—Cl 295
$[Cu(NH_3)_4(H_2O)_2]^{2+}$	Four Cu—N 205	One Cu—O 259
		One Cu—O 337
CuF_2	Four Cu—F 193	Two Cu—F 227

(i) Zinc

The metal zinc is not really a transition metal because it has filled 3d and 4s shells and does not show the typical properties of a transition metal, although it does form complexes. Some reactions of zinc are given in Fig. 12.33.

Calcium and zinc have the same outer electron configuration, namely $(4s)^2$, but zinc has the additional $3d^{10}$ electrons. There is a contraction of atomic size as we go across the first transition metal series. The Zn^{2+} ion is smaller than Ca^{2+} (74 and 99 pm radius respectively), and as a consequence the hydration energy of Zn^{2+} is greater than that of Ca^{2+} (2017 and 1564 kJ mol^{-1} respectively). Because of the $3d^{10}$ shell, the Zn^{2+} ion is more polarizable than Ca^{2+}, which leads to covalent character in its bonds.

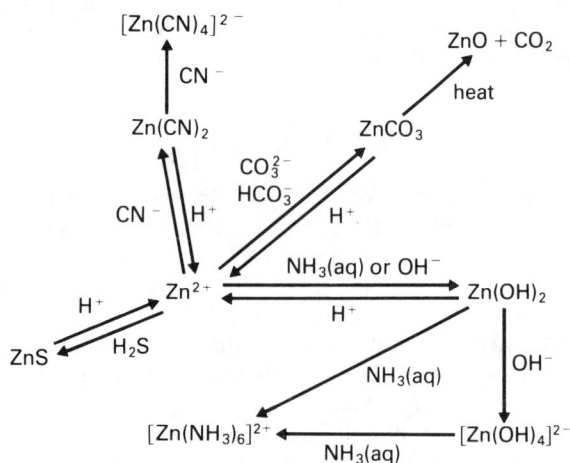

Fig. 12.33 Some reactions of the Zn^{2+} ion

The white oxide ZnO loses oxygen and becomes yellow on heating, with formation of a defect lattice. A small proportion of ions are lost from the solid structure, and generally more O^{2-} ions are missing than Zn^{2+}, i.e. the oxide is also non-stoichiometric, typically $Zn_{1.00033}O_{1.00}$. Some zinc ions in the structure must be reduced to Zn^+ to maintain electrical neutrality. The oxide is amphoteric:

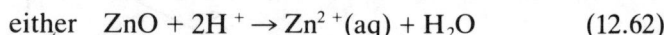

either $ZnO + 2H^+ \rightarrow Zn^{2+}(aq) + H_2O$ (12.62)

or $ZnO + H_2O \rightarrow Zn(OH)_2$ (12.63)

and $Zn(OH)_2 + 2OH^- \rightarrow Zn(OH)_4^{2-}$ (12.64)

Zinc is important in biological systems where it occurs in association with various enzymic proteins.

Exercises

12.1 Explain the trend in the metallic radius of the first row transition metals.

12.2 What effect does this trend have on the following p or s block elements?

12.3 Plot the atomization energies of the first row transition metals against atomic number. Discuss the reasons for the shape of the curve. Does atomization bear any relationship to hardness of the metal?

12.4 Explain why the transition metals have a wide range of oxidation states.

12.5 Why is the 6th ionization energy of vanadium so large?

12.6 W is used for the filaments of lamps, WC is used in the tips of drills, and Cr is used to coat iron. What characteristic of each of these materials is being made use of in these applications?

12.7 What happens when aqueous ammonia is added to: $FeSO_4$, $CuSO_4$, $NiSO_4$ and $Cr_2(SO_4)_3$? Write equations.

12.8 Write the electron configurations for Fe, Co(III), Cu(I), Zn(II), Cr, Mn(VII).

12.9 Write balanced equations for:

$$MnO_4^- + H_2O_2 \rightarrow$$
$$Cu^{2+}(aq) + Cu + I^-(aq) \rightarrow$$
$$Co^{2+}(aq) + NH_3(aq) + O_2 \rightarrow$$
$$Au + CN^-(aq) + O_2 \rightarrow$$
$$Fe(OH)_2 + O_2 \rightarrow$$

12.10 Discuss the nature of transition metal complexes, in particular the nature of the metal, the donor atoms, and the type of bonding.

12.11 What is a ligand? Give a range of examples.

12.12 What is the influence of six ligands symmetrically arranged around a metal ion on the d orbitals of the metal.

12.13 Discuss the origin of (a) colour and (b) paramagnetism in transition metal complexes.

12.14 List, by means of structural diagrams, the various isomers possible for: $[Co(L—L)_3]$, $[Co(L—L)_2X_2]$, $[PtX_2L_2]$, $[CoX_2L_4]$ ((L—L) is a bidentate ligand).

12.15 Discuss some of the more important industrial uses of Fe and Zn.

12.16 Name the following compounds: $K_3[Fe(CN)_6]$, $[Cr(H_2O)_6]Cl_3$, $[Co(NH_3)_4Cl_2]Cl$, Cs_2CuCl_4.

12.17 Discuss the industrial isolation of titanium metal. Give the chemical equations involved.

12.18 List the range of oxidation states of vanadium, and suggest chemical ways of reducing vanadium (V) to vanadium (II) including intermediate steps.

12.19 What is the structure of the dichromate ion?

12.20 Write the ion–electron equation for the reduction of dichromate in acid.

12.21 What use is made of Ag_2CrO_4 in chemical analysis?

12.22 Discuss the nature of chromium peroxy-compounds.

12.23 Which materials, in acid conditions, will be oxidised by dichromate? Cl^-, Fe^{2+}, H_2S (to S), Cu (to Cu^{2+}), H_2O; write equations

12.24 Giving your reasons list the *important* oxidation states of Cr and Mn, and give an example of each.

12.25 Write equations representing the reduction of MnO^{4-} in (a) acid conditions (b) neutral conditions.

12.26 Discuss the formation of the MnO_4^{2-} species and its stabilization.

12.27 Why is MnO_2 so stable?

12.28 What is the origin of the intermediate cherry red colour when MnO_4^- is reduced with the oxalate ion?

12.29 Is it possible to use HCl instead of H_2SO_4 when MnO_4^- is used as an oxidizing agent in acid condtions? Write the appropriate equations.

12.30 What are the structures of the iron oxides FeO, αFe_2O_3 and Fe_3O_4.

12.31 Discuss the formation of prussian blue. What is the origin of the intense blue colour?

12.32 Why is it necessary to stabilize Fe^{2+} in aqueous solution with acid?

12.33 Discuss the relative merits of the various ways of protecting iron from corrosion.

12.34 Why is Co(III) more stable than Co(II) in the presence of ligands such as NH_3.

12.35 Explain why $Co_2(CO)_8$ is dimeric and has a Co—Co bond.

12.36 Suggest what happens when, after sufficient concentrated HCl is added to a pink solution of Co^{2+} in water, the solution turns blue.

12.37 Describe an experimental way of distinguishing the complexes

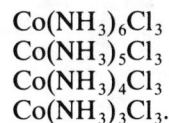

$$Co(NH_3)_6Cl_3$$
$$Co(NH_3)_5Cl_3$$
$$Co(NH_3)_4Cl_3$$
$$Co(NH_3)_3Cl_3.$$

Draw out the structures.

12.38 Why can Cu^+ be stabilized in the presence of I^- and Cl^-?

13 Compounds of Carbon and Hydrogen

13.1 Introduction to Organic Chemistry

Organic chemistry is the chemistry of the compounds of carbon. The term *organic* had its genesis in the beginnings of modern chemistry, at the start of the nineteenth century. Organic compounds were said to be those created by living organisms. At that time most scientists believed that all organic materials must have their origin in living matter; it was thought that some *vital force* was required in their production, and that this force could be provided only by a living organism. Today, after more than a century of development in the field of organic chemistry, it is generally accepted that it is possible to synthesize even the most complicated organic molecule in the laboratory. Of course, the task of synthesis may be a herculean one. It is convenient to continue to use the term 'organic' to refer specifically to carbon compounds partly because the number of such compounds is many times greater than the number of compounds which do not contain carbon. Over two million organic compounds are known and each year many thousands of new compounds are made or isolated from natural sources. There is a tremendous diversity of organic molecules ranging from very simple species, such as methane (CH_4), to giant molecules containing thousands of atoms. It can be argued that the diversity of the compounds which contain carbon is the most distinctive feature of organic chemistry.

Nature provides a spectacular array of unique carbon-containing compounds, whose identification is a challenge to the ingenuity of scientists. There is a continuous programme by which naturally occurring materials are screened for compounds of use to man, particularly for compounds of medicinal use. Each new organic compound, whether isolated from a natural source or synthesized in a laboratory, has its own peculiar shape and identity and exhibits subtleties of behaviour which invite detailed study. Life on earth is inextricably identified with the chemistry of carbon, and the study of the organic chemistry of living matter is still in its infancy.

As another aspect of organic chemistry we may note that the production of chemicals is the world's fifth largest manufacturing industry. Organic chemicals currently constitute one quarter of this industry. The organic component includes the production of dyes, paints, inks, plastics, perfumes, pharmaceutical and agriculture products. Ink sales alone amount to half a billion dollars yearly in the USA and more than two billion tons of synthetic rubber are produced each year. The two major reservoirs for industrial organic materials are the fossil fuels, coal and petroleum, which have been accumulated over millenia, and deposits of these raw materials are not replaceable. While there are alternatives to fossil fuels as sources of energy it is more difficult to find suitable alternative sources of basic raw materials. Ethanol produced by the fermentation of sugars is one possibility, but nevertheless the present concern with energy may turn out to be somewhat trivial compared with the inevitable future raw material shortage.

The work of organic chemists is directed primarily towards three major goals, namely:

1. structure determination—What is it?

2. synthesis—How do I make it?

3. molecular dynamics—How is it transformed?

The structure of even very complex molecules can be determined; for example the structures of chlorophyll *a,* a key substance in the process of photosynthesis by plants, and vitamin B_{12} are shown in Fig. 13.1. The elucidation of the structure of chlorophyll *a* was the culmination of years of work, using chemical techniques now regarded as classical; the structure of vitamin B_{12} was determined by modern X-ray crystallographic techniques. Both molecules have since been synthesized in the laboratory by R. B. Woodward and his colleagues. Despite the complexity of these molecules it is not difficult to recognize groupings of two or more atoms which recur frequently. Such aggregates of atoms are termed *functional groups* and their presence confers particular chemical properties on the molecule. The organic chemist understands complex molecules by considering the behaviour of their component functional groups. A number of examples of functional groups are shown in Table 13.1.

The synthesis of complex molecules requires the utiliza-

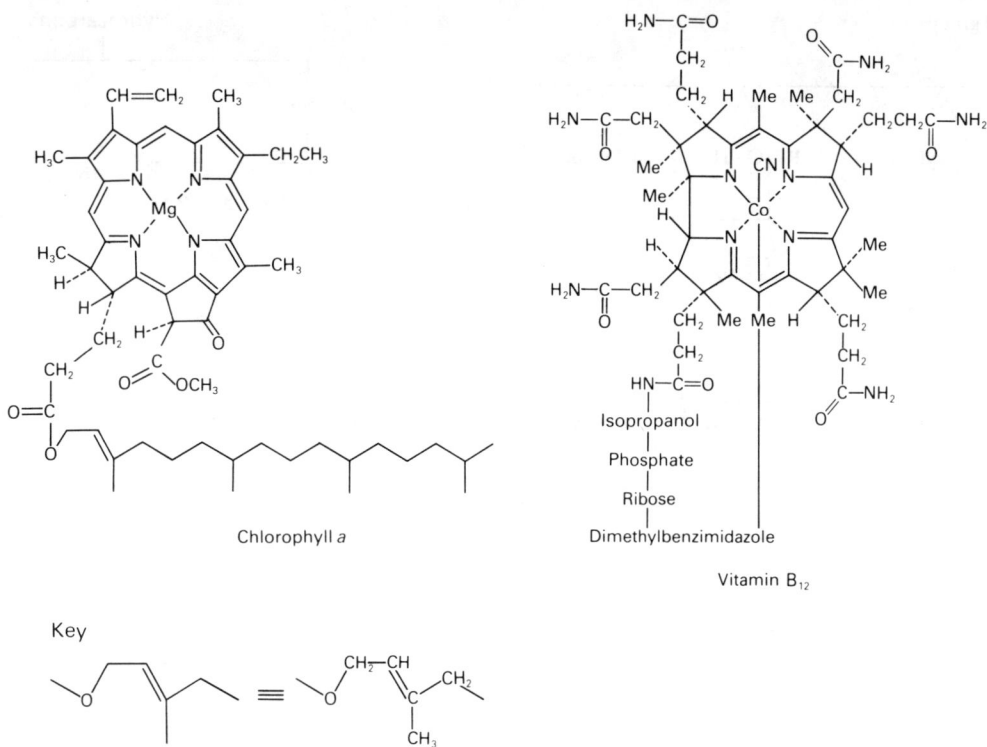

Chlorophyll *a*

Vitamin B$_{12}$

Key

Fig. 13.1 Structures of chlorophyll *a* and vitamin B$_{12}$

tion of the existing vast resources of chemical knowledge about reactions and reaction mechanisms: elucidation of the structure of complex molecules depends on resourceful and imaginative use of the many physical techniques available. X-ray crystallography provides essentially complete structural information, but takes a great deal of computer time and expensive facilities. The molecular formula and some information about a molecule's structure can be obtained from the high-resolution mass spectrum. The presence of a particular functional group in a molecule can often be inferred very rapidly from the molecule's infrared, ultraviolet or nuclear magnetic resonance spectrum. However, before we can begin to exploit the existing knowledge and techniques we must learn something of the basic chemistry of simple carbon compounds, as described in this and succeeding chapters. The assimilation of this information will allow plausible synthetic routes to simple compounds to be devised. This aspect of organic chemistry is emphasized in the problems found at the ends of chapters.

13.2 Alkanes: Structure and Nomenclature

Compounds which contain only the elements carbon and hydrogen are known as *hydrocarbons*. These compounds are described as either *aliphatic* or *aromatic*. Aliphatic compounds are further subdivided into families: namely *alkanes*, *alkenes* and *alkynes* (which exist both as open chains—*aliphatic* molecules—and rings—*alicyclic* molecules). Aromatic compounds have been described in Chapter 3 and have conjugated ring systems. The terminology is summarized in Fig. 13.2. Alkanes (sometimes called 'paraffins') are referred to as saturated hydrocarbons. In these compounds the tetracovalent nature of carbon is established by single bonds to carbon and hydrogen atoms. Alkenes contain a carbon–carbon double bond C=C and alkynes a carbon–carbon triple bond C≡C. Aromatic compounds are derivatives of the parent member, benzene.

The parent member of the alkane family is *methane*, CH$_4$ (Fig. 13.3). Replacement of one hydrogen of methane by a methyl (CH$_3$) group gives ethane, CH$_3$—CH$_3$, the second member of the alkane family. Propane, CH$_3$—CH$_2$—CH$_3$, is the third member of the series. Such a series of compounds is referred to as a homologous series. Alkanes have the general formula C$_n$H$_{2n+2}$ and any member differs from the next by one carbon and two hydrogen atoms.

The suffix for each of these compounds is the family ending 'ane' (Table 13.2).

Table 13.1 Functional groups

$CH_3-(CH_2)_n-$	alkyl

alkene

alkyne (R—C≡C—R)

R—OH   alcohol

R—O—R′   ether

R—NH$_2$ amine

R—X   halide

R—C≡N   nitrile

phenyl

aldehyde (R—C—H, ‖ O)

ketone (R—C—R, ‖ O)

carboxylic acid (R—C—OH, ‖ O)

ester (R—C—O—R′, ‖ O)

acid anhydride (R—C—O—C—R′, ‖ ‖ O O)

amide (R—C—NH$_2$, ‖ O)

acid halide (R—C—X, ‖ O)

X denotes F, Cl, Br or I
R, R′ denotes alkyl, phenyl or similiar grouping

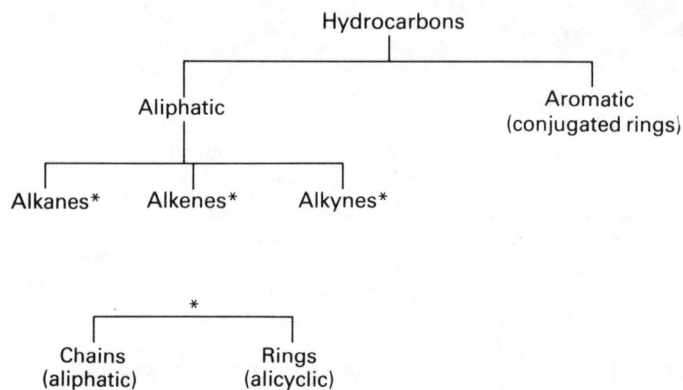

Fig. 13.2 The family of hydrocarbons

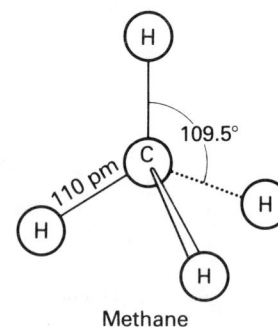

Fig. 13.3 3-dimensional representation of methane

Table 13.2 Alkanes

meth*ane*	CH_4	hex*ane*	C_6H_{14}
eth*ane*	C_2H_6	hept*ane*	C_7H_{16}
prop*ane*	C_3H_8	oct*ane*	C_8H_{18}
but*ane*	C_4H_{10}	non*ane*	C_9H_{20}
pent*ane*	C_5H_{12}	dec*ane*	$C_{10}H_{22}$

When a hydrogen of an alkane is replaced by a univalent substituent the compound is named by replacing the ending 'ane' of the hydrocarbon name with 'yl'. The class name for such groups is alkyl and several examples are given in Table 13.3. Numbering begins with the carbon having the free bond.

The hydrogen atoms surrounding the central carbon atom of methane are arranged at the corners of a regular tetrahedron, Fig. 13.3, corresponding to the sp^3 hybridization of the carbon valence orbitals (see Chapter 3). The arrangement provides maximum overlap of the bonding orbitals of carbon with the four 1s orbitals of the hydrogen atoms. The groups, bonded to each carbon in ethane remain at the

Table 13.3 Names of alkyl groups

CH_3-	methyl	$CH_3-CH_2-\overset{\displaystyle CH_3}{\overset{\displaystyle \vert}{CH}}-$	2-butyl (sec-butyl)
CH_3-CH_2-	ethyl		
$CH_3-CH_2-CH_2-$	1-propyl	$CH_3-\overset{\displaystyle CH_3}{\underset{\displaystyle CH_3}{\overset{\displaystyle \vert}{\underset{\displaystyle \vert}{C}}}}-$	2-methyl-2-propyl (tert-butyl)
$CH_3-\overset{\displaystyle CH_3}{\overset{\displaystyle \vert}{CH}}-$	2-propyl (isopropyl)		
$CH_3-CH_2-CH_2-CH_2-$	1-butyl		

corners of a tetrahedron and the molecules are free to rotate about the carbon–carbon bond. This rotation gives rise to different *conformations* of the molecule. Two conformations of the molecule are shown in Fig. 13.4 drawn end on (Newman projection) and on an angle (Saw horse projection). With all the hydrogens aligned the molecule is said to be in an *eclipsed* conformation, while with the hydrogens lying immediately between the adjacent hydrogens the conformation is said to be *staggered*. Conformations between these two extremes are said to be *skew* conformations. The hydrogen atoms on adjacent carbons are closest together in the eclipsed conformation, in which the potential energy of the system is at a maximum. The staggered conformation corresponds to a minimum of potential energy (Fig. 13.5). The energy difference between

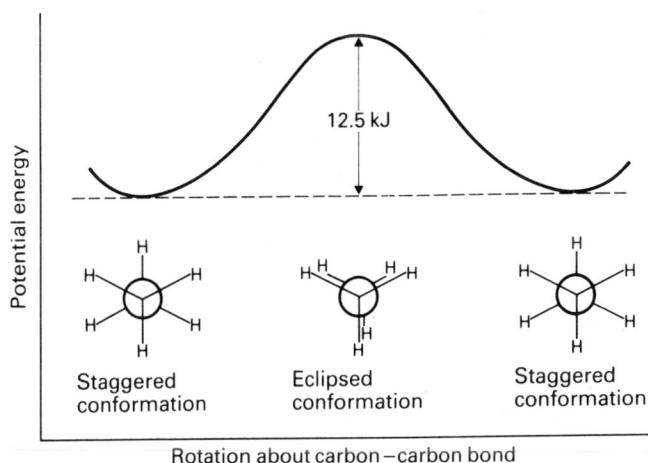

Fig. 13.5 Variation of the potential energy of ethane with rotation about the carbon–carbon bond

the staggered and eclipsed conformations is only 12.5 kJ mol^{-1} which at room temperature is too small a barrier to prevent free rotation about the carbon–carbon bond. However, the combined effect of many such barriers is commonly the major factor in deciding the shape of a large aliphatic or alicyclic molecule. Propane, the third member of the series, is often written $CH_3-CH_2-CH_3$ and referred to as a straight chain hydrocarbon. This is somewhat misleading, since with tetrahedral bond angles the carbon chain is bent, as shown in Fig. 13.6. As with ethane the potential energy of the molecule will depend on its rotational conformation. The barrier to rotation about a carbon–carbon bond is expected to be slightly higher for propane than for ethane, because a methyl group is larger than a hydrogen atom. In fact, the eclipsed conformation is 13.8 kJ mol^{-1} higher in energy than the staggered conformation.

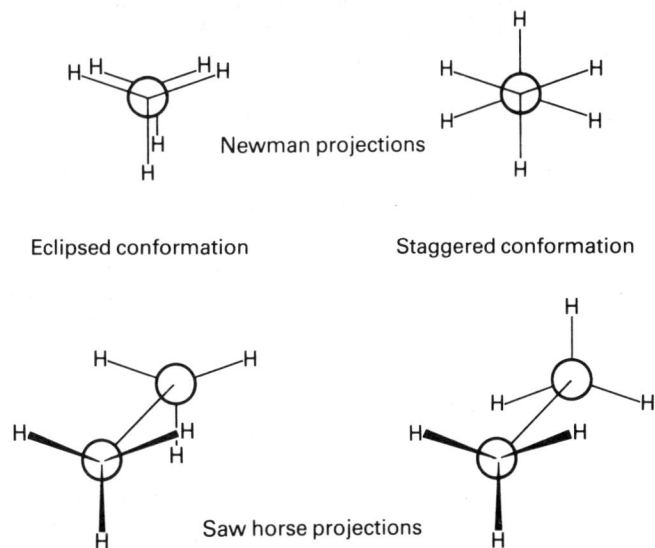

Fig. 13.4 Rotational conformations of ethane

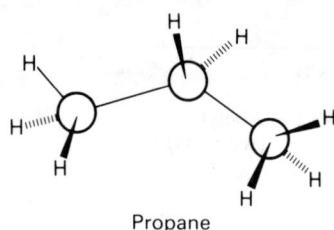

Propane

Fig. 13.6 3-dimensional representation of propane

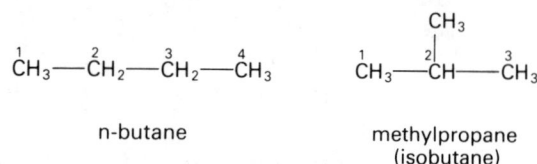

n-butane

methylpropane
(isobutane)

Fig. 13.7 Structural isomers of butane

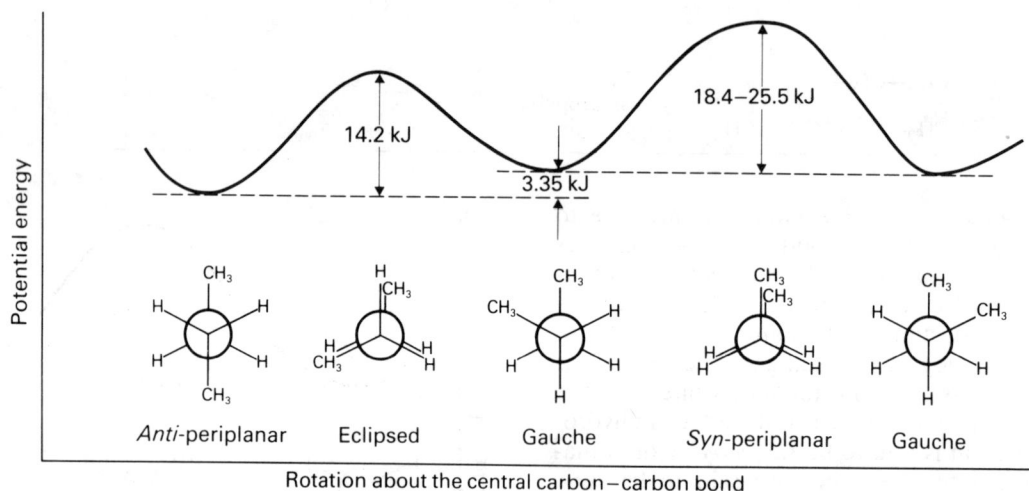

14.2 kJ

18.4–25.5 kJ

3.35 kJ

Potential energy

Anti-periplanar   Eclipsed   Gauche   *Syn*-periplanar   Gauche

Rotation about the central carbon–carbon bond

Fig. 13.8 Dependence of potential energy of butane on rotation about the carbon–carbon bond

Butane (C_4H_{10}), the next alkane, is interesting because there are two possible ways of arranging four carbons and ten hydrogen atoms which satisfy all the valence requirements of carbon and hydrogen. Consequently there are two compounds having the molecular formula C_4H_{10}, each with its own distinct physical properties. These compounds having the same formula but differing in their bonding arrangements, are termed *constitutional or structural isomers*. The two isomers of butane are shown in Fig. 13.7. The straight chain isomer, n-butane, has a considerable variety of possible conformations as shown in Fig. 13.8. The conformation is described using the terms; *anti*-periplanar, eclipsed, gauche, and *syn*-periplanar. The branched isomer methyl propane (Fig. 13.7) is named as a derivative of the longest or parent chain (propane); preceding this is the substituting methyl group (CH_3—) joined to C(2). It is not necessary to include a number to define the position of the methyl since the name methylpropane defines completely the constitutional isomer.

The next alkane, *pentane*, has the molecular formula C_5H_{12}. There are three possible ways to arrange five carbon and twelve hydrogen atoms and satisfy the bonding requirements of the atoms, giving the three constitutional isomeric pentanes shown in Fig. 13.9.

Hexane, the next homologue, has the molecular formula C_6H_{14} and exists as five different constitutional isomers (Fig. 13.10). As we ascend the series of alkanes we find that each successive member can exist as an increasing number of constitutional isomers, and each of these isomers must be given a distinctive, unambiguous name. For the straight chain hydrocarbons the prefix n- is used. To arrive at a systematic name for a more complicated structure we first select the longest chain of carbon atoms. This becomes the 'parent' chain, and is named according to the number of carbons in the chain. For example, in the molecule of Fig. 13.11 the longest chain has six carbon atoms and the molecule is therefore a derivative of the parent molecule hexane. The atoms of the chain are numbered from the end which will result in substituent groups being attached to carbon atoms which carry the lowest possible numbers. If the same alkyl group or other substituent occurs more than once at any position in the chain this is indicated by the prefix *di, tri* etc. The molecule in Fig. 13.11 is named 3,3,4-trimethyl-hexane since the longest chain is a hexane chain and there are three methyl(CH_3—) substituents, two at carbon 3 and one at position 4. If the substituents are different they are ordered alphabetically in the compound name. For hydrocarbons which contain a double or triple bond the molecule

$$CH_3-CH_2-CH_2-CH_2-CH_3$$

n-pentane

$$CH_3-CH_2-\underset{\underset{\displaystyle CH_3}{|}}{CH}-CH_3$$

methylbutane
(isopentane)

$$CH_3-\underset{\underset{\displaystyle CH_3}{|}}{\overset{\overset{\displaystyle CH_3}{|}}{C}}-CH_3$$

dimethylpropane
(neopentane)

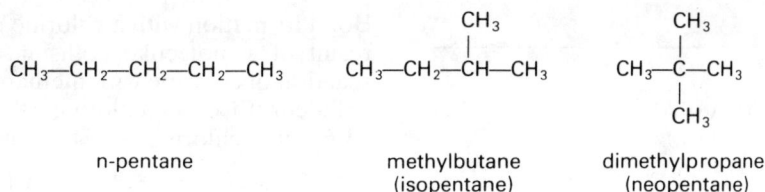

Fig. 13.9 Structural isomers of pentane

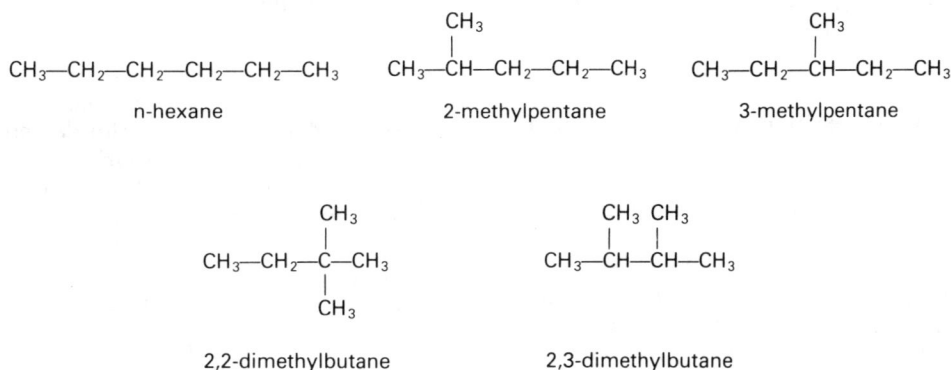

$$CH_3-CH_2-CH_2-CH_2-CH_2-CH_3$$

n-hexane

$$CH_3-\underset{\underset{\displaystyle CH_3}{|}}{CH}-CH_2-CH_2-CH_3$$

2-methylpentane

$$CH_3-CH_2-\underset{\underset{\displaystyle CH_3}{|}}{CH}-CH_2-CH_3$$

3-methylpentane

$$CH_3-CH_2-\underset{\underset{\displaystyle CH_3}{|}}{\overset{\overset{\displaystyle CH_3}{|}}{C}}-CH_3$$

2,2-dimethylbutane

$$CH_3-\underset{\underset{\displaystyle CH_3}{|}}{CH}-\underset{\underset{\displaystyle CH_3}{|}}{CH}-CH_3$$

2,3-dimethylbutane

Fig. 13.10 Structural isomers of hexane

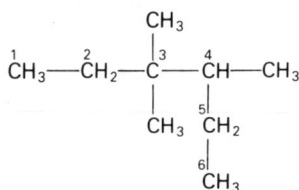

$$\overset{1}{CH_3}-\overset{2}{CH_2}-\underset{\underset{\displaystyle CH_3}{|}}{\overset{3}{\overset{\overset{\displaystyle CH_3}{|}}{C}}}-\overset{4}{\underset{\underset{\displaystyle \overset{5}{CH_2}}{|}}{CH}}-CH_3$$
$$\underset{\underset{\displaystyle \overset{6}{CH_3}}{|}}{}$$

Fig. 13.11 3,3,4-trimethylhexane

is named in the same way, but with the ending -ene or -yne in place of -ane (see Figs. 14.2 and 14.8). The longest chain containing the double or triple bond is referred to as the parent chain and the molecule is named accordingly as a derivative of that parent.

13.3 Properties and Reactions of Methane

Methane is the main component of natural gas and its availability makes it an important industrial raw material. For this reason we will examine the chemistry of methane in some detail. At room temperature methane is a gas. The melting point is $-183°C$, and boiling point $-161.5°C$. Liquid methane is colourless. Like all alkanes it is sparingly soluble in water, but is soluble in organic solvents. Methane

burns in air and once the reaction is initiated it gives off considerable heat (Eq. 13.1). Controlled partial oxidation of methane (reaction 13.2) provides acetylene, a valuable industrial chemical. High temperature catalytic reaction of methane (reaction 13.3) with water is a source of hydrogen. Methane reacts with chlorine, when heated to above 250°C in the dark and at room temperature in the presence of light, to yield hydrogen chloride and chloromethane CH_3Cl (Eq. 13.4).

$$CH_4 + Cl_2 \xrightarrow[\text{light}]{\text{heat or}} CH_3Cl + HCl \qquad (13.4)$$

In this reaction one of the hydrogens of methane has been replaced, or *substituted*, by a chlorine atom and such a reation is referred to as a *substitution* reaction. Chloromethane can itself react with chlorine to give dichloromethane which can react further to give trichloromethane (chloroform). The ultimate chlorinated product is tetrachloromethane (carbon tetrachloride) (scheme 13.5)

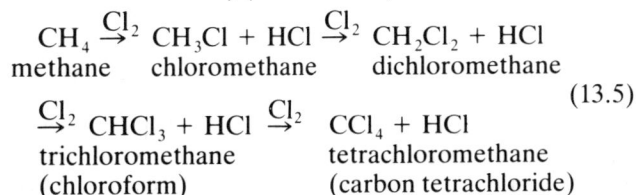

$$CH_4 \overset{Cl_2}{\rightarrow} CH_3Cl + HCl \overset{Cl_2}{\rightarrow} CH_2Cl_2 + HCl$$
methane   chloromethane   dichloromethane

$$\overset{Cl_2}{\rightarrow} CHCl_3 + HCl \overset{Cl_2}{\rightarrow} CCl_4 + HCl$$
trichloromethane   tetrachloromethane
(chloroform)   (carbon tetrachloride)

$$(13.5)$$

Table 13.4 Reactions of Methane

Oxidation

$$CH_4 + 2O_2 \xrightarrow{\text{burn}} CO_2 + 2H_2O + \text{heat (891 kJ)} \qquad (13.1)$$

$$6CH_4 + O_2 \xrightarrow{1500°C} 2CH\equiv CH + 2CO + 10H_2 \qquad (13.2)$$

$$CH_4 + H_2O \xrightarrow[\text{Ni}]{850°C} CO + 3H_2 \qquad (13.3)$$

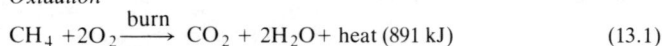

Halogenation

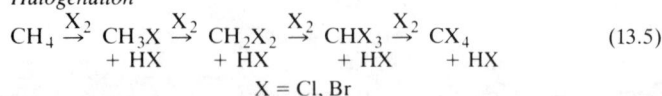

$$CH_4 \xrightarrow{X_2} CH_3X \xrightarrow{X_2} CH_2X_2 \xrightarrow{X_2} CHX_3 \xrightarrow{X_2} CX_4 \qquad (13.5)$$
$$\quad + HX \quad + HX \quad + HX \quad + HX$$
$$X = Cl, Br$$

It is not easy to arrange the reaction conditions in such a way as to obtain only a single *mono-*, *di-* or *tri*-chlorinated reaction product. The mono-chlorinated product cannot be obtained simply by reacting one mole of chlorine with one mole of methane, since in the early stages of the reaction chlorine will react with methane to form chloromethane, but as the concentration of chloromethane builds up, the reaction of chlorine with chloromethane, producing dichloro-methane, will compete with the reaction of chlorine with methane. Similarly, as the concentration of dichloromethane builds up, it will undergo reaction with chlorine to give trichloromethane. It is therefore usual to obtain a mixture of unreacted methane with all four possible chlorinated products. To obtain chloromethane free of other chlorinated compounds, chlorination must be carried out in a vast excess of methane and the chloromethane separated from unreacted methane. It is not difficult to obtain tetrachloro-methane free of the other chlorinated products since methane can be made to react with an excess of chlorine. Similar reactions occur between bromine and methane. Bromine is however less reactive than chlorine under the same reaction conditions.

The reaction of chlorine with methane is complex and we will now examine the *mechanism* of the reaction—that is, the detailed step by step changes which occur as the reaction proceeds. The first step in the chlorination of methane is the formation of chlorine atoms (Eq. 13.6) by thermal or photochemical fission of the chlorine–chlorine bond. Cleav-age of the chlorine–chlorine bond requires energy (243 kJ mol^{-1}) which can be provided in the form of heat or light. The energy required for such a process is equal to the bond dissociation energy.

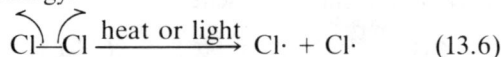

$$Cl \overgroup{\frown} Cl \xrightarrow{\text{heat or light}} Cl\cdot + Cl\cdot \qquad (13.6)$$

This is a *homolytic* process, i.e. one in which cleavage of the two-electron bond results in one electron being retained by each chlorine atom and is represented in Eq. 13.6 where a single headed arrow represents movement of an electron. The chlorine atoms, or radicals, are very reactive species.

Bond formation with a chlorine radical can only occur as a result of a molecular collision. At the beginning of the reaction of chlorine with methane the chlorine radicals can collide with other chlorine radicals (reverse of reaction 13.6), with chlorine (13.7) or with methane (13.8).

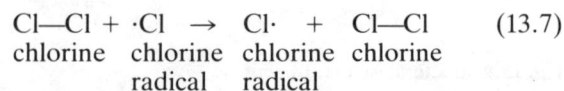

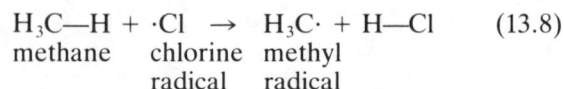

$$Cl—Cl + \cdot Cl \rightarrow Cl\cdot + Cl—Cl \qquad (13.7)$$
chlorine   chlorine   chlorine   chlorine
$\qquad$ radical $\quad$ radical

$$H_3C—H + \cdot Cl \rightarrow H_3C\cdot + H—Cl \qquad (13.8)$$
methane $\quad$ chlorine $\quad$ methyl
$\qquad\qquad$ radical $\quad$ radical

The reaction with chlorine (as represented in Eq. 13.7) is not productive. Reaction with methane as shown in Eq. 13.8 can produce hydrogen chloride and a methyl radical. The methyl radical formed in this process will collide most frequently with methane, the species present in highest concentration, but the reaction with methane is also not productive (Eq. 13.9). Collision of the methyl radical with chlorine as shown in Eq. 13.10 can produce methyl chloride and generate another chlorine radical.

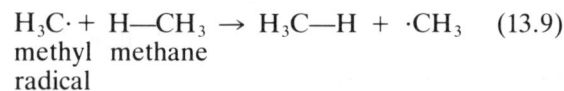

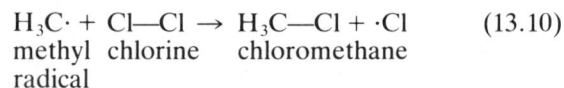

$$H_3C\cdot + H—CH_3 \rightarrow H_3C—H + \cdot CH_3 \qquad (13.9)$$
methyl $\quad$ methane
radical

$$H_3C\cdot + Cl—Cl \rightarrow H_3C—Cl + \cdot Cl \qquad (13.10)$$
methyl $\quad$ chlorine $\qquad$ chloromethane
radical

Thus the formation of methyl chloride involves the follow-ing processes:

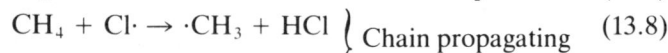

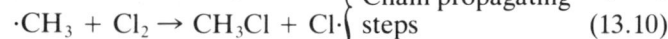

$$Cl_2 \rightarrow 2Cl\cdot \qquad\qquad \text{Initiation process} \qquad (13.6)$$

$$\left.\begin{array}{l} CH_4 + Cl\cdot \rightarrow \cdot CH_3 + HCl \\ \cdot CH_3 + Cl_2 \rightarrow CH_3Cl + Cl\cdot \end{array}\right\} \begin{array}{l}\text{Chain propagating} \quad (13.8)\\ \text{steps} \qquad\qquad\quad (13.10)\end{array}$$

In the formation of one molecule of chloromethane (Eq. 13.4) one chlorine radical has been required to produce the methyl radical (reaction 13.8) and a chlorine radical has been produced from the reaction of methyl radical with chlorine (reaction 13.10). The overall process can there-fore repeat itself since the chlorine radical used in reaction 13.8 is replaced in reaction 13.10. Such a self propagating reaction is referred to as a *chain reaction*. Steps 13.8 and 13.10 are repeated over and over again until such time as one of the following processes 13.11, 13.12 or 13.13 termi-nates the chain.

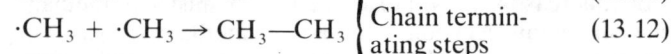

$$\left.\begin{array}{l} Cl\cdot + Cl\cdot \rightarrow Cl—Cl \\ \cdot CH_3 + \cdot CH_3 \rightarrow CH_3—CH_3 \\ \cdot CH_3 + Cl\cdot \rightarrow CH_3—Cl \end{array}\right\} \begin{array}{l}\text{Chain termin-}\\ \text{ating steps}\end{array} \begin{array}{l}(13.11)\\ (13.12)\\ (13.13)\end{array}$$

It is usual to find steps 13.8 and 13.10 being repeated on

the average several thousand times before the sequence is finally terminated.

While the overall reaction which results in formation of chloromethane from methane can be expressed by the equation

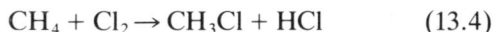

$$CH_4 + Cl_2 \rightarrow CH_3Cl + HCl \qquad (13.4)$$

the reaction in fact involves many steps. Each step other than initiation (13.6) requires a collision between reacting species.

In this overall reaction process (Eq. 13.4) a CH_3—H and a Cl—Cl bond have been broken, while a CH_3—Cl and an H—Cl bond have been formed. The energy change involved can be estimated from the bond dissociation energies which are given in Table 13.5.

Table 13.5 Bond dissociation energies in kJ mol^{-1}

CH_3—H	435		
Cl—Cl	243	Br—Br	193
CH_3—Cl	352	CH_3—Br	293
H—Cl	431	H—Br	368

A total of $435 + 243 = 678$ kJ of energy is required to break the CH_3—H and Cl—Cl bonds while a total of $352 + 431 = 783$ kJ of energy is gained by the system in the forming of the new CH_3—Cl and H—Cl bonds. The reaction is exothermic in that $783 - 678 = 105$ kJ of energy is released ($\Delta H = -105$ kJ) by the bond changes which occur in the reaction. Even though the overall reaction is exothermic it cannot start until chlorine is dissociated into chlorine radicals, and this process requires 243 kJ of energy. This explains why chlorine and methane do not react when kept in the dark at room temperature. However, once the reaction has been initiated by supplying to the system the energy required to produce chlorine radicals, the chain reaction will propagate and a considerable amount of energy in the form of heat will be dissipated to the surroundings.

13.4 Sources, Preparation and Reactions of Alkanes

Natural gas and petroleum are at present the chief sources of aliphatic hydrocarbons and are found trapped by rock structures near the surface of the earth. These products are believed to be the result of decay of the complex organic molecules derived from animal and plant material subjected to high pressure and the effects of bacteria and metal ions over a period of millions of years. Petroleum from various areas of the world varies in its constituents. Some are particularly rich in cycloalkanes (naphthenes) such as cyclohexane, methylcyclohexane, methylcyclopentane and dimethylcyclopentane. Coal is the major source of aromatic

hydrocarbons; these compounds can be reduced by the addition of hydrogen to form alicyclic compounds. Natural gas contains lower relative molecular mass alkanes: methane, ethane, propane and butane. Alkanes are non-polar covalent molecules, and whatever bond polarity does exist tends to be cancelled because of the symmetry of the molecules. The forces holding these non-polar molecules together are weak Van der Waals' forces which act only between portions of molecules in close contact. The larger molecules have a greater surface area and hence the forces between the molecules are greater. The lower relative molecular mass alkanes are gases while the intermediate ones are liquids and the higher relative molecular mass compounds are solids. Petroleum is separated by distillation into various fractions as shown in Table 13.6.

Table 13.6 Petroleum fractions

	Distillation temperature °C	Number of carbon atoms
Gas	20°	C_1–C_4
Petroleum ether	20–60°	C_5–C_6
Ligroin	60–100°	C_6–C_7
Natural gasoline	40–205°	C_5–C_{10} and cycloalkanes
Kerosene	175–300°	C_{10}–C_{16} and aromatics
Gas oil	275°	C_{16}–C_{25}
Lubricating oil and wax	Non-soluble liquids	Long chain and complex cyclic products
Asphalt and petroleum coke	Non-soluble solids	Polycyclic structures

Gas fractions are used for heating, petroleum ether and ligroin are used as solvents, and gasoline is used in internal combustion engines. Gasoline, a mixture of volatile hydrocarbons, is vapourized as it enters a motor cylinder. The straight chain alkanes are prone to pre-ignition in the cylinder, which leads to extra wear on the engine as a result of 'knocking' or 'pinking'. Highly branched alkanes, cyclic alkanes and aromatic hydrocarbons are less susceptible to knocking and are more suitable as engine fuels. Isomerization processes are therefore used to improve the properties of natural petroleum alkane fractions. In order to measure the suitability of a particular alkane mixture or compound for use as a fuel, an octane number scale has been established. Two arbitrary reference points are used. n-Heptane, a poor fuel, is given an octane number of 0, and 2,2,4-trimethyl-pentane (iso-octane) is given a number 100. The gasoline fraction of crude oil has an octane number less than 60. Since modern car engines function best on fuels having octane numbers approaching 100, this means that natural gasoline must be increased in octane rating by adding branched chain or cyclic alkanes and aromatic hydrocarbons or by introducing certain additives. Lead tetraethyl has been widely used as an additive for increasing octane number. In the combustion chamber it decomposes to give fine particles of lead or lead oxide which suppress chain reactions which would otherwise cause the mixture to begin to ignite before the spark. The atmospheric pollution caused by the added lead compound is considered undesirable and the allowable levels of lead additives in fuels are gradually being reduced.

Kerosene is used in jet engines, and gas oil in diesel engines. These latter two petroleum fractions are also used as furnace oil. Coke is mostly carbon, but can be converted into hydrocarbons. Calcium carbide is formed by heating coke with calcium oxide (Eq. 13.14) and reacts with water to form acetylene (reaction 13.15).

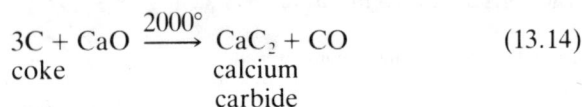

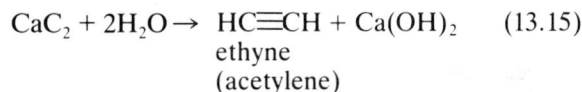

$$3C + CaO \xrightarrow{2000°} CaC_2 + CO \qquad (13.14)$$
$$\text{coke} \qquad\qquad \text{calcium}$$
$$\text{carbide}$$

$$CaC_2 + 2H_2O \rightarrow HC{\equiv}CH + Ca(OH)_2 \qquad (13.15)$$
$$\text{ethyne}$$
$$\text{(acetylene)}$$

This method of producing acetylene is virtually obsolete and it is now made from methane (reaction 13.16)

$$2CH_4 \xrightarrow{1500°C} HC{\equiv}CH + 3H_2 \qquad (13.16)$$

Acetylene is a particularly important industrial compound because it can be readily converted into many different organic compounds. One of the most famous uses is for the production of chloroethene (vinyl chloride) (reaction 13.17) the starting material for PVC (polyvinyl chloride) plastics.

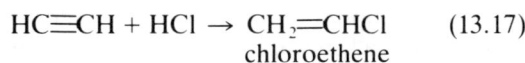

$$HC{\equiv}CH + HCl \rightarrow CH_2{=}CHCl \qquad (13.17)$$
$$\text{chloroethene}$$

In the laboratory alkanes can be prepared by reactions 13.18–13.23. Most of these reactions will be discussed later in the text and are listed at this stage for the sake of completeness in Table 13.7.

Hydrocarbons are used primarily as a source of energy, but are also the primary source of material for the synthesis of organic chemicals (Table 13.8). Petroleum fractions can be converted into other hydrocarbon compounds. *Cracking* processes convert higher alkanes to smaller alkanes and alkenes (reaction 13.26). Cracking of compounds involves pyrolysis or thermal degradation of the compound and is generally considered to involve the homolytic cleavage of C—C bonds. *Catalytic isomerization* changes straight chain alkanes into branched chain compounds and is shown in Eq. 13.28. *Catalytic reforming* converts alkanes and cyclo-alkanes into aromatic hydrocarbons.

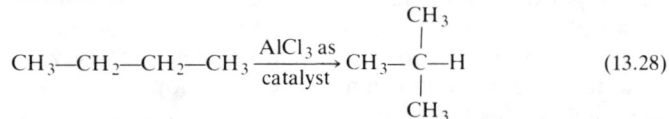

Table 13.7 Preparation of alkanes

Hydrogenation of alkenes
$$C_nH_{2n} \xrightarrow[\text{Pt or Pd}]{H_2} C_nH_{2n+2} \qquad (13.18)$$

Hydrolysis of a Grignard reagent
$$R{-}MgX \xrightarrow{H_2O} R{-}H \qquad (13.19)$$

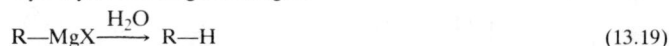

Reduction of a haloalkane
$$R{-}X \xrightarrow{Zn,H^+} R{-}H \qquad (13.20)$$

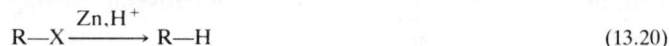

Coupling of haloalkanes with organometallic compounds
$$R{-}X \longrightarrow RLi \longrightarrow R_2CuLi \xrightarrow{R'{-}X} R{-}R' \qquad (13.21)$$
(p., sec., or   lithium   (primary)
tert., alkyl   dialkyl
lithium)   copper

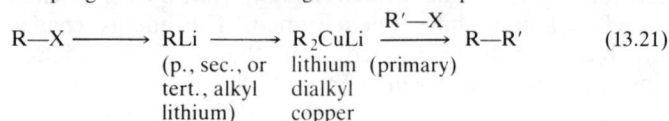

Reduction of ketones
$$R{-}\underset{\underset{O}{\|}}{C}{-}R' \xrightarrow{Zn/Hg/HCl} R{-}CH_2{-}R' \qquad (13.22)$$
$$\xrightarrow{N_2H_4/OH^-} R{-}CH_2{-}R' \qquad (13.23)$$

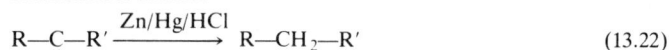

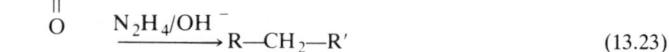

Table 13.8 Reactions of alkanes

Halogenation
$$R{-}\underset{\underset{R}{|}}{\overset{\overset{R}{|}}{C}}{-}H + X_2 \xrightarrow[\text{or light}]{\text{heat}} R{-}\underset{\underset{R}{|}}{\overset{\overset{R}{|}}{C}}{-}X + HX \qquad (13.24)$$
usually a mixture

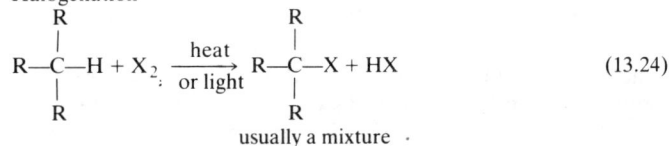

Reactivity $X_2 : Cl_2 > Br_2$
C—H : tertiary > secondary > primary > CH_3—H

Combustion
$$C_nH_{2n+2} + \frac{3n+1}{2} O_2 \text{ (excess)} \xrightarrow{\text{flame}} nCO_2 + (n+1)H_2O + \text{heat} \qquad (13.25)$$

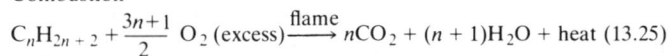

Pyrolysis or cracking
$$\text{alkane} \xrightarrow{400\text{–}900°} CH_2 + \text{smaller alkanes} + \text{alkenes} \qquad (13.26)$$

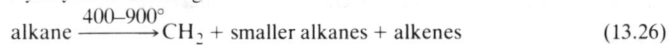

Alkylation
$$\underset{R}{\overset{R}{>}}C{=}C\underset{R}{\overset{R}{<}} + R{-}H \xrightarrow[25°C]{\text{conc. }H_2SO_4} R{-}\underset{\underset{H}{|}}{\overset{\overset{R}{|}}{C}}{-}\underset{\underset{R}{|}}{\overset{\overset{R}{|}}{C}}{-}R' \qquad (13.27)$$

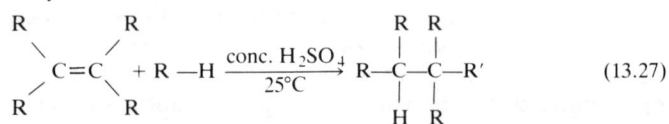

Isomerization
$$CH_3{-}CH_2{-}CH_2{-}CH_3 \xrightarrow[\text{catalyst}]{AlCl_3 \text{ as}} CH_3{-}\underset{\underset{CH_3}{|}}{\overset{\overset{CH_3}{|}}{C}}{-}H \qquad (13.28)$$

Methane is the source of hydrogen cyanide (reaction 13.29)

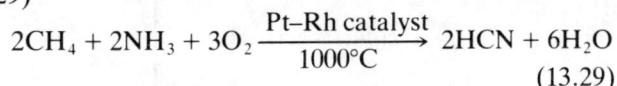

$$2CH_4 + 2NH_3 + 3O_2 \xrightarrow[1000°C]{\text{Pt–Rh catalyst}} 2HCN + 6H_2O$$

(13.29)

which is used extensively for the manufacture of cyanoethene (acrylonitrile) $CH_2{=}CHCN$, 1,4-dicyanobutane (adiponitrile) $NC(CH_2)_4CN$, and methyl 2-methyl-2-propenoate (methyl methacrylate) $CH_2{=}C(CH_3)CO_2CH_3$ which are used to make plastics and artificial fibres. Methane is used for production of chloromethane and dichloromethane. It is burnt to give carbon black, which is compounded in rubber to increase wear, and it is also converted into ethyne.

Ethane, propane, and butane are principal raw materials used for the manufacture of ethene (ethylene) (reaction 13.30), propene (13.31) and butadiene (13.32)

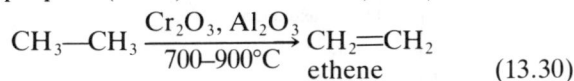

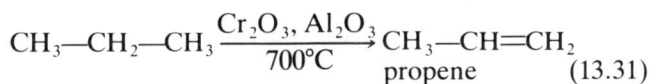

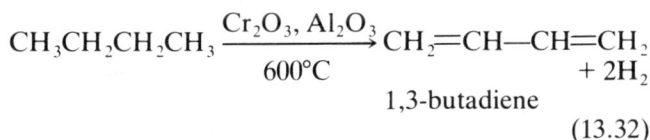

$$CH_3{-}CH_3 \xrightarrow[700-900°C]{Cr_2O_3, Al_2O_3} CH_2{=}CH_2 \quad \text{ethene}$$

(13.30)

$$CH_3{-}CH_2{-}CH_3 \xrightarrow[700°C]{Cr_2O_3, Al_2O_3} CH_3{-}CH{=}CH_2 \quad \text{propene}$$

(13.31)

$$CH_3CH_2CH_2CH_3 \xrightarrow[600°C]{Cr_2O_3, Al_2O_3} CH_2{=}CH{-}CH{=}CH_2 + 2H_2$$

1,3-butadiene

(13.32)

Ethene is the precursor of polyethene (polyethylene), polychloroethene (polyvinyl chloride), polyethanoxyethene (polyvinyl acetate) (see Chapter 14), ethanol, ethanal, ethanoic acid, ethanoic anhydride, oxirane (ethylene oxide), 1,2-ethanediol (ethylene glycol), chloroethane and lead tetraethyl. Propene is used to make propanone (acetone), polypropene (polypropylene) and polyurethane foams. Butadiene is essential to the synthetic rubber industry. The aromatic hydrocarbons, benzene, toluene, xylenes, and ethylbenzene, are converted to compounds like phenylethene (polystyrene), cyclohexane (an intermediate in the formation of nylon) and phenol (a precurser to cyclohexanol). Phenol is also used in the production of thermo-setting plastics, such as bakelite, in the production of substituted phenols used for synthesis of epoxy resins, and as precursors of weed killers such as 2,4-D. Xylene is oxidized to make terephthalic acid, which is used in the synthesis of terylene.

It is important not to underestimate the importance of petroleum both industrially and politically. The impact on world politics of countries rich in petroleum products is now well known. Over 80% of all organic materials are derived from petroleum and it is estimated that this figure will rise to 98% by 1985. It has been estimated that, even if it were possible to convert plant material to fuel products, it would require all the land in the USA planted in corn to supply the energy equivalent to the petroleum tonnage used at present in that country. It is possible to foresee the harnessing of alternative forms of energy, notably nuclear and solar energy, but it is not so easy to foresee an alternative raw material source. Coal will certainly become more important in the next decade and vast capital will be spent to explore coal technology. After the oil and coal era the need for an alternative source of organic material will be critical.

13.5 Chemical Environment in Alkanes

Methane is a symmetric molecule (Fig. 13.12) and each hydrogen is in an identical chemical environment, being joined to a carbon bearing three hydrogens. The same is true for ethane with each hydrogen joined to a carbon bearing two hydrogens and a methyl group (Fig. 13.13). For n-butane this is not the case. The hydrogens at C(1) and C(4) shown in Fig. 13.14 are in the same chemical environment but this environment is different from the hydrogens at C(2) and C(3) illustrated in Fig. 13.15. There are six hydrogens on C(1) and C(4) and four hydrogens on C(2) and C(3). Methylpropane has two chemically distinct groups of hydrogen atoms, namely the hydrogens on the methyl groups—nine hydrogens—(Fig. 13.16) and the hydrogen on C(2) (Fig. 13.17).

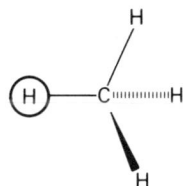

Fig. 13.12 Methane

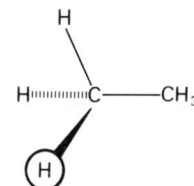

Fig. 13.13 Ethane

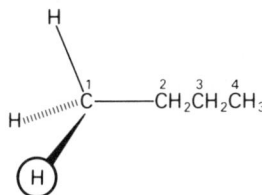

Fig. 13.14 n-butane

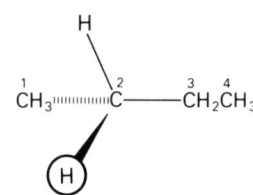

Fig. 13.15 n-butane

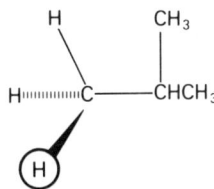

Fig. 13.16 Methylpropane

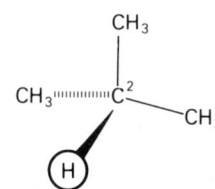

Fig. 13.17 Methylpropane

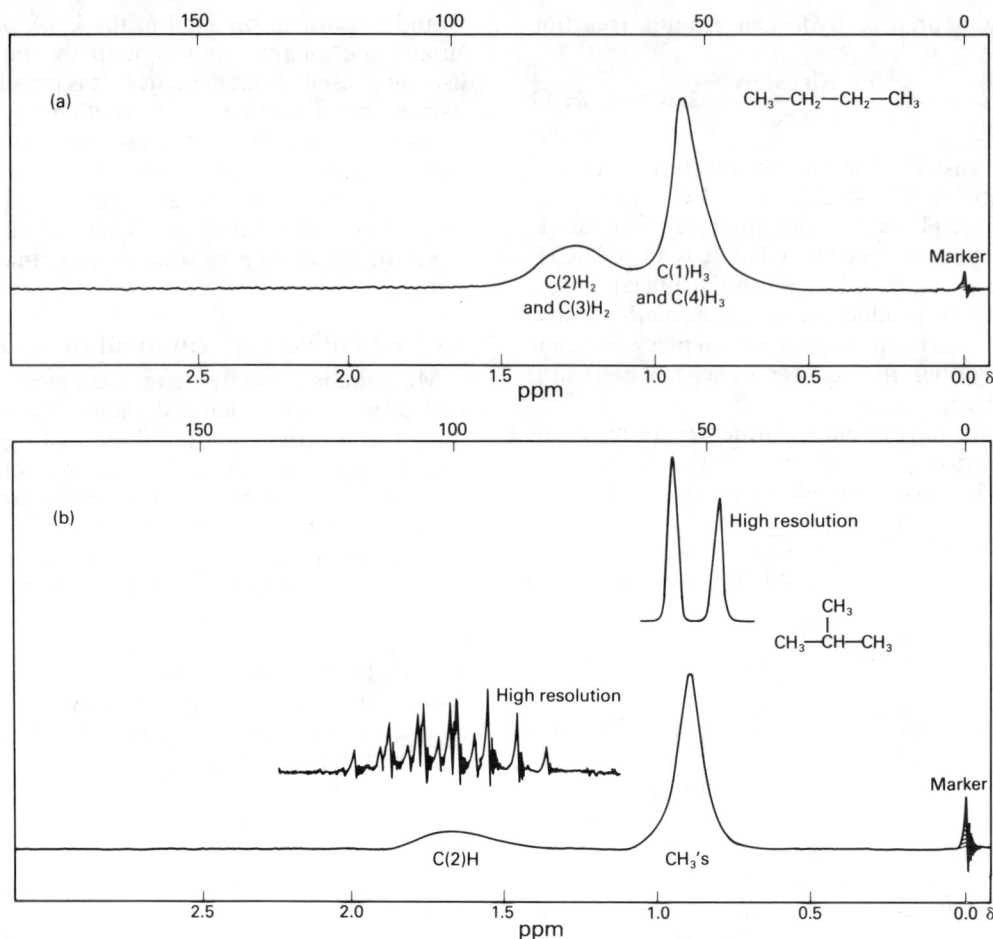

Fig. 13.18 Proton magnetic resonance spectra of (a) n-butane and (b) methylpropane

A very powerful tool which we have for investigating the environments of atoms in organic molecules is the nuclear magnetic resonance spectrometer. The most common type of nuclear magnetic resonance spectrometer is designed to detect the presence of hydrogen atoms in different chemical environments.

The nucleus of a hydrogen atom has a magnetic moment, i.e. acts as a little nuclear magnet, and when a magnetic field is applied the nucleus can be in either of two energy levels, depending on whether the proton's magnetic moment is parallel or antiparallel to the applied field. Any particular hydrogen atom experiences a magnetic field which depends upon the applied field and on the degree to which the applied field is diminished or augmented by the rest of the molecule. Therefore hydrogen atoms in different chemical environments experience slightly different magnetic fields, and there will be a characteristic energy difference between the parallel and antiparallel energy levels for hydrogen

atoms in each chemical environment. If electromagnetic radiation in the radiofrequency range is applied to the sample in a magnetic field it is possible to bring about transitions of the nuclear magnets between these energy states. The frequency ν of radiation required to effect this change is related to the energy difference ΔE by the Planck formula $\Delta E = h\nu$. Thus hydrogen atoms in different chemical environments when placed in a strong magnetic field absorb radiation of different frequency. The intensity of absorption at a particular frequency depends on the number of hydrogen atoms in that particular environment and so the ratio of numbers of hydrogen atoms in each of the different chemical environments are readily obtained (i.e. peak area is proportional to the number of hydrogens). The low resolution nmr spectra (Fig. 13.18) for n-butane and methylpropane show the two different types of atoms in each compound. For n-butane the ratio of the two types of hydrogen atoms is 3:2; there are six hydrogen atoms at C(1) and C(4) and four at C(2) and C(3). For methylpropane (isobutane) the ratio of hydrogen atoms in the different environments is 9:1 corresponding to the C(1), C(3) and C(4) hydrogens and the C(2) hydrogen. Analysis of the peaks of the high resolution spectra provides additional information about the environment of individual hydrogen atoms.

Carbon as found in nature is a mixture of two isotopes. The predominant isotope is ^{12}C; however, there is a small amount (1.1%) of ^{13}C in natural carbon. ^{12}C does not have a magnetic moment but ^{13}C does. This magnetic property of ^{13}C makes it possible to measure the carbon magnetic resonance spectra (cmr) of carbon containing compounds. The process is similar to proton magnetic resonance but it is more difficult experimentally because of the small amount of the ^{13}C isotope present, and because of the complicating effect of the proton magnetic moments. A ^{13}C nmr spectrum of 3-methylheptane is shown in Fig. 13.19. It can be seen that each chemically distinct carbon atom appears as a separate peak.

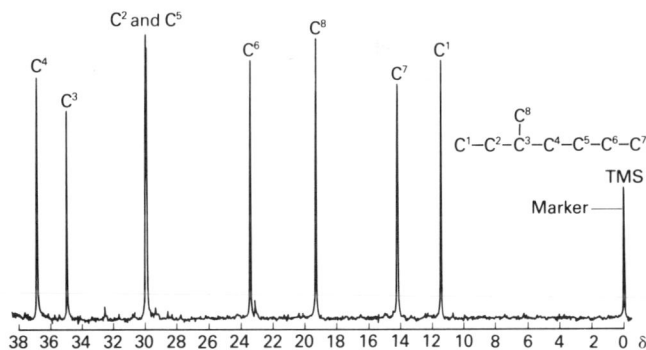

Fig. 13.19 Carbon magnetic resonance spectrum of 3-methylheptane

In contrast to the situation with proton magnetic resonance spectra, the area under the peaks is not simply related to the number of carbons in that particular environment. Carbon and proton magnetic resonance spectra facilitate the determination of the structure of organic compounds, and serve to illustrate the uniqueness of each atom in a molecule. More detailed analysis of the proton magnetic spectrum often allows us to determine the number of protons that are on neighbouring carbons. It is also possible from the ^{13}C spectrum to determine the number of hydrogens bonded to a particular carbon. A *primary* carbon is a carbon attached to only one carbon, a *secondary* carbon is one attached to two carbons, a *tertiary* carbon is one attached to three carbons, a *quaternary* carbon is one attached to four other carbon atoms. From cmr studies a carbon can be defined as primary, secondary, tertiary or quaternary. Three classes of carbon atoms are present in 3-methylheptane (Fig. 13.20). It should be remembered however that the develop-

ment of magnetic resonance spectra has only been widely used in the last decade and the structures of most of the compounds mentioned in this text were first worked out by considerably more laborious methods.

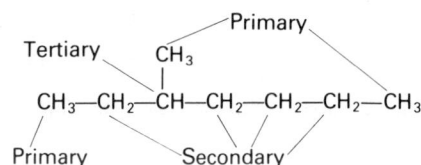

Fig. 13.20 3-methylheptane

13.6 Alicyclic Alkanes

Alicyclic compounds, a subgroup of the alkanes, are cyclic compounds and normally their names have the prefix *cyclo* before that of the appropriate alkane representing the number of carbon atoms in the ring. The ring carbons are numbered in such a way as to give substituents the lowest numbered positions. These molecules cannot really be adequately represented in two dimensions; in order to understand their chemical behaviour it is necessary to examine the orientation of atoms in space. It is helpful to examine models of these compounds. Consider cyclohexane, C_6H_{12}, which is a puckered molecule and exists preferentially in the *chair* conformation illustrated in Fig. 13.21. The

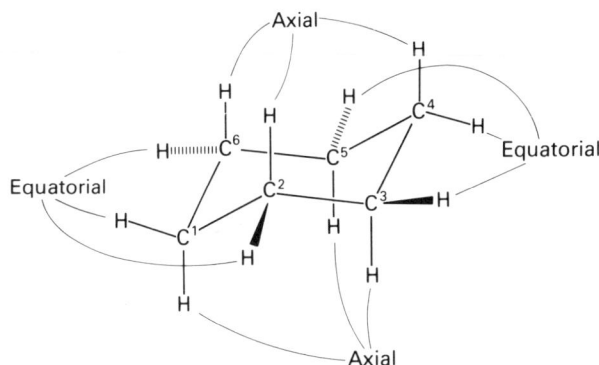

Fig. 13.21 Chair conformation of cyclohexane

chair form is the most stable conformation of cyclohexane and we will examine it by viewing it from several different directions. Atoms or groups on the cyclohexane ring which are directed up or down are referred to as axial and those that point outward as equatorial. Substituents on the same or opposite faces of the molecule are referred to as *cis* or *trans* respectively. The bond angles are all nearly tetrahedral and looking down the C(2),C(3) bond (Fig. 13.22)

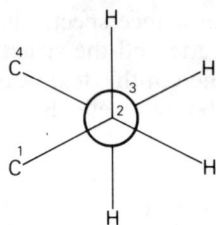

Fig. 13.22 Cyclohexane in chair conformation viewed down the C(2),C(3) bond

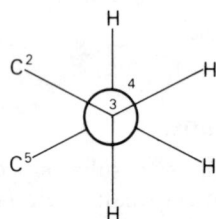

Fig. 13.23 Cyclohexane in chair conformation viewed down the C(3),C(4) bond

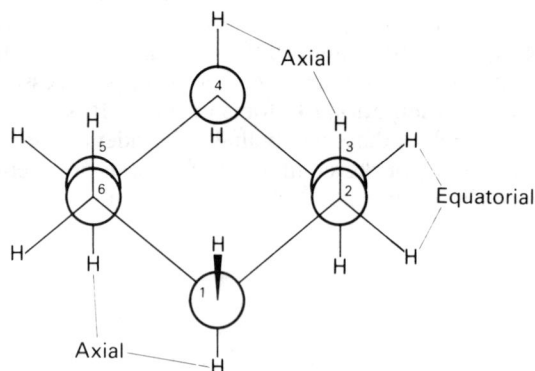

Fig. 13.24 Projection of cylcohexane in the chair conformation showing axial and equatorial orientation of hydrogens

we observe the staggered arrangement of groups and an interaction between C(1), C(2), C(3) and C(4) similar to that in gauche butane (p. 223). If we observe the molecule down the C(3),C(4) bond (Fig. 13.23), we see a similar relationship between the corresponding atoms and the same is true as we view the molecule down each carbon–carbon bond of the ring. A more complete view of the molecule can be obtained by looking down the C(2),C(3) and C(6),C(5) bonds (Fig. 13.24). An alternative configuration of cyclohexane is the boat conformation (Fig. 13.25). It contains some unfavourable interactions and is not as stable as the chair form. A three-dimensional representation is shown in Fig. 13.25 and the view down the C(1),C(6) bond in Fig. 13.26 shows the atoms are staggered. Viewed down the C(1) and C(2) bond (Fig. 13.27), C(3) and C(6) and the hydrogen atoms on C(1) and C(2) are seen to be eclipsed.

The skew butane interactions of the boat conformation can also be observed when the molecule is viewed down the C(1),C(2) and C(5),C(4) bonds as shown in Fig. 13.28. In addition to these unfavourable interactions the top hydrogen atoms on C(3) and C(6) are compressed (see Fig. 13.25). This is termed the flagpole interaction and results in a twisted boat (usually abbreviated to *twist boat*) conformation being somewhat more stable than a pure boat conformation. The relative potential energy of the different cyclohexane conformations is shown in Fig. 13.29.

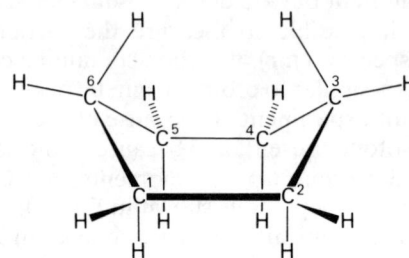

Fig. 13.25 Boat conformation of cyclohexane

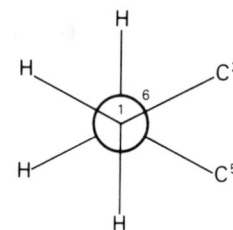

Fig. 13.26 Cyclohexane in boat conformation viewed down the C(1),C(6) bond

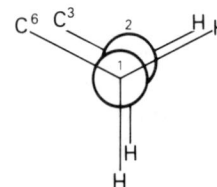

Fig. 13.27 Cyclohexane in boat conformation viewed down the C(1),C(2) bond

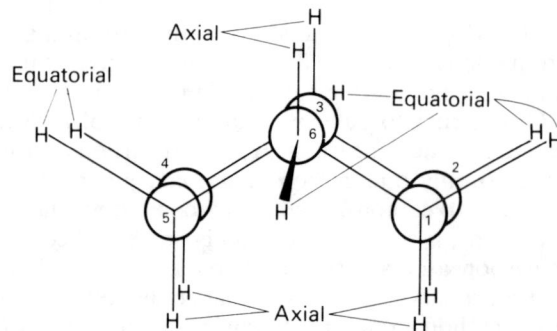

Fig. 13.28 Projection of cyclohexane in boat conformation

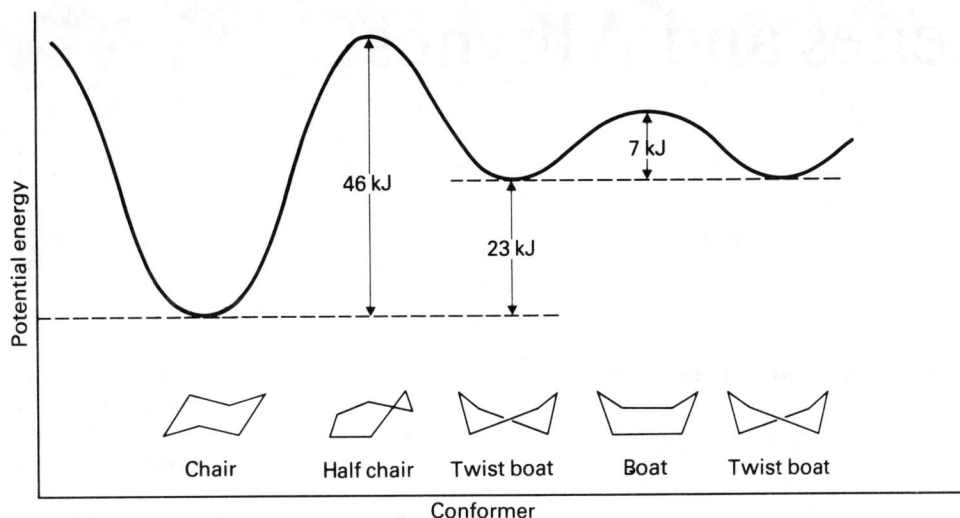

Fig. 13.29 Dependence of potential energy of cyclohexane on conformation

Exercises

13.1 Write and name the constitutional isomers of molecular formula C_7H_{16}. Indicate the hydrogen atoms attached to primary (p.) secondary (sec.) and tertiary (tert.) carbon atoms and for each compound determine the ratio of hydrogen atoms in the different chemical environments.

13.2 Methylbutane has hydrogen atoms in four different chemical environments. Write and name the structural formula of all possible mono-chlorinated methylbutane derivatives.

13.3 Draw in three dimensions the following cyclohexane derivatives.
   (1) *trans*-1,2-dimethylcyclohexane
   (2) *cis*-1,2-dimethylcyclohexane
   (3) *trans*-1,3-dimethylcyclohexane
   (4) *cis*-1,3-dimethylcyclohexane
   (5) *trans*-1,4-dimethylcyclohexane
   (6) *cis*-1,4-dimethylcyclohexane

14 Alkenes and Alkynes

14.1 Structure and Bonding in Alkenes and Alkynes

Alkenes and alkynes are unsaturated hydrocarbons, which differ from saturated hydrocarbons in that the carbon atom's desire to form four chemical bonds is satisfied by multiple bonding to other carbon atoms. Compounds with double bonds are called *alkenes*; those with triple bonds are called *alkynes*. The parent member of the alkene family is ethene (ethylene), C_2H_4, (Fig. 14.1) while that of the alkyne family is ethyne (acetylene), C_2H_2, (Fig. 14.2). Aromatic compounds form a separate class of unsaturated hydrocarbons—benzene, C_6H_6, (Fig. 14.3) is the parent member of this class.

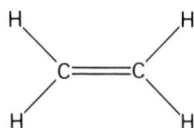

Fig. 14.1 Ethene (ethylene)   **Fig. 14.2** Ethyne (acetylene)

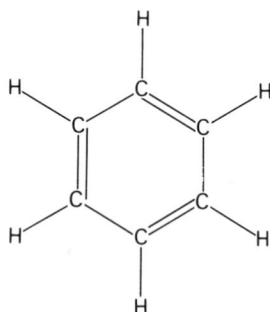

Fig. 14.3 Benzene

As we saw in Section 3.4, the carbon atoms in ethene are bonded to each other and to the two hydrogen atoms by three equivalent sp^2 hybrid orbitals, which form a planar arrangement of σ bonds with bond angles of 120°. The remaining p orbital on each carbon is occupied by a single electron. The p orbitals on each carbon overlap as shown in Fig. 14.4 to form a π bond, by overlap of the lobes both above and below the plane of the molecule. The double bond between two carbon atoms in ethene is markedly stronger (bond energy 682 kJ) than the single bond in ethane (367 kJ), and this is reflected in the reduced distance between the two carbons. The bond lengths of the parent hydrocarbons are shown in Fig. 14.5.

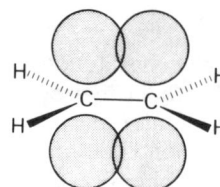

Fig. 14.4 Diagram showing overlap of p orbitals of ethene

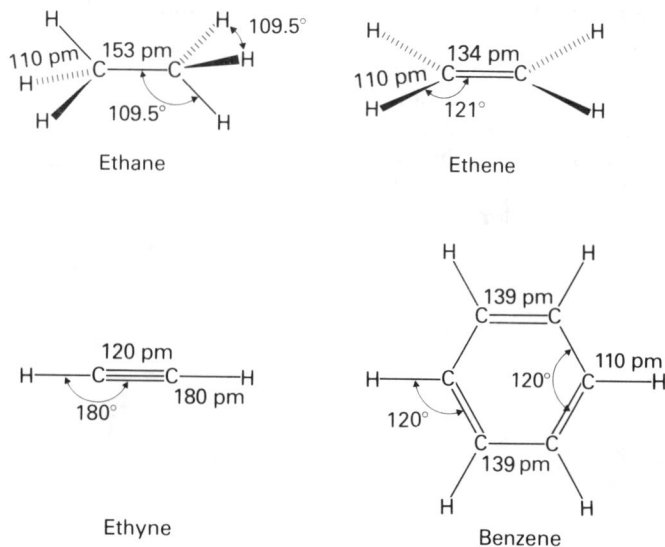

Fig. 14.5 Bond angles and bond lengths of parent hydrocarbons

In ethyne the σ bond framework of the molecule is formed from sp hybrid orbitals, which lie along a straight line. The remaining two p orbitals on each carbon atom can overlap with similar orbitals on the other carbon atom and thus strengthen the carbon–carbon linkage. Overlap of the p orbitals on each carbon, as shown in Fig. 14.6, allows pairing of electrons in each of two π bonds. The π bonds form a cylindrical sheath of electron density about the carbon–carbon σ bond. The triple bonding between the carbon atoms is reflected in the bond energy (829 kJ) and in the carbon–carbon bond length (Fig. 14.5). It is noteworthy that in both ethene and ethyne, as in ethane, the preferred bonding arrangement is that which leads to greatest separation of the atoms which are bound to a given atom.

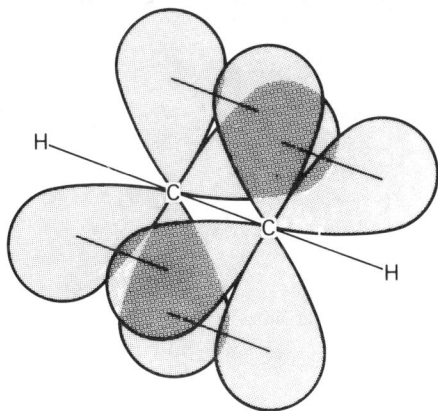

Fig. 14.6 Diagram showing overlap of p orbitals of ethyne

14.2 Alkenes: Structure and Nomenclature

The general formula of aliphatic alkenes containing one double bond is C_nH_{2n}. Ethene (Fig. 14.1) is the parent member of the family. The suffix *ene* is characteristic of the family. The next members of the series are propene (C_3H_6), butene (C_4H_8) and pentene (C_5H_{10}). Propene can be written either as in Fig. 14.7 or as in Fig. 14.8.

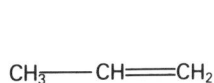

Fig. 14.7 Propene

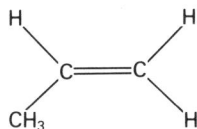

Fig. 14.8 Propene

Butene is more complicated, because the double bond can have two different locations. For the straight chain structure the double bond can occur between C(1) and C(2)

(Fig. 14.9) or between C(2) and C(3) (Fig. 14.10). With the double bond between C(3) and C(4) the molecule is identical with that drawn with the double bond between C(1) and C(2). It is usual to number the carbon chain so that the double bond carries the lowest number to indicate its position, i.e. numbering is begun from the end of the chain nearest the double bond. This number is generally placed in the name immediately before the parent chain containing the double bond. The double bond is always between the numbered carbon and that with the next higher number.

Fig. 14.9 1-butene

Fig. 14.10 2-butene

2-Methyl-1-propene, shown in Fig. 14.11, is another structural arrangement of atoms (*constitutional* or *structural isomer*) having the same molecular formula C_4H_8. In this case the numbers are not necessary since the structure is completely described by the name methylpropene. Branched molecules are named systematically as a derivative of the alkene corresponding to the longest carbon chain of the molecule, which in this case is propene.

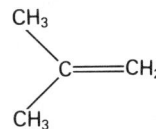

Fig. 14.11 Methylpropene (isobutylene)

2-Butene (Fig. 14.10) is interesting and warrants further attention. The energy barrier to rotation about a carbon–carbon double bond is of the order of 293 kJ and at ordinary temperatures rotation about a carbon–carbon double bond is not possible. There are therefore two unique structural arrangements of atoms in 2-butene: the atoms can be arranged so that the methyl groups are either on the same side (Fig. 14.12) or on different sides (Fig. 14.13) of the carbon–carbon double bond. These two structures are said to be *geometric isomers*. When there are two or more distinct ways in which atoms can be arranged in space with bonding occurring between the same atoms, the structures are referred to as *stereoisomers. Geometric isomers* are a special class of stereoisomers. When the two methyl groups are on the same side of the carbon–carbon double bond their relationship is

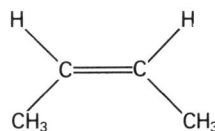

Fig. 14.12 *cis*-2-butene

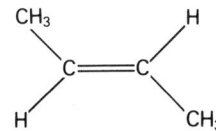

Fig. 14.13 *trans*-2-butene

referred to as *cis* (Latin—on this side). When they are on opposite sides of the molecule their relationship is said to be *trans* (Latin—across). For most purposes the structural relationship of substituents about a carbon–carbon double bond can be adequately described by the use of the prefix *cis* or *trans*. However, this nomenclature does require there to be one substituent group in common to each carbon atom of the double bond. Since this is not always the case a more general naming scheme has been introduced using the letters Z (on the same side—from German *zusammen*) and E (on the opposite—from German *entgegen*) to define the stereochemistry about a carbon–carbon double bond. The substituents on the double bond are arranged in a sequence which depends essentially on atomic number. Of the atoms immediately joined to the carbon of the double bond, that with the highest atomic number gets highest priority. Thus a chlorine substituent, with atomic number 17, has a higher sequence priority or preference than a methyl substituent, since the atomic number of carbon is only 6. If a sequence cannot be determined by this rule (e.g. methyl versus ethyl) then it is determined by a similar comparison of the next atoms in the substituent groups, and so on. The ethyl group has a higher sequence preference than the methyl group. Where the substituent contains a double or triple bond e.g. (—C—H or —C≡CR) the situa-

$$
\begin{array}{c}
\| \\
O
\end{array}
$$

tion is more complex and we will defer a discussion of this until later (p. 298). If the group of highest priority on one carbon of the double bond is on the same side as the group of highest priority on the other carbon the structure is described using the letter Z. If the group of higher priority on one carbon of the double bond is on the opposite side to the group of highest priority on the other carbon the structure is described using the letter E. The examples in Table 14.1 illustrate the use of this nomenclature.

Table 14.1 Nomenclature of alkenes

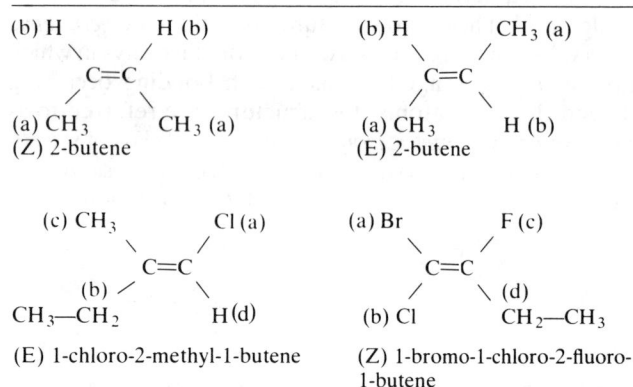

(Z) 2-butene   (E) 2-butene

(E) 1-chloro-2-methyl-1-butene   (Z) 1-bromo-1-chloro-2-fluoro-1-butene

14.3 Preparation and Reactions of Alkenes

The primary source of C_2 to C_5 alkenes is the petroleum industry. These lower boiling alkenes can be separated as pure compounds by distillation. The profusion of isomers of higher molecular weight alkenes makes separation of isomers by distillation impossible and they must be prepared by other methods. In the laboratory alkenes are prepared by reactions 14.1–14.6 (Table 14.2).

Table 14.2 Preparation of alkenes

Dehydrohalogenation of haloalkanes

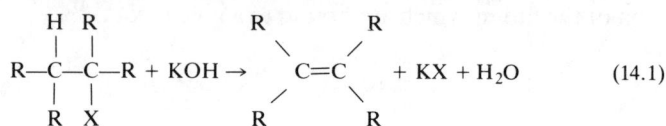

$$ (14.1) $$

Dehydration of alcohols

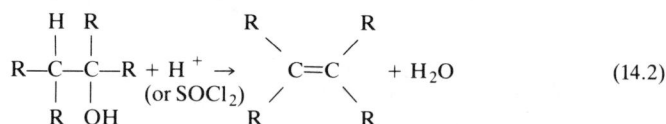

$$ (14.2) $$

Dehalogenation of vicinal dihalides

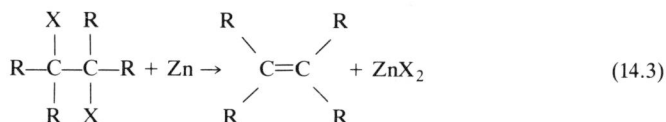

$$ (14.3) $$

Reduction of alkynes

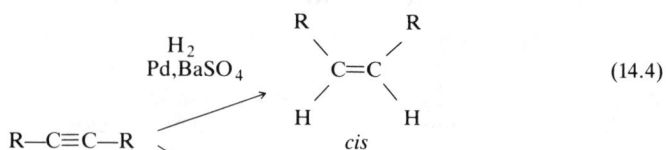

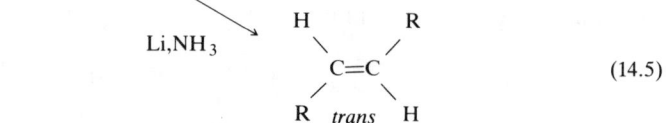

$$ (14.4) $$

$$ (14.5) $$

Replacement of carbonyl oxygen

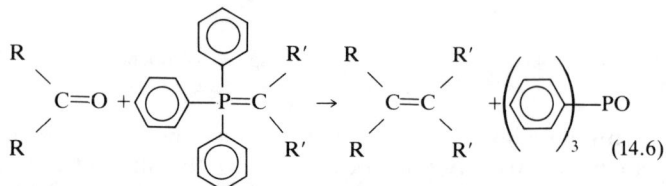

$$ (14.6) $$

Dehydrohalogenation (reaction 14.1) and dehydration (reaction 14.2) in structures where elimination of HX or H_2O can occur between more than one adjacent carbon atom leads to mixtures of alkenes. This is illustrated in Fig. 14.14 for the reaction of 2-chlorobutane with potassium hydroxide. Dehydration of alcohols by acids is often further complicated by acid catalysed isomerization of the initially formed alkene. Further discussion of these reaction processes will be deferred to Section 14.4.

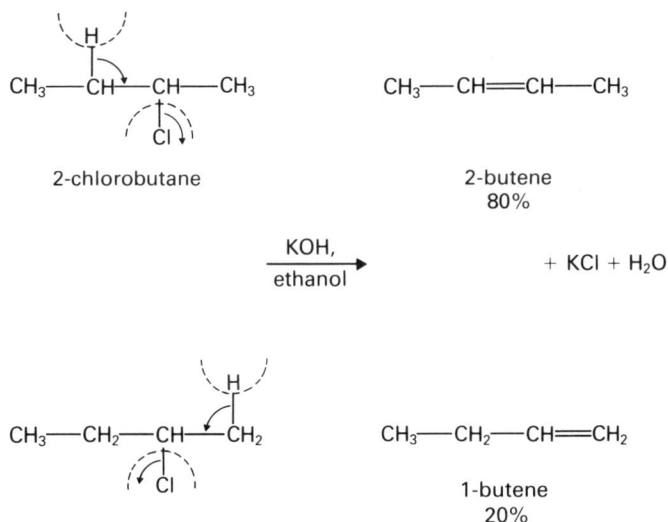

Fig. 14.14 Reaction of 2-chlorobutane with potassium hydroxide in ethanol

We have already indicated that the chemistry of alkenes is the chemistry of the π electron system. These electrons are held more loosely in the molecule than electrons in bonding σ orbitals. The double bond can therefore be regarded as a source of electrons and reaction can occur readily with species deficient in electrons. Electron deficient species are referred to as *electrophiles*. The type of reaction which is most characteristic of alkenes can be represented by the general equation:

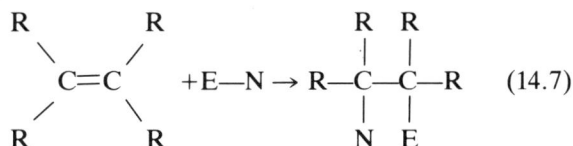

$$(14.7)$$

As this equation shows, the molecule E—N adds itself to the alkene, and so the reaction is refered to as an *addition reaction*. For reference, the different forms which the addition reaction can take are summarized in equations 14.8–14.23 (Table 14.3).

Table 14.3 Reactions of Alkenes

Addition of hydrogen

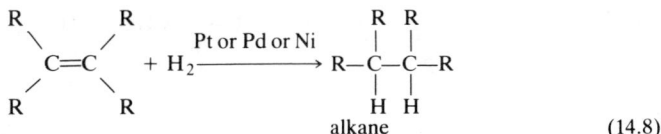

$$(14.8)$$

Addition of halogen

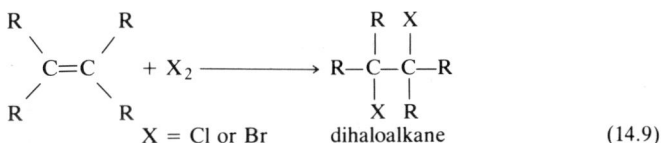

$$(14.9)$$

Addition of hydrogen halide

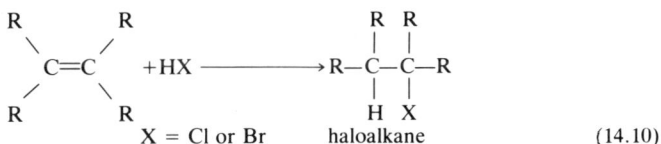

$$(14.10)$$

Addition of sulphuric acid

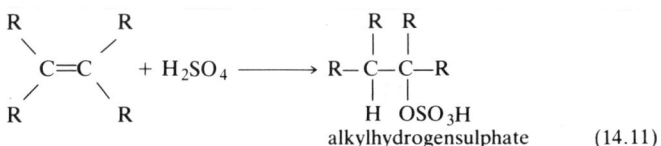

$$(14.11)$$

Addition of water

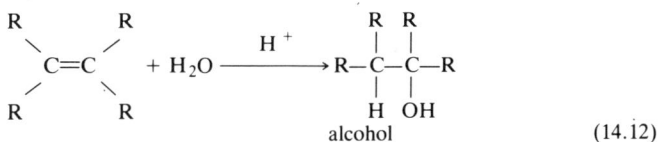

$$(14.12)$$

Halohydrin formation

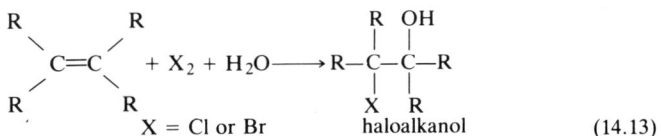

$$(14.13)$$

Table 14.3 continued

Oxymercuration–demercuration

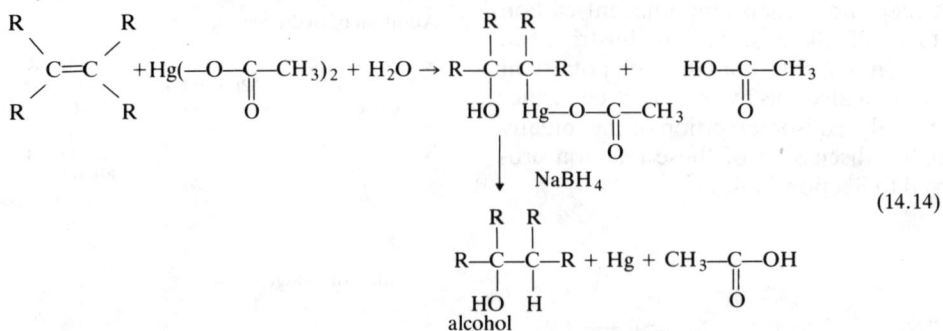

$$
\begin{array}{c}
\text{R} \quad\quad \text{R}\\
\diagdown \quad\;\; \diagup\\
\text{C}=\text{C} \quad +\text{Hg}(\!-\!\text{O}\!-\!\overset{\displaystyle \text{O}}{\underset{\displaystyle \parallel}{\text{C}}}\!-\!\text{CH}_3)_2 + \text{H}_2\text{O} \rightarrow \text{R}\!-\!\overset{\text{R}}{\underset{\text{HO}}{\text{C}}}\!-\!\overset{\text{R}}{\underset{\text{Hg}-\text{O}-\text{C}-\text{CH}_3}{\text{C}}}\!-\!\text{R} \quad + \quad \text{HO}\!-\!\overset{\displaystyle \text{O}}{\underset{\displaystyle \parallel}{\text{C}}}\!-\!\text{CH}_3\\
\diagup \quad\quad \diagdown\\
\text{R} \quad\quad \text{R} \quad\quad\quad \text{O}
\end{array}
$$

(14.14)

$$\downarrow \text{NaBH}_4$$

$$
\text{R}\!-\!\overset{\text{R}}{\underset{\text{HO}}{\text{C}}}\!-\!\overset{\text{R}}{\underset{\text{H}}{\text{C}}}\!-\!\text{R} + \text{Hg} + \text{CH}_3\!-\!\overset{\displaystyle\text{O}}{\underset{\displaystyle\parallel}{\text{C}}}\!-\!\text{OH}
$$
alcohol

Hydroboration–oxidation

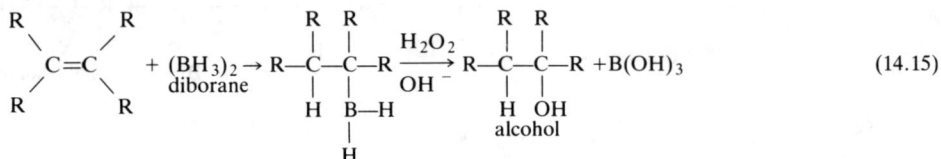

$$
\begin{array}{c}
\text{R} \quad\quad \text{R}\\
\diagdown \quad\;\; \diagup\\
\text{C}=\text{C} \quad + (\text{BH}_3)_2 \rightarrow \text{R}\!-\!\overset{\text{R}}{\underset{\text{H}}{\text{C}}}\!-\!\overset{\text{R}}{\underset{\underset{\text{H}}{\overset{|}{\text{B}-\text{H}}}}{\text{C}}}\!-\!\text{R} \xrightarrow[\text{OH}^-]{\text{H}_2\text{O}_2} \text{R}\!-\!\overset{\text{R}}{\underset{\text{H}}{\text{C}}}\!-\!\overset{\text{R}}{\underset{\text{OH}}{\text{C}}}\!-\!\text{R} +\text{B(OH)}_3\\
\diagup \quad\;\; \diagdown \quad \text{diborane}\\
\text{R} \quad\quad \text{R} \quad\quad\quad\quad\quad\quad\quad\quad\quad\quad \text{alcohol}
\end{array}
$$

(14.15)

Polymerization

(a) radical initiation

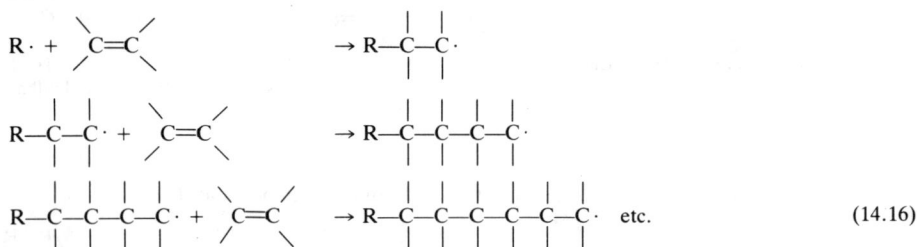

$$
\text{R}\cdot + \;\; \text{C}=\text{C} \quad\quad\quad\quad\quad \rightarrow \text{R}\!-\!\text{C}\!-\!\text{C}\cdot
$$

$$
\text{R}\!-\!\text{C}\!-\!\text{C}\cdot + \;\; \text{C}=\text{C} \quad\quad\quad \rightarrow \text{R}\!-\!\text{C}\!-\!\text{C}\!-\!\text{C}\!-\!\text{C}\cdot
$$

$$
\text{R}\!-\!\text{C}\!-\!\text{C}\!-\!\text{C}\!-\!\text{C}\cdot + \;\; \text{C}=\text{C} \quad \rightarrow \text{R}\!-\!\text{C}\!-\!\text{C}\!-\!\text{C}\!-\!\text{C}\!-\!\text{C}\!-\!\text{C}\cdot \;\; \text{etc.}
$$

(14.16)

(b) cation initiation

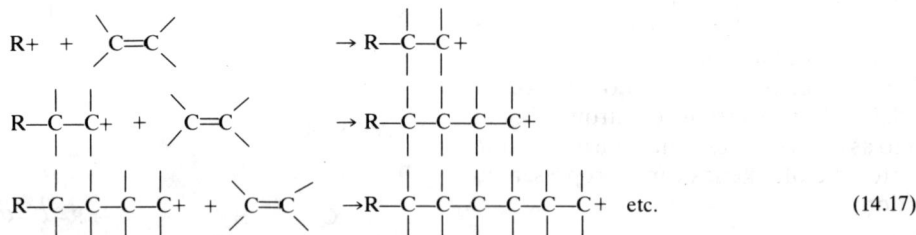

$$
\text{R}+ \;\; + \;\; \text{C}=\text{C} \quad\quad\quad\quad \rightarrow \text{R}\!-\!\text{C}\!-\!\text{C}+
$$

$$
\text{R}\!-\!\text{C}\!-\!\text{C}+ \;\; + \;\; \text{C}=\text{C} \quad\quad \rightarrow \text{R}\!-\!\text{C}\!-\!\text{C}\!-\!\text{C}\!-\!\text{C}+
$$

$$
\text{R}\!-\!\text{C}\!-\!\text{C}\!-\!\text{C}\!-\!\text{C}+ \;\; + \;\; \text{C}=\text{C} \quad \rightarrow \text{R}\!-\!\text{C}\!-\!\text{C}\!-\!\text{C}\!-\!\text{C}\!-\!\text{C}\!-\!\text{C}+ \;\; \text{etc.}
$$

(14.17)

(c) anion polymerization

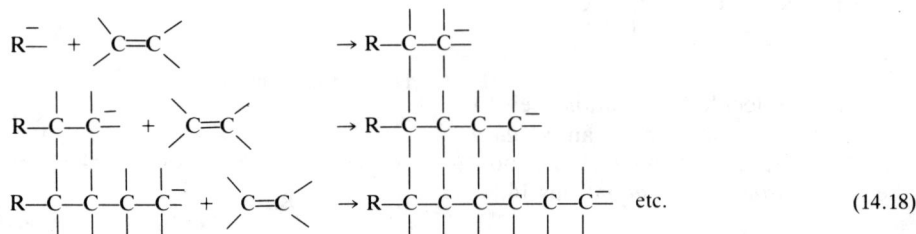

$$
\text{R}\!-\!\!^- \;\; + \;\; \text{C}=\text{C} \quad\quad\quad\quad \rightarrow \text{R}\!-\!\text{C}\!-\!\text{C}\!\!^-
$$

$$
\text{R}\!-\!\text{C}\!-\!\text{C}\!\!^- \;\; + \;\; \text{C}=\text{C} \quad\quad \rightarrow \text{R}\!-\!\text{C}\!-\!\text{C}\!-\!\text{C}\!-\!\text{C}\!\!^-
$$

$$
\text{R}\!-\!\text{C}\!-\!\text{C}\!-\!\text{C}\!-\!\text{C}\!\!^- \;\; + \;\; \text{C}=\text{C} \quad \rightarrow \text{R}\!-\!\text{C}\!-\!\text{C}\!-\!\text{C}\!-\!\text{C}\!-\!\text{C}\!-\!\text{C}\!\!^- \;\; \text{etc.}
$$

(14.18)

Table 14.3 continued

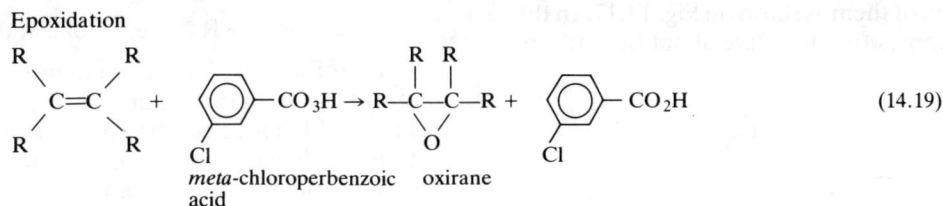

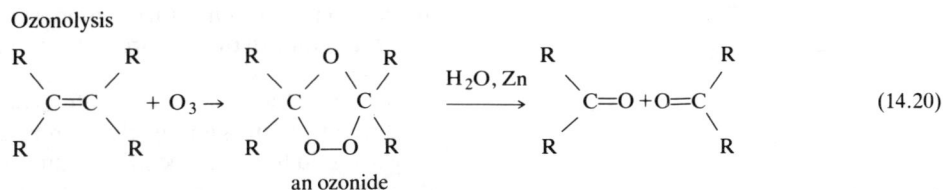

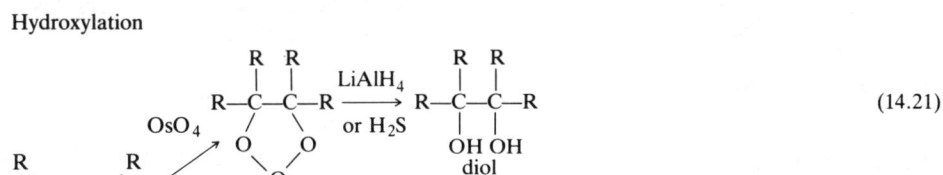

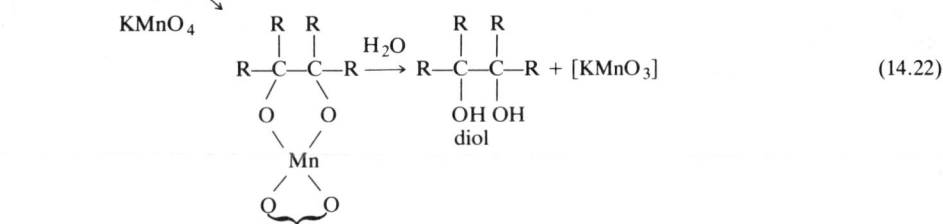

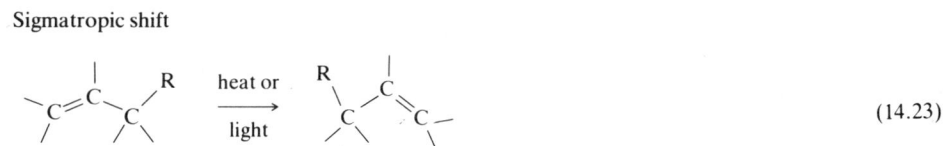

Epoxidation

$$R_2C=CR_2 + \text{(meta-chloroperbenzoic acid)}-CO_3H \rightarrow R-\underset{\underset{O}{\diagdown}\diagup}{\overset{R \; R}{C-C}}-R + \text{(Cl-benzene)}-CO_2H \qquad (14.19)$$

meta-chloroperbenzoic oxirane
acid

Ozonolysis

$$R_2C=CR_2 + O_3 \rightarrow \text{(an ozonide)} \xrightarrow{\text{H}_2\text{O, Zn}} R_2C=O + O=CR_2 \qquad (14.20)$$

an ozonide

Hydroxylation

$$R_2C=CR_2 \xrightarrow{\text{OsO}_4} R-\underset{\underset{O}{}\underset{Os}{}\underset{O}{}}{\overset{R \; R}{C-C}}-R \xrightarrow[\text{or H}_2\text{S}]{\text{LiAlH}_4} R-\underset{\underset{OH \; OH}{\text{diol}}}{\overset{R \; R}{C-C}}-R \qquad (14.21)$$

$$R_2C=CR_2 \xrightarrow{\text{KMnO}_4} R-\underset{\underset{Mn}{O \quad O}}{\overset{R \; R}{C-C}}-R \xrightarrow{\text{H}_2\text{O}} R-\underset{\underset{OH \; OH}{\text{diol}}}{\overset{R \; R}{C-C}}-R + [\text{KMnO}_3] \qquad (14.22)$$

Sigmatropic shift

$$\underset{\diagup}{\overset{\diagdown}{C}}=\underset{\diagdown}{\overset{\diagup}{C}}\underset{\diagdown}{\overset{\diagup}{C}}-R \xrightarrow[\text{light}]{\text{heat or}} R-\underset{\diagdown}{\overset{\diagup}{C}}\underset{\diagup}{\overset{\diagdown}{C}}=\underset{\diagdown}{\overset{\diagup}{C}}- \qquad (14.23)$$

14.4 Mechanism of Addition to Alkenes

Reactions of alkenes are generally initiated by an electrophilic species which embeds itself in the π electron cloud. The electrophile is often a positive ion, but polarizable molecules such as chlorine or bromine can also act as electrophiles. The electrophile may position itself symmetrically between the two carbon atoms as shown in Fig. 14.15, but more often it will not be symmetrically placed between

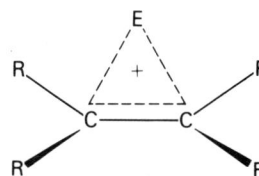

Fig. 14.15 Electrophile positioned symmetrically between two carbon atoms

the two carbon atoms (Fig. 14.16), and in an extreme case may bond to only one of them as shown in Fig. 14.17. In this case the system becomes free to rotate about the carbon–carbon bond.

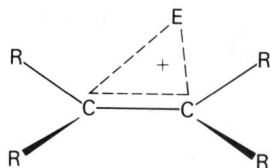

Fig. 14.16 Electrophile positioned unsymmetrically between two carbon atoms

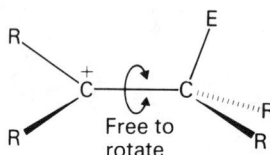

Fig. 14.17 Electrophile bonded to a single carbon atom

One of the most commonly encountered electrophiles is the positively charged hydrogen atom, or proton. (For simplicity we disregard the fact that the proton will normally be solvated.)

A proton can add to a double bond to form a *carbonium ion,* or carbon cation (*carbocation* for short) i.e. a cation with a formal positive charge on a carbon atom,

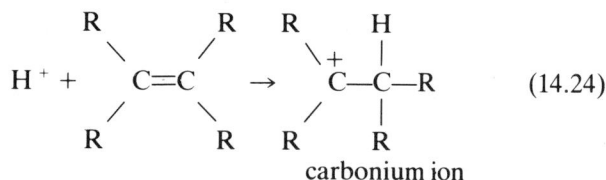

as shown in Eq. 14.24. Some chemists refer to this particular carbon cation as a *carbenium ion.* Carbonium or carbenium ions can be divided into classes depending on the number of substituents at the electron deficient carbon atom. This is illustrated in Table 14.4.

Table 14.4 Classes of carbonium ions

The stabilities of methyl, primary, secondary and tertiary radicals are known and the energy required to remove an electron from a radical (Eq. 14.25) can be determined.

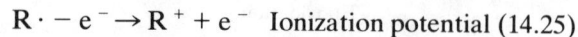

$$R\cdot - e^- \rightarrow R^+ + e^- \quad \text{Ionization potential} \quad (14.25)$$

It is therefore possible to compute the energies of the different classes of carbonium ion. This information is given in Fig. 14.18. These results apply to ions in the gas phase, but the stabilizing effect of solvation is not markedly dependent on the structure of the ion. It can be seen that the tertiary carbonium ion is more stable than the secondary carbonium ion, which in turn is more stable than the primary carbonium ion. In other words, the more alkyl groups there are about the positively charged atom, the more stable the species. Alkyl groups can therefore be regarded, in comparison with hydrogen, as electron releasing or donating. They are said to have a positive inductive effect. While this information has been determined by experiments with gaseous radicals and ions the stability trend is the same in solution.

The precise structure of the intermediate configurations in a reaction is very difficult to ascertain because of their transitory existence. However, under some conditions species which resemble transition states or intermediates can have more than transitory existence. For a carbon atom which carries three phenyl substituents, which have the ability to spread and stabilize a positive charge on an adjacent carbon atom, the cation is stable enough to exist as a salt at room temperature. The triphenylmethyl cation exists as a propeller-shaped structure, as shown in Fig. 14.19. Less stable carbocations can be formed at low temperature by reaction of alkenes with 'super acids', HF—SbF$_5$ and FSO$_3$H—SbF$_5$.

$$R—CH=CH—R \xrightarrow{FSO_3H—SbF_5} R—\overset{+}{C}H—CH_2—R$$
$$(14.26)$$

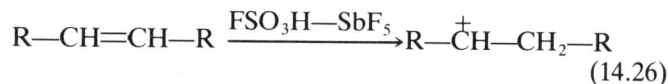

These carbocations can be studied by proton and carbon magnetic resonance techniques, and it has been possible from these studies to obtain proof that the central carbon atom of a carbonium ion is sp^2 hybridized and planar, with the carbon atom bearing a substantial positive charge. The fact that in strongly acid media at low temperature it is possible to observe carbocations spectroscopically does not prove for certain that such species exist as intermediates in reactions carried out at room temperature and in entirely different solvent media. Nevertheless there is strong evidence that addition reactions to olefins involve transition states and intermediates which have an excess of positive charge at the carbon to which the *nucleophile* (electron rich species) becomes attached.

We will examine the addition of HCl to the unsymmetrical olefin, propene (reaction 14.10). The reaction is a two step process. The electrophile, H$^+$, an electron deficient species, can add to the primary carbon atom (reaction 14.27) or the secondary carbon atom (reaction 14.28), with

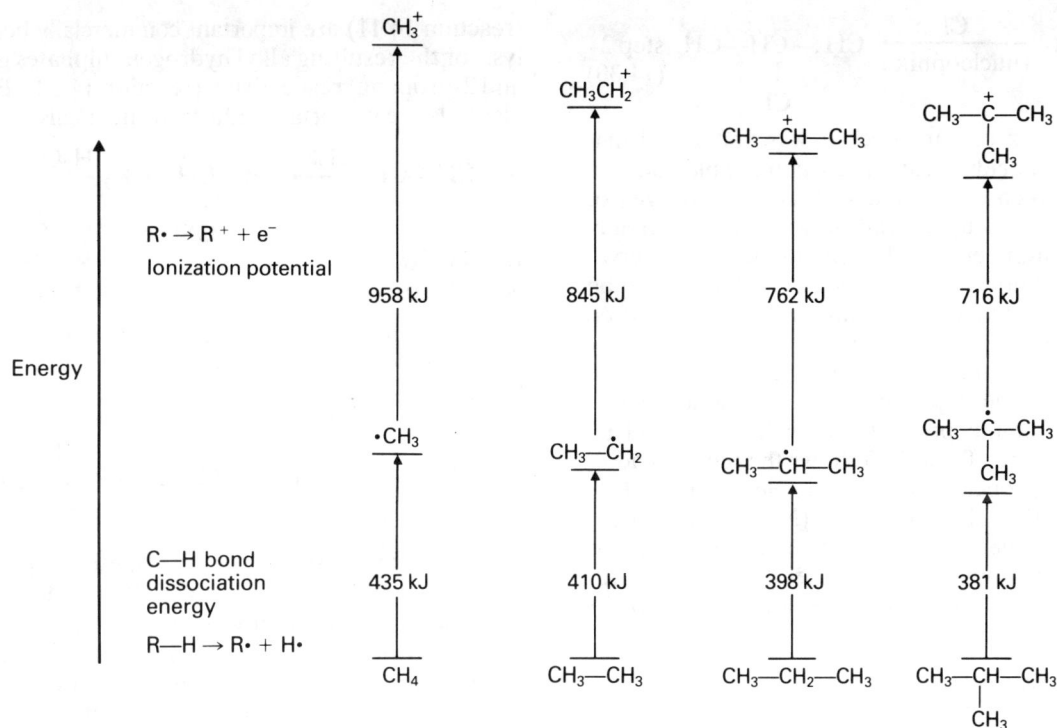

Fig. 14.18 Relative energies of radicals and carbonium ions

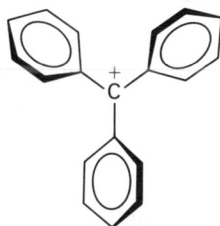

Fig. 14.19 Propeller shaped structure of triphenylmethyl carbocation

consequent development of positive charge on the secondary or primary carbon atom, respectively.

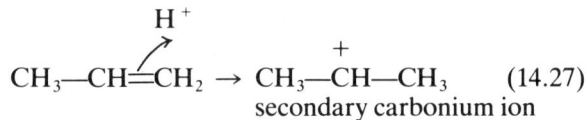

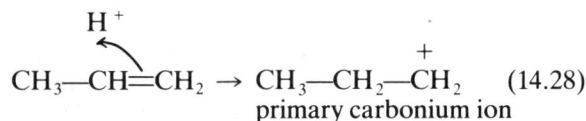

$$CH_3—CH{\overset{H^+}{\nearrow}}CH_2 \rightarrow CH_3—\overset{+}{CH}—CH_3 \quad (14.27)$$
secondary carbonium ion

$$CH_3—CH{\overset{H^+}{\nwarrow}}CH_2 \rightarrow CH_3—CH_2—\overset{+}{CH}_2 \quad (14.28)$$
primary carbonium ion

The formation of a carbocation intermediate in the reaction is the first step, and since the transition state leading to carbocation formation will reflect something of its nature and energy, protonation at the primary carbon with development of charge at the secondary carbon atom (reaction 14.27) will be favoured over protonation at the secondary carbon and consequent development of a positive charge on the primary carbon atom (reaction 14.28). This is illustrated by this reaction of propene with HCl, since only 2-chloropropane (14.29) is formed.

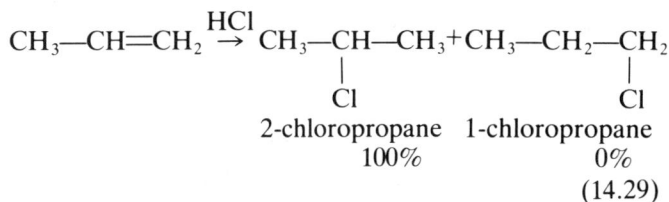

$$CH_3—CH{=}CH_2 \overset{HCl}{\rightarrow} CH_3—\underset{Cl}{CH}—CH_3 + CH_3—CH_2—\underset{Cl}{CH_2}$$

2-chloropropane   1-chloropropane
100%   0%
(14.29)

The rate-determining step here is protonation, to give the more stable, lower energy, secondary carbonium ion. The second step of the reaction is rapid and involves attack of the chloride nucleophile at the secondary carbon (reaction 14.30).

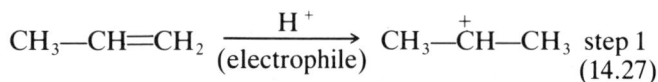

$$CH_3—CH{=}CH_2 \xrightarrow[\text{(electrophile)}]{H^+} CH_3—\overset{+}{CH}—CH_3 \text{ step 1}$$
(14.27)

$$CH_3—\overset{+}{C}H—CH_3 \xrightarrow[\text{(nucleophile)}]{Cl^-} CH_3—CH—CH_3 \text{ step 2}$$
$$\underset{Cl}{\vert} \quad (14.30)$$

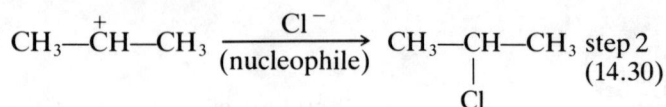

It is more than a century since it was first noted that addition of proton-acids to carbon–carbon double bonds of unsymmetrical alkenes occurred so that the hydrogen of the acid was attached to the carbon atom which already held the greater number of hydrogen atoms. This observation has become known as *Markovnikov's Rule*. It can be paraphrased as 'To him who has, more will be given'. Often addition reactions give almost exclusively one addition product as shown in reaction 14.29, and when this occurs the reaction is said to be *regiospecific*. It is now in general possible to predict the product that will be formed by addition to an alkene. Table 14.5 shows the scope of addition reactions to propene, and lists the electrophilic (E^+, electron deficient) and nucleophilic (N^-, electron rich) species involved in each reaction. The electrophile always adds to the carbon atom with the greater number of hydrogen atoms attached; the nucleophile adds to the other carbon atom.

Table 14.5 Electrophilic addition to alkenes

$CH_3—CH{=}CH_2$	E^+ $\rightarrow CH_3—\underset{\vert}{\overset{E}{C}H}—\overset{+}{C}H_2$	N^- $\rightarrow CH_3—\underset{\vert}{\overset{N}{C}H}—\overset{E}{\underset{\vert}{C}}H_2$	
E-N	E^+	N^-	
Cl_2	Cl^{+*}	Cl^-	(14.9)
Br_2	Br^{+*}	Br^-	(14.9)
HCl	H^+	Cl^-	(14.10)
HBr	H^+	Br^-	(14.10)
H_2SO_4	H^+	$HOSO_3^-$	(14.11)
H_2O, H^+	H^+	OH^- or H_2O	(14.12)
Br_2, H_2O	Br^+	OH^- or H_2O	(14.13)
$Hg(OAc)_2, H_2O$	$Hg^+(OAc)$	OH^- or H_2O	(14.14)
$(BH_3)_2$	H_2B^+	H^-	(14.15)

 * Note: the species Cl^+ and Br^+ do not exist free in solutions, but as parts of a single polarized halogen molecule. The positive charge is less than one electron charge.

There are exceptions to Markovnikov's rule which occur for certain alkenes (see reaction 17.11) and when addition does not occur by an ionic mechanism but instead involves free radicals. For example, the addition of HBr to methylpropene in the presence of peroxides gives 1-bromomethylpropane, an anti-Markovnikov product (Eq. 14.31).

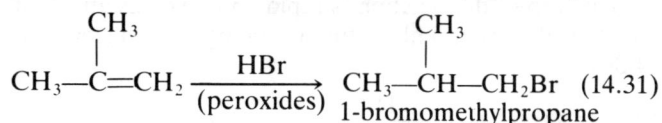

$$\underset{CH_3}{\overset{CH_3}{\vert}}CH_3—C{=}CH_2 \xrightarrow[\text{(peroxides)}]{HBr} CH_3—\underset{\vert}{\overset{CH_3}{C}}H—CH_2Br \quad (14.31)$$
1-bromomethylpropane

Reactions of ethene and propene with sulphuric acid (reaction 14.11) are important commercially because hydrolysis of the resulting alkyl hydrogen sulphates gives ethanol and 2-propanol respectively (reaction 14.32). Both of these alcohols are important industrial chemicals.

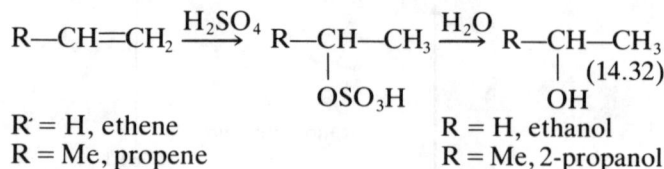

$$R—CH{=}CH_2 \xrightarrow{H_2SO_4} R—\underset{OSO_3H}{\overset{\vert}{C}}H—CH_3 \xrightarrow{H_2O} R—\underset{OH}{\overset{\vert}{C}}H—CH_3 \quad (14.32)$$

R = H, ethene
R = Me, propene

R = H, ethanol
R = Me, 2-propanol

With more reactive alkenes (e.g. methylpropene) it is possible to add water directly to the double bond in acid solution (reaction 14.33).

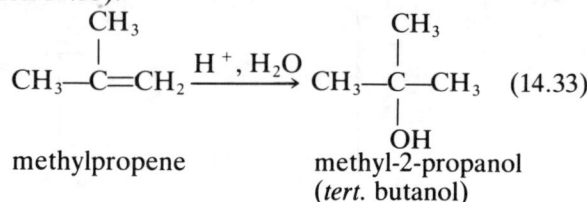

$$CH_3—\underset{}{\overset{CH_3}{\vert}}C{=}CH_2 \xrightarrow{H^+, H_2O} CH_3—\underset{OH}{\overset{CH_3}{\vert}}C—CH_3 \quad (14.33)$$

methylpropene

methyl-2-propanol
(*tert.* butanol)

Because of the orientation inherent in acid catalysed addition to an olefin there are some alcohols that cannot be made in this way, e.g. 1-propanol from propene. However, 1-propanol can be made by the addition of diborane to propene (Eq. 14.34), where the B—H bond of diborane is polarized so that boron is electrophilic and hydrogen nucleophilic. The boron adduct is conveniently oxidized by peroxide to give the alcohol. Overall this reaction sequence represents anti-Markovnikov addition of water, but the initial step, addition of diborane, is consistent with the mechanistic principles outlined earlier bearing in mind that the boron is the electrophile and hydrogen the nucleophile.

Alkenes commonly undergo rearrangement in acid solution. For example, 3,3-dimethyl-1-butene in sequence 14.35 gives both 3,3-dimethyl-2-butanol and 2,3-dimethyl-2-butanol by reaction with sulphuric acid, followed by hydrolysis of the intermediate alkyl hydrogen sulphates. The latter product arises by a methyl shift from C(3) in the secondary carbocation to C(2), forming the more stable tertiary cation. This shift of methyl group, with its bonding electrons, to an adjacent carbon is referred to as a 1,2-methyl shift.

Oxymercuration–demercuration (Eq. 14.14 and also shown in sequence 14.35) can be used in circumstances where Markovnikov addition of water is required and where products of rearrangement are not wanted.

Addition of halogens to alkenes (Eq. 14.9) involves a reaction process very similar to the addition of hydrogen halide. The reaction is initiated by the positive end of a halogen dipole reacting with the π electrons of the alkene. The second step of the reaction involves attack by the halide nucleophile. If sodium chloride is added to a reaction

$$CH_3—CH=CH_2 \xrightarrow{(BH_3)_2} CH_3—\underset{\underset{H}{|}}{CH}—\underset{\underset{\underset{H}{|}}{B—H}}{CH_2} \xrightarrow[OH^-]{H_2O_2} CH_3—CH_2—\underset{\underset{OH}{|}}{CH_2}$$

1-propanol (14.34)

$$\underset{\underset{CH_3}{|}}{\overset{\overset{CH_3}{|}}{CH_3—CH}}—CH=CH_2 \xrightarrow{H_2SO_4} CH_3—\overset{\overset{CH_3}{|}}{\underset{\underset{CH_3}{|}}{C}}{}^+—CH—CH_3 \underset{\text{shift}}{\overset{\text{1,2-methyl}}{\rightleftharpoons}} CH_3—\overset{+}{\underset{\underset{CH_3}{|}}{C}}—\overset{\overset{CH_3}{|}}{CH}—CH_3$$

3,3-dimethyl-1-butene   secondary carbonium ion   tertiary carbonium ion

oxymercuration Hg(OAc)₂ H₂O   ↓ H₂SO₄   ↓ H₂SO₄

$$CH_3—\overset{\overset{CH_3}{|}}{\underset{\underset{CH_3}{|}}{C}}—\overset{}{\underset{\underset{OSO_3H}{|}}{CH}}—CH_3$$ $$CH_3—\overset{\overset{OSO_3H}{|}}{\underset{\underset{CH_3}{|}}{C}}—\overset{}{\underset{\underset{CH_3}{|}}{CH}}—CH_3$$

↓ H₂O   ↓ H₂O

$$CH_3—\overset{\overset{CH_3}{|}}{\underset{\underset{CH_3\ OH}{|\ \ |}}{C}}—\overset{}{\underset{\underset{Hg—O—C—CH_3}{|\ \ \ \ \ \ \ \ \|}}{CH}}—CH_2 \xrightarrow{NaBH_4} CH_3—\overset{\overset{CH_3}{|}}{\underset{\underset{CH_3\ OH}{|\ \ |}}{C}}—CH—CH_3$$ $$CH_3—\overset{\overset{OH}{|}}{\underset{\underset{CH_3\ CH_3}{|\ \ \ |}}{C}}—CH—CH_3$$

O   3,3-dimethyl-2-butanol   2,3-dimethyl-2-butanol (14.35)

$$CH_3—CH=CH_2 \xrightarrow{Br_2,\ Cl^-} CH_3—\overset{\overset{Br}{|}}{\underset{\underset{H}{|}}{C}}—CH_2Br + CH_3—\overset{\overset{Cl}{|}}{\underset{\underset{H}{|}}{C}}—CH_2Br$$

1,2-dibromopropane   1-bromo-2-chloropropane (14.36)

mixture of propene and bromine the added chloride ions can compete with bromide as nucleophiles and both 1,2-dibromopropane and 1-bromo-2-chloropropane are formed (reaction 14.36).

Propene fails to react with sodium chloride in the absence of bromine, thereby demonstrating that the reaction is initiated not by the chloride ion, a nucleophile, but by an electrophile. The formation of 1-bromo-2-chloropropane in reaction 14.36 demonstrates that the nucleophile becomes attached to the carbon with the least number of hydrogens,

and these observations are consistent with a two step mechanism. The first step is addition of electrophile ($Br^{\delta+}$—$Br^{\delta-}$ or Br^+) and the second is attack by the nucleophile ($Br^{\delta-}$—$Br^{\delta+}$ or Br^-).

Hydrogen adds to alkenes (reaction 14.8) in the presence of metal catalysts. The reaction occurs at the surface of the metal, and is exothermic. The metal catalyst lowers the energy of activation and thereby increases the rate of reaction, but does not affect the course of the reaction, the position of equilibrium, or the net energy change in the overall process (see Fig. 14.20). The enthalpy difference between the product alkane and the starting alkene is referred to as the heat of hydrogenation.

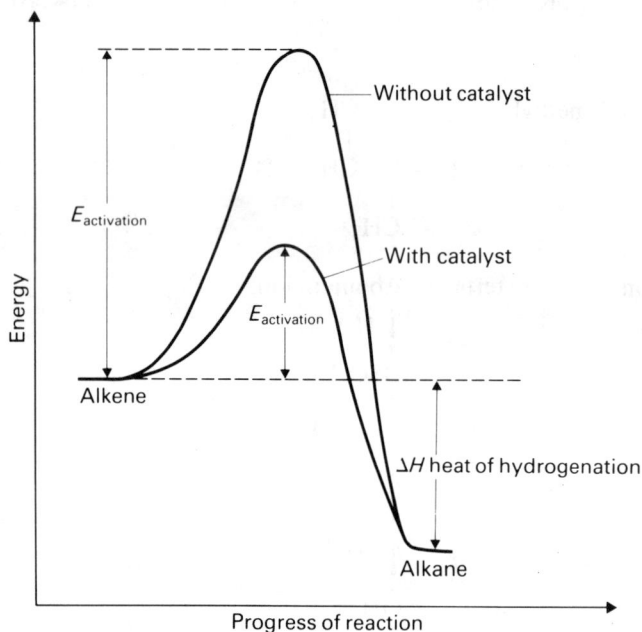

Fig. 14.20 Effect of catalyst on the addition of hydrogen to an alkene

Alkenes react with peroxy-acids to give oxiranes (epoxides) (reaction 14.19). A particularly convenient reagent for this purpose is *meta*-chloroperbenzoic acid, which is stable at room temperature and is available commercially. Oxiranes are valuable intermediates in organic synthesis. Reaction of an alkene with ozone (reaction 14.20) results in formation of an ozonide, which on reaction with water and zinc gives a doubly bonded oxygen attached to each carbon atom of the initial double bond. The overall reaction is shown for methylpropene in Eq. 14.37. By isolating and identifying the carbonyl products (aldehydes or ketones) the structure of the starting alkene can be deduced. In the past this method has been used to establish the identity of a number of alkenes.

14.5 Stereochemistry of Selected Addition Reactions

Oxidation of an alkene by carefully controlled reaction with permanganate produces a diol (Eq. 14.22). The same diol can be obtained by allowing the alkene to react with osmium tetroxide followed by reduction of the resulting osmate ester (Eq. 14.21). We will examine the stereochemistry of products of these reactions in detail. The reagents approach the alkene perpendicular to the plane of the molecule and form cyclic intermediates (sequence 14.38). Both oxygens of the resulting diol are therefore on the same face of the molecule—either both above or both below—and such reactions are often referred to as *cis*-addition reactions.

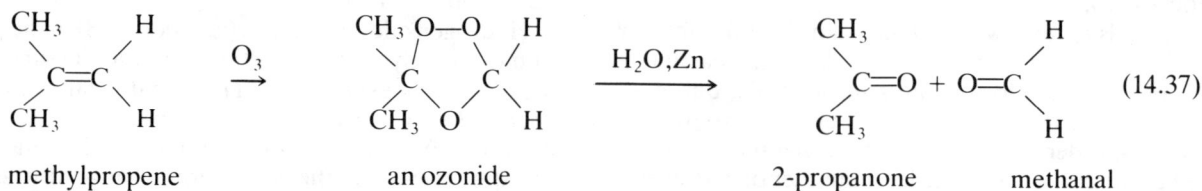

Fig. 14.21 *meso*-2,3-butanediol

In the reaction of *cis*-2-butene with permanganate or osmium tetroxide the reagent can attack from either the top or the bottom face. In either case the product is the same and this is shown in sequence 14.39. If we turn structure II so that the hydroxyls point upward, as shown at III, and rotate this structure so that the back carbon comes to the front, structure I is produced. It can be therefore seen that structure II is the same as structure I. This diol has a symmetry such that each end of the molecule is the mirror image of the other, as shown in Fig. 14.21. Compounds which have this symmetry are referred to as having a *meso* structure. Reaction of *trans*-2-butene with permanganate produces two diols, depending on whether attack is from

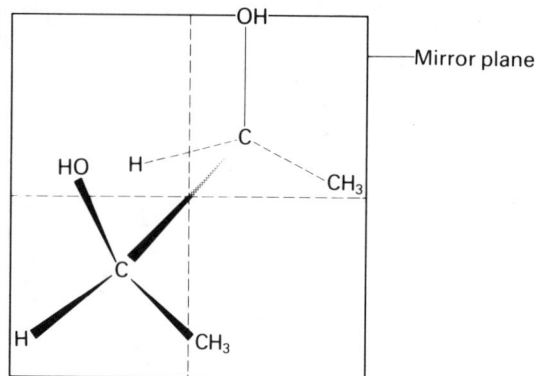

methylpropene   an ozonide   2-propanone   methanal

$$\begin{array}{c}
\text{CH}_3 \quad\quad \text{H} \\
\diagdown \quad \diagup \\
\text{C}=\text{C} \\
\diagup \quad \diagdown \\
\text{CH}_3 \quad\quad \text{H}
\end{array} \xrightarrow{\text{O}_3} \begin{array}{c}
\text{CH}_3 \; \text{O}-\text{O} \; \text{H} \\
\diagdown \;|\quad\quad|\; \diagup \\
\text{C}\quad\quad\text{C} \\
\diagup \;|\quad\quad|\; \diagdown \\
\text{CH}_3 \; \text{O} \quad\; \text{H}
\end{array} \xrightarrow{\text{H}_2\text{O,Zn}} \begin{array}{c}
\text{CH}_3 \quad\quad\quad\quad \text{H} \\
\diagdown \quad\quad\quad\quad \diagup \\
\text{C}=\text{O} + \text{O}=\text{C} \\
\diagup \quad\quad\quad\quad \diagdown \\
\text{CH}_3 \quad\quad\quad\quad \text{H}
\end{array} \quad (14.37)$$

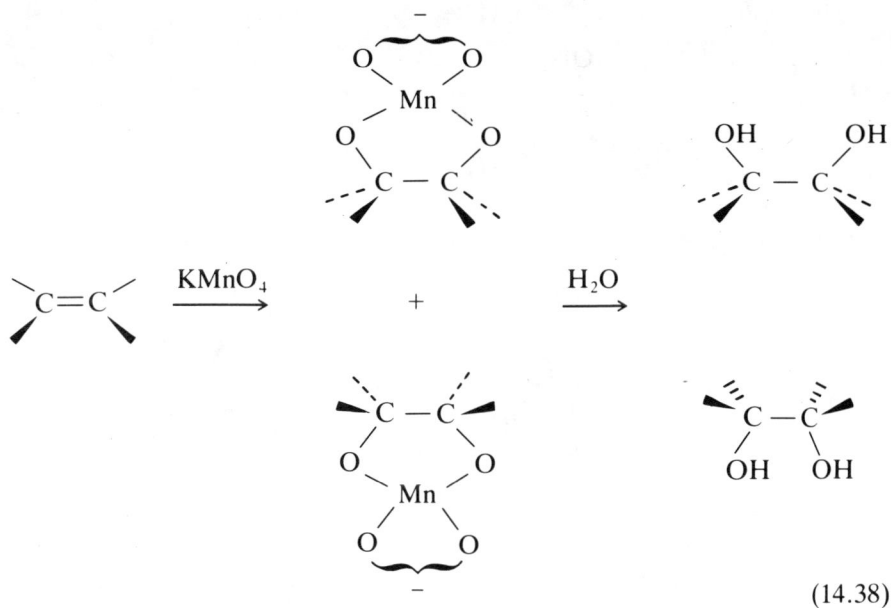

$$(14.38)$$

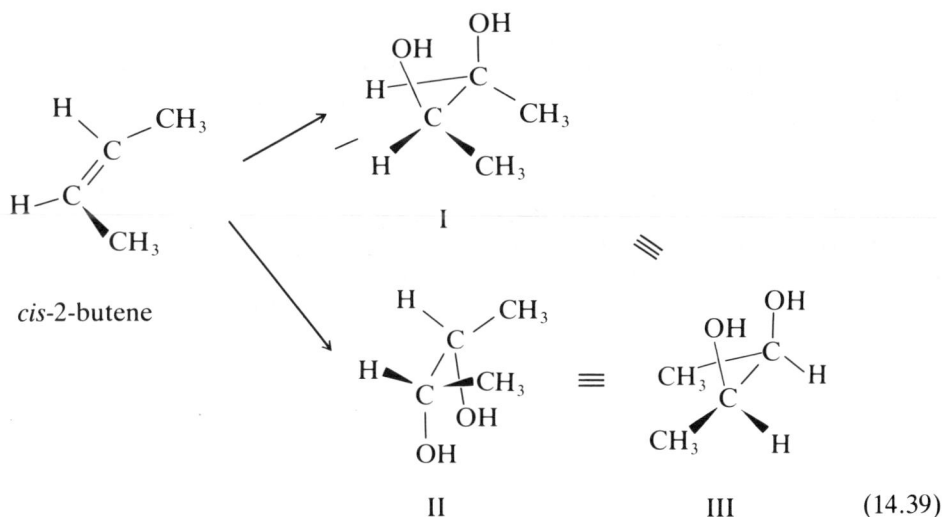

$$(14.39)$$

above or below the plane of the molecule, and this is shown in sequence 14.40. Since the alkene is symmetrical with respect to a plane passing through the carbon atoms the reagent will approach with equal probability from above or below the plane, to produce a 50:50 mixture of the diols I and II. Diol II can be viewed from another perspective, as III. There is, however, no way in which the products I and II can be reoriented to make them appear identical. They each have the same structural formula, in that the same

pairs of atoms are joined together, but they are unique and different stereoisomers.

To help our understanding of this aspect of stereochemistry we will consider a compound containing one carbon atom joined to four different substituent groups. There are two unique ways of joining these four different groups to the central carbon atom; the two structures are related in that one is a mirror image of the other, and the two forms, which are non-superimposable, are termed

CH₃ ... H
C
C
H ... CH₃

→

I

II ≡ III (14.40)

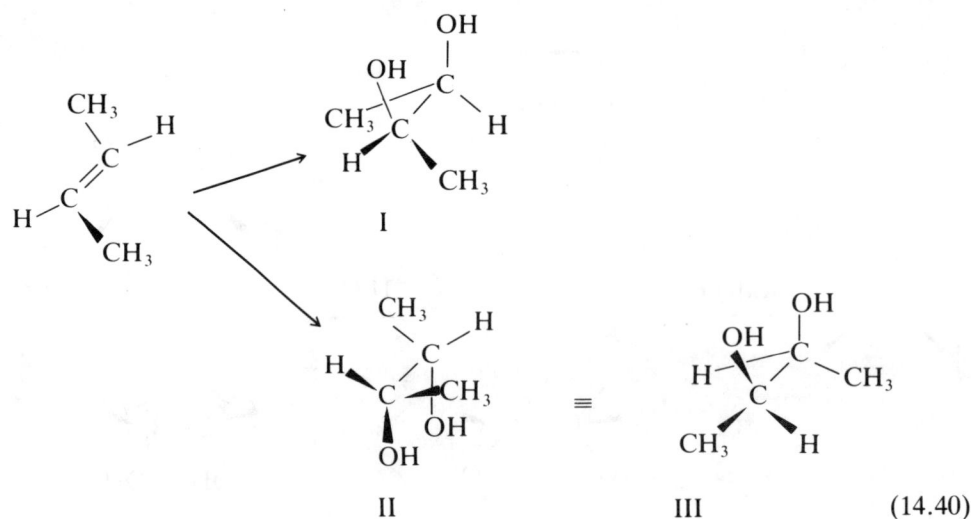

stereoisomers (Fig. 14.22). The compounds have the same structural formula, and differ only in the way the groups are oriented in space. Structures which are related as mirror images are referred to as *enantiomers*. A single carbon atom in a molecule must carry four different groups before the molecule can exist as two distinct enantiomers. A carbon atom which is joined to four different groups is referred to as an *asymmetric* or *chiral* carbon. Enantiomers have identical physical and chemical properties, but differ in the direction in which they rotate the plane of polarization of polarized light. A one to one mixture of enantiomers is termed a *racemic* mixture or *racemate*; such a mixture does not rotate the plane of polarized light because the effect of one optically active species is exactly cancelled by the other.

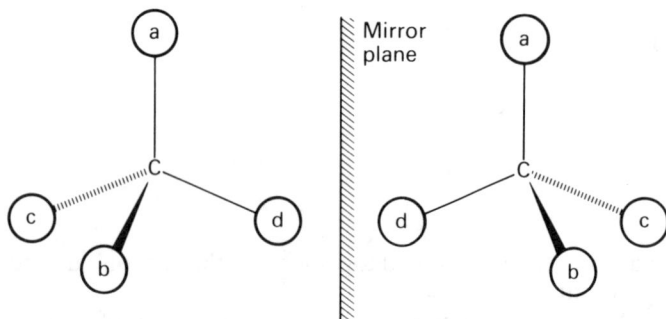

Fig. 14.22 Enantiomers

The measure of optical activity is the *molecular rotation*, the amount by which plane polarized light travelling a unit optical path through a standard solution of an optically active compound has its plane of polarization rotated. The molecular rotation in degrees is determined with an instrument called a polarimeter. The magnitude of the rotation varies with the wavelength of the polarized light. The size of the angle through which the plane is rotated is proportional to the number of molecules of the substance in the path of the light beam, i.e. to the product of the concentration and the optical path length through the cell containing the solution. The *specific rotation* of any compound is the number of degrees of rotation observed if a 1 decimetre tube is used, and the compound is present in a concentration of $1 \, g \, cm^{-3}$.

The arrangement in space of atoms about a chiral centre is referred to as the *configuration*. The configuration of a chiral centre can be defined as R or S according to the following rules. The four atoms attached to the chiral centre are arranged in sequence according to atomic number. The atom, joined to the carbon, with the highest atomic number is then given the highest sequence priority. If the sequence cannot be determined by this rule then it is determined by similar comparison of the next atoms on the group (and so on). The molecule is then visualized as being oriented with the group of lowest priority (commonly hydrogen) directed away from the observer. If, looking from the carbon atom towards this atom or group, the direction from the group of first to second to third sequence priority is in a clockwise direction the configuration is classified as R (Latin: rectus, right) and if anti-clockwise the configuration is classified as S (Latin: sinister, left). The examples in Figs. 14.23–14.25 illustrate these rules of nomenclature (see also Chapter 18). For the structure shown in Fig. 14.25, to decide the order of preference of the two carbon substitutents —CH₂OH and —CH(CH₃)₂, the atoms attached are placed

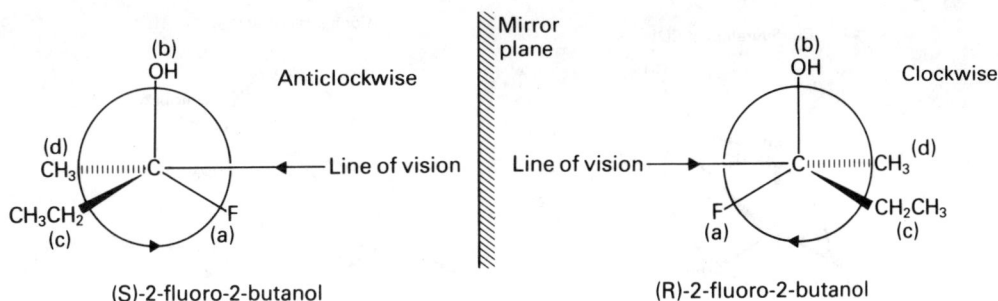

(S)-2-fluoro-2-butanol   (R)-2-fluoro-2-butanol

Fig. 14.23

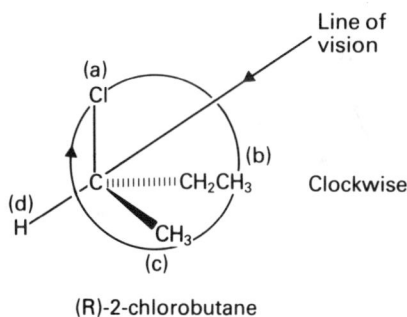

(R)-2-chlorobutane

Fig. 14.24

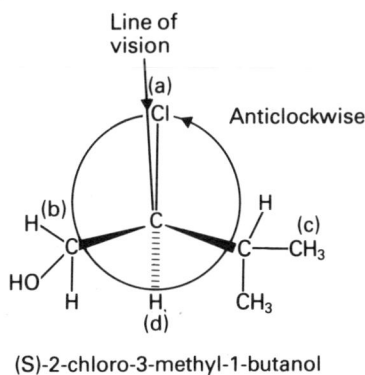

(S)-2-chloro-3-methyl-1-butanol

Fig. 14.25

in order of atomic number (—C(O,H,H) and —C(C,C,H). We then compare the preferred atom of one set with that of the other, i.e. O with C. The group containing the atom of highest atomic number (i.e. —CH_2OH) has the higher priority.

The 2,3-butanediols obtained from reaction of *cis* and *trans*-2-butene with potassium permanganate (reactions 14.39 and 14.40 respectively) have two chiral carbon atoms

and are shown in Fig. 14.26. The structures are drawn several times in order to illustrate the definition of configuration at each chiral carbon atom. Because the substituents at the chiral carbons (C(2) and C(3)) are the same, namely H, CH_3, OH and —CH(OH)CH_3 (see also scheme 14.39) the (2R,3S)-configuration is identical to the (3S,2R)-configuration, and the structure is referred to as *meso* (see Fig. 14.21). The (2R,3R)- and (2S,3S)-diols are mirror images, or enantiomers, as shown in Fig. 14.27 and therefore have the same physical and chemical properties and differ only in their reactions with optically active reagents and in the manner in which they rotate plane polarized light. Stereoisomers of this type which are not mirror images or enantiomers of each other are called *diastereoisomers*. The (R,R)- and (S,S)-structures are diastereoisomers of the (R,S)-structure. Diastereoisomers have different physical properties and can be chemically distinguished.

The plane of symmetry of a *meso*-compound (Fig. 14.21) is lost if substituents on each chiral carbon atom are different. For example C(2) and C(3) in 2,3-pentanediol differ in being bonded to a methyl and an ethyl group respectively. For this structure the (2R,3S)- and (2S,3R)-configurations are different, but related as mirror images (Fig. 14.28). A compound which contains two chiral carbon atoms can exist in as many as four unique stereoisomers; (R,R), (S,S), (R,S) and (S,R).

The addition of bromine to alkenes (reaction 14.9) is a two step process. The first step of the reaction involves attack by the electrophilic bromine on the π electrons to form the bromonium ion, which is subsequently attacked by bromide ion (sequence 14.41). In the bromonium ion the bromine atom is bonded more or less equally between the two carbon atoms.

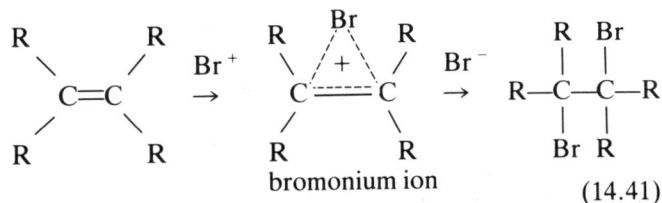

$$(14.41)$$

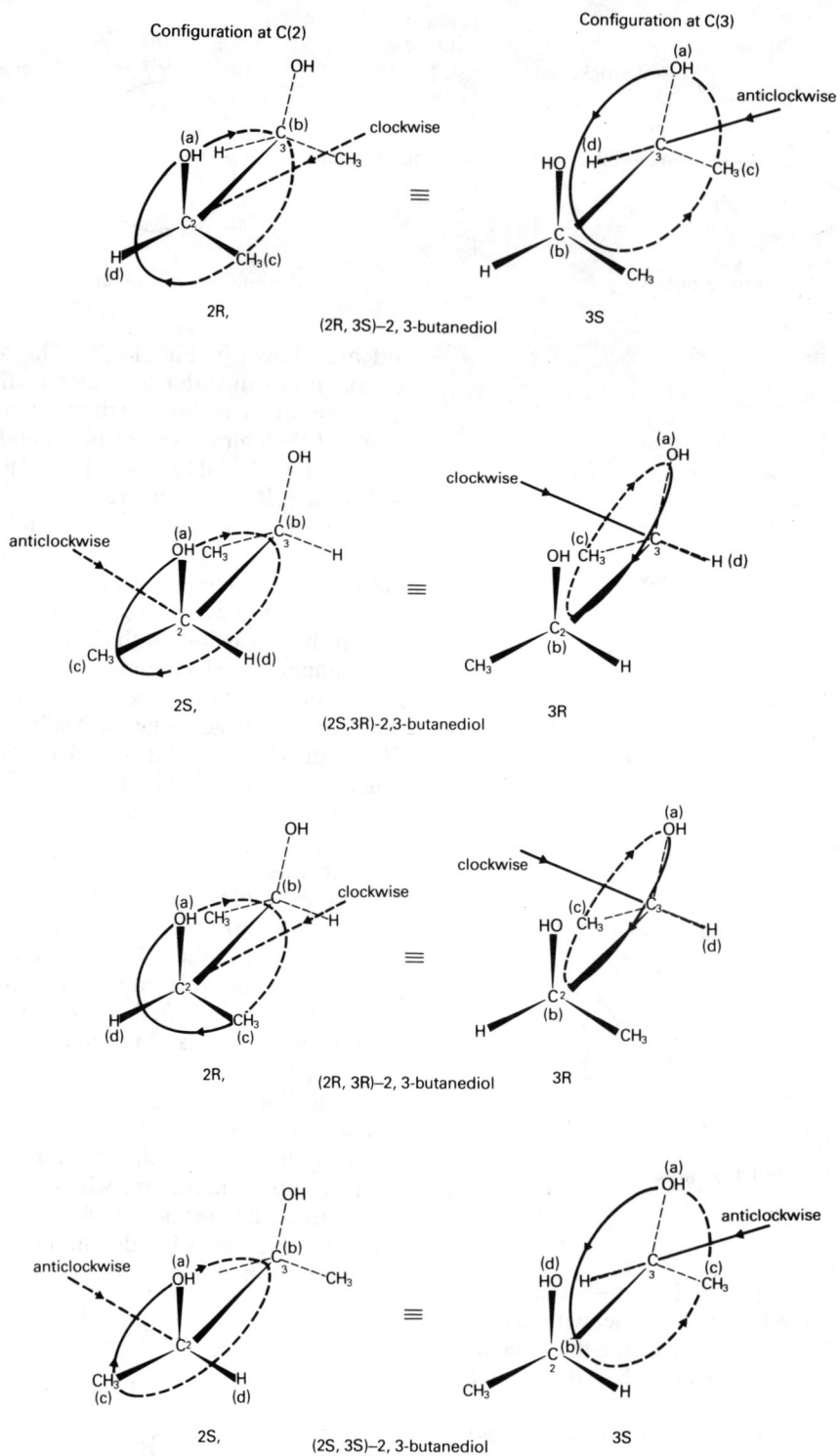

Fig. 14.26 Configurations of 2,3-butanediols

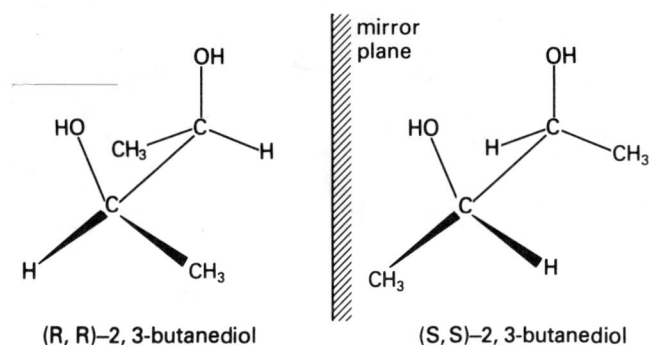

(R, R)–2, 3-butanediol   (S, S)–2, 3-butanediol

Fig. 14.27

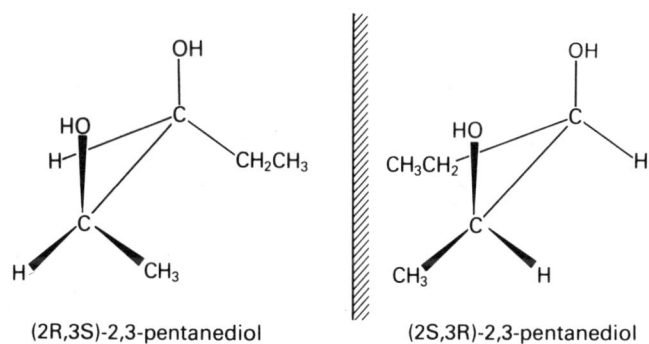

(2R,3S)-2,3-pentanediol   (2S,3R)-2,3-pentanediol

Fig. 14.28

Bromine addition to alkenes generally occurs with complete stereospecificity, with the bromine atoms approaching the alkene one from each face of the molecule (*anti*-addition). Reaction of *cis*-2-butene with bromine gives a 50:50 mixture of (2S,3S)- and (2R,3R)-dibromobutanes (Eq. 14.42), while the similar reaction with *trans*-2-butene gives only *meso* or (2R,3S)-dibromobutane (Eq.14.43). The marked stereospecificity of reaction of bromine with *cis*- and *trans*-2-butene cannot be accounted for if a species with the bromine atom attached to only one carbon atom is a reaction intermediate. Both alkenes would give rise to classical carbonium ions related by rotation about the C(2),C(3) bond as shown in Fig. 14.29. Since rotation

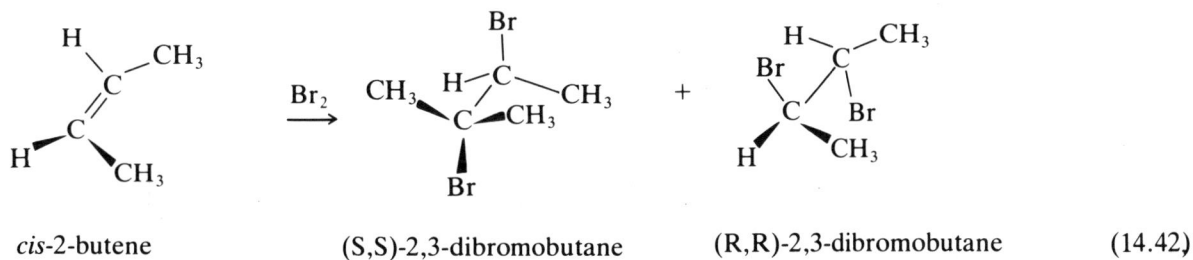

trans-2-butene   (R,S)-2,3-dibromobutane

(14.43)

about this bond would be expected at room temperature, the product mixture of dibromobutanes obtained from *cis*- and *trans*-2-butene would be independent of the configuration of the starting alkene. The bromonium ion intermediate however is not free to rotate about the carbon–carbon bond and hence the integrity of the starting alkene is maintained. The second step of the reaction involves attack by bromide ion from the reverse side of the molecule and the overall process is represented in Fig. 14.30.

14.6 Sigmatropic Reaction

A *sigmatropic reaction* (e.g. reaction 14.44) is one which involves the migration to a new position of a σ bond adjacent to a π electron system.

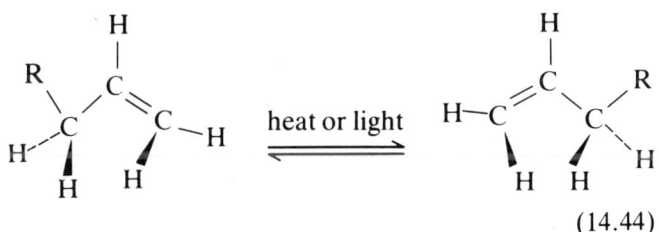

(14.44)

The process involves an uncatalysed intramolecular reaction, and can be brought about by either heat or light. The reaction may go undetected if, as in the above example, the product is the same as the reactant. A reaction where the product is the same as the reactant is termed a *degenerate reaction*. Sigmatropic reactions often occur in quite complex molecular systems. One of the simplest known examples is the conversion of 3,3,4-trimethly-4-penten-2-one by light

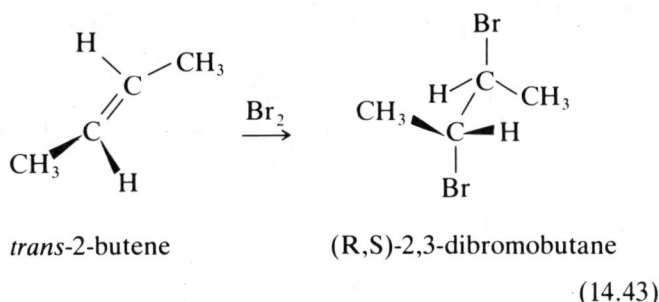

cis-2-butene   (S,S)-2,3-dibromobutane   (R,R)-2,3-dibromobutane   (14.42)

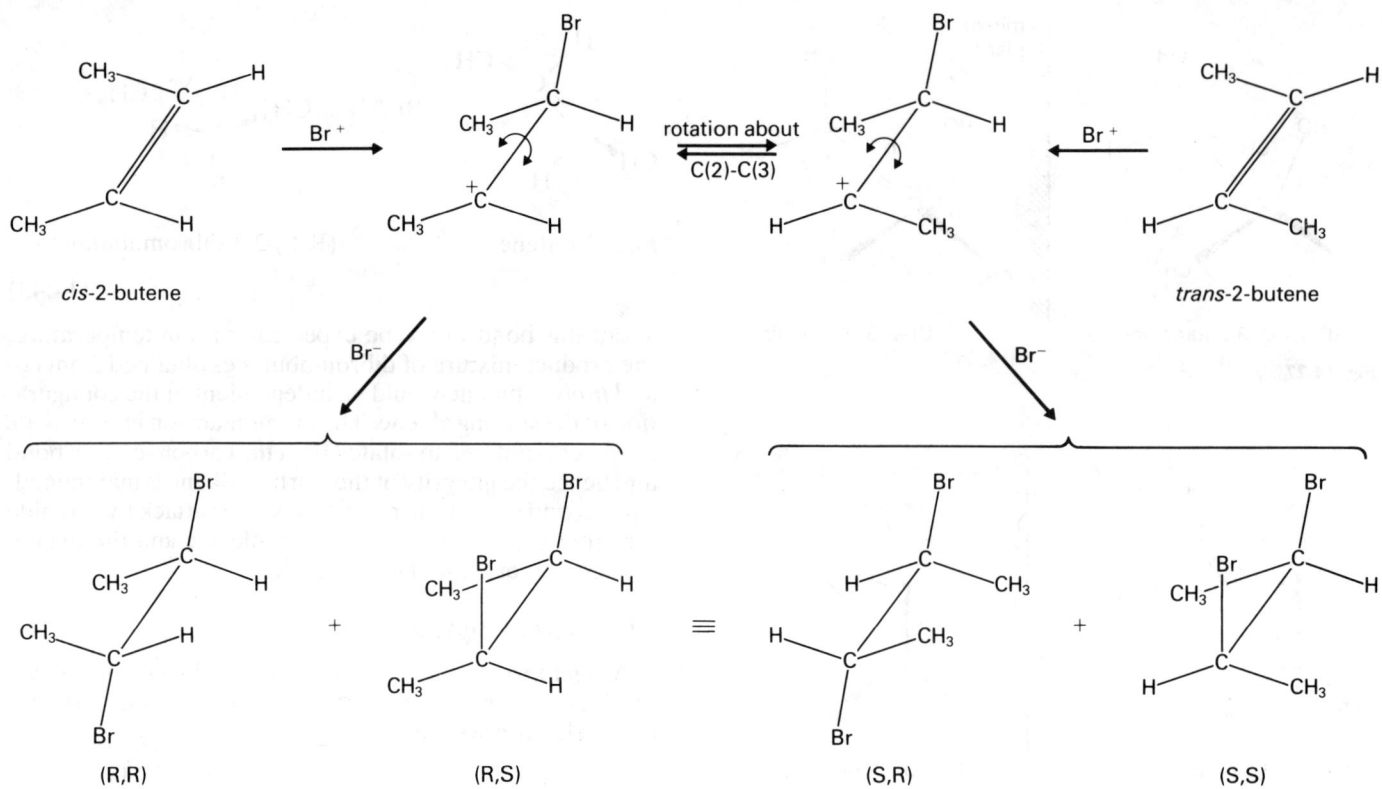

Fig. 14.29 Non-stereospecific addition of bromine to *cis*- and *trans*-2-butene

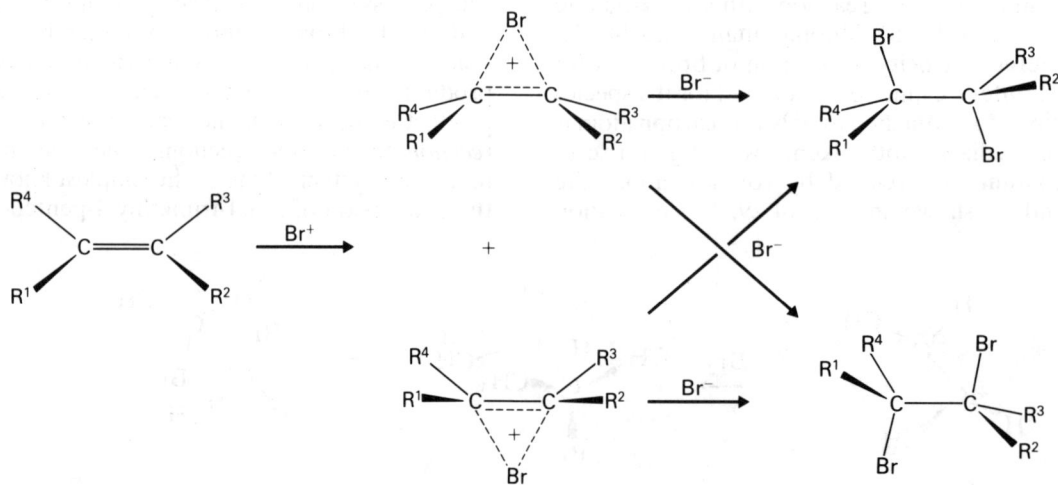

Fig. 14.30 Stereospecific addition of bromine to an alkene

to 4,5-dimethyl-4-hexen-2-one (Eq. 14.45). A detailed examination of reactions of this type is outside the scope of this text.

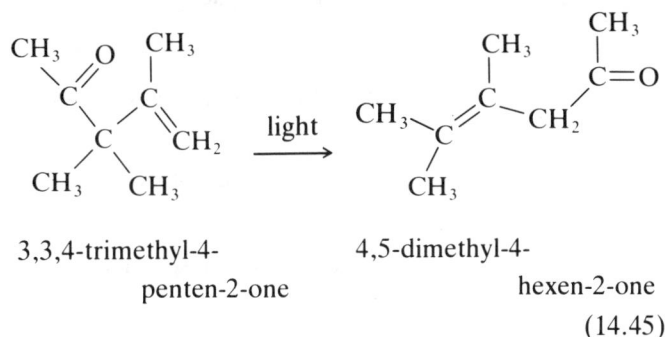

3,3,4-trimethyl-4-
penten-2-one

4,5-dimethyl-4-
hexen-2-one

(14.45)

14.7 Polymers

Polymerization (reactions 14.16–14.18) is an important industrial process and a vast number of products are now manufactured from polymeric materials. Polymers are built up by joining together monomer units. There are two main types of polymers: an *addition polymer* is one formed from unsaturated hydrocarbons, while a *condensation polymer* is one in which a small molecule, often water, is eliminated in a condensation reaction between two monomer units. Polymers can be prepared which contain more than one type of monomer unit—proteins are an example. They are described in Chapter 19.

Polymerization of alkenes can be initiated by free radicals (reaction 14.16), by cations (reaction 14.17) or by anions (reaction 14.18). A large variety of polymers made from alkene derivatives exist—some are shown in Table 14.6 (page 250). Polymers of importance include synthetic elastomers, polyphenylethene (polystyrene), polyethene (polyethylene), polyethanoxyethene (polyvinylacetate), polychloroethene (polyvinyl chloride), polymethylmethacrylate (perspex), polytetrafluoroethene (teflon), orlon, nylon, terylene (or dacron), and most of the ion exchange resins. Natural rubber is also an alkene polymer. Polyester and polyamide polymers are examples of condensation polymers in which two types of monomer units are copolymerized. Teflon and Nylon 66 are examples of this type. The properties of polymers depend on the nature of any attached substituents and side chains, the degree of cross-linking between chains, the number of monomer units in each polymer molecule, the regularity (or lack of it) of the stereochemistry along the polymer chain, and the strengths of the bonds within the chain.

14.8 Alkynes: Structure, Nomenclature, Preparation and Reactions

The parent member of the alkyne family is ethyne (acetylene), Fig. 14.2, which can be made from readily available raw materials—coal, limestone and water, and by the controlled partial oxidation of methane at high temperatures. This compound is of particular industrial importance because it can be converted into a wide variety of more valuable chemicals.

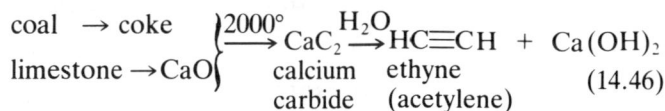

$$coal \rightarrow coke \left. \begin{cases} \xrightarrow{2000°} CaC_2 \xrightarrow{H_2O} HC{\equiv}CH + Ca(OH)_2 \\ \quad\; calcium \quad\; ethyne \\ \quad\; carbide \quad (acetylene) \end{cases} \right. \quad (14.46)$$
$$limestone \rightarrow CaO$$

$$6CH_4 + O_2 \xrightarrow{1500°} 2CH{\equiv}CH + 2CO + 10H_2 \quad (14.47)$$

In the laboratory alkynes can be prepared from dihalides (reaction 14.48, page 251). Terminal alkynes can be modified by reaction of the sodium salt with haloalkanes (reaction 14.49). The naming of alkynes is similar to that for alkenes except that the suffix -ene is replaced by -yne. The parent structure in the name is the longest continuous chain that contains the triple bond, and the position of the multiple bond is given by the number of the first triply-bonded carbon. Numbering is from the end of the chain nearest the triple bond.

The chemistry of alkynes resembles the chemistry of alkenes in that it is intimately linked with the π electrons of the multiple bond. Addition reactions are the most common reactions that occur, but alkynes, which have a —C≡C—H group, show appreciable acidity. This facilitates their use as nucleophiles. The carbon–carbon triple bond is more electronegative than a carbon–carbon single or double bond which helps to stabilize the anion. Ethyne, for example reacts with sodium to give sodium acetylide and hydrogen (reaction 14.50).

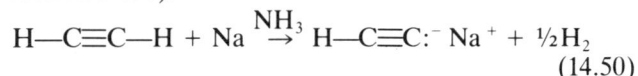

$$H{-}C{\equiv}C{-}H + Na \xrightarrow{NH_3} H{-}C{\equiv}C{:}^- Na^+ + \tfrac{1}{2}H_2$$
$$(14.50)$$

These acetylides are capable of acting as nucleophiles in displacing halogens from haloalkanes (Eq. 14.49). It is possible to reduce disubstituted alkynes to either *cis* or *trans* alkenes by careful choice of reaction conditions (reactions 14.51 and 14.52, page 251). The stereospecific formation of a particular geometrical isomer is a particularly useful synthetic reaction. Alkenes can be further reduced to alkanes by using a more reactive catalyst (Eq. 14.53).

The addition of water to alkynes (reaction 14.56, page 252), catalysed by proton acids or Hg^{2+} ions, is an important industrial process. Addition of water to ethyne (reaction 14.58) gives ethenol (vinyl alcohol) which is the tautomeric or enol form of ethanal.

Table 14.6 Polymers

(1) Addition polymers

monomer	polymer	
$CH_2\!\!=\!\!CH_2$ ethene	$-\!(CH_2\!\!-\!\!CH_2)\!-\!_n$	polyethene (polyethylene)

$CH_2\!\!=\!\!CHCl$
chloroethene
	$-\!(CH_2\!\!-\!\!CH)\!-\!_n$
	    |
	    Cl
		polychloroethene (polyvinyl chloride)

$CH_2\!\!=\!\!CHPh$
phenylethene
	$-\!(CH_2\!\!-\!\!CH)\!-\!_n$
	    |
	    Ph
		polyphenylethene (polystyrene)

$CH_2\!\!=\!\!CHOAc$
ethanoxyethene
	$-\!(CH_2\!\!-\!\!CH)\!-\!_n$
	    |
	    OAc
		polyethanoxyethene (polyvinyl acetate)

$CF_2\!\!=\!\!CF_2$ tetrafluoroethene	$-\!(CF_2\!\!-\!\!CF_2)\!-\!_n$	teflon

$CH_2\!\!=\!\!CHCN$
cyanoethene
	$-\!(CH_2\!\!-\!\!CH)\!-\!_n$
	    |
	    CN
		orlon

CH_3
   |
$CH_2\!\!=\!\!C\!\!-\!\!CH\!\!=\!\!CH_2$
2-methyl-1,3-butadiene
(isoprene)
	CH_3
	   |
	$-\!(CH_2\!\!-\!\!C\!\!=\!\!CH\!\!-\!\!CH_2)\!-\!_n$
		natural rubber

(2) Condensation polymers

$HO\!\!-\!\!CH_2\!\!-\!\!CH_2\!\!-\!\!OH$ 1,2-ethanediol (ethylene glycol) + $CO_2H\!\!-\!\!(CH_2)_4\!\!-\!\!CO_2H$ hexanedioic acid (adipic acid)	$-\!(O\!\!-\!\!(CH_2)_2\!\!-\!\!O\!\!-\!\!\underset{\displaystyle O}{\overset{\displaystyle \|}{C}}\!\!-\!\!(CH_2)_4\!\!-\!\!\underset{\displaystyle O}{\overset{\displaystyle \|}{C}})\!-\!_n$	Dacron
$NH_2\!\!-\!\!CH_2)_6\!\!-\!\!NH_2$ 1,6-diaminohexane (hexamethylenediamine) + $CO_2H\!\!-\!\!(CH_2)_4\!\!-\!\!CO_2H$ hexanedioic acid (adipic acid)	$-\!(NH\!\!-\!\!(CH_2)_6\!\!-\!\!NH\!\!-\!\!\underset{\displaystyle O}{\overset{\displaystyle \|}{C}}\!\!-\!\!(CH_2)_4\!\!-\!\!\underset{\displaystyle O}{\overset{\displaystyle \|}{C}})\!-\!_n$	Nylon 66

$CH_2\!\!-\!\!CH_2$
 | $\backslash$
CH_2 $C\!\!=\!\!O$
 | |
CH_2 NH
  $\backslash$ |
   CH_2
azacyloheptan-2-one
(γ-caprolactam)
	$-\!(\underset{\displaystyle O}{\overset{\displaystyle \|}{C}}\!\!-\!\!(CH_2)_5\!\!-\!\!NH)\!-\!_n$	Nylon 6

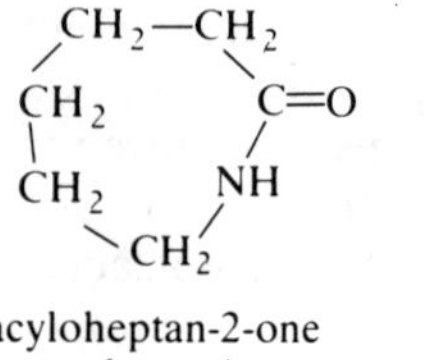

Table 14.7 Preparation of alkynes

Dehydrohalogenation of 1,2-dihaloalkanes

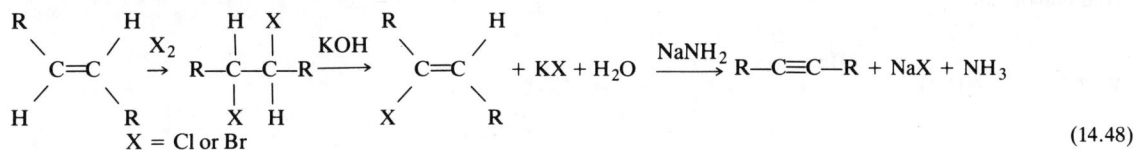

$$\underset{\substack{/\ \ \backslash \\ \text{H} \qquad \text{R}}}{\overset{\substack{\text{R} \qquad \text{H} \\ \backslash \quad /}}{\text{C}=\text{C}}} \xrightarrow{\text{X}_2} \underset{\substack{| \ | \\ \text{X} \ \text{H}}}{\overset{\substack{\text{H} \ \text{X} \\ | \ |}}{\text{R}-\text{C}-\text{C}-\text{R}}} \xrightarrow{\text{KOH}} \underset{\substack{/\ \ \backslash \\ \text{X} \qquad \text{R}}}{\overset{\substack{\text{R} \qquad \text{H} \\ \backslash \quad /}}{\text{C}=\text{C}}} + \text{KX} + \text{H}_2\text{O} \xrightarrow{\text{NaNH}_2} \text{R}-\text{C}\equiv\text{C}-\text{R} + \text{NaX} + \text{NH}_3$$

X = Cl or Br

(14.48)

Reaction of sodium acetylides with primary haloalkanes

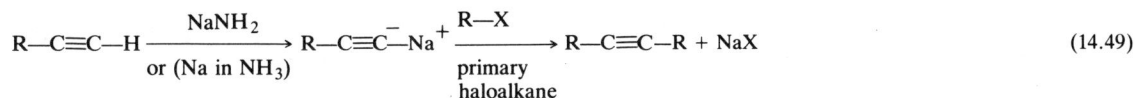

$$\text{R}-\text{C}\equiv\text{C}-\text{H} \xrightarrow[\text{or (Na in NH}_3)]{\text{NaNH}_2} \text{R}-\text{C}\equiv\overset{-}{\text{C}}-\text{Na}^+ \xrightarrow[\substack{\text{primary} \\ \text{haloalkane}}]{\text{R}-\text{X}} \text{R}-\text{C}\equiv\text{C}-\text{R} + \text{NaX}$$

(14.49)

Table 14.8 Reactions of alkynes

Addition of hydrogen

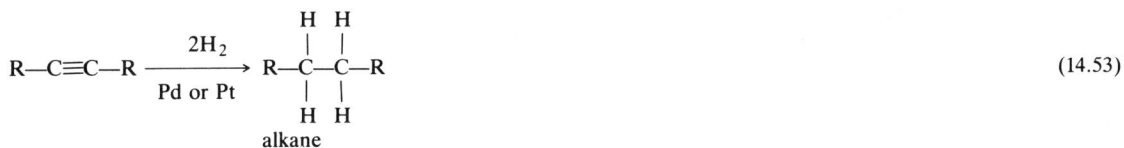

$$\text{R}-\text{C}\equiv\text{C}-\text{R} \xrightarrow{\substack{\text{H}_2,\text{Pd}, \\ \text{BaSO}_4}} \underset{\substack{/\ \ \backslash \\ \text{H} \qquad \text{H}}}{\overset{\substack{\text{R} \qquad \text{R} \\ \backslash \quad /}}{\text{C}=\text{C}}}$$

cis-alkene

(14.51)

$$\xrightarrow{\text{Li},\text{NH}_3} \underset{\substack{/\ \ \backslash \\ \text{H} \qquad \text{R}}}{\overset{\substack{\text{R} \qquad \text{H} \\ \backslash \quad /}}{\text{C}=\text{C}}}$$

trans-alkene

(14.52)

$$\text{R}-\text{C}\equiv\text{C}-\text{R} \xrightarrow[\text{Pd or Pt}]{2\text{H}_2} \underset{\substack{| \ | \\ \text{H} \ \text{H}}}{\overset{\substack{\text{H} \ \text{H} \\ | \ |}}{\text{R}-\text{C}-\text{C}-\text{R}}}$$

alkane

(14.53)

Addition of halogens

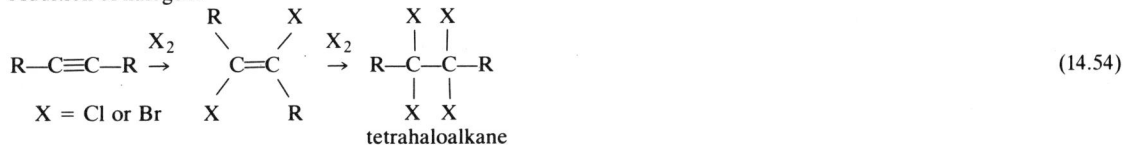

$$\text{R}-\text{C}\equiv\text{C}-\text{R} \xrightarrow{\text{X}_2} \underset{\substack{/\ \ \backslash \\ \text{X} \qquad \text{R}}}{\overset{\substack{\text{R} \qquad \text{X} \\ \backslash \quad /}}{\text{C}=\text{C}}} \xrightarrow{\text{X}_2} \underset{\substack{| \ | \\ \text{X} \ \text{X}}}{\overset{\substack{\text{X} \ \text{X} \\ | \ |}}{\text{R}-\text{C}-\text{C}-\text{R}}}$$

X = Cl or Br

tetrahaloalkane

(14.54)

Addition of hydrogen halides

$$\text{R}-\text{C}\equiv\text{C}-\text{R} \xrightarrow{\text{HX}} \underset{\substack{/\ \ \backslash \\ \text{H} \qquad \text{R}}}{\overset{\substack{\text{R} \qquad \text{X} \\ \backslash \quad /}}{\text{C}=\text{C}}} \xrightarrow{\text{HX}} \underset{\substack{| \ | \\ \text{H} \ \text{X}}}{\overset{\substack{\text{H} \ \text{X} \\ | \ |}}{\text{R}-\text{C}-\text{C}-\text{R}}}$$

X = Cl or Br

dihaloalkane

(14.55)

Table 14.8 Reactions of alkynes (continued)

Addition of water

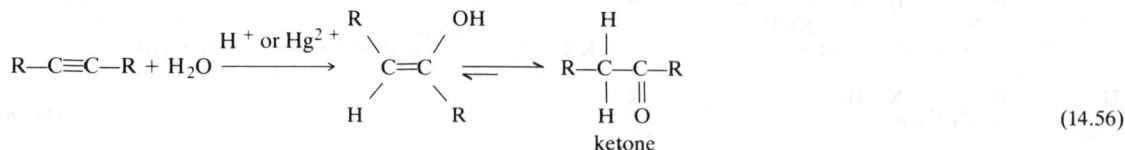

$$R—C{\equiv}C—R + H_2O \xrightarrow{H^+ \text{ or } Hg^{2+}} \begin{matrix} R \\ \diagdown \\ C{=}C \\ \diagup \quad \diagdown \\ H \qquad R \end{matrix} \begin{matrix} OH \\ \diagup \end{matrix} \rightleftarrows \underset{\text{ketone}}{R—\overset{H}{\underset{H}{C}}—\overset{\parallel}{\underset{O}{C}}—R}$$

(14.56)

Formation of metal salt

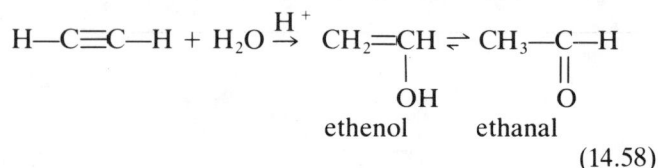

$$R—C{\equiv}C—H + Na \xrightarrow{\text{liq. NH}_3} R—C{\equiv}C—\overset{-}{N}\overset{+}{a} + \tfrac{1}{2}H_2$$

(14.57)

$$H—C{\equiv}C—H + H_2O \xrightarrow{H^+} \underset{\underset{\text{ethenol}}{OH}}{CH_2{=}\overset{|}{CH}} \rightleftarrows \underset{\underset{\text{ethanal}}{O}}{CH_3—\overset{\parallel}{C}—H}$$

(14.58)

Tautomerism is the name given to the property of molecules which exist as a rapidly equilibrating mixture of two isomers. The interconversion normally involves the removal of a proton from a molecule and the addition of it or another proton at a different site in the molecule. Compounds which are related by a facile equilibrium of this sort are referred to as *tautomers*. In the case of ethenol (vinyl alcohol) and ethanal the proton on the hydroxyl group of the alcohol is lost, and another proton is added at C(2). The equilibrium is very much in favour of ethanal. This is a particular example of aldo-enol tautomerism. In aldo-enol tautomerism the species which are in equilibrium are an aldehyde (aldo form) and an unsaturated alcohol (enol form). Ethanal is the carbonyl form of ethenol (vinyl alcohol), ethenol is the enol form of ethanal. Ethanal produced from ethyne by the addition of water is oxidized to ethanoic acid (acetic acid) in a commercial process. Addition of water to propyne produces the enol of 2-propanone (acetone) which rapidly changes to the more stable ketone tautomer (reaction 14.59).

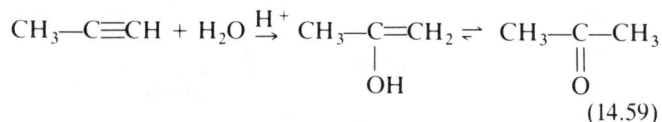

$$CH_3—C{\equiv}CH + H_2O \xrightarrow{H^+} \underset{\underset{OH}{|}}{CH_3—C{=}CH_2} \rightleftarrows \underset{\underset{O}{\parallel}}{CH_3—C—CH_3}$$

(14.59)

Alkynes react with proton acids (reaction 14.55) and with halogens (reaction 14.54) in a similar manner to alkenes.

Exercises

14.1 Write and name the constitutional and geometrical alkene isomers having the molecular formula C_5H_{10}.

14.2 How would you prepare:
 (a) 1-butanol from 1-butene
 (b) 2-butanol from 1-butene
 (c) 2-butyne from 2-butene

14.3 Give the structural formula of the alkenes which on ozonolysis and subsequent hydrolysis will give
 (1) Methanal (formaldehyde) and 2-methylpropanal
 (2) 2-propanone (acetone)

14.4 For the constitutional isomers of molecular formula C_7H_{16} (problems 13.1) which could be resolved into optically active enantiomers? Assign the configuration (R or S) to the enantiomers.

14.5 Which constitutional isomers of monochloromethylbutane (problem 13.2) could be resolved into optical enantiomers?

14.6 Draw and name the diols obtained by oxidation of (E)- and (Z)-2-pentene with potassium permanganate.

14.7 How would you synthesize 2-butanone from 2-butyne.

14.8 How could $CH_3—C{\equiv}C—CD_3$ be prepared from propene and CD_3Cl?

15 Aromatic Compounds

15.1 Aromatic Compounds and their Reactions

Benzene C_6H_6 (Fig. 15.1) can be regarded as the parent of the aromatic class of compounds. In benzene, as in ethene, the carbon framework of the molecule is formed by overlap of sp^2 hybrid orbitals. Benzene is a planar, symmetrical molecule, with each carbon atom lying at the corner of a regular hexagon. Each carbon atom has the remaining electron in the p orbital perpendicular to the plane of the molecule. As noted in Section 3.4, these orbitals overlap both above and below the molecular plane to form three bonding π orbitals, (Fig. 15.2), each of which is able to be occupied by a pair of electrons. The electrons in these π orbitals form a symmetrical cloud of electron density above and below the plane of the carbon atoms (Fig. 15.3).

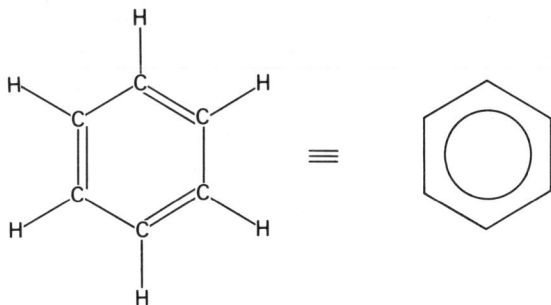

Fig. 15.2 Bonding π orbitals of benzene

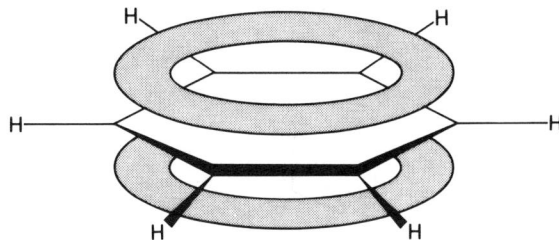

Fig. 15.1 Benzene: the double bonds are not fixed

Electrons in π orbitals are more accessible to electron-seeking reagents (*electrophiles*) than are electrons in σ bonds, and so it is to be expected that the chemistry of unsaturated hydrocarbons with electrophiles will be inextricably linked with the behaviour of these π electrons. The behaviour of π electrons in benzene and other aromatic compounds is significantly different from that of π electrons in alkenes and alkynes.

The distinguishing feature of aromatic molecules is the π electron cloud containing $4n + 2$ electrons surrounding

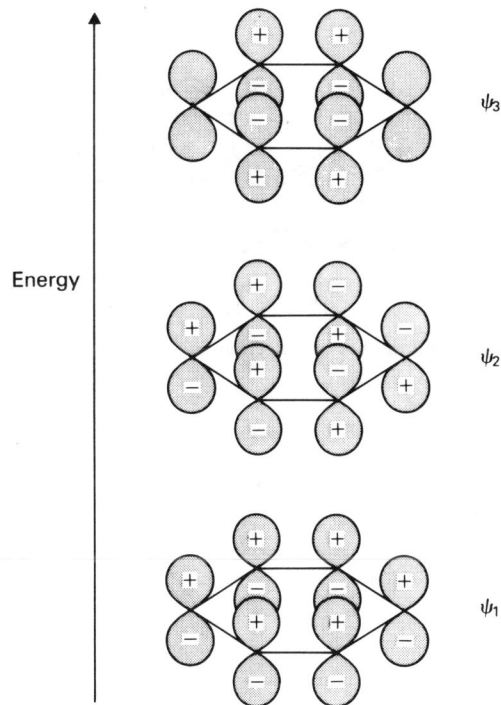

Fig. 15.3 π electron cloud of benzene

both faces of the cyclic molecule. Figures 15.4–9 show some molecules which have 'aromatic character'. The cyclopentadienyl anion (Fig. 15.6) and cycloheptatrienyl cation (Fig. 15.7) are cyclic charged species which contain six

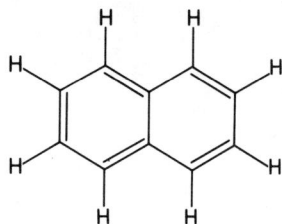

Fig. 15.4 Naphthalene, 10 π electrons, $n = 2$

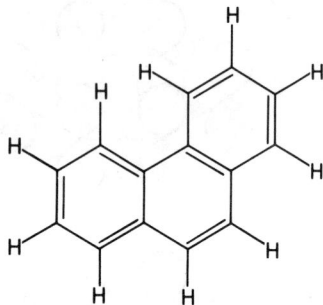

Fig. 15.8 Pyrrole, 6 π electrons, $n = 1$

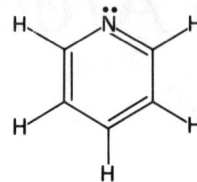

Fig. 15.9 Pyridine, 6 π electrons, $n = 1$

Fig. 15.5 Phenanthracene, 14 π electrons, $n = 3$

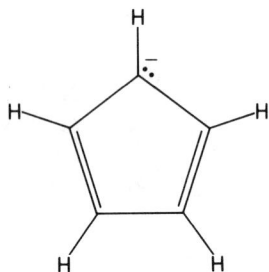

Fig. 15.6 Cyclopentadienyl anion, 6 π electrons, $n = 1$

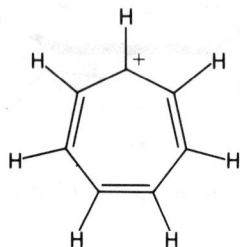

Fig. 15.7 Cycloheptatrienyl cation, 6 π electrons, $n = 1$

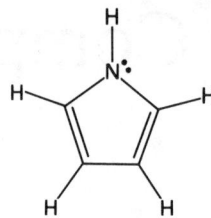

π electrons. These charged species have a number of chemical properties which characterize them as aromatic compounds, notably unusual stability and an unwillingness to undergo typical alkene reactions.

Benzene does not react at all with HCl or HBr and this is at first sight surprising, because alkenes react so readily with proton acids. Bromine *adds* to ethene, which is converted to a compound of molecular formula $C_2H_4Br_2$, but the reaction of bromine with benzene is of an entirely different nature. Benzene (C_6H_6) is converted into a compound with a molecular formula C_6H_5Br. In this reaction, shown in Eq. 15.1, one hydrogen atom of benzene has been replaced by a bromine atom. This is a *substitution* reaction

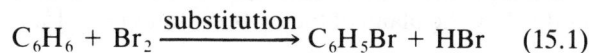

$$C_6H_6 + Br_2 \xrightarrow{\text{substitution}} C_6H_5Br + HBr \quad (15.1)$$

and is characteristic of reactions undergone by aromatic compounds. In summary, alkenes undergo *addition*— aromatic compounds undergo *substitution* reactions.

The most common reactions of benzene and aromatic compounds are initiated by an attacking electrophile and be referred to as *electrophilic substitution reactions*. Such reactions include nitration (reaction 15.2), sulphonation (reaction 15.3), halogenation (reaction 15.4), Friedel– Crafts alkylation (reaction 15.5) and acylation (reaction 15.6). These reactions form the basis of the aromatic chemical industry.

Table 15.1 Reactions of benzene

Nitration

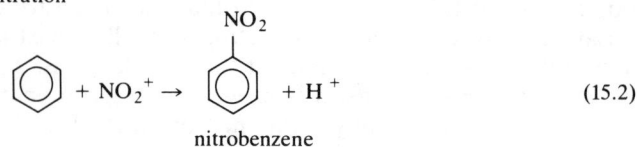

$$\text{electrophile} = NO_2^+$$

$$HNO_3 + H_2SO_4 \rightleftharpoons NO_2^+ + H_3O^+ + 2HSO_4^-$$

Sulphonation

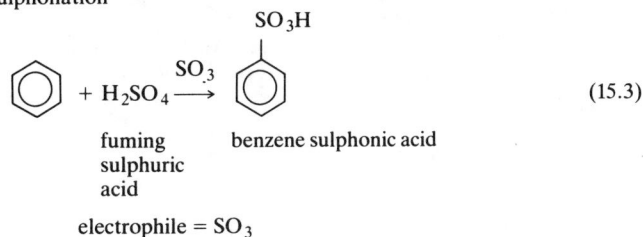

$$\text{benzene} + H_2SO_4 \xrightarrow{SO_3} \text{benzene sulphonic acid (SO}_3H) \qquad (15.3)$$

fuming
sulphuric
acid

benzene sulphonic acid

electrophile = SO_3

Halogenation

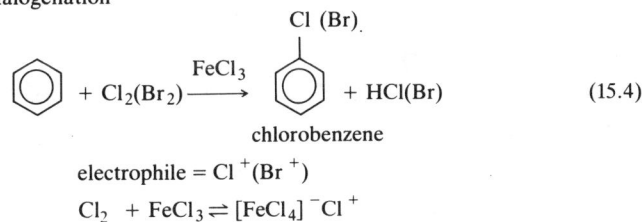

$$\text{benzene} + Cl_2(Br_2) \xrightarrow{FeCl_3} \text{Cl (Br)} + HCl(Br) \qquad (15.4)$$

chlorobenzene

electrophile = $Cl^+ (Br^+)$

$$Cl_2 + FeCl_3 \rightleftharpoons [FeCl_4]^- Cl^+$$

Friedel–Crafts alkylation

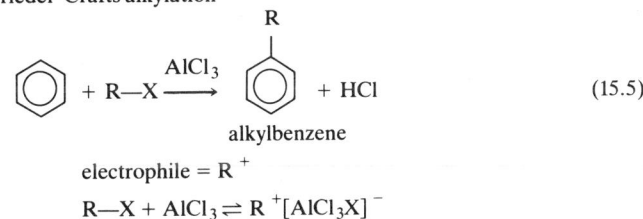

$$\text{benzene} + R{-}X \xrightarrow{AlCl_3} R\text{-alkylbenzene} + HCl \qquad (15.5)$$

alkylbenzene

electrophile = R^+

$$R{-}X + AlCl_3 \rightleftharpoons R^+[AlCl_3X]^-$$

Friedel–Crafts acylation

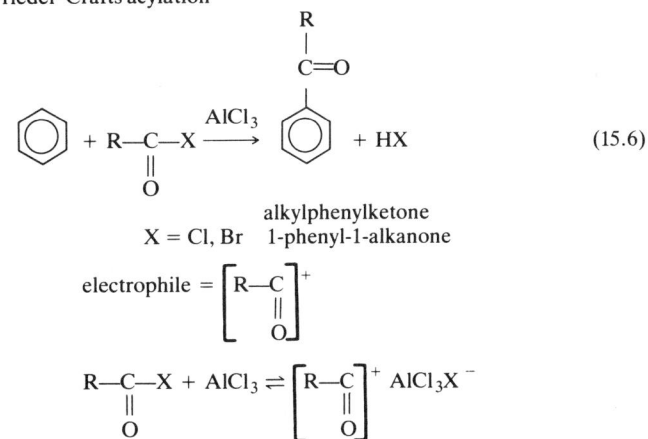

$$\text{benzene} + R{-}C{-}X \xrightarrow{AlCl_3} \underset{\substack{O}}{R-C=O} + HX \qquad (15.6)$$

alkylphenylketone
1-phenyl-1-alkanone

X = Cl, Br

$$\text{electrophile} = \left[R{-}\underset{O}{\overset{\|}{C}} \right]^+$$

$$R{-}\underset{O}{\overset{\|}{C}}{-}X + AlCl_3 \rightleftharpoons \left[R{-}\underset{O}{\overset{\|}{C}} \right]^+ AlCl_3X^-$$

15.2 The Mechanism of Electrophilic Aromatic Substitution

To understand why electrophilic substitution occurs with aromatic compounds, whereas similar reagents react with alkenes to give addition products, it is necessary to examine the detailed mechanisms of the reactions. Electrophilic aromatic substitution is a two step process. The first step involves the electrophile (e.g. NO_2^+, SO_3, X^+, R^+ or $R{-}\underset{O}{\overset{\|}{C}}{}^+$) becoming embedded in the π electron cloud and forming a bond to one of the carbons as shown in scheme 15.7. The positive charge is spread over the remaining five carbons of the benzene ring. Such a species is often called a 'Wheland' intermediate.

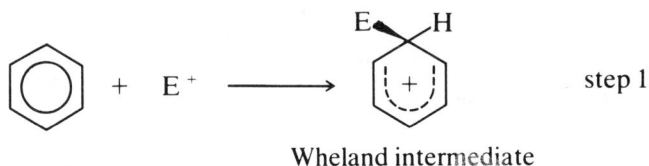

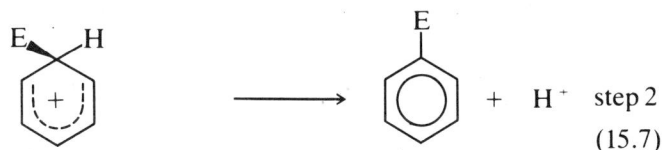

step 1

Wheland intermediate

step 2

(15.7)

These species can commonly be observed at low temperature by nmr techniques, and their presence as intermediates in electrophilic substitution reactions is consistent with kinetic studies. For example, if we replace one hydrogen atom on the benzene ring by deuterium, the rate of substitution at that carbon is almost the same as at any other carbon atom. If cleavage of the carbon–hydrogen or carbon–deuterium bond were involved in the rate-determining step we would expect a significant difference (isotope effect) in reaction rate. Almost all electrophilic substitution reactions reveal no such rate difference so cleavage of a carbon–hydrogen bond is not involved in the rate-determining step. The reaction is therefore a two step process where the transition state for the step involving proton loss is lower in potential energy than the transition state for formation of the Wheland intermediate (Fig. 15.10). If the reaction involved only a single step, where the Wheland species represented a transition state and not an intermediate as shown in scheme 15.8 (page 258), the breaking of the carbon–hydrogen bond would necessarily be involved in the rate-determining step and a kinetic isotope effect would be observed.

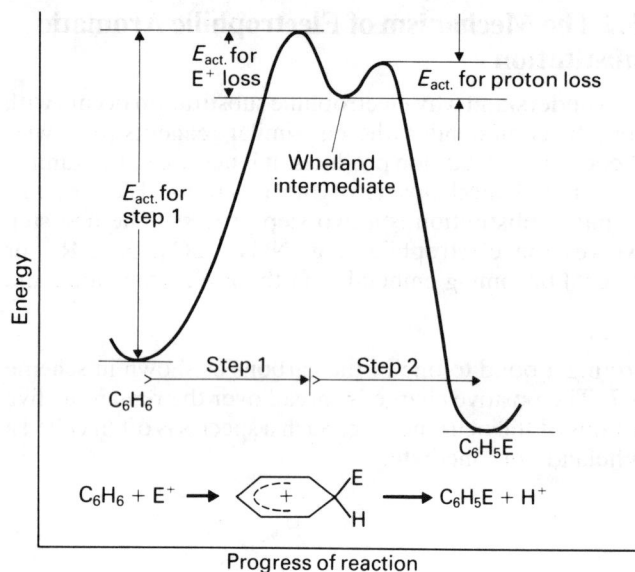

Fig. 15.10 Potential energy changes in electrophilic aromatic substitution

15.3 Modification of Monosubstituted Benzene Derivatives

From the products of nitration (reaction 15.2), sulphonation (reaction 15.3), halogenation (reaction 15.4), Friedel–Crafts alkylation (reaction 15.5) and acylation (reaction 15.6) a number of other benzene derivatives can be prepared by modification of the substituent groups. A series of such reactions are listed in Table 15.2. These include side chain halogenation (15.9–15.11) initiated by light. Halogenation of methylbenzene (15.36 and 15.37) is analogous to the reaction of methane with chlorine discussed in section 13.3. The phenyl group stabilizes an unpaired electron (radical) on the adjacent carbon, and since the formation of this radical is endothermic the activation energy for its formation is lowered compared to the activation energy for formation of a methyl radical. Methylbenzene is therefore even more reactive with chlorine radicals than methane.

Table 15.2 Modification of substituents in monosubstituted benzenes

Side chain halogenation

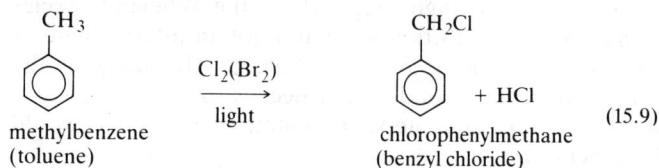

$$\text{(15.9)}$$

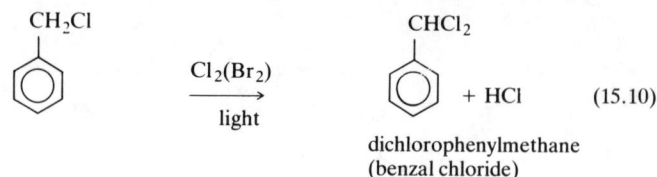

$$\text{(15.10)}$$

dichlorophenylmethane
(benzal chloride)

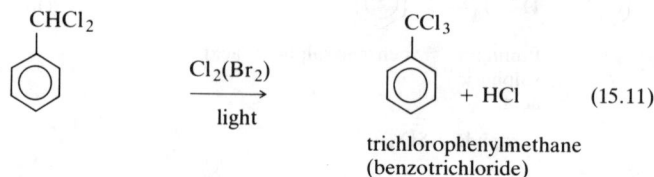

$$\text{(15.11)}$$

trichlorophenylmethane
(benzotrichloride)

Oxidation

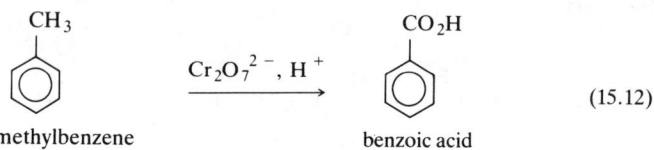

$$\text{(15.12)}$$

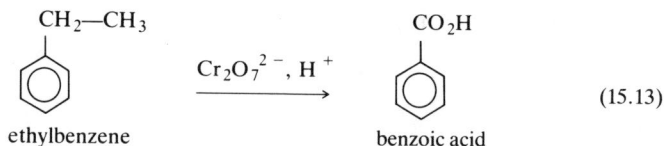

$$\text{(15.13)}$$

Reduction

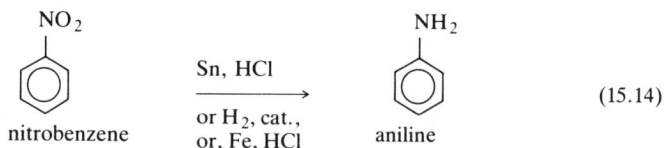

$$\text{(15.14)}$$

Preparation of aryl magnesium halide

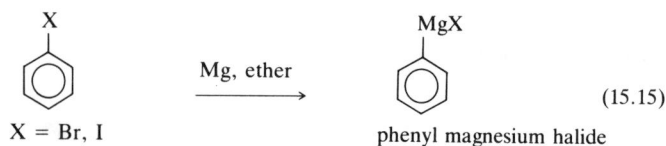

$$\text{(15.15)}$$

Reduction

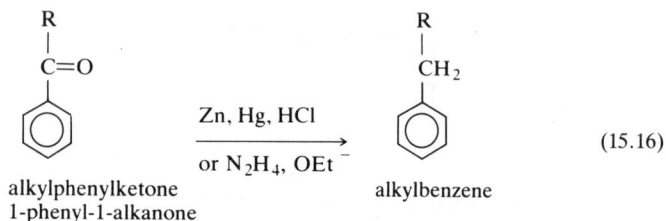

$$\text{(15.16)}$$

Displacement

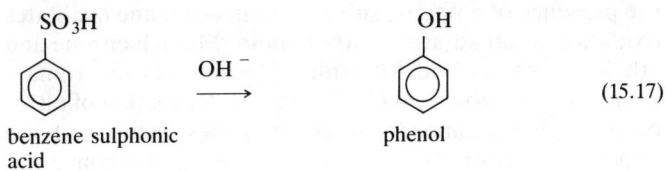

benzene sulphonic acid → phenol   (15.17)

Hydrolysis

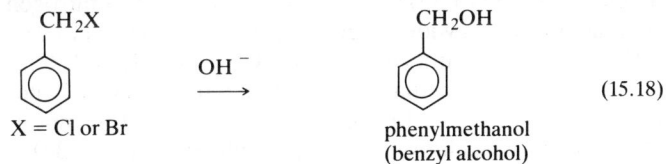

CH_2X ($X = Cl$ or Br) → CH_2OH phenylmethanol (benzyl alcohol)   (15.18)

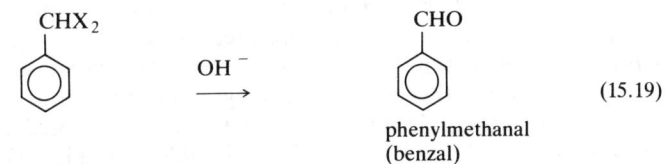

CHX_2 → CHO phenylmethanal (benzal)   (15.19)

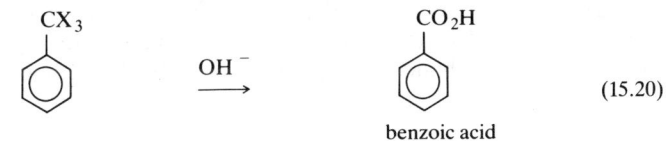

CX_3 → CO_2H benzoic acid   (15.20)

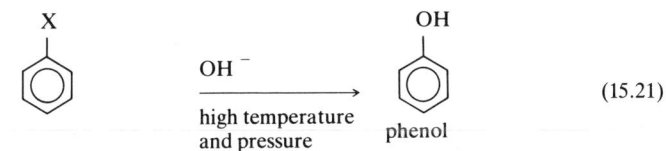

X → OH phenol, high temperature and pressure   (15.21)

Reactions of phenyl magnesium halide

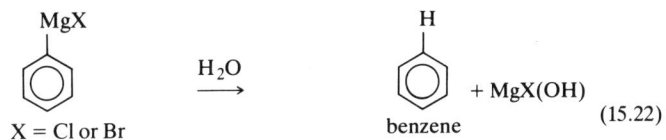

MgX ($X = Cl$ or Br) → H benzene $+ MgX(OH)$ (15.22)

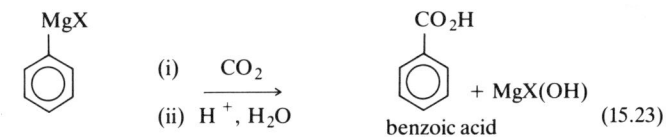

MgX (i) CO_2 (ii) H^+, H_2O → CO_2H benzoic acid $+ MgX(OH)$ (15.23)

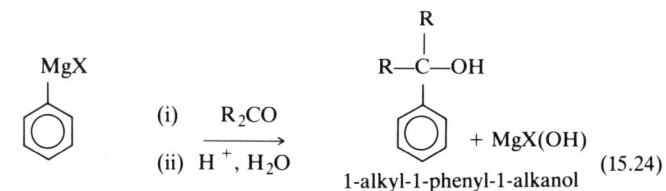

MgX (i) R_2CO (ii) H^+, H_2O → R—C—OH 1-alkyl-1-phenyl-1-alkanol $+ MgX(OH)$ (15.24)

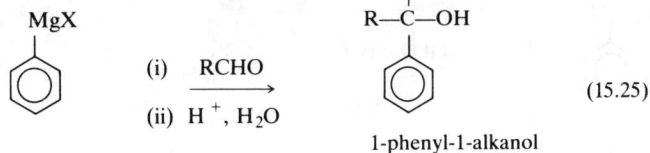

MgX (i) RCHO (ii) H^+, H_2O → R—CH—OH 1-phenyl-1-alkanol   (15.25)

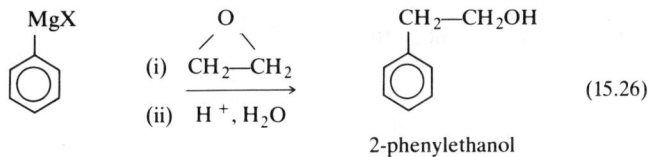

MgX (i) CH_2—CH_2 O (ii) H^+, H_2O → CH_2—CH_2OH 2-phenylethanol   (15.26)

Formation of diazonium salts

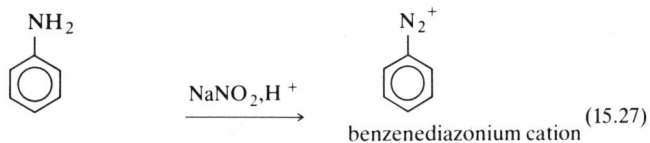

NH_2 $NaNO_2, H^+$ → N_2^+ benzenediazonium cation   (15.27)

Reactions of diazonium cation

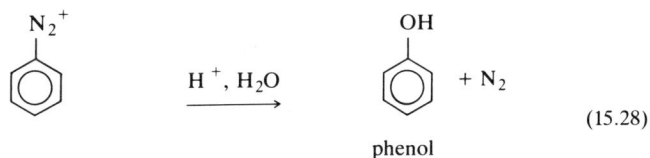

N_2^+ H^+, H_2O → OH phenol $+ N_2$ (15.28)

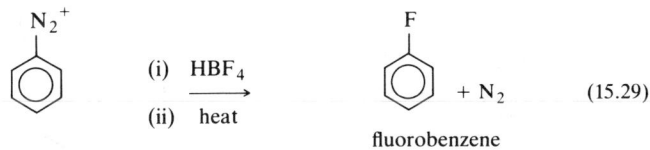

N_2^+ (i) HBF_4 (ii) heat → F fluorobenzene $+ N_2$ (15.29)

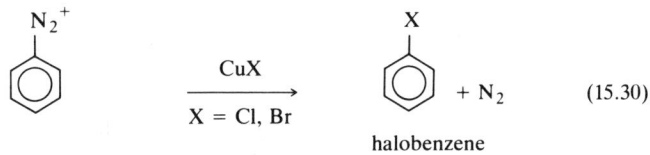

N_2^+ CuX ($X = Cl, Br$) → X halobenzene $+ N_2$ (15.30)

N_2^+ CuCN → CN cyanobenzene $+ N_2$ (15.31)

N_2^+ ROH → OR alkylphenylether $+ N_2$ (15.32)

Table 15.2 (continued)

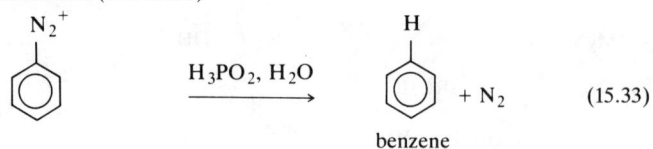

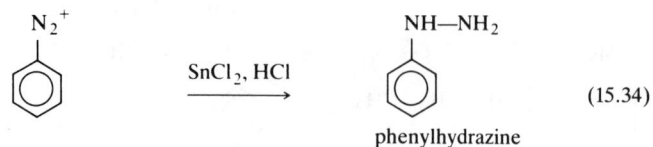

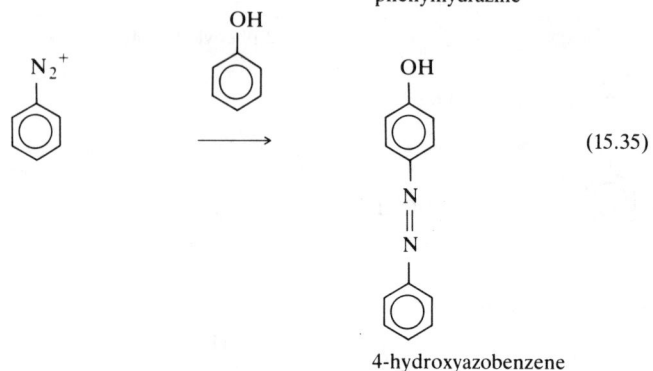

$$\text{N}_2^+ \text{benzene} \xrightarrow{\text{H}_3\text{PO}_2,\ \text{H}_2\text{O}} \text{benzene} + \text{N}_2 \quad (15.33)$$

$$\text{N}_2^+ \xrightarrow{\text{SnCl}_2,\ \text{HCl}} \text{NH—NH}_2 \quad (15.34)$$

phenylhydrazine

$$\text{N}_2^+ + \text{OH} \longrightarrow \text{4-hydroxyazobenzene} \quad (15.35)$$

4-hydroxyazobenzene

While alkanes are unreactive to normal oxidizing conditions the presence of a phenyl substituent in an alkane facilitates oxidation at an adjacent carbon atom. Methylbenzene and ethylbenzene can both be oxidized by an acidic dichromate solution to benzoic acid (15.12, 15.13). Reduction of nitrobenzene gives aniline (15.14) and both these compounds are important industrial chemicals. Aniline can be converted by reaction with nitrous acid to the benzenediazonium cation (15.27). This cation is highly reactive, and reacts as an electrophile (15.35) in attacking phenol or loses nitrogen on reaction with a wide range of nucleophiles; H_2O (15.28), BF_4^- (15.29) Cl^- and Br^- (15.30), CN^- (15.31) and ROH (15.32). The benzenediazonium cation is reduced with hypophosphorous acid to benzene (15.33) and with stannous chloride in hydrochloric acid to phenylhydrazine (15.34).

Halobenzenes do not undergo nucleophilic displacement reactions (15.21) under conditions easily attained in a laboratory (see section 16.3), but are nevertheless important in synthesis since they react with magnesium to give aryl magnesium halides (15.15) which react with water (15.22), carbon dioxide (15.23), ketones (15.24), aldehydes (15.25) and oxirane (ethylene oxide) (15.26). For the side chain halogenated substrates chlorophenylmethane (15.18), dichlorophenylmethane (15.19) and trichlorophenylmethane

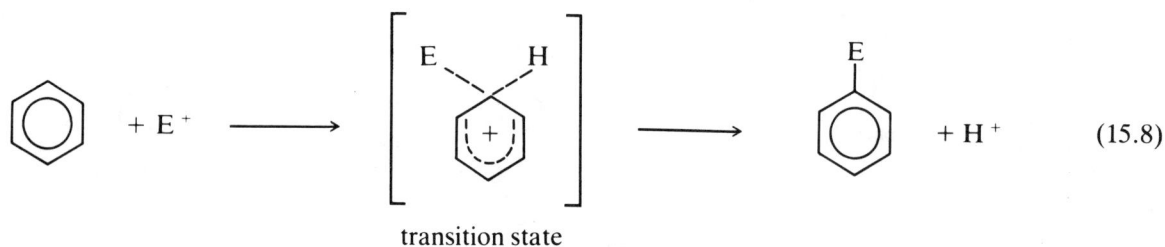

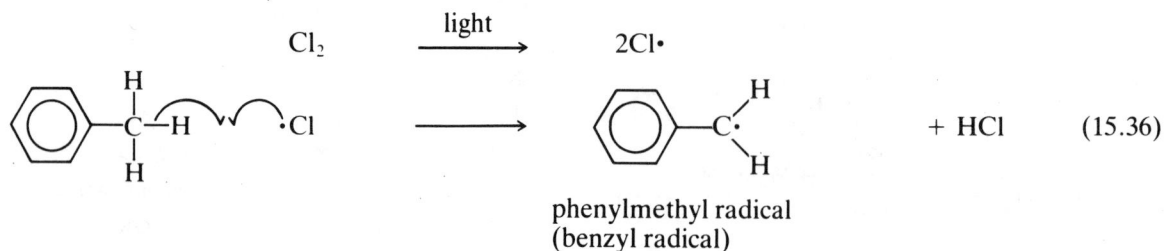

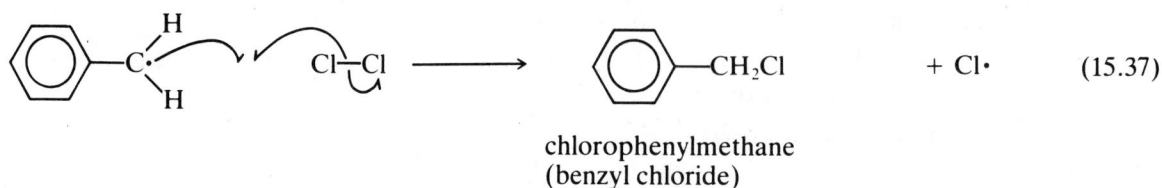

$$\text{benzene} + \text{E}^+ \longrightarrow \left[\text{transition state} \right] \longrightarrow \text{E} + \text{H}^+ \quad (15.8)$$

transition state

$$\text{Cl}_2 \xrightarrow{\text{light}} 2\text{Cl}\cdot$$

$$\text{C}_6\text{H}_5\text{CH}_3 + \cdot\text{Cl} \longrightarrow \text{phenylmethyl radical (benzyl radical)} + \text{HCl} \quad (15.36)$$

phenylmethyl radical
(benzyl radical)

$$\text{radical} + \text{Cl—Cl} \longrightarrow \text{—CH}_2\text{Cl} + \text{Cl}\cdot \quad (15.37)$$

chlorophenylmethane
(benzyl chloride)

(15.20) displacement of halogen by hydroxide occurs to give phenylmethanol, phenylmethanal and benzoic acid respectively. These processes are shown in schemes 15.38–15.40. The first step in each sequence involves replacement of Cl⁻ by OH⁻. For reactions 15.39 and 15.40 the intermediate chlorohydroxy species decompose to phenylmethanal and benzoyl chloride. Benzoyl chloride reacts further with hydroxide to give benzoic acid. Benzene sulphonic acid also undergoes a nucleophilic displacement reaction with hydroxide to give phenol (15.17).

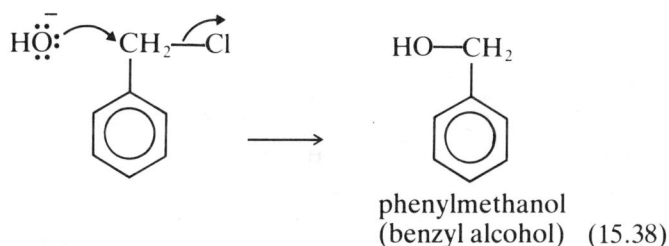

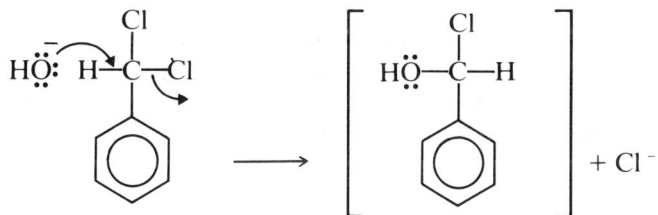

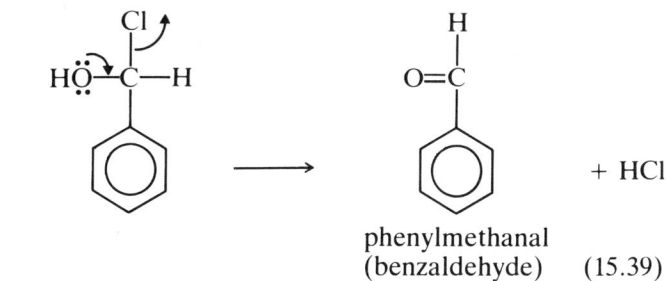

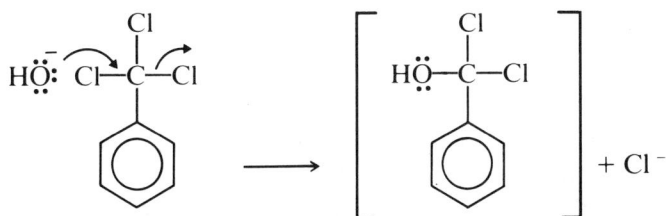

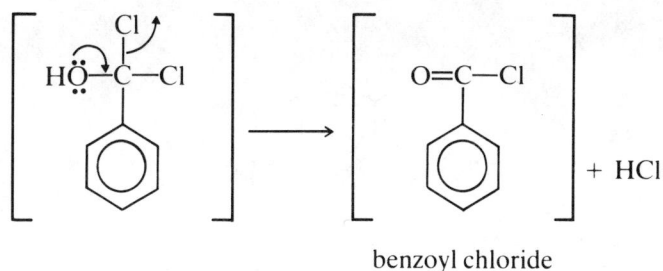

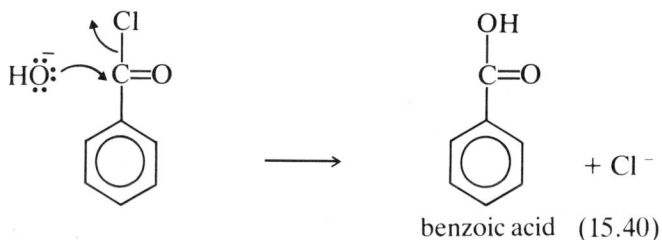

benzoyl chloride

phenylmethanol
(benzyl alcohol)   (15.38)

+ HCl

benzoic acid   (15.40)

15.4 Disubstituted Benzene Derivatives

Benzene can obviously carry more than one substituent —there are altogether six hydrogen atoms which can be replaced. For a monosubstituted benzene derivative (Fig. 15.11) there are three distinguishable positions in the ring where a second substituent can be placed. The substituents can be in positions 1 and 2 (*ortho*) (Figs. 15.12), 1 and 3 (*meta*) (Fig. 15.13) or 1 and 4 (*para*) (Fig. 15.14). For any monosubstituted benzene derivative there are two *ortho* and two *meta* carbon atoms that can be attacked by an electrophile, but only one *para* position and this is shown in Fig. 15.15. On statistical grounds it would be expected that substitution of a monosubstituted benzene derivative would give *ortho*, *meta* and *para* disubstituted products in the ratio 2:2:1. In fact, this statistical ratio of products is never observed in practice and we must therefore examine in more detail the effect of substituents on electrophilic substitution.

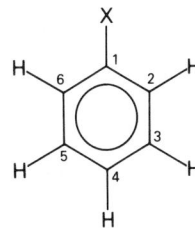

+ Cl⁻

phenylmethanal
(benzaldehyde)   (15.39)

+ HCl

+ Cl⁻

Fig. 15.11 Monosubstituted benzene

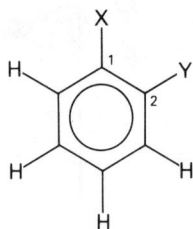

Fig. 15.12 1,2- or *ortho* disubstituted benzene

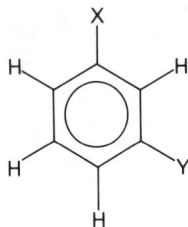

Fig. 15.13 1,3- or *meta* disubstituted benzene

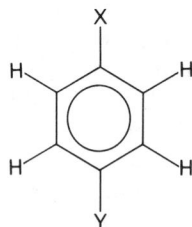

Fig. 15.14 1,4 or *para* disubstituted benzene

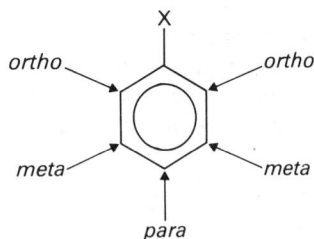

Fig. 15.15 *Ortho, meta* and *para* positions of a monosubstituted benzene

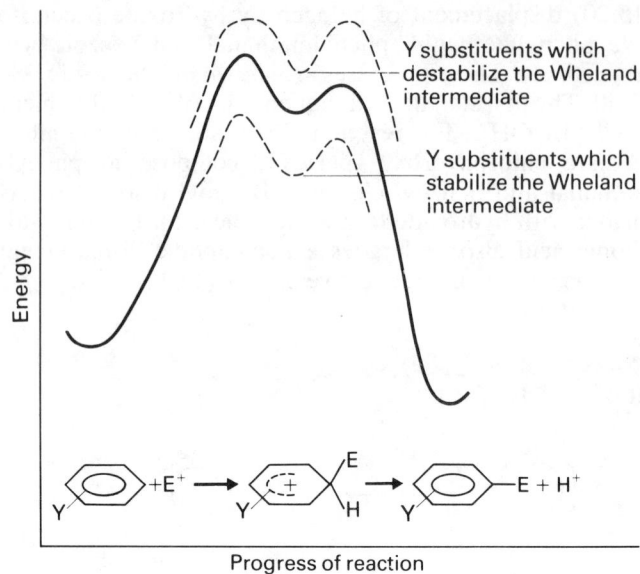

Fig. 15.16 Effect of substituents on the reaction profile for electrophilic aromatic substitution

A substituent whose presence has the effect of lowering the potential energy of the reaction intermediate relative to the parent molecule will also influence the energy of the transition state leading to that intermediate (Fig. 15.16). The result is lowering of the effective energy of activation for the reaction, which causes an increase in the rate of reaction. The substituent is said to activate the benzene ring. In a similar manner, substituents which destabilize the carbonium ion intermediate relative to that formed by the unsubstituted benzene molecule will decrease the rate of reaction, because they increase the activation energy of the process leading to the Wheland intermediate. For electrophilic substitution in benzene, *deactivating* substituents must cause a reduction in electron density at the ring. They are therefore electron withdrawing relative to the parent substituent, hydrogen. Such substituents, shown in Table 15.3, destabilize an already positively charged reaction intermediate. It is noticeable that most groups which are deactivating involve multiple bonds.

Table 15.3 Deactivating substituents

—halogen	halo
—NO$_2$	nitro
—N(CH$_3$)$_3$$^+$	quaternary ammonium cation
—CN	cyano
—CO$_2$H	carboxy
—SO$_3$H	sulfo
—CHO	formyl
—CRO	oxo (keto)

Substituents which increase the electron density in the ring relative to that when only hydrogen is present stabilize the carbonium ion intermediate in an electrophilic substitution. This lowers the energy of the transition state for the rate-determining step of the reaction, and so increases the rate of reaction. These substituents are referred to as activating substituents and a list is given in Table 15.4. It is noticeable that with the exception of the alkyl group they all have a lone pair of electrons on one atom adjoining the benzene ring.

Table 15.4 Activating substituents

—OH	hydroxy
—OR	alkoxy
—NH$_2$	amino
—NHCOR	N-acetyl
—R	alkyl

Substituents not only effect the overall rate of reaction of a benzene derivative, they also influence the relative rates of reaction at the *ortho, meta* and *para* positions around the ring. The intermediate shown in Fig. 15.17 can be described in valence bond terms as a resonance hybrid of three canonical forms (Fig. 15.18). The excess positive charge of this ion is seen to be very unequally distributed over the carbon atoms. The charge is able to be located only at the *ortho* and *para* carbon atoms as shown in Fig. 15.19. Substituents which increase electron density in the aromatic ring either by an inductive mechanism (e.g. methyl) or by overlap of electron-rich orbitals with the π electron system (e.g. the p orbital electrons on oxygen in —OH or —OR or of nitrogen in —NH$_2$ or —NHCOR) will therefore favour *ortho-para* product formation more than *meta* since they release electron density to the most electron deficient carbons

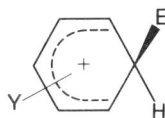

Fig. 15.19 Diagram showing centres of maximum electron deficiency in Wheland intermediate

(Fig. 15.20). Electron withdrawing substituents (e.g. —N(CH$_3$)$_3$$^+$, —NO$_2$, —CN, —CO$_2$H, —COR, —SO$_3$H) markedly decrease the rate of reaction at all carbons of the ring but have their greatest deactivating effect in the *ortho* and *para* positions, i.e. they will direct substituents to the *meta* positions. Halogen substituents fall into a separate category, in that while they are electronegative and deactivate all positions to electrophilic substitution relative to benzene, the lone pair electrons overlap with the π system in the Wheland intermediate and favour *ortho-para* product formation more than *meta*. Table 15.5 classifies various substituents as *ortho-para* directing or *meta* directing in aromatic electrophilic substitution.

Table 15.5 Directing effect of substituents in aromatic electrophilic substitution

ortho-para directing	*meta* directing
—OH	—NO$_2$
—OR	—N(CH$_3$)$_3$$^+$
—NH$_2$	—CN
—NHCOCH$_3$	—CO$_2$H
—R	—SO$_3$H

—halogen	—C—H
	   ‖
	   O

	—C—R
	   ‖
	   O

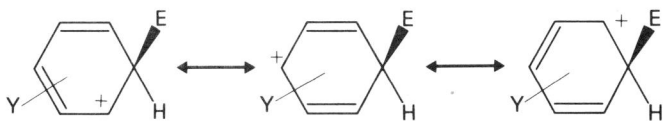

Fig. 15.17 Wheland intermediate in electrophilic substitution of a monosubstituted benzene derivative

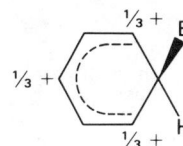

Fig. 15.18 Canonical forms of a monosubstituted Wheland intermediate

The examples of electrophilic substitution of some monosubstituted benzene derivatives given in Tables 15.6 and 15.7 illustrate the type of results which form the basis of the generalizations made. In interpreting these tables it should be remembered that the rates of formation of *ortho, meta* and *para* products do vary with reaction conditions such as solvent and temperature. A knowledge of the reactivity and directing influence of substituents obtained both from rate data and product analyses allows synthetic schemes for a number of aromatic compounds to be devised. Some examples of such schemes appear in the exercises given at the end of this chapter.

ortho-para substituents   *meta* substituents

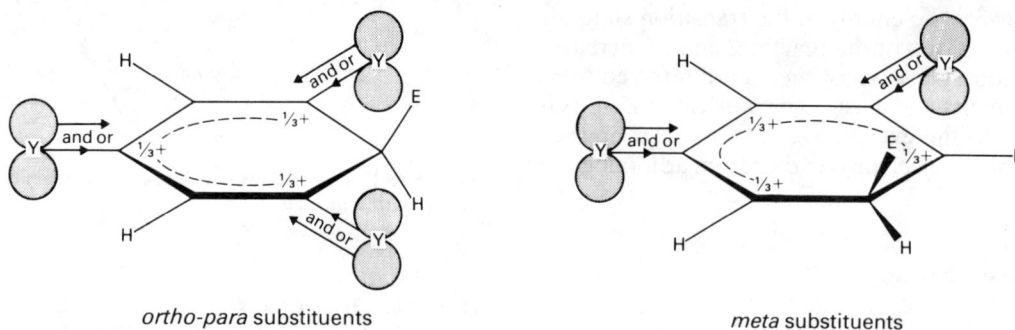

Fig. 15.20 Diagram showing how electron releasing substituents stabilize the Wheland intermediate for *ortho-para* substitution as compared with *meta* substitution

Table 15.6 Isomer distribution (%) for chlorination of monosubstituted benzene

—Y	*ortho*	*meta*	*para*
meta directing substituents —NO$_2$	17.6	80.9	1.5
—CN	23.2	73.9	2.9
—CHO	30.7	63.5	5.8
ortho-para directing substituents —Br	39.7	3.4	56.9
—Cl	36.4	1.3	62.3
—Me	74.7	2.2	23.1
—OMe	34.9	—	65.1

Table 15.7 Isomer distribution (%) for nitration of monosubstituted benzene

Y	*ortho*	*meta*	*para*
meta directing substituents —NO$_2$	6.4	93.2	0.3
—CN	17.1	80.7	2.0
—Br	36.5	1.2	62.4
ortho-para directing substituents —Cl	29.6	0.9	69.5
—Me	58.5	4.0	37.1
—OMe	44.0	—	55.0
—NHCOCH$_3$	19.0	2.0	79.0

Exercises

15.1   How would you undertake the synthesis of the following compounds starting from benzene.
(1) *meta*-chloronitrobenzene
(2) *ortho*-chloronitrobenzene
(3) *para*-methylbenzene sulphonic acid
(4) *meta*-nitrobenzoic acid
(5) phenol
(6) phenylmethanol (benzyl alcohol)
(7) *para*-bromoaniline
(8) 1-phenyl-1-ethanamine

16 Haloalkanes, Alcohols, Amines and Ethers

16.1 Structure and Nomenclature of Haloalkanes, Alcohols, Amines and Ethers

In Chapter 13 we examined the chemistry of alkanes, in which carbon has a coordination number of four, and there is a notable absence of lone pairs, π orbitals, or vacant orbitals of suitable energy for bonding with other chemical species. The four atoms about each carbon atom form a formidable protective sheath, and consequently the reactions of alkanes are invariably initiated at a peripheral hydrogen atom. For example, halogenation involves attack on hydrogen by the highly reactive halogen atom. Alkenes, which were examined in Chapter 14, have no lone pairs or vacant orbitals of suitable energy for bonding, but they do have a framework which exposes to attack the carbon atoms attached by double bonds. This exposure, coupled with the presence of relatively diffuse π bonding electrons, facilitates reaction at the carbon atom, and so most reactions of alkenes are initiated by direct attachment of an electrophilic species to carbon. The distinctive feature of compounds to be discussed in this chapter is the presence of a hetero atom (i.e. non-carbon atom) joined by a sigma bond to a carbon atom. The hetero atoms all carry one or more lone pairs of electrons, and are all larger than hydrogen—often larger than carbon. The differences in size and electron configuration in comparison with carbon and hydrogen are reflected in differences in bond length and bond energy. Some selected examples are shown in Table 16.1. With the exception of the C—F bond, these bonds are all weaker than C—H bonds. The weaker C—X bonds more readily take part in chemical reactions. In the early stages of a reaction an electrophile can become attached by the lone pair of electrons to the hetero atom, further weakening the C—X bond. This lowers the activation energy for C—X bond cleavage. Hetero atoms are therefore sites of enhanced chemical reactivity, where chemical modifications can readily be performed. Thus they are a key to chemical synthesis.

The high electronegativity of most hetero atoms relative to carbon results in a polarization of the σ bond between carbon and the hetero atom. This *inductive effect* causes the carbon

Table 16.1 Bond energies (kJ mol^{-1})

C—F	485
C—H	413
C—O	351
C—C	348
C—Cl	328
C—N	292
C—Br	276
C—I	248

atom to become slightly electron deficient and thereby subject to attack by electron rich species or *nucleophiles*.

Alcohols and haloalkanes can be classified into three groups, depending upon whether the hydroxyl or halide group is attached to a primary, secondary or tertiary carbon atom (Fig. 16.1). Amines are also classified as primary, secondary and tertiary, but in this case the classification is

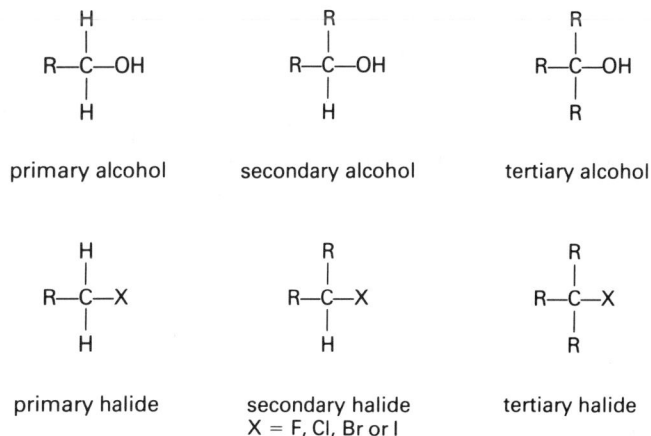

Fig. 16.1 Classification of alcohols and halides

based on the number of carbon atoms attached to the nitrogen atom (Fig. 16.2). Aliphatic amines are named by adding the suffix *-amine* to the name of the chain or ring system to which nitrogen is attached. The final 'e' of the hydrocarbon is omitted. Methanamine (methylamine), shown in Fig. 16.3, is the parent member of the family. Aromatic amines are generally named as derivatives of the simplest aromatic amine, aniline (Fig. 16.5 and Fig. 16.6). Salts of amines are named by replacing the suffix *-amine* by *ammonium*; tetramethylammonium iodide is shown in Fig. 16.7. Monohydric alcohols are named by adding the suffix *ol* to the name of the corresponding hydrocarbon, the final 'e' of the hydrocarbon being elided (Fig. 16.8). The parent aromatic alcohol is phenol (Fig. 16.9). Polyhydric alcohols are named by adding the appropriate suffix diol, triol, tetrol, etc. to the name of the corresponding hydrocarbon system (see Fig. 14.28). Haloalkanes are named as halo

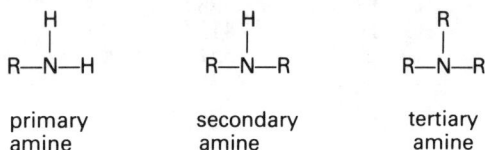

Fig. 16.2 Classification of amines

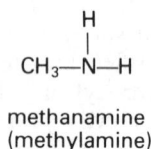

methanamine
(methylamine)

Fig. 16.3

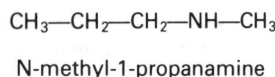

N-methyl-1-propanamine

Fig. 16.4

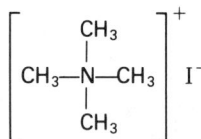

aniline   N-methylaniline   tetramethylammonium iodide

Fig. 16.5   **Fig. 16.6**   **Fig. 16.7**

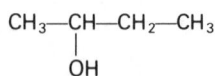

2-butanol
(2-hydroxybutane)

phenol

Fig. 16.8   **Fig. 16.9**

derivatives of the longest carbon aliphatic chain, or, in a cyclic molecule, by the position of the substituent in the ring. Two selected examples are shown in Figs 16.10 and 16.11. Where more than one substituent is joined to a parent chain and the substituents are named as prefixes

then they are recorded alphabetically (e.g. Fig. 16.10) (see also section 18.1). Simple ethers are named according to the groups attached to the ether oxygen (Figs 16.12 and 16.13). Alicyclic OR groups can be referred to as alkoxy when the alkyl group contains less than five carbons (Fig. 16.14). For example an alternative name for ethylmethyl-ether is methoxyethane (Fig. 16.12).

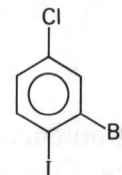

2-bromo-4-chloro-1-iodobenzene

Fig. 16.10

CH_3—CH—CH_2Cl
|
OH

1-chloro-2-propanol

Fig. 16.11

CH_3—O—CH_2—CH_3

ethylmethylether
(methoxyethane)

Fig. 16.12

CH_3CH_2—O—CH_2CH_3

diethylether
(ethoxyethane)

Fig. 16.13

CH_3—CH_2—O—CH_2—CH_2—Br

1-bromo-2-ethoxyethane

Fig. 16.14

16.2 Alcohols: Preparation and Reactions

Alcohols are derivatives of water in which one hydrogen is replaced by a carbon-containing group. The hydroxyl group confers water solubility on alcohols with small alkyl groups. As the size of the attached alkyl or aryl group increases, the physical properties of the hydrocarbon component of the molecule become dominant and the molecule becomes less and less soluble.

Alcohols are prepared industrially by addition of water to alkenes, as shown in reaction 16.1. This is a process known as hydration. Ethanol is obtained by fermentation of sugars (reaction 16.2), a process which forms the basis of the brewing industry. In the laboratory alcohols are prepared in a variety of ways; some of the more common preparations are listed in Table 16.2. Discussion of most of these reactions will be found in other sections of the text.

Table 16.2 Preparation of alcohols

Hydration of alkenes

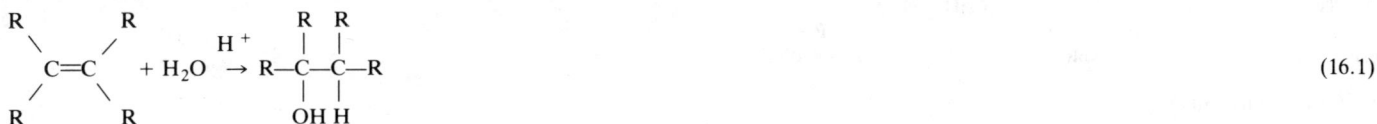

$$\begin{array}{c} R \quad\quad R \\ \backslash \quad\quad / \\ C{=}C \quad + H_2O \xrightarrow{H^+} R{-}\underset{\underset{OH}{|}}{C}{-}\underset{\underset{H}{|}}{C}{-}R \\ / \quad\quad \backslash \\ R \quad\quad R \end{array} \qquad\qquad (16.1)$$

Fermentation of sugars

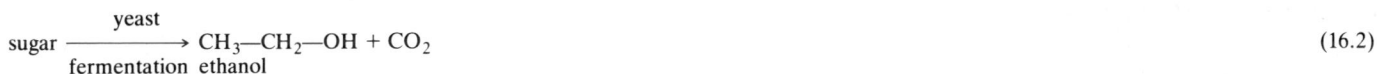

$$\text{sugar} \xrightarrow[\text{fermentation}]{\text{yeast}} \underset{\text{ethanol}}{CH_3{-}CH_2{-}OH} + CO_2 \qquad\qquad (16.2)$$

Oxymercuration–demercuration

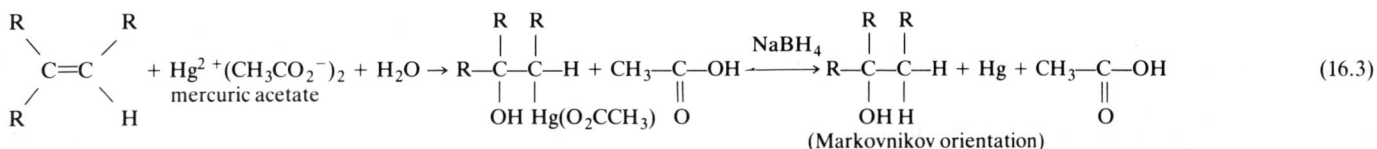

$$\begin{array}{c} R \quad\quad R \\ \backslash \quad\quad / \\ C{=}C \quad + Hg^{2+}(CH_3CO_2{}^-)_2 + H_2O \rightarrow R{-}C{-}C{-}H + CH_3{-}C{-}OH \xrightarrow{NaBH_4} R{-}C{-}C{-}H + Hg + CH_3{-}C{-}OH \\ / \quad\quad \backslash \qquad\quad \text{mercuric acetate} \\ R \quad\quad H \end{array}$$

(Markovnikov orientation)   (16.3)

Hydroboration–oxidation

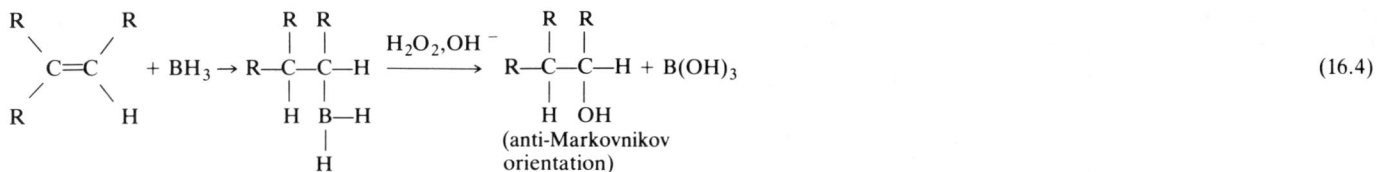

$$\begin{array}{c} R \quad\quad R \\ \backslash \quad\quad / \\ C{=}C \quad + BH_3 \rightarrow R{-}C{-}C{-}H \xrightarrow{H_2O_2, OH^-} R{-}C{-}C{-}H + B(OH)_3 \\ / \quad\quad \backslash \\ R \quad\quad H \end{array}$$

(anti-Markovnikov orientation)   (16.4)

Action of Grignard reagent on carbon–oxygen bonds

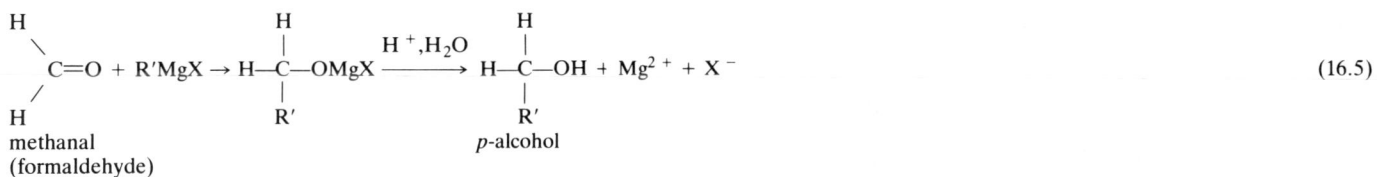

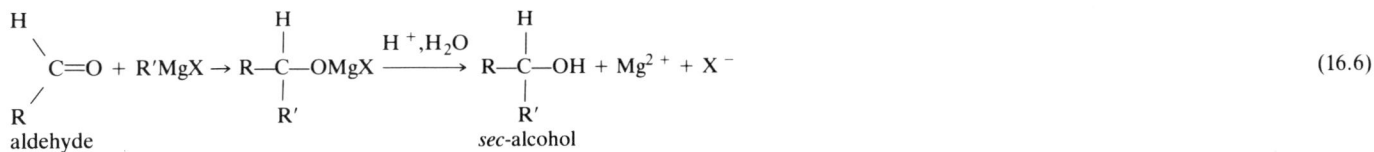

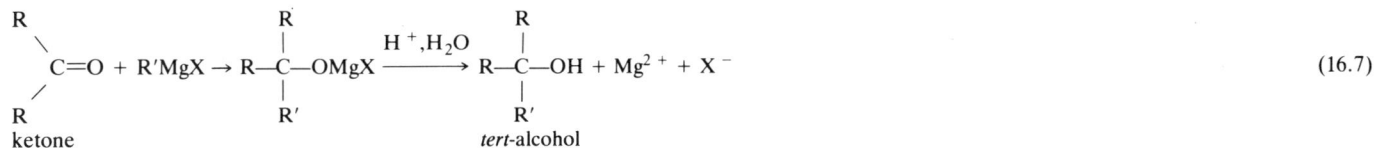

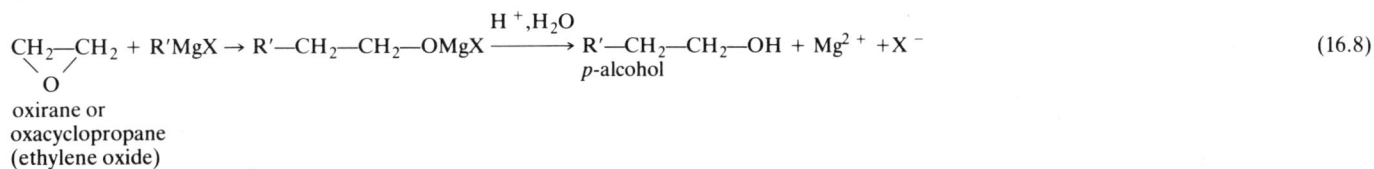

$$\begin{array}{c} H \\ \backslash \\ C{=}O + R'MgX \rightarrow H{-}\underset{\underset{R'}{|}}{C}{-}OMgX \xrightarrow{H^+, H_2O} \underset{\underset{R'}{|}}{H{-}C{-}OH} + Mg^{2+} + X^- \\ / \\ H \end{array}$$

methanal (formaldehyde)   *p*-alcohol   (16.5)

$$\begin{array}{c} H \\ \backslash \\ C{=}O + R'MgX \rightarrow R{-}\underset{\underset{R'}{|}}{C}{-}OMgX \xrightarrow{H^+, H_2O} \underset{\underset{R'}{|}}{R{-}C{-}OH} + Mg^{2+} + X^- \\ / \\ R \end{array}$$

aldehyde   *sec*-alcohol   (16.6)

$$\begin{array}{c} R \\ \backslash \\ C{=}O + R'MgX \rightarrow R{-}\underset{\underset{R'}{|}}{\overset{\overset{R}{|}}{C}}{-}OMgX \xrightarrow{H^+, H_2O} \underset{\underset{R'}{|}}{\overset{\overset{R}{|}}{R{-}C{-}OH}} + Mg^{2+} + X^- \\ / \\ R \end{array}$$

ketone   *tert*-alcohol   (16.7)

$$\underset{\underset{O}{\diagdown\diagup}}{CH_2{-}CH_2} + R'MgX \rightarrow R'{-}CH_2{-}CH_2{-}OMgX \xrightarrow{H^+, H_2O} \underset{\text{*p*-alcohol}}{R'{-}CH_2{-}CH_2{-}OH} + Mg^{2+} + X^- \qquad (16.8)$$

oxirane or
oxacyclopropane
(ethylene oxide)

Table 16.2 Preparation of alcohols (continued)

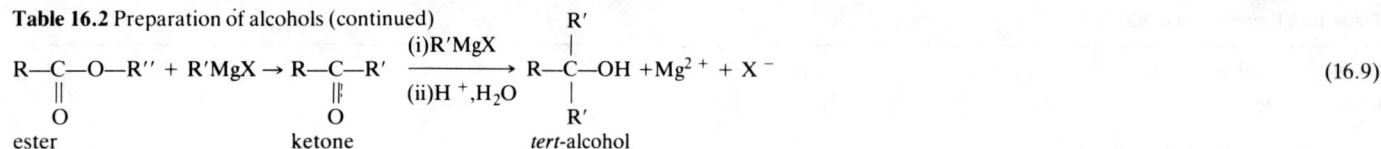

$$R-\underset{\underset{O}{\parallel}}{C}-O-R'' + R'MgX \rightarrow R-\underset{\underset{O}{\parallel}}{C}-R' \xrightarrow[\text{(ii)H}^+,H_2O]{\text{(i)}R'MgX} R-\underset{\underset{R'}{\vert}}{\overset{\overset{R'}{\vert}}{C}}-OH + Mg^{2+} + X^- \qquad (16.9)$$

ester　　　　　　　　　　ketone　　　　　　　　　　*tert*-alcohol

Hydrolysis of haloalkanes

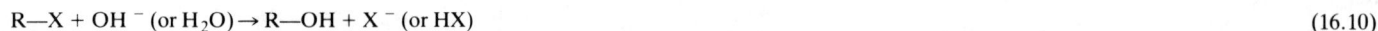

$$R-X + OH^- \text{ (or } H_2O) \rightarrow R-OH + X^- \text{ (or HX)} \qquad (16.10)$$

Reduction of carbonyl compounds

$$\underset{\text{(H)}R'}{\overset{R}{\diagdown}}C{=}O \xrightarrow[\text{(ii) H}^+,H_2O]{\text{(i) LiAlH}_4} R'-\underset{\underset{H}{\vert}}{\overset{\overset{R}{\vert}}{C}}-OH \qquad (16.11)$$

Reduction of esters

$$R'-\underset{\underset{O}{\parallel}}{C}-O-R \xrightarrow[\text{(ii) H}^+, H_2O]{\text{(i) LiAlH}_4} R'-CH_2-OH + R-OH \qquad (16.12)$$

Oxidation of alkenes

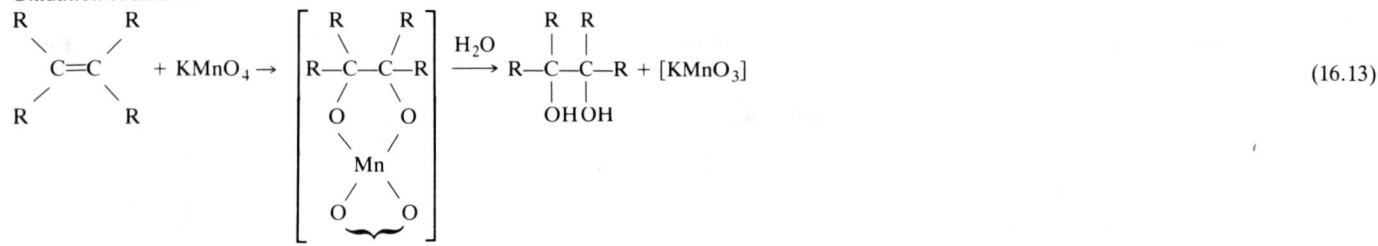

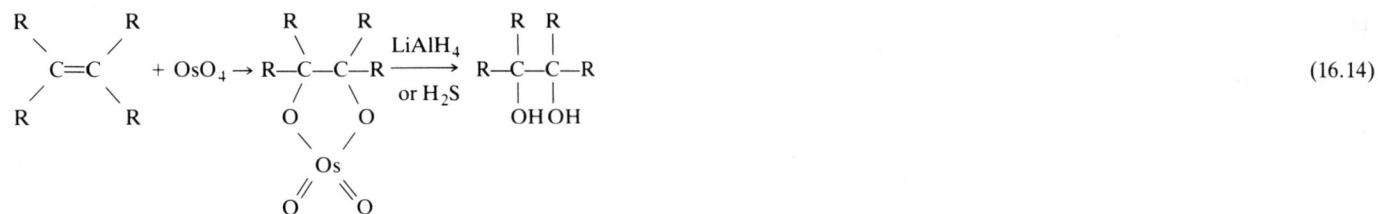

$$\underset{\underset{R}{\diagup}}{\overset{\overset{R}{\diagdown}}{}}C{=}C\overset{\overset{R}{\diagup}}{\underset{\underset{R}{\diagdown}}{}} + KMnO_4 \rightarrow \left[\begin{array}{c} \underset{\underset{O}{\diagdown}}{\overset{\overset{R}{\diagdown}}{R}}-C-C-\underset{\underset{O}{\diagup}}{\overset{\overset{R}{\diagup}}{R}} \\ Mn \\ O \diagup \diagdown O \\ \overbrace{} \\ - \end{array}\right] \xrightarrow{H_2O} R-\underset{\underset{OH}{\vert}}{\overset{\overset{R}{\vert}}{C}}-\underset{\underset{OH}{\vert}}{\overset{\overset{R}{\vert}}{C}}-R + [KMnO_3] \qquad (16.13)$$

$$\underset{\underset{R}{\diagup}}{\overset{\overset{R}{\diagdown}}{}}C{=}C\overset{\overset{R}{\diagup}}{\underset{\underset{R}{\diagdown}}{}} + OsO_4 \rightarrow R-\underset{\underset{O}{\diagdown}}{\overset{\overset{R}{\diagdown}}{C}}-\underset{\underset{O}{\diagup}}{\overset{\overset{R}{\diagup}}{C}}-R \xrightarrow[\text{or } H_2S]{LiAlH_4} R-\underset{\underset{OH}{\vert}}{\overset{\overset{R}{\vert}}{C}}-\underset{\underset{OH}{\vert}}{\overset{\overset{R}{\vert}}{C}}-R \qquad (16.14)$$

$$\underset{\underset{O}{\diagup}}{\overset{}{}}Os\overset{}{\underset{\underset{O}{\diagdown}}{}}$$

Displacement

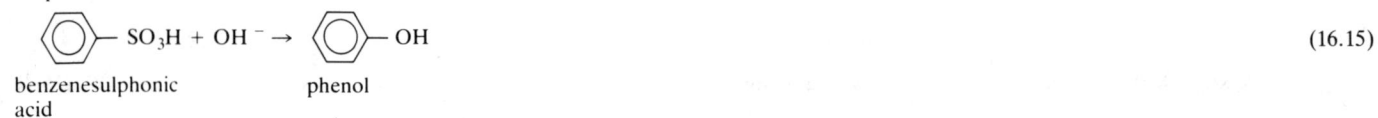

$$\text{C}_6\text{H}_5-SO_3H + OH^- \rightarrow \text{C}_6\text{H}_5-OH \qquad (16.15)$$

benzenesulphonic　　　　phenol
acid

Hydrolysis of diazonium salts

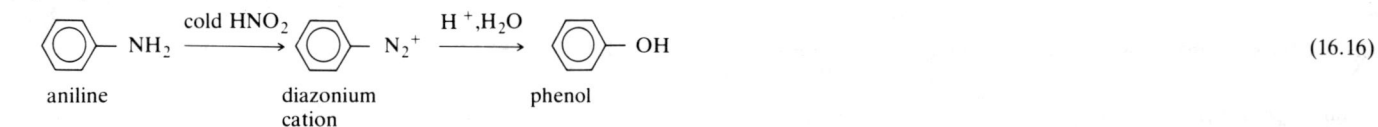

$$\text{C}_6\text{H}_5-NH_2 \xrightarrow{\text{cold HNO}_2} \text{C}_6\text{H}_5-N_2^+ \xrightarrow{H^+,H_2O} \text{C}_6\text{H}_5-OH \qquad (16.16)$$

aniline　　　　　　　diazonium　　　　　　phenol
　　　　　　　　　　cation

Table 16.3 Reactions of alcohols

Reaction with hydrogen halides

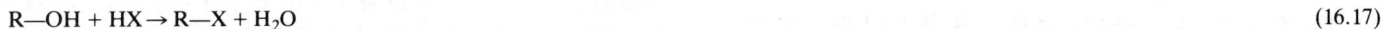

$$R—OH + HX \rightarrow R—X + H_2O \qquad\qquad (16.17)$$

Reactions with phosphorus trihalides and thionyl chloride

$$R—OH + PX_3 \ \ \rightarrow R—X + H_3PO_3$$
$$R—OH + SOCl_2 \rightarrow R—X + SO_2 + HCl \qquad\qquad (16.18)$$
$$X = Br, Cl$$

Dehydration

$$(16.19)$$

Reaction with active metals

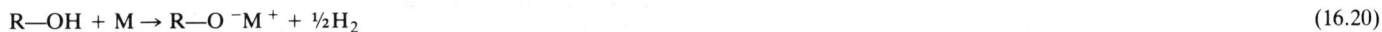

$$R—OH + M \rightarrow R—O^-M^+ + \tfrac{1}{2}H_2 \qquad\qquad (16.20)$$
$$M = Na, K, Mg, Al$$

Ester formation

$$(16.21)$$

$$(16.22)$$

Oxidation

(i)

$$(16.23)$$

primary alcohol — aldehyde — carboxylic acid

(ii)

$$(16.24)$$

secondary alcohol — ketone

(iii)

$$(16.25)$$

tertiary alcohol

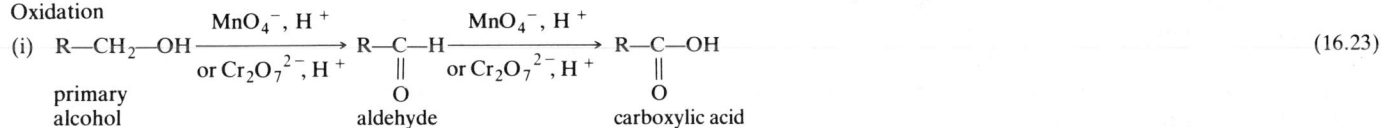

Reaction with sulfuric acid

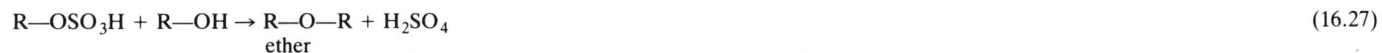

$$R—OH + H_2SO_4 \rightarrow R—OSO_3H + H_2O \qquad\qquad (16.26)$$
primary alcohol — alkyl hydrogen sulphate

$$R—OSO_3H + R—OH \rightarrow R—O—R + H_2SO_4 \qquad\qquad (16.27)$$
ether

The chemistry of alcohols (Table 16.3) is essentially the chemistry of the hydroxyl group. The C—O bond of the alcohol can be broken and the OH replaced (e.g. reaction 16.17). Alternatively water can be removed from the molecule, a process known as dehydration or elimination (reaction 16.19). Alcohols are slightly acidic, and react with the most reactive metals to liberate hydrogen gas and form alkoxide salts of the metal (Eq. 16.20). Alcohols react with carboxylic acids, in the presence of a trace of mineral acid as catalyst, to give esters (reaction 16.22). The reaction of methanol with ethanoic acid (plus a small quantity of mineral acid) is shown in Eq. 16.28:

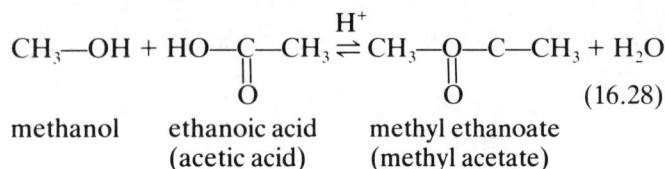

$$CH_3—OH + HO—\overset{\overset{\displaystyle H^+}{}}{\underset{\underset{\displaystyle O}{\|}}{C}}—CH_3 \rightleftharpoons CH_3—O—\underset{\underset{\displaystyle O}{\|}}{C}—CH_3 + H_2O \quad (16.28)$$

methanol   ethanoic acid   methyl ethanoate
            (acetic acid)   (methyl acetate)

The reaction is reversible, and the equilibrium can be driven to the right by removal of water or to the left by adding water. This is an example of a *condensation* reaction, involving the loss of a molecule of water from two reacting species. An ester can also be prepared by reaction of an alcohol with an acid halide (reaction 16.21). Thus propanoyl chloride reacts with methanol to give methyl propanoate (reaction 16.29).

A similar reaction occurs between an alcohol and 4-toluenesulphonyl chloride to form a 4-toluenesulphonic acid ester (reaction 16.30) often referred to as a tosylate. The tosylate group is fairly easy to dislodge with another reagent (nucleophile)—we say that it is an excellent leaving group—and for this reason it is used extensively in nucleophilic substitution and elimination reactions. Alcohols undergo a condensation reaction with sulphuric acid to give alkyl hydrogen sulphates (Eq. 16.26). Methyl hydrogen sulphate is shown being formed from methanol and H_2SO_4 in Eq. 16.31. Alkyl hydrogen sulphates react further with

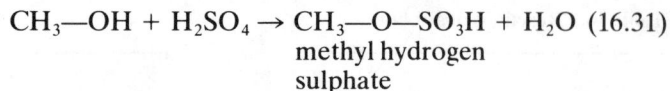

$$CH_3—OH + H_2SO_4 \rightarrow CH_3—O—SO_3H + H_2O \quad (16.31)$$
methyl hydrogen
sulphate

excess alcohol to give ethers (Eq. 16.27) and this is shown for the reaction of methyl hydrogen sulphate with methanol in reaction 16.32.

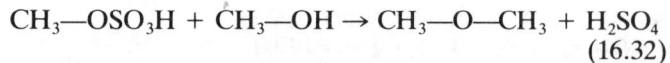

$$CH_3—OSO_3H + CH_3—OH \rightarrow CH_3—O—CH_3 + H_2SO_4 \quad (16.32)$$

Primary and secondary alcohols can be oxidized with aqueous acidic potassium dichromate or potassium permanganate. Primary alcohols are oxidized to aldehydes, which can be further oxidized to carboxylic acids (reaction 16.23). Secondary alcohols are oxidized to ketones, (16.24) but tertiary alcohols do not react under these conditions (16.25). The difference in chemical reactivity of the primary, secondary and tertiary alcohols provides a means of distinguishing between them.

We now examine in some detail the reaction of mineral acids with alcohols as shown in Eq. 16.17. The overall reaction involves replacement or substitution of a hydroxyl group by a nucleophile, and is illustrated in Eq. 16.33 for the reaction of ethanol with hydrogen bromide.

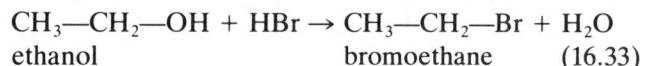

$$CH_3—CH_2—OH + HBr \rightarrow CH_3—CH_2—Br + H_2O$$
ethanol   bromoethane   (16.33)

The first step of the reaction involves protonation of the oxygen atom (reaction 16.34). This is an equilibrium reaction and the equilibrium favours the free alcohol. It is however the protonated species that undergoes further reaction.

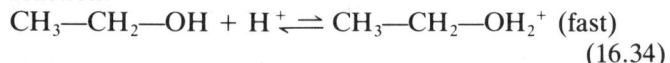

$$CH_3—CH_2—OH + H^+ \rightleftharpoons CH_3—CH_2—OH_2^+ \text{ (fast)}$$
$$(16.34)$$

The nature of the next step depends on the structure of the alkyl group. For primary alcohols the nucleophilic portion of the acid starts to attach itself to the reverse side of the carbon as the protonated hydroxyl group departs, as

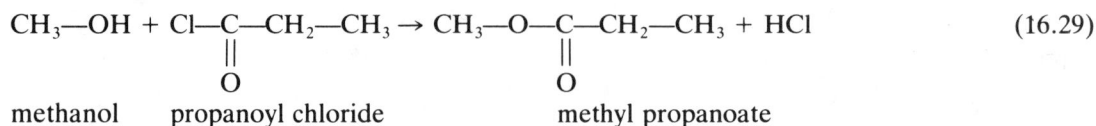

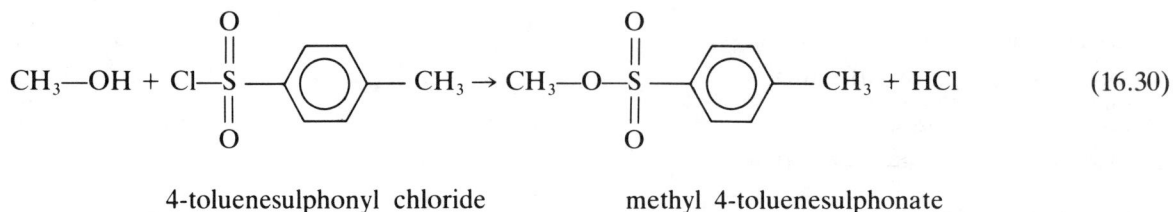

$$CH_3—OH + Cl—\underset{\underset{\displaystyle O}{\|}}{C}—CH_2—CH_3 \rightarrow CH_3—O—\underset{\underset{\displaystyle O}{\|}}{C}—CH_2—CH_3 + HCl \quad (16.29)$$

methanol   propanoyl chloride   methyl propanoate

$$CH_3—OH + Cl—\overset{\overset{\displaystyle O}{\|}}{\underset{\underset{\displaystyle O}{\|}}{S}}—\bigcirc\!\!\!-CH_3 \rightarrow CH_3—O—\overset{\overset{\displaystyle O}{\|}}{\underset{\underset{\displaystyle O}{\|}}{S}}—\bigcirc\!\!\!-CH_3 + HCl \quad (16.30)$$

4-toluenesulphonyl chloride   methyl 4-toluenesulphonate

illustrated in Eq. 16.35. This is the slow step, i.e. the rate-determining step, of the reaction. Only the protonated alcohol and the bromide ion, are involved in this step and the process is therefore *bimolecular,* obeying second-order kinetics—the rate of formation of bromoethane equals $k[CH_3—CH_2—OH_2^+][Br^-]$—and is given the abbreviation SN_2. (S = substitution, N = nucleophilic, 2 = number of species involved in the rate-determining step). The Gibbs free energy changes which occur during the progress of the reaction are shown in Fig. 16.15.

For tertiary alcohols the overall reaction mechanism and reaction kinetics are different but the initial step, the reversible protonation of the oxygen atom (Eq. 16.36) is the same. The difference is that for a tertiary alcohol, heterolysis (splitting into ions) of the carbon–oxygen bond occurs before the nucleophile starts to bond with the carbon, as shown in reaction 16.37. This heterolysis of the carbon–oxygen bond to form a discrete carbonium ion is the slow or rate-determining step of the reaction and involves only one chemical species, namely the protonated alcohol. This step of the reaction obeys the rate equation: reaction rate = $k[R—OH_2^+]$. Thus the reaction is a *unimolecular substitution reaction,* and is given the abbreviation SN_1. The carbonium ion intermediate can be attacked from either face by the nucleophile, as shown in Eq. 16.38.

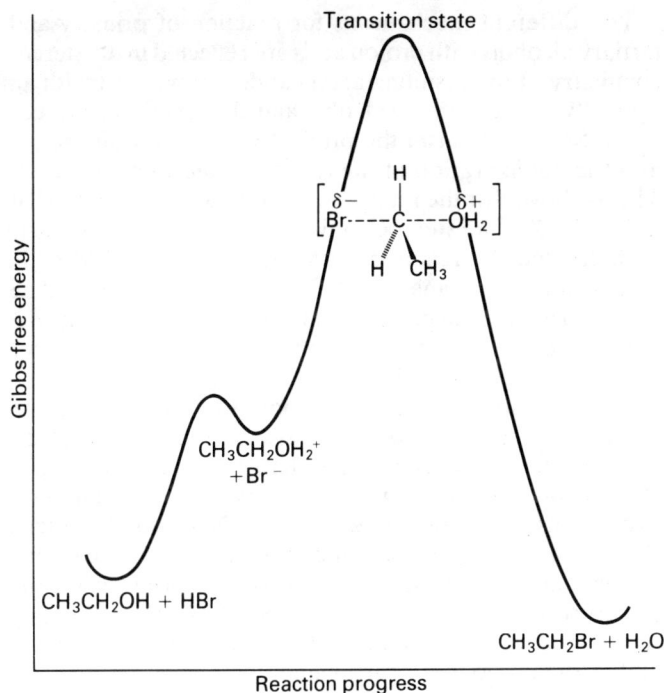

Fig. 16.15 Gibbs free energy changes during progress of the substitution reaction $CH_3CH_2OH + HBr \rightarrow CH_3CH_2Br + H_2O$

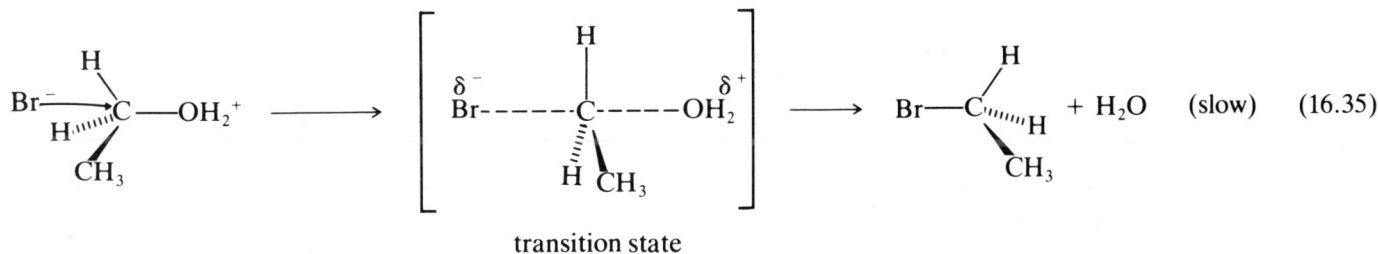

transition state   (slow)   (16.35)

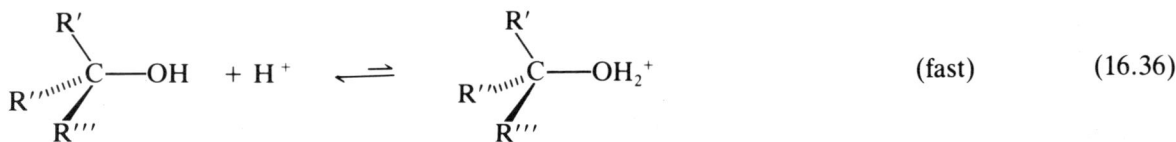

(fast)   (16.36)

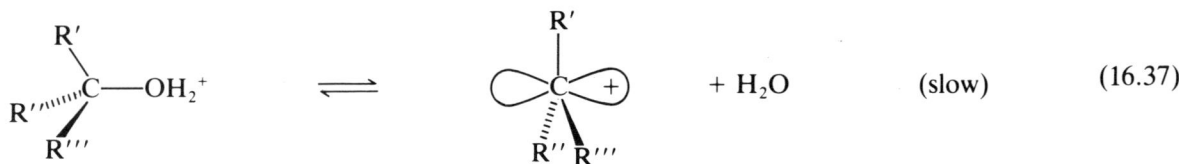

(slow)   (16.37)

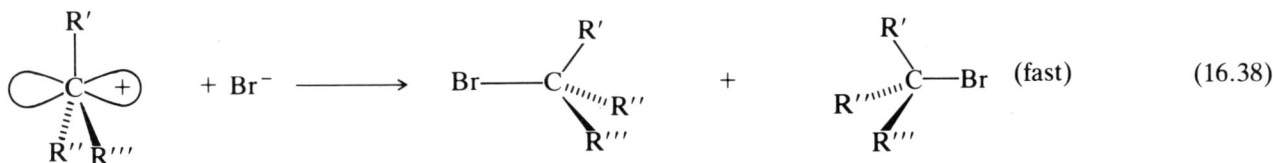

(fast)   (16.38)

The different mechanisms for reaction of primary and tertiary alcohols with proton acids are reflected in the stereochemistry of the resulting alkyl halides. If we start with an optically active primary alcohol and the reaction proceeds by an SN_2 mechanism the product will be optically active and the configuration of the carbon is said to be inverted. This is shown for the reaction of (S)-ethanol-1-D with HBr in Eq. 16.39. The stereochemistry of the reaction contrasts with that seen for reaction of a tertiary alcohol with HBr. In this instance the planar carbonium ion intermediate will be attacked with equal probability from either face, producing a racemic mixture of bromoalkanes. This is shown for (S)-3-methyl-3-hexanol in Eq. 16.40.

To summarize, there are two points of difference in the reactions of primary and tertiary alcohols with proton acids. For a primary alcohol positive charge does not develop at the carbon atom. For a tertiary alcohol a carbonium ion is present as a reaction intermediate. Nucleophilic attack then occurs on this planar *tert*-carbonium ion, whereas for a primary alcohol attack on the carbon atom occurs while the structure is still essentially tetrahedral. We have already noted (section 14.4) that tertiary carbonium ions are more stable than secondary carbonium ions, which in turn are more stable than primary carbonium ions. It is therefore not surprising that tertiary alcohols undergo substitution by a mechanism which involves development of charge on a carbon atom. The more bulky groups of a tertiary alcohol also offer steric hindrance to an attacking nucleophile, which discourages the synchronous type of displacement of the protonated hydroxyl by nucleophile that occurs with primary alcohols. Heterolysis of the C—O bond of a primary alcohol to form a carbonium ion would have a large activation energy; reaction can occur at ordinary temperatures only because there is little steric hindrance to nucleophilic attack, allowing a bimolecular reaction to take place.

The reaction of a secondary alcohol with a proton acid is more complex. The mechanism is often said to be intermediate between those found for primary and tertiary alcohols. The reaction of both secondary and tertiary alcohols with proton acids can lead to rearrangement of the skeleton of the molecule. For example, 3-methyl-2-butanol reacts with HCl as shown in Eq. 16.41 to give 2-chloro-2-methylbutane as well as 2-chloro-3-methylbutane. Rearrangement is often found in reactions which involve carbonium ion intermediates. These intermediates rearrange by hydrogen and alkyl shifts which occur at rates comparable to nucleophilic attack (see section 18.2). Haloalkanes can also be prepared from alcohols by reaction with phosphorus

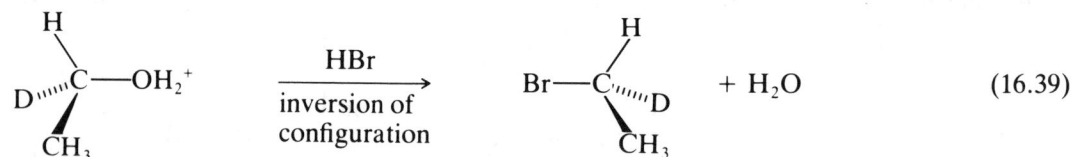

protonated (S)-ethanol-1-D
(primary alcohol)

(R)-bromoethane-1-D

(16.39)

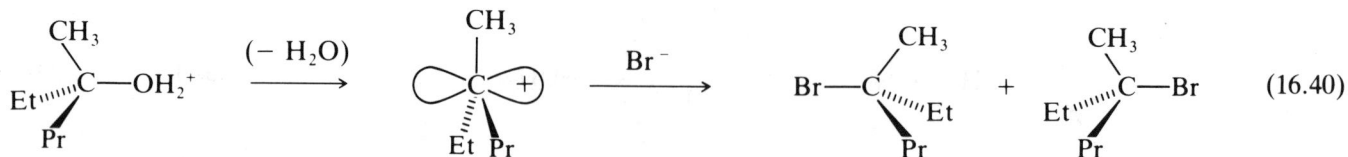

protonated (S)-3-methyl-
3-hexanol
(tertiary alcohol)

reation intermediate
(tertiary carbonium ion)

(R)-   (S)-
racemic mixture of
3-bromo-3-methylhexane

(16.40)

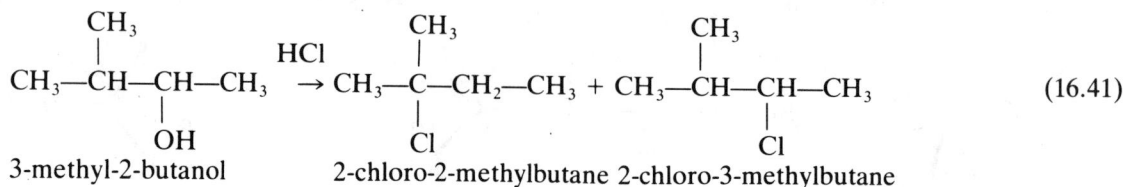

3-methyl-2-butanol   2-chloro-2-methylbutane   2-chloro-3-methylbutane

(16.41)

trihalides (reaction 16.18). This is a good method because less rearrangement occurs than during reaction with hydrogen halides.

Reaction of an alcohol with acid can also result in the formation of alkene products. This is known as an *elimination* reaction. Elimination often competes with nucleophilic substitution, and both reactions have a common first step involving protonation of the oxygen atom of the alcohol (Eq. 16.42). The protonated alcohol can undergo C—O bond heterolysis to give a carbonium ion (reaction 16.43) and this is the slowest step in the reaction sequence. The reaction is therefore unimolecular and given the abbreviation E_1 since only one species, the protonated alcohol, is involved in the rate-determining step. Proton loss from the α carbon, as shown in Eq. 16.44, can occur rapidly to give alkene provided the conformation of the ion allows the C(α)—H bond to align itself with the vacant p orbital (in Eq. 16.44; B: is any base).

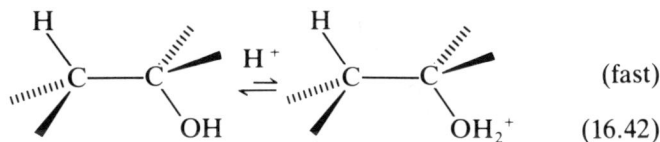

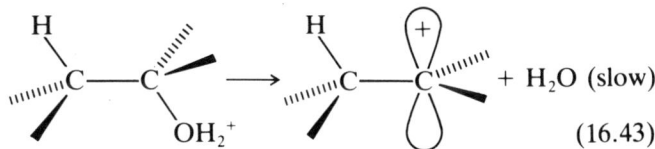

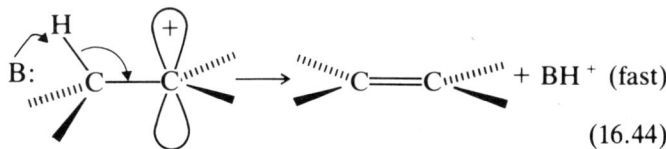

(fast)   (16.42)

+ H_2O (slow)   (16.43)

+ BH^+ (fast)   (16.44)

An alternative mechanism of elimination involves first protonation of the alcohol (reaction 16.42) followed by the simultaneous loss of the adjacent proton and C—O bond cleavage (reaction 16.45). This is a bimolecular elimination reaction because two species, the base and the protonated alcohol are involved in the rate-determining step. The reaction mechanism is given the abbreviation E_2.

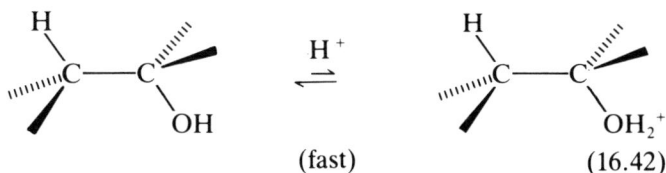

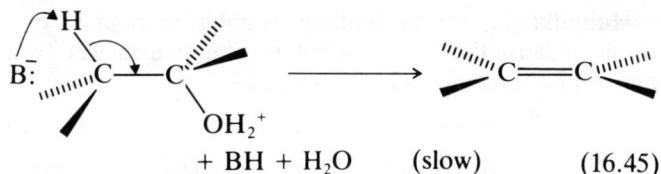

(fast)   (16.42)

+ BH + H_2O (slow)   (16.45)

16.3 Haloalkanes: Preparation and Reactions

We have already considered in some detail the preparation of haloalkanes by free radical halogenation of alkanes (Eq. 13.3), by addition of hydrogen halides to alkenes (Eq. 14.3) and from reaction of alcohols with HX or PX_3 (Eq. 16.18). Equations 16.46–16.52 summarize the reactions used in the laboratory to prepare haloalkanes.

Table 16.4 Preparation of haloalkanes

Halogenation of hydrocarbons

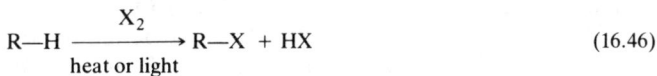

$$R—H \xrightarrow[\text{heat or light}]{X_2} R—X + HX \qquad (16.46)$$

From alcohols

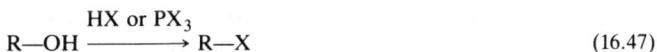

$$R—OH \xrightarrow{HX \text{ or } PX_3} R—X \qquad (16.47)$$

Addition to alkenes and alkynes

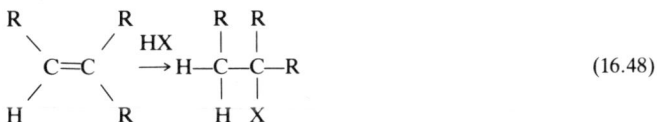

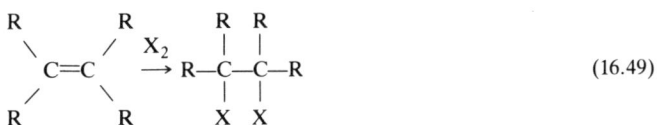

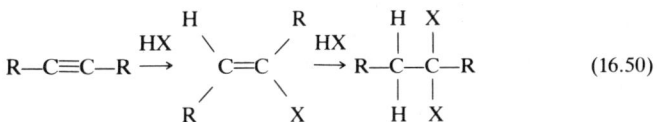

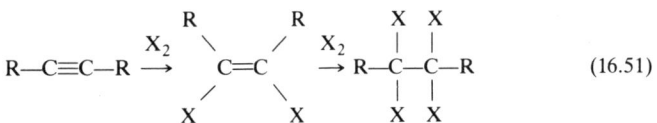

(16.48)

(16.49)

(16.50)

(16.51)

Nucleophilic displacement

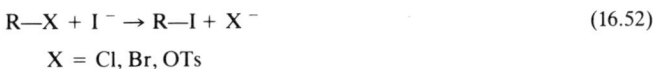

$$R—X + I^- \rightarrow R—I + X^- \qquad (16.52)$$

X = Cl, Br, OTs

Haloalkanes are particularly valuable in organic synthesis because the halide ion can be readily displaced by a nucleophile (reactions 16.53 and 16.54).

$$R—X + :Z^- \rightarrow R—Z + X^- \tag{16.53}$$

$$R—X + :ZH \rightarrow R—Z + HX \tag{16.54}$$

Reactions of this type are referred to as *nucleophilic substitution* reactions. The nucleophile may be a negative ion or a neutral molecule with one or more lone pairs of electrons. The great majority of reactions of haloalkanes are nucleophilic displacement reactions; the scope of such reactions is shown by Eqs 16.55–16.65. Other important reactions of haloalkanes are given in Eqs 16.66–16.68. In order to become familiar with these reactions the student should complete the second problem at the end of this chapter.

The mechanism of reaction of an alcohol with a proton acid depends on the structure of the alcohol (section 16.2). Similarly the mechanism of displacement of a halogen from a haloalkane by HO^- (reaction 16.55), RO^- (reaction 16.57), $RC{\equiv}C^-$ (reaction 16.58), RCO_2^- (reaction 16.60). CN^- (reaction 16.61), ammonia (reaction 16.62) and amines (reactions 16.63 and 16.64), for example, depends on the structure of the haloalkane. The reaction between bromoethane and hydroxide ion (Eq. 16.69) is second-order i.e. reaction rate $= k[CH_3—CH_2—Br][OH^-]$ which implies that the slow step in the reaction involves a collision between hydroxyl ion and the bromoethane. To be effective in

$$CH_3—CH_2—Br + OH^- \xrightarrow[\text{solvent}]{\text{ethanol}} CH_3—CH_2—OH + Br^- \tag{16.69}$$

displacing the bromide ion the hydroxide ion must approach from the opposite side of the carbon atom from that to which the bromine is attached, as illustrated in scheme 16.70. This is an example of a bimolecular nucleophilic substitution reaction (SN_2).

The reaction of 2-bromo-2-methylpropane (a tertiary haloalkane) with hydroxide ion follows a different reaction mechanism; the rate law is reaction rate $= k[R—Br]$. The slow step (Eq. 16.71) in the reaction involves heterolysis of the tertiary C—Br bond, and this is followed by rapid capture of a hydroxyl ion by the carbonium ion (Eq. 16.72).

Table 16.5 Reactions of haloalkanes

Nucleophilic substitution

$$R—X + :OH^- \rightarrow \underset{\text{alcohol}}{R—OH} + X^- \tag{16.55}$$

$$R—X + H_2O \rightarrow \underset{\text{alcohol}}{R—OH} + HX \tag{16.56}$$

$$R—X + :\overset{-}{O}—R^1 \rightarrow \underset{\text{ether}}{R—O—R^1} + X^- \tag{16.57}$$

$$R—X + :C{\equiv}CR^1 \rightarrow \underset{\text{alkyne}}{R—C{\equiv}C—R^1} + X^- \tag{16.58}$$

$$R—X + :I^- \rightarrow \underset{\substack{\text{iodoalkane}\\\text{(haloalkane)}}}{R—I} + X^- \tag{16.59}$$

$$R—X + :\overset{-}{O}—\underset{\underset{O}{\|}}{C}—R \rightarrow R—O—\underset{\underset{O}{\|}}{C}—R + X^- \tag{16.60}$$
ester

$$R—X + :CN^- \rightarrow \underset{\text{alkanenitrile}}{R—CN} + X^- \tag{16.61}$$

$$R—X + :NH_3 \rightarrow \underset{\textit{primary} \text{ amine}}{R—NH_2} + HX \tag{16.62}$$

$$R—X + :NH_2R \rightarrow \underset{\textit{sec.} \text{ amine}}{R—NHR} + HX \tag{16.63}$$

$$R—X + :NHR_2 \rightarrow \underset{\textit{tert.} \text{ amine}}{R—NR_2} + HX \tag{16.64}$$

$$R—X + Ar—H \xrightarrow{AlCl_3} Ar—R + HX \tag{16.65}$$

Dehydrohalogenation: elimination

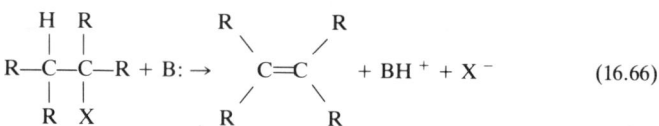

$$\tag{16.66}$$

Preparation of an organo magnesium halide (Grignard reagent)

$$R—X \xrightarrow[\text{ether}]{Mg} R—MgX \tag{16.67}$$

Reduction

$$R—X \xrightarrow{LiAlH_4} R—H \tag{16.68}$$

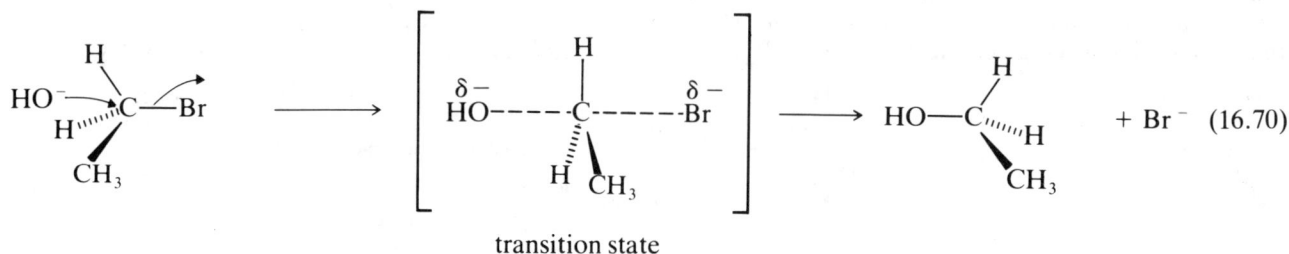

transition state   (16.70)

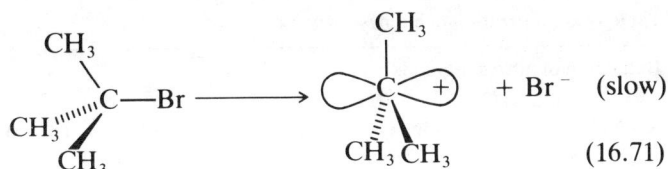

$$+ \text{ Br}^- \quad \text{(slow)} \quad (16.71)$$

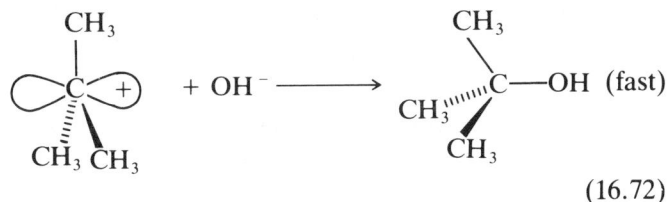

$$(16.72)$$

This is an example of a unimolecular nucleophilic substitution reaction, an SN_1 reaction. The energy required to effect the cleavage of the C—Br bond is supplied mainly by ion–solvent interactions. The carbonium ion produced as an intermediate in reaction 16.71 can be attacked from either face; consequently an optically active haloalkane undergoing nucleophilic substitution by an SN_1 mechanism will give a racemic mixture of products. Nucleophilic substitution reactions of optically active haloalkanes which follow first-order kinetics not only result in racemization but often give products of rearrangement. Both these observations are consistent with the presence of carbonium ions as intermediates in the reaction. The reactivity sequence for unimolecular substitution (SN_1) is tertiary halides > secondary halides > primary halides. With bimolecular nucleophilic substitution reactions racemization and rearrangement are not observed and the order of reactivity for such reactions is primary > secondary > tertiary. Reactions are, however seldom purely SN_1 or SN_2 in character. The contribution of SN_1 character to a reaction is least for a primary halide and greatest for a tertiary halide while the contribution of SN_2 character is greatest for reaction of a primary halide.

Elimination often competes with nucleophilic substitution in haloalkanes, with the result that alkenes are frequent by-products of such reactions. There are two extreme types of reaction mechanism for elimination of HX from haloalkanes. Elimination can occur via a carbonium ion intermediate, as shown in Eq. 16.73, and such a reaction is a unimolecular (E_1) process. The second step (Eq. 16.74) which is rapid, involves abstraction of a proton by a base with formation of the alkene. Unimolecular

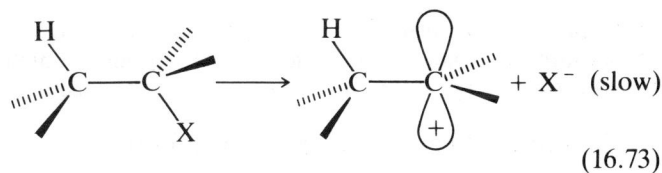

$$+ \text{ X}^- \quad \text{(slow)} \quad (16.73)$$

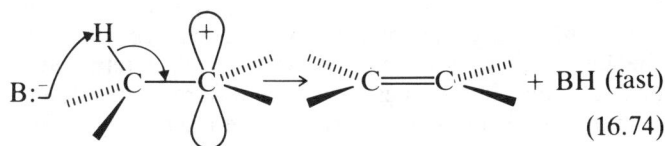

$$+ \text{ BH} \quad \text{(fast)} \quad (16.74)$$

elimination reactions follow first-order kinetics, show the same effect of structure on reactivity as SN_1 reactions, and have the same tendency to be accompanied by rearrangement. Alternatively, elimination can occur by a bimolecular mechanism (E_2) where both haloalkane and base are involved in the rate-determining step, and where rearrangement does not occur (Eq. 16.75).

Nucleophilic substitution reactions of aromatic halides are high-energy processes and do not occur under the mild conditions effective in bringing about substitution reactions of aliphatic halides. It is not possible for the attacking nucleophile to attack the carbon atoms from the side opposite to the halogen (Fig. 16.16) as required for an SN_2 mechanism and unimolecular displacement is extremely slow because the carbonium ion that would be formed by heterolysis of the C—X bond, as shown in Fig. 16.17, is a particularly high-energy species. The reaction therefore occurs only under vigorous conditions of high temperature and pressure. So, for example, phenol (Fig. 16.11) cannot

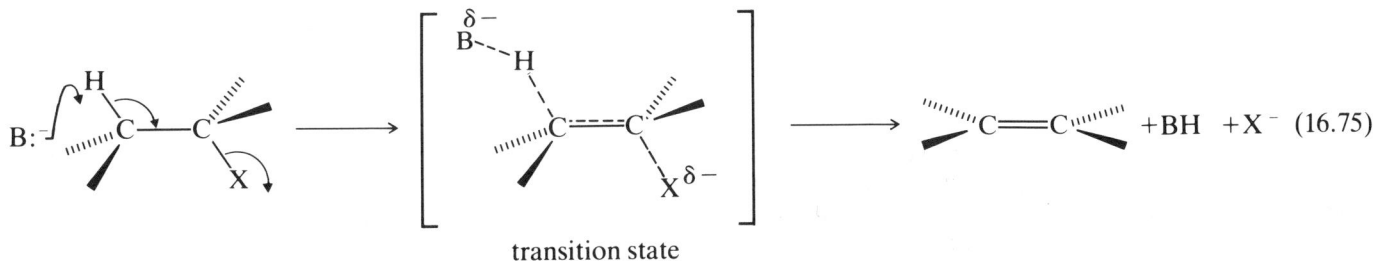

transition state

$$+\text{BH} \ +\text{X}^- \quad (16.75)$$

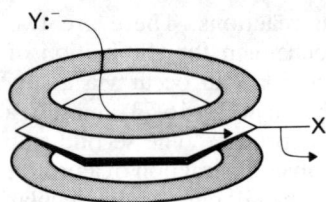

Fig. 16.16   **Fig. 16.17**

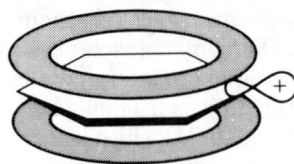

be prepared using normal laboratory procedures by reaction of sodium hydroxide with chlorobenzene, but at high temperature and pressure such displacement does occur.

16.4 Amines: Preparation and Reactions

Amines can be regarded as derivatives of ammonia in which one, two, or all three of the hydrogen atoms have been replaced by alkyl or aryl groups. The three bonds and the lone pair of the nitrogen atom are in a tetrahedral arrangement, but amines with three different substituents do not exhibit optical activity because the energy barrier between enantiomers (the barrier to inversion) is low enough for interconversion to be rapid at room temperature (scheme 16.76).

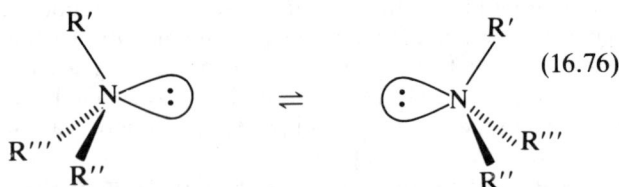

$$(16.76)$$

The reduction of nitro compounds by metal and acid or with lithium aluminium hydride are important methods of preparing primary amines (reactions 16.77 and 16.78). Aromatic amines are important as precursors of diazonium salts (reaction 16.86) which are intermediates in the synthesis of a number of aromatic compounds (see reactions 15.28–15.35).

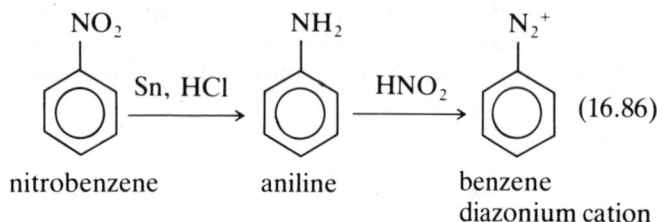

$$(16.86)$$

nitrobenzene   aniline   benzene diazonium cation

Haloalkanes react with ammonia to give primary amines as shown in Eq. 16.87 but the primary amine formed will react further to give secondary and tertiary amines and quaternary ammonium salts (scheme 16.79).

Table 16.6 Preparation of amines

Reduction of nitro componds

$$R—NO_2 \xrightarrow[\text{or Sn, HCl}]{\text{LiAlH}_4} R—NH_2 \qquad (16.77)$$

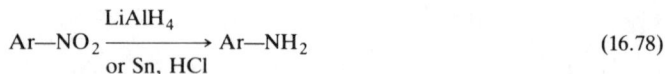

$$Ar—NO_2 \xrightarrow[\text{or Sn, HCl}]{\text{LiAlH}_4} Ar—NH_2 \qquad (16.78)$$

Reaction of haloalkanes with ammonia or amines

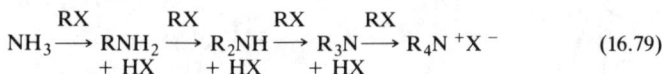

$$NH_3 \xrightarrow{RX} \underset{+\ HX}{RNH_2} \xrightarrow{RX} \underset{+\ HX}{R_2NH} \xrightarrow{RX} \underset{+\ HX}{R_3N} \xrightarrow{RX} R_4N^+X^- \qquad (16.79)$$

Reduction of nitriles

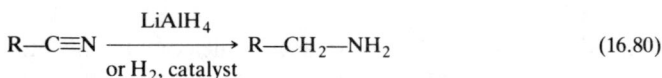

$$R—C≡N \xrightarrow[\text{or } H_2,\ \text{catalyst}]{\text{LiAlH}_4} R—CH_2—NH_2 \qquad (16.80)$$

Reduction of oximes

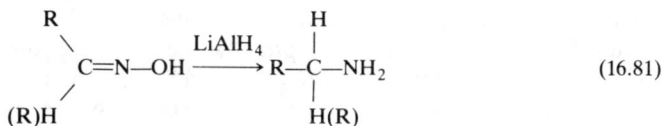

$$(16.81)$$

Hofmann degradation of amides

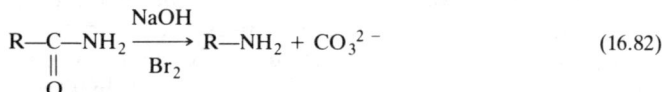

$$R—\underset{\underset{O}{\|}}{C}—NH_2 \xrightarrow[Br_2]{NaOH} R—NH_2 + CO_3^{2-} \qquad (16.82)$$

Reductive amination

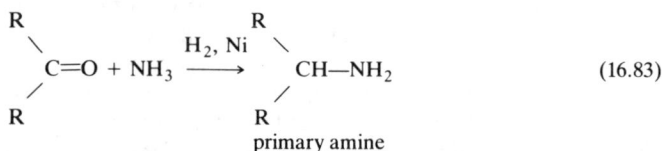

$$(16.83)$$
primary amine

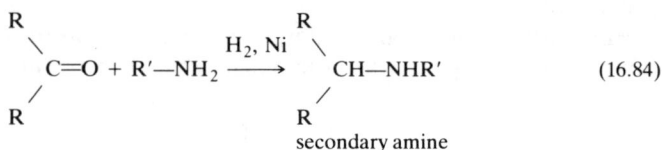

$$(16.84)$$
secondary amine

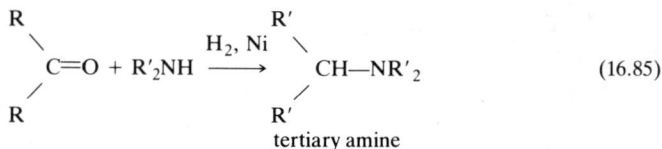

$$(16.85)$$
tertiary amine

$$CH_3-CH_2-CH_2-Br \xrightarrow{NH_3} CH_3-CH_2-CH_2-NH_2 + HBr$$

1-bromopropane   1-propanamine (1-propylamine)

(16.87)

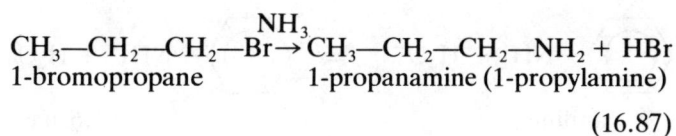

The further reaction of primary amines with haloalkanes limits the usefulness of this procedure as a method of synthesis of primary amines. The usefulness of the reaction is limited still further by the fact that tertiary halides on reaction with ammonia tend to give alkenes, which are products of elimination rather than substitution. This is illustrated in Eq. 16.88 by the formation of methylpropene from 2-bromomethylpropane.

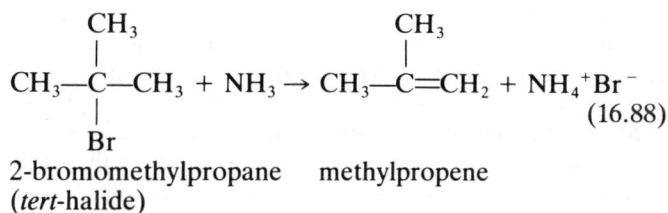

$$\underset{\underset{Br}{|}}{CH_3-\overset{\overset{CH_3}{|}}{C}-CH_3} + NH_3 \rightarrow CH_3-\overset{\overset{CH_3}{|}}{C}=CH_2 + NH_4^+Br^-$$

(16.88)

2-bromomethylpropane   methylpropene
(*tert*-halide)

Primary amines can also be prepared by reduction of alkyl or aryl nitriles (reaction 16.80) and oximes (16.81). These reactions involve the addition of hydrogen to the multiple C—N bonds, as is shown for ethanenitrile in Eq. 16.89 and for the oxime of ethanal in Eq. 16.90.

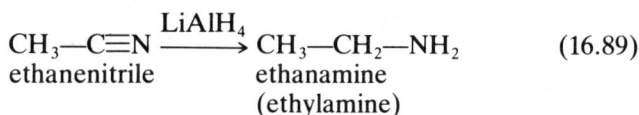

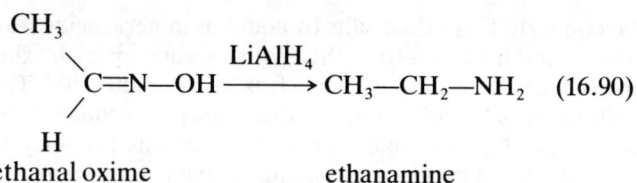

$$CH_3-C\equiv N \xrightarrow{LiAlH_4} CH_3-CH_2-NH_2$$

(16.89)

ethanenitrile   ethanamine
(ethylamine)

$$\underset{\underset{H}{|}}{\overset{\overset{CH_3}{\diagdown}}{}C}=N-OH \xrightarrow{LiAlH_4} CH_3-CH_2-NH_2$$

(16.90)

ethanal oxime   ethanamine

Primary amines can also be prepared by reaction of amide with bromine and base (Eq. 16.82). This reaction is known as the Hofmann degradation reaction, and is important because it provides a means of reducing the number of carbon atoms in the molecule by one. Thus the reaction of propanamide with bromine and base gives ethanamine, which has one carbon less than the starting amide (Eq. 16.91).

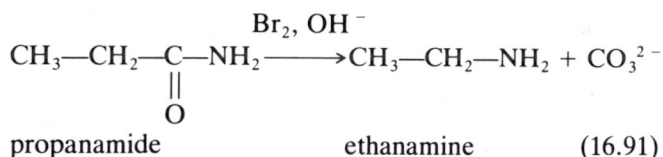

$$CH_3-CH_2-\overset{\overset{}{\underset{\underset{O}{||}}{C}}}{}-NH_2 \xrightarrow{Br_2, \, OH^-} CH_3-CH_2-NH_2 + CO_3^{2-}$$

propanamide   ethanamine   (16.91)

Aldehydes and ketones can be converted into primary, secondary and tertiary amines by reduction in the presence of ammonia, a primary amine, or a secondary amine respectively (reactions 16.83, 16.84 and 16.85). Imines are believed to be intermediates in these reactions and examples are shown in Eq. 16.92, 16.93 and 16.94.

The reactions of amines are dictated mainly by the lone pair of electrons on the nitrogen atom. An amine is able to act as a nucleophile or as a base. Amines, like ammonia, can

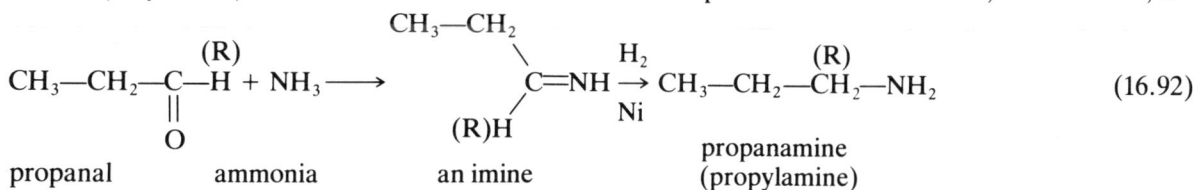

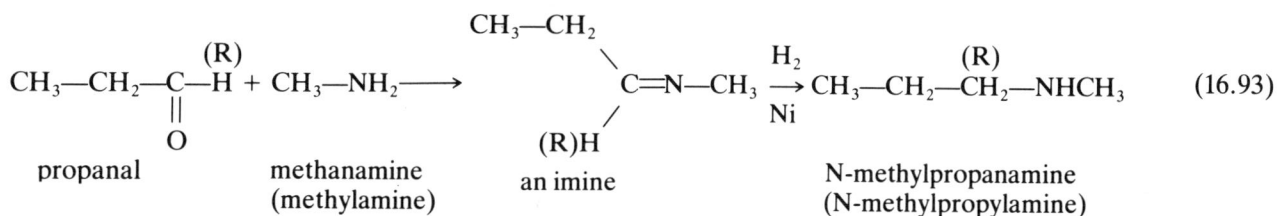

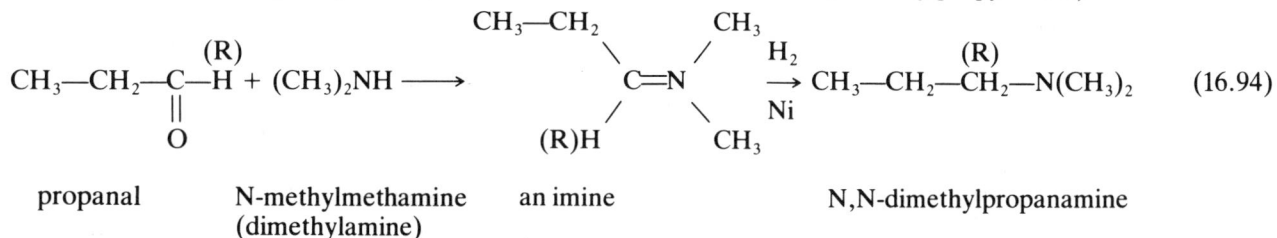

$$CH_3-CH_2-\overset{(R)}{\underset{\underset{O}{||}}{C}}-H + NH_3 \longrightarrow \underset{(R)H}{\overset{CH_3-CH_2}{\diagdown}}C=NH \xrightarrow[Ni]{H_2} CH_3-CH_2-\overset{(R)}{CH_2}-NH_2$$

(16.92)

propanal   ammonia   an imine   propanamine
(propylamine)

$$CH_3-CH_2-\overset{(R)}{\underset{\underset{O}{||}}{C}}-H + CH_3-NH_2 \longrightarrow \underset{(R)H}{\overset{CH_3-CH_2}{\diagdown}}C=N-CH_3 \xrightarrow[Ni]{H_2} CH_3-CH_2-\overset{(R)}{CH_2}-NHCH_3$$

(16.93)

propanal   methanamine
(methylamine)   an imine   N-methylpropanamine
(N-methylpropylamine)

$$CH_3-CH_2-\overset{(R)}{\underset{\underset{O}{||}}{C}}-H + (CH_3)_2NH \longrightarrow \underset{(R)H}{\overset{CH_3-CH_2}{\diagdown}}C=N\overset{\diagup CH_3}{\diagdown CH_3} \xrightarrow[Ni]{H_2} CH_3-CH_2-\overset{(R)}{CH_2}-N(CH_3)_2$$

(16.94)

propanal   N-methylmethamine
(dimethylamine)   an imine   N,N-dimethylpropanamine

be converted into their salts by aqueous mineral acids (Eq. 16.95) and liberated from their salts by aqueous hydroxide ion. This is shown for aniline in Eqs 16.106 and 16.107. The salts are more soluble in water than the amines themselves, and it is often possible to make use of this property to extract an amine from an organic solvent to an aqueous solution, and then back to an organic solvent by adding acid or base to the aqueous phase. (See also p. 61 onwards.)

Table 16.7 Reactions of amines

Basicity

(16.95)

Alkylation

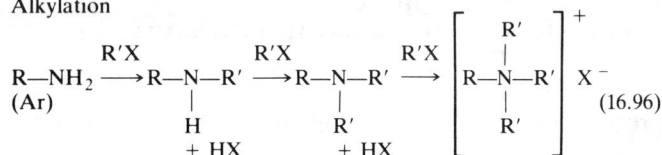

(16.96)

Conversion to amides

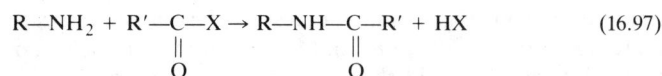

(16.97)

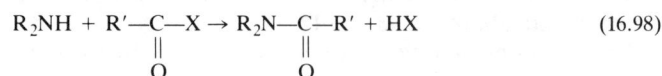

(16.98)

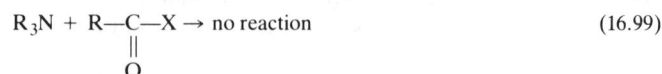

(16.99)

Reaction with nitrous acid Ar = aromatic *aryl* group

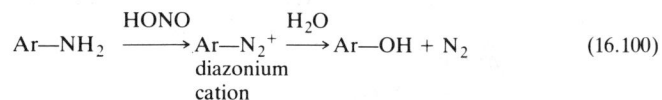

(16.100)

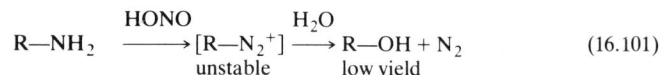

(16.101)

(16.102)

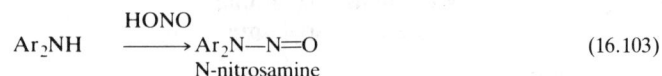

(16.103)

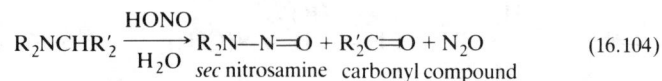

(16.104)

(16.105)

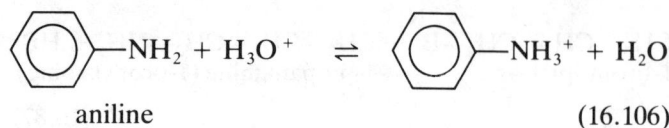

aniline

(16.106)

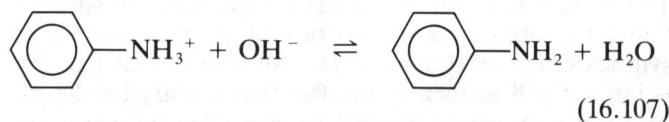

(16.107)

Alkyl groups attached to the nitrogen atom increase the electron density at the nitrogen atom, and also stabilize the protonated amine. Thus the basicity of amines generally increases with alkyl substitution. Aromatic amines such as aniline are weaker bases than ammonia because the nitrogen lone pair becomes involved with the aromatic π electrons and is less available for bonding with electrophiles. One consequence of this involvement is that the aromatic ring of aniline is extremely reactive to electrophilic substitution. Direct bromination of aniline (reaction 16.108) occurs so readily that it cannot be stopped at a monobrominated product and the 2,4,6-tribromoaniline is produced despite the deactivating effect of each added bromine.

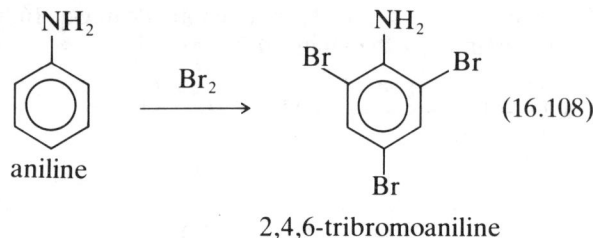

(16.108)

2,4,6-tribromoaniline

Ortho and *para* monobromoanilines are prepared by bromination of acetanilide (reaction 16.109) followed by hydrolysis (see section 17.5) of the bromoacetanilides. The amide substituent (Fig. 16.18), activates the aromatic ring to electrophilic substitution to a much lesser extent than does an amine group. This can be rationalized by saying that the carbonyl group adjacent to the nitrogen atom makes competing demands for a lone pair of electrons on nitrogen which is therefore less available to the aromatic ring. This explains why the monobrominated products can be isolated when acetanilide is brominated.

Aromatic primary amines readily form diazonium salts (reaction 16.100), on reaction with cold aqueous mineral acid and sodium nitrite. The active reagent is nitrous acid, HONO. Diazonium salts, as we have already noted, are important in synthesis but decompose readily and are therefore used as soon as they are prepared. As well as undergoing nucleophilic displacement reactions (see reactions 15.28–15.33), they undergo a coupling process important in

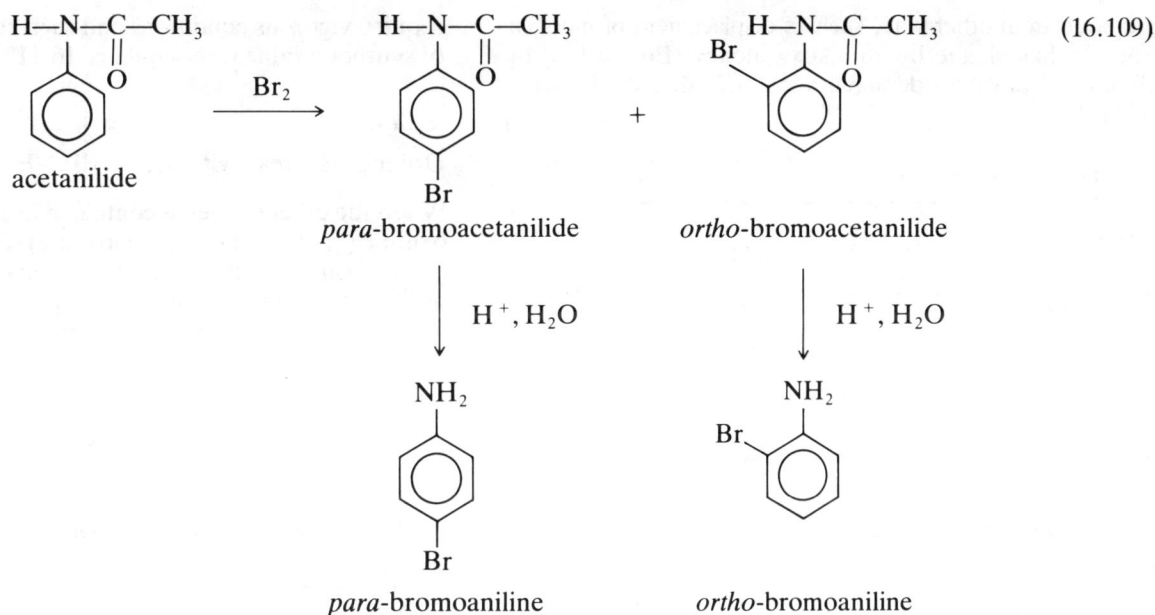

$$(16.109)$$

acetanilide → *para*-bromoacetanilide + *ortho*-bromoacetanilide

para-bromoaniline   *ortho*-bromoaniline

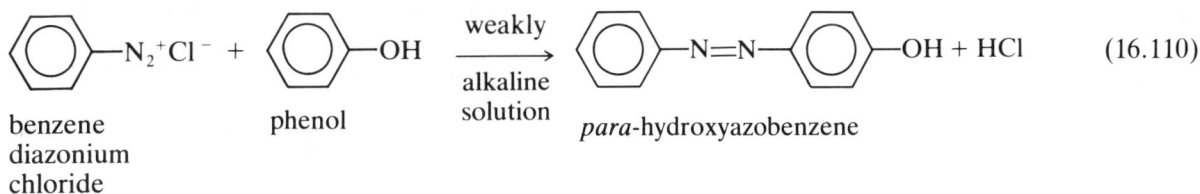

$$(16.110)$$

benzene diazonium chloride + phenol → *para*-hydroxyazobenzene + HCl

the preparation of azo dyes. This latter process involves the positively charged diazonium species acting as an electrophile and attacking activated benzene derivatives such as phenol (reaction 16.110).

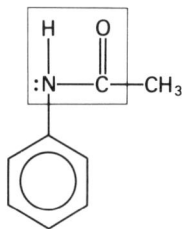

Fig. 16.18 Amide grouping in acetanilide

Primary aliphatic amines react with nitrous acid by liberating nitrogen and forming alcohol and alkenes (Eq. 16.101). The product mixture is generally complex and of no synthetic value but it does provide a useful qualitative test for primary amines. Secondary aliphatic and aromatic amines give N-nitroso products (reactions 16.102 and 16.103), while tertiary aliphatic amines, contrary to common belief, undergo dealkylation to form a carbonly compound, a *sec.* nitrosamine and nitrous oxide (Eq. 16.104). Tertiary aromatic amines have been reported to give *para*-nitroso derivatives (reaction 16.105).

16.5 Ethers: Preparation and Reactions

Ethers are widely used as solvents, Diethylether (ethoxyethane) is the most commonly used and is often known simply as 'ether'. Ethers are prepared by reaction of alcohols with sulphuric acid under carefully controlled conditions to avoid formation of alkene by-products (reaction 16.111). Diethylether (ethoxyethane) is prepared from ethanol by heating with concentrated sulphuric acid at 140 °C, care being taken to ensure that there is a considerable excess of ethanol at all times. At higher temperatures, in the vicinity of 180 °C, ethene is the major product. Unsymmetrical ethers such as methylethylether (methoxyethane) must be prepared in another way, such as displacement of halogen from an haloalkane by an alkoxy moiety (Eq. 16.112) or alkoxymercuration—demercuration of alkenes (scheme 16.114).

Table 16.8 Preparation of ethers

1 From alcohols

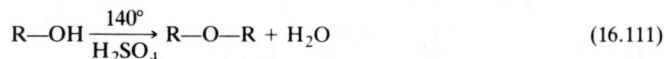

$$R—OH \xrightarrow[H_2SO_4]{140°} R—O—R + H_2O \qquad (16.111)$$

2 Alkoxy displacement of halogen from haloalkanes

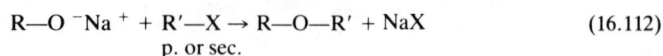

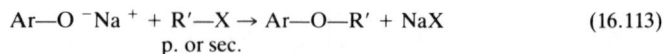

$$R—O^-Na^+ + R'—X \rightarrow R—O—R' + NaX \qquad (16.112)$$
$$\text{p. or sec.}$$

$$Ar—O^-Na^+ + R'—X \rightarrow Ar—O—R' + NaX \qquad (16.113)$$
$$\text{p. or sec.}$$

3 Alkoxymercuration–demercuration

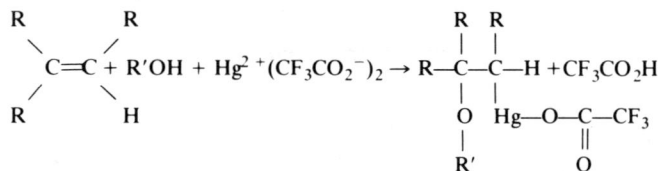

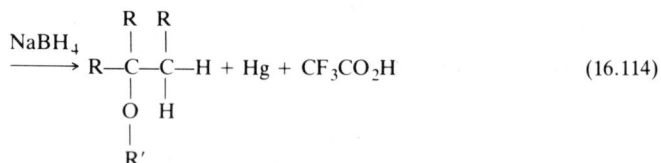

$$(16.114)$$

Ethers are useful as solvents because numerous organic and inorganic compounds are soluble in them and because they are stable towards oxidizing agents, reducing agents and bases. They do react with concentrated acids, but

require vigorous conditions and such reactions are seldom of synthetic value (see sequence 16.115).

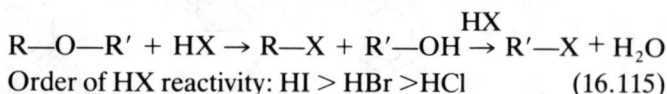

$$R—O—R' + HX \rightarrow R—X + R'—OH \xrightarrow{HX} R'—X + H_2O$$
Order of HX reactivity: HI > HBr > HCl $\qquad (16.115)$

When the ether oxygen is contained in a small ring, as in an oxirane (epoxide or oxacyclopropane) (Fig. 16.19), cleavage of a carbon–oxygen bond readily occurs to relieve the strain of the small ring. Oxiranes are prepared by oxidation of an alkene with a peracid such as *meta*-chloroperbenzoic acid (Eq. 16.116). Reaction of oxiranes can be initiated by acid (reactions 16.117 and 16.118) or base (reaction 16.119).

Fig. 16.19 Oxirane or oxacylcopropane (ethylene oxide)

Acid catalysed cleavage of oxiranes involves as the initial step rapid protonation of the basic oxirane oxygen, as shown in scheme 16.120. Cleavage of the C—O bond then occurs between the oxygen atom and the more substituted carbon atom to give, for example, (Eq. 16.121) a tertiary carbonium ion intermediate. This carbonium ion is then attacked by a nucleophile (Eq. 16.122) to give the final product.

Because the epoxide carbon–oxygen bond breaks towards the more substituted carbon the nucleophile becomes attached to the carbon atom bearing the smaller number of hydrogen atoms. For example, the oxirane of methylpropene, 2,2-dimethyloxacyclopropane gives, on reaction with HCl, 2-chloro-2-methyl-1-propanol (Eq. 16.123). This is in contrast to base-catalysed cleavage of epoxides where the nucleophile (H$^-$ or CH$_3$—CH$_2$—O$^-$) becomes attached to the least hindered carbon (Eqs 16.124 and 16.125).

Table 16.9 Preparation of oxiranes (epoxides)

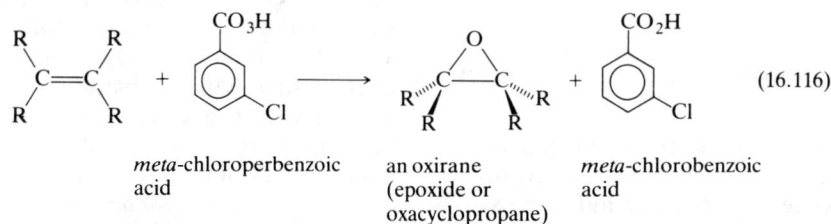

meta-chloroperbenzoic acid   an oxirane (epoxide or oxacyclopropane)   *meta*-chlorobenzoic acid (16.116)

Table 16.10 Reaction of oxiranes

1. Acid catalysed cleavage

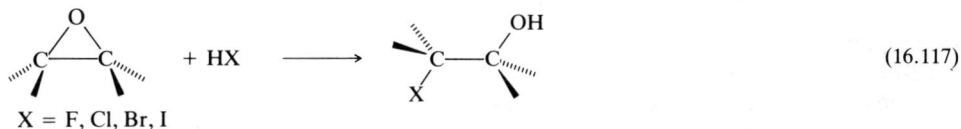

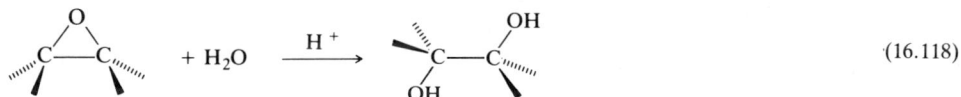

(16.117)

X = F, Cl, Br, I

(16.118)

2. Base catalysed cleavage

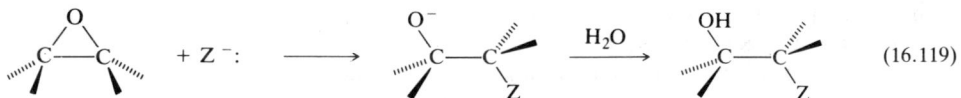

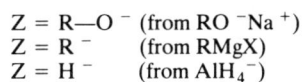

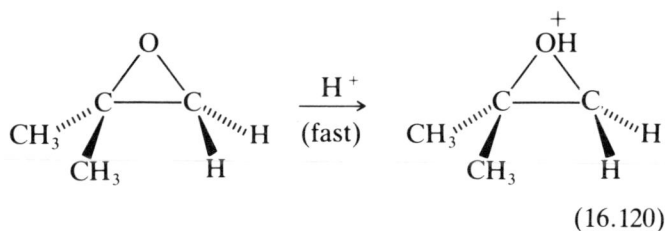

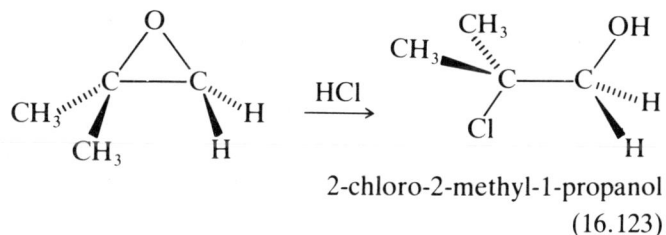

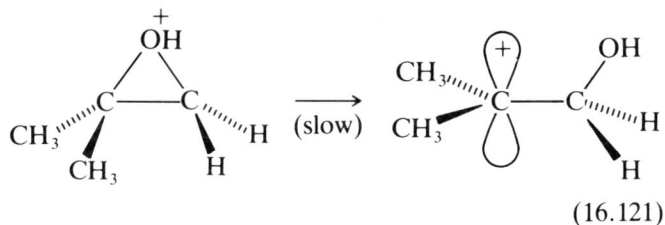

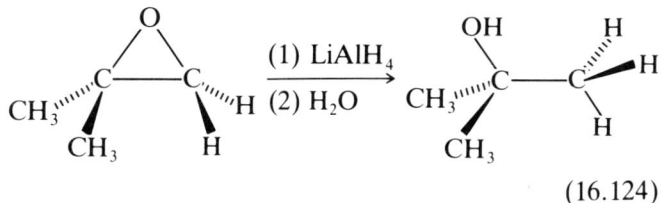

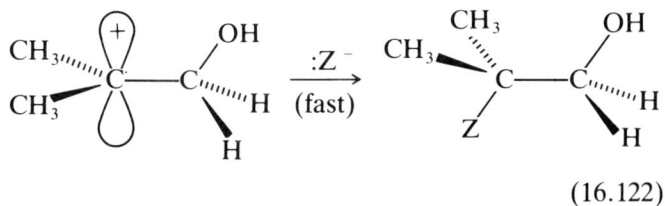

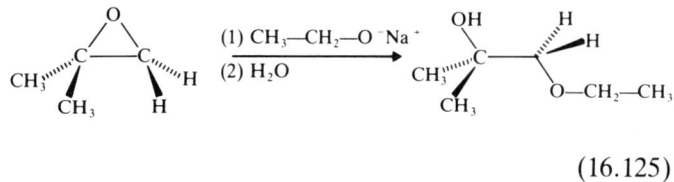

(16.119)

Z = R—O⁻ (from RO⁻Na⁺)
Z = R⁻ (from RMgX)
Z = H⁻ (from AlH₄⁻)

(16.120)

2-chloro-2-methyl-1-propanol
(16.123)

(16.121)

(1) LiAlH₄
(2) H₂O
(16.124)

:Z⁻
(fast)
(16.122)

(1) CH₃—CH₂—O⁻Na⁺
(2) H₂O
(16.125)

Exercises

16.1 Show, by means of structural formulae, how the following transformations may be carried out. Name all reagents used.

(a) ethanol to 2-propanol
(b) 1-butanol to 2-butanol
(c) 2-bromopropane to methyl-1-propanol
(d) bromoethane to propanoic acid
(e) chloroethane to 1-propanamine
(f) ethanol to 1-butanol
(g) ethanol to ethylmethylether (methoxyethane)
(h) 1-butene to 1-ethoxy-2-butanol

16.2 Name the reagents required to prepare the following compounds from bromoethane:(a) ethanol, (b) ethylmethyl-ether (methoxyethane), (c) 2-pentyne, (d) ethyl propanoate, (e) propanenitrile ($CH_3CH_2C{\equiv}N$), (f) N,N-dimethyl-ethanamine, (g) ethylbenzene, (h) ethyl magnesium bromide, (i) ethane

16.3 Show how an organo magnesium halide reagent can be used in the synthesis of the following compounds:

(a) 1-phenyl-1-ethanone (acetophenone, methylphenyl-ketone) from benzal
(b) methyl-2-propanol from 2-propanol
(c) 1-propanol from ethyl magnesium bromide
(d) 1-butanol from ethyl magnesium bromide
(e) 2-butanol from ethyl magnesium bromide
(f) 1-methyl-1-cyclohexanol from methyl magnesium bromide
(g) 1-buten-3-ol from methyl magnesium bromide
(h) 1-phenyl-1-propanol from phenyl magnesium bromide
(i) 2-methyl-2-butanol from methyl magnesium bromide.

17 Aldehydes, Ketones, Carboxylic Acids, Acid Halides, Amides, Anhydrides and Esters

17.1 Introduction to Carbonyl Compounds

This chapter is concerned with the chemistry of the compounds shown in Table 17.1, all of which have the general formula R—C—Z. Carbon forms single σ bonds to

$$R-\underset{\overset{\|}{O}}{C}-Z$$

many elements, and is known to form multiple bonds to oxygen, nitrogen, sulphur, phosphorus and itself. The carbon–oxygen and carbon–nitrogen multiple bonds are stronger than carbon–sulphur multiple bonds. This is because of the similarity in size of the carbon, oxygen and nitrogen atoms, which facilitates overlap of the electron in the p orbitals of the carbon atom and hetero atom. By comparison, sulphur is a large atom, and the p orbital overlap between sulphur and carbon is correspondingly reduced. All carbonyl compounds show a characteristic infrared absorption at the carbonyl ($>$C$=$O) stretching frequency (i.e. *ca* 1700 cm^{-}, Figs 17.1, 17.2 and 17.3, pages 282 and 283). Infrared spectroscopy is a very powerful tool for investigating molecular structure; the features of the molecule's spectrum for carbonyl containing compounds will often indicate the nature of Z quite precisely. Thus, for example, the O—H group of ethanoic acid is responsible for the absorption peak at 3200 cm^{-1} in Fig. 17.3.

17.2 Aldehydes and Ketones

Aldehydes are named by adding the suffix *al* to the name of the corresponding hydrocarbon, the final '*e*' of the hydro-

carbon name being omitted (Figs 17.4–17.7). Positions of substituents on the chain are numbered in such a way that the carbonyl carbon is always considered as C(1).

$$H-\underset{\overset{\|}{O}}{C}-H \qquad CH_3-\underset{\overset{\|}{O}}{C}-H \qquad$$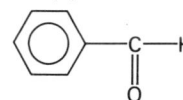

methanal   ethanal   benzal
(formaldehyde)   (acetaldehyde)   (benzaldehyde)

Fig. 17.4   **Fig. 17.5**   **Fig. 17.6**

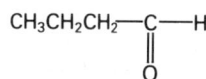

butanal

Fig. 17.7

Ketones are named by adding the suffix *one* to the name of the corresponding hydrocarbon with the final '*e*' of the hydrocarbon name omitted.

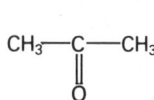 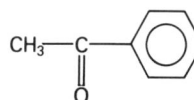 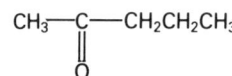

propanone   1-phenyl-1-ethanone   2-pentanone
(acetone)   (acetophenone)

Fig. 17.8   **Fig. 17.9**   **Fig. 17.10**

The position of the carbonyl group is given by a number which can be placed before the parent name as in Figs 17.9 and 17.10 or immediately before the suffix (e.g. pentan-2-one).

Table 17.1 Carbonyl compounds

R—C—R (O)	R—C—H (O)	R—C—OH (O)	R—C—X (O)	R—C—OR (O)	R—C—NH$_2$ (O)	R—C—O—C—R (O)(O)
Z = R	H	OH	X	OR	NH$_2$	O—C—R (O)
ketone	aldehyde	carboxylic acid	acid halide	ester	amide	anhydride

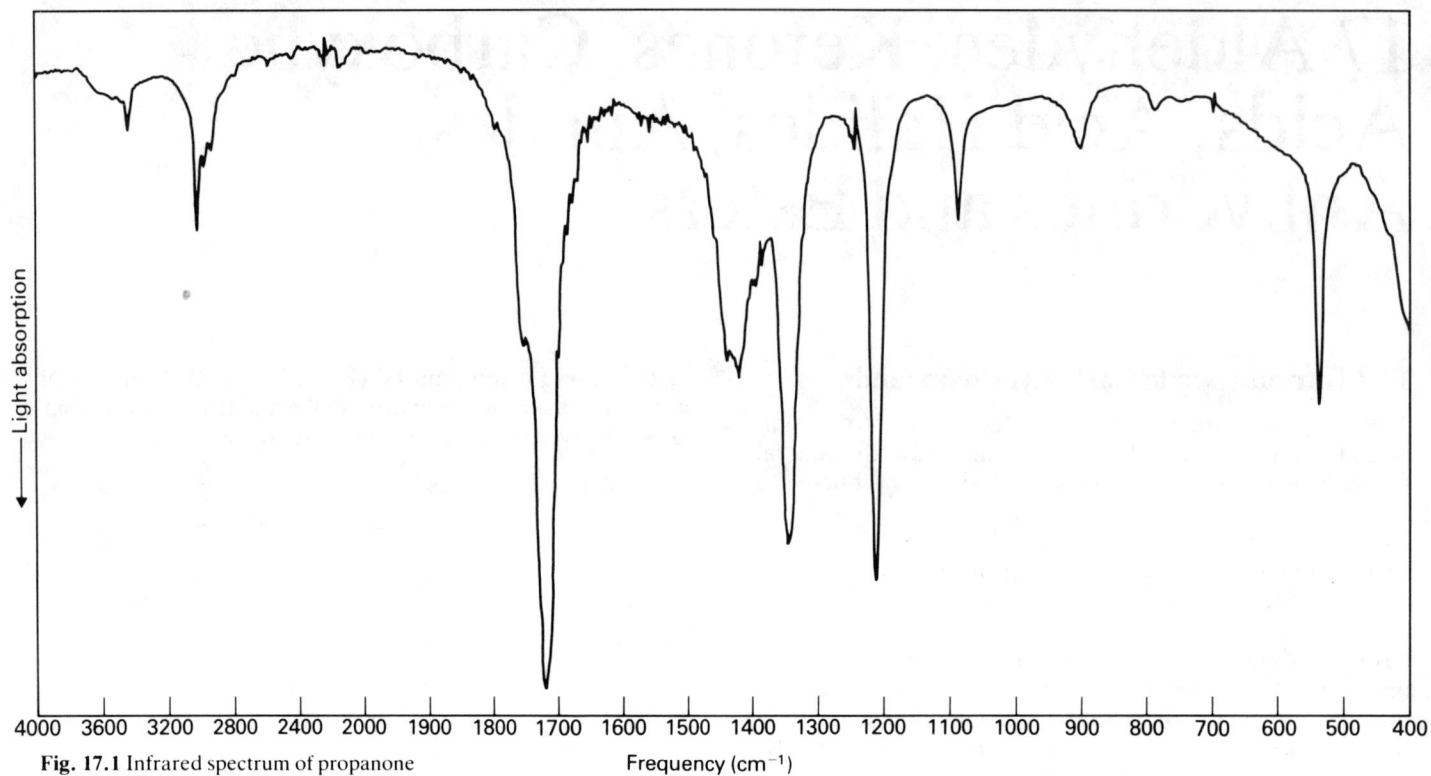

Fig. 17.1 Infrared spectrum of propanone   Frequency (cm^{-1})

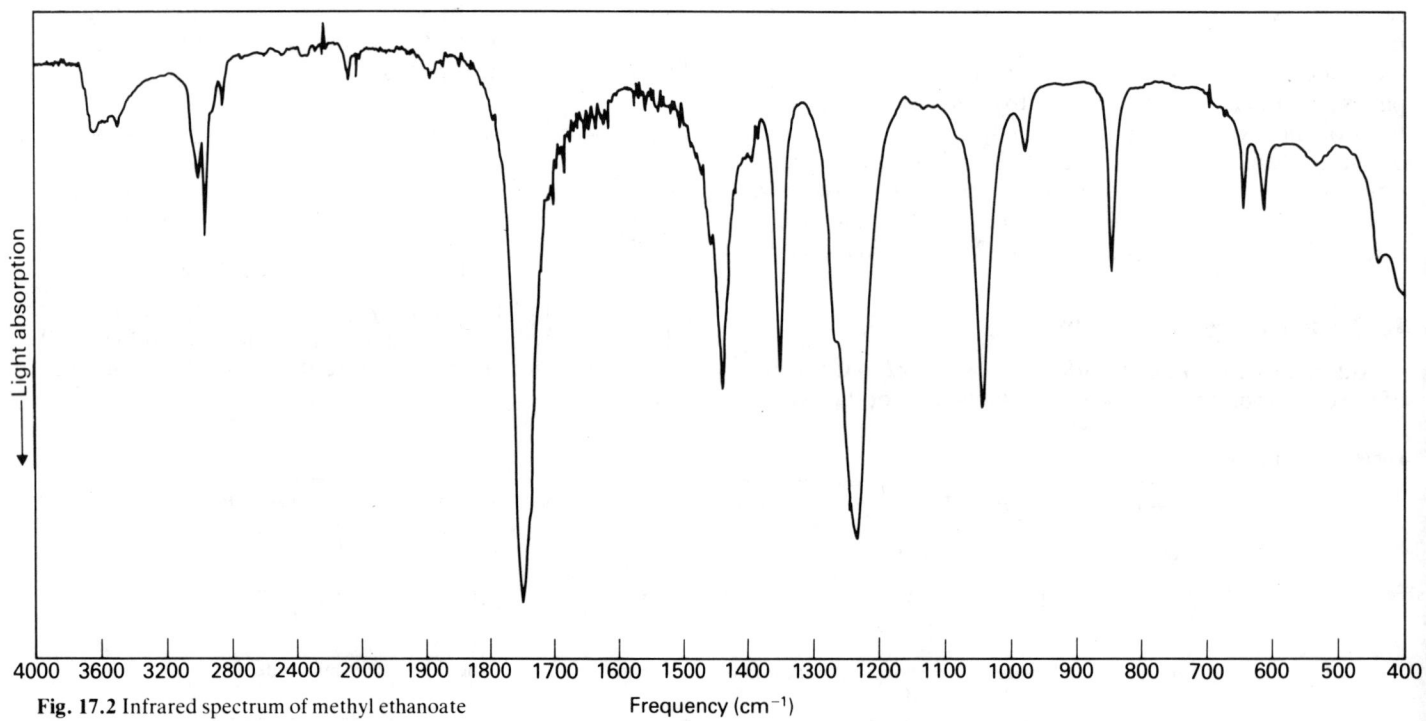

Fig. 17.2 Infrared spectrum of methyl ethanoate   Frequency (cm^{-1})

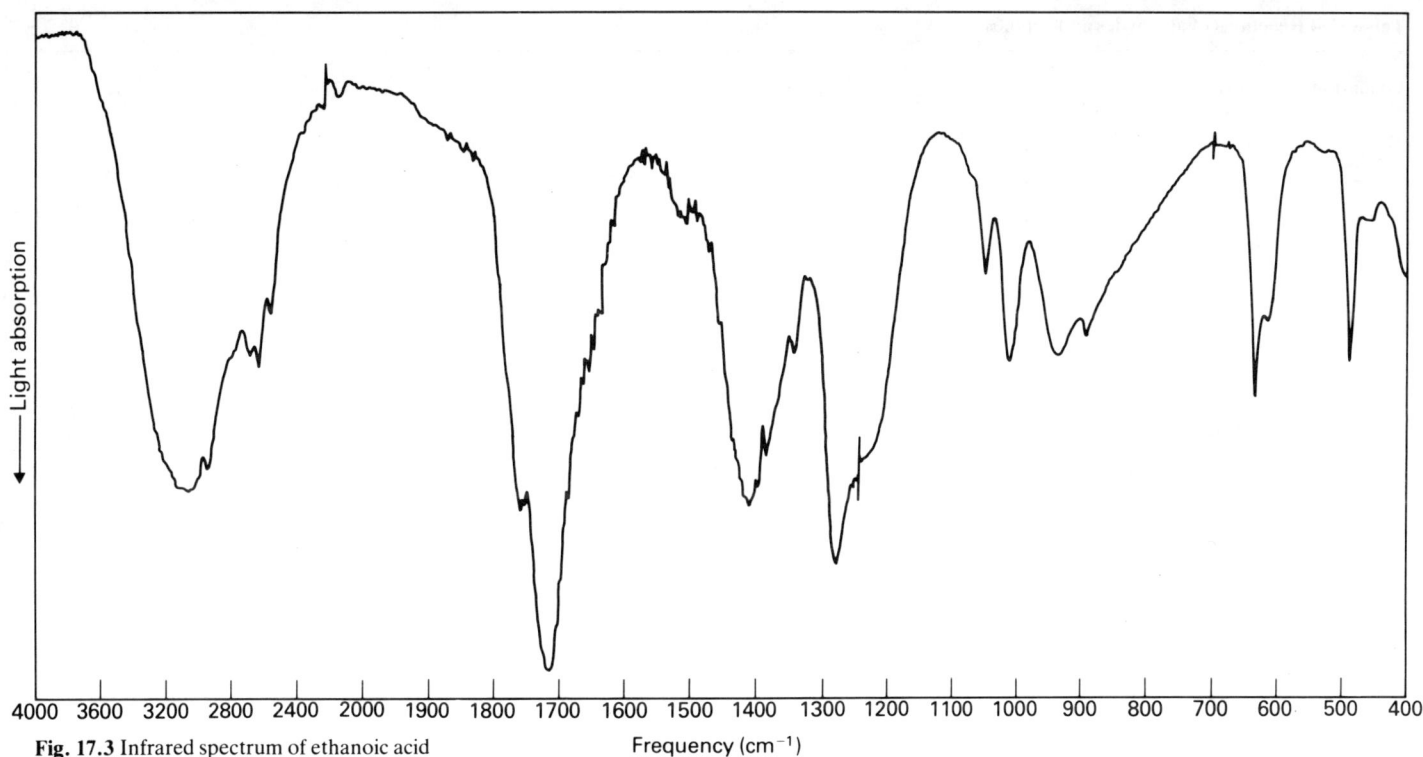

Fig. 17.3 Infrared spectrum of ethanoic acid

Frequency (cm^{-1})

Most of the procedures for the preparation of aldehydes and ketones have been met earlier (Tables 17.2 and 17.3).

Table 17.2 Preparation of aldehydes

Oxidation of primary alcohol

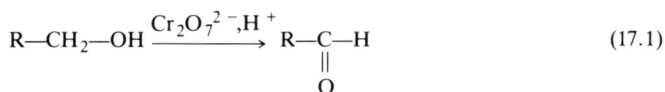

$$R—CH_2—OH \xrightarrow{Cr_2O_7^{2-},H^+} R—\underset{\underset{O}{\|}}{C}—H \qquad (17.1)$$

Oxidation of methyl benzenes

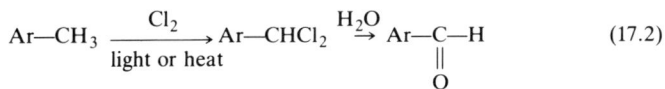

$$Ar—CH_3 \xrightarrow[\text{light or heat}]{Cl_2} Ar—CHCl_2 \xrightarrow{H_2O} Ar—\underset{\underset{O}{\|}}{C}—H \qquad (17.2)$$

Reduction of acid chlorides

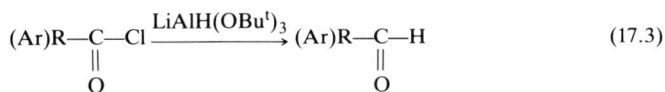

$$(Ar)R—\underset{\underset{O}{\|}}{C}—Cl \xrightarrow{LiAlH(OBu^t)_3} (Ar)R—\underset{\underset{O}{\|}}{C}—H \qquad (17.3)$$

Table 17.3 Preparation of ketones

Oxidation of secondary alcohols

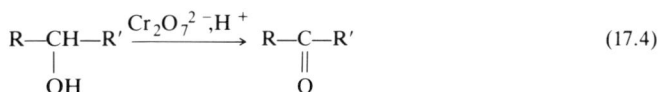

$$R—\underset{\underset{OH}{|}}{CH}—R' \xrightarrow{Cr_2O_7^{2-},H^+} R—\underset{\underset{O}{\|}}{C}—R' \qquad (17.4)$$

Friedel-Crafts alkylation

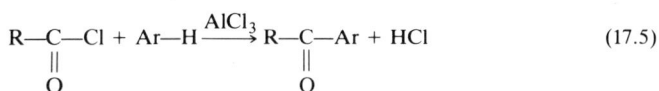

$$R—\underset{\underset{O}{\|}}{C}—Cl + Ar—H \xrightarrow{AlCl_3} R—\underset{\underset{O}{\|}}{C}—Ar + HCl \qquad (17.5)$$

The chemistry of aldehydes and ketones is dominated by the carbonyl (C=O) group. A list of reactions which aldehydes and ketones undergo is shown in Table 17.4. Aldehydes differ from ketones in that they are readily oxidized to carboxylic acids (reaction 17.6), whereas ketones require vigorous conditions before oxidation can

Table 17.4 Reactions of aldehydes and ketones

Oxidation of aldehydes

$$R\text{—}\underset{\underset{O}{\|}}{C}\text{—}H \xrightarrow{Cr_2O_7{}^{2-},\ H^+} R\text{—}\underset{\underset{O}{\|}}{C}\text{—}OH$$

$$Ar\text{—}\underset{\underset{O}{\|}}{C}\text{—}H \xrightarrow{Cr_2O_7{}^{2-},\ H^+} Ar\text{—}\underset{\underset{O}{\|}}{C}\text{—}OH$$

(17.6)

Oxidation of methylketones

$$R\text{—}\underset{\underset{O}{\|}}{C}\text{—}CH_3 \xrightarrow{X_2,\ OH^-} R\text{—}\underset{\underset{O}{\|}}{C}\text{—}O^- + CHX_3$$

$$X = Cl,\ Br,\ I$$

(17.7)

Reduction to alcohols

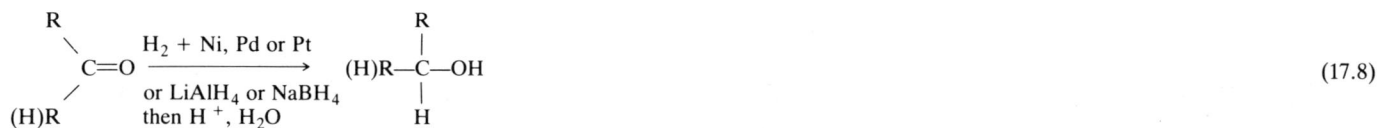

$$\underset{(H)R}{\overset{R}{\diagdown}}C{=}O \xrightarrow[\substack{\text{or } LiAlH_4 \text{ or } NaBH_4 \\ \text{then } H^+,\ H_2O}]{H_2 + Ni,\ Pd \text{ or } Pt} (H)R\text{—}\underset{\underset{H}{|}}{\overset{\overset{R}{|}}{C}}\text{—}OH$$

(17.8)

Reduction to hydrocarbons

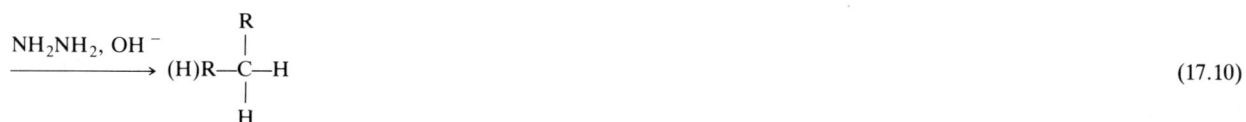

$$\underset{(H)R}{\overset{R}{\diagdown}}C{=}O \xrightarrow[\text{conc. } HCl]{Zn,\ Hg,} (H)R\text{—}\underset{\underset{H}{|}}{\overset{\overset{R}{|}}{C}}\text{—}H$$

(17.9)

$$\xrightarrow{NH_2NH_2,\ OH^-} (H)R\text{—}\underset{\underset{H}{|}}{\overset{\overset{R}{|}}{C}}\text{—}H$$

(17.10)

Addition of an organo magnesium halide

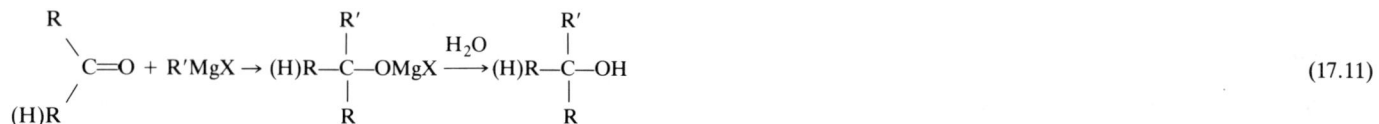

$$\underset{(H)R}{\overset{R}{\diagdown}}C{=}O + R'MgX \rightarrow (H)R\text{—}\underset{\underset{R}{|}}{\overset{\overset{R'}{|}}{C}}\text{—}OMgX \xrightarrow{H_2O} (H)R\text{—}\underset{\underset{R}{|}}{\overset{\overset{R'}{|}}{C}}\text{—}OH$$

(17.11)

Addition of hydrogen cyanide

$$\underset{(H)R}{\overset{R}{\diagdown}}C{=}O + CN^- \xrightarrow{H^+} (H)R\text{—}\underset{\underset{R}{|}}{\overset{\overset{CN}{|}}{C}}\text{—}OH$$

(17.12)

Addition of bisulphite

$$\underset{(H)R}{\overset{R}{\diagdown}}C{=}O + Na^+HSO_3^- \rightleftharpoons (H)R{-}\underset{\underset{OSO_2^-Na^+}{|}}{\overset{\overset{R}{|}}{C}}{-}OH \qquad (17.13)$$

bisulphite addition product

Addition of alcohols

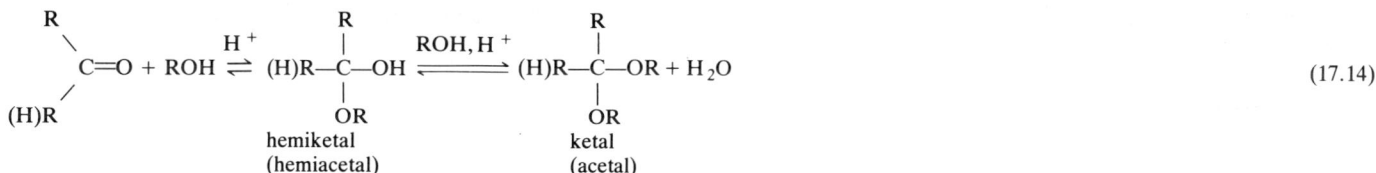

$$\underset{(H)R}{\overset{R}{\diagdown}}C{=}O + ROH \overset{H^+}{\rightleftharpoons} (H)R{-}\underset{\underset{OR}{|}}{\overset{\overset{R}{|}}{C}}{-}OH \overset{ROH, H^+}{\rightleftharpoons} (H)R{-}\underset{\underset{OR}{|}}{\overset{\overset{R}{|}}{C}}{-}OR + H_2O \qquad (17.14)$$

hemiketal   ketal
(hemiacetal)   (acetal)

Addition of ammonia derivatives

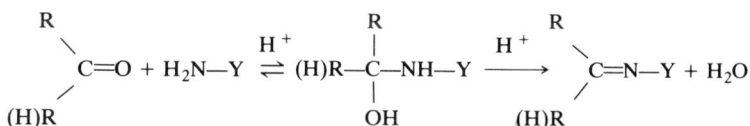

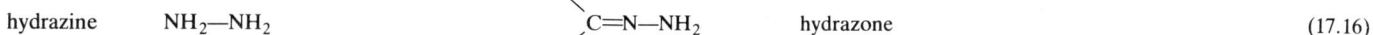

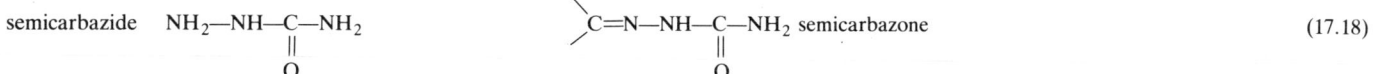

$$\underset{(H)R}{\overset{R}{\diagdown}}C{=}O + H_2N{-}Y \overset{H^+}{\rightleftharpoons} (H)R{-}\underset{\underset{OH}{|}}{\overset{\overset{R}{|}}{C}}{-}NH{-}Y \overset{H^+}{\longrightarrow} \underset{(H)R}{\overset{R}{\diagdown}}C{=}N{-}Y + H_2O$$

hydroxylamine	$H_2N{-}OH$	$\diagdown C{=}N{-}OH$	oxime	(17.15)
hydrazine	$NH_2{-}NH_2$	$\diagdown C{=}N{-}NH_2$	hydrazone	(17.16)
phenylhydrazine	$NH_2{-}NHPh$	$\diagdown C{=}N{-}NHPh$	phenylhydrazone	(17.17)
semicarbazide	$NH_2{-}NH{-}\underset{\overset{\|}{O}}{C}{-}NH_2$	$\diagdown C{=}N{-}NH{-}\underset{\overset{\|}{O}}{C}{-}NH_2$	semicarbazone	(17.18)

α-Halogenation of ketones

$$R'{-}\underset{\underset{H}{|}}{\overset{\overset{R'}{|}}{C}}{-}\underset{\overset{\|}{O}}{C}{-}R + X_2 \xrightarrow[\text{or } OH^-]{H^+} R'{-}\underset{\underset{X}{|}}{\overset{\overset{R'}{|}}{C}}{-}\underset{\overset{\|}{O}}{C}{-}R \qquad (17.19)$$

Aldol condensation

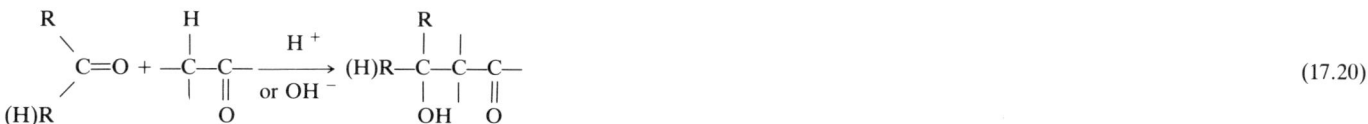

$$\underset{(H)R}{\overset{R}{\diagdown}}C{=}O + {-}\underset{\overset{\|}{O}}{C}{-}\overset{\overset{H}{|}}{C}{-} \xrightarrow[\text{or } OH^-]{H^+} (H)R{-}\underset{\underset{OH}{|}}{\overset{\overset{R}{|}}{C}}{-}\overset{\overset{|}{|}}{C}{-}\underset{\overset{\|}{O}}{C}{-} \qquad (17.20)$$

be effected. Aldehydes can be oxidized using reagents as mild as the silver ion (reaction 17.21).

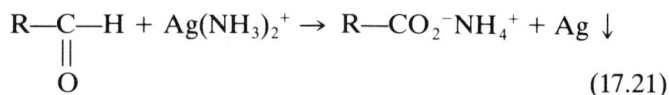

$$R{-}\underset{\overset{\|}{O}}{C}{-}H + Ag(NH_3)_2^+ \rightarrow R{-}CO_2^-NH_4^+ + Ag \downarrow \qquad (17.21)$$

Aldehydes also reduce Fehling's solution, which is an alkaline solution of cupric ion complexed with tartrate ion, in a reaction in which the deep blue colour of the solution is discharged and a red precipitate of cuprous oxide is formed. These reactions have been used in the past to distinguish aldehydes from ketones; however, today infrared and

$$CH_3-CH_2-\overset{\displaystyle O}{\underset{\displaystyle \|}{C}}-CH_3 \xrightarrow{I_2,\ OH^-} CH_3-CH_2-\overset{\displaystyle }{C}H_2-\overset{\displaystyle O}{\underset{\displaystyle \|}{C}}-O^- + CHI_3 \qquad (17.22)$$

iodoform

nuclear magnetic resonance spectroscopy would be used to solve such problems.

Oxidation of a ketone requires the cleavage of a C—C bond. As a rule this takes place only under vigorous conditions with the single exception that methyl ketones undergo reaction with hypohalite (X_2, OH^-) to give the carboxylic acid anion and haloform (Eq. 17.7). When iodine is used the reaction produces triiodomethane (iodoform) and is known as the iodoform test for methyl ketones. This is shown in reaction 17.22 for 2-butanone.

Reduction of aldehydes gives primary alcohols, and reduction of ketones gives secondary alcohols, as shown in reactions 17.23 and 17.24 for ethanal and propane.

$$CH_3-\overset{\displaystyle O}{\underset{\displaystyle \|}{C}}-H \xrightarrow{LiAlH_4} CH_3-CH_2-OH \qquad (17.23)$$

ethanal   ethanol
(acetaldehyde)

$$CH_3-\overset{\displaystyle O}{\underset{\displaystyle \|}{C}}-CH_3 \xrightarrow{LiAlH_4} CH_3-\overset{\displaystyle OH}{\underset{\displaystyle |}{C}}H-CH_3 \qquad (17.24)$$

propanone   2-propanol
(acetone)

Reduction (Eq. 17.8) can be accomplished either by use of lithium aluminium hydride, $LiAlH_4$, sodium borohydride, $NaBH_4$, or by catalytic hydrogenation. The first two reagents can be regarded as sources of H^- ions. The reaction involves 1,2 addition of the reagent $LiAlH_4$ to the carbonyl group as shown in scheme 17.25. Hydrolysis of the resulting aluminium complex results in formation of alcohol product. Reactions of aldehydes and ketones with R^- as

RMgX (reaction 17.11), CN^- (reaction 17.12), HSO_3^- (reaction 17.13) and alcohols (reaction 17.14) also involve attack by a nucleophile at the carbonyl carbon and this is shown with three-dimensional representations in schemes 17.26–17.29.

$$(17.27)$$

$$(17.28)$$

$$(17.29)$$

When aldehydes and ketones react with derivatives of ammonia the first step of the reaction (scheme 17.30) involves attack by the nitrogen nucleophile on the carbonyl carbon. Proton exchange gives a neutral species which can lose water to give a compound containing a carbon–nitrogen

$$+ Al(OH)_3 + H_2 \qquad (17.25)$$

$$+ Mg(OH)_2 + HX \qquad (17.26)$$

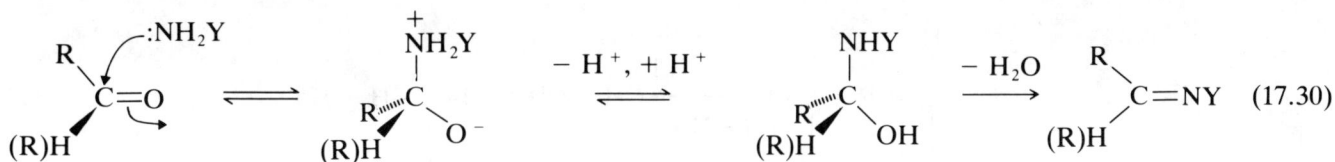

$$- H^+, + H^+ \qquad - H_2O \qquad (17.30)$$

double bond. This mechanism accounts for the formation of oximes, hydrazones, phenylhydrazones and semicarbazones (reactions 17.15–17.18).

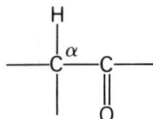

Fig. 17.11

Ketones having a hydrogen atom on the carbon adjacent to the carbonyl group as shown in Fig. 17.11 can undergo substitution reactions in which that hydrogen atom is replaced. Two reactions which involve substitution of an α-hydrogen atom, the first replacement by halogen and the second known as an 'aldol condensation', are shown in reactions 17.19 and 17.20. The propensity of carbonyl compounds to undergo reactions of this nature is a result of the acidity (Eq. 17.31) of the proton on the carbon adjacent (α) to the carbonyl group which enables that carbon atom to act as a nucleophile. The negative charge of the carbon

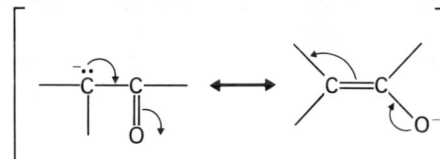

Fig. 17.12

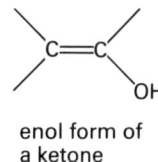

enol form of
a ketone

Fig. 17.13

anion or carbanion is spread onto the carbonyl oxygen which in terms of valence bond theory is a resonance hybrid of two structures which contribute to the overall structure (Fig. 17.12). Protonation of this anion produces either ketone or the enol of the ketone shown in Fig. 17.13. The carbanion α to a carbonyl group is able to react with another carbonyl group or with electro-positive halogen. Reaction with another carbonyl compound is catalysed by either acid or base and the mechanism generally accepted for the base catalysed condensation of ethanal to give 3-hydroxybutanal (an aldol) is shown in Eqs 17.32–17.34. The purpose of the hydroxide ion catalyst is to produce the carbanion (Eq. 17.32) which attacks the carbonyl group of another molecule of ethanal (Eq. 17.33). The anion produced is protonated by water to give 3-hydroxybutanal, as in Eq. 17.34. 3-Hydroxybutanal readily eliminates water on warming with

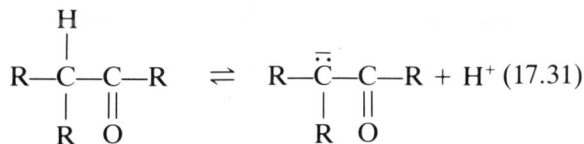

$$R-\underset{\substack{| \\ R}}{\overset{\substack{H \\ |}}{C}}-\underset{\substack{|| \\ O}}{C}-R \;\rightleftharpoons\; R-\underset{\substack{| \\ R}}{\overset{\ddot{C}}{C}}-\underset{\substack{|| \\ O}}{C}-R + H^+ \quad (17.31)$$

dilute acid to produce 2-butenal (Eq. 17.35). This facile loss of water is related to the presence of the carbonyl group, but we will not consider this further.

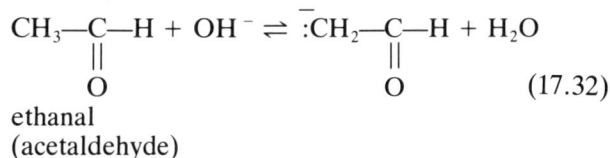

ethanal
(acetaldehyde)

$$(17.32)$$

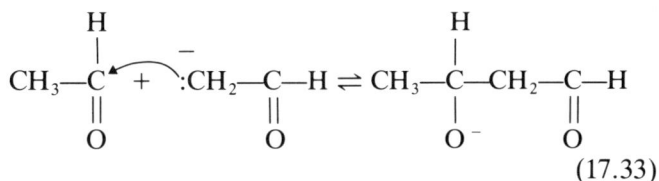

$$(17.33)$$

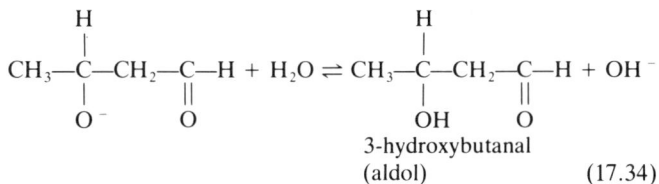

3-hydroxybutanal
(aldol)

$$(17.34)$$

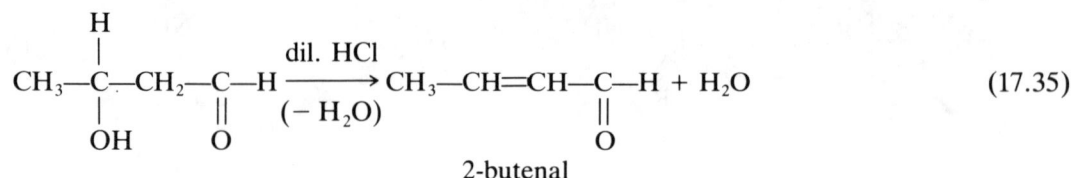

$$CH_3\!-\!\underset{\underset{\textstyle OH}{|}}{\overset{\overset{\textstyle H}{|}}{C}}\!-\!CH_2\!-\!\underset{\underset{\textstyle O}{\|}}{C}\!-\!H \xrightarrow[(-H_2O)]{dil.\ HCl} CH_3\!-\!CH\!=\!CH\!-\!\underset{\underset{\textstyle O}{\|}}{C}\!-\!H + H_2O \qquad (17.35)$$

2-butenal

Halogenation α to a carbonyl group (Eq. 17.19) is similar to the aldol condensation and is also catalysed by either acid or base. In the presence of base the reaction involves the carbanion produced by loss of a proton from the α-carbon atom. This is shown for propanone in Eq. 17.36 and 17.37. In the presence of a basic catalyst the slowest step, that is the rate-determining step, is the formation of the carbanion (Eq. 17.36) which then reacts readily with halogen (Eq. 17.37). The acid catalysed reaction is discussed on page 109.

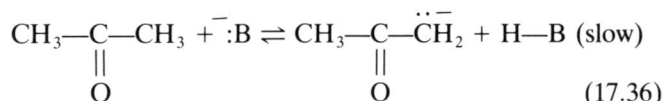

$$CH_3\!-\!\underset{\underset{\textstyle O}{\|}}{C}\!-\!CH_3 + {}^-\!:\!B \rightleftharpoons CH_3\!-\!\underset{\underset{\textstyle O}{\|}}{C}\!-\!\overset{..}{C}\overline{H}_2 + H\!-\!B \text{ (slow)}$$

$$(17.36)$$

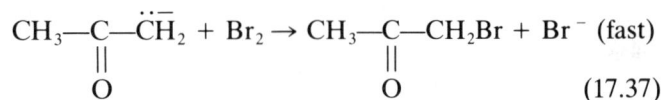

$$CH_3\!-\!\underset{\underset{\textstyle O}{\|}}{C}\!-\!\overset{..}{C}\overline{H}_2 + Br_2 \rightarrow CH_3\!-\!\underset{\underset{\textstyle O}{\|}}{C}\!-\!CH_2Br + Br^- \text{ (fast)}$$

$$(17.37)$$

17.3 Carboxylic Acids: Structure, Preparation and Reactions

Since carboxylic acids have been known for a long time many of them possess trivial or common names. In the systematic naming of a carboxylic acid the suffix *oic acid* is added to the name of the corresponding hydrocarbon, the final '*e*' of the hydrocarbon being omitted. Substituents carry the number given by their position on the chain with the carboxylic acid carbon given the number C(1). Selected examples are given in Figs 17.14–17.16. The carboxylic acid function is polar and boiling points of these compounds are high because of extensive intermolecular hydrogen bonding. In non-polar solvents carboxylic acids tend to form dimers as a result of this hydrogen bonding (Fig. 17.17). Salts of carboxylic acids are readily formed by reaction with cold dilute aqueous sodium hydroxide (reaction 17.38).

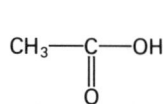

ethanoic acid
(acetic acid)

Fig. 17.14

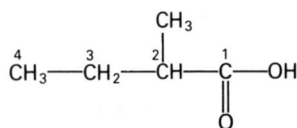

2-methylbutanoic acid

Fig. 17.15

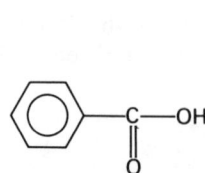

benzoic acid

Fig. 17.16

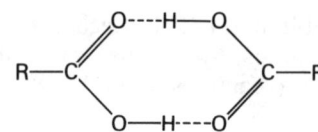

H-bonding between
carboxylic acid
molecules

Fig. 17.17

These salts are extremely soluble in aqueous media but are insoluble in non-polar organic solvents. Most carboxylic acids are themselves relatively insoluble in cold aqueous solutions. The ready interconversion of carboxylic acids and their alkali metal salts facilitates their identification and separation. Carboxylic acids can be transferred from an organic solution into an aqueous alkaline solution, which facilitates purification and separation from neutral or basic molecules, which are generally insoluble in such a medium.

$$R\!-\!CO_2H + NaOH \rightleftharpoons R\!-\!CO_2^-Na^+ + H_2O \quad (17.38)$$

soluble in   soluble in
organic solvent,   water,
insoluble in   insoluble in
water   organic solvent

The negative charge on the carboxylic acid anion is spread equally over the two oxygen atoms. In terms of valence bond theory the anion is a resonance hybrid of two equivalent structures which contribute equally to the overall structure (Fig. 17.18). From a molecular orbital viewpoint two electrons are in both the ψ_1 and ψ_2 molecular orbitals (Fig. 17.19). It is clear from the symmetry of the ψ_1 and ψ_2

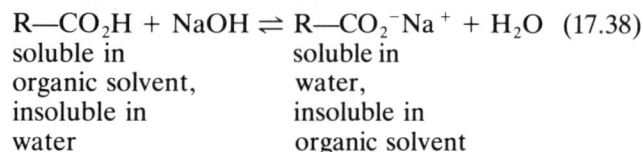

carboxylic acid anion

Fig. 17.18

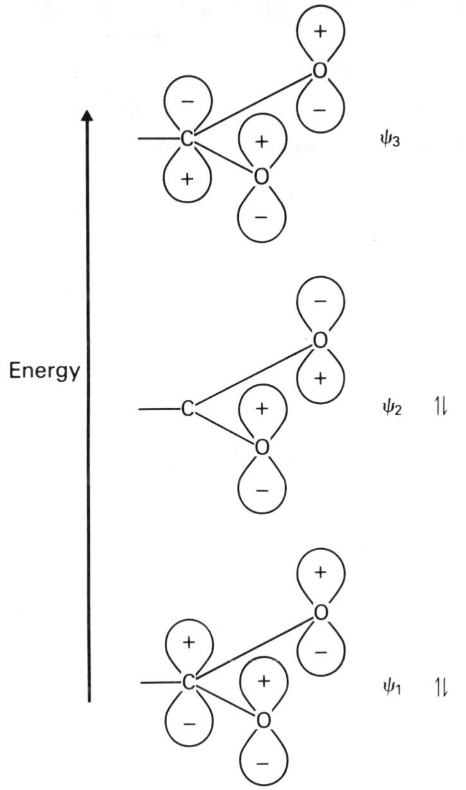

Fig. 17.19

π molecular orbitals of
carboxylic anion

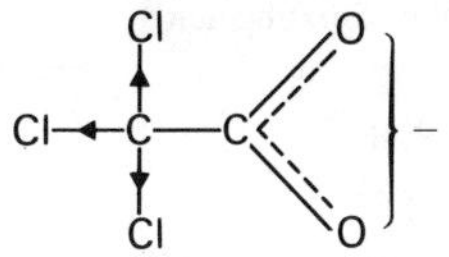

Fig. 17.20   trichloroethanoic acid anion

Ethanoic acid is by far the most important aliphatic carboxylic acid. It is prepared industrially from ethyne by hydration to give ethanal, followed by air oxidation (sequence 17.41).

$$HC\equiv CH \xrightarrow[H_2SO_4]{H_2O,} CH_3-\underset{\underset{O}{\|}}{C}-H \xrightarrow[oxidation]{air} CH_2-\underset{\underset{O}{\|}}{C}-OH$$

ethyne   ethanal   ethanoic acid

(17.41)

Benzoic acid is the most important aromatic carboxylic acid and is produced on an industrial scale by oxidation of alkylbenzenes obtained from coal tar (reaction 17.42).

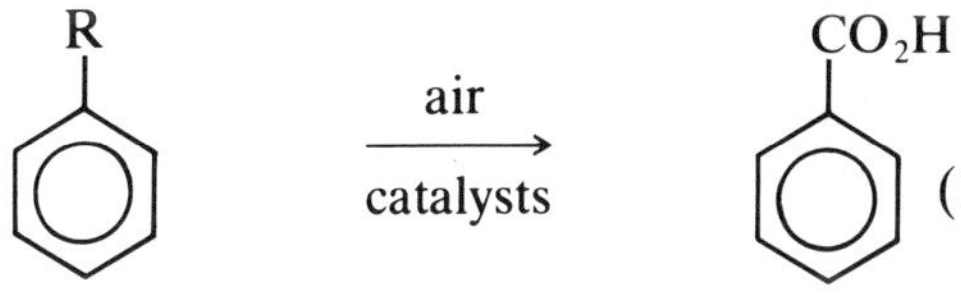

(17.42)

alkylbenzenes   benzoic acid

molecular orbitals that the electron distribution is the same on both oxygen atoms. The structural consequence of this electron delocalization is that in the anion both C—O bonds are the same length. Of course, in the parent carboxylic acid the bond lengths of the C—O bonds are different. The acid strength, measured by the pK_a, of a carboxylic acid depends on the stability of the acid anion. The electron withdrawing nature of chlorine causes trichloroethanoic acid (Eq. 17.39) to be a much stronger acid ($pK_a = 0.63$) than ethanoic acid (acetic acid) ($pK_a = 4.76$, Eq. 17.40) since the anion of trichloroethanoic acid is stabilized by the inductive effect of the three chlorine substituents (Fig. 17.20).

$$Cl_3C-\underset{\underset{O}{\|}}{C}-OH + H_2O \rightleftharpoons Cl_3C-CO_2^- + H_3O^+ \quad (17.39)$$

trichloroethanoic acid

$$CH_3-\underset{\underset{O}{\|}}{C}-OH + H_2O \rightleftharpoons CH_3-CO_2^- + H_3O^+ \quad (17.40)$$

ethanoic acid

The methods used for the industrial preparation of chemicals in large quantities are not often suited to laboratory synthesis. Carboxylic acids are produced in the laboratory by oxidation of primary alcohols (reaction 17.43) or alkyl benzenes (reaction 17.44), carbonation of organo magnesium halides (reaction 17.45) or hydrolysis of nitriles (reaction 17.46), amides, esters, acid halides and anhydrides (reaction 17.47). The preparation of carboxylic acids from organo magnesium halides shown in sequence 17.45, is an important method of extending the skeleton of a molecule by one carbon atom. Nitriles, which are generally prepared from haloalkanes, are readily hydrolysed to carboxylic acids as in Eq. 17.46 and this overall reaction sequence also adds a carbon to the framework of the molecule.

In most reactions of carboxylic acids and their derivatives the carbonyl group is retained; however, the reactivity of these molecules is a consequence of the presence of the C=O group. Carboxylic acids are readily converted to acid halides, esters, anhydrides or amides. Conversion of a carboxylic acid into an acid chloride (reaction 17.49) occurs with replacement of the OH by Cl. Reagents commonly

Table 17.5 Preparation of carboxylic acids

Oxidation of primary alcohols

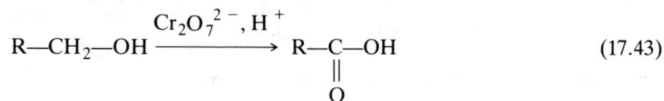

$$R-CH_2-OH \xrightarrow{Cr_2O_7^{2-},\ H^+} R-\underset{\underset{O}{\|}}{C}-OH \qquad (17.43)$$

Oxidation of alkylbenzenes

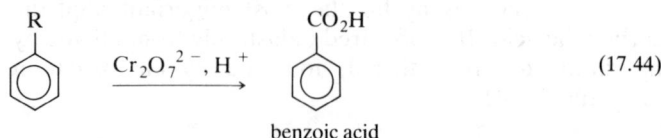

$$\qquad (17.44)$$

benzoic acid

Carbonation of organo magnesium halides

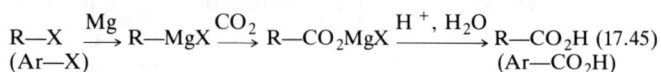

$$\underset{(Ar-X)}{R-X} \xrightarrow{Mg} R-MgX \xrightarrow{CO_2} R-CO_2MgX \xrightarrow{H^+,\ H_2O} \underset{(Ar-CO_2H)}{R-CO_2H}\ (17.45)$$

Hydrolysis of nitriles

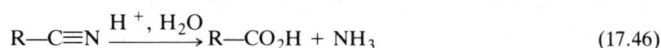

$$R-C\equiv N \xrightarrow{H^+,\ H_2O} R-CO_2H + NH_3 \qquad (17.46)$$

Hydrolysis of amides, esters, anhydrides and acid halides

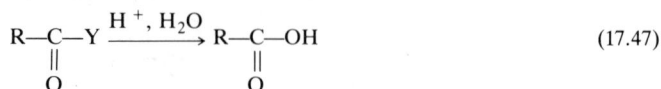

$$R-\underset{\underset{O}{\|}}{C}-Y \xrightarrow{H^+,\ H_2O} R-\underset{\underset{O}{\|}}{C}-OH \qquad (17.47)$$

Y = NH$_2$, OR, X, OCOR

Table 17.6 Reactions of carboxylic acids

Salt formation

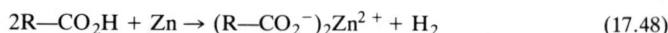

$$R-CO_2H + NaOH \rightarrow R-CO_2^-\ Na^+ + H_2O \qquad (17.38)$$

$$2R-CO_2H + Zn \rightarrow (R-CO_2^-)_2Zn^{2+} + H_2 \qquad (17.48)$$

Conversion to acid chloride

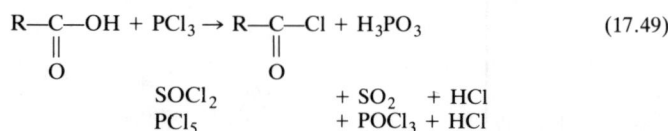

$$R-\underset{\underset{O}{\|}}{C}-OH + PCl_3 \rightarrow R-\underset{\underset{O}{\|}}{C}-Cl + H_3PO_3 \qquad (17.49)$$

$$\begin{array}{ll} SOCl_2 & + SO_2\ + HCl \\ PCl_5 & + POCl_3 + HCl \end{array}$$

Conversion to esters

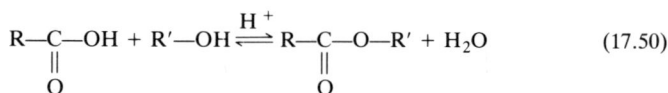

$$R-\underset{\underset{O}{\|}}{C}-OH + R'-OH \underset{}{\overset{H^+}{\rightleftharpoons}} R-\underset{\underset{O}{\|}}{C}-O-R' + H_2O \qquad (17.50)$$

Conversion to anhydride

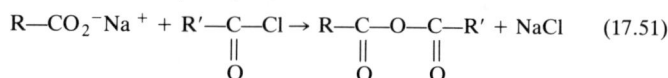

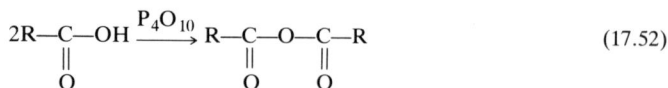

$$R-CO_2^-\ Na^+ + R'-\underset{\underset{O}{\|}}{C}-Cl \rightarrow R-\underset{\underset{O}{\|}}{C}-O-\underset{\underset{O}{\|}}{C}-R' + NaCl \qquad (17.51)$$

$$2R-\underset{\underset{O}{\|}}{C}-OH \xrightarrow{P_4O_{10}} R-\underset{\underset{O}{\|}}{C}-O-\underset{\underset{O}{\|}}{C}-R \qquad (17.52)$$

Conversion to amides

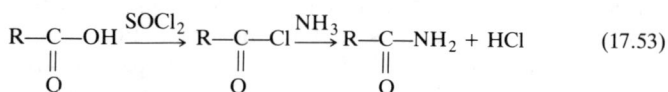

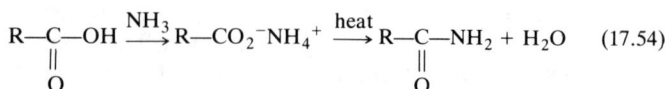

$$R-\underset{\underset{O}{\|}}{C}-OH \xrightarrow{SOCl_2} R-\underset{\underset{O}{\|}}{C}-Cl \xrightarrow{NH_3} R-\underset{\underset{O}{\|}}{C}-NH_2 + HCl \qquad (17.53)$$

$$R-\underset{\underset{O}{\|}}{C}-OH \xrightarrow{NH_3} R-CO_2^-NH_4^+ \xrightarrow{heat} R-\underset{\underset{O}{\|}}{C}-NH_2 + H_2O \qquad (17.54)$$

Reduction

$$R-CO_2H \xrightarrow{LiAlH_4} R-CH_2-OH \qquad (17.55)$$
primary alcohols

Substitution at the α-carbon

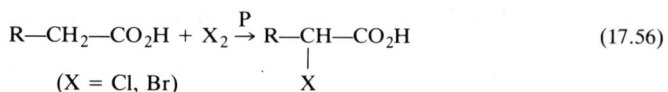

$$R-CH_2-CO_2H + X_2 \xrightarrow{P} R-\underset{\underset{X}{|}}{CH}-CO_2H \qquad (17.56)$$
(X = Cl, Br)

used for this purpose are thionyl chloride SOCl$_2$, phosphorus trichloride PCl$_3$, and phosphorus pentachloride PCl$_5$. Thionyl chloride has the advantage that the other products of reaction are all gases and are therefore easily removed or separated from the product acid halide. This is shown for reactions of ethanoic acid (Eq. 17.57) and benzoic acid (reaction 17.58).

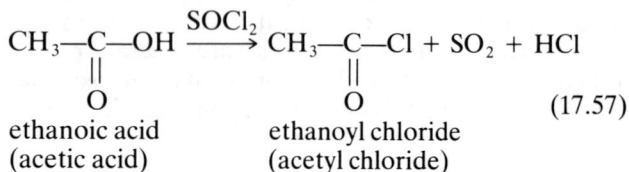

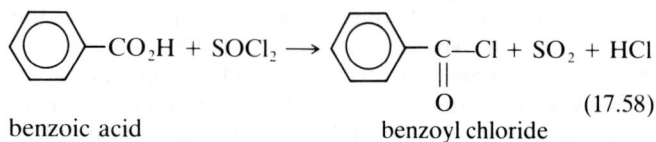

$$CH_3-\underset{\underset{O}{\|}}{C}-OH \xrightarrow{SOCl_2} CH_3-\underset{\underset{O}{\|}}{C}-Cl + SO_2 + HCl$$

$$\qquad (17.57)$$

ethanoic acid   ethanoyl chloride
(acetic acid)   (acetyl chloride)

$$\qquad (17.58)$$

benzoic acid   benzoyl chloride

Conversion of a carboxylic acid to an ester (reaction 17.50) involves the removal of water by a condensation reaction between the carboxylic acid and alcohol as shown in Eq. 17.59 for ethanoic acid and ethanol. This is an equilibrium reaction and the position of equilibrium can be displaced towards the ester side by continuous removal of water from the reaction mixuture as fast as it is formed. The reaction is catalysed by mineral acids.

Formation of an anhydride (reaction 17.52) from two carboxylic acid molecules involves the removal of a water molecule, and is another example of a condensation reaction. Only symmetrical anhydrides (reaction 17.60) can be prepared in reasonable yield in this manner; unsymmetrical anhydrides are prepared by nucleophilic displacement of halide from an acid halide by a carboxylic acid anion (reaction (17.51), as shown in reaction 17.61 for the formation of benzoic ethanoic anhydride.

Amides can be prepared by heating the ammonium salt formed by reaction of a carboxylic acid with ammonia (reaction 17.54). The preparation of ethanamide is shown in sequence 17.62. Amides are however more often prepared by nucleophilic displacement of the halide of an acid halide by ammonia (reaction 17.53). Substituted amides can similarly be prepared by using a primary or secondary amine. The preparation of N-methylethanamide in this way is shown in reaction 17.63. Lithium aluminium hydride

$$CH_3-\overset{\underset{\|}{O}}{C}-Cl + NH_2-CH_3 \rightarrow CH_3-\overset{\underset{\|}{O}}{C}-NHCH_3 + HCl \quad (17.63)$$

ethanoyl   methanamine   N-methylethanamide
chloride

is one of the few reagents that will reduce a carboxylic acid directly to a primary alcohol, (Eq. 17.55) but the reagent is too expensive for its use to be practical on an industrial scale.

Substitution at the carbon atom adjacent to a carboxylic acid (Eq. 17.56) can be brought about, in a carboxylic acid containing an α-hydrogen, by reaction with red phosphorus plus chlorine or bromine. The formation of 2-bromopropanoic acid from propanoic acid is shown in Eq. 17.64. The reagent

$$CH_3-CH_2-CO_2H \xrightarrow{P, Br_2} CH_3-\underset{\underset{Br}{|}}{CH}-CO_2H \quad (17.64)$$

propanoic acid   2-bromopropanoic acid

mixture produces traces of PX_3 (reaction 17.65) which converts the carboxylic acid to the acid halide (reaction

$$CH_3-\overset{\underset{\|}{O}}{C}-OH + HO-CH_2-CH_3 \overset{H^+}{\rightleftharpoons} CH_3-\overset{\underset{\|}{O}}{C}-O-CH_2-CH_3 + H_2O \quad (17.59)$$

ethanoic acid   ethanol   ethyl ethanoate
(acetic acid)   (ethyl acetate)

$$CH_3-\overset{\underset{\|}{O}}{C}-OH + HO-\overset{\underset{\|}{O}}{C}-CH_3 \xrightarrow{P_2O_5} CH_3-\overset{\underset{\|}{O}}{C}-O-\overset{\underset{\|}{O}}{C}-CH_3 \quad (17.60)$$

ethanoic acid   ethanoic anhydride
(acetic acid)   (acetic anhydride)

$$CH_3-CO_2^-Na^+ + Cl-\overset{\underset{\|}{O}}{C}-\bigcirc \rightarrow CH_3-\overset{\underset{\|}{O}}{C}-O-\overset{\underset{\|}{O}}{C}-\bigcirc + NaCl \quad (17.61)$$

sodium ethanoate   benzoyl chloride   benzoic ethanoic anhydride
(sodium acetate)

$$CH_3-CO_2H + NH_3 \rightarrow CH_3-CO_2^-NH_4^+ \xrightarrow{heat} CH_3-\overset{\underset{\|}{O}}{C}-NH_2 + H_2O \quad (17.62)$$

ethanoic acid   ethanamide
(acetamide)

$$P + X_2 \rightarrow PX_3 \qquad (17.65)$$

$$R-CH_2-CO_2H + PX_3 \rightarrow R-CH_2-\underset{\substack{\| \\ O}}{C}-X \qquad (17.66)$$

$$R-CH_2-\underset{\substack{\| \\ O}}{C}-X + X_2 \rightarrow R-\underset{\substack{| \\ X}}{CH}-\underset{\substack{\| \\ O}}{C}-X + HX \qquad (17.67)$$

$$R-\underset{\substack{| \\ X}}{CH}-\underset{\substack{\| \\ O}}{C}-X + R-CH_2-\underset{\substack{\| \\ O}}{C}-OH \rightarrow$$

$$R-\underset{\substack{| \\ X}}{CH}-\underset{\substack{\| \\ O}}{C}-OH + R-CH_2-\underset{\substack{\| \\ O}}{C}-X \qquad (17.68)$$

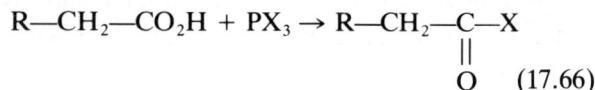

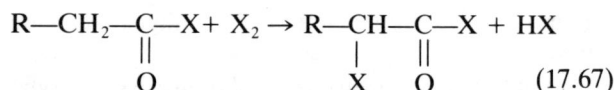

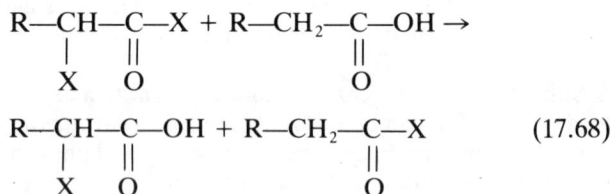

17.66). It is the acid halide which undergoes α-substitution (reaction 17.67). The α-substituted acid halide is converted to the α-substituted carboxylic acid by reaction with the carboxylic acid used as starting material, to produce another acid halide molecule (reaction 17.68) which in turn can undergo substitution. The reaction proceeds until substitution at the α-position is complete. These α-halocarboxylic acids are important in synthesis. The halogen can be displaced with a nucleophile to produce α-amino acids (reaction 17.69) and α-hydroxy acids (reaction 17.70) and loss of the β-hydrogen atom and the halide gives α, β-unsaturated acids (sequence 17.71). All of these compounds are important biologically.

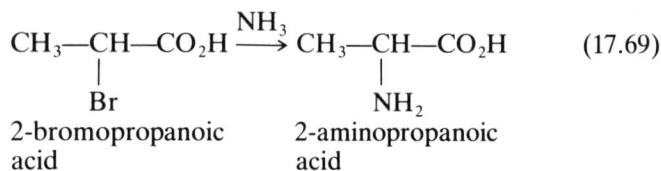

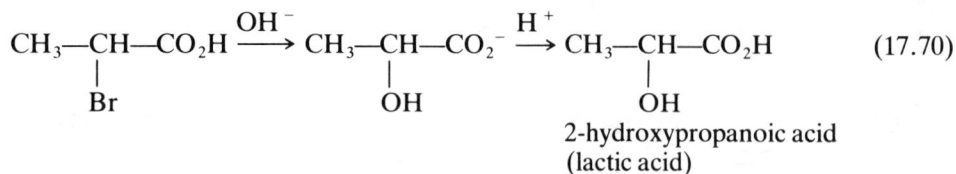

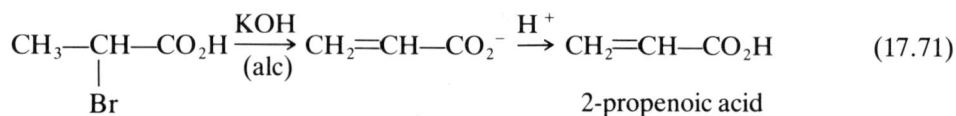

$$CH_3-\underset{\substack{| \\ Br}}{CH}-CO_2H \xrightarrow{NH_3} CH_3-\underset{\substack{| \\ NH_2}}{CH}-CO_2H \qquad (17.69)$$

2-bromopropanoic acid   2-aminopropanoic acid

$$CH_3-\underset{\substack{| \\ Br}}{CH}-CO_2H \xrightarrow{OH^-} CH_3-\underset{\substack{| \\ OH}}{CH}-CO_2^- \xrightarrow{H^+} CH_3-\underset{\substack{| \\ OH}}{CH}-CO_2H \qquad (17.70)$$

2-hydroxypropanoic acid (lactic acid)

$$CH_3-\underset{\substack{| \\ Br}}{CH}-CO_2H \xrightarrow[\text{(alc)}]{KOH} CH_2{=}CH-CO_2^- \xrightarrow{H^+} CH_2{=}CH-CO_2H \qquad (17.71)$$

2-propenoic acid

17.4 Acid Halides: Structure, Preparation and Reactions

Acid halides (or acyl halides) take their name from the parent carboxylic acids by replacing the ending *ic acid* with the suffix *oyl* or *yl* and adding the appropriate halide (Figs 17.21 and 17.22). Acyl halides derived from acids having names ending in carboxylic acid are named using the suffix *carbonyl halide*. Acid halides are generally prepared directly from carboxylic acids as shown in Table 17.7.

$$CH_3-\underset{\substack{\| \\ O}}{C}-Cl$$

ethanoyl chloride (acetyl chloride)

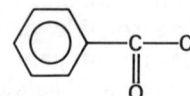

Fig. 17.21

benzoyl chloride

Fig. 17.22

Table 17.7 Preparation of acid halides

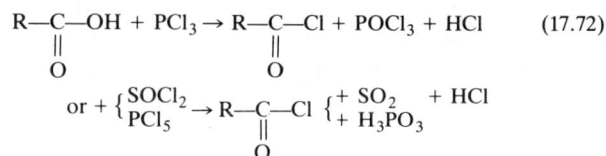

$$R-\underset{\substack{\| \\ O}}{C}-OH + PCl_3 \rightarrow R-\underset{\substack{\| \\ O}}{C}-Cl + POCl_3 + HCl \qquad (17.72)$$

$$\text{or} + \left\{ \begin{matrix} SOCl_2 \\ PCl_5 \end{matrix} \right. \rightarrow R-\underset{\substack{\| \\ O}}{C}-Cl \left\{ \begin{matrix} + SO_2 & + HCl \\ + H_3PO_3 \end{matrix} \right.$$

Acid halides are particularly reactive due to the inductive effect of the halogens and of all the acid derivatives they most readily undergo nucleophilic substitution (reaction 17.73).

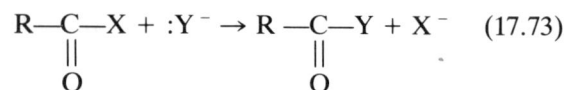

$$R-\underset{\substack{\| \\ O}}{C}-X + :Y^- \rightarrow R-\underset{\substack{\| \\ O}}{C}-Y + X^- \qquad (17.73)$$

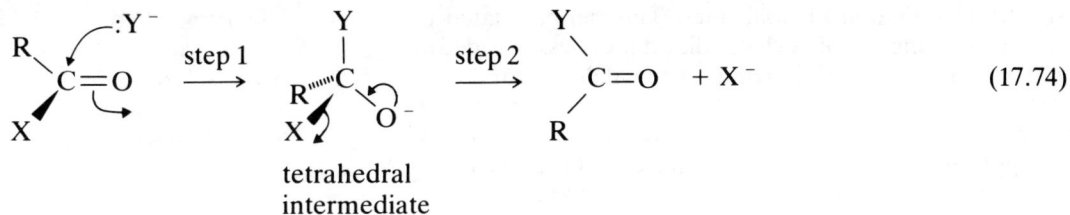

(17.74)

Substitution involves direct nucleophilic attack on the carbonyl carbon to give a tetrahedral anion intermediate (Eq. 17.74). If a mineral acid is present the proton will become attached to the carbonyl oxygen, making the carbonyl group even more susceptible to nucleophilic attack. This is shown in the sequence 17.75. A series of reactions which involve nucleophilic displacement of halogen are shown in Table 17.8.

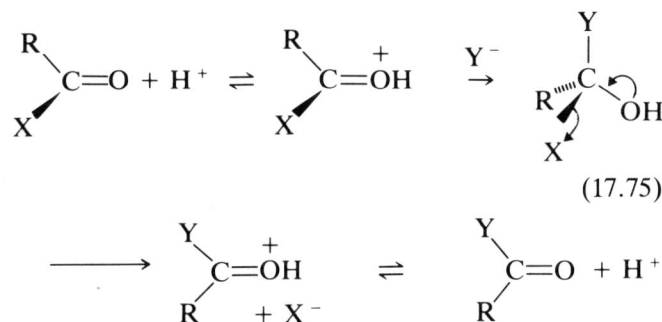

(17.75)

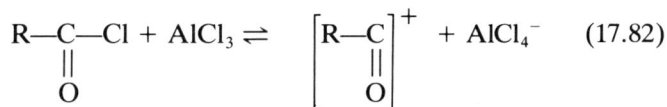

Friedel–Crafts acylation (reaction 17.80 and 15.6) involves the formation of the electrophilic $R-\underset{\underset{O}{\parallel}}{C}{}^{+}$ species from the reaction of acid halide with aluminium chloride (reaction 17.82). Such a species can effect electrophilic substitution on an aromatic molecule as is shown for reaction with toluene in Eq. 17.83. Acid halides are more reactive to nucleophilic

$$R-\underset{\underset{O}{\parallel}}{C}-Cl + AlCl_3 \rightleftharpoons \left[R-\underset{\underset{O}{\parallel}}{C}\right]^{+} + AlCl_4^{-}$$ (17.82)

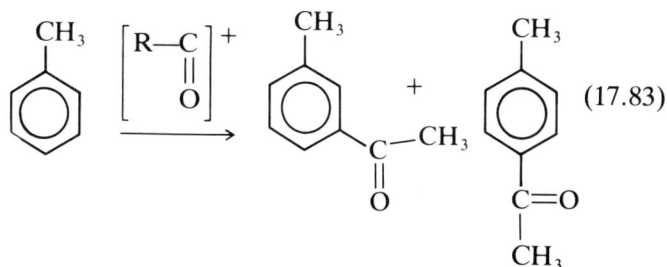

(17.83)

Table 17.8 Reactions of acid chlorides

Conversion into carboxylic acid—hydrolysis

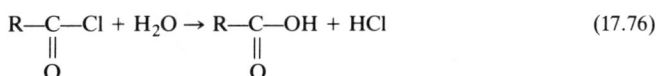

(17.76)

Conversion to amides

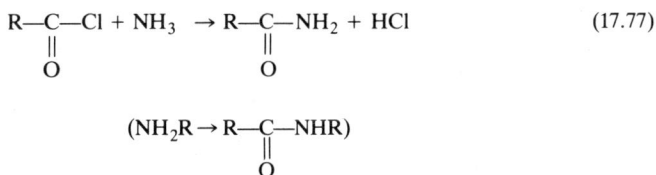

(17.77)

Conversion to esters

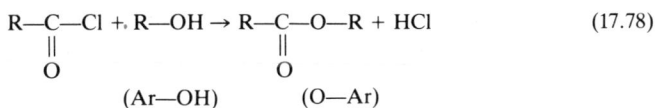

(17.78)

Conversion to anhydrides

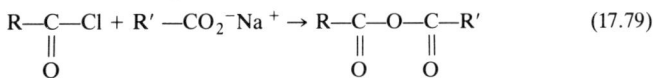

(17.79)

Friedel–Crafts acylation

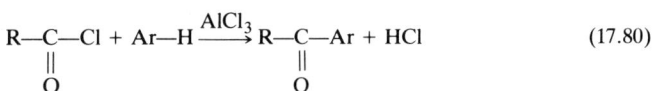

(17.80)

Reduction

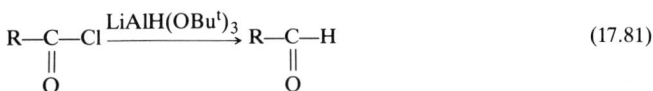

(17.81)

substitution than are haloalkanes. This can be related to the fact that the site of nucleophilic attack is less crowded at the sp^2 carbon (Fig. 17.23) of the acid halide than at the sp^3 carbon (Fig. 17.24) of the haloalkane. In addition the carbon atom of the acid halide is more electron deficient because of the polarity of the C—O bond. This electron deficiency is often further increased by proton attachment to the carbonyl

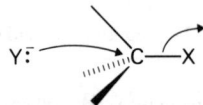
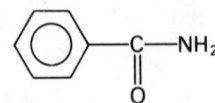

ethanamide
(acetamide)

benzamide

Fig. 17.25   **Fig. 17.26**

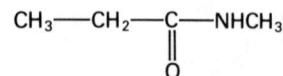

Fig. 17.27   N-methylpropanamide

oxygen. Bimolecular nucleophilic displacement of halogen in a haloalkane involves the formation of a five-coordinated carbon atom in the transition state, as shown in sequence 17.84, whereas nucleophilic displacement of acid halides involves the formation of a four-coordinated species which is expected to be inherently lower in energy (sequence 17.85). The attacking nucleophile can be H_2O, CN^-, NH_3, NH_2R, ROH, RO^-, RCO_2^- or RCO_2H. The contrast in reactivity between acid halides and haloalkanes is shown, for example, by the fact that ethanoyl chloride undergoes vigorous hydrolysis when mixed with water whereas chloroethane reacts only slowly under such conditions. For similar reasons, amides are more reactive than amines, and esters more reactive than ethers.

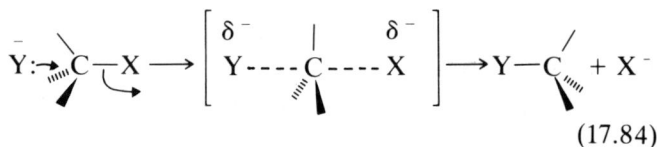

$$(17.84)$$

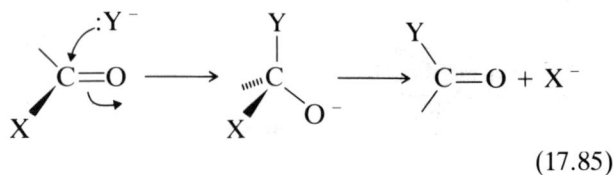

$$(17.85)$$

17.5 Amides: Structure, Preparation and Reactions

Amides, as derivatives of carboxylic acids, are named by replacing the suffix *oic* of the parent acid with *amide* (see Figs 17.25, 17.26 and 17.27). They are prepared by reac-

tion of acid halides with ammonia (reaction 17.86), primary (equation 17.87) or secondary amines (reaction 17.88), or industrially by heating the carboxylic acid with ammonia (reaction 17.89). Amides undergo acid or base catalysed

Table 17.9 Preparation of amides

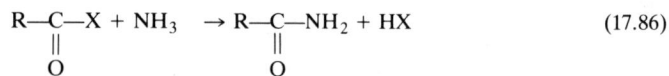

$$(17.86)$$

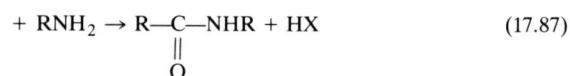

$$(17.87)$$

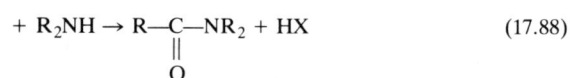

$$(17.88)$$

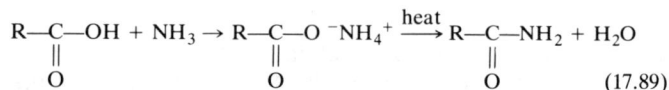

$$(17.89)$$

Table 17.10 Reactions of amides

Hydrolysis

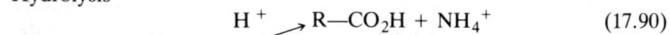

$$(17.90)$$

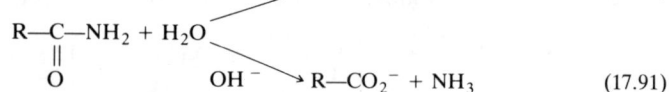

$$(17.91)$$

Degradation to amines

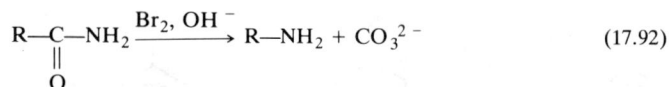

$$(17.92)$$

Dehydration to a nitrile

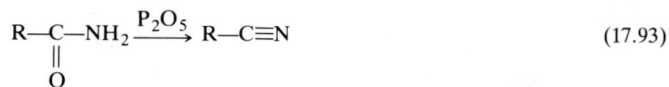

$$(17.93)$$

hydrolysis (reactions 17.90 and 17.91) to the parent carboxylic acid, and are degraded to amines by reaction with bromine and strong alkali (reaction 17.92). The latter reaction is synthetically useful because it involves the removal of a carbon atom from the molecules. This is shown in scheme 17.94 for propanamide which is degraded to ethanamine.

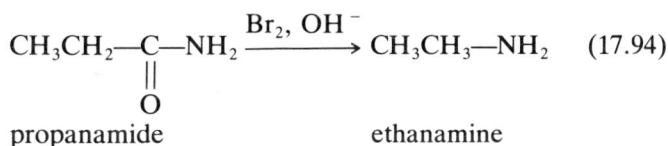

$$CH_3CH_2-\overset{\underset{||}{O}}{C}-NH_2 \xrightarrow{Br_2,\ OH^-} CH_3CH_3-NH_2 \quad (17.94)$$

propanamide   ethanamine

A water molecule can be abstracted from an amide, with phosphorus pentoxide, to form a nitrile (reaction 17.93).

17.6 Acid Anhydrides

Symmetrical anhydrides are named by replacing the suffix *acid* of the parent carboxylic acid with *anhydride* (Fig. 17.28). Unsymmetrical anhydrides derived from two different monocarboxylic acids are given three-word names (in alphabetical order) representing the two acids, the third word being anhydride (Fig. 17.29). They are prepared by reaction of an acid anion with an acid chloride (reaction 17.95) or by removal of water from acid molecules (reaction 17.96).

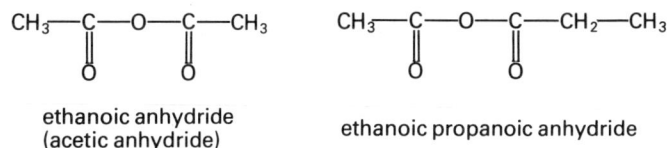

$$CH_3-\overset{\underset{||}{O}}{C}-O-\overset{\underset{||}{O}}{C}-CH_3 \qquad CH_3-\overset{\underset{||}{O}}{C}-O-\overset{\underset{||}{O}}{C}-CH_2-CH_3$$

ethanoic anhydride
(acetic anhydride)

ethanoic propanoic anhydride

Fig. 17.28   **Fig. 17.29**

Table 17.11 Preparation of anhydrides

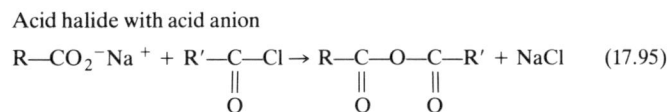

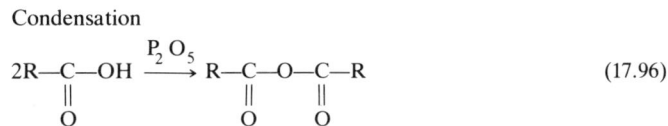

Acid halide with acid anion

$$R-CO_2^-Na^+ + R'-\overset{\underset{||}{O}}{C}-Cl \rightarrow R-\overset{\underset{||}{O}}{C}-O-\overset{\underset{||}{O}}{C}-R' + NaCl \quad (17.95)$$

Condensation

$$2R-\overset{\underset{||}{O}}{C}-OH \xrightarrow{P_2O_5} R-\overset{\underset{||}{O}}{C}-O-\overset{\underset{||}{O}}{C}-R \quad (17.96)$$

Anhydrides undergo the same reactions as acid halides, but are less reactive (Table 17.12). Ethanoic anhydride (acetic anhydride), the most common anhydride is an inexpensive and convenient reagent for converting alcohols to their ethanoate (acetate) esters. This is shown for the

reaction with ethanol in Eq. 17.101. The ester product must be seperated from the carboxylic acid produced in the reaction; this is readily accomplished by converting the carboxylic acid to the salt and removing it in the aqueous phase.

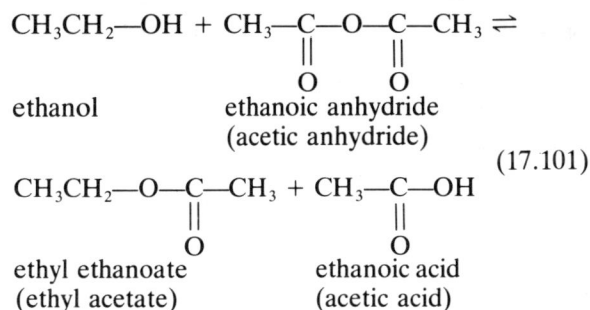

$$CH_3CH_2-OH + CH_3-\overset{\underset{||}{O}}{C}-O-\overset{\underset{||}{O}}{C}-CH_3 \rightleftharpoons$$

ethanol   ethanoic anhydride
                     (acetic anhydride)

$$\quad (17.101)$$

$$CH_3CH_2-O-\overset{\underset{||}{O}}{C}-CH_3 + CH_3-\overset{\underset{||}{O}}{C}-OH$$

ethyl ethanoate   ethanoic acid
(ethyl acetate)   (acetic acid)

Table 17.12 Reactions of anhydrides

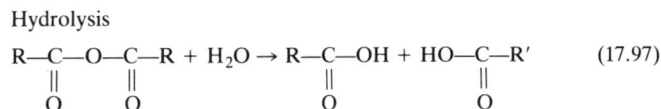

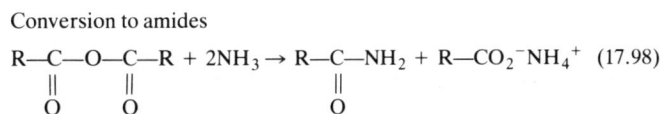

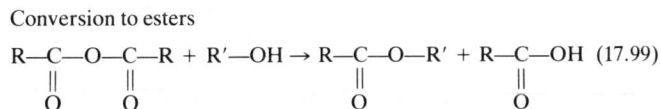

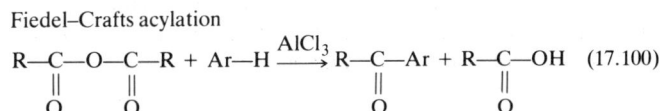

Hydrolysis

$$R-\overset{\underset{||}{O}}{C}-O-\overset{\underset{||}{O}}{C}-R + H_2O \rightarrow R-\overset{\underset{||}{O}}{C}-OH + HO-\overset{\underset{||}{O}}{C}-R' \quad (17.97)$$

Conversion to amides

$$R-\overset{\underset{||}{O}}{C}-O-\overset{\underset{||}{O}}{C}-R + 2NH_3 \rightarrow R-\overset{\underset{||}{O}}{C}-NH_2 + R-CO_2^-NH_4^+ \quad (17.98)$$

Conversion to esters

$$R-\overset{\underset{||}{O}}{C}-O-\overset{\underset{||}{O}}{C}-R + R'-OH \rightarrow R-\overset{\underset{||}{O}}{C}-O-R' + R-\overset{\underset{||}{O}}{C}-OH \quad (17.99)$$

Fiedel–Crafts acylation

$$R-\overset{\underset{||}{O}}{C}-O-\overset{\underset{||}{O}}{C}-R + Ar-H \xrightarrow{AlCl_3} R-\overset{\underset{||}{O}}{C}-Ar + R-\overset{\underset{||}{O}}{C}-OH \quad (17.100)$$

17.7 Esters

Esters are named as derivatives of carboxylic acids by replacing the ending *ic acid* by *ate*, the acid stem being preceded by the name of the other organic group as a separate word. Esters can be prepared by reaction of an alcohol with a carboxylic acid (reaction 17.102), an acid halide (Eq. 17.103) or with an anhydride (reaction 17.104).

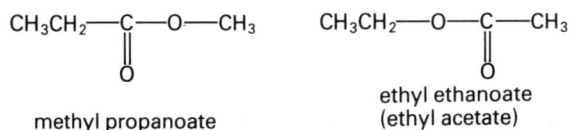

$$CH_3CH_2-\overset{\underset{||}{O}}{C}-O-CH_3 \qquad CH_3CH_2-O-\overset{\underset{||}{O}}{C}-CH_3$$

methyl propanoate

ethyl ethanoate
(ethyl acetate)

Fig. 17.30   **Fig. 17.31**

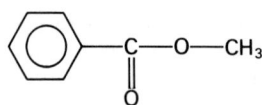

methyl benzoate

Fig. 17.32

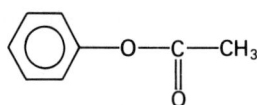

phenyl ethanoate

Fig. 17.33

Table 17.13 Preparation of esters

From an acid

$$R-\underset{\underset{O}{\|}}{C}-OH + R'-OH \overset{H^+}{\rightleftharpoons} R-\underset{\underset{O}{\|}}{C}-O-R' + H_2O \qquad (17.102)$$

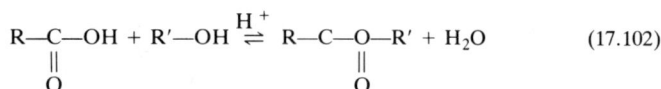

From acid halide

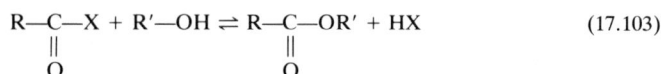

$$R-\underset{\underset{O}{\|}}{C}-X + R'-OH \rightleftharpoons R-\underset{\underset{O}{\|}}{C}-OR' + HX \qquad (17.103)$$

From an anhydride

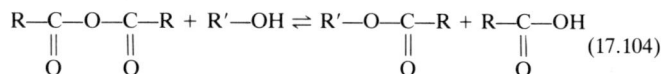

$$R-\underset{\underset{O}{\|}}{C}-O-\underset{\underset{O}{\|}}{C}-R + R'-OH \rightleftharpoons R'-O-\underset{\underset{O}{\|}}{C}-R + R-\underset{\underset{O}{\|}}{C}-OH \qquad (17.104)$$

Esters undergo the nucleophilic displacement reactions that by now can be regarded as typical of carboxylic acid derivatives; namely displacement of the OR' group by OH (reaction 17.105), OR'' (Eq. 17.107) or NH$_2$ (reaction 17.106), and so on. Hydrolysis of esters (reaction 17.105), a process which involves cleavage of the bond between the oxygen and the acyl group, (Fig. 17.34), is catalysed by either acid or base. Under basic conditions the reaction is bimolecular, the rate depending on both ester and hydroxide ion concentration. The hydroxide ion attacks the carbonyl carbon rather than the alkyl carbon. The reaction is a two stage one, involving the reversible formation of an intermediate tetrahedral anion (reaction 17.110). The anion is capable of undergoing proton exchange at a rate which is comparable with the rate of formation of carboxylic acid and alkoxy ion. Information of this type is obtained by using esters produced from optically active alcohols and carboxylic acids labelled at the oxygen atoms by ^{18}O.

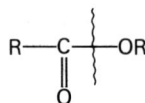

$$R-\underset{\underset{O}{\|}}{C} \vdash OR$$

Fig. 17.34

The mechanism of acid catalysed hydrolysis of an ester (scheme 17.111) is the reverse of the mechanism for acid catalysed formation of an ester. Since all the steps in the reaction are reversible, addition of water will displace the equilibrium towards carboxylic acid and alcohol, while removal of water will drive the equilibrium to the ester side.

Table 17.14 Reactions of esters

Hydrolysis

$$R-\underset{\underset{O}{\|}}{C}-O-R' + H_2O \rightarrow R-\underset{\underset{O}{\|}}{C}-OH + R'-OH \qquad (17.105)$$

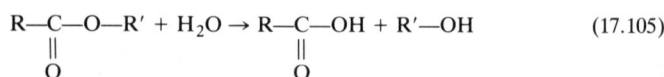

Conversion to amides

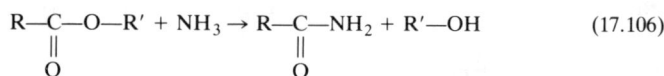

$$R-\underset{\underset{O}{\|}}{C}-O-R' + NH_3 \rightarrow R-\underset{\underset{O}{\|}}{C}-NH_2 + R'-OH \qquad (17.106)$$

Conversion into other esters

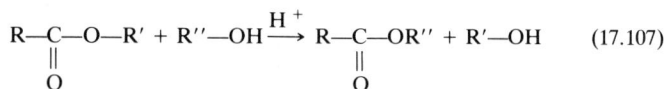

$$R-\underset{\underset{O}{\|}}{C}-O-R' + R''-OH \overset{H^+}{\rightarrow} R-\underset{\underset{O}{\|}}{C}-OR'' + R'-OH \qquad (17.107)$$

Reduction

$$R-\underset{\underset{O}{\|}}{C}-O-R' \overset{LiAlH_4}{\longrightarrow} R-CH_2-OH + R'-OH \qquad (17.108)$$

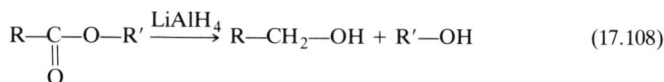

Reaction with organo magnesium halides

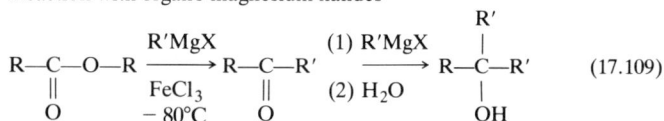

$$R-\underset{\underset{O}{\|}}{C}-O-R \overset{R'MgX}{\underset{FeCl_3 \atop -80°C}{\longrightarrow}} R-\underset{\underset{O}{\|}}{C}-R' \overset{(1)\ R'MgX}{\underset{(2)\ H_2O}{\longrightarrow}} R-\underset{\underset{OH}{|}}{\overset{\overset{R'}{|}}{C}}-R' \qquad (17.109)$$

Tertiary alcohols can be prepared by reaction of an organo magnesium halide with an ester (reaction 17.109). By carefully regulating the conditions it is sometimes possible to isolate the ketone intermediate. Under normal circumstances this ketone will react further to give tertiary alcohol, as is shown for the reaction of methyl propanoate with methyl magnesium halide in reaction sequence 17.112. Esters can be reduced with LiAlH$_4$ (reaction 17.108), or by catalytic hydrogenation at elevated temperatures under high pressure to give the primary alcohol derived from the acid part of the ester. Methyl propanoate is for example reduced to 1-propanol and methanol (reaction 17.113).

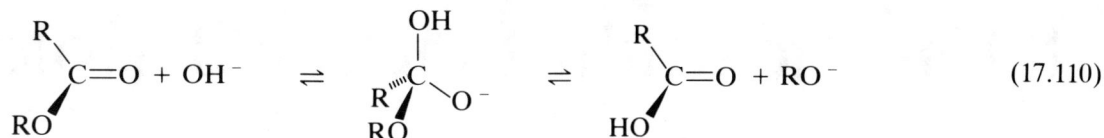

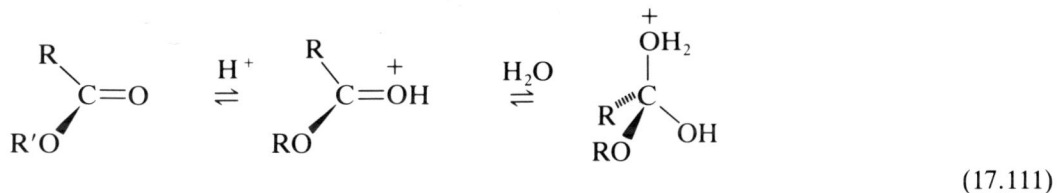

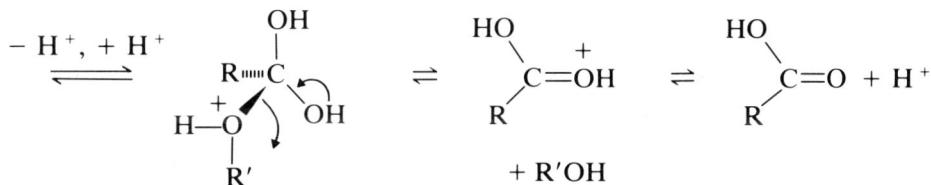

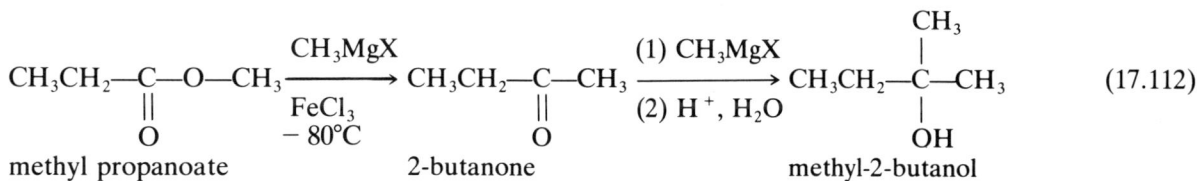

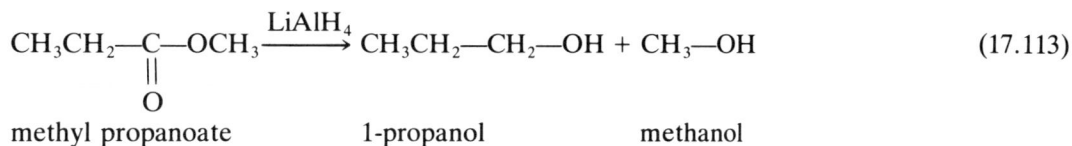

$$\underset{RO}{\overset{R}{>}}C=O + OH^- \; \rightleftharpoons \; \underset{RO}{\overset{OH}{\underset{R\,''''}{|}}}C\!-\!O^- \; \rightleftharpoons \; \underset{HO}{\overset{R}{>}}C=O + RO^- \tag{17.110}$$

$$\underset{R'O}{\overset{R}{>}}C=O \; \overset{H^+}{\rightleftharpoons} \; \underset{RO}{\overset{R}{>}}C\!\!\overset{+}{=}\!OH \; \overset{H_2O}{\rightleftharpoons} \; \underset{RO}{\overset{\overset{+}{O}H_2}{\underset{R\,''''}{|}}}C\!\!\underset{OH}{} \tag{17.111}$$

$$\overset{-H^+, +H^+}{\rightleftharpoons} \; R\,''''\overset{OH}{\underset{\underset{R'}{|}}{\underset{\overset{+}{H-O}}{\overset{|}{C}}}}OH \; \rightleftharpoons \; \underset{R}{\overset{HO}{>}}C\!\!\overset{+}{=}\!OH \; \rightleftharpoons \; \underset{R}{\overset{HO}{>}}C=O + H^+$$

$$+ R'OH$$

$$\underset{\underset{\text{methyl propanoate}}{O}}{CH_3CH_2\!-\!\overset{\|}{C}\!-\!O\!-\!CH_3} \xrightarrow[\substack{FeCl_3 \\ -80°C}]{CH_3MgX} \underset{\underset{\text{2-butanone}}{O}}{CH_3CH_2\!-\!\overset{\|}{C}\!-\!CH_3} \xrightarrow[\substack{(2)\ H^+,\ H_2O}]{(1)\ CH_3MgX} \underset{\underset{\text{methyl-2-butanol}}{OH}}{CH_3CH_2\!-\!\overset{\overset{CH_3}{|}}{C}\!-\!CH_3} \tag{17.112}$$

$$\underset{\underset{\text{methyl propanoate}}{O}}{CH_3CH_2\!-\!\overset{\|}{C}\!-\!OCH_3} \xrightarrow{LiAlH_4} \underset{\text{1-propanol}}{CH_3CH_2\!-\!CH_2\!-\!OH} + \underset{\text{methanol}}{CH_3\!-\!OH} \tag{17.113}$$

Exercises

17.1 By means of structural formula devise synthetic schemes for the following conversions:

(a) 1-propanol to methyl propanoate
(b) 1-propanol to 2-aminopropanoic acid
(c) propene to methylpropanoic acid
(d) ethanol to ethanamide (acetamide)
(e) propanoic acid to N,N-dimethylpropanamide
(f) ethanol to ethanoic anhydride (acetic anhydride)
(g) methylpropanoic acid to 2-propanamine
(h) 1-propanol to prop-1-yl ethanoate
(i) ethyl pentanoate to (i) ethanol and pentanoic acid, (ii) 3-ethyl-3-heptanol, (iii) heptanoic acid
(j) 2-butanone to 2-butanamine
(k) ethanamine to N-ethanoylethanamine (N-ethyl-ethanamide).

18 Di- and Multi-functional Molecules

18.1 Nomenclature of Di- and Multi-functional Molecules

In previous chapters we have examined a number of the more important functional groups which can occur in an organic molecule and have noted some of the modifications that can be carried out to a functional group or at the site of a functional group. In the main the compounds considered contained only one reactive site or functional group so that their chemical behaviour was relatively simple. However, by far the greater proportion of organic molecules are more complex, and contain two or more functional groups.

We consider first the naming of some simple compounds that have more than one functional group. First it is necessary to specify the carbon chain. In the name of the molecule as many of the functional groups as possible are included in the main chain. If one or more of the functional groups have suffix names, the group highest in the conventional order of precedence (see Table 18.1) is used in deriving the name of the parent compound. The name is then written so that this principal group becomes the suffix to the parent whereas all other substituents are prefixes. The priority rating in Table 18.1 also governs which substituent is given the lowest possible position number. The number associated with the principal group may be placed immediately before the suffix (e.g. butan-2-ol) or, as in this text, before the parent chain (e.g. 2-butanol). A suffix receives the lowest possible number unless there is a functional group present of higher priority. If more than one simple (and unsubstituted) prefix name is required these are arranged alphabetically. Multiplying prefixes such as *di, tri,* and *tetra* do not affect the alphabetical arrangement of simple prefix names. Examples of the nomenclature of multi-functional molecules are given in Table 18.2. The best way for the student to become familiar with the system is to practise naming a variety of hypothetical polyfunctional molecules.

18.2 Rearrangement Reactions

The chemistry of di-functional, tri-functional, and even polyfunctional molecules can usually still be understood in

Table 18.1 Order of precedence of functional groups–Prefix and/or Suffix

Class of compound	Functional group	Prefix	Suffix
carboxylic acid	$-CO_2H$	carboxy	oic acid carboxylic acid
sulphonic acid	$-SO_2OH$	sulfo	sulfonic acid
acid halide	$-\overset{\displaystyle \|}{\underset{\displaystyle O}{C}}-X$	haloformyl	yl halide carbonyl halide
amide	$-\overset{\displaystyle \|}{\underset{\displaystyle O}{C}}-NH_2$	—amide —carboxamide	carbamoyl
nitrile	$-C\equiv N$	cyano	nitrile carbonitrile
aldehyde	$-CHO$	formyl	al carboxaldehyde
ketone	$C-\overset{\displaystyle \|}{\underset{\displaystyle O}{C}}-C$	oxo (keto)	one
alcohol	$-\overset{\displaystyle \|}{\underset{\displaystyle \|}{C}}-OH$	hydroxy	ol
amine	$-\overset{\displaystyle \|}{N}-$	amino	amine
ether	$-O-$	alkoxy phenoxy	ether
alkene	$C=C$	—	ene
alkyne	$C\equiv C$	—	yne
alkane	$C-C$	—	ane
halo substituent	$-F$ $-Cl$ $-Br$ $-I$	fluoro chloro bromo iodo	— — — —
alkyl or aryl substituent	$-CH_3$ $-C_6H_5$	methyl phenyl	— —
nitro substituent	$-NO_2$	nitro	—

Table 18.2 Examples of multi-function nomenclature

CH$_2$=CH—CH$_2$—C≡CH
1-penten-4-yne
not 4-penten-1-yne

CH$_3$—CH—CH$_2$—CH—CH$_3$
    | |
  OH   NH$_2$
4-amino-2-pentanol
not 4-hydroxy-2-pentylamine

CH$_3$—CH$_2$—CH$_2$—CH—C—CH$_3$
3-[prop-1-yl]-4-penten-2-one

3-bromo-2-fluorobenzoyl chloride

CH$_3$—⟨ ⟩—CN
3-chloro-4-methylbenzonitrile

Br—CH$_2$—CH—CH$_2$—CH—CH$_3$
1-bromo-4-chloro-2-methylpentane
not 5-bromo-2-chloro-4-methylpentane

CH$_3$—CH—CH$_2$—CO$_2$H
    |
  OH
3-hydroxybutanoic acid

CH$_3$—CH—CHO
    |
  Br
2-bromopropanal

CH$_3$O—CH$_2$—CH—C—NH$_2$
2-chloro-3-methoxypropanamide

CH$_2$—C—CH$_3$
2,2-dibromo-1-chloropropane

terms of the behaviour of the individual functional groups, provided it is realized that the possibility exists for their behaviour to be modified as a result of the interplay of functional groups within the molecule. For most of the reactions we have examined so far, reaction is limited to the site of the functional group or to an adjacent carbon atom. For example, on reaction with a hydroxyl nucleophile 2-bromopropane gives 2-propanol (reaction 18.1), a product of nucleophilic displacement at the site of the bromo group. In addition propene (reaction 18.2) is formed by loss of the bromo group from carbon(2) and a proton from an adjacent carbon atom.

Reactions can also involve rearrangement of atoms adjacent to the functional group. An example we have

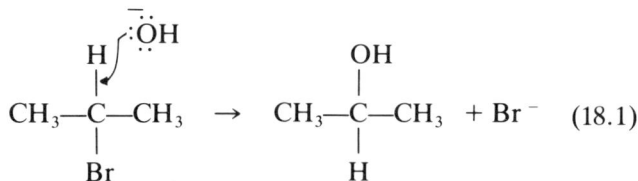

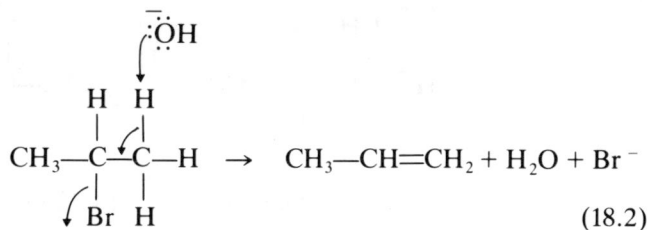

$$CH_3—\overset{H}{\underset{Br}{C}}—CH_3 \rightarrow CH_3—\overset{OH}{\underset{H}{C}}—CH_3 + Br^- \quad (18.1)$$

$$CH_3—C—C—H \rightarrow CH_3—CH=CH_2 + H_2O + Br^- \quad (18.2)$$

already considered (reaction 16.41) is the reaction of 3-methyl-2-butanol with HCl to give 2-chloro-3-methylbutane, a product of nucleophilic displacement, and 2-chloro-2-methylpentane (reaction 18.3). Under the reaction conditions an equilibrium is immediately established between the alcohol and the protonated alcohol. Water can be lost from the protonated species to give a secondary carbonium (carbenium) ion or, alternatively, can be displaced by a chloride ion in an SN$_2$ reaction to give 2-chloro-3-methylbutane. The secondary carbonium ion can itself react with chloride ion to give 2-chloro-3-methylbutane, an SN$_1$ process, or undergo rearrangement by migration of a hydride (a hydrogen atom with both bonding electrons) to give the more stable tertiary carbonium ion. Reaction of this ion with chloride ion gives 2-chloro-2-methylbutane. The reaction leading to formation of this product is referred to as a rearrangement reaction and for this substrate involves migration of the C(3)H of 3-methyl-2-butanol to the adjacent carbon atom, the one initially bonded to oxygen.

The above example involved migration of a hydride along a carbon skeleton. Rearrangement of the carbon skeleton itself can occur, and an example of this was shown in reaction 14.35, where 3,3-dimethyl-1-butene reacted with H$_2$SO$_4$ and water to give 2,3-dimethyl-2-butanol as well as 3,3,-dimethyl-2-butanol. A more complex type of rearrangement reaction, where extensive skeletal rearrangement is observed, occurs with several steroid substrates. Steroids have a tetracyclic skeleton made up of three six-membered rings and one five-membered ring (Fig. 18.1). They are important in nature and the class of steroids contains many

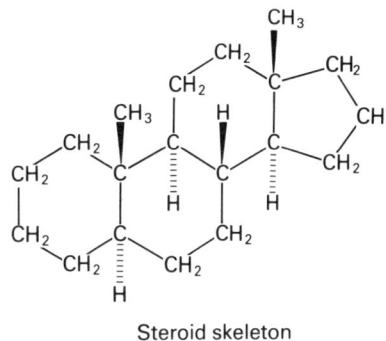

Steroid skeleton

Fig. 18.1

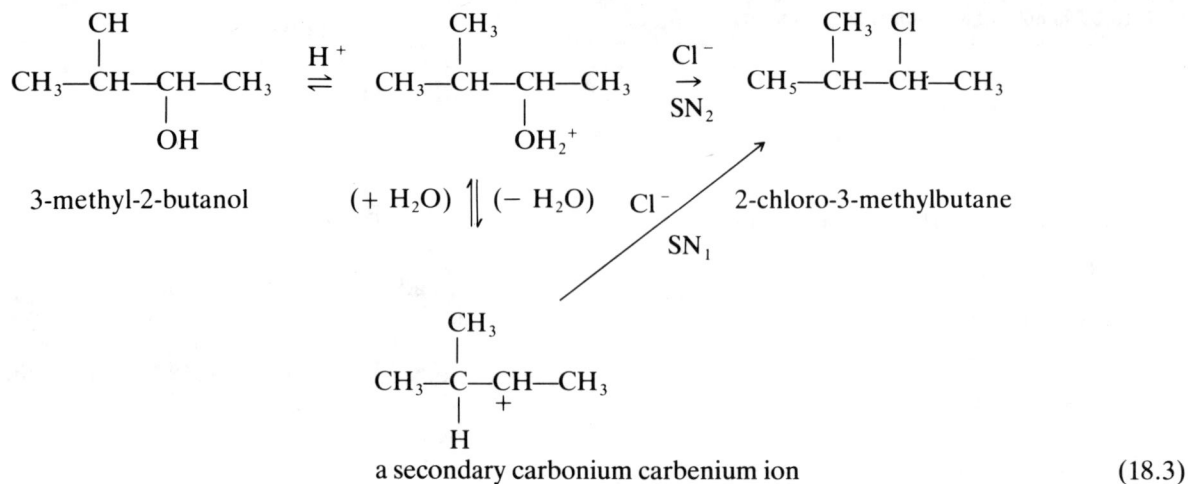

a secondary carbonium carbenium ion

(18.3)

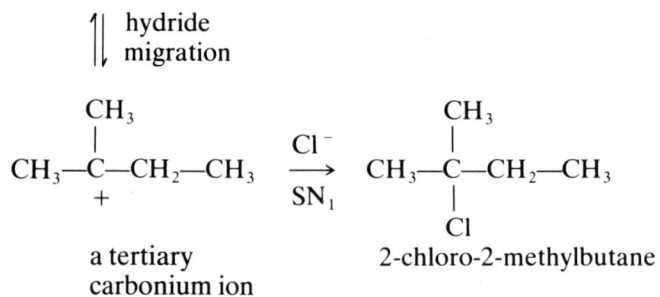

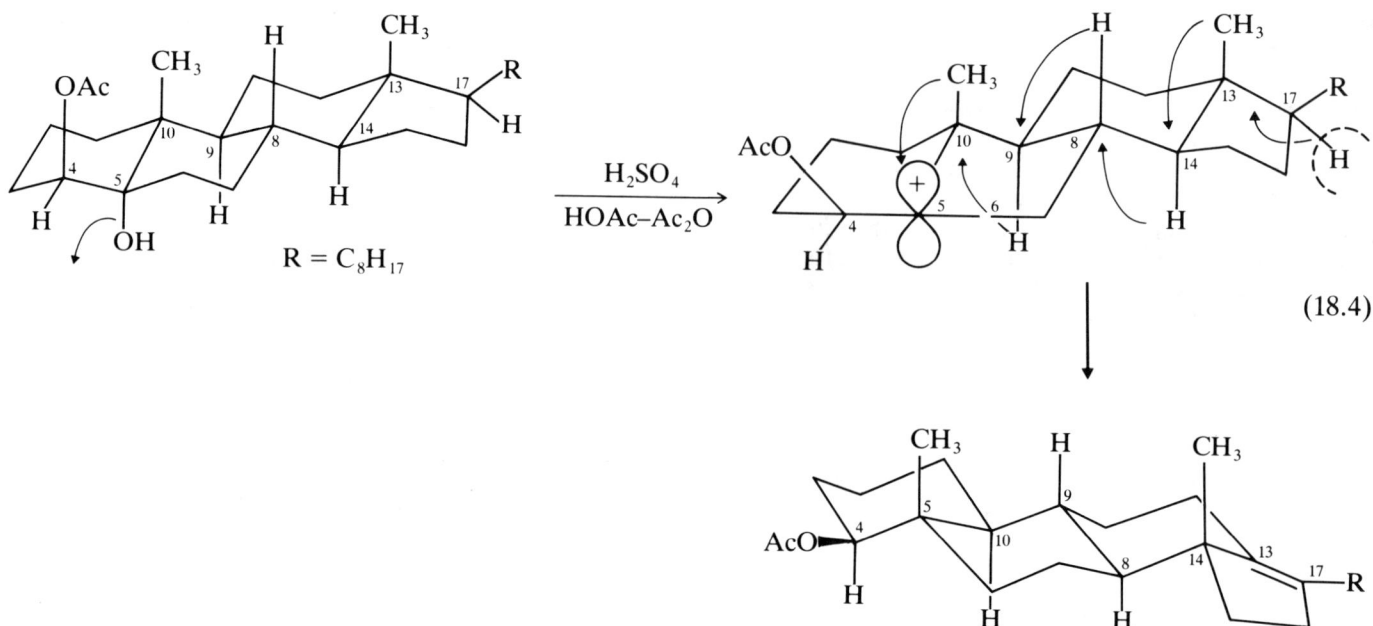

(18.4)

important hormones, including the sex hormones. Elimination of water from the steroid 4β-acetoxy-5α-cholestan-5-ol, shown in reaction 18.4, is an example where a product is formed by extensive rearrangement of the carbon skeleton—a rearrangement which extends to the end of the molecule quite remote from the site where the reaction was initiated. The reaction is initiated by loss of the hydroxyl to give a carbonium ion with the positive charge centred at C(5). The methyl at C(10) migrates to C(5) and its place at C(10) is taken by the C(9)-hydrogen, which must migrate to do so. The hydrogen atom on C(9) is replaced by the C(8)-hydrogen, which migrates in turn to take its place. The series of migrations continues as shown until a proton is finally lost from C(17), with consequent formation of a double bond in the five-membered ring, at a point far removed from the site of the hydroxyl group in the original alcohol.

18.3 Neighbouring Group Effects

The presence of a functional group can effect the reactivity of an adjacent or 'proximate' group. This is well illustrated in the electrophilic substitution reactions of mono-substituted benzenes discussed in section 15.4. The substituents —OH, —OR, —NH$_2$, —NHCOR, —R and —X direct an electrophile to the *ortho* and *para* positions,

whereas —NO$_2$, —N(CH$_3$)$_3$$^+$, —CN, —CO$_2$H, —SO$_3$H, —CHO and —CRO direct an electrophile to the *meta* carbon. These substituents affect the rate of reaction at each carbon atom of the benzene ring. In Table 18.3 the rate of reaction at the different ring carbons for several monosubstituted benzenes are compared with the rate of reaction of a position in benzene under the same reaction conditions. These rate comparisons are known as *partial rate factors*. Because of the wide range of reactivity of monosubstituted benzenes it is often necessary to use different reaction conditions for different compounds if the reaction is to be observed on a measurable time scale. Thus, for example nitration of nitrobenzene in HNO$_3$—CH$_3$NO$_2$ mixtures is too slow to be measured, and more vigorous conditions have to be employed. Because of this wide range of reactivity, numerical comparisons for a wide range of substituents on benzene are difficult to make. The substituents on benzene which increase the electron density of the ring carbons by an *inductive* effect (e.g. —CH$_3$) or by overlap of lone pair electrons with the π electrons of the benzene ring (e.g. —O—CH$_3$) (a *resonance* effect) have their greatest effect at the *ortho* and *para* carbons. Electron withdrawing substituents such as nitro (—NO$_2$) decrease the reactivity of all carbon positions and similarly have their greatest effect at the *ortho* and *para* positions, resulting in electro-

Table 18.3 Partial rate factors for chlorination and nitration of some monosubstituted benzenes in comparison with benzene

philic attack occurring slowly at the least deactivated positions, the *meta* positions. Halogen substituents are unusual in that the inductive withdrawing effect through the C—X σ bond decreases the rate of reaction at all carbons relative to the carbon of benzene, but the resonance effect involving overlap of the lone pair electrons with the π electrons favours reaction occurring at the *ortho* and *para* positions.

Some of the most dramatic effects of a functional group proximate to a reaction site are shown for reactions where chloro, bromo or tosylate groups are displaced by a nucleophilic solvent such as water (reaction 18.5) or ethanoic

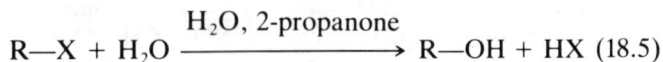

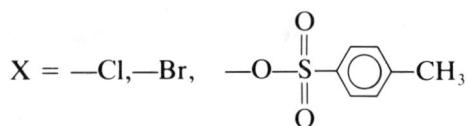

$$R—X + H_2O \xrightarrow{\text{H}_2\text{O, 2-propanone}} R—OH + HX \quad (18.5)$$

$$X = —Cl, —Br, \quad —O—\overset{\displaystyle O}{\underset{\displaystyle O}{\overset{\|}{\underset{\|}{S}}}}—\langle\bigcirc\rangle—CH_3$$

acid. The tosylate group, abbreviated as —OTs, is often used as a leaving group. It is readily prepared by reaction

of an alcohol with tosyl chloride (CH$_3$—$\langle\bigcirc\rangle$—SO$_2$Cl).

In nucleophilic displacement reactions, or when the only nucleophile present is solvent and the reaction is referred to as a solvolysis reaction, elimination to give alkene products is frequently competitive with attack at carbon by a nucleophile. A dramatic illustration of the effect of a proximate double bond and cyclopropyl group on the rate of solvolysis of some bi- and tri-cyclic tosylate derivatives is shown in Table 18.4. The double bond and cyclopropyl groups are so positioned as to be able to provide electron density to the C(7) carbon joined initially to the tosylate.

The double bond and cyclopropyl groups are said to *participate*, that is they provide *anchimeric* assistance, which they do by stabilizing the intermediate carbonium ion and lowering the activation energy for C—O bond heterolysis. This results in an immense acceleration in the rate of reaction, as is shown diagrammatically in Fig. 18.2. For each reaction as the C—O bond is stretched as a result of interaction of the substrate with the solvent molecule, rehybridization at C(7) from sp^3 to sp^2 occurs. The developing, electron deficient p orbital interacts with the π electrons of the double bond or, in the case of the cyclopropyl derivative, with the electrons in the σ bond. The electron density of a cyclopropyl σ bond lies outside a line joining the carbon atoms—one could say that the bond is shaped like a banana. The rich area of electron density of the 'banana bond' is suitably positioned to overlap with the developing electron deficient centre, and as a result the cyclopropyl electrons provide for an even greater intramolecular acceleration (*anchimeric* assistance) than the alkene moiety.

The previous examples illustrate the effect a functional group can have on the rate at which reaction occurs at a nearby site in the molecule. The participation of a functional group at a reaction centre may affect not merely the reaction rate, but even the nature of the products formed. Alkene participation is perhaps nowhere more extensive than for the reactions of squalene oxide. Squalene (Fig. 18.3) is a long chain polyalkene which, both in biological systems and in a laboratory, can be selectively oxidized at the terminal double bond to give squalene oxide (Fig. 18.4). Despite the fact that squalene can fold in an almost infinite number of conformations, its reactions are often very stereospecific. This indicates that reaction occurs to the molecule when it is folded in a very specific manner. It is now generally accepted that the typical plant triterpenoids are built up in nature through an all-chair folding and

Table 18.4 Relative rate of solvolysis

| Relative rate | 1 | 10^{11} | 10^{14-16} |

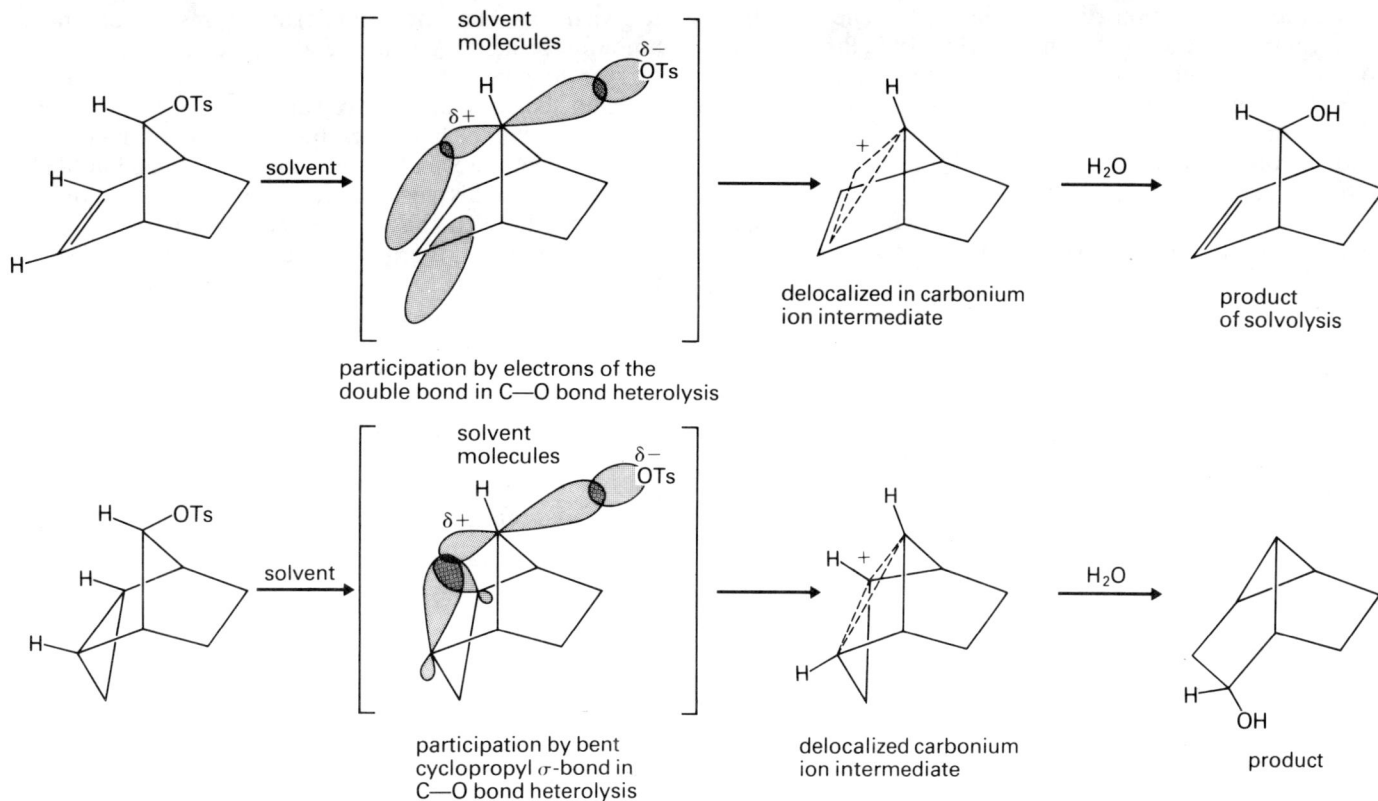

Fig. 18.2 Schematic diagram showing alkene and cyclopropane participation in solvolysis reactions

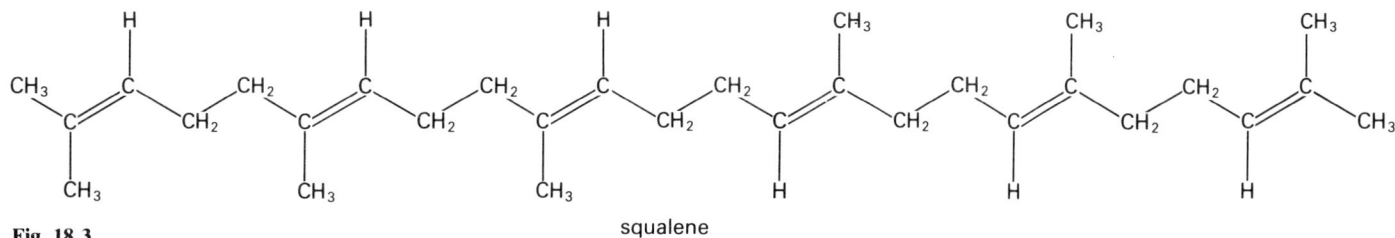

Fig. 18.3 squalene

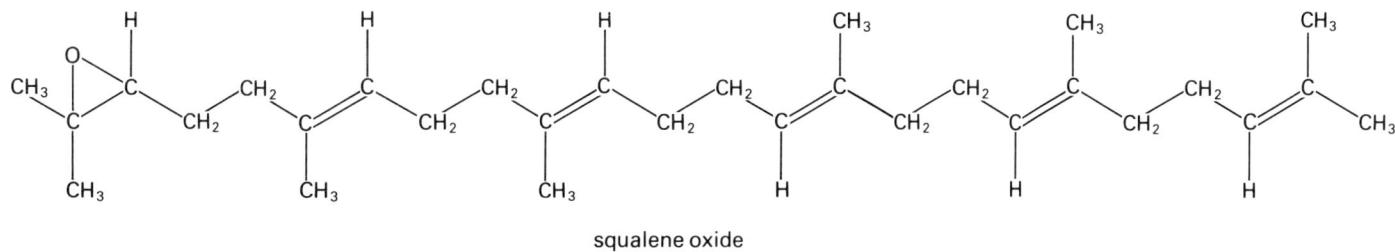

squalene oxide

Fig. 18.4

subsequent cyclization of squalene oxide. The essential features of this reaction are shown in Fig. 18.5. Protonation of the oxide occurs in acid media (H$^+$ or H$\pm$enzyme) and the alkene group participates with C—O bond cleavage. With formation of a new C—C bond the process is repeated with involvement of the adjacent alkene moiety until a tetra- or penta-cyclic molecule is formed.

The steroid precursor, lanosterol, results from an alternative way of folding squalene oxide, as is shown in Fig. 18.6. The tetracyclic carbonium ion so produced then undergoes a series of methyl-hydrogen migrations and, the sequence is terminated by loss of a proton from C(9). There is a similarity between the cyclization reactions of squalene oxide and the polymerization of alkenes (section 14.7) for the case where polymerization is initiated by a cation.

Another important neighbouring group interaction involves the intramolecular interaction of aldehydes (and ketones) with alcohols. Aldehydes (and ketones) react with alcohols

to give hemiacetals (hemiketals) in the presence of catalytic amounts of acid. A hemiacetal (or hemiketal) will react further to give acetals (or ketals) (reaction 18.6a). If a molecule contains an aldehyde and a hydroxyl group separated by a flexible chain of carbon atoms it may be possible to achieve intramolecular acetal formation (reaction 18.6b). It happens that this particular reaction is of great importance in carbohydrate chemistry (see Chapter 19).

Ester and alcohol groups are also able to participate in reactions occurring to other groups nearby. An interesting example which demonstrates a remarkable amount of molecular reorganization is shown for the reaction of the 1-ethanoxy-*trans*-3,4-epoxypentane with acid to give *cis*-2-methyl-3-ethanoxytetrahydrofuran (reaction 18.7). In the first step of the reaction in the presence of acid the already strained C(3)–oxygen bond is cleaved and nucleophilic attack with inversion by the carbonyl (C=O) of the ethanoate group gives an intermediate ethanonium ion. After rota-

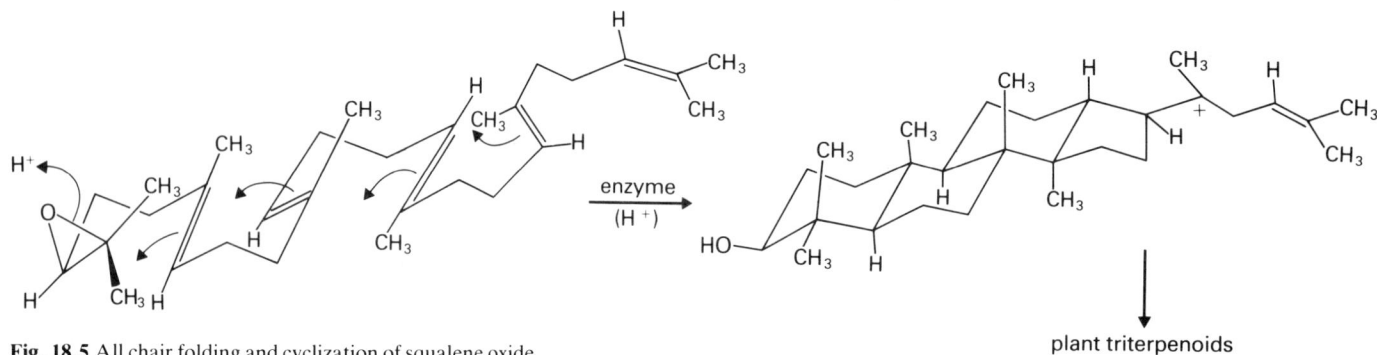

Fig. 18.5 All chair folding and cyclization of squalene oxide

plant triterpenoids

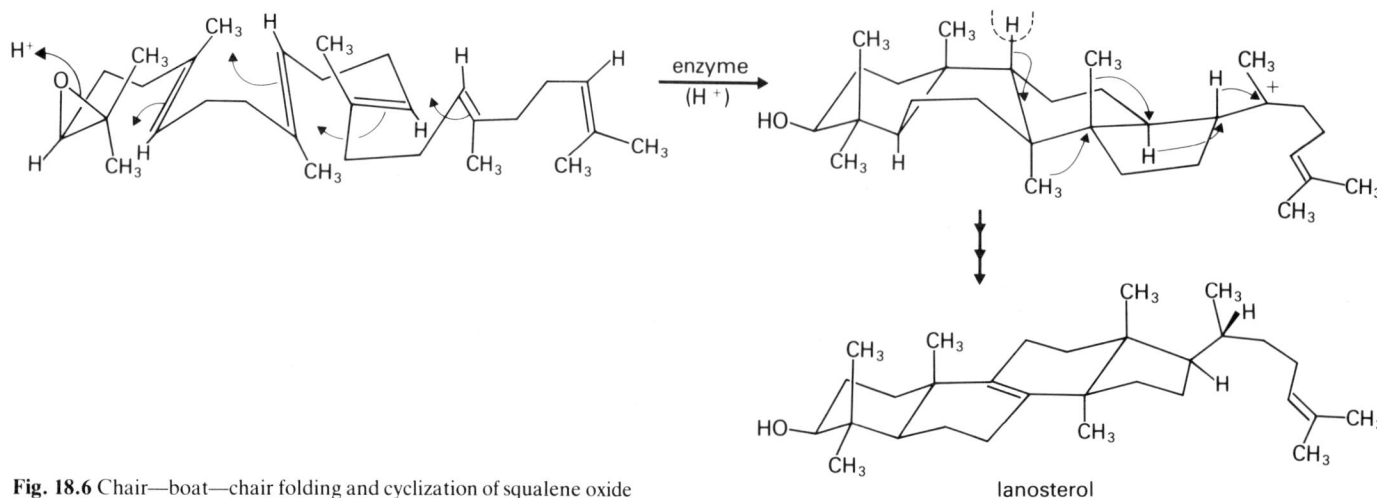

Fig. 18.6 Chair—boat—chair folding and cyclization of squalene oxide

lanosterol

$$\text{(18.6a)}$$

hemiacetal
(hemiketal)

acetal
(ketal)

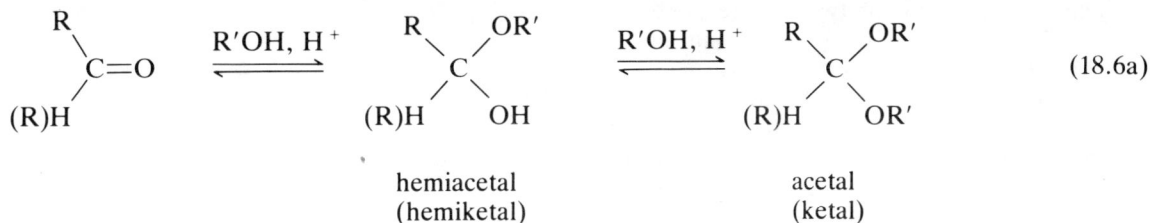

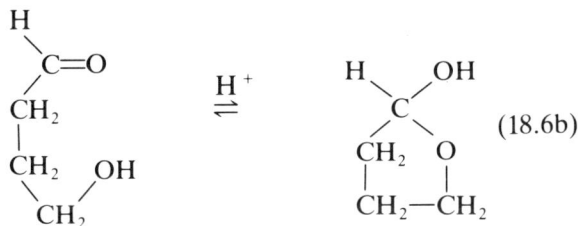

nucleophilic attack by the hydroxyl, with inversion of con-figuration occurring at the site of the attack. To determine the finer detail of such a reaction, labelling experiments are often performed. In this instance the carbonyl oxygen of the starting ethanoate was labelled with ^{18}O and this is indicated by an asterisk. The ^{18}O in the product is found to be linked directly to the tetrahydrofuran ring, which is consistent with the molecular transformation occurring as shown in the reaction scheme.

$$\text{(18.6b)}$$

tion about the C(2)—C(3) bond (step 2) the C(2)-hydroxyl is suitably positioned to attack the carbon atom of the ethanonium ion to give the neutral *ortho*-ester. This ester rearranges in the third step of the reaction to give a second ethanonium ion. The final step (4) in the reaction involves

18.4 Conjugated Systems

If two functional groups are proximate, the resulting molecular system may undergo reactions which would not occur with either of the functional groups in isolation. A

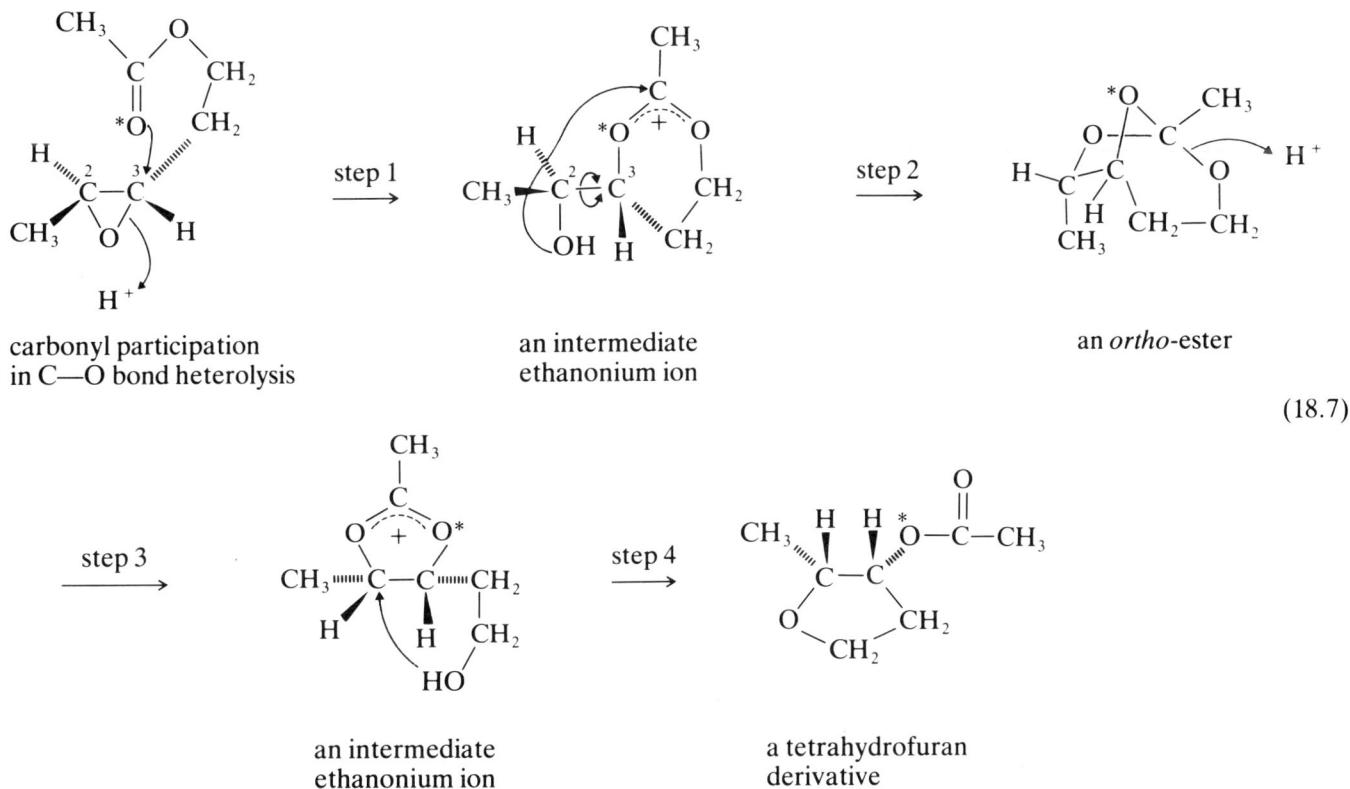

carbonyl participation
in C—O bond heterolysis

an intermediate
ethanonium ion

an *ortho*-ester

$$\text{(18.7)}$$

an intermediate
ethanonium ion

a tetrahydrofuran
derivative

molecule which contains two double bonds is termed a diene. The positions of the double bonds in the chain are indicated by numbering atoms in the usual manner. When the double bonds are joined directly by a single bond, the diene is referred to as a conjugated diene. A conjugated diene undergoes several reactions which are not possible for a non-conjugated diene. 1,3-Butadiene, shown in Fig. 18.7, is the simplest such compound. The molecule can rotate freely about the C(2)–C(3) bond; however, the most stable conformation of the molecule is achieved when the four carbon atoms lie in a plane. The *transoid* conformation is more stable than the *cisoid* conformation (Fig. 18.7).

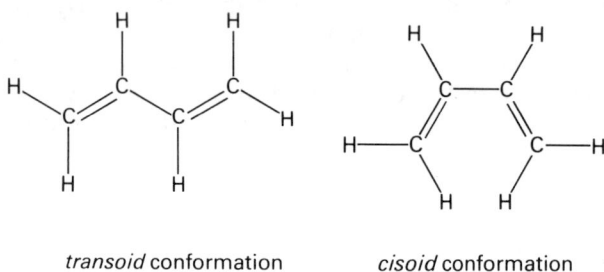

transoid conformation   *cisoid* conformation

Fig. 18.7 1,3-butadiene

When 1,3-butadiene exists in either of these planar conformations it is no longer possible to regard the double bonds as independent of one another. As discussed in section 3.5, the p orbitals on each carbon atom overlap to form a new set of molecular orbitals X_1, X_2, X_3 and X_4 as shown in Fig. 18.8, with an overall reduction in the total energy of the molecule. Because of this interaction the bond between C(2) and C(3) acquires some double bond character.

1,3-Butadiene can be transformed into cyclobutene (reaction 18.8), which is an example of an *electrocyclic*

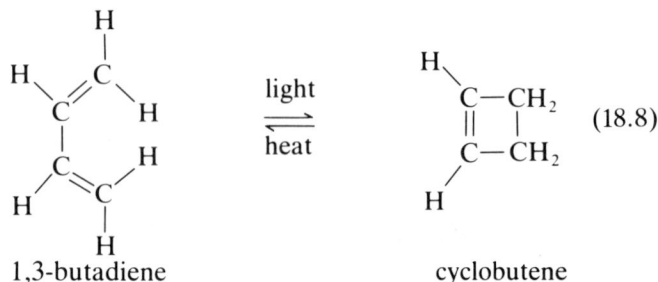

1,3-butadiene   cyclobutene

reaction. An electrocyclic reaction involves either the formation of a single bond between the ends of a linear system containing conjugated π electrons or the reverse process. The full elegance of such a reaction only becomes apparent when the substrate diene carries substituents, as in Fig. 18.9. (E,E)-2,4-Hexadiene is interconvertible under thermal

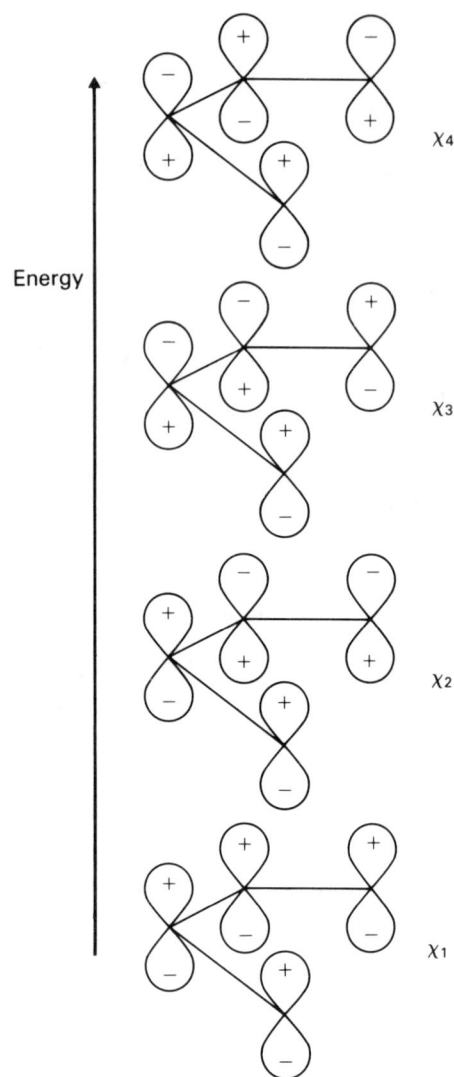

Fig. 18.8 Molecular orbitals of butadiene

conditions with *trans*-3,4-dimethyl-1-cyclobutene and *cis*-3,4-dimethyl-1-cyclobutene is interconvertible under similar conditions with (Z,E)-2,4-hexadiene. These processes are remarkably stereospecific. Rotation about the C(2)–C(3) and C(4)–C(5) bonds of the dienes occurs in the same direction (termed *conrotation*) as a new bond is formed between C(2) and C(5). In a similar way conrotation occurs for the reverse process, the conversion of cyclobutene to diene and these processes are also shown in Fig. 18.9. When rotation of the terminal carbons of a conjugated system occurs in opposite directions the process is termed *disrotation*. This is observed when 1,3-dienes are photolysed and transformed to cyclobutenes.

(E,E)-2,4-hexadiene

conrotation

trans-3,4-dimethyl-1-cyclobutene

conrotation

heat

cis-3,4-dimethyl-1-cyclobutene

conrotation

heat

(Z,E)-2,4,- hexadiene

conrotation

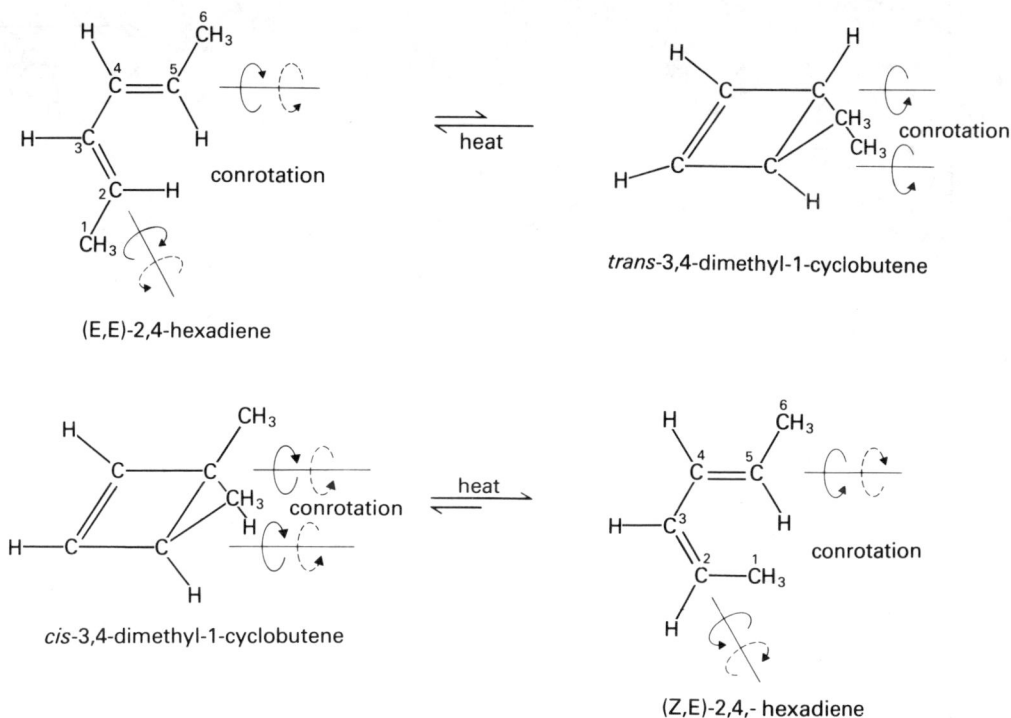

Fig. 18.9 2,4-Hexadiene—3, 4-dimethyl-1-cyclobutene interconversion

In recent years a single principle has been shown to be widely applicable for many reactions initiated by heat and light. It is known as the principle of 'conservation of orbital symmetry.' Simply stated the principle is that the symmetry of molecular orbitals is retained in reactions where bond making occurs at the same time as bond breaking. A reaction of this type is referred to as a *symmetry-allowed concerted reaction*. The following example shows how the principle arises.

For the conversion of a 1,3-butadiene to a cyclobutene (or the converse) the predominant modification to the molecular orbitals for the diene occurs to the π orbitals. These are the χ_1, χ_2, χ_3 and χ_4 molecular orbitals shown both in Figs 18.8 and 18.10. They are transformed to give the C(3)–C(4) σ and σ^* orbitals and the C(1)–C(2) π and π^* orbitals of the cyclobutene (see Fig. 18.10). As these molecular orbitals are transformed there are, of course, changes to all the other carbon–carbon and carbon–hydrogen bonds. These latter changes contribute to the energy barrier, that is the activation energy, between the butadiene and the cyclobutene, but are not relevant to the present discussion.

Butadiene contains four electrons in its π orbitals, these electrons being spin-paired and contained in the χ_1 and χ_2 molecular orbitals. During the transformation to cyclobutene these electrons are used to form the C(3)–C(4) σ

bond and the C(1)–C(2) π bond of ground state cyclobutene. Since bonding will only occur when there is overlap of orbitals of like sign, the formation of cyclobutene requires that both end carbon atoms rotate in the same direction (conrotation) to allow for σ bond formation between the terminal lobes of the highest occupied molecular orbital (χ_2). Fig. 18.10 illustrates how the χ_2 molecular orbital is transformed into the σ bond and the χ_1 molecular orbital is transformed into the π orbital of cyclobutene.

If rotation of the end carbon atoms were to occur in a disrotatory manner (Fig. 18.11) the χ_1 molecular orbital of butadiene would transform to the C(3)–C(4) σ bond of cyclobutane. The χ_2 molecular orbital would then have to transform into the π^* orbital, since the sign of the wave function C(2) and C(3) of the χ_2 orbital is the same as the sign of the π^* orbital wave function. The formation of a cyclobutene from a 1,3-butadiene by disrotatory ring closure would therefore require the product to be generated in the doubly excited state where two electrons occupy the π^* molecular orbital. The converse, disrotatory ring opening of a cyclobutene to a butadiene, would require the product to be generated in the doubly excited state with two electrons in each of the χ_1 and χ_3 molecular orbitals. In either case, the excitation energy would have to be added to the activation energy of the reaction. For this reason disrota-

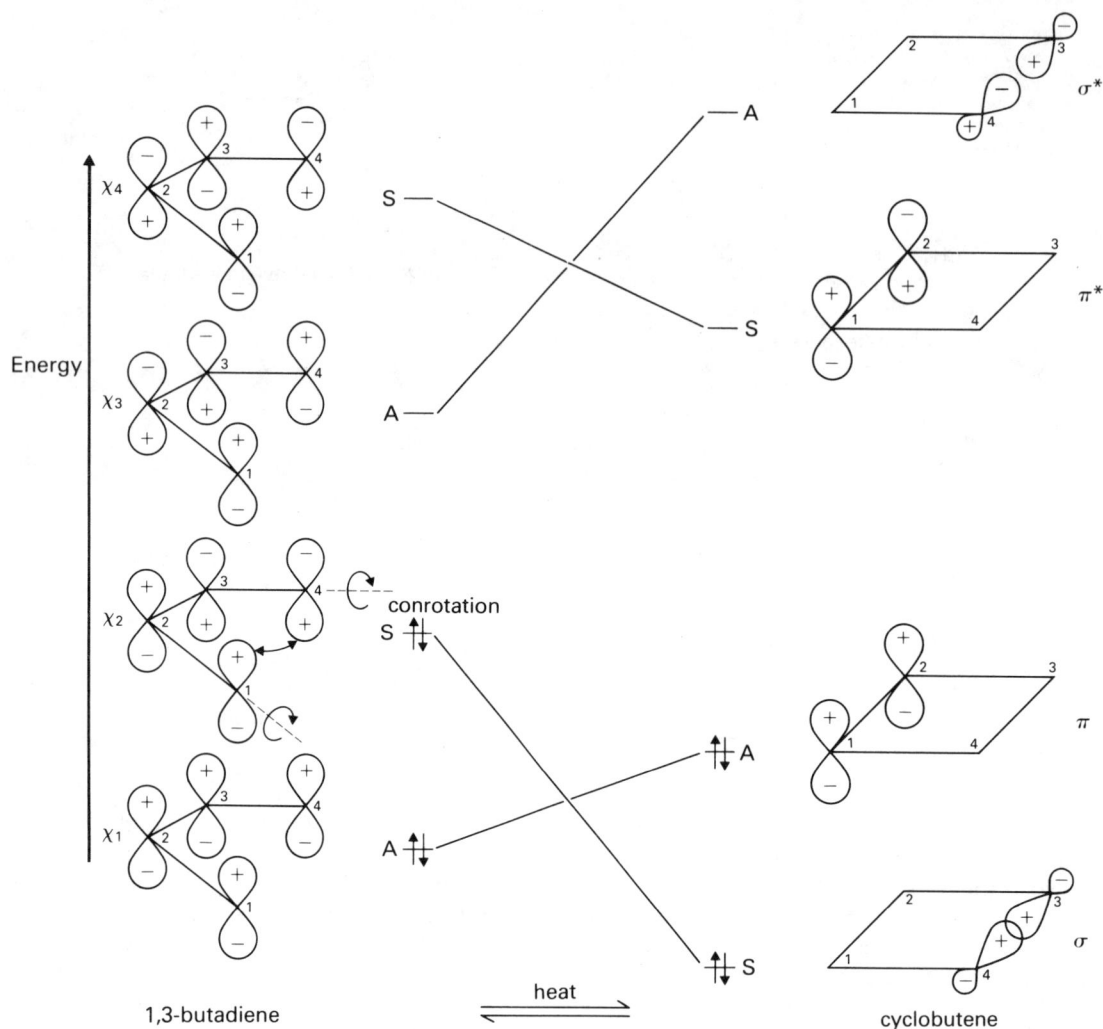

Fig. 18.10 Correlation diagram showing conrotation interconversion of butadiene–cyclobutene

tory interconversion between butadiene and cyclobutene is markedly less favoured than conrotatory interconversion.

For the orbitals being transformed in the conrotatory butadiene–cyclobutene interconversion a two-fold axis of symmetry is preserved throughout and this is shown in Fig. 18.12. For disrotatory interconversion a plane of symmetry is preserved throughout the reaction. For any symmetry transformation of the molecule the orbitals involved in the reaction can be classified as either symmetric 'S' or *anti*symmetric 'A' with respect to the particular transformation, or symmetry element. For example the χ_1 molecular orbital of butadiene is symmetric with respect to reflection in the plane of symmetry of the molecule, and *antisym*-

metric with respect to a two-fold axis of symmetry. The symmetries of the orbitals involved in the butadiene–cyclobutene interconversion are assigned in Figs 18.10 and 18.11, with respect to the appropriate symmetry elements. When the orbitals of butadiene are compared with orbitals of like symmetry in cyclobutene, for conrotatory and disrotatory cyclization, a *correlation diagram* can be constructed (Figs 18.10 and 18.11). The quantum mechanical non-crossing rule, which states that only levels of unlike symmetry are allowed to cross one another as we go from one side to the other, must be observed in constructing these diagrams. The correlation diagrams show that for conrotatory closure the bonding levels of butadiene (χ_1, χ_2) and cyclobutene

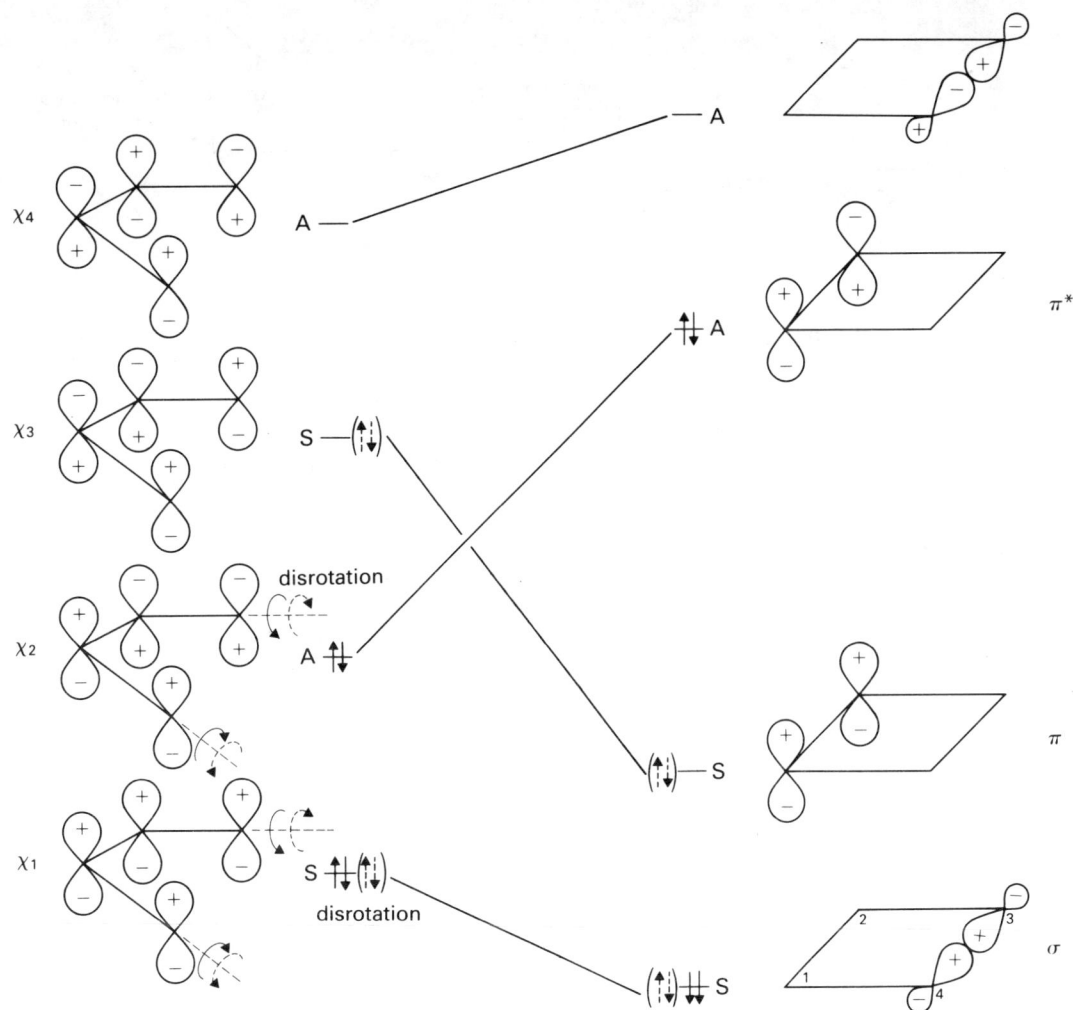

Fig. 18.11 Correlation diagram showing disrotatory interconversion of butadiene–cyclobutene

(σ, π) correlate directly, whereas for disrotatory closure this is not the case. Thermal cyclization of butadiene to cyclobutene, and thermal cleavage of cyclobutene to butadiene, therefore proceed by conrotation and are symmetry allowed processes.

A different situation occurs in the photochemistry of this system. If light is absorbed by butadiene an electron can be promoted from the χ_1 molecular orbital to the χ_3 molecular orbital. The resulting molecular state is referred to by photochemists (chemists who study the interaction of light with matter) as the first excited state of butadiene. The electron configuration in this state can be abbreviated as $\chi_1{}^2, \chi_2{}^1, \chi_3{}^1$, which shows that two electrons occupy the χ_1

molecular orbital and the χ_2 and χ_3 molecular orbitals each contain one electron. The first excited state of butadiene $(\chi_1{}^2, \chi_2{}^1, \chi_3{}^1)$ correlates directly with the first excited state of cyclobutene $(\sigma^2, \pi^1, \pi^{*1})$ for disrotatory (Fig. 18.13) but not conrotatory closure (Fig. 18.14). We would therefore predict that photocylization of butadiene to cyclobutene, and photocleavage of cyclobutene to butadiene, would proceed by disrotation, the symmetry allowed process in this case, and experiment confirms the correctness of this prediction for this and many similar examples.

Conjugated dienes add to alkenes in a reaction which is known by the names of the men who discovered it, the Diels–Alder reaction. An example is the addition of ethene

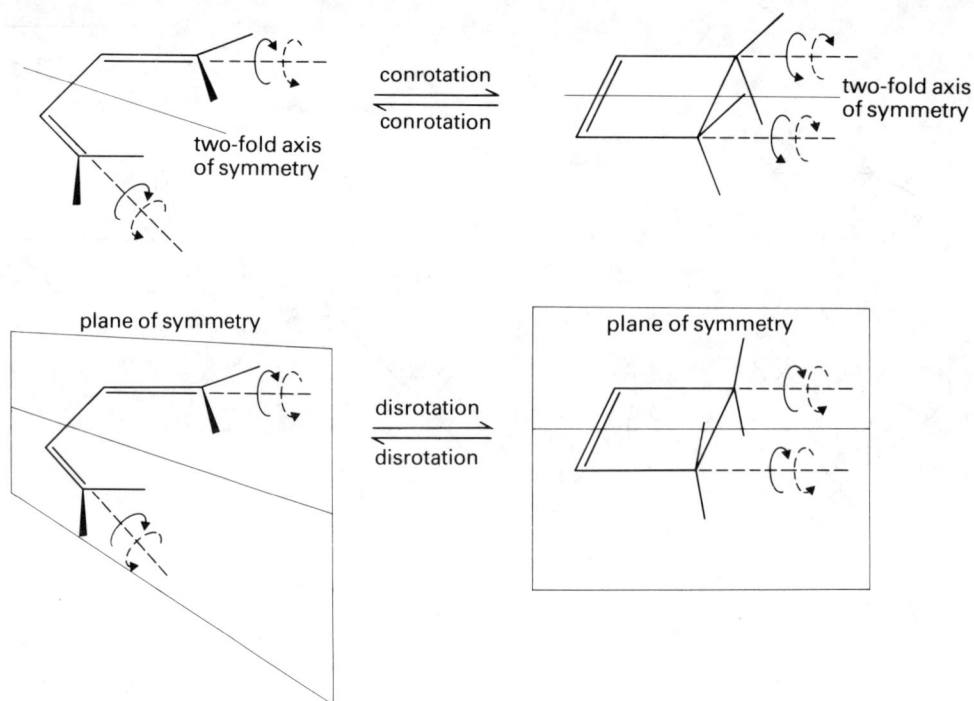

Fig. 18.12 Diagram showing the symmetry of conrotatory and disrotatory interconversion of a 1,3 butadiene and cyclobutene

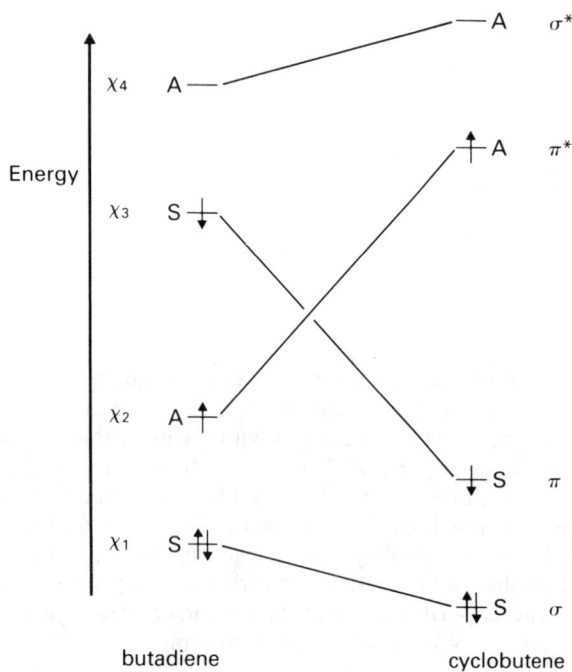

Fig. 18.13 Correlation diagram showing the first excited state disrotatory interconversion butadiene–cyclobutane

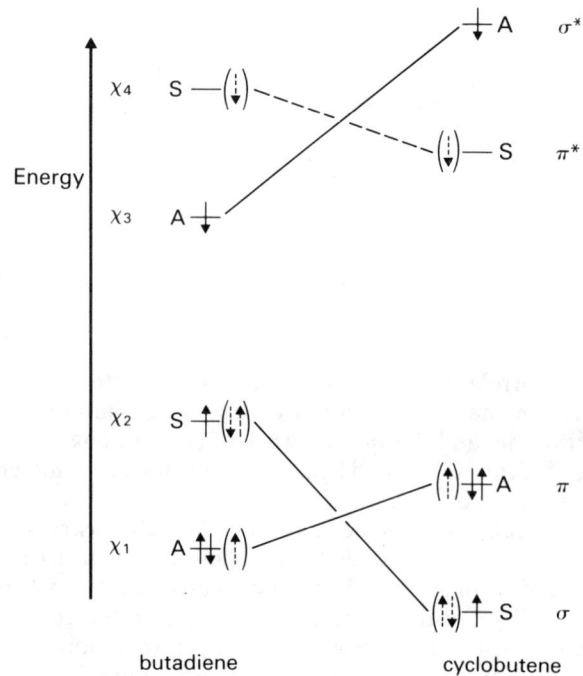

Fig. 18.14 Correlation diagram showing the first excited state conrotatory interconversion butadiene–cyclobutene

to 1,3-butadiene to give cyclohexene (reaction 18.9). This is one of a general class of reactions referred to as *cyclo-addition* reactions and for which the principle of 'conservation of orbital symmetry' is useful in predicting reaction products and their stereochemistry. These reactions provide a useful route to the synthesis of cyclic molecules. The

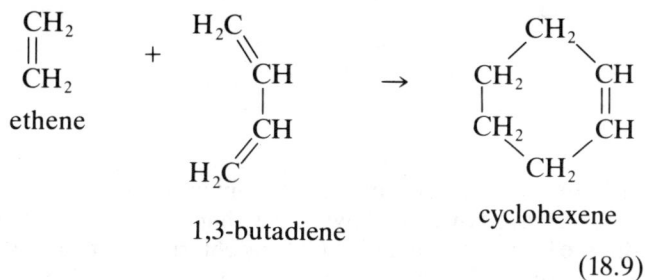

1,3-butadiene   cyclohexene

(18.9)

Diels–Alder reaction involves the addition of a 2π electron system (a π system with two electrons) to the ends of a conjugated 4π electron system and can be described as a $[\pi 2 + \pi 4]$ cycloaddition. Two other examples are shown in reactions 18.10 and 18.11.

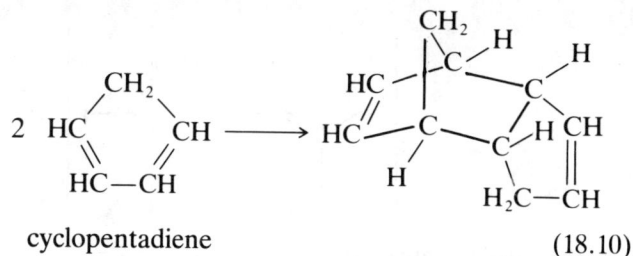

cyclopentadiene

(18.10)

Reaction of butadiene with bromine (reaction 18.12) results in the formation of two dibromobutenes, 1,2-dibromo-3-butene and 1,4-dibromo-2-butene. In a similar way addition of HBr (reaction 18.13) gives not only 3-bromo-1-butene but also 1-bromo-2-butene. The reaction shown in sequence 18.14 involves attack by an electrophile (H^+ or Br^+) to form an allylic carbonium ion intermediate in which the positive charge is spread over three carbon atoms. The intermediate is attacked by a nucleophile (N^-) at C(2) or at C(4) to give the 1,2- or 1,4-addition products respectively. The allylic carbonium ion can be written as a resonance hybrid of two equivalent canonical structures

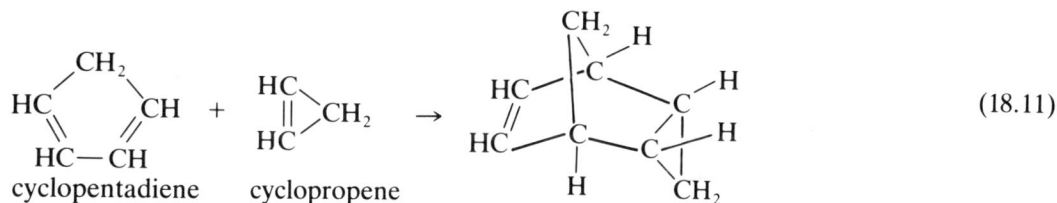

cyclopentadiene   cyclopropene

(18.11)

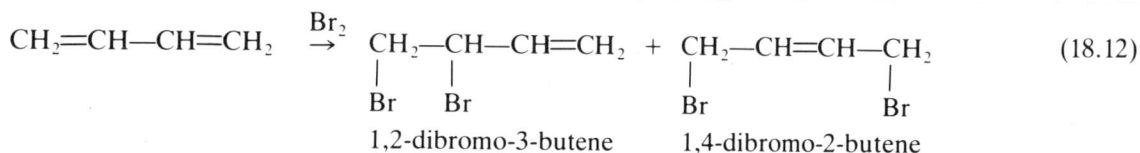

(18.12)

1,2-dibromo-3-butene   1,4-dibromo-2-butene

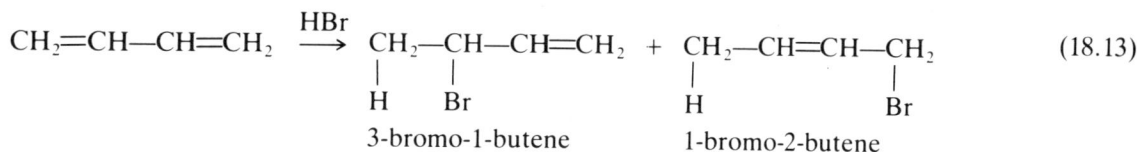

(18.13)

3-bromo-1-butene   1-bromo-2-butene

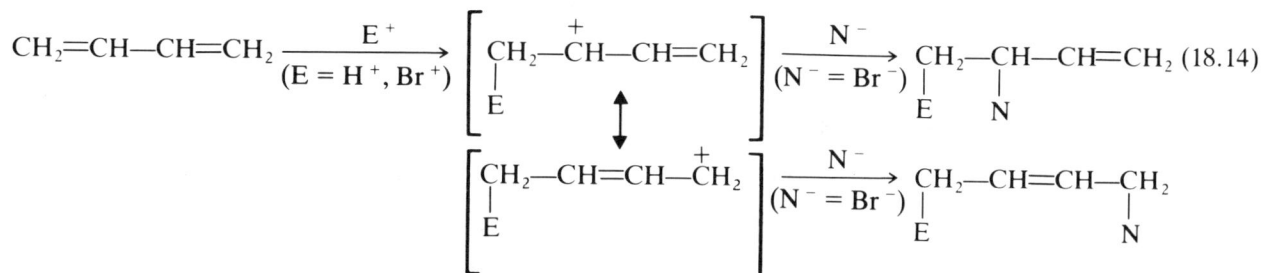

(18.14)

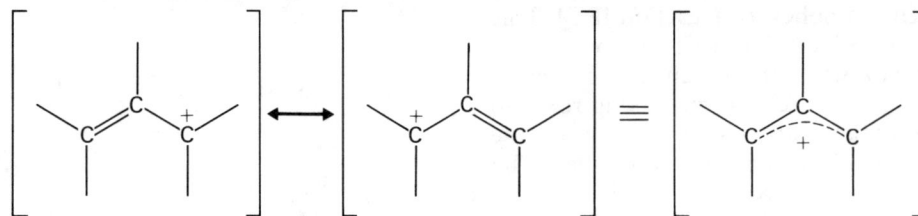

Fig. 18.15 An allylic carbonium ion

(Fig. 18.15) which differ only in the positions of electrons, i.e. the electrons are delocalized, and as a consequence of the delocalization of electrons an ion of this type is as stable as a tertiary carbonium ion.

When the double bonds of a diene are separated by a methylene group, as in 1,4-pentadiene (Fig. 18.16), the effects of conjugation and delocalization disappear. The double bonds become relatively independent of each other and the molecule undergoes the reactions characteristic of alkenes.

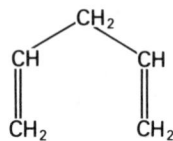

1,4-pentadiene

Fig. 18.16

When an alkene is positioned adjacent to a carbonyl group (Fig. 18.17), to give what is frequently referred to as a conjugated α,β-unsaturated carbonyl function, 1,4-addition can compete with 1,2-addition in the same way as we saw with 1,3-butadiene. For example, HCl adds to 2-propenoic acid to give 3-chloropropanoic acid (reaction 18.15) rather than 2-chloropropanoic acid. The first step of the

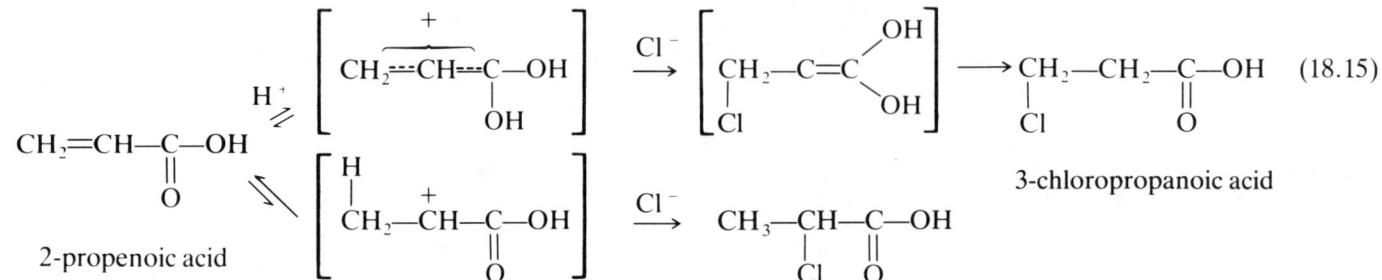

reaction involves addition of a proton to the most electron rich site—the carbonyl oxygen. The intermediate carbonium ion, stabilized by orbital overlap with the π electrons of the double bond, is attacked at C(3) by chloride ion to give the enol form of the carboxylic acid, which reverts by proton and electron shifts to the more stable 3-chloropropanoic acid. 2-Chloropropanoic acid, a 'Markovnikov' product, is not formed because the carbonium ion formed by protonation at C(2) is destabilized by the electron withdrawing carboxylic group and is therefore higher in energy than the allylic carbonium ion formed by protonation at the carbonyl oxygen.

Fig. 18.17 An α,β-unsaturated carbonyl moiety

The α,β-unsaturated-carbonyl functional group is important in synthesis since it is possible to add a carbon nucleophile, in a process known as a Michael reaction (reaction 18.16). The carbon nucleophile is generated by the removal of a proton, by a base, from a CH or CH_2 group which is

2-propenoic acid

3-chloropropanoic acid   (18.15)

2-chloropropanoic acid
not observed

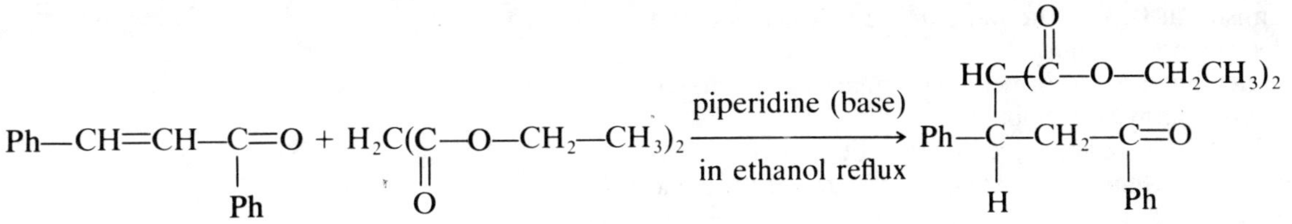

$$Ph—CH=CH—\underset{\underset{Ph}{|}}{C}=O + H_2C(\overset{\overset{O}{||}}{C}—O—CH_2—CH_3)_2 \xrightarrow[\text{in ethanol reflux}]{\text{piperidine (base)}} Ph—\underset{\underset{H}{|}}{\overset{\overset{HC-(C—O—CH_2CH_3)_2}{|}}{C}}—CH_2—\underset{\underset{Ph}{|}}{C}=O$$ (18.16)

1,3-diphenyl-2-propen-1-one   diethyl malonate
(benzalacetophenone)

$$H_2C(CO_2Et)_2 + \text{Base} \rightleftharpoons H—\text{Base}^+ + :\bar{C}H(CO_2Et)$$ (18.17)

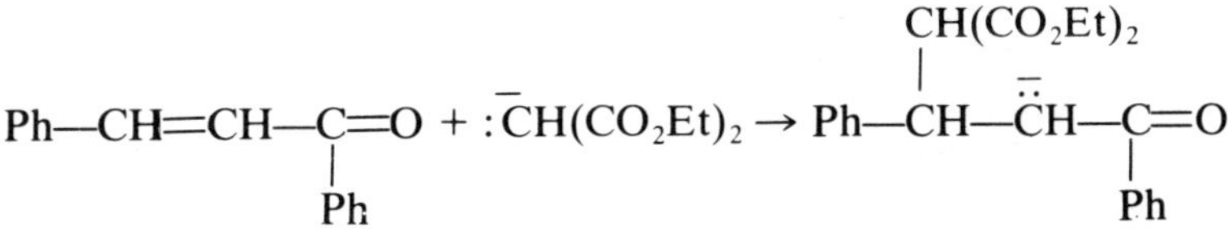

$$Ph—CH=CH—\underset{\underset{Ph}{|}}{C}=O + :\bar{C}H(CO_2Et)_2 \rightarrow Ph—CH—\underset{\underset{Ph}{|}}{\overset{\overset{CH(CO_2Et)_2}{|}}{C}H}—C=O$$ (18.18)

$$H—\text{Base}^+ + Ph—CH—\underset{\underset{Ph}{|}}{\overset{\overset{CH(CO_2Et)_2}{|}}{C}H}\!\!=\!\!\!=\!\!\!C\!\!=\!\!O \rightarrow Ph—CH—\underset{\underset{Ph}{|}}{\overset{\overset{CH(CO_2Et)_2}{|}}{C}}H_2—C=O + :\text{Base}$$ (18.19)

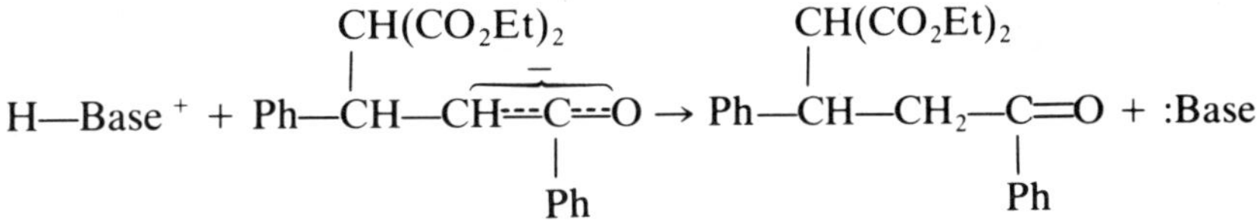

activated by being adjacent to one or more carbonyl groups. The carbonyl group has the effect of stabilizing the adjacent negative charge (reaction 18.17). The carbanion, or carbon nucleophile, generated in this way attacks the conjugated system (reaction 18.18), and this is followed by proton transfer from the base (reaction 18.19), which completes the sequence. A Michael reaction is represented schematically in reaction 18.20, which shows the overall process to involve addition, to the conjugated unsaturated carbonyl, of a hydrogen at C(2) and a carbon at C(3).

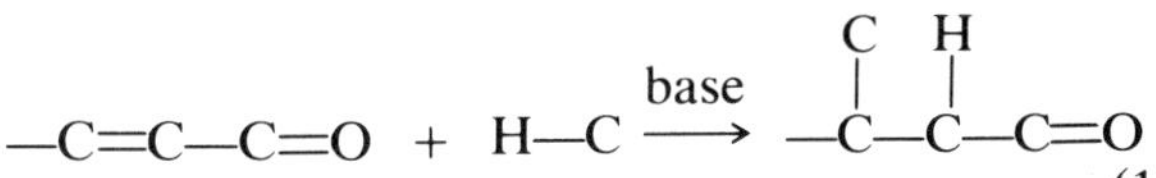

$$—C=C—C=O + H—C \xrightarrow{\text{base}} —\underset{\underset{}{|}}{\overset{\overset{C}{|}}{C}}—\underset{\underset{}{|}}{\overset{\overset{H}{|}}{C}}—C=O$$ (18.20)

Direct 1,2-addition to the carbonyl of a conjugated carbonyl functional group can also occur. For example, reaction of 4-methyl-3-penten-2-one with ethyl magnesium halide, followed by hydrolysis, gives 2,4-dimethyl-2-hexen-4-ol (reaction 18.21). The mode of addition to a conjugated system is influenced by many factors, such as the nature of the solvent and the temperature, but for any particular substrate there is a tendency for what are regarded as the stronger nucleophiles to undergo 1,2-addition reactions and for the weaker nucleophiles to undergo 1,4-additions.

Exercises

18.1 Suggest a mechanism to explain the conversion of 2,3-dimethyl-2,3-butanediol by sulphuric acid into 3,3-dimethyl-2-butanone.

18.2 Show how tetrahydrofuran

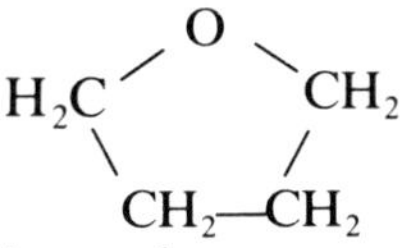

$$\underset{CH_2—CH_2}{H_2C \diagup \overset{\displaystyle O}{} \diagdown CH_2}$$

could be formed from 4-chloro-1-butanol.

18.3 Solvolysis of (2R,3S)-2-ethanoxy-3-bromobutane in ethanoic acid gives (R,S)-2,3-diethanoxybutane where the

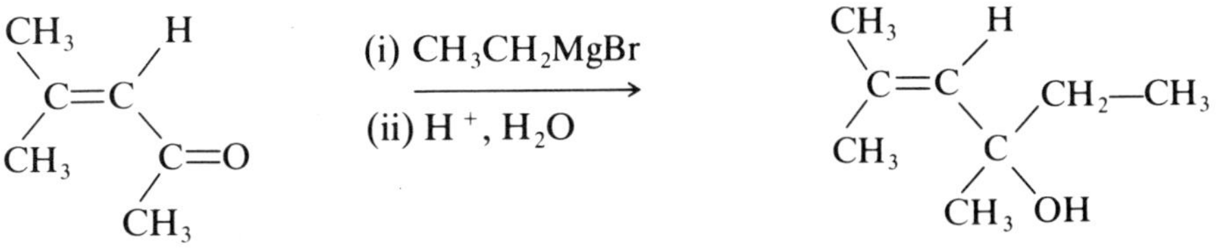

$$\underset{CH_3}{\overset{CH_3}{\diagdown}}C=C\underset{\underset{CH_3}{|}}{\overset{\overset{H}{\diagup}}{C}=O} \xrightarrow[\text{(ii) } H^+, H_2O]{\text{(i) } CH_3CH_2MgBr}$$

4-methyl-3-penten-2-one

$$\underset{CH_3}{\overset{CH_3}{\diagdown}}C=C\underset{\underset{CH_3 \quad OH}{\diagup \diagdown}}{\overset{\overset{H}{\diagup}}{C}}—CH_2—CH_3$$ (18.21)

2,4-dimethyl-2-hexen-4-ol

stereochemistry at C(3) is the same in the product as in the starting compound. Explain.

18.4   Predict the product formed by thermal ring opening of *trans*-3-ethyl-4-methyl-1-cyclobutene.

18.5   (E,E)-2,4-Hexadiene is converted by light to a 3,4-dimethyl-1-cyclobutene—an example of an electrocyclic reaction occurring in a disrotatory manner. What is the stereochemistry of the product?

18.6   When HCl is added to 2-methyl-1,3-butadiene the only 1,4-addition product obtained is 1-chloro-3-methyl-2-butene. 1-Chloro-2-methyl-2-butene is not formed. Why?

18.7   HBr adds to $CH_2{=}CHCO_2Et$ under both ionic and radical-chain conditions to give only $BrCH_2CH_2CO_2Et$. Write mechanisms to account for these observations.

18.8   Write equations for the Michael condensations;

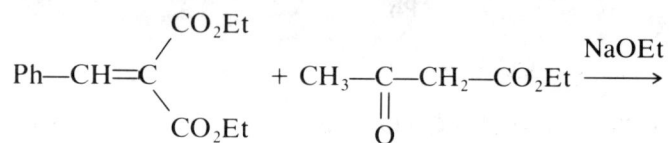

and

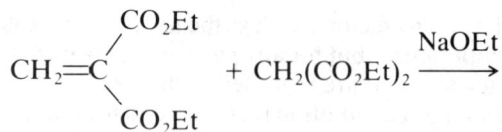

19 The Chemistry of Living Things

19.1 Introduction

The initial impetus for the development of organic chemistry that has occurred during the last 150 years was provided by the discovery that organic chemicals could be synthesized in the laboratory without the involvement of a living organism, other than the experimenter. The direction of the development has been such that most of the two million or more organic compounds that have been characterized during this period have only the most tenuous connection with living matter. Consequently, it is very interesting to observe that the wheel has now come full circle, in that the area of organic chemistry which promises the most exciting developments for the future is the area which is most intimately involved with the chemistry of life. In this chapter we shall discuss several groups of compounds which are of vital importance to living things, namely carbohydrates, proteins, haemoglobin and myoglobin, and nucleic acids.

19.2 Carbohydrates

Carbohydrates are compounds which have the general formula $C_n(H_2O)_m$—i.e. hydrated carbon—hence the name carbohydrate. Carbohydrates are eaten in large amounts as bread, potatoes and rice, worn as cotton and linen, and written on as paper. They are produced in nature from carbon dioxide and water by the action of sunlight in the presence of chlorophyll, the green pigment of plants (Eq. 19.1).

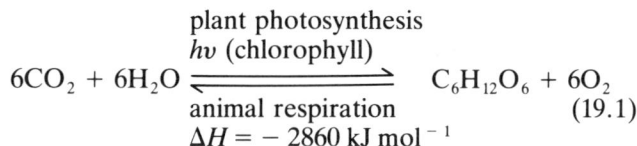

$$6CO_2 + 6H_2O \underset{\substack{\text{animal respiration} \\ \Delta H = -2860 \text{ kJ mol}^{-1}}}{\overset{\substack{\text{plant photosynthesis} \\ h\nu \text{ (chlorophyll)}}}{\rightleftharpoons}} C_6H_{12}O_6 + 6O_2 \quad (19.1)$$

The driving force for the reaction is supplied by sunlight. Animals eat the carbohydrates and convert them back to carbon dioxide and water, in the process releasing and using the stored energy of the sunlight.

In terms of their functional groups, carbohydrates are polyhydroxy aldehydes, polyhydroxy ketones or molecules which yield polyhydroxy aldehydes and ketones on hydrolysis. The smallest carbohydrates, those containing three, four, five or six carbon atoms, are called monosaccharides. The simplest is glyceraldehyde $C_3(H_2O)_3$ (Fig. 19.1).

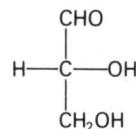

Fig. 19.1 Glyceraldehyde

Glyceraldehyde contains one asymmetric or chiral carbon atom, and can therefore exist in two unique forms which are related to each other as mirror images. Before the chirality at the asymmetric carbon can be defined (R or S) the sequence rule (p. 244) must be extended to include substituent groups containing multiple bonds (e.g. —CHO). Where there is a double bond both atoms of the double bond are considered, for sequence purposes, to be duplicated.

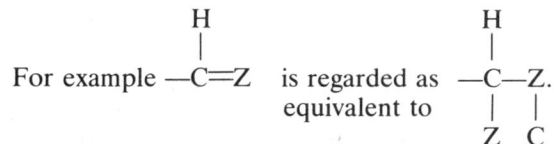

For example —C=Z is regarded as equivalent to —C—Z.

For glyceraldehyde the —OH group has the highest priority, while the aldehyde group (—CHO) for sequence purposes is regarded as

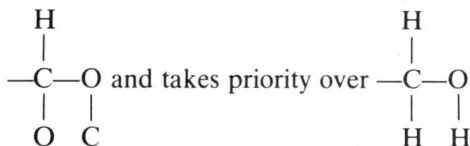

—C—O and takes priority over —C—O

The sequence is therefore —OH, —CHO, —CH$_2$OH, —H and the asymmetry of glyceraldehyde can be assigned* as

* (R)-glyceraldehyde has in the past been referred to as D-glyceraldehyde and (S)-glyceraldehyde as L-glyceraldehyde.

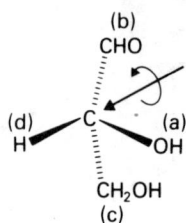

Fig. 19.2 (R)-glyceraldehyde

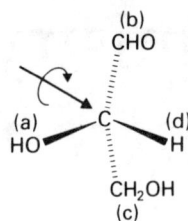

Fig. 19.3 (S)-glyceraldehyde

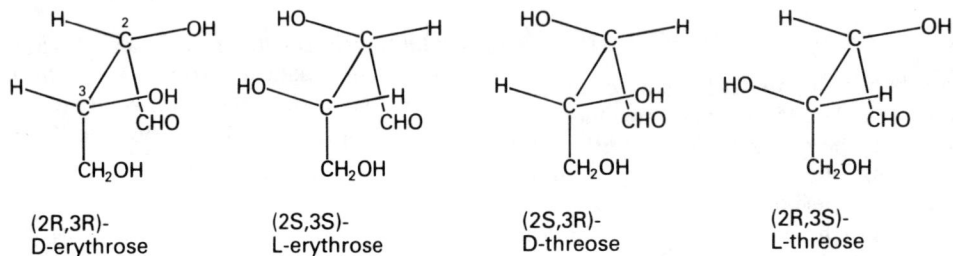

(2R,3R)-
D-erythrose

(2S,3S)-
L-erythrose

(2S,3R)-
D-threose

(2R,3S)-
L-threose

Fig. 19.4 Stereoisomers of $C_4(H_2O)_4$

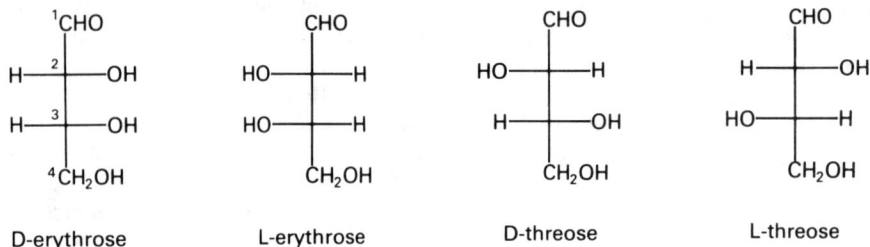

D-erythrose

L-erythrose

D-threose

L-threose

Fig. 19.5 Fischer projections of $C_4(H_2O)_4$ stereoisomers

shown in Figs 19.2 and 19.3. There is an increasing number of stereoisomers as the homologous series is ascended and the next carbohydrate homologue, $C_4(H_2O)_4$, with two asymmetric carbon atoms, can exist as four possible stereoisomers (Fig. 19.4). The carbon atoms are numbered beginning with the aldehyde carbon atom as C(1). These compounds can be drawn as *Fischer projections* (Fig. 19.5) where bonds drawn horizontally are above the plane of the page and those drawn vertically are below the plane of the page.*

Monosaccharides with five carbon atoms, $C_5(H_2O)_5$, are called pentoses and those with six carbon atoms, $C_6(H_2O)_6$, are called hexoses. Six-carbon monosaccharides which con-

tain an aldehyde group are known as aldohexoses, and those which contain a ketone group are known as ketohexoses. D-fructose, a ketohexose, is shown in Fig. 19.6.

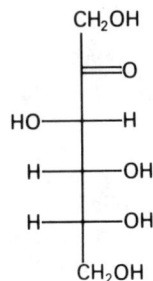

Fig. 19.6 D-fructose

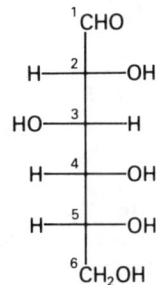

Fig. 19.7 D-glucose

* By definition the carbon atom farthest from the aldehyde group determines the configuration as D or L (cf. glyceraldehyde). The most important naturally occurring sugars are all in the D-series i.e. the carbon farthest from the carbonyl has an R-configuration.

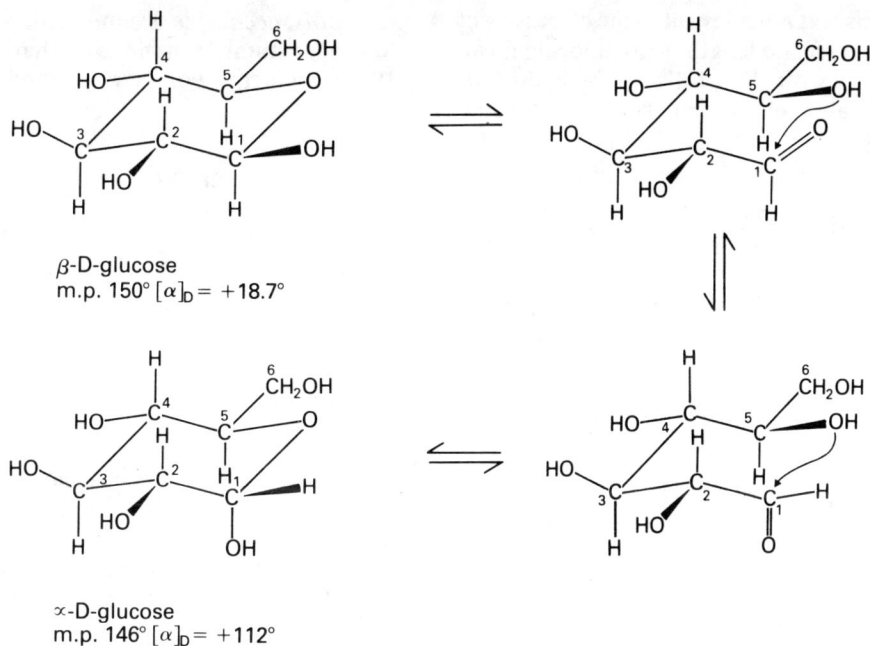

β-D-glucose
m.p. 150° $[\alpha]_D = +18.7°$

α-D-glucose
m.p. 146° $[\alpha]_D = +112°$

Fig. 19.8 Equilibrium between the glucose forms

Of all sugars, D-glucose (Fig. 19.7) is the most common. It is also thermodynamically the most stable of the aldohexoses. D-glucose exists in two crystalline forms, α- and β-, which can be obtained separately (Fig. 19.8). Neither form contains an aldehyde group. β-D-glucose, with all the ring substituents equatorial, is thermodynamically more stable than α-D-glucose, where the C(1)-hydroxyl is axial. At equilibrium in aqueous solution D-glucose exists as a mixture of 64% β-, 36% α- and 0.02% free aldehyde form. The C(5) hydroxyl in the α- and β- forms has formed a hemiacetal with the aldehyde, thereby producing a new asymmetric carbon atom and giving rise to two diastereoisomers. The aldehyde of D-glucose is free to rotate about the C(1)—C(2) bond and can be attacked by the C(5) hydroxyl when the carbonyl is above or below the plane of the molecule. Thus two distinct products, can be formed, one with the oxygen axial and the other with the oxygen equatorial. The essential features of this process are represented schematically in Fig. 19.9. The interconversion of hemiacetals through the open chain aldehyde form when dissolved in water is generally accompanied by a change in optical rotation and is known as *mutarotation*.

Because the functional groups of sugars can interact intramolecularly, it is often advisable to protect one part of a molecule, (e.g. with a temporary substituent) before

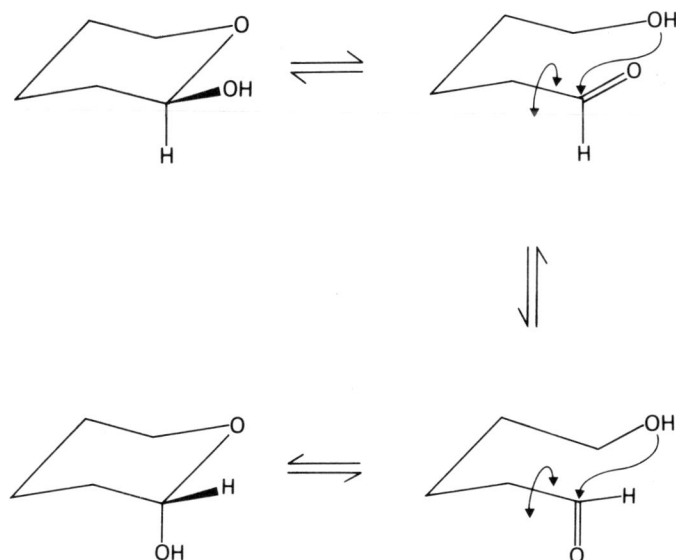

Fig. 19.9 Schematic representation showing the essential features of the equilibrium between the glucose forms

effecting reaction at another part. With this proviso, the reactions of the alcohol, ketone and aldehyde functions examined earlier in this text are relevant to the chemistry of sugars. For example, the chain length of an aldo-sugar can be extended by reaction with HCN followed by hydrolysis to the carboxylic acid and subsequent reduction (scheme 19.2).

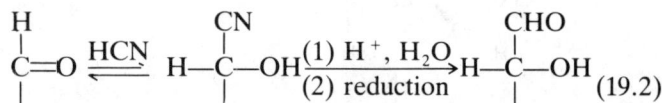

$$
\underset{\substack{|\\ \text{C}=\text{O}\\|}}{\overset{\text{H}}{}} \quad \xrightarrow{\text{HCN}} \quad \underset{\substack{|\\ \text{H}-\text{C}-\text{OH}\\|}}{\overset{\text{CN}}{}} \xrightarrow[\text{(2) reduction}]{\text{(1) H}^+,\ \text{H}_2\text{O}} \underset{\substack{|\\ \text{H}-\text{C}-\text{OH}\\|}}{\overset{\text{CHO}}{}} \quad (19.2)
$$

Using reactions discussed earlier the chain length of an aldo-sugar can also be reduced by one carbon atom (sequence 19.3).

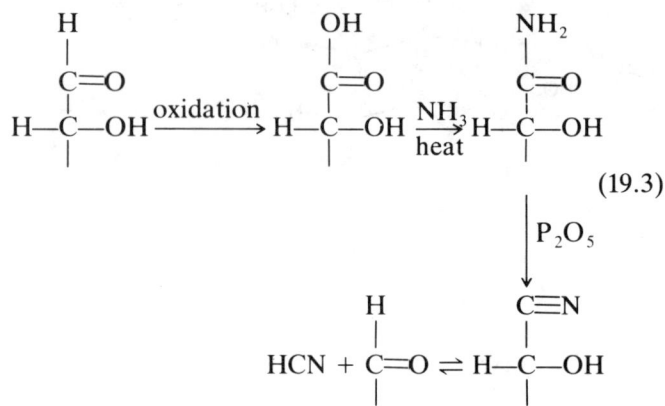

$$
\underset{\substack{|\\ \text{H}-\text{C}-\text{OH}\\|}}{\overset{\text{H}}{\text{C}=\text{O}}} \xrightarrow{\text{oxidation}} \underset{\substack{|\\ \text{H}-\text{C}-\text{OH}\\|}}{\overset{\text{OH}}{\text{C}=\text{O}}} \xrightarrow[\text{heat}]{\text{NH}_3} \underset{\substack{|\\ \text{H}-\text{C}-\text{OH}\\|}}{\overset{\text{NH}_2}{\text{C}=\text{O}}}
$$

(19.3)

$$\downarrow \text{P}_2\text{O}_5$$

$$
\text{HCN} + \underset{\substack{|\\ \text{C}=\text{O}\\|}}{\overset{\text{H}}{}} \rightleftharpoons \underset{\substack{|\\ \text{H}-\text{C}-\text{OH}\\|}}{\overset{\text{C}\equiv\text{N}}{}}
$$

Monosaccharide units can be linked together to give disaccharides, where a *disaccharide* is a compound that can be hydrolysed to two monosaccharide units. Sucrose, obtained from sugar beet and sugar cane, and lactose (milk sugar) are both disaccharides. Their hydrolyses are shown in reactions 19.4 and 19.5.

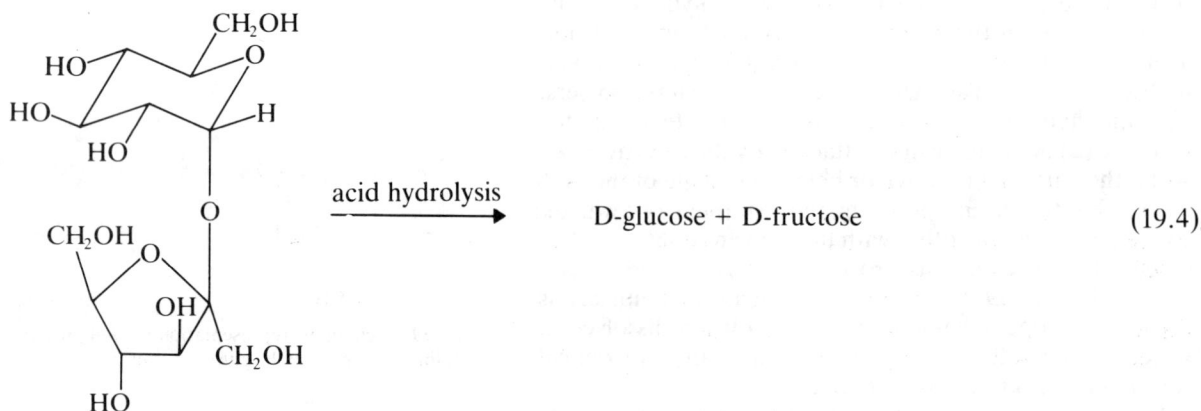

A *polysaccharide* is a molecule which can be hydrolysed to a large number of monosaccharide units. Cellulose (Fig. 19.10) a glucose polymer, is found in most plants. Wood is

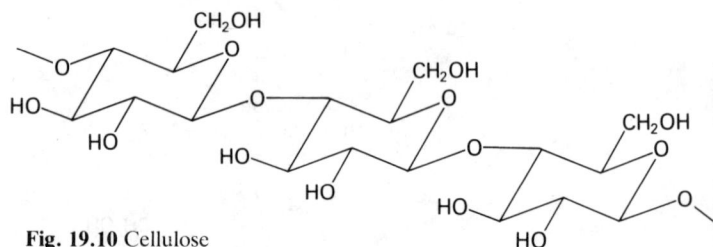

Fig. 19.10 Cellulose

about 50% cellulose. Each cellulose molecule contains about 3000 glucose units. For those species which possess the enzymes necessary to break down cellulose to glucose this is a rich source of energy. Ungulates (hoofed mammals) lack these enzymes, but are able to digest cellulose with the aid of symbiotic bacteria which live in their gut.

19.3 Proteins

Proteins in life provide protection as skin, hair and nails, and motive power as muscles. As enzymes, or organic catalysts, they sustain life by speeding up otherwise slow chemical reactions, often by factors of 10^6 or more. They are large molecules, with molecular weights ranging from 13 000 to several million and are therefore particularly likely to possess complicated structures. The sheer number of atoms in the molecule makes the determination of protein structure very difficult. Proteins are built up of long linear

$$\xrightarrow{\text{acid hydrolysis}} \quad \text{D-glucose} + \text{D-fructose} \qquad (19.4)$$

sucrose: α-D-glucopyranosyl-β-D-fructofuranoside

acid hydrolysis

D-galactose   (19.5)

+

D-glucose

lactose: 4-O-β-D-galactopyranosyl-1-O-β-D-glucopyranose

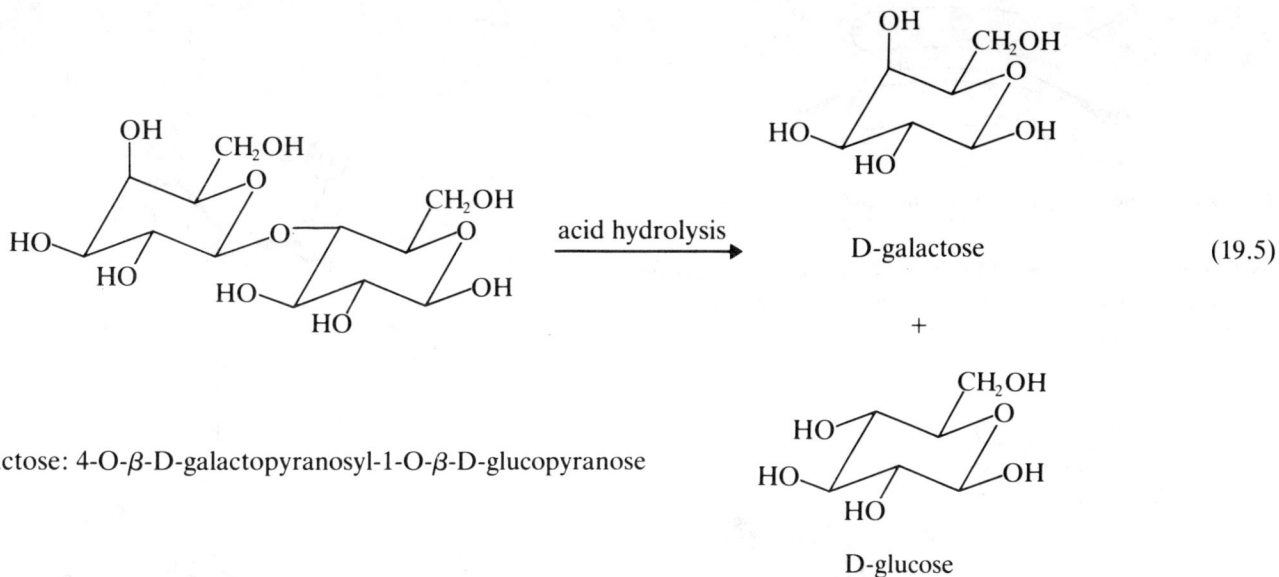

chains of α-amino carboxylic acids (Fig. 19.11). Twenty different amino acids are known to take part in protein

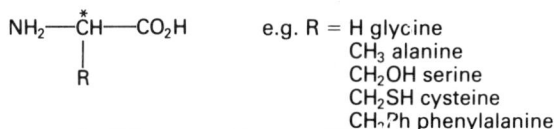

linked together in a chain by a condensation reaction involving the loss of a molecule of water between each joined unit. The hydroxyl group of the carboxylic acid and the proton from the amine end of another α-amino acid are removed in the formation of an amide or *polypeptide* linkage (reaction 19.6).

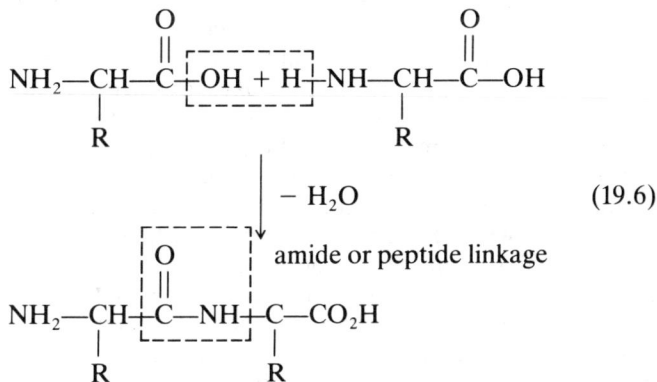

$$NH_2-CH-C\!\!\overset{O}{\underset{R}{\|}}\!\!-OH + H-NH-CH-C\!\!\overset{O}{\underset{R}{\|}}\!\!-OH$$

$$\downarrow \quad -H_2O \qquad\qquad (19.6)$$

$$NH_2-CH-C\!\!\overset{O}{\underset{R}{\|}}\!\!-NH-C-CO_2H$$

amide or peptide linkage

NH₂—CH—CO₂H   e.g. R = H glycine
       | CH₃ alanine
       R   CH₂OH serine
                                        CH₂SH cysteine
                                        CH₂Ph phenylalanine

Fig. 19.11 Examples of amino acids found in nature

construction. Each of these twenty amino acid units has a different R group, but all have the same absolute configuration and are often referred to as L-amino acids. With the exception of cysteine and cystine the remainder belong to the (S) chiral family (Fig. 19.12). Amino acid units are

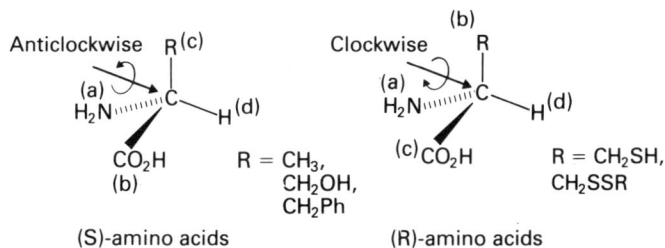

When only a few amino acids are linked together in this way the resulting molecule is referred to as a polypeptide. The amide linkage is essentially planar, since the nitrogen orbital containing a lone pair of electrons overlaps with the carbonyl π electrons, and the electron density is spread over the O—C—N atoms as shown in Fig. 19.13. This overlap imposes a barrier to rotation about the carbonyl C—N bond. Protein chains are, however, able to twist, curl and entwine as a consequence of the geometry about the asymmetric carbon atom between the nitrogen atom and

Anticlockwise R(c)   Clockwise R (b)
(a)   (a)
H₂N''''''C—H(d)   H₂N''''''C—H(d)
    | |
CO₂H   R = CH₃,   (c)CO₂H   R = CH₂SH,
(b)   CH₂OH,   CH₂SSR
              CH₂Ph

(S)-amino acids   (R)-amino acids

Fig. 19.12 Amino acid configuration

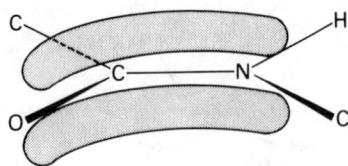

Fig. 19.13 Amide: lone pair overlap with π electrons

carbonyl carbon atom (Fig. 19.14). Association between amide linkages can occur by hydrogen bonding between the hydrogen attached to the nitrogen of one group and the

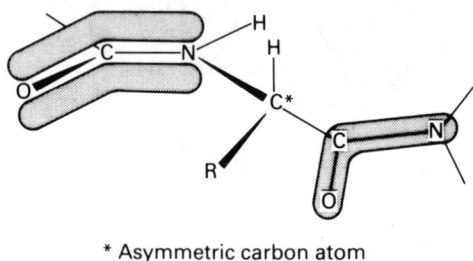

* Asymmetric carbon atom

Fig. 19.14 Twisting of protein chain

oxygen attached to a carbon of another amide group. Such intramolecular association within a chain results in a quite rigid helical structure for the polypeptide (Fig. 19.15). The configuration of the helix is maintained by the hydrogen bond between the nitrogen and the oxygen attached to the carbon atom three amino acid groups along the chain. Alternatively, intermolecular hydrogen bonding can occur between two different polypeptide chains, a type of configuration which occurs in fibrous proteins such as silk and which is shown in Fig. 19.16. It is thought that the polypeptide chains in muscle and other contractible fibres may function by a reversible transition from the helical to the chain configuration. A third structural modification of protein is found in the non-stretch connective tissues in ligaments, and in the strengthening networks of skin. This structure is shown in Fig. 19.17 and involves three left handed single chain helices wrapped around with a right handed twist. A protein chain can also be joined from one cysteine residue (Fig. 19.11) to another by a disulphide linkage ($—CH_2—S—S—CH_2—$) formed by oxidative coupling of the thiol ($—SH$) groups.

In order to establish the structure of a protein it is necessary first to determine the number and kind of amino acid units present, then to determine the sequence, or order, in which they occur in the chain, and finally to determine the stereochemistry and gross structure of the

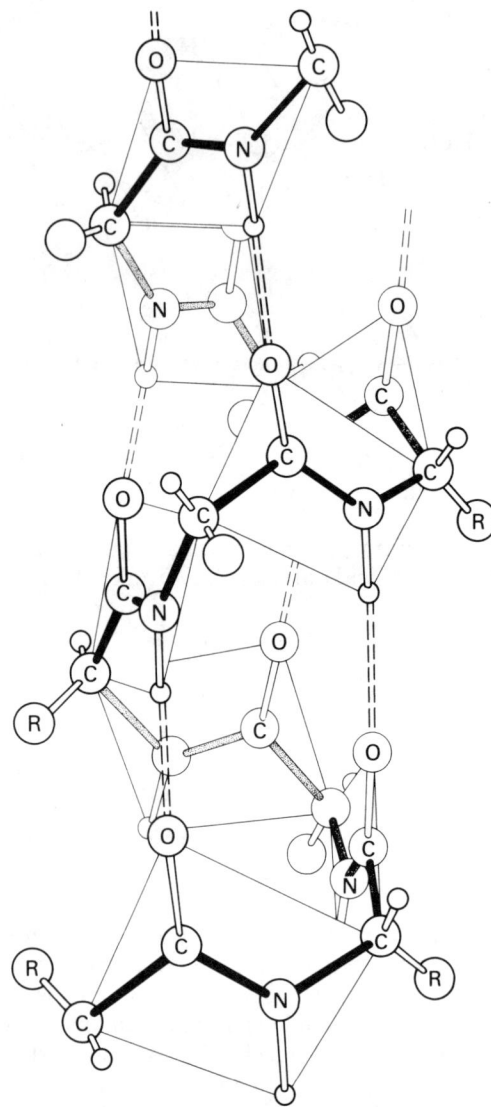

Fig. 19.15 Intramolecular hydrogen bonding in the α-helix of a protein chain

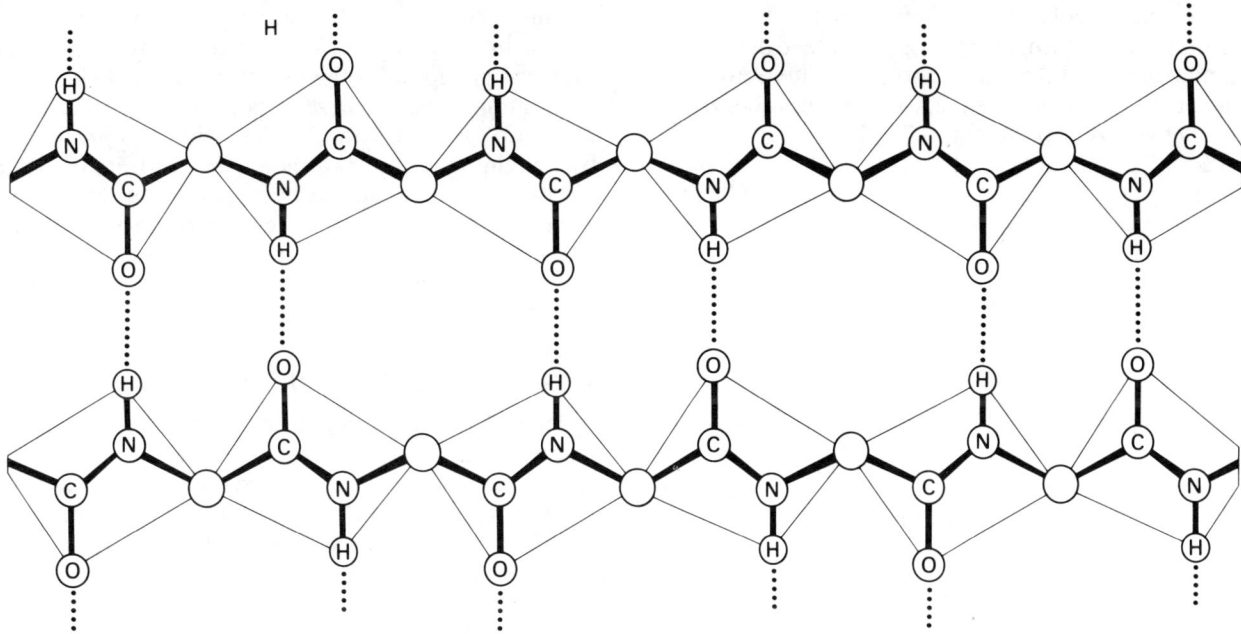

Fig. 19.16 Intermolecular hydrogen bonding between protein chains

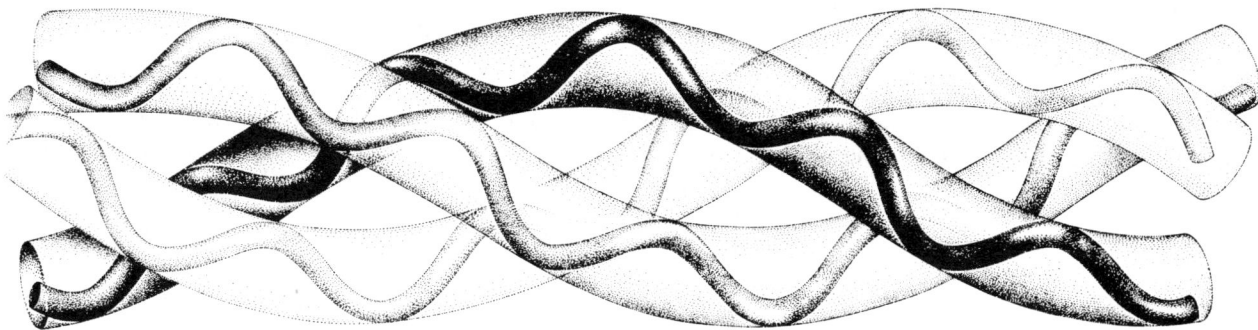

Fig. 19.17 Non-stretch connective tissue protein of ligaments

protein. The amide links of proteins are readily hydrolysed; a protein, therefore, can be broken down to give a mixture of different α-amino acids. It is then possible to determine the relative amounts of different amino acids in the protein by quantitative chromatography. It is possible to determine which amino acid is at the amino end of a protein chain by reaction of the protein with 2,4-dinitrofluorobenzene, followed by hydrolysis (reaction 19.7).

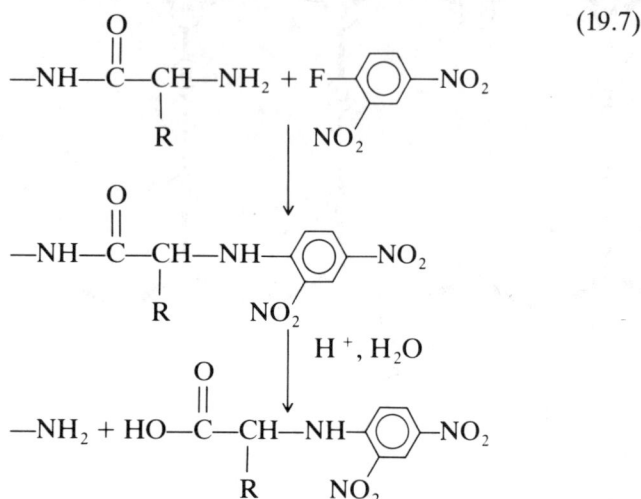

$$(19.7)$$

The amino acid which is found to contain the dinitrophenyl group can therefore be established as the terminal amino acid. It is also possible, by an analogous procedure, to determine the amino acid which is at the carboxylic acid end of the molecule.

A protein chain can be broken up into smaller polypeptide fragments by specific enzymes which cleave amide bonds between specific pairs of amino acids. The terminal groups of these smaller polypeptide chains can then be determined as above. In this way, using careful logic and making maximum use of the information gained at each stage, it is possible to determine the complete sequence of the amino acid units in a protein molecule.

The three-dimensional structure of a protein molecule, like that of any other molecule which can be obtained in a crystalline form, can be determined by X-ray diffraction, but the large number of atoms in the protein molecule makes this an enormous task. Merely obtaining suitable crystalline samples can be extremely difficult. However, steady advances are being made in this area and the structures of several important proteins are now known in detail. Before we can begin to understand the manner in which a protein is involved in the chemistry of living systems, it is necessary to know both the sequence of the amino acids in the molecule and also the overall three-dimensional structure of the

molecule. Thus, for example, many proteins function as enzymes, which as noted above can often speed up an otherwise slow reaction by a factor of several million. This is accomplished by the cooperative interaction of suitably placed functional groups within the protein, in such a way that one group holds the substrate molecule in the correct orientation while another functional group effects the desired reaction. One can visualize the action of a protein enzyme by imagining that it takes hold of a substrate molecule and operates on it, with one hand holding it down while the other hand wields the knife. Enzymes are easily denatured, i.e. rendered inactive, by heating, the effect of the heat being to alter the gross shape of the molecule without breaking any covalent bonds.

19.4 Haemoglobin and Myoglobin

Two of the most important proteins in vertebrates are haemoglobin (Fig. 19.18) and myoglobin (Fig. 19.19), whose structures are now known in detail as a result of Nobel-Prize winning X-ray diffraction studies. Both of these molecules have metal ions in association with the protein *globin*. Haemoglobin in red blood corpuscles binds molecular oxygen to give oxyhaemoglobin, and carries it through the arteries to the tissues where oxygen is required. In the tissues oxyhaemoglobin gives up its oxygen to myoglobin. Myoglobin has a high affinity for oxygen under the conditions in which oxyhaemoglobin is giving up its supply. Myoglobin contains a polypeptide chain of molecular weight 17 800 which can be regarded as a protective sheath for a heme group. The eight right handed helices of the molecule are folded back upon themselves to form a pocket which contains the heme molecule. The iron atom of the heme group of both oxyhaemoglobin and oxymyoglobin is octahedrally coordinated, as it usually is in inorganic complexes. (Its six nearest neighbour atoms are at the corners of an octahedron and each donates an electron pair to the orbitals surrounding the iron.) In the deoxy forms the octahedral site normally occupied by oxygen is vacant and the iron atoms have a somewhat less common five-fold coordination.

In Chapter 15 we examined some important aspects of the chemistry of benzene and its derivatives. One or more of the carbon atoms in benzene can be replaced by a hetero atom as for example in pyridine (Fig. 19.20), where a carbon

Fig. 19.20 Pyridine

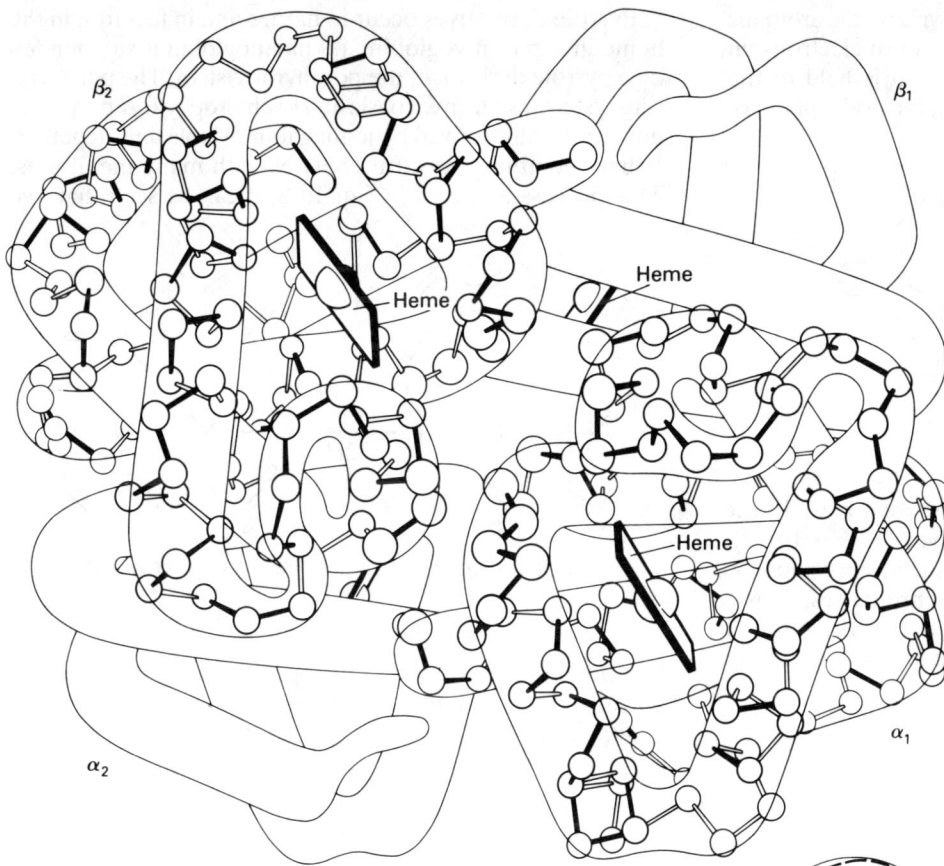

Fig. 19.18 Haemoglobin

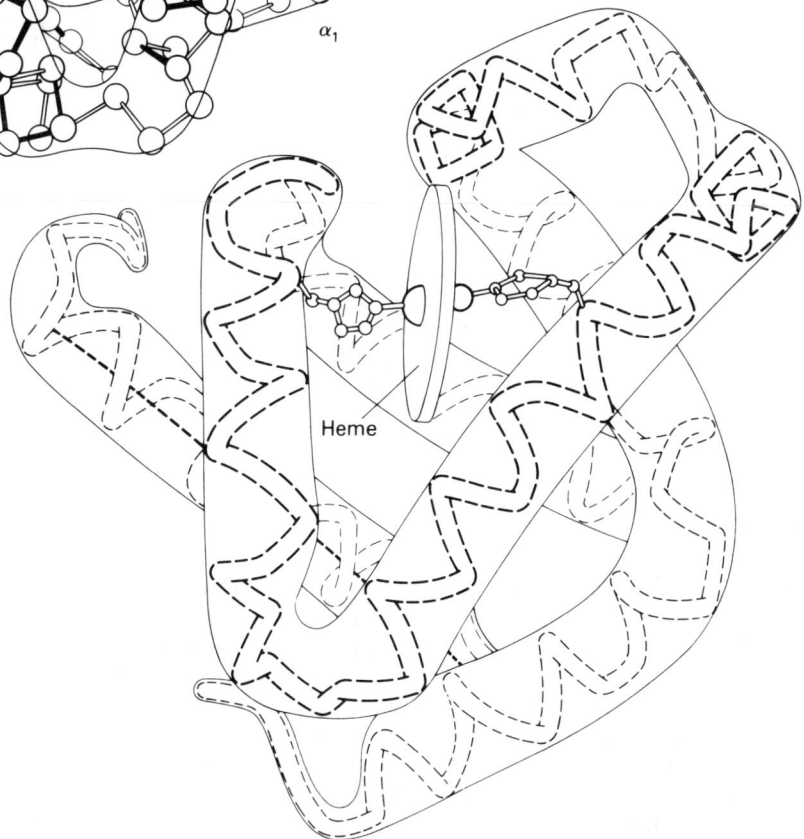

Fig. 19.19 Myoglobin

atom has been replaced by nitrogen. In pyridine the aromatic nature of the ring is retained. The lone pair of electrons on the nitrogen atom retains its identity, and is held in the plane of the molecule (Fig. 19.21) i.e. in the nodal plane of

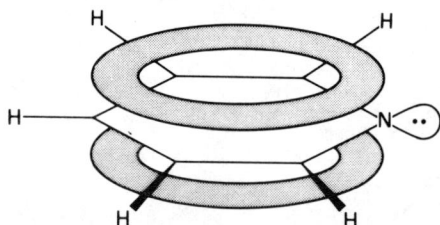

Fig. 19.21 Pyridine showing electron density

the π electron system. Pyridine is therefore a base, in the same way as a normal amine is a base. Pyrrole, (Fig. 19.22), is a five-membered unsaturated nitrogen containing compound that does not have any significant basic properties. This is because the lone pair of electrons on the nitrogen atom of pyrrole (Fig. 19.23) is involved with the four π electrons of

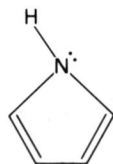

Fig. 19.22 Pyrrole

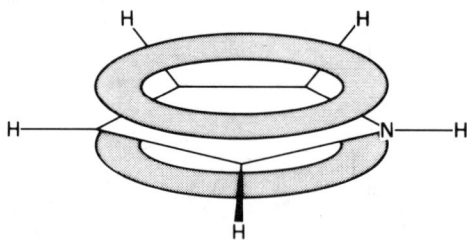

Fig. 19.23 Pyrrole showing electron density

the diene in forming a set of molecular orbitals rather similar to those of benzene, and also containing six electrons. The lone pair of electrons on the nitrogen atom is not free to bond with hydrogen ions, and pyrrole is therefore only weakly basic, undergoing reactions which resemble those of benzene, rather than those of 1,3-butadiene and of a normal amine. (This illustrates the pronounced effect that one functional group can have on another when they are in close proximity.)

Pyrrole derivatives occur in nature and in fact iron in the heme group of myoglobin and haemoglobin is surrounded by a pyrrole derivative, the porphyrin system. The porphyrin ring system also forms a basic unit of chlorophyll. A porphyrin unit is formally derived by joining four pyrrole units together. It has the ability to form chelates with many metal ions. This process is shown in Eq. 19.8; it can be seen that two

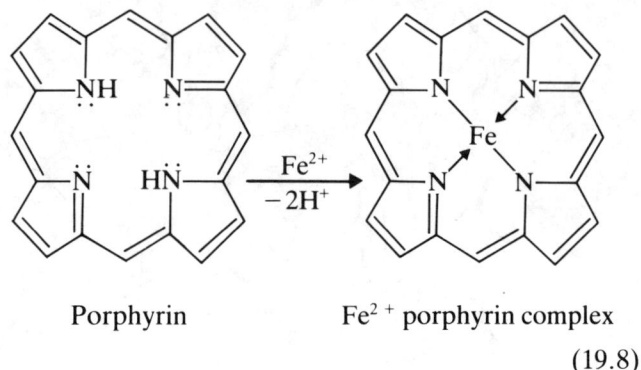

Porphyrin Fe^{2+} porphyrin complex

$$(19.8)$$

protons are lost from the nitrogen atoms in forming the chelate complex. It is the iron-porphyrin complex or 'heme' group (Fig. 19.24) in haemoglobin that gives blood its red

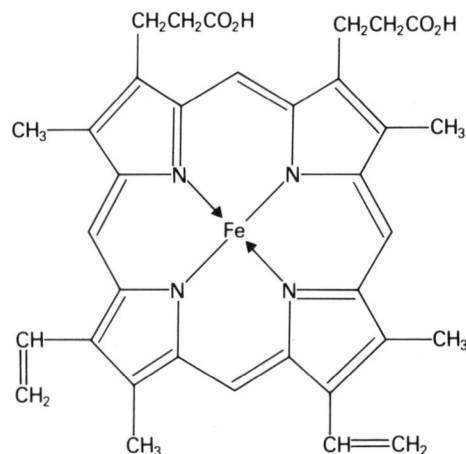

Fig. 19.24 Heme group

colour. Each haemoglobin molecule contains four heme groups enfolded in one of the four chains of amino acid units that constitutes globin. The four chains of globin consist of two identical pairs known as α and β chains. The heme group is planar, and at its centre is an iron atom in the +2 (ferrous) oxidation state. The oxygen molecule is bound reversibly to the ferrous iron. This ferrous iron is thought to acquire its capacity for bonding molecular oxygen through the combination with heme and globin. Chemists have been interested to determine the structure of haemoglobin, and to understand the factors which allow the Fe^{2+} ion to bind oxygen reversibly without producing permanent oxidation of the iron to the ferric state. The gross molecular structure of haemoglobin has been determined by X-ray diffraction techniques, but the orientation of the molecular oxygen in the molecule is not yet known for certain. The precise structure of the slightly less complicated but closely related oxygen containing molecule, myoglobin is known, but only for the case when it is not bound with oxygen. The

role of the protein in these systems appears to be to provide protection for the potentially unstable FeO_2 entity.

Recently a number of 'model' compounds have been synthesized which behave in a similar manner to haemoglobin and myoglobin as far as the ability to coordinate oxygen is concerned. A synthetic compound which exhibits similar chemical properties to natural myoglobin has been prepared and the protein part of the molecule replaced by an imidazole ring (imidazole is a five-membered ring compound containing two nitrogen atoms). This molecule has been prepared from chlorophyll *a* by reactions which in the main have been met in earlier sections of this book (cf. sequence 19.9). The first step is the most complicated, among other things involving

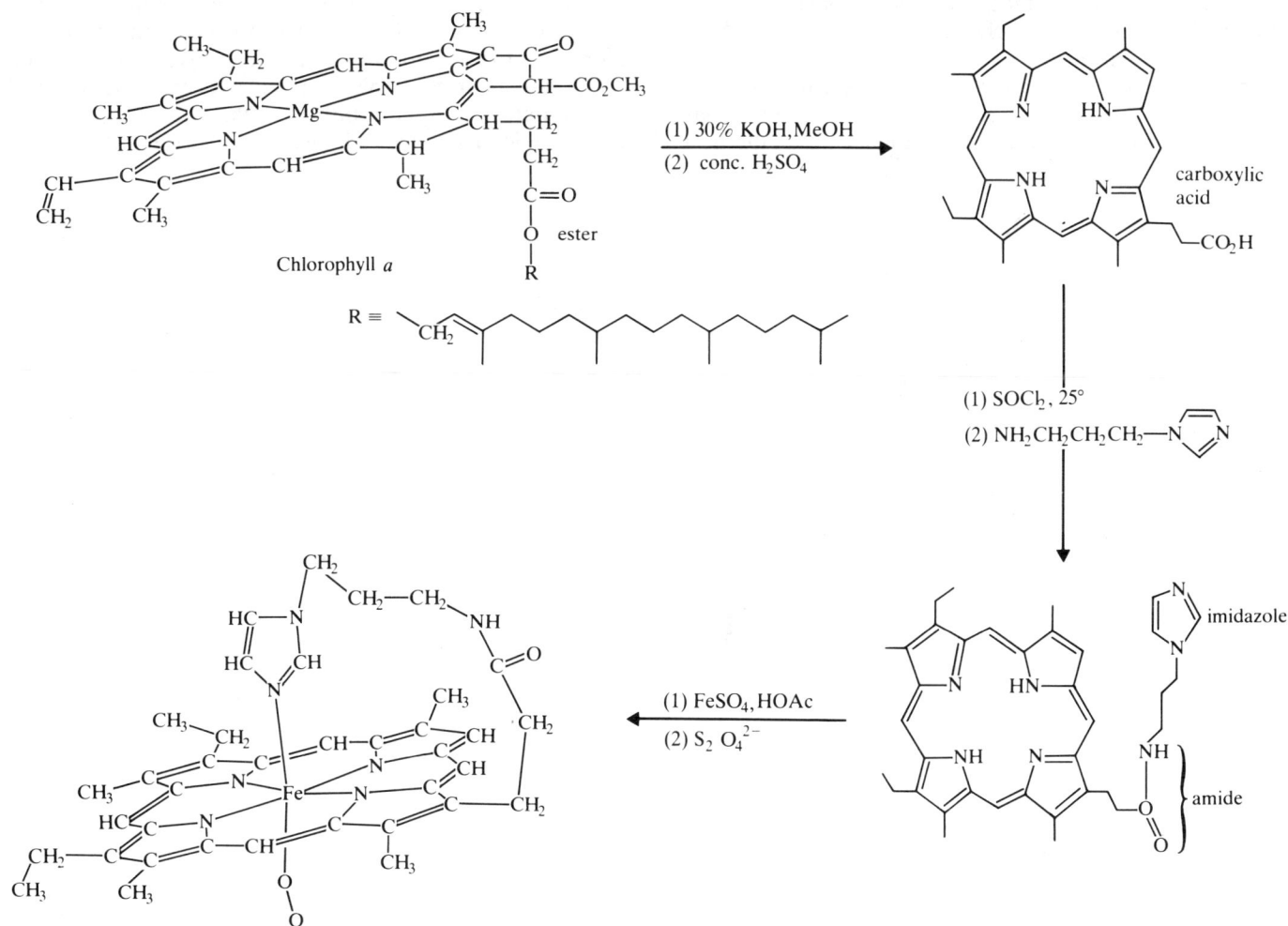

(19.9)

hydrolysis of an ester group to a carboxylic acid. The carboxylic acid is converted to the acid chloride (reaction 19.10)

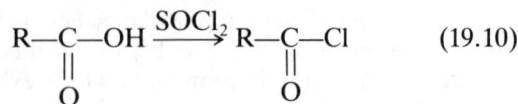

$$R—\underset{\underset{O}{\|}}{C}—OH \xrightarrow{SOCl_2} R—\underset{\underset{O}{\|}}{C}—Cl \qquad (19.10)$$

with thionyl chloride, and the chloride atom is then displaced by a reaction with a primary amine (reaction 19.11) to give

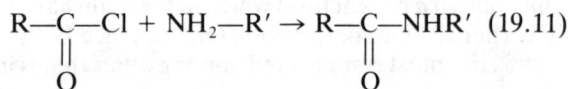

$$R—\underset{\underset{O}{\|}}{C}—Cl + NH_2—R' \rightarrow R—\underset{\underset{O}{\|}}{C}—NHR' \quad (19.11)$$

an amide. Ferrous ion is then introduced into the molecule, and the product molecule binds reversibly with molecular oxygen at $-40°C$. This is the first indication that the complex protein ligand is not entirely necessary for iron to bind molecular oxygen reversibly. In another study ferrous ion has been chelated by a modified porphyrin ring and complexed with 1-methylimidazole. The resulting complex has a structure which resembles a picket fence about the iron atom, and creates a cavity within which molecular oxygen binds (Fig. 19.25). This complex is stable at room temperature and it has been possible, by X-ray diffraction, to determine the stereochemical relationship of the molecular oxygen to the iron. This stereochemistry has been a matter of controversy for many years. It is now fairly well established that the Fe—O—O system is bent at about 120° in haemoglobin, but the position of the Fe—O—O triangle relative to the porphyrin group is still uncertain.

The availability of synthetic oxygen-carrying compounds could possibly lead to the development of improved artificial lung machines; more important, it should lead to a better understanding of the mechanism of oxygen transport in mammals.

19.5 Nucleic Acids

One of the basic questions of life is the mechanism of heredity, i.e. the manner in which genetic information is stored and transmitted within an organism, and between parent and offspring. This problem has been partially solved by the determination of the molecular structure of deoxyribonucleic acid. Deoxyribonucleic acid (DNA) is found

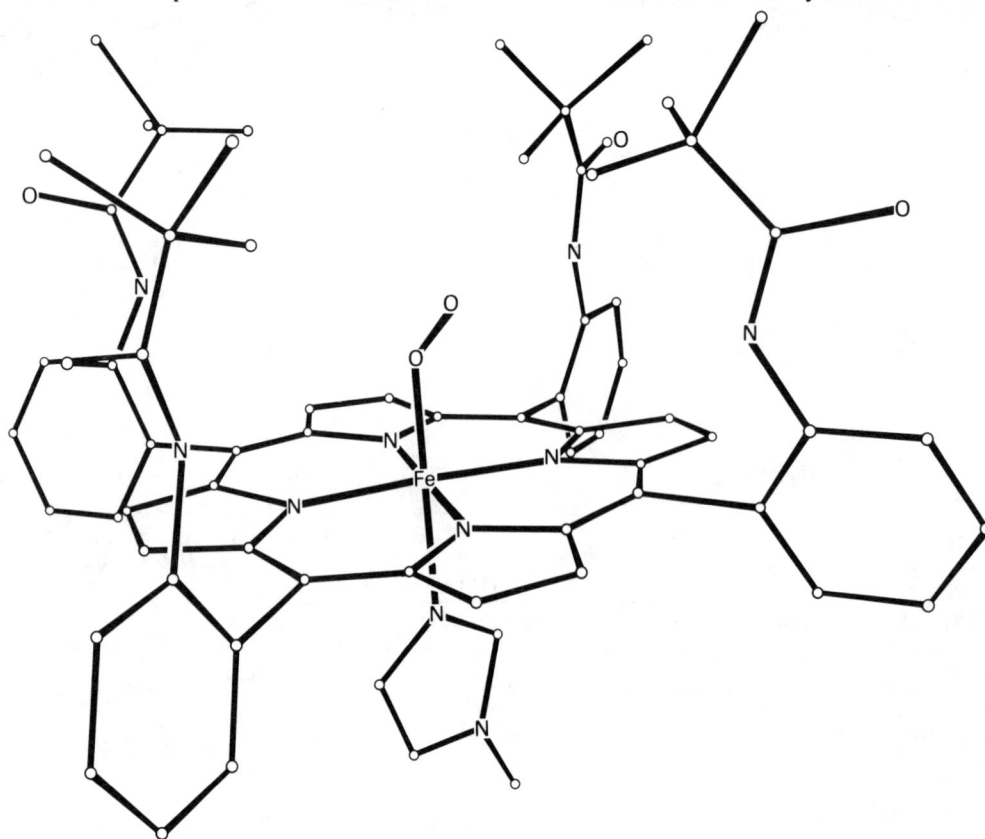

Fig. 19.25 Three dimensional structure of ferrous complex containing molecular oxygen

with ribonucleic acid (RNA) in cells, and both are essential for the biosynthesis of proteins. Both compounds are long chain proteins, known as nucleoproteins. They are, however, chemically quite distinct from proteins, and are made up from nucleic acid units (Fig. 19.26). The backbone of the

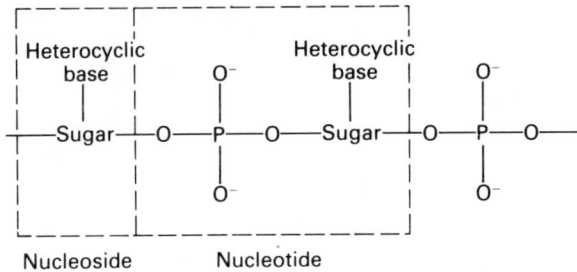

Fig. 19.26 Nucleic acid

molecule is a polyester chain. The acid part of the ester is phosphoric acid, and the alcohol part is a sugar. Each sugar is linked (at C (1) through a β-linkage) to a heterocyclic base. Mild degradation of a nucleic acid yields a mixture of acids known as *nucleotides,* each containing one phosphate, one sugar and one base unit. The phosphate can be removed

to give a *nucleoside* containing one sugar–base unit. Complete hydrolysis gives phosphate, sugar and bases. In RNA the sugar is D-ribose and in DNA it is D-2-deoxyribose. The bases of DNA are adenine, guanine, cytosine and thymine (Fig. 19.27) and those of RNA are adenine, guanine, cytosine and uracil (Fig. 19.28). A segment of a DNA chain is shown in Fig. 19.29. It consists of D-2-deoxyribose units combined with bases and linked by phosphate units. The sequence of the bases in the chain differs from one nucleic acid to another, and provides the means by which genetic information is stored. The 'genetic code' relates the base sequence to the nature of the proteins whose synthesis a particular DNA strand controls. The DNA molecule consists of two complementary chains which are thought to be wound about each other to form a helix or spiral. Both spirals are right handed and both have ten nucleotides per turn. An alternative configuration for DNA has recently been proposed which involves 'side by side' meshing of the complementary chains rather than intertwining of them. The chains are held together by hydrogen bonding between base pairs. Adenine of one chain is hydrogen bonded to thymine of the opposite chain (Fig. 19.30), and guanine is bonded similarly to cytosine (Fig. 19.31). When the spiral splits during cell division each chain can act as a template to guide the synthesis of a matching chain (Fig. 19.32).

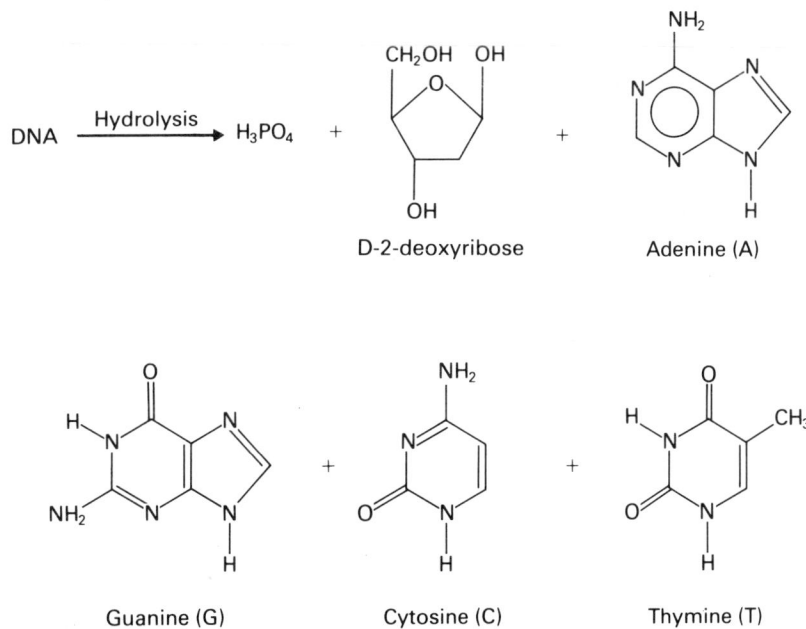

Fig. 19.27 Hydrolysis of DNA

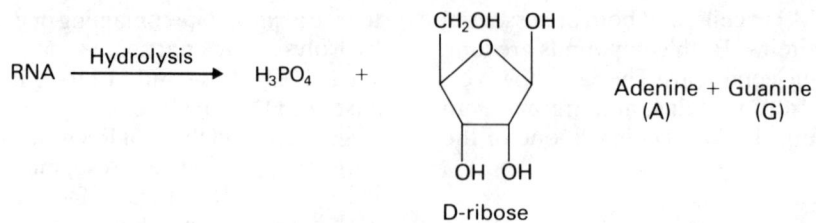

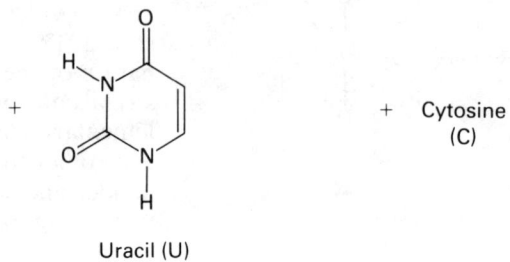

Fig. 19.28 Hydrolysis of RNA

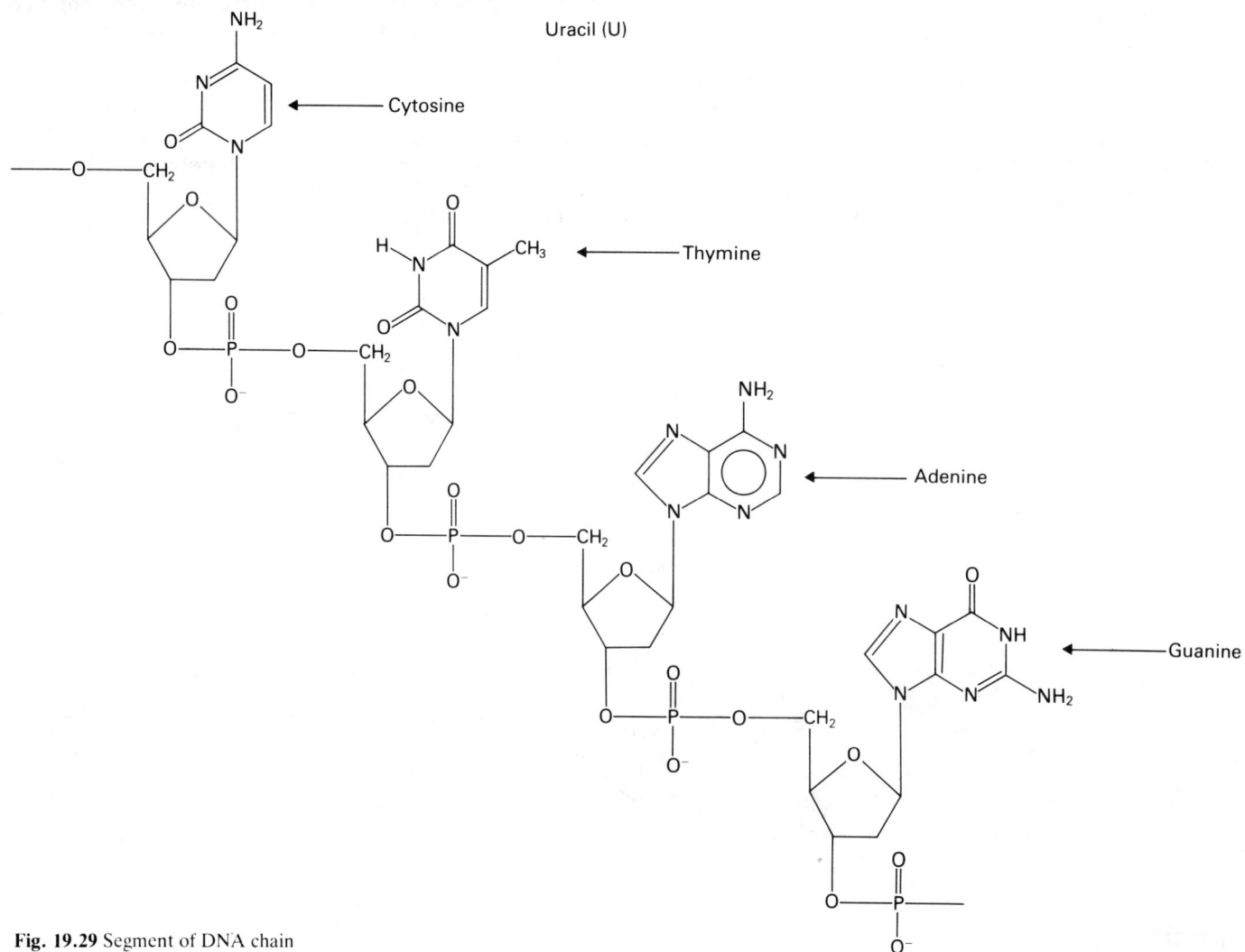

Fig. 19.29 Segment of DNA chain

In a biological system, the DNA molecule can duplicate itself, and can control the synthesis of the specific proteins that are characteristic of each organism. It is the facility for hydrogen bonding between specific base pairs that is responsible for the accuracy of self-duplication of DNA. After the two strands of the DNA molecule dissociate (an 'un-zipping' process) free nucleotides form hydrogen bonds with the complementary nucleotides of the separated strand. The nucleotides held by hydrogen bonds to the DNA strand are then polymerized by an enzyme catalyst to form the complementary nucleic acid chain. As noted above, the sequence of base pairs of DNA control the synthesis of specific proteins in an organism. The process involves the synthesis of RNA using DNA as a template. As the DNA unzips, strands of RNA are formed in the same manner as when DNA reproduces itself. However, the new chains contain ribose instead of deoxyribose and have a different base sequence, which is determined by the base sequence of the DNA chain. Opposite each adenine of DNA a uracil appears in the RNA strand, opposite each guanine is cytosine, opposite each thymine is adenine, and opposite each cytosine is guanine. Such a molecule of RNA, termed messenger RNA, takes the information from DNA to the ribosome where the protein is synthesized. The messenger RNA is then able to call up transport RNA molecules, each of which transports an amino acid to the site of protein synthesis. The order in which the amino acids are built into a protein chain depends on the sequence of bases of the messenger RNA. Three base units are required to call up a specific amino acid. There are 64 possible combinations or three base units (codons) and only 20 amino acid units, so that the same amino acid can be called up by more than one codon. Current work in this field includes studies aimed at relating specific codons to specific amino acid units, at understanding the details of protein synthesis in the ribosomes, and at understanding the control mechanisms which govern whether or not a particular DNA molecule is active in protein synthesis.

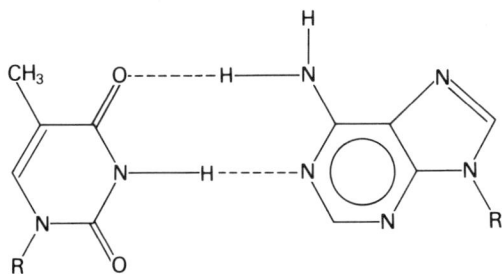

Fig. 19.31 Cytosine – guanine hydrogen bonding

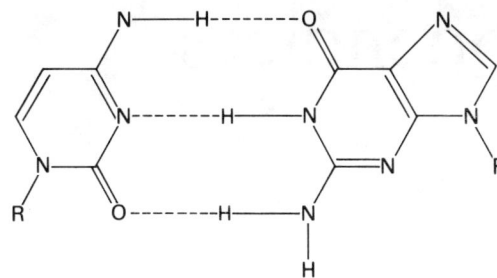

Fig. 19.32 Replication of DNA

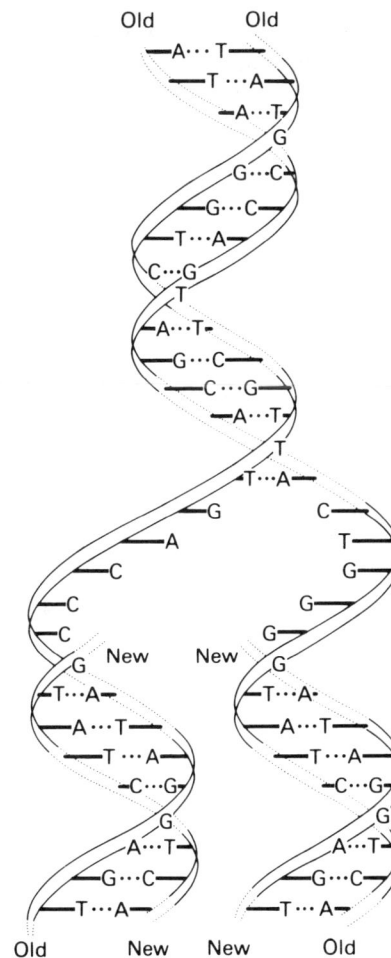

Fig. 19.30 Thymine – adenine hydrogen bonding

Appendices

I Fundamental constants and conversion factors

c Speed of light $2.997\,925 \times 10^8 \text{ m s}^{-1}$
h Planck's constant $6.626\,18 \times 10^{-34} \text{ J s}$
e Charge on proton $1.602\,19 \times 10^{-19} \text{ C}$
N_0 Avogadro's number $6.022\,05 \times 10^{23} \text{ mol}^{-1}$
F Faraday $9.648\,46 \times 10^4 \text{ C mol}^{-1}$
R Gas constant $8.314\,41 \text{ J K}^{-1} \text{ mol}^{-1}$
k Boltzmann's constant $1.380\,66 \times 10^{-231} \text{ J K}^{-1}$
μ_0 Permeability of vacuum $4\pi \times 10^{-7} \text{ J s}^2 \text{ C}^{-2} \text{ m}^{-1}$

Pressure: 1 atmosphere = 760 torr = 760 mm Hg = 101.325 kPa = $1.013\,25 \times 10^5$ N m^{-2}; 1 torr = 133.322 Pa.

Gas constant: $R = 8.314\,41$ J K^{-1} mol^{-1} = 1.98717 cal K^{-1} mol^{-1} = $0.082\,0575$ litre atm K^{-1} mol^{-1}.

Energy: 1 calorie = 4.184 J (Note: the 'Calorie' used in nutrition is actually the kilocalorie); 1 eV (electron-volt) = $1.602\,189 \times 10^{-19}$ J = 96.486 kJ mol^{-1} = 23.060 kcal mol^{-1} = 8065.5 cm^{-1}.

Mass: Atomic mass unit = $1.660\,57 \times 10^{-27}$ kg; electron mass, $m_e = 9.109\,53 \times 10^{-31}$ kg; proton mass, $m_p = 1.672\,65 \times 10^{-27}$ kg; neutron mass, $m_n = 1.674\,95 \times 10^{-27}$ kg.

II Electron configurations of the elements

Atomic number	Element	Electron configuration	Atomic number	Element	Electron configuration
1	H	$1s$	19	K	$[\text{Ar}]\,4s$
2	He	$1s^2$	20	Ca	$—4s^2$
3	Li	$[\text{He}]\,2s$	21	Sc	$—3d4s^2$
4	Be	$—2s^2$	22	Ti	$—3d^24s^2$
5	B	$—2s^22p$	23	V	$—3d^34s^2$
6	C	$—2s^22p^2$	24	Cr	$—3d^54s$
7	N	$—2s^22p^3$	25	Mn	$—3d^54s^2$
8	O	$—2s^22p^4$	26	Fe	$—3d^64s^2$
9	F	$—2s^22p^5$	27	Co	$—3d^74s^2$
10	Ne	$—2s^22p^6$	28	Ni	$—3d^84s^2$
11	Na	$[\text{Ne}]3s$	29	Cu	$—3d^{10}4s$
12	Mg	$—3s^2$	30	Zn	$—3d^{10}4s^2$
13	Al	$—3s^23p$	31	Ga	$—3d^{10}4s^24p$
14	Si	$—3s^23p^2$	32	Ge	$—3d^{10}4s^24p^2$
15	P	$—3s^23p^3$	33	As	$—3d^{10}4s^24p^3$
16	S	$—3s^23p^4$	34	Se	$—3d^{10}4s^24p^4$
17	Cl	$—3s^23p^5$	35	Br	$—3d^{10}4s^24p^5$
18	Ar	$—3s^23p^6$	36	Kr	$—3d^{10}4s^24p^6$

II Electron configurations of the elements (continued)

Atomic number	Element	Electron configuration	Atomic number	Element	Electron configuration
37	Rb	[Kr] $5s$	71	Lu	$—4f^{14}5d6s^2$
38	Sr	$—5s^2$	72	Hf	$—4f^{14}5d^26s^2$
39	Y	$—4d5s^2$	73	Ta	$—4f^{14}5d^36s^2$
40	Zr	$—4d^25s^2$	74	W	$—4f^{14}5d^46s^2$
41	Nb	$—4d^45s$	75	Re	$—4f^{14}5d^56s^2$
42	Mo	$—4d^55s$	76	Os	$—4f^{14}5d^66s^2$
43	Tc	$—4d^55s^2$	77	Ir	$—4f^{14}5d^76s^2$
44	Ru	$—4d^75s$	68	Pt	$—4f^{14}5d^96s$
45	Rh	$—4d^85s$	79	Au	$—4f^{14}5d^{10}6s$
46	Pd	$—4d^{10}$	80	Hg	$—4f^{14}5d^{10}6s^2$
47	Ag	$—4d^{10}5s$	81	Tl	$—4f^{14}5d^{10}6s^26p$
48	Cd	$—4d^{10}5s^2$	82	Pb	$—4f^{14}5d^{10}6s^26p^2$
49	In	$—4d^{10}5s^25p$	83	Bi	$—4f^{14}5d^{10}6s^26p^3$
50	Sn	$—4d^{10}5s^25p^2$	84	Po	$—4f^{14}5d^{10}6s^26p^4$
51	Sb	$—4d^{10}5s^25p^3$	85	At	$—4f^{14}5d^{10}6s^26p^5$
52	Te	$—4d^{10}5s^25p^4$	86	Rn	$—4f^{14}5d^{10}6s^26p^6$
53	I	$—4d^{10}5s^25p^5$	87	Fr	[Rn] $7s$
54	Xe	$—4d^{10}5s^25p^6$	88	Ra	$—7s^2$
55	Cs	[Xe] $6s$	89	Ac	$—6d7s^2$
56	Ba	$—6s^2$	90	Th	$—6d^27s^2$
57	La	$—5d6s^2$	91	Pa	$—5f^26d7s^2$
58	Ce	$—4f^26s^2$	92	U	$—5f^36d7s^2$
59	Pr	$—4f^36s^2$	93	Np	$—5f^46d7s^2$
60	Nd	$—4f^46s^2$	94	Pu	$—5f^67s^2$
61	Pm	$—4f^56s^2$	95	Am	$—5f^77s^2$
62	Sm	$—4f^66s^2$	96	Cm	$—5f^76d7s^2$
63	Eu	$—4f^76s^2$	97	Bk	$—5f^97s^2$
64	Gd	$—4f^75d6s^2$	98	Cf	$—5f^{10}7s^2$
65	Tb	$—4f^96s^2$	99	Es	$—5f^{11}7s^2$
66	Dy	$—4f^{10}6s^2$	100	Fm	$—5f^{12}7s^2$
67	Ho	$—4f^{11}6s^2$	101	Md	$—5f^{13}7s^2$
68	Er	$—4f^{12}6s^2$	102	No	$—5f^{14}7s^2$
69	Tm	$—4f^{13}6s^2$	103	Lw	$—5f^{14}6d7s^2$
70	Yb	$—4f^{14}6s^2$	104	Unq	$—5f^{14}6d^27s^2$

Answers

Chapter 1

1.1 $\dfrac{96\,847}{6.0225 \times 10^{23}} = 1.608 \times 10^{-19}\,\text{C}$

1.2 $\dfrac{1}{1840}\,g = 5.435 \times 10^{-7}\,\text{kg}$

1.3 $\dfrac{6.02 \times 10^{23} \times 6.627 \times 10^{-34} \times 3 \times 10^{8}}{500 \times 10^{-9}} = 239.4\,\text{kJ}$

1.4 $m = E/C^2 = 239.4 \times 10^{3}/(3 \times 10^{8})^2 = 2.66 \times 10^{-12}\,\text{kg}$

1.5 $6.02 \times 10^{23} \times 1 \times 10^{-10} = 6 \times 10^{13}\,\text{kg} = 6 \times 10^{10}\,\text{tonnes}$

1.6 $\lambda = 6.627 \times 10^{-34}/\text{mass} \times 10^{-3}$; wavelengths are $0.734\,\text{m}, 2.2 \times 10^{-5}\,\text{m}, 6.63 \times 10^{-21}\,\text{m}$, and $9.1 \times 10^{-32}\,\text{m}$. (Masses must be expressed in kilograms, since h is given in J s.)

1.7 Substitute the expression for Ψ into the one-dimensional Schroedinger equation, Eq. 1.10, and verify that the equation is satisfied.

1.8 We require $\Psi = 0$ at $x = 0$; hence the cosine term in Ψ disappears. We also require $\Psi = 0$ at $x = a$; hence the argument of the sine term must be of the form $2\pi nx/a$, where n is a positive whole number. We therefore have $2\pi nx/a = 2mvx/h$ so that $m^2v^2 = n^2h^2/a^2$, and the kinetic energy $= \frac{1}{2}mv^2 = n^2h^2/2ma^2$. Since there is no potential energy, this is the required formula for the energy levels, E_n. The first level E_1 has $n = 1$, the second $n = 2$, and so on.

1.9 The kinetic energy is $\frac{1}{2}mv^2$, which is $1.5RT$ per mole. The mass per mole, m is 32 g, or 0.032 kg. Hence

$$v^2 = \frac{3 \times 8.314 \times 300}{0.032}$$

and $v_{rms} = 484\,\text{m s}^{-1}$

Chapter 2

2.1 Use Eq. 2.2 with $h = 6.627 \times 10^{-34}$ J s, and $p = mv$ where $m = 9.1 \times 10^{-31}$ kg and $v = 10^{-3}$ ms^{-1}. The uncertainty in p is $10^{-2} \times 9.11 \times 10^{-34}$ kg m s^{-1}. Hence

$\Delta x \geq \dfrac{6.627 \times 10^{-34}}{2\pi \times 9.1 \times 10^{-34}} = 0.116\,\text{m} = 11.6\,\text{cm}$

2.2 Use Eq. 2.7 with $n_2 = 2$ and $n_1 = 1$. The result is $\lambda = 121.6\,\text{nm}$.

2.3 From Eq. 2.6, the energy levels are proportional to Z^2. Hence for He$^+$, with $Z = 2$, the transition will have four times the frequency, i.e $\lambda = 121.6/4 = 30.4\,\text{nm}$.

2.4 The angular momentum is given by mvr where r is the radius of a circular orbit. Bohr set $mvr = nh/2\pi$. Using $\lambda = h/mv$, this gives $n\lambda = 2\pi r$, the circumference of the orbit. Thus the orbit contains an integral number of wavelengths, i.e. the waves must be standing waves.

2.5 (a) $L = 1$ only; (b) $L = 2, 1$ or 0; (c) $L = 4, 3, 2, 1$ or 0; (d) $L = 3, 2$, or 1.

2.6 (a) -1(rule 5); (b) $+6$ (rules 3, 5 and 7); (c) $+7$ (rules 5 and 7); (d) $+4\frac{1}{2}$ (rules 5 and 7). This implies that two of the I atoms are in the $+5$ state and two in the $+4$ state. (e) $+2\frac{2}{3}$. As in (d) with two Fe^{3+} ions and one Fe^{2+} ion per Fe$_3$O$_4$ unit.

2.8 If the atom is ionized in the upper state n_1 must be infinite, i.e. $1/n_1$ is zero. Putting $n_2 = 1$ in Eq. 2.7 gives $\lambda = 91.2\,\text{nm}$ as the wavelength corresponding to a photon which can just ionize H.

2.9 The surface area of a sphere of radius r is $4\pi r^2$, so the volume of a spherical shell of thickness δr is $4\pi r^2\,\delta r$. The probability δP of finding the electron between r and $r + \delta r$ is therefore $\delta P = 4\pi r^2\Psi^2\delta r$, and the probability per unit δr is $P = \delta p/\delta r = 4\pi r^2\,(\pi\alpha^3)^{-1}\text{e}^{-2r/\alpha}$. Differentiating this with respect to r and setting $dP/dr = 0$ gives $r = \alpha$ at the most probable distance.

Chapter 3

3.1 4.803 debye corresponds to a distance of 100 pm and a charge of 1 electron. Hence 1.85 debye at 91.7 pm corresponds to a charge of

$$\frac{1.85}{4.803} \times \frac{100}{91.7} = 0.42 \text{ electron.}$$

3.2 An energy decrease can be predicted directly from the effect of increasing the value of a in the energy formula.

3.3 Three hydrogen atoms in a plane can be bonded through the sp^2 hybrids. Two other atoms are held by overlap of their 1s orbitals with the p orbital lobes above and below the plane of the sp^2 hybrids, to form a three-centre molecular orbital that holds two electrons.

3.4 (a) LiH^{2+} is isoelectronic with H_2, but there is a significant dipole moment because the molecular orbitals are concentrated around the lithium nucleus with its higher positive charge. (b) CO is isoelectronic with N_2. Perhaps surprisingly, it has only a very small dipole moment (0.2 debye). (c) NF is isoelectronic with O_2, and so has an analogous triplet ground state.

Chapter 4

4.1 Use $w = P\Delta V$; $P = 101.3$ kN m^{-2} = 1.01×10^5 N m^{-2}, $\Delta V = 2 l = 2 \times 10^{-3}$ m^3. Hence $w = 202.6$ J. $q = 250$ J, so $\Delta U = +47.4$ J $= C_v \Delta T$. Hence $\Delta T = 3.8$ K.

4.2 It is best to calculate w first. $w = $ force $\times$ distance $= mgh = 10^4 \times 9.81 \times 10^{-2} = 981$ J. As well as heating the bar to 250°C, the heat source also has to supply 981 J of energy to raise the heavy weight. To merely heat the bar requires $C\Delta T = 10^4 \times 1.1 \times 230 = 2.530 \times 10^6$ J. This is ΔU, and $q = \Delta U + w = 2.531 \times 10^6$ J.

4.3 The power dissipated in the resistance R is $EI = E^2/R = I^2R$. The energy dissipated in 10 minutes is $I^2Rt = 10^{-6} \times (5 \times 10^3) \times (10 \times 60) = 3$ J. This is $w = -\Delta U$. No temperature change is involved here: stored chemical energy is converted to external heat.

4.4 1.5 J $= C\Delta T$; $\Delta T = 1.5/50 = 3 \times 10^{-2}$ K.

4.5 $-w = 5 \times 10^3 \times 9.81 \times 4 \times 10^{-2} = 1.962$ J $= \Delta U$. $\Delta U = C\Delta T$; $\Delta T = 1.962/(700 \times 1.1) = 2.55 \times 10^{-3}$ K.

4.6 $\Delta H = \Delta U + \Delta(PV) = \Delta U + \Delta n (RT)$ where $\Delta n = -2$. Hence $\Delta U = -92.38 + 2 \times 8.314 \times 298 \times 10^{-3} = -87.42$ kJ.

4.7 For the reaction $C_6H_6(l) + 7\frac{1}{2}O_2(g) \rightarrow 6CO_2(g) + 3H_2O(l)$ Table 4.1 gives $\Delta H = -3.268 \times 10^3$ kJ. Δn is -1.5, so $\Delta U = \Delta H - \Delta n (RT) = 3.271 \times 10^3$ kJ $= q$.

4.8 For $C_2H_2(g) \rightarrow 2C(g) + 2H(g)$, $\Delta H = 1.646$ kJ. From Table 4.2 we would predict $\Delta H = 817 + 2 \times 413 = 1.643$ kJ.

4.9 For $2NH_3(g) + 1\frac{1}{2}O_2(g) \rightarrow N_2 + 3H_2O$, Table 4.1 gives $\Delta H = 3 \times (-241.8) - 2 \times (-46.19) = -633.0$ kJ. Table 4.2 gives $\Delta H = 6 \times 391 + 3 \times 247.5 - 2 \times 472.7 - 6$

$\times 463$ (where $D_{N\equiv N}$ and $D_{O=O}$ must be found from Table 4.1) $= -634.9$ kJ.

4.10 (a) -315.6 kJ; (b) -41.8 kJ; (c) -191.0 kJ; (d) -198.8 kJ, (e) -597.2 kJ; (f) -518.5 kJ.

4.11 The reaction $2H_2(g) + O_2(g) \rightarrow 2H_2O(g)$ has $\Delta H = -483.6$ kJ, from Table 4.1. This heat has to raise the products from 298 K to the flame temperature. For the products (including nitrogen) $C_p = 2 \times 47 + 4 \times 35 = 234$ J K^{-1}. Hence $\Delta T = -\Delta H/C_p = 2067$ K and $T_{\text{flame}} = 2365$ K. In practice this would over-estimate T by about 100 K because of the presence of significant amounts (ca. 5%) of H and OH in the flame gases.

4.12 From Table 4.1 we find $\Delta H = -633.02$ kJ. Because the temperature range is large we use Eq. 4.14, with $\Delta C_p = \Delta a + \Delta b T + \Delta c T^2$, and from Table 4.3 $\Delta a = 27.16$, $\Delta b = -5.072 \times 10^{-2}$, $\Delta c = 1.4754 \times 10^{-5}$. Integrating gives $\Delta H(1500) = -633.02 \times 10^3 + 27.16 (1500 - 298) - \frac{1}{2} \times 5.072 \times 10^{-2} \times (1500^2 - 298^2) + \frac{1}{3} \times 1.4754 \times 10^{-5} \times (1500^3 - 298^3)$ J $= -638.71$ kJ. Care is needed in evaluating $\Delta a, \Delta b$ and Δc and, in the final integral, not to forget the factors of $\frac{1}{2}$ and $\frac{1}{3}$ that result from integrating T and T^2.

4.13 From Table 4.1, ΔH at 298 K is -184.60 kJ. We use Eq. 4.14 with $\Delta a = -4.43$, $\Delta b = -5.687 \times 10^{-3}$, $\Delta c = 5.120 \times 10^{-6}$. Hence $\Delta H (1000) = -1.846 \times 10^5 - 4.43 \times 702 - \frac{1}{2} \times 5.687 \times 10^{-3} \times (10^6 - 298^2) + \frac{1}{3} \times 5.120 \times 10^{-6} \times (10^9 - 298^3)$ J $= -188.64$ kJ.

4.14 For an ideal monatomic gas $C_v = \frac{3}{2} R$ and $C_p = \frac{5}{2} R = 20.785$ J K^{-1} per mole. Hence ΔH between ambient air and combustion chamber is 42.19 kJ mol^{-1} and between combustion chamber and exhaust is -16.63 kJ mol^{-1}. (The high pressure in the combustion chamber does not affect H, of course).

4.15 (a) For the reversible isothermal case we use Eq. 4.27. $w = 8.314 \times 243 \times \ln 2 = 1.573$ kJ. Since it is isothermal, $\Delta U = \Delta H = 0$ and $q = w$. (b) $w = 0$ (no external pressure to work against); $\Delta H = \Delta U = 0$ also, so $q = 0$. (c) $w = P\Delta V = 101.3 \times 10^3 \times 10^{-3} \times 11.2 = 1.135$ kJ, also $\Delta U = \Delta H = 0$ and $q = w$. (d) $w = 0$ as in (b), $q = 0$ by definition, so $\Delta U = 0$, therefore $\Delta T = 0$ and $\Delta H = 0$. (e) $q = 0$, $w = P\Delta V$, where P is the fixed final pressure and $\Delta V = 11.2 \times 10^{-3}$ cubic metres. Also $w = -\Delta U = -C_v\Delta T$, where $C_v = \frac{3}{2} R = 12.471$ J K^{-1}. Now $PV = RT$ where P is the final pressure, $V = 22.4 \times 10^{-3}$ and $T = 273 + \Delta T$. We have two equations: $P \times 22.4 \times 10^{-3} = 8.314 \times (273 + \Delta T)$ and $P \times 11.2 \times 10^{-3} = -12.471 \times \Delta T$. Solving gives $P = 76.0$ kPa and $\Delta T = -68.25$ K. Hence $w = 851.2$ J, $\Delta U = -w = -851.2$ J, and $\Delta H = C_p\Delta T = -1.419$ kJ. (f) Using Eq. 4.35 with $\gamma - 1 = 0.6667$, we find the final temperature to be 172 K. Hence $\Delta T = -101$ K, $\Delta U = -1.260$ kJ $= -w$, $\Delta H = -2.099$ kJ, and $q = 0$ by definition.

Chapter 5

5.1   The boiling point elevation of 0.51°C is equal to that predicted for a one molar solution on the basis of the data in Table 5.1. Hence acetic acid is not signficantly ionized, nor appreciably dimerized in a one molar aqueous solution.

5.2 $P = x_1 P^0$; $P_{methanol} = 0.4 \times 11.83 = 4.73$ kPa; $P_{ethanol} = 0.6 \times 5.93 = 3.56$ kPa; $P_{total} = 8.29$; x(methanol) in the vapour $= 4.73/8.29 = 0.57$.

5.3 $P_{naphthalene} = 0.964 - 0.930 = 0.034$ atmospheres in the vapour. Mole fraction in condensate $= 0.034/0.964 = 0.035$.

5.4 P_{CO_2} in air $= (0.033/100) \times 101.3 = 33.4$ Pa. Hence x_{CO_2} in water $= 33.4/(1.42 \times 10^5) = 2.35 \times 10^{-4}$ and x_{CO_2} in benzene $= 33.4/(9.71 \times 10^3) = 3.44 \times 10^{-3}$. (Using 1 kg litre^{-1} and 0.90 kg litre^{-1} as the densities of water, relative molecular mass 18, and benzene, relative molecular mass 78, respectively, we find that these mole fractions correspond to concentrations of 4.23×10^{-6} and 2.98×10^{-4} mol litre^{-1}). The distribution coefficient is $k_{H_2O}/k_{C_6H_6} = 14.6$.

5.5   Yes (there is no azeotrope). The number of theoretical plates is four.

5.6   (a) P_{CO}^2/P_{CO_2}
(b) $P_{SO_2}/P_{S_2O}^2$
(c) $a_{I_3^-}/[I_2]a_{I^-} = [I_3^-]/[I_2][I^-]$ in dilute solution
(d) $1/P_{Cl_2}$
(e) $P_{NO} \cdot P_{NO_2}/[HNO_2]^2$
(f) $a_{H_3O^+} a_{OH^-} = [H_3O^+][OH^-]$ in very dilute solution.

5.7 $P_{Na} = 7.13$ atm., $P_{Na_2} = 2.87$ atm., hence $K_p = 2.87/(7.13)^2 = 5.65 \times 10^{-2}$ atm^{-1}.

5.8   Let [Ethyl Acetate] $= x$ (number of moles), $[H_2O]_x = x$, [Ethanol] $= 4 - x$, [Acetic Acid]$_x = 1 - x$, $K_C = 4 = x^2/(4 - x)(1 - x)$. Hence $x = 5.74$ or 0.930. We can discard the first answer; $x = 0.930$ and [Acetic Acid] $= 0.070$ moles. The mole fraction is then 9070/5 = 0.014.

5.9 $N_2 + 3H_2 \rightleftharpoons 2NH_3$; total pressure $= P_{N_2} + 3P_{H_2} + 2P_{NH_3}$. For 90% decomposition $P_{NH_3} = 0.01\ P$, $P_{N_2} = 0.45\ P$, and $P_{H_2} = 2.70\ P$, giving a total pressure of $3.25\ P$, where P is the unknown initial NH_3 pressure. Hence $P^2 \times 1.6 \times 10^{-4} = 0.10^2/0.45 \times (2.70)^3$ which gives $P = 7.06$ atm., and total pressure $= 3.25\ P = 22.9$ atm.

5.10   As in 5.9, $P_{NH_3} = 0.05\ P$, $P_{N_2} = 0.725\ P$, and $P_{H_2} = 1.35\ P$, where P is the initial $NH_3 + N_2$ pressure. The total pressure is now $0.05\ P + 0.725\ P + 1.35\ P = 2.125\ P$, and $P^2 \times 1.6 \times 10^{-4} = 0.05^2/0.725 \times (1.35)^3$. Hence $P = 8.76$ atm. and the total pressure is 18.61 atm. (As a further exercise, consider the effect of adding N_2 in terms of Le Chatelier's Principle).

5.11 $P_{Cl_2} + P_{BrCl} + P_{Br_2} = 1.0$; hence $P_{Cl_2} = 0.719 - P_{BrCl}$ $P_{BrCl}^2/P_{Cl_2} = 2.032$. Calculate $P_{BrCl}x$. Then $x^2/(0.719 - x) = 2.032$, giving $x = +0.563$ or -2.595. Ignore the second answer, so that $x = 0.563$ and $0.719 - x = 0.156$ atmospheres.

5.12   (a) 0.3 M; (b) 0.2 M; (c) 0.1 M; (d) 0.003 M

5.13   (a) $\gamma(Zn^{2+}) = 0.190$, $\gamma(Cl^-) = 0.660$; $\gamma \pm = 0.436$
(b) $\gamma(K^+) = 0.696 = \gamma(NO_3^-) = \gamma \pm$
(c) $\gamma(Na^+) = 0.755 = \gamma(HSO_4^-) = \gamma \pm$
(d) $\gamma(Na^+) = 0.941 = \gamma(H^+)$; $\gamma(SO_4^{2-}) = 0.784$, $\gamma \pm = 0.885$.

5.14   The pink crystallites would tend to dissolve and the green crystallites to grow, because of the difference in solubility, until all the pink material had gone.

5.15   (a) We use $K_{PbBr_2} = 9 \times 10^{-6} = a_{Pb^{2+}} a_{Br^-}^2 = 4m^3 \gamma\pm^3$ where m is the concentration in moles per litre. Putting $\gamma = 1$ gives $m = 1.31 \times 10^{-2}$, so that $I = 3.93 \times 10^{-2}$ and γ is actually 0.679, giving $m = 1.93 \times 10^{-2}$. The new γ is 0.635, and new $m = 2.06 \times 10^{-2}$. The solubility product is not given very precisely so we settle for $m = 2.1 \times 10^{-2}$.
(b) The ionic strength of 0.05 molar lead nitrate is 0.15, so $\gamma_{Pb^{2+}} = 0.270$ and $\gamma_{Br^-} = 0.721$. The activity of Pb^{2+} is $(0.05 + m) \times 0.270$ and that of Br^- is $2 \times 0.721 \times m$. Hence, by successive approximations, keeping activity coefficients constant, we find $m = 1.6 \times 10^{-2}$ molar. The ionic strength of the solution is now 0.198, so $\gamma_{Pb^{2+}} = 0.236$ and $\gamma_{Br^-} = 0.697$, and m may be adjusted slightly, to 1.7×10^{-2} molar.
(c) We begin with the activity coefficients found in the last problem, since the ionic strength is expected to be almost the same. This gives $m = 2.7 \times 10^{-2}$, so the ionic strength is actually 0.231. The new $\gamma\pm$ is 0.467, so $m = 1.31 \times 10^{-2}/0.467 = 2.8 \times 10^{-2}$ molar.

5.16 K for the reaction $Cu^{2+} + 5NH_3 \rightleftharpoons Cu(NH_3)_5^{2+}$ is equal to $K_1 \times K_2 \times K_3 \times K_4 \times K_5$, i.e. log $K = 12.6$ and $K = 4 \times 10^{12} = [Cu(NH_3)_5^{2+}]/[NH_3]^5[Cu^{2+}]$. Hence $[Cu(NH_3)_5^{2+}]/[Cu^{2+}] = 4 \times 10^7$.

5.17   pH = $pK_a - \log \gamma^2 = pK_a - 2 \times 0.509 \times (0.1)^{1/2}/(1 + (0.1)^2)$ i.e. pH = $pK_a + 0.24$ (pH units).

5.18   pH = $pK_a + \log_{10}$ (conjugate acid/base), $\log_{10}$ (acid/base) $= -0.74 = \log_{10}([NH_4^+]/[NH_3])$. Hence $[NH_4^+]/[NH_3] = 0.182$. We have, therefore, $V/(10 - V) = 0.182$, and the volume of HCl, V, works out to be 1.54 ml.

5.19   Table 5.8 shows that thymol blue, with a pK of 1.7, is quite a strong acid. To neutralize 0.5 ml of 0.05 molar indicator will require 0.25 ml of 0.1 molar base. This is the titration error. In practice it would be measured by doing a 'blank' titration, with just the indicator in 10 ml of distilled water.

5.20 $S_{vap} = \Delta H_{vap}/T$. Hence the values are: 94, 102, 98, 108, 88, 96 and 91 J mol^{-1} K^{-1}. Trouton's rule says that for all liquids the entropy of vaporization at the boiling point is about 92 J mol K^{-1}. Hydrogen bonded liquids such as H$_2$O and H$_2$SO$_4$ have anomalously high entropies of vaporization.

5.21 The reversible path is water $(-20°)$ $\overset{(1)}{\longrightarrow}$ water $(0°)$ $\overset{(2)}{\longrightarrow}$ ice $(0°)$ $\overset{(3)}{\longrightarrow}$ ice $(-20°)$. For step (1) $\Delta S_1 = +80 \ln 273/253$; for step (2) $\Delta S_2 = -6 \times 10_3/273$; for step (3) $\Delta S_3 = -40 \ln 273/253$. Hence $\Delta S = -18.9$ J K^{-1}.

5.22 The entropy change in the surroundings is $+\Delta H_f/253$. ΔH_f at 253 K must be calculated using the Kirchhoff equation. $\Delta H_f = 6 \times 10^3 + (80-40) \times (-20) = 5.2 \times 10^3$. Hence $\Delta S_{surroundings}$ is $+20.6$ J K^{-1}. (Note that the total ΔS for system plus surroundings is positive.)

5.23 $\Delta S = nR \ln (V_2/V_1) = 1.5 \times 8.315 \times \ln 2 = 8.65$ J K^{-1}. The temperature is irrelevant.

5.24 Eq. 5.103 gives $\Delta \bar{S} = 4.85$ J K^{-1}.

5.25 Hess's law, plus data from Table 5.9, gives $\Delta S = 160.4$ J K^{-1} (all substances are in their standard states).

5.26 As in 5.25, all substances being in their standard states, $\Delta S = -309.0$ J K^{-1} mol^{-1}.

5.27 Using data from Table 5.10, we get $\Delta G^{\oplus} = +130.6$ kJ for the calcium carbonate dissociation and $\Delta G^{\oplus} = -134.4$ kJ for the reaction of SO$_3$ with water.

5.28 Using Eq. 5.123 we obtain $K_a = 1.3 \times 10^{-23}$ atm, the dissociation pressure of calcite at 298 K, and $K_a = 3.6 \times 10^{23}$ mol^3 l^{-3} atm^{-1} for the reaction of SO$_3$ with water to form H$^+$ and SO$_4^{2-}$.

5.29 From Table 5.10, $\Delta G^{\oplus}$ is $+5.40$ kJ. Hence, using Eq. 5.122, K_p is 0.113 atm.

5.30 At 298 K $\Delta G^{\oplus}$ is $+130.6$ kJ, from exercise 5.27; $\Delta S^{\oplus}$, from exercise 5.25, is $+160.4$ J K^{-1}. (Hence, using Eq. 5.108, $\Delta H^{\oplus}$ is $+178$ kJ, which agrees with Table 4.1). Using Eq. 5.113 we find $\Delta G^{\oplus} = 13.06 - 0.1604 \times (1000-298) = +18.0$ kJ at 1000 K. Hence, using Eq. 5.122, K_p is 0.115 atm. at 1000 K.

Chapter 6

6.1   (a) $Cu^{2+} + Pb \rightarrow Pb^{2+} + Cu$
(b) $Cl_2 + 2I^- \rightarrow 2Cl^- + I_2$
(c) $Mn^{2+} + 2H_2O + Cl_2 \rightarrow MnO_2(s) + 4H^+ + 2Cl^-$
(d) $2MnO_4^- + 16H^+ + 10Cl^- \rightarrow 2Mn^{2+} + 8H_2O + 5Cl_2$

The $E^{\ominus}$ values are: (a) 0.48 V; (b) 0.82 V; (c) 0.13 V (in practice this reaction can be driven in reverse by heating to drive off Cl_2 and it provides a convenient laboratory preparation of chlorine gas); (d) 0.15 V.

6.2   (a) $Fe/FeSO_4(m_1)/KCl$ bridge$/Pb(NO_3)_2(m_2)/Pb$
(b) $Pt/FeSO_4(m_1), Fe_2(SO_4)_3(m_2)/KCl$ bridge$/KI(m_3), I_2(s)/Pt$
(c) $Ag, AgCl(s)/HCl(m)/Hg_2Cl_2(s), Hg(l)$
(d) $Zn/ZnCl_2(m_1)/KCl$ bridge$/HCl(m_2)/H_2(p), Pt$

6.3   (a) $E_{cell} = E^{\ominus}_{Fe^{2+}/Fe} - E^{\ominus}_{Pb^{2+}/Pb} - \dfrac{RT}{2F} \ln(a_{Fe^{2+}}/a_{Pb^{2+}})$

(b) $E_{cell} = E^{\ominus}_{Fe^{3+}/Fe^{2+}} - E^{\ominus}_{I_2/I^-} - \dfrac{RT}{F} \ln(a_{Fe^{2+}}/a_{Fe^{3+}} \, a_{I^-})$

(c) $E_{cell} = E^{\ominus}_{AgCl/Ag} - E^{\ominus}_{Hg_2Cl_2/Hg}$

(d) $E_{cell} = E^{\ominus}_{Zn^{2+}/Zn} - \dfrac{RT}{2F} \ln (P_{H_2} . a_{Zn^{2+}}/a^2_{H^+})$

6.4   (a) The half-cell reaction is
$$MnO_4^- + 4H^+ + 3e \rightarrow MnO_2(s) + 2H_2O(l)$$
Hence $E = E^{\ominus}_{MnO_4^-/MnO_2} + \dfrac{RT}{3F} \ln(a_{MnO_4^-} . a^4_{H^+})$

(b) The half-cell reaction is
$$Pb^{2+} + 2e \rightarrow Pb(s)$$
with $Pb^{2+} + 2Cl^- \rightleftharpoons PbCl_2(s)$

$E = E^{\ominus}_{Pb^{2+}/Pb} + \dfrac{RT}{2F} \ln a_{Pb^{2+}}$

where $a_{Pb^{2+}} . a^2_{Cl^-} = K_{PbCl_2}$

Hence $E = E^{\ominus}_{Pb^{2+}/Pb} + (RT/2F) \ln K_{PbCl_2} - (RT/F) \ln a_{Cl^-}$

(c)   We have $Cu^{2+} + 2e \rightarrow Cu(s)$
with $Cu^{2+} + 4NH_3 \rightleftharpoons Cu(NH_3)_4$
So $E = E^{\ominus}_{Cu^{2+}/Cu} + \dfrac{RT}{2F} \ln a_{Cu^{2+}}$

where $K = K_{Cu(NH_3)_4^{2+}} = a^4_{NH_3} . a_{Cu^{2+}}/a_{Cu(NH_3)_4^{2+}}$

$= P^4_{NH_3} . a_{Cu^{2+}}/a_{Cu(NH_3)_4^{2+}}$

So $E = E^{\ominus}_{Cu^{2+}/Cu} + \dfrac{RT}{2F} \ln K + \dfrac{RT}{2F} \ln (a_{Cu(NH_3)_4^{2+}}/P^4_{NH_3})$

(d) $E = E^{\ominus}_{Q/H_2O} - \dfrac{RT}{2F} \ln (a_{H_2O}/a_Q a^2_{H^+})$

$= E^{\ominus}_{Q/H_2O} + \dfrac{RT}{F} \ln a_{H^+}$

(since $a_Q = a_{H_2O}$). Hence $E = E^{\ominus}_{Q/H_2O} - \dfrac{2.303 \, RT}{F} . pH$

Chapter 7

7.1 (a) $t_{1/2} = 0.6931/k$, hence $k = 9.23 \times 10^{-3}$ hr^{-1} = 2.56×10^{-6} s^{-1}.

(b) $N_2O \rightarrow N_2 + \frac{1}{2}O_2$. Hence when 0.5 atm. of N_2O has decomposed, 0.5 atm. of N_2 and 0.25 atm. of O_2 have been formed, so $P = 1.25$ atm.

7.2 (a) $-d[CH_3CO_2CH_3]/dt = k[CH_3CO_2CH_3][H^+]$
(b) $[H^+]$ is constant so $k[H^+]$ amounts to a pseudo first-order rate constant whose value is $k' = 1.3 \times 10^{-4}$ s^{-1}. Hence $t_{1/2} = 0.693/k' = 5.33 \times 10^4$ s, or almost 89 minutes.

7.3 The reaction 2 Butadiene → Dimer produces a pressure decrease equal to half the pressure of butadiene that has reacted. Call the initial amount of butadiene a and the amount that has reacted x, where $x = 2\Delta P$. This gives the first three columns of the table below. The next two columns are used to test whether the reaction is first-order. We find k_1 is not constant (a graph of $\ln[a/(a-x)]$ versus t is curved), so it is not first-order. From the data in the last two columns, a graph of $1/(a-x)$ versus t is a straight line and the calculated k_2 is reasonably constant. Hence the reaction is second-order, with a mean rate constant from the table, of 1.74×10^{-4} kPa^{-1} min^{-1}. (The graph weights the values at long times more highly than those at short times and gives 1.72×10^{-4}).

t	x	$(a-x)$	$\ln a/(a-x)$	k_1	$1/(a-x)$	k_2
0	0	84.26	0.0	—	1.187×10^{-2}	—
20.78	20.00	64.26	0.2710	1.30×10^{-2}	1.556×10^{-2}	1.78×10^{-4}
49.50	35.70	48.56	0.5511	1.11×10^{-2}	2.059×10^{-2}	1.76×10^{-4}
77.57	44.58	39.68	0.7531	9.71×10^{-3}	2.520×10^{-2}	1.72×10^{-4}
103.58	50.50	33.76	0.9146	8.83×10^{-3}	2.962×10^{-2}	1.71×10^{-4}

7.4 (a) A graph of $\ln$ [sucrose] versus time is linear with negative slope 3.57×10^{-3} min^{-1}. This is the rate constant for the first-order reaction. (b) Since H_2O is present in large excess, its concentration is constant and may be incorporated in the actual second-order rate constant to give the observed first-order one.

7.5 The result for $t = \infty$ shows the size of the thiosulphate excess. Hence the initial $[C_3H_7Br] = 0.0395$ and to evaluate k we use the formula $(b-a)kt = \ln\{a(b-x)/b(a-x)\}$.

t/s	$a/(a-x)$	$b/(b-x)$	$\ln\{a(b-x)/b(a-x)\}$	$k(1 \text{ mol}^{-1}\text{s}^{-1})$
1110	1.186	1.0686	0.1044	1.65×10^{-3}
2010	1.353	1.119	0.1894	1.65×10^{-3}
3192	1.593	1.179	0.3004	1.65×10^{-3}

7.6 The rate-determining step must involve either a reaction of the methyl acetate with a proton, or a short-lived intermediate formed by such a reaction (cf. the iodine–acetone reaction). Hence we arrive at:

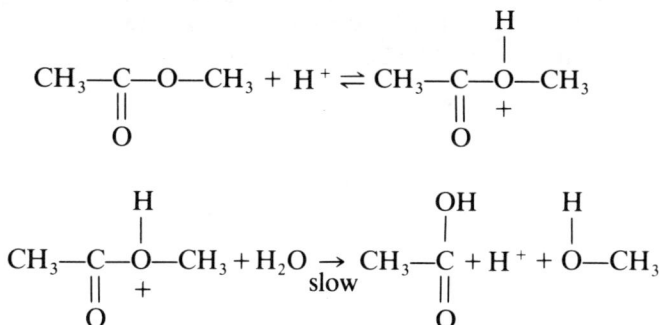

$$CH_3-\overset{\displaystyle O}{\underset{\|}{C}}-O-CH_3 + H^+ \rightleftharpoons CH_3-\overset{\displaystyle \overset{H}{|}}{\underset{\| +}{\underset{O}{C}}}-O-CH_3$$

$$CH_3-\overset{\displaystyle \overset{H}{|}}{\underset{\| +}{\underset{O}{C}}}-O-CH_3 + H_2O \xrightarrow{\text{slow}} CH_3-\overset{\displaystyle \overset{OH}{|}}{\underset{\|}{\underset{O}{C}}} + H^+ + \overset{H}{\underset{}{O}}-CH_3$$

where the H^+ comes from the H_2O molecule.

7.7 (a) $[O]_{\text{steady-state}} = k_1[O_3]/(k_2[O_3] + k_3[O_2][M])$
(b) Reaction (2) removes *two* O_3 molecules—the one in reaction (2) and the one that would have been reformed by reaction (3). Hence

$$\frac{-d[O_3]}{dt} = 2k_2[O][O_3] = \frac{2k_2k_1[O_3]^2}{k_2[O_3] + k_3[O_2][M]}$$

$$= 2k_1[O_3] \text{ when } [O_2] \text{ is very small.}$$

7.8 Using Eq. 7.62, $k_{65}/k_0 = \exp\{-\Delta H^*/R(1/273 - 1/338)\}$. Hence, using $R = 8.314$, $\Delta H^* = 103.4$ kJ. The experimental $k = 2k_{7.15}k_{7.18}/k_{7.17}$ (Eq. 7.24). Hence $\Delta H^* = \Delta H^*_{7.15} + \Delta H^*_{7.18} - \Delta H^*_{7.17}$.

7.9 At 0°C we have $7.8 \times 10^{-7} = Ae^{-\Delta H^*/273 R}$. Hence $A = 4.9 \times 10^{13}$. $kT/h = 1.38 \times 10^{-23} \times 273 \div 6.627 \times 10^{-34} = 5.7 \times 10^{12}$. ($A$ exceeds kT/h; hence ΔS^* must be positive, if $k = 1$.)

7.10 (a) The $T^{1/2}$ factors are there because the experimental data has been fitted to a collision-theory model, where $T^{1/2}$ occurs in the expression for Z_{AB}. (b) $19\,600 \times R = 163.0$ kJ; $21\,400 \times R = 177.9$ kJ. (c) $K = k_1/k_2 = 1.65 \exp(1800/T) = 690$ at 298 K. $K = k_1/k_2$ follows because at equilibrium the two reaction rates are equal. (d) $\Delta H^* = -1800 \times R = -15.0$ kJ.

Chapter 13

13.1

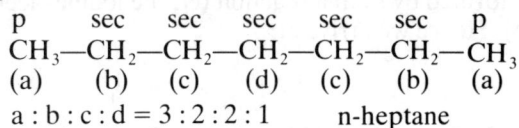

```
   p    sec   sec   sec   sec   sec    p
CH₃—CH₂—CH₂—CH₂—CH₂—CH₂—CH₃
  (a)   (b)   (c)   (d)   (c)   (b)   (a)
```
a : b : c : d = 3 : 2 : 2 : 1   n-heptane

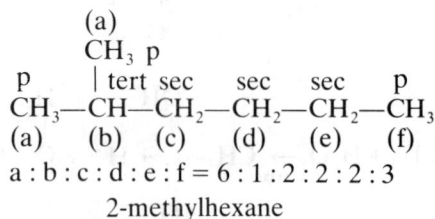

```
         (a)
         CH₃ p
   p      | tert sec   sec   sec    p
CH₃—CH—CH₂—CH₂—CH₂—CH₃
  (a)    (b)   (c)    (d)   (e)   (f)
```
a : b : c : d : e : f = 6 : 1 : 2 : 2 : 2 : 3

   2-methylhexane

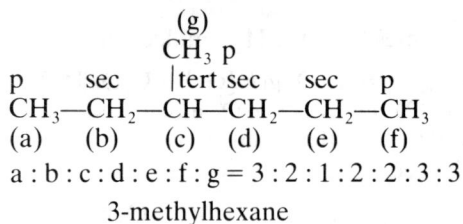

```
                (g)
                CH₃ p
   p    sec     |tert sec   sec    p
CH₃—CH₂—CH—CH₂—CH₂—CH₃
  (a)   (b)    (c)   (d)   (e)   (f)
```
a : b : c : d : e : f : g = 3 : 2 : 1 : 2 : 2 : 3 : 3

   3-methylhexane

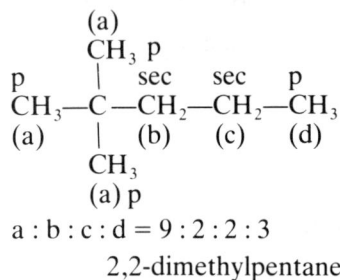

```
        (a)
        CH₃ p
   p     |  sec   sec    p
CH₃—C—CH₂—CH₂—CH₃
  (a)   |  (b)   (c)   (d)
        CH₃
        (a) p
```
a : b : c : d = 9 : 2 : 2 : 3

   2,2-dimethylpentane

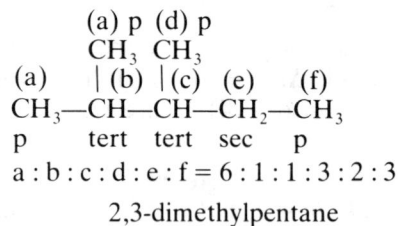

```
        (a) p (d) p
        CH₃ CH₃
  (a)    | (b) |(c) (e)     (f)
CH₃—CH—CH—CH₂—CH₃
   p    tert  tert  sec    p
```
a : b : c : d : e : f = 6 : 1 : 1 : 3 : 2 : 3

   2,3-dimethylpentane

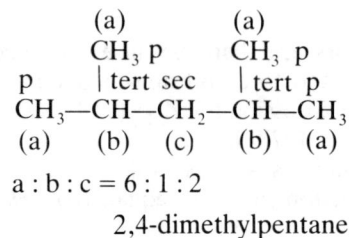

```
       (a)          (a)
       CH₃ p        CH₃ p
  p     | tert sec    | tert p
CH₃—CH—CH₂—CH—CH₃
  (a)   (b)   (c)    (b)   (a)
```
a : b : c = 6 : 1 : 2

   2,4-dimethylpentane

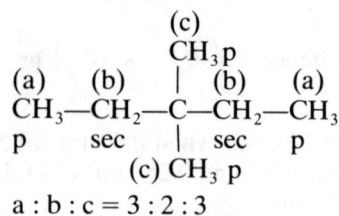

```
                (c)
                CH₃ p
  (a)   (b)      |   (b)   (a)
CH₃—CH₂—C—CH₂—CH₃
   p    sec     |  sec    p
               (c) CH₃ p
```
a : b : c = 3 : 2 : 3

   3,3-dimethylpentane

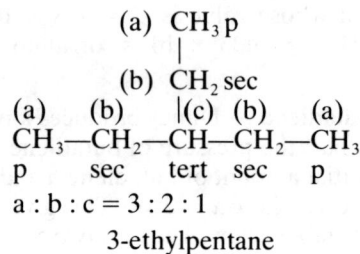

```
         (a)  CH₃ p
               |
         (b)  CH₂ sec
  (a)    (b)   |(c)  (b)   (a)
CH₃—CH₂—CH—CH₂—CH₃
   p    sec   tert  sec    p
```
a : b : c = 3 : 2 : 1

   3-ethylpentane

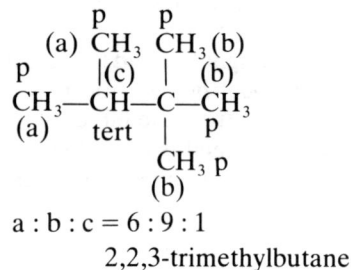

```
             p    p
      (a)  CH₃  CH₃ (b)
   p        |(c)  |    (b)
CH₃—CH—C—CH₃
  (a)    tert  |   p
              CH₃ p
              (b)
```
a : b : c = 6 : 9 : 1

   2,2,3-trimethylbutane

13.2

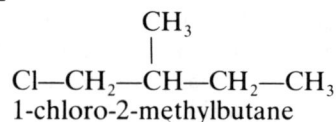

```
          CH₃
           |
Cl—CH₂—CH—CH₂—CH₃
1-chloro-2-methylbutane
```

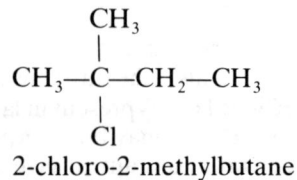

```
          CH₃
           |
CH₃—C—CH₂—CH₃
           |
          Cl
2-chloro-2-methylbutane
```

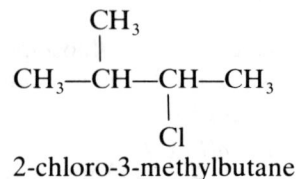

```
          CH₃
           |
CH₃—CH—CH—CH₃
           |
          Cl
2-chloro-3-methylbutane
```

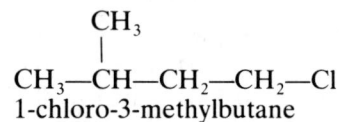

```
          CH₃
           |
CH₃—CH—CH₂—CH₂—Cl
1-chloro-3-methylbutane
```

13.3

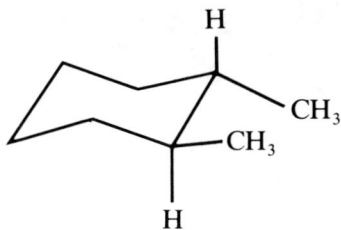

(1) *trans*-1,2-dimethylcyclohexane

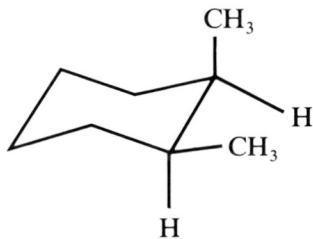

(2) *cis*-1,2-dimethylcyclohexane

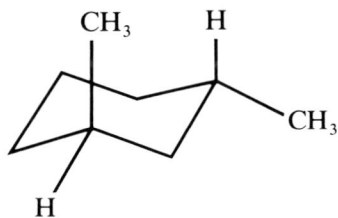

(3) *trans*-1,3-dimethylcyclohexane

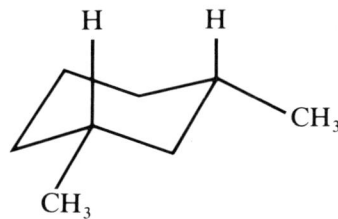

(4) *cis*-1,3-dimethylcyclohexane

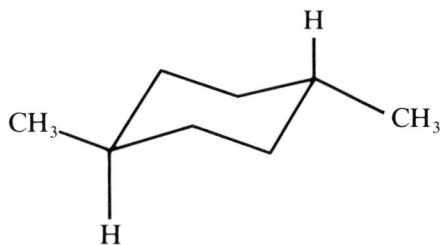

(5) *trans*-1,4-dimethylcyclohexane

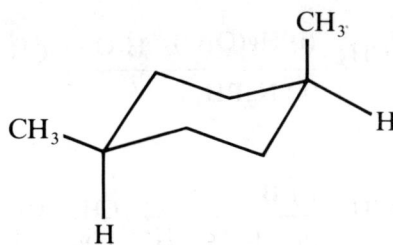

(6) *cis*-1,4-dimethylcyclohexane

Chapter 14
14.1

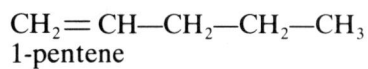

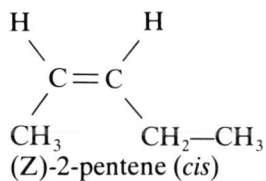

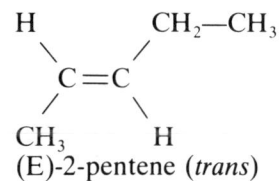

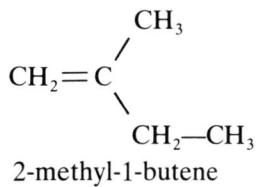

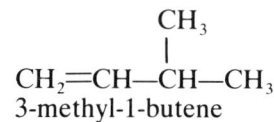

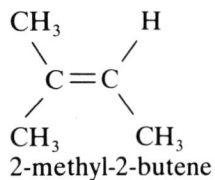

$$CH_2\!=\!CH\!-\!CH_2\!-\!CH_2\!-\!CH_3$$
1-pentene

(Z)-2-pentene (*cis*)

(E)-2-pentene (*trans*)

2-methyl-1-butene

$$CH_2\!=\!CH\!-\!\underset{\underset{CH_3}{|}}{CH}\!-\!CH_3$$
3-methyl-1-butene

2-methyl-2-butene

14.2

(a)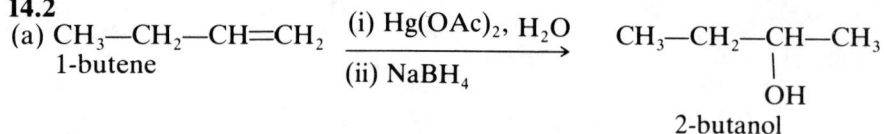

(b)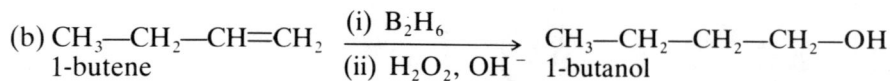

(c)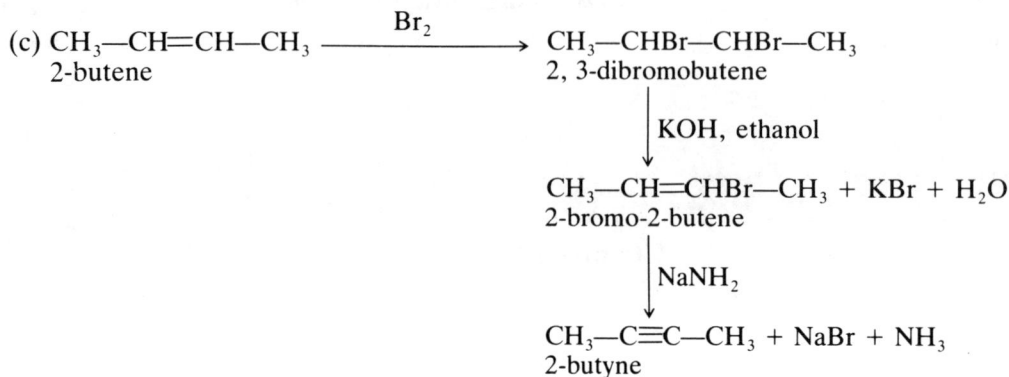

14.3

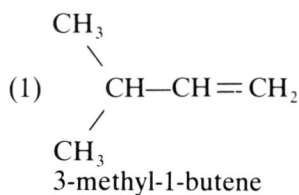

(1) 3-methyl-1-butene

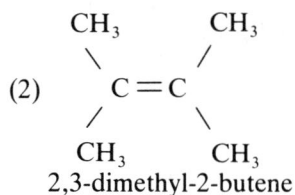

(2) 2,3-dimethyl-2-butene

14.4

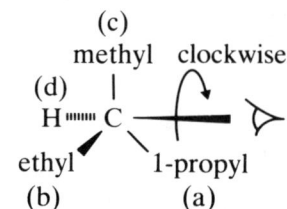

(R)-3-methylhexane

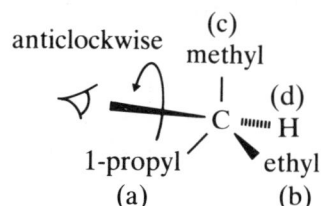

(S)-3-methylhexane

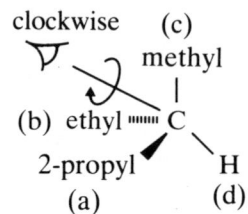

(R)-2,3-dimethylpentane

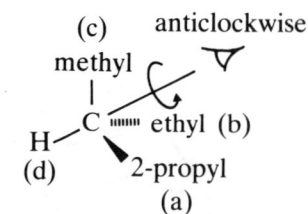

(S)-2,3-dimethylpentane

14.5

1-chloro-2-methylbutane and 2-chloro-3-methylbutane

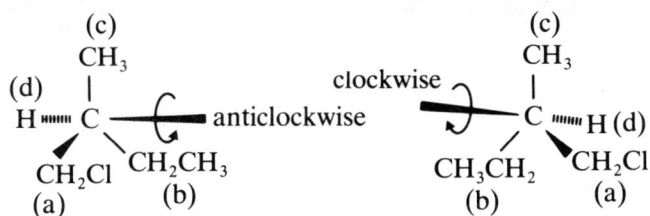

(S)-1-chloro-2-methylbutane   (R)-1-chloro-2-methylbutane

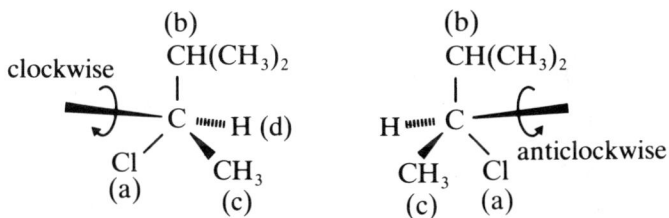

(R)-2-chloro-3-methylbutane   (S)-2-chloro-3-methylbutane

14.7

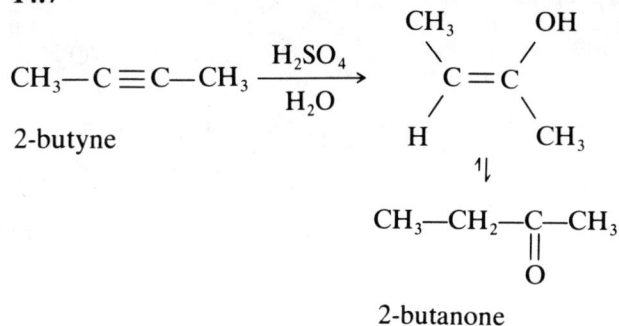

2-butyne

2-butanone

14.8

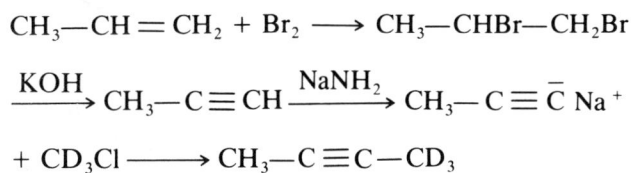

$$CH_3—CH=CH_2 + Br_2 \longrightarrow CH_3—CHBr—CH_2Br$$

$$\xrightarrow{KOH} CH_3—C\equiv CH \xrightarrow{NaNH_2} CH_3—C\equiv \bar{C}\ Na^+$$

$$+ CD_3Cl \longrightarrow CH_3—C\equiv C—CD_3$$

14.6

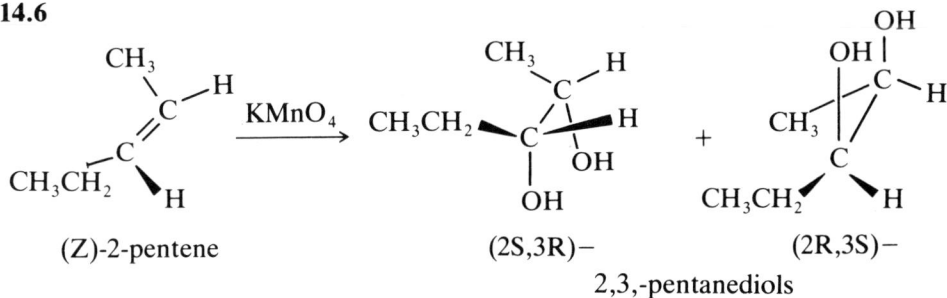

(Z)-2-pentene   (2S,3R)−   (2R,3S)−

2,3,-pentanediols

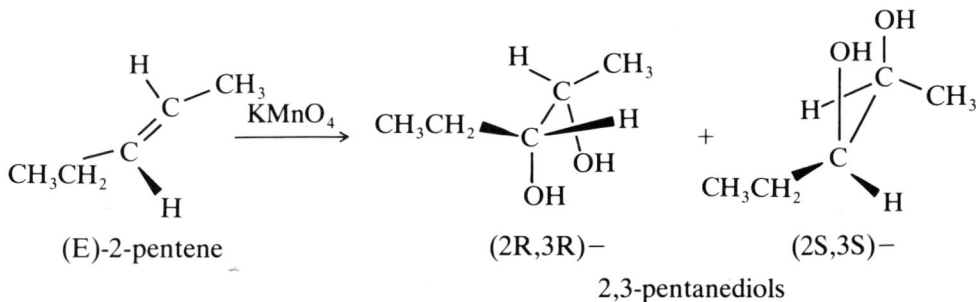

(E)-2-pentene   (2R,3R)−   (2S,3S)−

2,3-pentanediols

Chapter 15
15.1

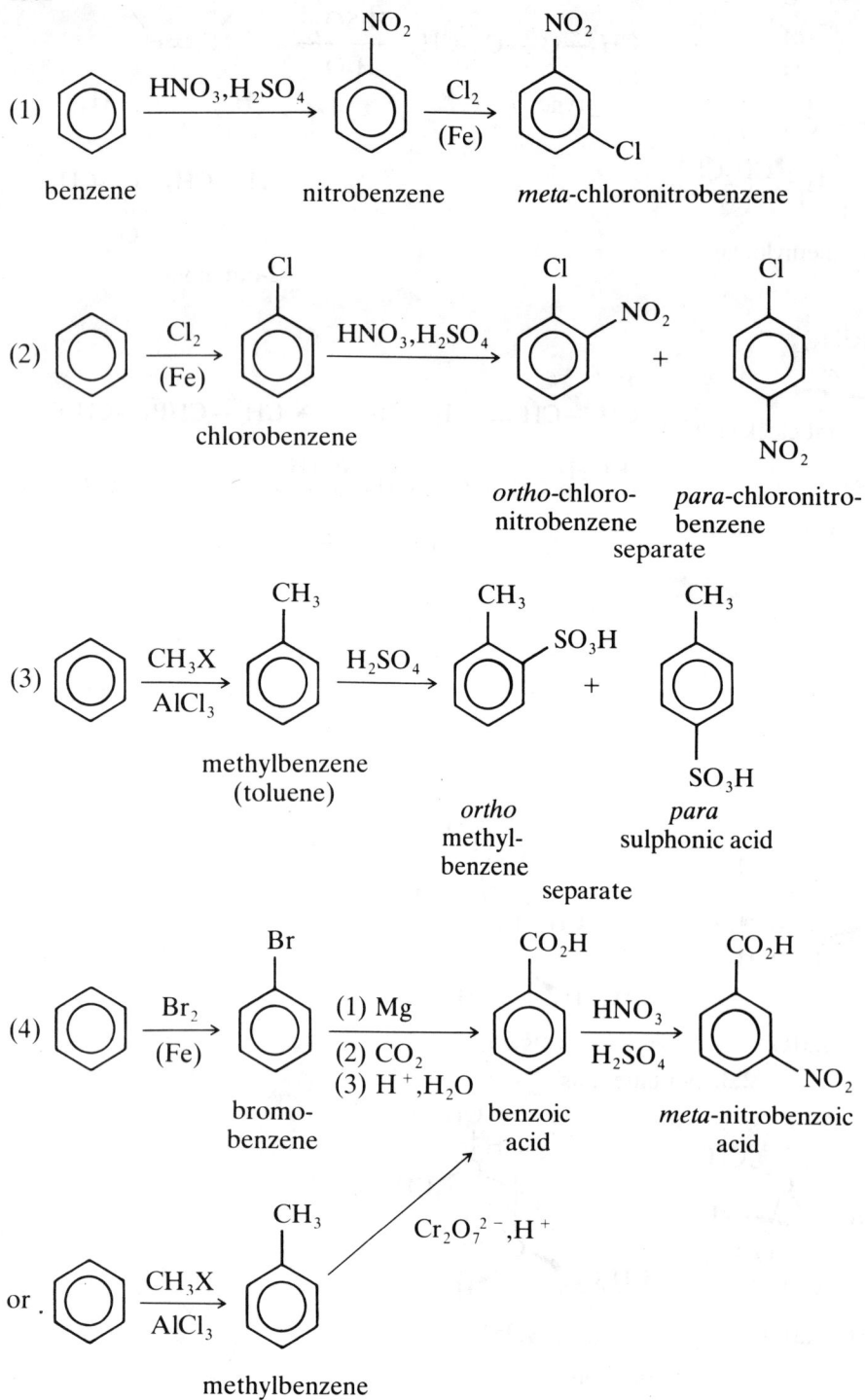

(1)

benzene   nitrobenzene   *meta*-chloronitrobenzene

HNO_3, H_2SO_4 Cl_2 (Fe)

(2)

benzene   chlorobenzene

Cl_2 (Fe) HNO_3, H_2SO_4

ortho-chloro-nitrobenzene   *para*-chloronitro-benzene

separate

(3)

benzene   methylbenzene (toluene)

CH_3X $AlCl_3$ H_2SO_4

ortho methyl-benzene   *para* sulphonic acid

separate

(4)

benzene   bromo-benzene   benzoic acid   *meta*-nitrobenzoic acid

Br_2 (Fe)   (1) Mg (2) CO_2 (3) H^+, H_2O HNO_3 H_2SO_4

or .

benzene   methylbenzene

CH_3X $AlCl_3$ $Cr_2O_7^{2-}, H^+$

15.1 (continued)

(5)

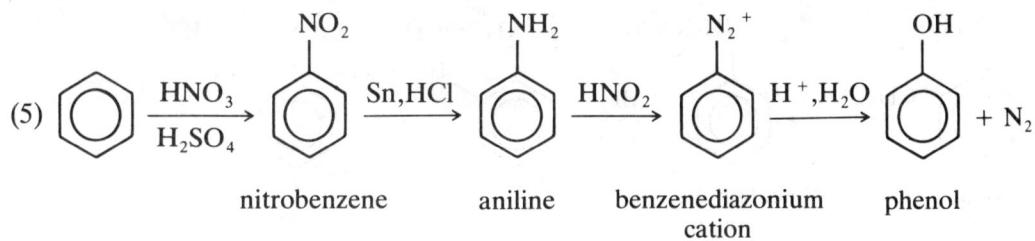

nitrobenzene   aniline   benzenediazonium cation   phenol

or

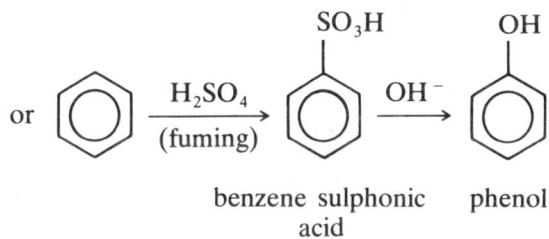

benzene sulphonic acid   phenol

(6)

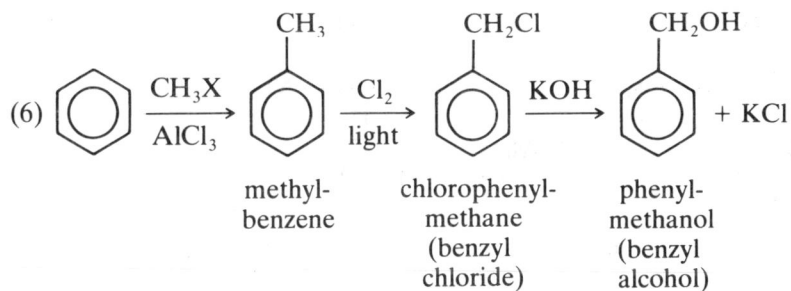

methyl-benzene   chlorophenyl-methane (benzyl chloride)   phenyl-methanol (benzyl alcohol)

or

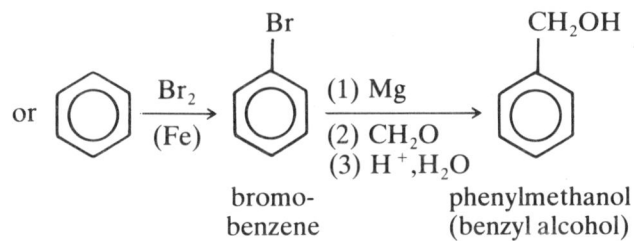

bromo-benzene   phenylmethanol (benzyl alcohol)

15.1 (continued)

(7)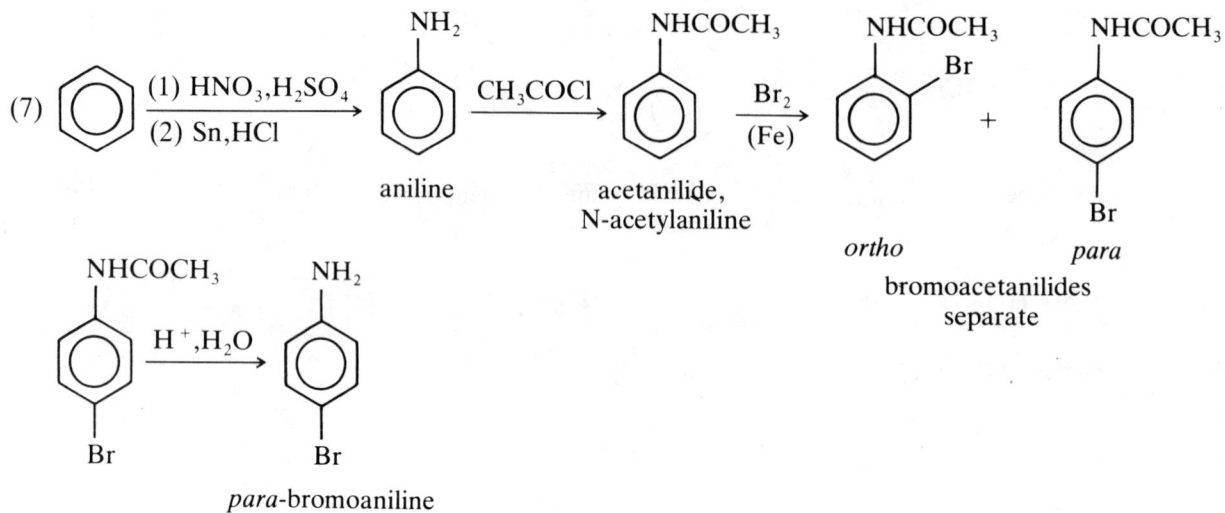

aniline → acetanilide, N-acetylaniline → ortho + para bromoacetanilides separate

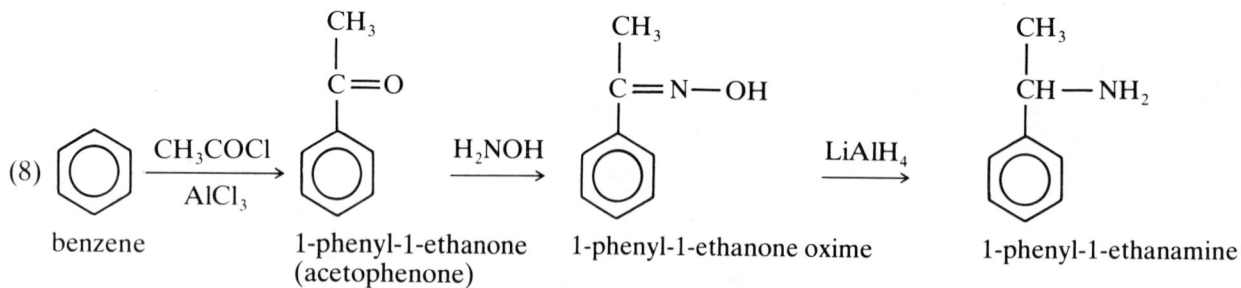

para-bromoaniline

(8)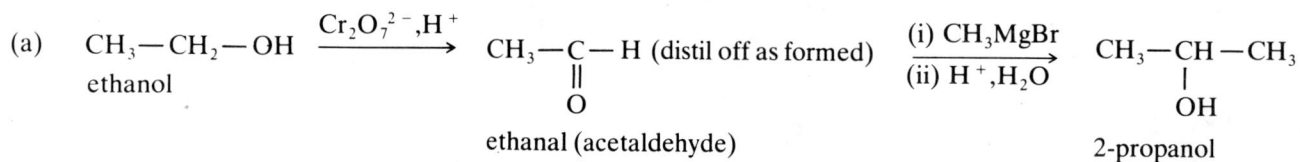

benzene → 1-phenyl-1-ethanone (acetophenone) → 1-phenyl-1-ethanone oxime → 1-phenyl-1-ethanamine

Chapter 16

16.1

(a) CH_3-CH_2-OH $\xrightarrow{Cr_2O_7^{2-},H^+}$ $CH_3-\overset{\overset{\displaystyle O}{\|}}{C}-H$ (distil off as formed) $\xrightarrow[\text{(ii) } H^+,H_2O]{\text{(i) } CH_3MgBr}$ $CH_3-\underset{\underset{\displaystyle OH}{|}}{CH}-CH_3$

ethanol   ethanal (acetaldehyde)   2-propanol

(b)

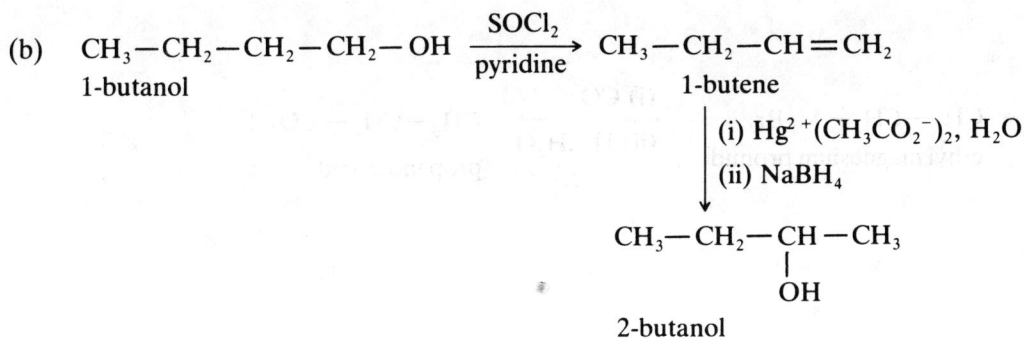

CH$_3$—CH$_2$—CH$_2$—CH$_2$—OH $\xrightarrow[\text{pyridine}]{\text{SOCl}_2}$ CH$_3$—CH$_2$—CH=CH$_2$
1-butanol   1-butene

(i) Hg^{2+}(CH$_3$CO$_2^-$)$_2$, H$_2$O
(ii) NaBH$_4$

CH$_3$—CH$_2$—CH—CH$_3$
                    |
                    OH
2-butanol

(c)

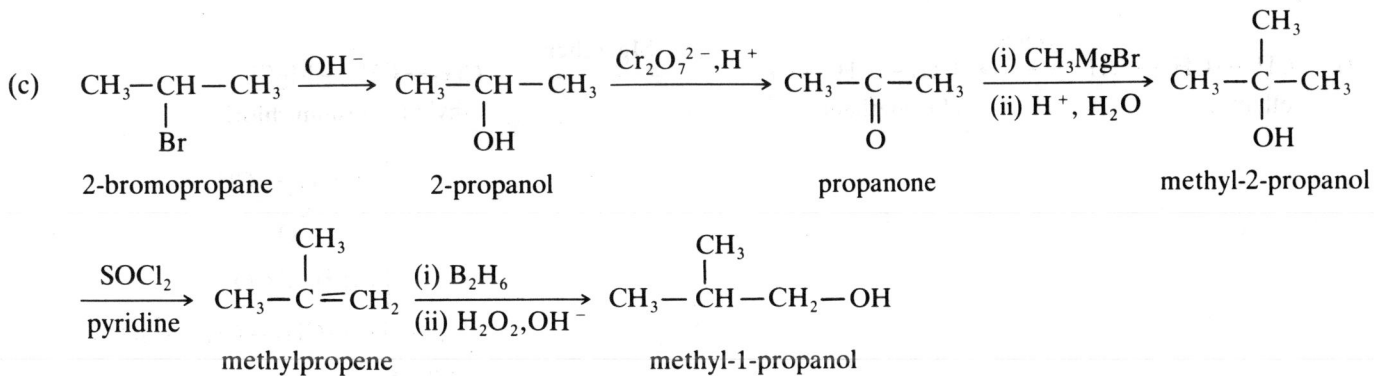

CH$_3$—CH—CH$_3$ $\xrightarrow{\text{OH}^-}$ CH$_3$—CH—CH$_3$ $\xrightarrow{\text{Cr}_2\text{O}_7^{2-},\text{H}^+}$ CH$_3$—C—CH$_3$ $\xrightarrow[\text{(ii) H}^+,\text{ H}_2\text{O}]{\text{(i) CH}_3\text{MgBr}}$ CH$_3$—C—CH$_3$
         | | || |
         Br   OH   O   OH
2-bromopropane   2-propanol   propanone   methyl-2-propanol

$\xrightarrow[\text{pyridine}]{\text{SOCl}_2}$ CH$_3$—C=CH$_2$ $\xrightarrow[\text{(ii) H}_2\text{O}_2,\text{OH}^-]{\text{(i) B}_2\text{H}_6}$ CH$_3$—CH—CH$_2$—OH
                                  | |
                                  CH$_3$ CH$_3$
                         methylpropene   methyl-1-propanol

or

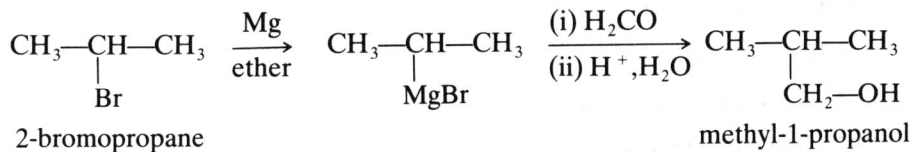

CH$_3$—CH—CH$_3$ $\xrightarrow[\text{ether}]{\text{Mg}}$ CH$_3$—CH—CH$_3$ $\xrightarrow[\text{(ii) H}^+,\text{H}_2\text{O}]{\text{(i) H}_2\text{CO}}$ CH$_3$—CH—CH$_3$
         | | |
         Br   MgBr   CH$_2$—OH
2-bromopropane   methyl-1-propanol

16.1 (continued)

(d) CH_3-CH_2-Br $\xrightarrow[\text{ether}]{\text{Mg}}$ CH_3-CH_2-MgBr $\xrightarrow[\text{(ii) H}^+,\text{H}_2\text{O}]{\text{(i) CO}_2}$ $CH_3-CH_2-CO_2H$
      bromoethane   ethyl magnesium bromide   propanoic acid

(e) CH_3-CH_2-Cl $\xrightarrow{\text{CN}^-}$ $CH_3-CH_2-C\equiv N$ $\xrightarrow{\text{LiAlH}_4}$ $CH_3-CH_2-CH_2-NH_2$
      chloroethane   propanenitrile   1-propanamine

(f) CH_3-CH_2-OH $\xrightarrow{\text{SOCl}_2}$ CH_3-CH_2-Cl $\xrightarrow{\text{Mg, ether}}$ CH_3-CH_2-MgCl
      ethanol   chloroethane   ethyl magnesium chloride

(i) CH_2-CH_2
        $\diagdown\diagup$
           O

(ii) H^+,H_2O

$CH_3-CH_2-CH_2-CH_2-OH$
1-butanol

(g) $CH_3-CH_2-OH + Na \longrightarrow CH_3-CH_2-O^-Na^+ + \frac{1}{2}H_2$
      ethanol   sodium ethoxide

$CH_3-CH_2-O^-Na^+ + CH_3-Br \longrightarrow CH_3-CH_2-O-CH_3$
                          bromomethane   ethylmethylether
                                         methoxyethane

16.1 (continued)

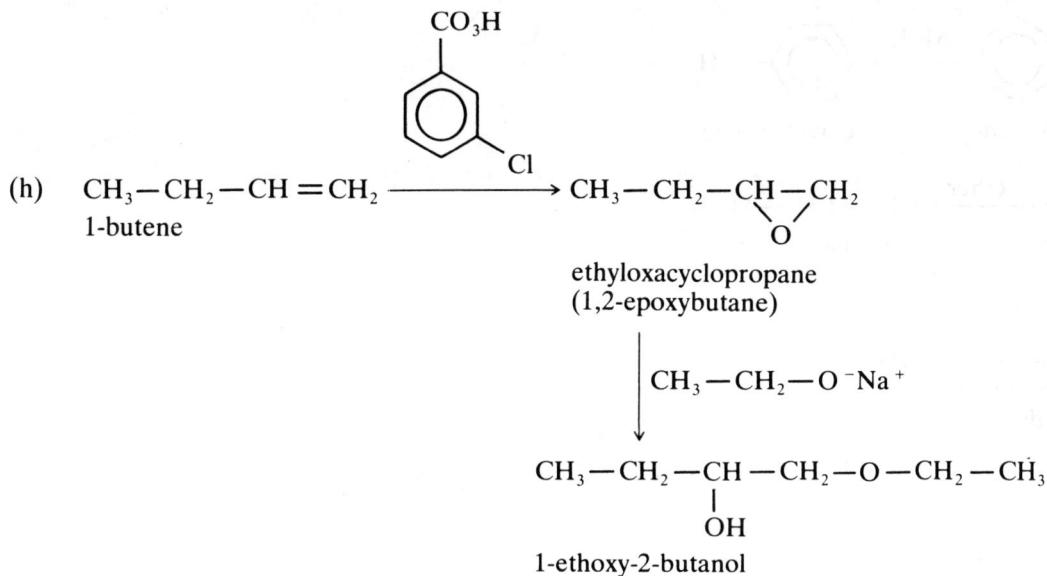

(h) $CH_3-CH_2-CH=CH_2$ $\xrightarrow{\text{[3-chlorobenzoic acid]}}$ $CH_3-CH_2-\underset{\underset{O}{\diagup\diagdown}}{CH}-CH_2$

1-butene

ethyloxacyclopropane
(1,2-epoxybutane)

$\downarrow CH_3-CH_2-O^-Na^+$

$CH_3-CH_2-\underset{\underset{OH}{\mid}}{CH}-CH_2-O-CH_2-CH_3$

1-ethoxy-2-butanol

16.2

(a) $CH_3-CH_2-Br + OH^- \longrightarrow CH_3-CH_2-OH$

bromoethane   aqueous   ethanol
          hydroxide

(b) $CH_3-CH_2-Br + CH_3-O^-Na^+ \longrightarrow CH_3-CH_2-O-CH_3 + NaBr$

         sodium methoxide   ethylmethylether

(c) $CH_3-CH_2-Br + CH_3-C\equiv C^-Na^+ \longrightarrow CH_3-CH_2-C\equiv C-CH_3$

        sodium   2-pentyne
        methylacetylide

(d) $CH_3-CH_2-Br + CH_3-CH_2-CO_2^-Na^+ \longrightarrow CH_3-CH_2-O-\underset{\underset{O}{\parallel}}{C}-CH_2-CH_3$

         sodium propanoate

                            ethyl propanoate

(e) $CH_3-CH_2-Br + CN^- \longrightarrow CH_3-CH_2-C\equiv N$

              cyanide   propanenitrile

(f) $CH_3-CH_2-Br + (CH_3)_2NH \longrightarrow CH_3-CH_2-\underset{\underset{CH_3}{\mid}}{N}-CH_3$

      (excess)

        N-methylmethanamine
         (dimethylamine)   N,N-dimethylethanamine

16.2 (continued)

(g)   CH₃—CH₂—Br +

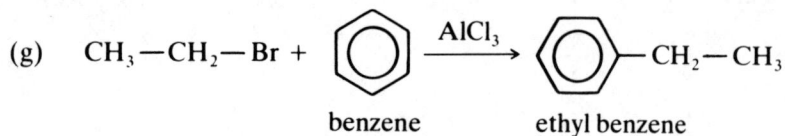

benzene   ethyl benzene

(h)   CH₃—CH₂—Br + Mg $\xrightarrow{\text{ether}}$ CH₃—CH₂—MgBr

magnesium   ethyl magnesium bromide

(i)   CH₃—CH₂—Br + LiAlH₄ ⟶ CH₃—CH₃

lithium   ethane
aluminium
hydride

16.3

(a)

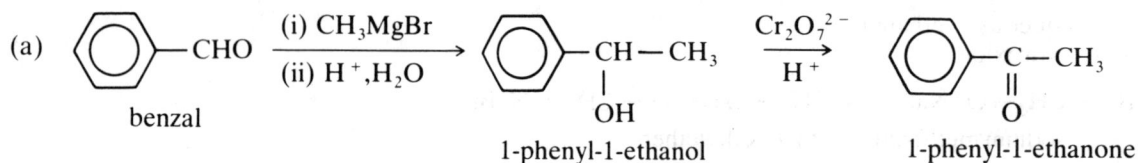

benzal

1-phenyl-1-ethanol

1-phenyl-1-ethanone

(b)

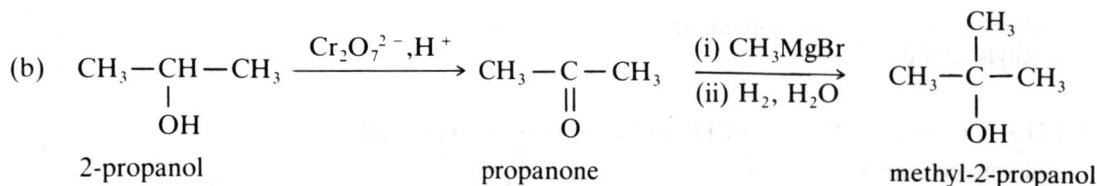

2-propanol   propanone   methyl-2-propanol

(c)   CH₃—CH₂—MgBr $\xrightarrow[\text{(ii) H}^+,\text{ H}_2\text{O}]{\text{(i) H}_2\text{CO}}$ CH₃—CH₂—CH₂—OH

ethyl magnesium   1-propanol
bromide

16.3 (continued)

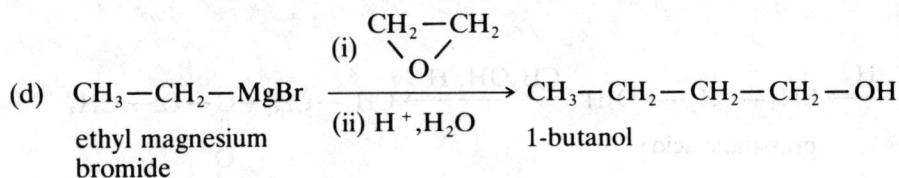

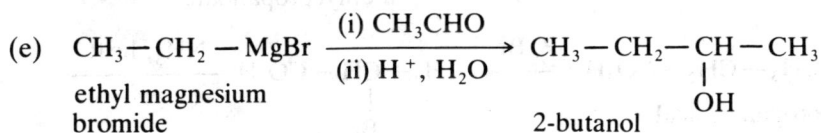

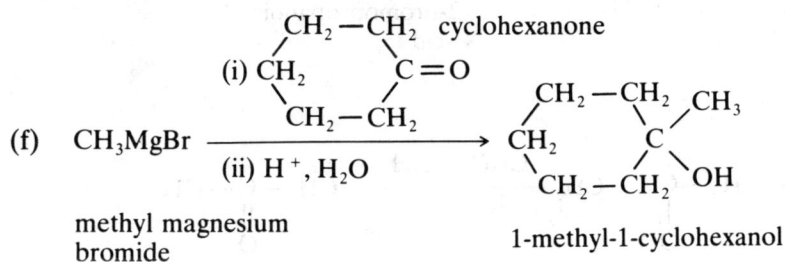

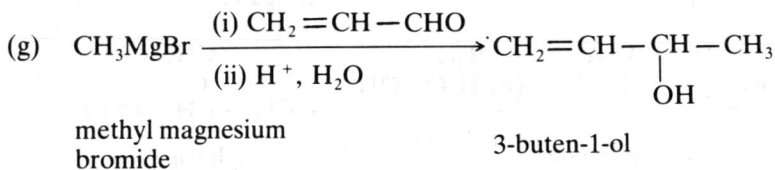

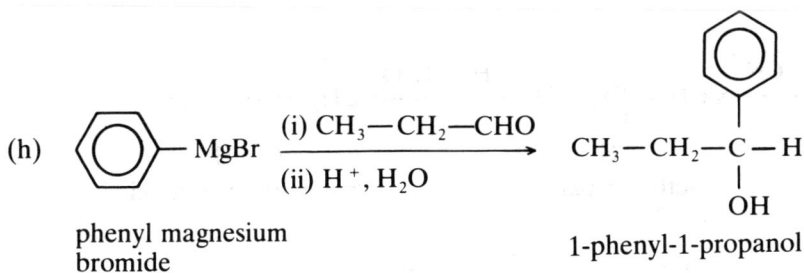

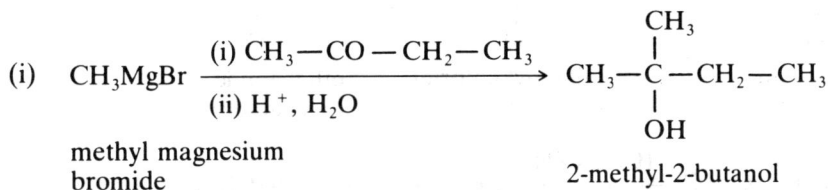

(d) CH_3-CH_2-MgBr $\xrightarrow[\text{(ii) } H^+, H_2O]{\text{(i)} \quad \overset{CH_2-CH_2}{\underset{O}{\diagdown \diagup}}}$ $CH_3-CH_2-CH_2-CH_2-OH$

ethyl magnesium bromide   1-butanol

(e) CH_3-CH_2-MgBr $\xrightarrow[\text{(ii) } H^+, H_2O]{\text{(i) } CH_3CHO}$ $CH_3-CH_2-\underset{\underset{OH}{|}}{CH}-CH_3$

ethyl magnesium bromide   2-butanol

(f) CH_3MgBr $\xrightarrow[\text{(ii) } H^+, H_2O]{\text{(i)} \quad \text{cyclohexanone}}$

methyl magnesium bromide

1-methyl-1-cyclohexanol

(g) CH_3MgBr $\xrightarrow[\text{(ii) } H^+, H_2O]{\text{(i) } CH_2=CH-CHO}$ $CH_2=CH-\underset{\underset{OH}{|}}{CH}-CH_3$

methyl magnesium bromide   3-buten-1-ol

(h) $\langle O \rangle-MgBr$ $\xrightarrow[\text{(ii) } H^+, H_2O]{\text{(i) } CH_3-CH_2-CHO}$ $CH_3-CH_2-\underset{\underset{OH}{|}}{C}-H$

phenyl magnesium bromide

1-phenyl-1-propanol

(i) CH_3MgBr $\xrightarrow[\text{(ii) } H^+, H_2O]{\text{(i) } CH_3-CO-CH_2-CH_3}$ $CH_3-\underset{\underset{OH}{|}}{\overset{\overset{CH_3}{|}}{C}}-CH_2-CH_3$

methyl magnesium bromide   2-methyl-2-butanol

Chapter 17

17.1

Suggested solutions

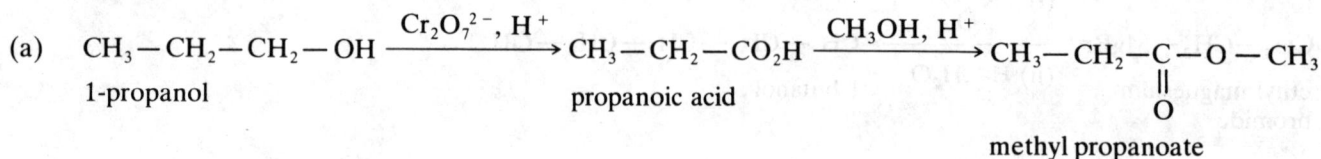

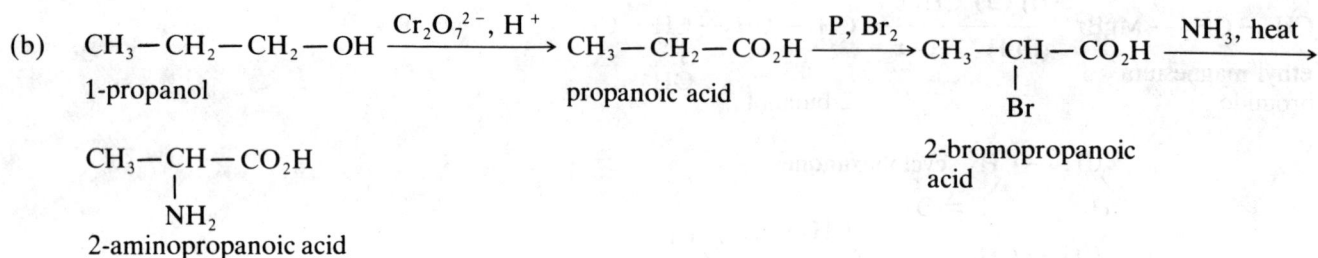

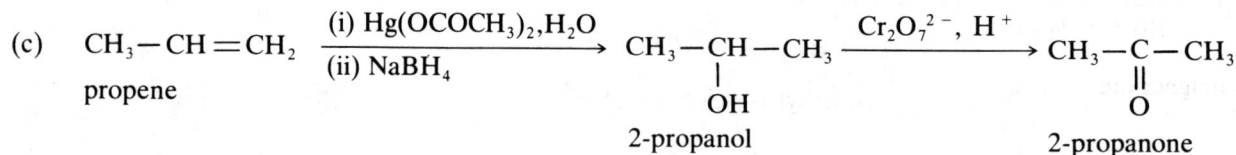

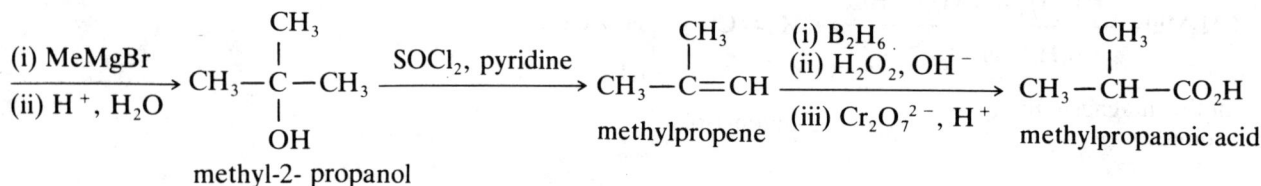

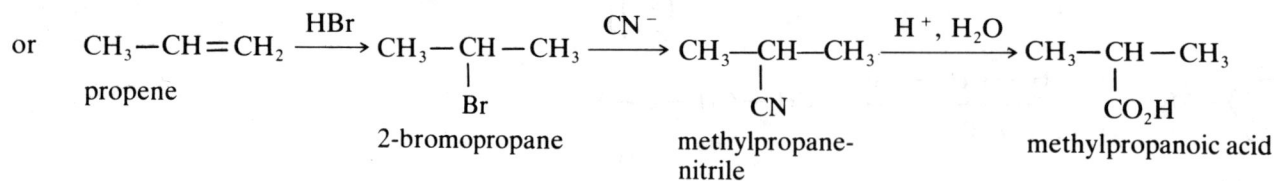

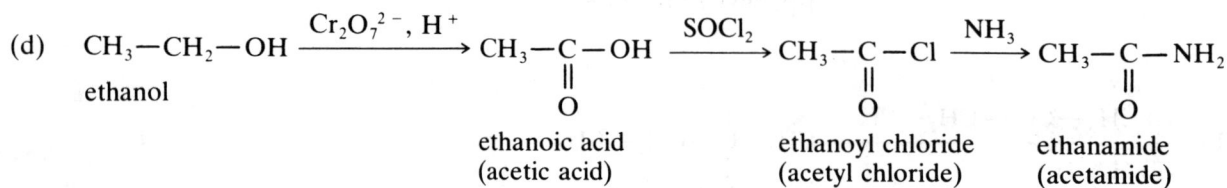

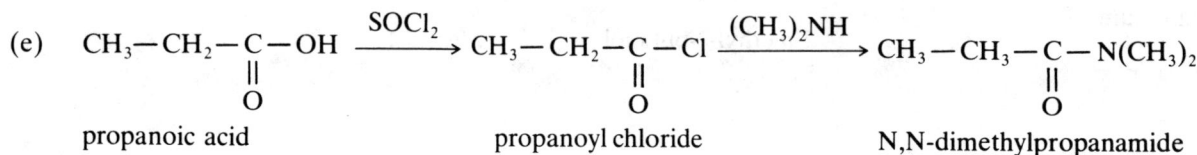

(a) $CH_3-CH_2-CH_2-OH \xrightarrow{Cr_2O_7^{2-},\ H^+} CH_3-CH_2-CO_2H \xrightarrow{CH_3OH,\ H^+} CH_3-CH_2-\overset{\displaystyle O}{\overset{\displaystyle \|}{C}}-O-CH_3$

1-propanol   propanoic acid   methyl propanoate

(b) $CH_3-CH_2-CH_2-OH \xrightarrow{Cr_2O_7^{2-},\ H^+} CH_3-CH_2-CO_2H \xrightarrow{P,\ Br_2} CH_3-\underset{\underset{\displaystyle Br}{|}}{CH}-CO_2H \xrightarrow{NH_3,\ heat}$

1-propanol   propanoic acid   2-bromopropanoic
acid

$CH_3-\underset{\underset{\displaystyle NH_2}{|}}{CH}-CO_2H$

2-aminopropanoic acid

(c) $CH_3-CH=CH_2 \xrightarrow[\text{(ii) NaBH}_4]{\text{(i) Hg(OCOCH}_3)_2, H_2O} CH_3-\underset{\underset{\displaystyle OH}{|}}{CH}-CH_3 \xrightarrow{Cr_2O_7^{2-},\ H^+} CH_3-\overset{\displaystyle O}{\overset{\displaystyle \|}{C}}-CH_3$

propene   2-propanol   2-propanone

$\xrightarrow[\text{(ii) H}^+, \text{H}_2\text{O}]{\text{(i) MeMgBr}} CH_3-\underset{\underset{\displaystyle OH}{|}}{\overset{\overset{\displaystyle CH_3}{|}}{C}}-CH_3 \xrightarrow{SOCl_2,\ pyridine} CH_3-\overset{\overset{\displaystyle CH_3}{|}}{C}=CH \xrightarrow[\text{(iii) Cr}_2\text{O}_7^{2-}, \text{H}^+]{\substack{\text{(i) B}_2\text{H}_6 \\ \text{(ii) H}_2\text{O}_2, \text{OH}^-}} CH_3-\underset{\underset{\displaystyle CH_3}{|}}{CH}-CO_2H$

methyl-2- propanol   methylpropene   methylpropanoic acid

or $CH_3-CH=CH_2 \xrightarrow{HBr} CH_3-\underset{\underset{\displaystyle Br}{|}}{CH}-CH_3 \xrightarrow{CN^-} CH_3-\underset{\underset{\displaystyle CN}{|}}{CH}-CH_3 \xrightarrow{H^+,\ H_2O} CH_3-\underset{\underset{\displaystyle CO_2H}{|}}{CH}-CH_3$

propene   2-bromopropane   methylpropane-   methylpropanoic acid
nitrile

(d) $CH_3-CH_2-OH \xrightarrow{Cr_2O_7^{2-},\ H^+} CH_3-\overset{\displaystyle O}{\overset{\displaystyle \|}{C}}-OH \xrightarrow{SOCl_2} CH_3-\overset{\displaystyle O}{\overset{\displaystyle \|}{C}}-Cl \xrightarrow{NH_3} CH_3-\overset{\displaystyle O}{\overset{\displaystyle \|}{C}}-NH_2$

ethanol   ethanoic acid   ethanoyl chloride   ethanamide
(acetic acid)   (acetyl chloride)   (acetamide)

(e) $CH_3-CH_2-\overset{\displaystyle O}{\overset{\displaystyle \|}{C}}-OH \xrightarrow{SOCl_2} CH_3-CH_2-\overset{\displaystyle O}{\overset{\displaystyle \|}{C}}-Cl \xrightarrow{(CH_3)_2NH} CH_3-CH_3-\overset{\displaystyle O}{\overset{\displaystyle \|}{C}}-N(CH_3)_2$

propanoic acid   propanoyl chloride   N,N-dimethylpropanamide

17.1 (continued)

(f) $CH_3-CH_2-OH \xrightarrow{Cr_2O_7^{2-},\ H^+} CH_3-\underset{\underset{O}{\|}}{C}-OH \xrightarrow{P_2O_5} CH_3-\underset{\underset{O}{\|}}{C}-O-\underset{\underset{O}{\|}}{C}-CH_3$

ethanol   ethanoic acid   ethanoic anhydride
                                    (acetic acid)   (acetic anhydride)

(g) $CH_3-\underset{}{\overset{\overset{\displaystyle CH_3}{|}}{CH}}-\underset{\underset{O}{\|}}{C}-OH \xrightarrow{SOCl_2} CH_3-\underset{}{\overset{\overset{\displaystyle CH_3}{|}}{CH}}-\underset{\underset{O}{\|}}{C}-Cl \xrightarrow{NH_3} CH_3-\underset{}{\overset{\overset{\displaystyle CH_3}{|}}{CH}}-\underset{\underset{O}{\|}}{C}-NH_2 \xrightarrow{Br_2,OH^-}$

methylpropanoic acid   methylpropanoyl   methylpropanamide
                          chloride

$CH_3-\underset{}{\overset{\overset{\displaystyle CH_3}{|}}{CH}}-NH_2$

2-propanamine

(h) $CH_3-CH_2-CH_2OH \xrightarrow{CH_3CO_2H,\ H^+} CH_3-CH_2-CH_2-O-\underset{\underset{O}{\|}}{C}-CH_3$

1-propanol   prop-l-yl ethanoate

(i) $CH_3-CH_2-CH_2-CH_2-\underset{\underset{O}{\|}}{C}-O-CH_2-CH_3 \xrightarrow{H^+,\ H_2O} CH_3-CH_2-CH_2-CH_2-CO_2H \ +$

ethyl pentanoate   pentanoic acid CH_3-CH_2-OH
                                                      ethanol

(ii) $CH_3(CH_2)_3\underset{\underset{O}{\|}}{C}-O-CH_2CH_3 \xrightarrow[\text{(ii) } H^+,\ H_2O]{\text{(i) } CH_3CH_2MgBr} CH_3-CH_2-CH_2-CH_2-\underset{\underset{\displaystyle OH}{|}}{\overset{\overset{\displaystyle CH_2-CH_3}{|}}{C}}-CH_2-CH_3$

                                                3-ethyl-3-heptanol

(iii) $CH_3(CH_2)_3\underset{\underset{O}{\|}}{C}-O-CH_2CH_3 \xrightarrow{LiAlH_4} CH_3-CH_2-CH_2-CH_2-CH_2-OH + CH_3-CH_2-OH$

                                         1-pentanol   ethanol

$CH_3(CH_2)_3CH_2OH \xrightarrow[\substack{\text{(iii) } CH_2-CH_2 \\ \text{(iv) } H^+,\ H_2O \\ \text{(v) } Cr_2O_7^{2-},\ H^+}]{\substack{\text{(i) } SOCl_2 \\ \text{(ii) Mg, ether} \\ \overset{\displaystyle O}{\diagup\ \diagdown}}} CH_3-CH_2-CH_2-CH_2-CH_2-CH_2-CO_2H$

1-pentanol   heptanoic acid

17.1 (continued)

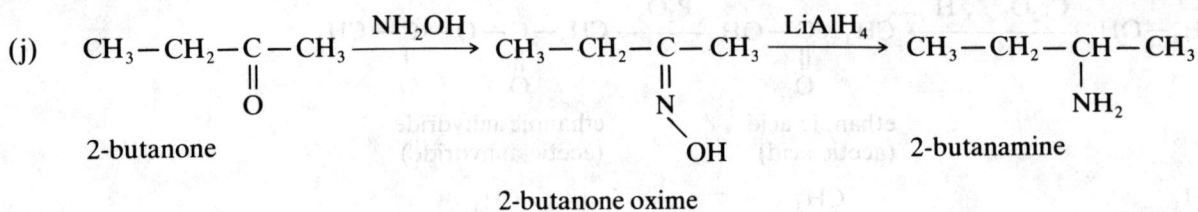

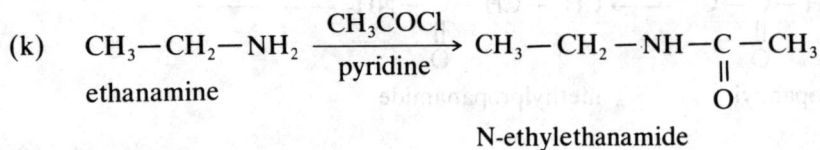

(j) $CH_3-CH_2-C-CH_3$ $\xrightarrow{NH_2OH}$ $CH_3-CH_2-C-CH_3$ $\xrightarrow{LiAlH_4}$ $CH_3-CH_2-CH-CH_3$

$\quad\quad\quad\quad\quad\quad\overset{||}{O}$ $\quad\quad\quad\quad\quad\quad\overset{||}{\underset{OH}{N}}$ $\quad\quad\quad\quad\quad\quad\overset{|}{NH_2}$

2-butanone $\quad\quad\quad\quad\quad\quad\quad$ 2-butanone oxime $\quad\quad\quad$ 2-butanamine

(k) $CH_3-CH_2-NH_2$ $\xrightarrow[\text{pyridine}]{CH_3COCl}$ $CH_3-CH_2-NH-C-CH_3$

ethanamine $\quad\quad\quad\quad\quad\quad\quad\quad\quad\quad\quad\quad\overset{||}{O}$

N-ethylethanamide

Chapter 18

18.1

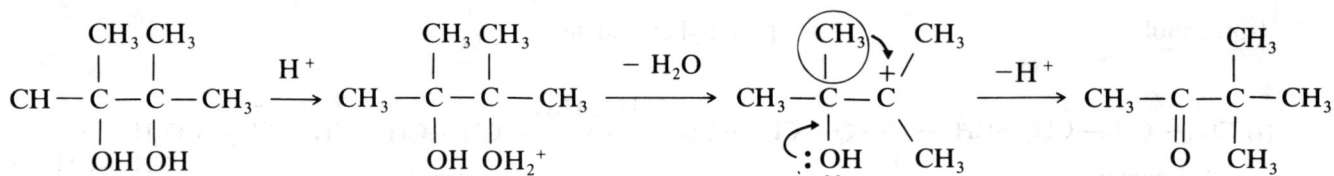

2,3-dimethyl-2,3-butanediol $\quad\quad\quad\quad\quad\quad\quad\quad\quad\quad\quad$ 3,3-dimethyl-2-butanone

18.2

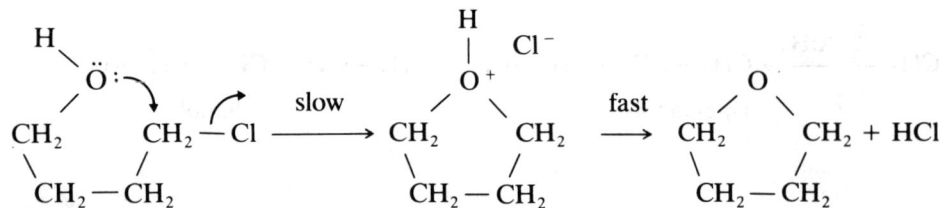

4-chloro-1-butanol $\quad\quad\quad\quad\quad\quad\quad\quad\quad$ tetrahydrofuran

18.3

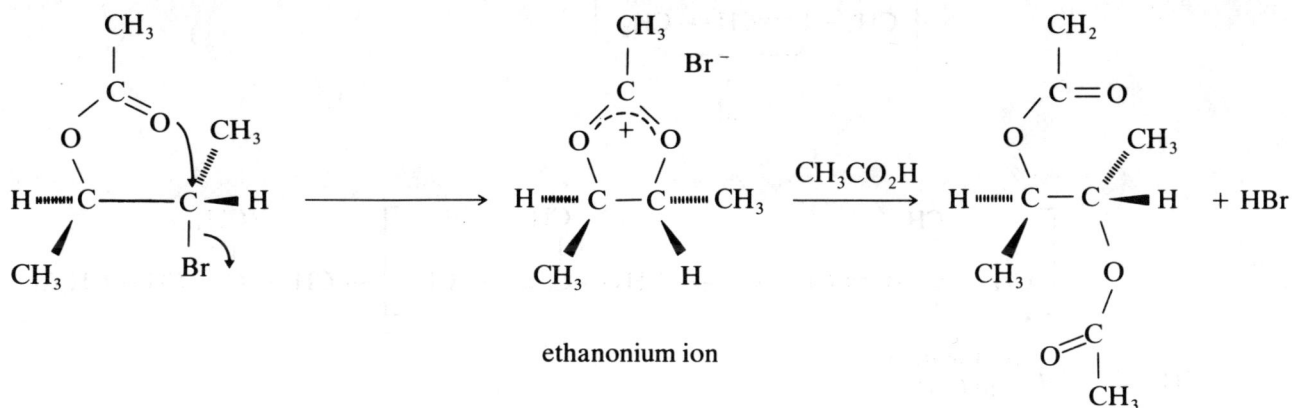

(2R,3S)-2-ethanoxy-3-bromobutane   (R,S)-2,3-diethanoxybutane

The stereochemistry at C(3) is retained by neighbouring group participation by the ethanoxy function to form the ethanonium ion.

18.4

trans-3-ethyl-4-methyl-1-cyclobutene   (E,E)-2,4-heptadiene

18.5

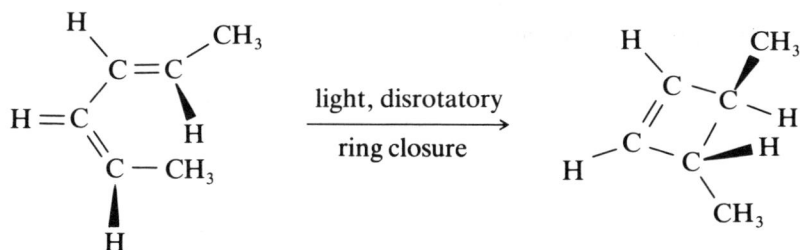

(Z,E)-2,4-hexadiene   *trans*-3,4-dimethyl-1-cyclobutene

18.6

$$\left[\begin{array}{c} CH_3 \\ | \\ CH_3-C\!=\!\!=\!CH\!=\!\!=\!CH_2 \\ \underbrace{\qquad\qquad}_{+} \\ ||| \end{array} \right]$$

$$\left[\begin{array}{c} CH_3 \\ | \\ CH_3-\overset{+}{C}-C\!=\!CH_2 \end{array} \longleftrightarrow \begin{array}{c} CH_3 \\ | \\ CH_3-C\!=\!\overset{+}{C}-CH_2 \end{array} \right] \rightarrow \begin{array}{c} CH_3 \\ | \\ CH_3-C\!=\!CH-CH_2Cl \end{array}$$

$$\begin{array}{c} CH_3 \\ | \\ CH_2\!=\!C-CH\!=\!CH_2 \end{array}$$

protonation at C(1)

protonation at C(4)

$$\left[\begin{array}{c} CH_3 \\ | \\ CH_2\!=\!C-\overset{+}{C}H-CH_3 \\ ||| \end{array} \longleftrightarrow \begin{array}{c} CH_3 \\ | \\ \overset{+}{C}H_2-C\!=\!CH-CH_3 \end{array} \right] \rightarrow \begin{array}{c} CH_3 \\ | \\ ClCH_2-C\!=\!CHCH_3 \end{array}$$

not formed

$$\left[\begin{array}{c} CH_3 \\ | \\ CH_2\!=\!\!=\!C\!=\!\!=\!C-CH_3 \\ \underbrace{\qquad\qquad}_{+} \end{array} \right]$$

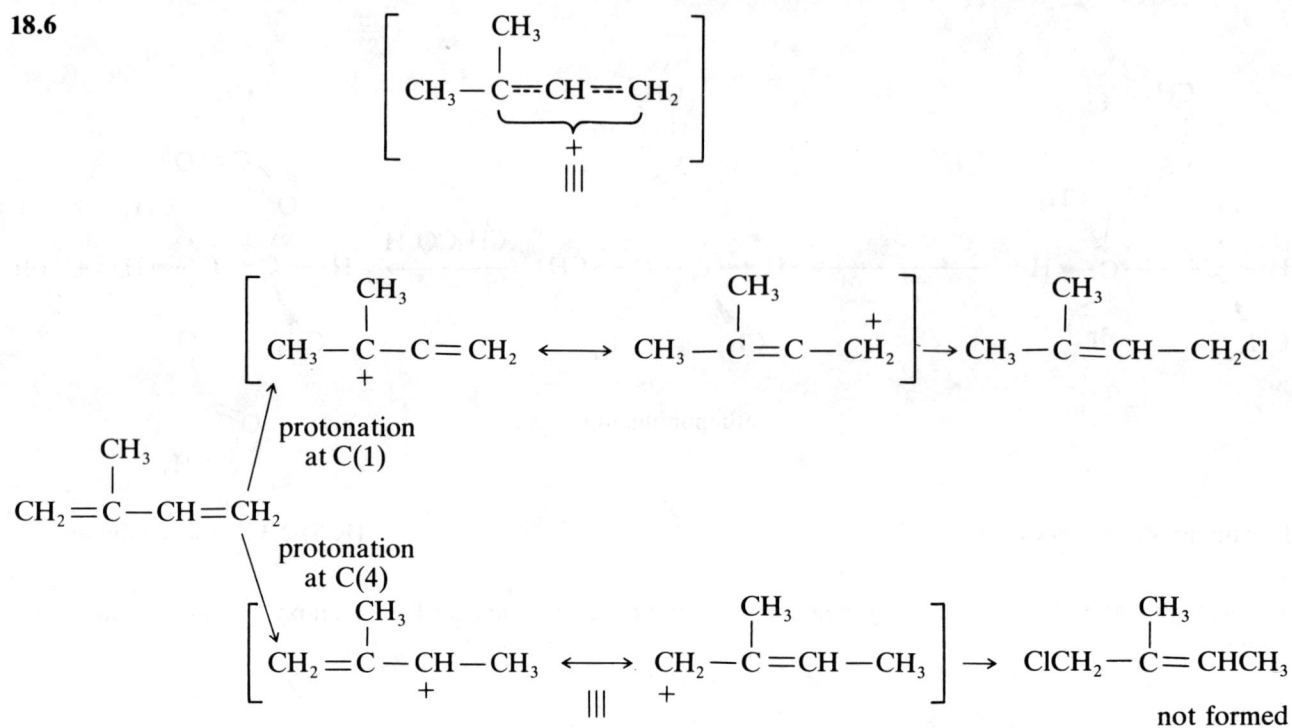

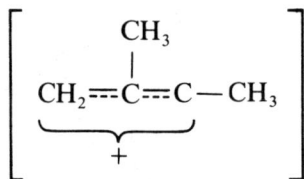

The allylic ion formed by protonation at C(1) is more stable than that formed by protonation at C(4) due to the contribution of the canonical structure

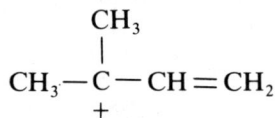

$$\begin{array}{c} CH_3 \\ | \\ CH_3-\overset{+}{C}-CH\!=\!CH_2 \end{array}$$

to the resonance stabilization of the former allylic carbonium ion.

18.7

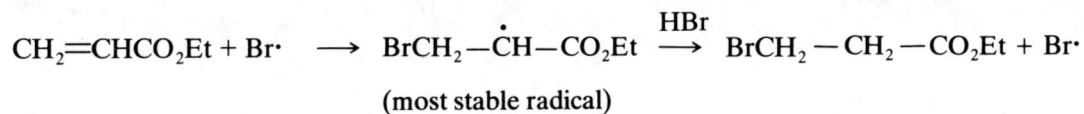

CH₂=CHCO₂Et + Br· ⟶ BrCH₂—ĊH—CO₂Et $\xrightarrow{\text{HBr}}$ BrCH₂—CH₂—CO₂Et + Br·

(most stable radical)

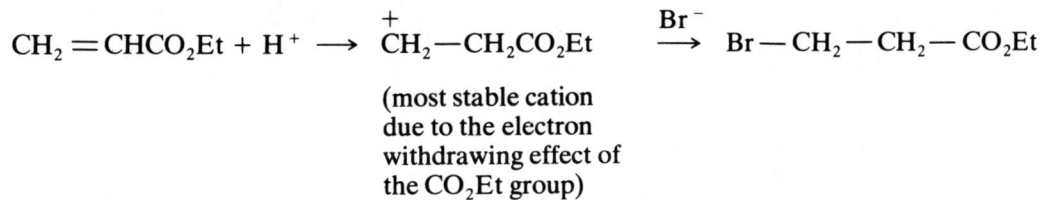

CH₂=CHCO₂Et + H⁺ ⟶ ⁺CH₂—CH₂CO₂Et $\xrightarrow{\text{Br}^-}$ Br—CH₂—CH₂—CO₂Et

(most stable cation
due to the electron
withdrawing effect of
the CO₂Et group)

18.8

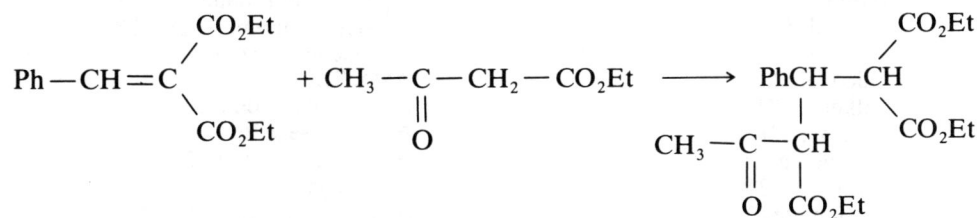

18.8

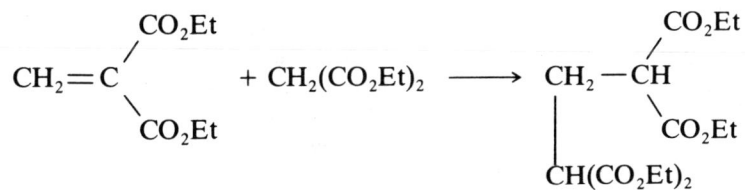

Index